AF902870

Springer Geology

More information about this series at http://www.springer.com/series/10172

Jai Krishna

The Indian Mesozoic Chronicle

Sequence Stratigraphic Approach

Springer

Jai Krishna
Former Professor and Head
Centre of Advanced Study in Geology
Institute of Science
Banaras Hindu University
Varanasi, Uttar Pradesh
India

ISSN 2197-9545 ISSN 2197-9553 (electronic)
Springer Geology
ISBN 978-981-10-2476-4 ISBN 978-981-10-2477-1 (eBook)
DOI 10.1007/978-981-10-2477-1

Library of Congress Control Number: 2016952005

© Springer Science+Business Media Singapore 2017
This work is subject to copyright. All rights are reserved by the Publisher, whether the whole or part of the material is concerned, specifically the rights of translation, reprinting, reuse of illustrations, recitation, broadcasting, reproduction on microfilms or in any other physical way, and transmission or information storage and retrieval, electronic adaptation, computer software, or by similar or dissimilar methodology now known or hereafter developed.
The use of general descriptive names, registered names, trademarks, service marks, etc. in this publication does not imply, even in the absence of a specific statement, that such names are exempt from the relevant protective laws and regulations and therefore free for general use.
The publisher, the authors and the editors are safe to assume that the advice and information in this book are believed to be true and accurate at the date of publication. Neither the publisher nor the authors or the editors give a warranty, express or implied, with respect to the material contained herein or for any errors or omissions that may have been made.

Printed on acid-free paper

This Springer imprint is published by Springer Nature
The registered company is Springer Nature Singapore Pte Ltd.
The registered company address is: 152 Beach Road, #22-06/08 Gateway East, Singapore 189721, Singapore

To the memory of my loving revered parents, Shyama and Parashuram, whose human values, upbringing, and guidance transformed me to what I am today at the youthful age of 71, and whatever little substance, endurance, and positivity I could realize in academics and personal life.

Foreword

No other era in the geological history of the earth was more significant and exciting than the Mesozoic. The Permian end witnessed the most pronounced extinction of life, when approximately 90–95 % of marine life and 70 % of terrestrial life perished. Thus in a way, life once again evolved almost from scratch during the Triassic, the oldest period of the Mesozoic. There was proliferation and rapid evolution of conodont animals and of ammonoids, appearance of dinosaurs and coral reefs. Initially, the Triassic sea is believed to have been shallow; it rapidly deepened in Early Triassic, followed by shallowing, particularly after the Carnian.

During the Jurassic, the conodont animals disappeared though preponderance of ammonoids continued. The first bird *Archaeopteryx* appeared, *Pterosaurs* ruled the sky, and other dinosaurs dominated the land, rise in sea level caused transgression in several parts of the peninsula, the sea level kept on fluctuating. There was lush growth of forests.

The Cretaceous marked breakup of the Gondwana land, rise in the sea level and incursion of sea in the peninsular and the Lesser Himalayan parts, major volcanic activity, and another event of extinction of life which included majestic dinosaurs. India, like a Noah's Ark, with its southern fauna and flora, embarked on its long northward journey.

Paleontologic, sedimentologic, paleogeographic, volcanic, basinal, and tectonic aspects of the Indian Mesozoic succession have been studied by various researchers in fairly great detail, but mostly in isolation. What has lacked is identification of mega-sequences and their further differentiation, integration of fauna within the ambit of sequence stratigraphy and erection of regional sequence stratigraphic framework by utilizing the ammonoids in précising the surfaces and correlating these with the regional surfaces and events. An assimilation of updated various geologic elements and their unbiased synthesis to seek whether the evolutionary changes, tectonics, sea level fluctuations, and evolving sedimentologic framework were independent or closely linked and formed part of one story. In other words, erecting the sequence and event stratigraphy based on detailed lithology and paleontology to reconstruct a holistic view of the geological history of the Indian

Mesozoic. This exercise could be attempted only by someone who has an intimate knowledge of various aspects of the Mesozoic history, not only of India but also of the world, particularly the Gondwana Land.

There could be no person better than Prof. Jai Krishna to attempt this kind of synthesis. He has spent more than five decades meticulously studying the Jurassic sequences in India along with his dedicated team. He often educated me with his views regarding the regional evolution of basins inter-alia tectonics and evolution of life. Most impressive was his effort to correlate poorly fossiliferous/ unfossiliferous sequences on the basis of regional sequence stratigraphic context. I am glad that after relinquishing the sponsored teaching and research, Prof. Jai Krishna decided to synthesize various aspects of the Indian Mesozoic inclusive of comparison with other basins between Arabia and Australia. The book "The Indian Mesozoic Chronicle: Sequence Stratigraphic Approach" would be encyclopedic in nature and useful to the students, teachers, researchers, and petroleum geoscientists in the educational institutions and industries.

O.N. Bhargava

Preface

Time to time in the past two decades, my research collaborators in India and abroad, also others passionately involved in the progress of Indian geology, even students in India, particularly at BHU, desired of me a book on my comprehension of the Indian Mesozoic sequences. Near my formal retirement in 2010 came 'The Making of India', an immensely inspiring magnum opus on Indian Geology, written by my revered guru (teacher) Prof. K.S. Valdiya. This also kindled the fire in me to go for an exciting post-retirement innings. Once formally free from teaching and sponsored research in 2014, I began scribbling my comprehension of the Kachchh Mesozoic. As I continued mostly in early morning sessions in Varanasi, Gurgaon, and Bangalore, the canvas of the project went on assuming yet larger dimensions than originally contemplated to engulf the entire Gondwanian Tethyan Margin from Arabia to Australia.

The whole array of integrated progress in the Indian Mesozoic geology, realized since my initiation in the Indian Mesozoic in the mid-1960s, is interwoven with the innovative threadwork of sequence stratigraphy. The key to the study is development of ammonoid-based high-resolution zonal scales in the Indian Triassic and Jurassic comparable to the best of the world, while the Cretaceous is considerably reinforced with the composite records of ammonoids, foraminifers, dinocysts, and nannoplanktons.

Emphasis is placed primarily to high-quality lithostratigraphic differentiation along with meticulous recording of the sections, bed by bed ammonoid collection in the field, and the resultant ammonoid stratigraphic range charts based upon comprehensive taxonomic studies in the laboratory. The cardinal requisite of good stratigraphic refinement has been the differentiation of a large number of stratigraphically precise ammonoid bearing levels, in several fold larger number of beds in the sedimentary column, to ensure the development of near-optimum number of zones/subzones/horizons as the best measure of quality refinement. The ammonoid-based ~25 zone succession in the ~25 my long Bathonian–Tithonian scale of Kachchh is uniquely developed in a single superfamily—Perisphinctacaea, that too near exclusively in the Kachchh basin.

The author's attraction to sequence stratigraphy stemmed primarily from it being an effective powerful tool to unify the highly kaleidoscopic array of apparently disjunct geological facets strewn over large regions of the like of Indian subcontinent. His understanding of sequence stratigraphy has been an integrated refined chronostratigraphically controlled realistic comprehension of rock relationships of genetically related cyclic strata, units, events, and geological phenomena based on sedimentology, paleontology, eustatics, tectonism, magmatism, and many other such interrelated subdisciplines.

The eventful Mesozoic has been the game changer of the earth's Phanerozoic history, especially in the Indian context. Indian Mesozoic studies have been done mostly in isolation at basinal level, only occasionally extending to specific sectors. Compilations lack the very element of life and dynamism. The Cretaceous in Spiti is seldom related to Cauvery, or to that of Kachchh, or Triassic of Kashmir to that in Myanmar. It is this holistic subcontinental approach that has been endeavoured through the application of sequence surface timelines across the subcontinent and even beyond on the Gondwanian Tethyan margin.

The Indian Late Precambrian–Neogene record is organized into five mega-sequences. The fourth among them—also the most important one—spans through the intra-Permian–intra-Paleocene interval from the origin to the closure of the Neotethys. It is further differentiated into three first-order sequences, 35 second-order sequences, and then several second-order sequences into third-, fourth-, and fifth-order ones. The first-order sequences are developed in Spiti Himalaya (intra-Permian to Pliensbachian), in Kachchh (Toarcian to Barremian), and compositely in Cauvery, Spiti, and Kachchh (Aptian–Paleocene), and subsequently extended on either side of India from Arabia to Australia.

The mega-sequence began with outpouring of the Panjal and coeval volcanics. The first-order sequences are found punctuated by major Gondwana dismemberment extensional tectonics and magmatism at the ~183 ma intra-Jurassic Karoo and coeval volcanism with principal manifestation in the west sector, intra-Cretaceous ~126 ma Rajmahal and coeval volcanism with cardinal expression in the east sector as the intervening first-order SBs. The first-order MFSs, respectively, signal the initiation of oceanization in the northwest spreading away of the Aargo block in northeast at the ~159 ma intra-Oxfordian MFS, and farther up the initiation of spreading between India and Madagascar at the ~92 ma intra-Turonian MFS. The mega-sequence formally closed soon after the termination of Deccan magmatic event, yet notional closure of the Neotethys stretched farther into Paleogene collision of India and Asia.

The validation of the developed first/second-order sequence framework in east and west of India on the Gondwanian Tethyan Margin is realized through comparison with Arabia, East Africa, Madagascar, Pakistan, High Himalaya, Indonesia, Papua New Guinea, Timor, West Irian, and NW Australia. The first- and second-order sequence surfaces are consequences of intra-basinal to inter-regional tectonics and geographically restricted to broad tectono-stratigraphically homogeneous regions of the Gondwanian Tethyan Margin in the present study. The resultant sequence surfaces are often found discordant to those of other such regions, for example, to the Eurasian Tethyan Margin framework. On the contrary, the third- and finer order sequences shorter than ~1.5–2.0 my are governed by earth's orbital dynamics, and thus found globally isochronous.

Comparison of the Indian sequence framework also includes important hydrocarbon-producing regions and basins across the Indian divergent margins both in east and west; and based there upon, a highly positive scenario of the Indian Mesozoic hydrocarbon perspective of source/reservoir rocks is outlined in sequence stratigraphic backdrop as an edifice for elaborate evaluation in near future.

The Indian Mesozoic geological developments irrespective of location in east, north, south, or west sectors, whether in intra-cratonic or peri-cratonic rift basins, frontier or foreland basins, marine or non-marine, guide-fossil bearing or devoid, in carbonate or clastic facies, exposed, onshore, or offshore, deep or shallow, in Indian ocean or South Tibetan collision arena are considered genetically related resultants of the same regional tectonics, allowing correlation of units and chronicle of regional events through the all-pervading sequence surface timelines. So time corresponded the magnetic anomalies, origin, climax, and termination of igneous activities, eustatics, climatic and anoxic events, breakup, dispersal, and collision episodes, rifting, transform sliding, and spreading phases in an alternative innovative sequence stratigraphic context. The single all unifying sequence surface ages are found echoed through highly precise SHRIMP zircon, U/Pb, and Ar40/Ar39 radiometric ages, mega/micro faunal/floral records, and magnetic anomalies through the span of the mega-sequence almost equally strong from all the sectors and varied geological settings of the region as unanticipated strong support to interdisciplinary manifestations of a single spreading framework encircling and engulfing the Indian plate.

In the precision chronicle of the multifaceted happenings at sub-zonal resolution and multiples in east, west, north, and south, the plethora of litho-stratigraphic unit names has been largely done away with and rather made redundant. The large intra-Jurassic stratigraphic gap in Kachchh with increase in duration from margin to basin, has been précised in different sections, and its long held interpretation radically revised from subaerial to submarine all over the GTM from Arabia to Australia. Other major gaps of the Indian geological record are also differentiated as subaerial or submarine. All the formations and members throughout the basin have been precisely age ranged. Ammonoid heterochronic evolution, geography, expansion, and migration events back and forth Indian subcontinent during the Jurassic are also outlined in first- and second-order sequence stratigraphic context.

Ammonoids as guide fossils provide a singular advantage to its researchers. Even a single incomplete fragmented specimen when found, much like the aircraft black-box, starts unfolding the geological history like a time machine on the outcrop itself as if the specimen itself is in a narrative conversation with the discoverer. It is these conversations with ammonoids on the outcrops, and discourses on the ammonoid bearing and coeval Indian Mesozoic sequences in the classrooms, seminars, and conferences, that I, in the form of this book, present unto the learned Mesozoic geoscience and allied scholars, researchers, teachers, and professionals across the world.

Varanasi, India Jai Krishna

Acknowledgements

In the backdrop of not so creativity-conducive Indian post-retirement scenario, the book quite unwillingly has been singly written, yet it incorporates and heavily dwells on the published and unpublished data of joint works with the former research scholars and other collaborators.

At the outset, I salute with immense gratitude all my teachers at the Department of Geology, University of Lucknow, where I learnt the alphabets of geology, in particular late Prof. S.N. Singh who sowed in me the first seeds of paleontology and stratigraphy. Professor Singh not only supervised my Ph.D. thesis but also provided me exemplary lessons to meet the challenges of the worldly sphere. I express my indebtedness to the Banaras Hindu University, my colleagues and students there, amidst whom I could progress in the academic arena. I offer my reverence to the land of Kachchh where I carried on my professional field studies, also my thankfulness to Mr. P.H. Bhatti and family of Bhuj for their diverse help over the decades during my numerous stays there.

I thank in particular Profs. Gerd Westermann, James Howard, Raymond Enay, Elie Cariou, Jacques Thierry, Eric Bueffteau, Guillermo Melendez, Indrabir Singh, and Surendra Kumar, Dr. Syed Jafar, Dr. K.P. Jain, Dr. Rahul Garg, Dr. M. Venkateshwarlu, and many others with whom I coauthored publications. I profusely thank my former research students, Dr. Deobrat Pathak, Dr. Bindhyachal Pandey, Dr. Jairam Ojha, and Dr. Nageshwar Dubey who immensely helped me in field and laboratory in our long yet continuing association. Innumerable colleagues in India and abroad are also thanked with whom I have had fruitful interaction during my umpteen sojourns across India, and outside.

During my travels, particularly, in those young impressionable years, I found myself singularly fortunate to have met and interacted with stalwarts of the world geological arena who may have made impressions and germinated seeds on my then young mind, and to whom, of the like of Profs. C. Roy, D.N. Wadia, W.D. West, M.R. Sahni, B.P. Radhakrishna, Rajnath, G.W. Chiplonkar, F.J. Pettijohn, G. Middleton, Tuzo Wilson, A. Gansser, Dennis Shaw, John Talent, T. Matsumato,

C. Egeler, Henry Tintant, B. Ziegler, A. Seilacher, and many others, I very pridefully remember at this juncture.

I am grateful to all the reviewers who in spite of their engaging schedules painstakingly perused the MS and provided valuable comments that helped me improve upon the content, organization, and the MS at large.

Besides being appreciative of the help from several colleagues for consulting them on specific points, or making available valuable reprints, I am pleased to record here with gratitude the exemplary support and multifarious help constantly over the past 3 years from two senior fellow geoscientists, Dr. O.N. Bhargava and Prof. N.C. Ghose. They earnestly desired to see the venture fructify, and I fail to find words in expressing my indebtedness to them.

I appreciate with thankfulness the role played by Mr. Aninda Bose, the first Springer executive that I had the privilege to meet. Only his coolness and endurance let me surge through the time-taking review process. I also appreciate the help received from a few other Springer personnel at different stages of the publication of the book.

I myself, not a computer suave to the desired level, always necessitated assistance of a computer graphic specialist. This work and my many other publications would not have been complete but for the constant support of my Corel graphic specialist Abhishek Jaiswal who has over the last 10 years drawn all the figures/tables of this book and of other publications.

All the Jaikrishnas now numbering ten inclusive of my grandsons and granddaughters have constantly encouraged me in spite of my cutting on the time normally spent with them. Words fail me in acknowledging my wife Neera's exemplary coolness, patience, endurance, positive energy, and servicefulness, and what not that singularly kept me marching forward, particularly in intervals of slow progress in this venture. Timely help by Neeharika, Mohini, and Vladimir in cross-checking the references is gratefully appreciated.

Jai Krishna

Contents

About the Author

Prof. Jai Krishna has more than 50 years of experience in teaching and research in geology. He commenced university teaching and research in 1964, and formally retired in 2011 as Professor and Head, Centre of Advanced Study in Geology, Banaras Hindu University, India. Professor Krishna completed his doctoral studies from the University of Lucknow in 1968 on the West Indian ammonoid bearing Callovian–Albian succession. Professor Krishna carried out postdoctoral studies at McMaster, Canada, jointly with Prof. Gerd Westermann in 1979, 1982, and 1985 on the Jurassics of Kachchh, Jaisalmer, Spiti, and Kumaon Himalaya. During the 1980s and 1990s, Prof. Krishna traveled extensively across the world as Visiting Professor/Scientist/Fellow, inclusive of the coveted Marie Curie Fellowship of the European Economic Community. He realized collaborative researches, participated and chaired conferences, delivered invited lectures, visited stratigraphic sections, consulted type collections in Canada, USA, Argentina, Brazil, France, Germany, Switzerland, Netherland, UK, Spain, Belgium, Italy, Russia, Poland, Austria, Japan, China, Australia, Singapore, Malaysia, Indonesia, Thailand, and Myanmar.

His focus has been ammonoid and sequence stratigraphy based applicative Mesozoic geology, biogeography, and integrated dynamics. Professor Krishna has published about 100 research papers in peer-reviewed reputed journals inclusive of NATURE, more than 50 alone on the Kachchh Jurassic; developed a highly resolved ammonoid zonal scale in the Kachchh Jurassic; formulated comprehensive first- to third-order sequence stratigraphic framework for the Gondwanian Tethyan Margin in India; Spitiian (Late Permian–Pliensbachian), Kachchhian (Toarcian–Barremian), and Dravidian (Aptian–intra-Paleogene) respectively in the north, west, and east sectors. Professor Krishna at present continues to be actively focused on the riddles of the Indian Mesozoics and related hydrocarbon exploratory geological aspects.

Abbreviations

AAP	Aargo Abyssal Plain
ADm	Adhoi member
Am	Member A of Jhurio formation
Ar40/Ar39	Argon 40/Argon39
BCSm	Babia Cliff Sandstone member
BHU	Banaras Hindu University
BLGOm	Badi Lower Golden Oolite member
BM	Basal Member
Bm	B member of Jhurio formation
BNO	Bangong-Co Nujiang ocean
BNSZ	Bangong-Co Nujian Suture Zone
BSm	Bambhanka Shale member
BWLm	Badi White Limestone member
C	Cretaceous
C	Carbon
C10, 20	Cretaceous MFSs of the Arabian plate
CAP	Cuvier Abyssal Plain
CAR	Central Aargo Region
CBCm	Cheria Bet Conglomerate member
C-I to C-XII	Cretaceous second-order sequences
CLm	Canyon Limestone member
Cm	Member of Jhurio formation
CO_2	Carbon Dioxide
COB	Continent Ocean Boundary
DGH	Directorate General of Hydrocarbons
Dm	Dingy member
Dm	Member of Jhurio formation
DOM	Dhosa Oolite Member
DS	Depositional Sequence
DSDP	Deep Sea Drilling Project

DSM	Dhosa Sandstone Member
EAR	East Aargo Region
EECO	Early Eocene Climatic Optimum
Em	E Member of Jhurio formation
EP	Exmouth Plateau
ETM	Eurasian Tethyan Margin
FAD	First Appearance Datum
Fm	Member of Jhurio formation
FSm	Fort Sandstone member
FT	Fission Track
GAP	Gascoyne Abyssal Plain
GBm	Gangta Bet member
GM	Ghuneri Member
Gm	G member of Jhurio formation
GSI	Geological Survey of India
GSM	Gypseous Shale Member
GSm	Gadadha Sandstone member in Khadir
GSm	Gadputa Sandstone member in Patcham
GSm	Gamadu Sandstone member in Wagad
GSSP	Global Boundary Stratotype Section and Points
GTM	Gondwanian Tethyan Margin
GYFm	Goradongar Yellow Flagstone member
H	Habo locality of Kachchh
HG	Hard Ground
HSHm	Hadibhadang Shale member
HSm	Hadibhadang Sandstone member
HST	High-Stand Systems Tract
ICS	International Commission of Stratigraphy
IEAP	Indo-East-African Province
ISJS	International Sub-commission of Jurassic Stratigraphy
ISZ	Indus Suture zone
ITSZ	Indus Tsangpo Suture Zone
J	Jurassic
J	Jhura locality of Kachchh
J 10, 20	Jurassic MFSs on the Arabian plate
J/C	Jurassic/Cretaceous
Ja	Jara locality of Kachchh
JAIS	Jaisalmer
Jf	Jhurio formation
J-I to J-XIV	Jurassic second-order sequences
JLm	Jumara Limestone member
JM	Jhadasa Member
Ju	Jumara locality of Kachchh
K	Keera locality of Kachchh
K/Ar	Potassium/Argon

KAb	Kantkote Ammonoid bed
K-G	Krishna–Godavari
KHF	Katrol Hill fault
KKSSZ	Kohistan-Karakorum-Shyok-Suture-Zone
KLA	Kohistan Ladakh Arc
KM	Kantkote Member
KMF	Kachchh Mainland fault
KMU	Kachchh Mainland Uplift
KSm	Kaladongar Sandstone member
LAB	Lower Astarte Bed
LAD	Last Appearance Datum
LH	Lesser Himalaya
LIP	Large Igneous Province
LM	Lower Member
LPR	Leptosphinctes Rudstone
LSM	Lower Shale Member
LSM	Lower Sandstone Member
LYFm	Lower Yellow Flagstone member
M	Macroconch
m	Microconch
Ma	Million years ago
MECO	Middle Eocene Climatic Optimum
MFI	Maximum Flooding Interval
MFS	Maximum Flooding Surface
MH	Median High
MHm	Modar Hill member
MI 10 to MI 80	Mesozoic first-order SBs of ONGC
MII	Mesozoic second-order SBs of ONGC
MIII	Mesozoic third-order SBs of ONGC
MLJT	Middle/Late Jurassic Transition
MSm	Middle Sandstone member
My	Million years by way of duration
NARM	Narmada
NC	Nannoplankton zones in Cretaceous
NJ	Nannoplankton zones in Jurassic
NLm	Nara Limestone member
NSm	Nara Shale member in Wagad
OAE	Oceanic Anoxic Event
ODP	Ocean Drilling Program
ONGC	Oil and Natural Gas Corporation
P/PF	Planktonic Foraminfer Zone
P/T	Permian/Triassic
PAP	Perth Abyssal Plain
PSm	Patasar Shale member
PSm	Purple Sandstone member

RLm	Raimalro Limestone member
RSM	Ridge Sandstone Member
RST	Regressive Systems Tract
SAUR	Saurashtra
SB	Sequence Boundary
SBZ	Shallow Benthic Zones
SCLm	Sadhara Coral Limestone member
T–I to T–X	Triassic second-order sequences
T/J	Triassic/Jurassic
T/Tr	Triassic
TOC	Total Organic Content
Tr 10, 20	Triassic MFSs of the Arabian plate
TST	Transgressive Systems Tract
UAb	Upper Astarte bed
UM	Ukra Member
U/Pb	Uranium/Lead
ULM	Upper Limestone Member
WAR	West Aargo Region
WBZP	Walby Zenith Plateau
WIMP	West Irian Micro Plate
WP	Wallby plateau
WZFZ	Walby Zenith Fracture Zone
YTSZ	Yarlung Tsangpo Suture Zon
ZP	Zenith plateau

List of Figures

List of Tables

Contents of Field Photographs

Chapter 1
Introduction and Paleogeographic Context, Previous Work, High-Resolution Scale, Magnetochronologic Perspective, Radiometric Scenario, Igneous Activities, Anoxic Events and Eustatic Fluctuations

Abstract The chapter outlines the multidisciplinary geodynamic transformations in the Mesozoic with focus on the evolution of India as part of the Gondwana dismemberment. The rationale of approaching the chronicle through the innovative sequence stratigraphic approach is emphasized upon through formulation of depositional sequences in the key basins of Spiti in north, Kachchh in west and Cauvery in southeast, respectively, for the Triassic, Jurassic and Cretaceous intervals between the origin to the closure of the Neotethys. Paramount significant is given to ammonoids in developing and updating the integration towards high-resolution chronicle. Updates are also outlined on important aspects of tectonics, volcanism, tectono-sedimento-stratigraphy, eustatics, anoxicism, geomagnetism, radiometry, organic evolution, and others, along with their interrelationships.

Keywords Mesozoic · Geodynamics · Paleogeography · Gondwana · Neotethys · Indian Ocean · Gondwanian tethyan margin (GTM) · Eurasian tethyan margin (ETM) · Gondic and Dravidian corridors · Igneous events · Oceanic anoxic events (OAEs) · Sequence stratigraphy · Spiti · Kachchh · Cauvery · Magnetostratigraphy · Radiometry · Ammonoids · Indo-East-African province (IEAP) · Eustatics · Tectonics

1.1 Introduction and Paleogeographic Context

The Mesozoic interval holds paramount significance in Earth history. Important geological concepts developed during the study of the Mesozoic record across the globe, particularly in West Europe. The science of stratigraphy grew with formulation of the concept of the order of superposition, stages, zones, horizons, paleogeographic reconstructions and many others in context of the study of the Mesozoic. The varied progress of the human civilization for well over a century has

© Springer Science+Business Media Singapore 2017
J. Krishna, *The Indian Mesozoic Chronicle*, Springer Geology,
DOI 10.1007/978-981-10-2477-1_1

been motored by the hydrocarbon energy that is principally sourced in the Mesozoic sediments. Stupendous geological and geographical changes occurred during the Mesozoic.

The geodynamic changes in turn transformed the globe quite close to what we find it today (Fig. 1.1a–e).

The transformations contextual to India were even more spectacular. As a Gondwana constituent in the south hemisphere at the start, India was neighboured by land, practically, from three sides except for its Gondwanian Tethyan Margin (GTM, Fig. 1.2) in the north, which, since the start of the Late Permian, faced the onslaught of the Neotethys. The GTM extended from Arabia in west to Australia in east with East Africa, Madagascar, Pakistan, West India, Higher Himalaya, New Guinea and Sula in between the two extremes, while the north margin of the Neotethys represented the Eurasian Tethyan Margin (ETM).

Soon after the close of the Mesozoic disappeared the Neotethys, and India became a distinguished constituent of Asia. During the Mesozoic, there developed the transient shallow marine Gondic and Dravidian corridor seaways (Krishna 1987a, b, 1996, 2009) on account of prolonged rifting and eustatic rise in the east and west divergent sectors as precursors to the subsequent origin of the Indian Ocean (Fig. 1.3).

Igneous events punctuated the Mesozoic dismemberment history in the region as Panjal, Lodhika (Gujarat), Myanmar, Rajmahal, St. Marys and Deccan volcanism. The peaking of the Panjal volcanism heralded the Mesozoic which closed with the climaxing of the Deccan event. The intervening Lodhika and Rajmahal events punctuated the major rift/drift events in the west and east divergent sectors in India's struggle to emancipation from the Gondwana. The Lodhika (Gujarat) and Myanmar igneous events respectively signalled the initiation of localized geographically restricted oceanization in the east and west. The Rajmahal volcanism symbolized complete oceanic separation from Australo-Antarctica, while the St. Marys volcanism marked the final oceanic isolation of India in all the four directions, and its eventual rapid escape from the clutches of its Gondwanian captors. The tectonic and igneous events guided the entire course of India's geodynamic evolution, so also of the Gondwana, and of the Pangea at large, through sea-level changes, intricate tectono-sedimento-stratigraphic mosaic, oceanic anoxism, energy resources, etc. that are inscripted in the corresponding geological record. The said Gondwanian age also witnessed the transformation of the Neotethyan ocean into the snow clad majestic Himalaya—the highest mountain system of the world in the north at the expense of the erstwhile equally high Gondwanian mountain chain in the south and east of India into the seamless and ever growing Indian Ocean. The Indian coast lines developed as persistent features during the fragmentation process. The Gondwana fragmentation culminated in India's liberation and its subsequent voyage for several thousand kilometres across the equator. Medially evolved terrestrial and marine biota flourished during the

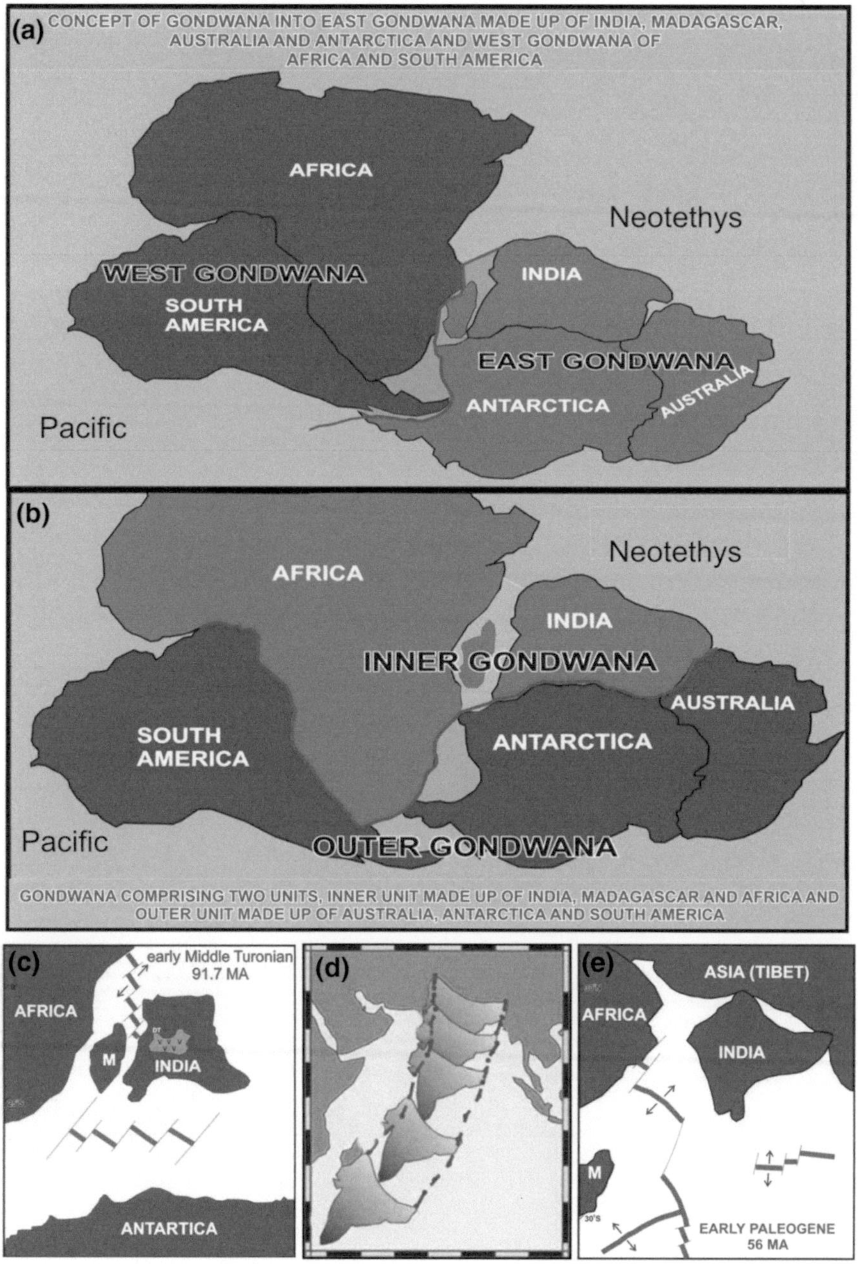

Fig. 1.1 **a**, **b** Gondwana at the intra-Permian start of the mega-sequence, **c** oceanic separation of Madagascar from India **d** schematic drift of India after separation of Madagascar and union with Asia, **e** union of India in Asia in Paleogene (modified after Krishna 1986, 1987a, b)

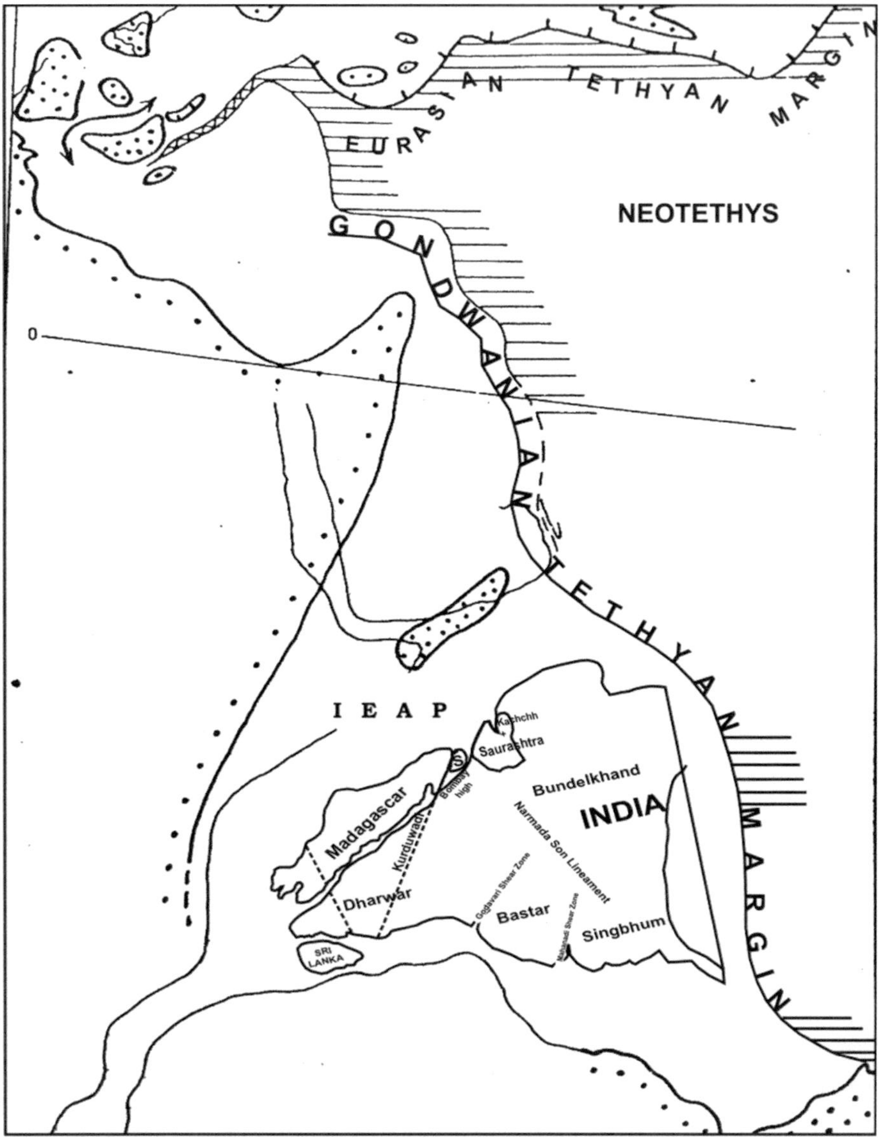

Fig. 1.2 Neotethys with GTM and ETM during the Jurassic (modified after Krishna and Cariou 1990)

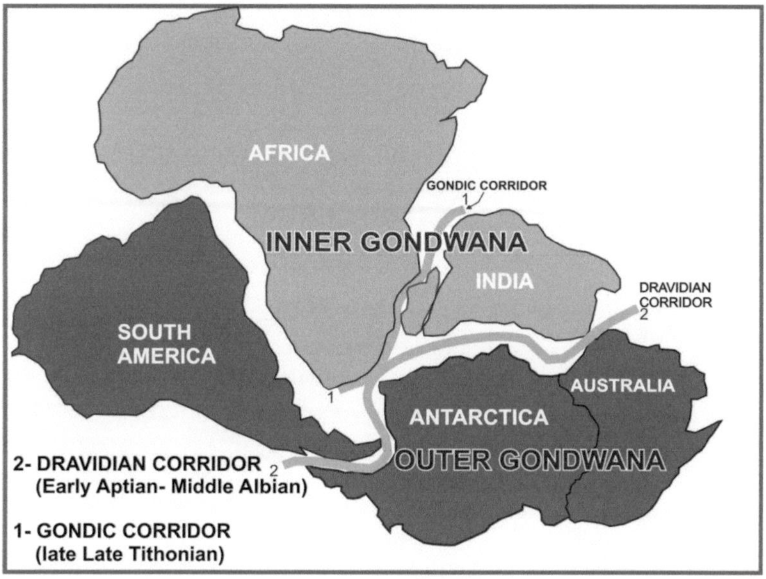

Fig. 1.3 Gondic and Dravidian shallow marine corridors sea-ways (modified after Krishna 1987a, b)

Mesozoic, and then mostly perished at the close of the interval to pave way for more evolved and better sustainable life patterns. Dinosaurs—the giant rulers of the land across the globe—could not survive the sudden and drastic environmental degradation at the close of the Mesozoic, so also vanished the ammonoids that commanded the world's shallow seas.

1.2 Rationale of the Study

The rationale of the refinement of the rich Indian Mesozoic geological record and geodynamic chronicle, as also its regional comprehension undertaken here based on the concept of sequence stratigraphy, cannot be over emphasized. The interdisciplinary events have been to some extent time précised, yet not genetically linked to the finely layered sedimentary record in context of the succession of depositional sequences. The basins of Spiti in north, Kachchh in west, and Cauvery in the east/south located close to the respective action theatres witnessed many of the above major regional and global geological events. The signatures and evidences of the said events are well imprinted in the said three and other Indian sedimentary basins. The author has over the past five decades primarily focused on Kachchh and

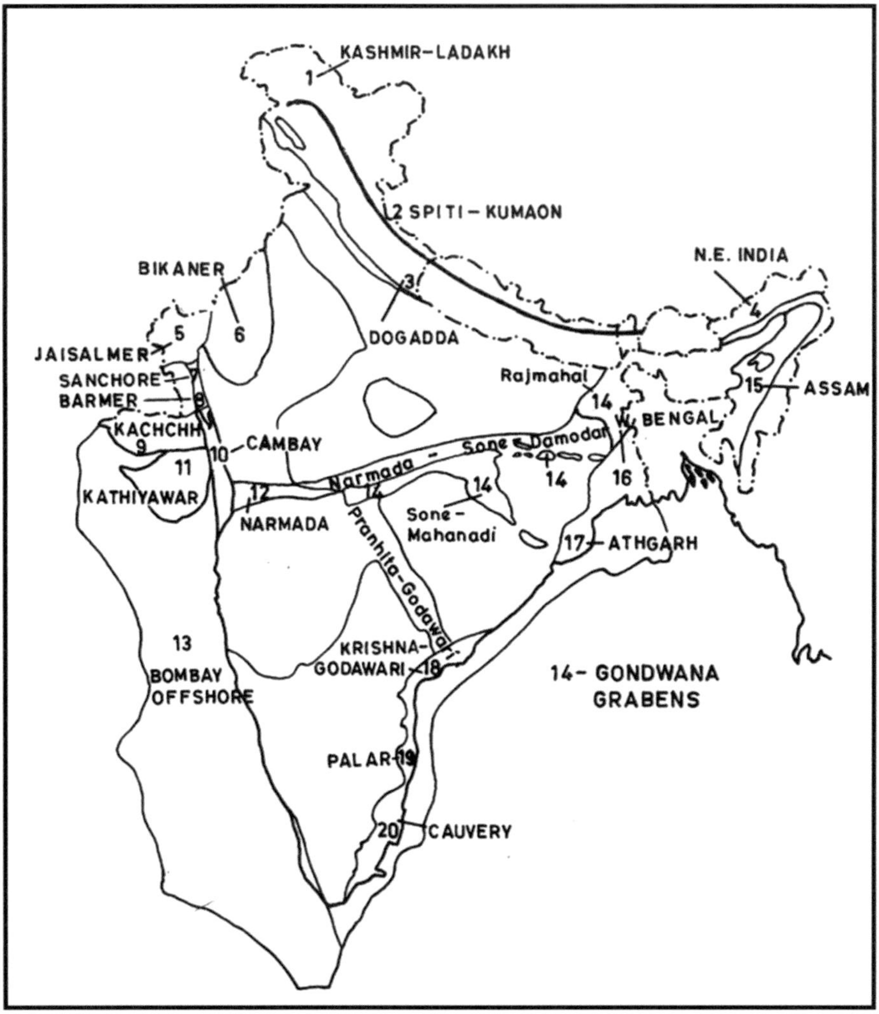

Fig. 1.4 Indian sedimentary basins (modified after Bhandari 1983)

neighbourhood, yet also remained intermittently and secondarily interested in Spiti and Cauvery basins (Figs. 1.4, 1.5, 1.6 and 1.7). The cumulative studies in association of his research group at the Banaras Hindu University (BHU) make the foundation of the present work.

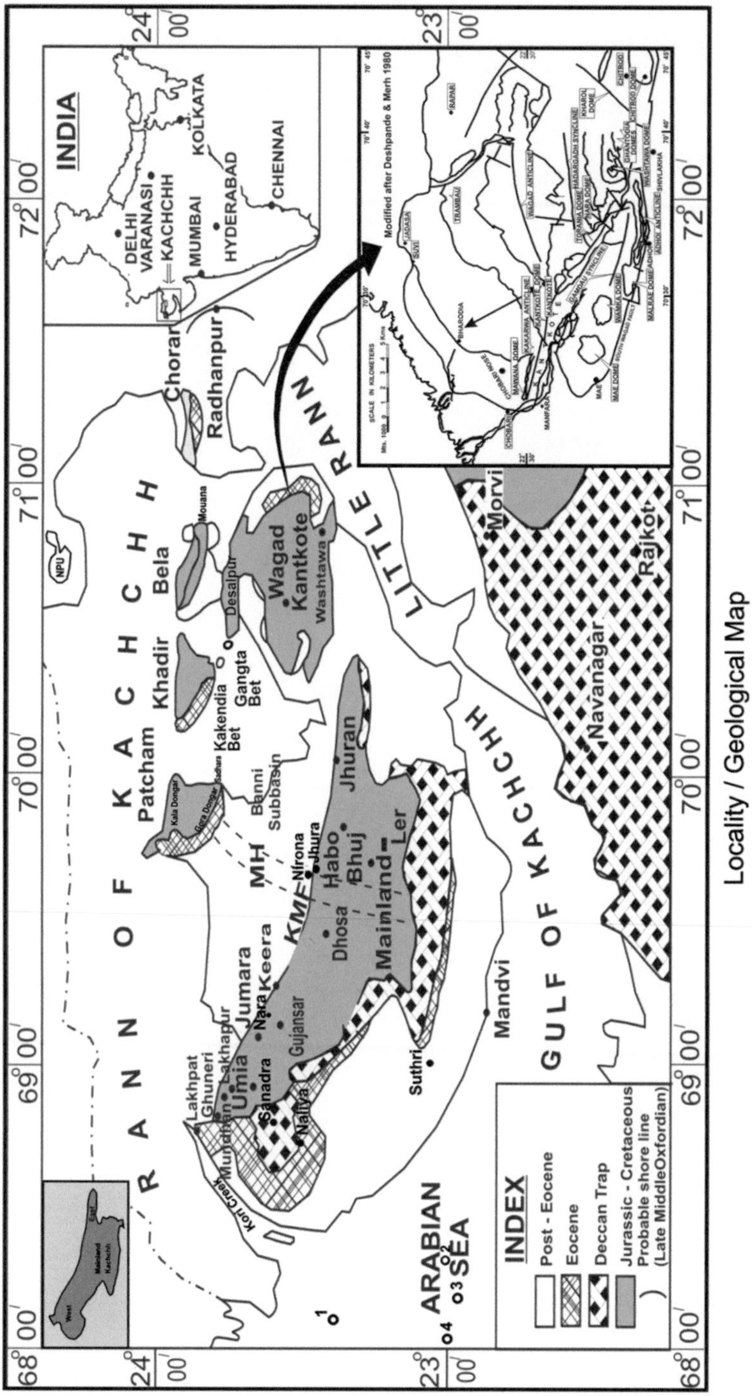

Fig. 1.5 Schematic geological map of Kachchh with principal localities

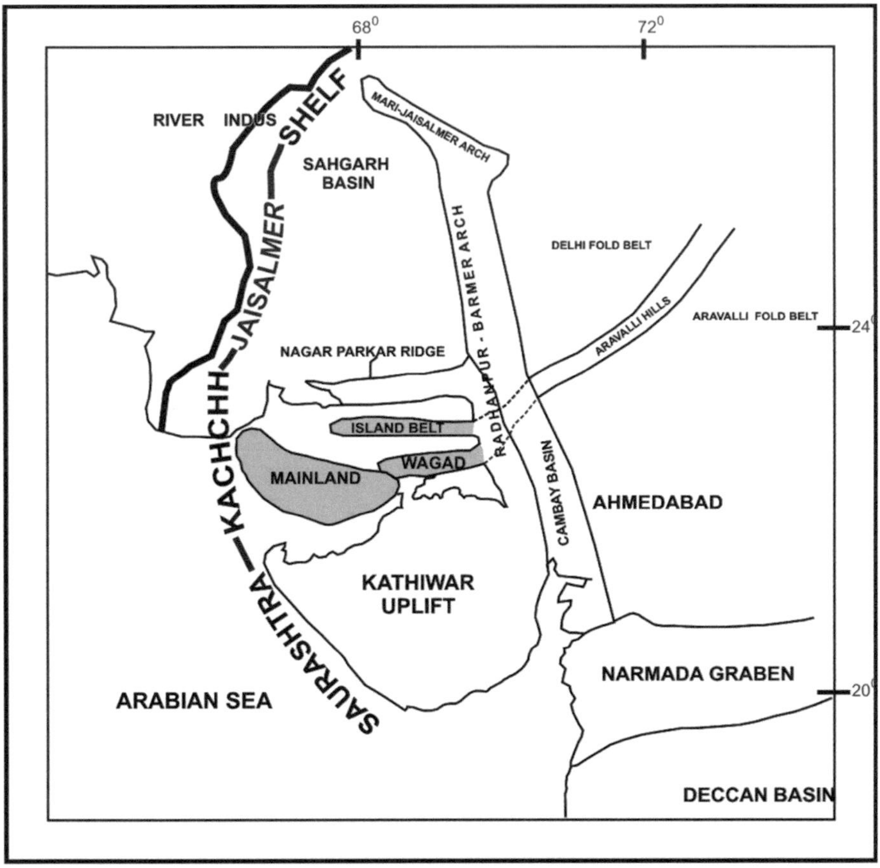

Fig. 1.6 Elements of Saurashtra–Kachchh shelf inclusive of Kachchh basin modified after Biswas (1993)

1.3 Previous Work

Inspite of enveloping a rather large spatio-temporal canvas, the principal focus in this work is on the Kachchh Mesozoic sedimentary record, and previous work is reviewed only in Kachchh, that too briefly, as an example for other Indian basins. Further, the review centres on the development of the ammonoid-based high-resolution scale in the Jurassic, while the associated equally important chronology of igneous events is developed in the subsequent chapters.

Reference to Kachchh existed in several thousand year old ancient Indian scriptures, so named for its turtle-like broad shape (inset in the Fig. 1.5 in the Kachchhi language tortoise means Kachchh). The geoscience community in Europe took notice of Kachchh after the 1819 major earthquake in context of the significant geo-changes that the region underwent. Accounts of the said earthquake have been

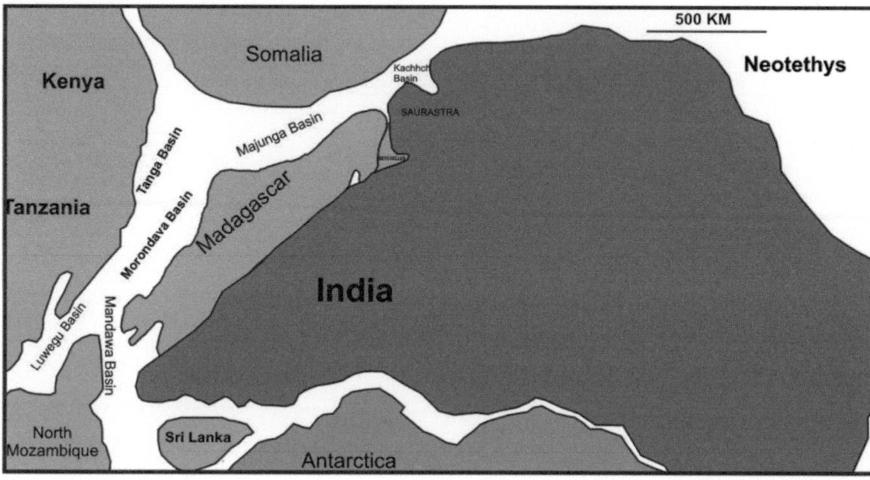

Fig. 1.7 Gondwana constituents excluding Australia and South America at the base of Toarcian (~183 ma) (modified after Reeves 2014)

given by Lyell (1853). Sykes was the first to report the Mesozoic fossil finds from Kachchh in the 1930s that were collected by Captain Smee of British Army. Later Sowerby (1840) systematically described these fossils, which included the Jurassic ammonoids as well. Grant (1837) provided the first geological account of Kachchh along with a map. After three decades, Blanford (1869) provided the first meaningful description of Kachchh Geology. Wynne and Fedden during 1868–72 systematically mapped the entire Kachchh, which included a detailed geological framework by Wynne (1872). The publication formed a good edifice for later studies. Wynne (1872) differentiated the older Lower Jurassic marine fossiliferous rocks from the younger Upper Jurassic plant fossil bearing rocks in his text, and also delineated these in the map. He considered his Lower Group as the equivalent of the European Dogger, since the same did include the full Dogger in addition of the youngest Lias and full Malm (the early four Jurassic stages grouped as Lias—Hettangian, Sinemurian, Pliensbachian and Toarcian; Dogger made up of Aalenian, Bajocian, Bathonian and Callovian; and Malm comprising Oxfordian, Kimmeridgian and Tithonian)), while his Upper Group turned out to be of Early Cretaceous age.

In 1867, Stoliczka, a geologist in his thirties of the then famous Vienna school, visited Kachchh. As per his unpublished field diary, Stoliczka, based on his field studies, distinguished four stratigraphic units in the Kachchh Mesozoic as Patcham Group, Chari (earlier spelled as Charee) Group, Katrol Group and Umia (earlier spelled as Oomea) Group in ascending order (Table 1.1). The scheme was later incorporated in Waagen (1871).

In his fourfold differentiation, Stoliczka was apparently guided by the spectacular limestone development in the upper part of the Patcham Group in the 'Island

Table 1.1 Lithostratigraphic framework of Stoliczka and Biswas in the Mesozoic of Kachchh

Stoliczka units with ages revised here	Age	BISWAS (1971)		
		E. KUTCH	PACHHAM	MAIN LAND
UMIA FN.	ALBIAN TO ? SANTONIAN	Not Exposed		BHUJ FORMATION: UKRA MEMBER
	APTIAN			
	BERRIASIAN TO BARREMIAN			JHURAN FORMATION
KATROL FN.	TITHONIAN TO KIMMERIDGIAN	WAGED SANDSTONE KANTHKOT MEMBER		
CHARI FN.	OXFORDIAN TO CALLOVIAN	WASHTAWA FORMATION	Not Exposed	DHOSA OOLITE MEMBER JUMARA FORMATION
		KHADIR FORMATION		
PATCHAM FN.	BATHONIAN		GORADONGAR FORMATION	JHURIO FORMATION
			KALADONGAR FORMATION	Not Exposed
		PRECAMBRIAN BASEMENT EXPOSED AT MERUDA HILL	Not Exposed	

belt', particularly, at Patcham that was followed up by thick richly fossiliferous shales all over the Mainland particularly so in the neighbourhood of Charee–Fulae as Chari Group with marker Dhosa Oolite at the top. Farther up, he was presumably influenced by the poorly fossiliferous shale/sand alternations in the Katrol hill region, while the differentiation of the youngest sand dominating Umia Group was, besides gross lithology, additionally, constrained by the green glauconitic marker bed in the west Mainalnd, being so typified at Umia at the base. Due to his untimely death in 1868 while working in High Himalaya, his preliminary studies remained unpublished. The above-mentioned units were essentially lithostratigraphic in character, although in those days differentiation into litho, bio and chronostratigraphic units was not practiced. The practice commenced only in the latter half of the twentieth century. Later, Waagen (1871) incorporated Stoliczka's fourfold stratigraphic framework. The present author and numerous other workers in India and abroad have continued to follow the stratigraphic units of Stoliczka with minor changes/clarifications. However, many others including geologists of the ONGC, instead of following Stolickza's classification, erected separate and new units for the three sectors of the basin, namely Mainland, 'Island belt' and Wagad (Table 1.1—Biswas 1977), which have been found untenable as discussed in sequel. The succession in Mainland Kachchh comprised the Jhurio Formation, Jumara Formation, Jhuran Formation and Bhuj Formation. In the opinion of the present author, the fundamental dent to Kachchh Mesozoic stratigraphy was made by Stoliczka as a solid foundation for all time to come, which would be impossible to erase. Waagen (1873–75), without ever visiting Kachchh, based on Stoliczka's collections, published ammonoid systematic and stratigraphic results. He commented that Stoliczka's stratigraphic units did not exactly correspond with the units then recognized in Europe, yet he confirmed the presence in the Kachchh Jurassic of most of the European ammonoid evolutionary succession and assemblages. Waagen (1873–75) organized the ammonoid succession of Kachchh Chari and Katrol units into nine European zonal intervals from the start of the Kelloway Group, through the Oxford Group to the close of Kimmeridge Group. In addition, he corresponded the Patcham Group, the oldest of Kachchh to Bath Group of Europe, while the youngest Umia Group was correlated to the Tithon Group of Europe and also to Early Cretaceous. Spath (1924, 1927–33) revised and furthered Waagen's work using Stoliczka's units. Spath (1924, 1927–33) developed a succession of 16 ammonoid faunas in the Kachchh Jurassic in comparison to 9 of Waagen (in Pandey and Dave 1993). Besides studying the collections of Blake and Smeeth that were available to him in the British Museum, Spath also used Rajnath's bed-by-bed ammonoid collections in Jumara that he had carried along for doctoral studies in London. Blake also had made collections, which due to his untimely death could not be studied. Meanwhile, comprehensive accounts on bivalves (Cox 1940), brachiopods (Kitchin 1893–1900) and corals (Gregory 1893) and echinoids (Gregory 1900) were also published.

In the late 1920s began researches on the Kachchh Mesozoic in far off BHU (Banaras Hindu University). BHU was founded in 1916 as a nationalist university to scuttle the British out of India. Its Department of Geology was established a few years later in 1923, while the first M.Sc. in Geology came out in 1925. What prompted such strange initiative in a time of acute resource constraints is difficult to surmise. The then ruler of Kachchh made a special request to the founder Vice Chancellor of BHU—Madan Mohan Malaviya, one of the foremost champions of India's freedom movement against the British rule. The request was for sending geological expertise towards geological evaluation of the mineral and fossil wealth of Kachchh, which the then ruler of Kachchh thought the British were fast plundering. Rajnath initiated comprehensive researches on the Kachchh Mesozoic in 1927 at BHU. Since then the researches on Kachchh at BHU not only have continued but have far outweighed any other Indian or foreign research endeavours on the Kachchh Mesozoic. Rajnath, then merely a fresh M.Sc. pass out, also the first Indian, at the direction of the then Vice Chancellor of BHU, proceeded to Kachchh for comprehensive studies on the Mesozoic stratigraphy of Kachchh. Rajnath in his very first visit travelled widely in Mainland of the marshy and desert country of Kachchh for several months in 1927. He was the first to work out a detailed succession of beds in a stratigraphic section (Rajnath 1932—Table 1.2) in Kachchh

Table 1.2 Rajnath's 26 beds succession in Jumara—first effort of details differentiation in Kachchh Jurassic units

Rajnath 1932			
Kantkote	sandstone, not visited		
	Bed No.1	Green (Upper) Dhosa Oolite Brown (Lower) Dhosa oolite	Argovian
	" 1 a	Dhosa oolite	
	" 2		
	" 3		
	" 4		
	" 5		
	" 6		
	" 7	Zeilleria zone	
	" 7a		
Upper Shales greenish	" 8		
	" 9		
	" 10		Callovian+
	" 11		
	" 12		
	" 13		
	" 14		
Lower Shales greenish	" 15		
	" 16		
	" 17		
	" 18		
	" 19		
	" 20		
	" 21		
	Bed No. 22		
	" 23		
	" 24		
	" 25		Bathonian
	" 26		
	(Lower beds exposed at Patcham not examined)		

with thicknesses of beds of Patcham Formation and Chari Formation in Jumara dome in West Mainland ~100 km west of Bhuj—the capital town of the Kachchh state, along with a detailed map of the Jumara dome as also of West Mainland. Rajnath's famous succession of beds comprised 26 beds, the older 5 of Patcham Formation and the younger 21 of Chari Formation.

By demonstrating the mappability of the Stoliczka'units (Table 1.1), Rajnath provided ample credence to Stoliczka's stratigraphic scheme. Rajnath also collected ammonoids and other mega-invertebrates from most of his 26 bed succession. Opportunity came forth soon to Rajnath for doctoral studies in London. There in the name of urgency of bringing out significant results from the Kachchh Jurassic, Spath, at the British Museum of Natural history, who had already established himself as the leading Mesozoic ammonoid worker of the world as also of Kachchh, used Rajnath's field notes and ammonoid material in his monograph on the Kachchh Jurassic ammonoids (Spath 1927–33). Rajnath had to satisfy himself with a D.Sc. on his geologically not so significant brachipod material. Apart from many of his significant contributions on the Kachchh Mesozoic stratigraphy, he inspired several younger colleagues and students at BHU for studies in Kachchh. Five generations have passed, yet researches at BHU on the Kachchh Mesozoic have continued uninterruptedly for about nine decades (1927 onward). Pascoe (1959) reinterpreted and summarized the Kachchh Mesozoic sedimentary succession into 11 sediment intervals; two in Patcham unit, five in Chari unit and four in major part of Katrol unit, and ranged them from Bathonian to Kimmeridgian. Post independence, many other Indian research centres also joined efforts to resolve and refine the Kachchh Mesozoic stratigraphy. Apart from ammonoids, bivalves and brachiopods, studies on micro-faunal/floral groups have also come up, in particular context of the hydrocarbon exploratory geological studies both on surface and wells, onshore and offshore. However, these works apart from vast amount of data generation in numerous sections could not make the much desired progress in the refinement of Kachchh Jurassic stratigraphy. It was primarily due to neglect of any comprehensive program of guide fossil (ammonoids, dinoflagellate and coccoliths) studies in the lithostratigraphically well-differentiated sections.

Biswas (1971, 1977, 1993, 2012) proposed a new nomenclature of stratigraphic units for Mainland, Island belt and Wagad (Table 1.2), which the ONGC (Oil and Natural Gas Corporation) continues to follow in its exploration activities. Biswas' work realized quantum progress in the Kachchh Jurassic lithostratigraphy, structure, and tectonics. However, not much integrated progress could be made in spite of the 3-D facies analysis, architecture and correlations. In absence of even stage level time planes, these studies resulted in faulty ages, correlations and unrealistic geological modeling as (discussed in Chaps. 2 and 4). Unpublished and published skeletal sequence stratigraphic frameworks have also emerged (ONGC 1993, 2008; Dave 2011, 2012), commented upon later in Chap. 3 on sequence stratigraphy.

The ONGC geologists inclusive of Biswas did collect ammonoids and other fossils (personal oral communication) which, however, could never be studied. Numerous microfossil studies on foraminifers, ostracods, dinoflagellates, coccoliths, spores and pollens in outcrop sections and wells have been made. These in view of invariably long and poorly known ranges of the benthic foraminifers, dinoflagellates, ostracods, pollen/spores fell short of refining the Kachchh Jurassic strtigraphy (discussed in detail later in Chap. 4) while the coccolith studies undertaken only in bits. The said studies also lacked integration with ammonoid biostratigraphy. This is crucially needed in present times as already accomplished in Europe (summarized in Cariou and Hantzpergue 1997).

The present author undertook his first field visit to Kachchh only in 1980. Thus began his endeavour towards bringing the Kachchh Jurassic stratigraphy to international standard. He started with fresh detailed measurement and record of the undeclared stratotype sections of Stoliczka's four stratigraphic units (Krishna 1987c), which Stoliczka's untimely death had earlier prevented from formal proposition and description. Similar studies were extended in many other sections

(Krishna and coworkers since 1982). Emphasis was laid on the stratigraphical differentiation of maximum possible number of stratigraphcally precise ammonoid levels. Most of the sections were meticulously measured under sponsored researches with the assistance of the project team members (Drs. Pathak, Pandey, Ojha, and Dubey, who still continue as research associates). It resulted in several fold better litho/biostraigraphic differentiation. Krishna's exposure over the last three decades to European centres of Jurassic studies greatly facilitated his understanding of the usefulness of the sequence stratigraphic studies in his modern multidisciplinary integrative studies in Kachchh Mesozoic towards objective intra-basinal to regional geological comprehension. Spanning over two decades, indigenous ammonoid zonal schemes for the younger five ammonoid bearing stages were established (Krishna 1983a, b, 1991, 2002, 2004a, b, c, 2005, 2009; unpublished reports 1994, 2014; Krishna et al. 1994a, b, c, 1995a, b, 1996a, b, c, 1998, 2000, 2009a, b and in the present work on the Oxfordian and early part of Kimmeridgian, Ojha 1996 and Krishna and Ojha 1996, 2000 on the Callovian, Pathak 1989; Krishna and Pathak 1989, 1991, 1993, 1994; Pandey 1993 on the Kimmerdgian, and Tithonian stages; Pandey and Pathak 2015; Pandey et al. 2013a, b on the Bathonian stage, and others). It is hoped that parallel schemes based on coccoliths and dinoflagellates and others fossils would soon emerge with integration to ammonoid biostrtigraphy. Dubey (1992 unpublished Ph.D thesis, Dubey and Chatterjee 1997, and others) focused on the sedimentology, facies and depositional framework of the Cretaceous Umia Formation. The ammonoid-based high-resolution scale facilitated development of mega- to third-order sequence stratigraphic schemes as a model for the entire GTM (Krishna 2002, 2005; Krishna and Pathak 1994; Krishna et al. 2010, 2011) to in turn facilitate the spatio-temporal constraints and exploitation of hydrocarbon source and reservoir rocks in progress of the nation.

Another group headed by Fursich (numerous valuable publications since the mid-1990s listed later under individual chapters) started benthic invertebrate-based paleoenvironmental and paleoecological studies in the Kachchh Jurassic. The group has made rapid progress in diverse directions. However, in some works in absence of precise age assignments, the models proposed could not achieve the precise temporal control that the models deserved. Fursich's group later also added ammonoid workers to revise and refine their earlier age frameworks. Ages to Kachchh units assigned by Fursich and his group have been discussed at length later in Chap. 4.

Raju (2008, 2011) summarized the Kachchh Cretaceous stratigraphic framework. Table 6 of Raju (2011) ranges the Umia Formation from Late Berriasian to close of Albian, Bhuj Formation of Biswas (1971, 1977, 1993) from Cenomanian to Campanian, Naliya Formation to Santonian, Kori Formation from Santonian to Campanian and Anjar Formation to Middle Maestrichtian. In the text (Raju 2011), the Bhuj Formation is ranged from Neocomian to Santonian, Naliya Formation from Coniacian to Santonian, Kori Formation from Coniacian to Maestrichtian and Mundra Formation to Santonian. In an earlier table Raju (2008) showed the Anjar Formation to range from Early to Middle Maestrichtian, Bhuj Formation from

Cenomanian to Campanian and Kori Formation from Santonian to Campanian. In Raju's successive publications, discrepancies in ages of the Cretaceous units and stratigraphic gaps in other Indian basins are present. Creation of new Jurassic and Cretaceous stages in Kachchh and elsewhere in India (Pandey and Dave 1993; Prasad et al. 1996, and others) is unnecessary and has been commented upon in Chap. 4.

Inclusive of the numerous wells drilled, the Jurassic stratigraphic record of Kachchh is now almost complete. The first three stages without intra or interstage differentiation are present in non-marine palynofossil bearing clastic facies in wells, while the younger eight stages are known in outcrops through guide fossils. The presence of Toarcian and Aalenian has been confirmed through recent coccolith records (Rai and Jain 2012, 2013). The younger six stages from Bajocian to Tithonian are known through succession of ammonoids.

In recent years, two books on Indian geology (Ramkrishna and Vaidyanathan 2008; Valdiya 2010, 2016) have comprehensively reviewed Indian geology inclusive of the Mesozoics.

1.4 Development of a Scale

The prime objective of the present work has been the integration of the interdisciplinary data towards high-resolution objective geological comprehension. True measure of any such attempt is the resolution of the scale used. Independent high-resolution ammonoid-based zonal scales could not be developed for long for any Mesozoic geological system in India. It took nearly two decades to develop and perfect a refined time scale in the Kachchh Jurassic (Fig. 1.8) based on ammonoids that are considered the all time fastest evolving organisms. Incidentally, the developed Kachchh Jurassic scale is also biochronologically the highest resolved of the entire exposed Indian geological record from Cambrian to Neogene. Excellently resolved ammonoid zonal scale for the Triassic in India has been developed in Spiti basin (Bhargava et al. 2004; Krystyn et al. 2004). The progress and development in the Indian Cretaceous stratigraphy are also summarized here, using the composite ammonoid and planktonic-based state of the art (Ayyasami 2011; Raju 2008, 2011; Singh 2006 and others), yet short of a formal scale. Nevertheless, the geological integration here is principally focused and in a large measure based on the Kachchh basin, through further extension and integration to other basins.

1.5 Magnetostratigraphic Perspective

During the Mesozoic, the magnetic polarity scale even in global context continues to be far from satisfactory. Only some information is available for the terminal and early part of Cretaceous and Late Jurassic (Figs. 1.9, 1.10, 1.11, 1.12 and 1.13). Information from Triassic, Early and Middle Jurassic is meager and fragmentary,

Fig. 1.8 Ammonoid zonal scale for the Jurassic in IEAP developed in Kachchh

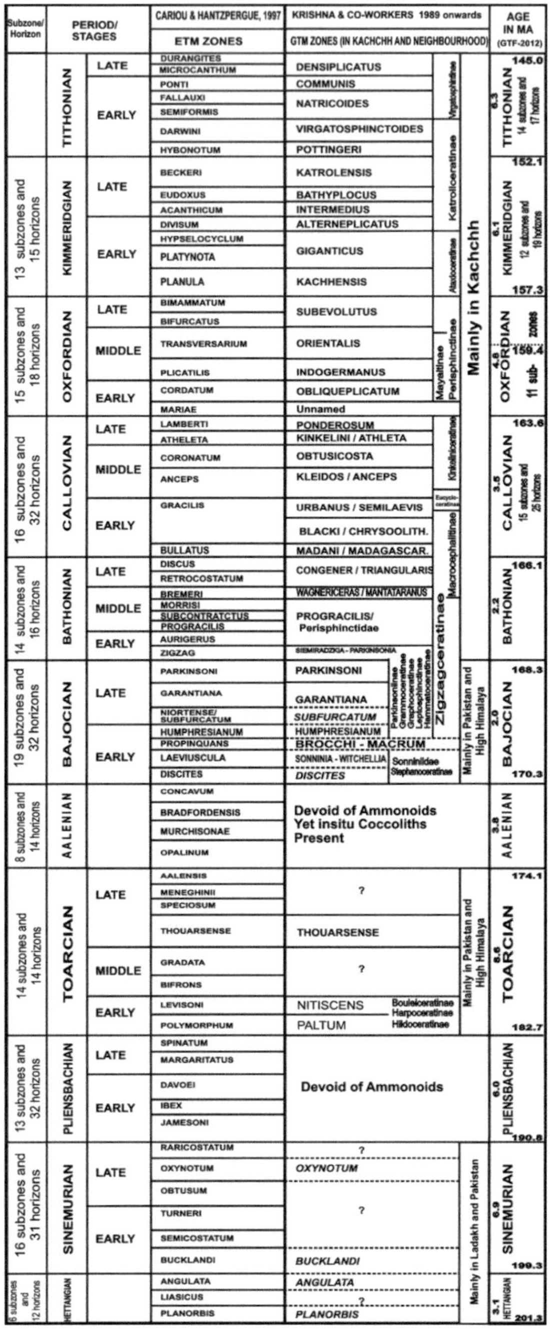

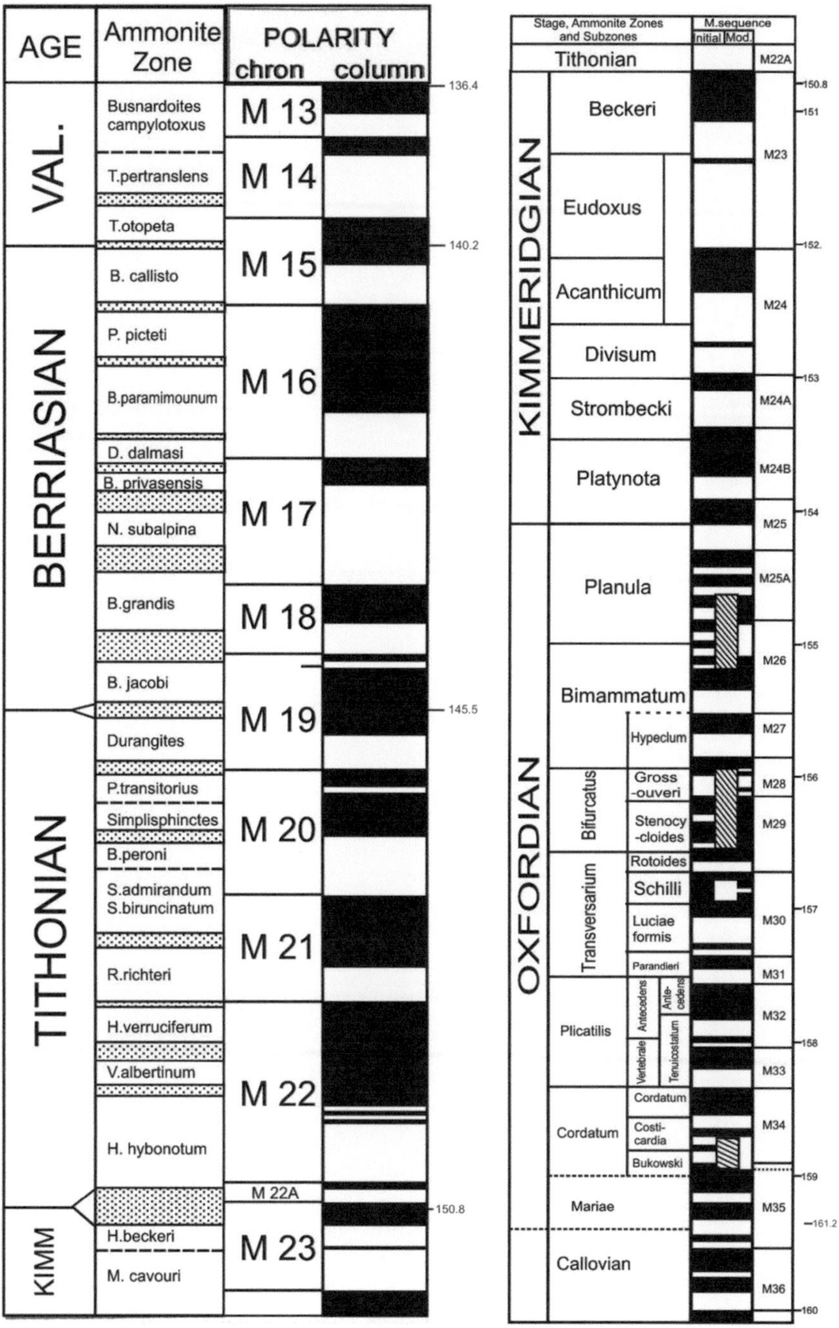

Fig. 1.9 International magnetic polarity chrons from Callovian to Valanginian interval modified after Gradstein et al. (2005)

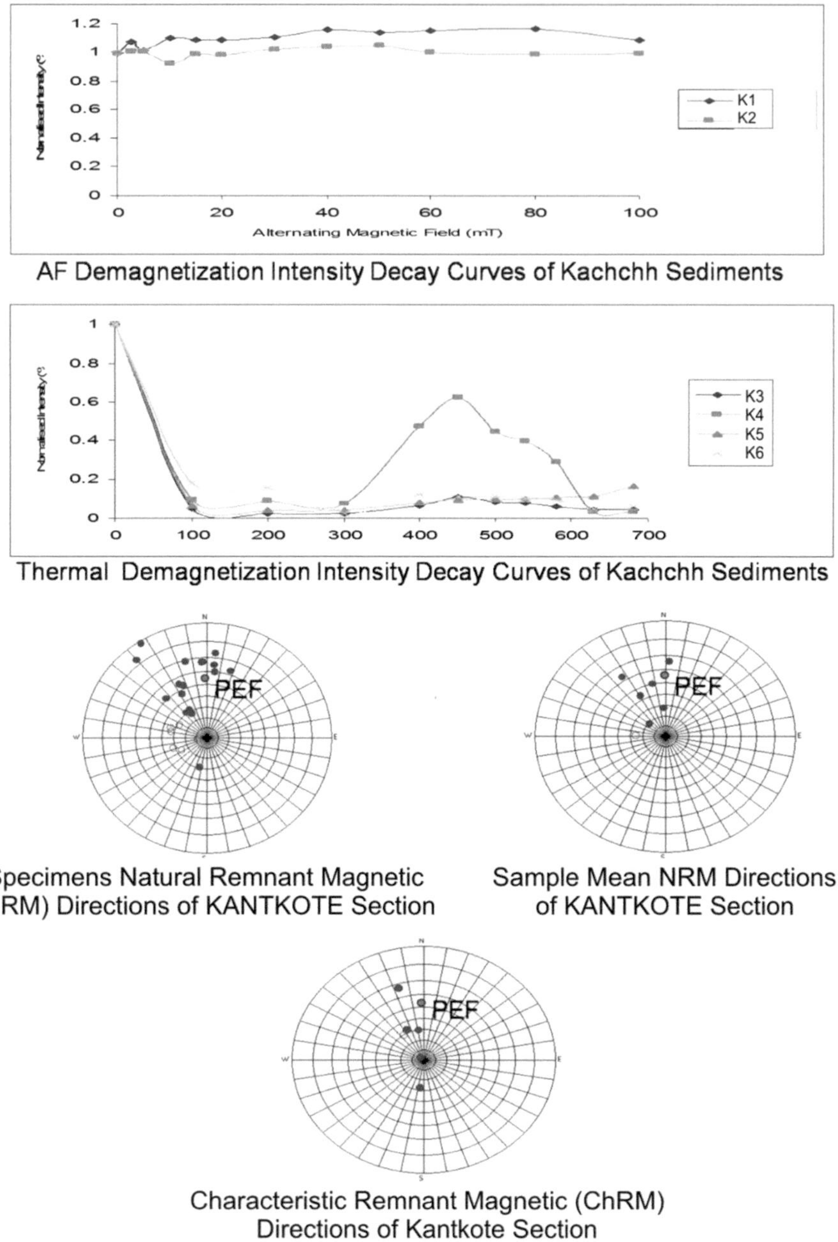

Fig. 1.10 Demonstration of appropriate retention of magnetism in Kachchh samples for geomagnetic investigation (after Venkateshwarlu et al. 2008)

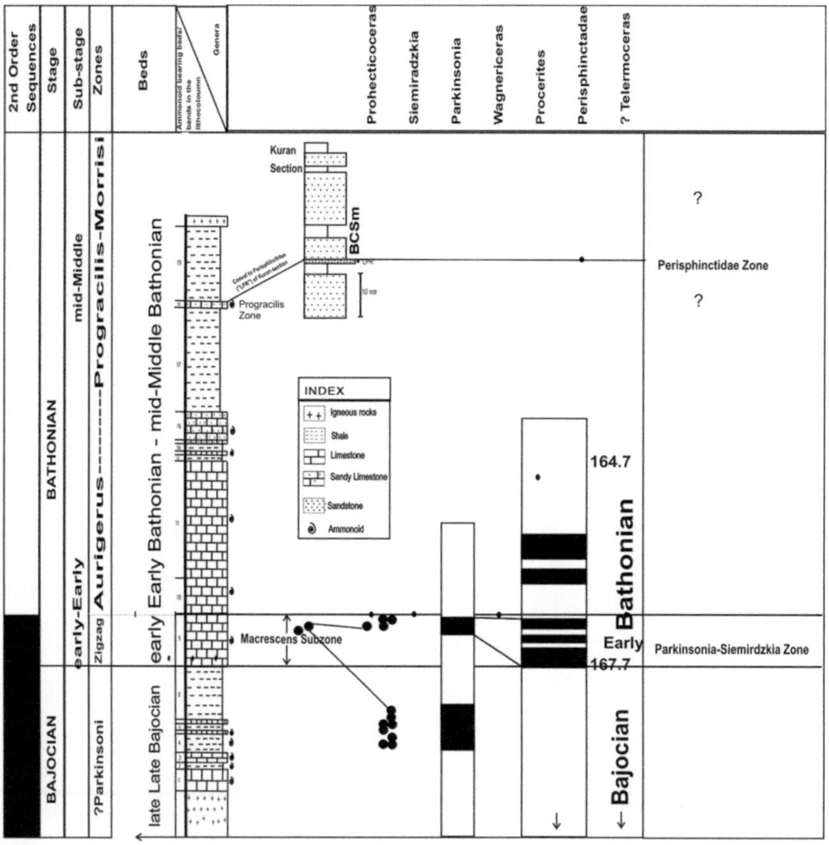

Fig. 1.11 Preliminary/tentative zonation in the newly discovered late Late Bajocian to mid Middle Bathonian stratigraphic column at Nara vis-a-vis second-order sequences and magnetic polarity intervals (modified after Ogg and Hinnov 2012 and Pandey and Pathak 2015)

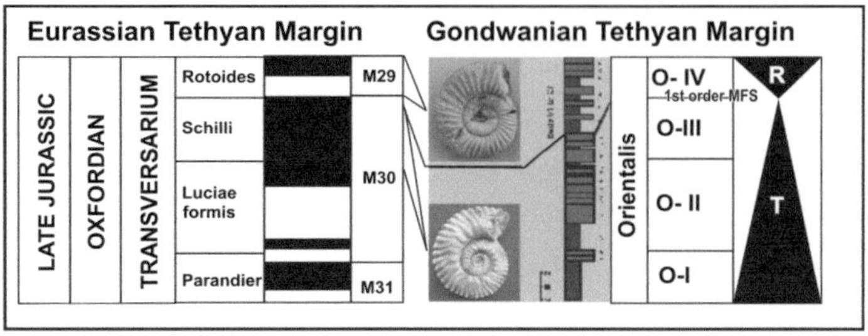

Fig. 1.12 Integration of magnetic polarity chrons, ammonoid zones/subzones and sequence surfaces in the late Middle Oxfordian Tranversarium Zone/Orientalis Zone in Kachchh (after Krishna et al. 2011)

SAMPLE	STRATIGRAPIC UNIT	SECTION	FT AGE AS PER 1988 TIME SCALE AND REVISED AGES AS PER 2012 SCALE	AMMONOID CONSTRAINED AGE	REMARK
Glauconite	Glauconite Bed, Amarsar Member	Ghuneri-Amarsar Section	89 MA / 92 MA	early Middle Turonian	Concordant
Glauconite	Ukra Member of Umia Formation	Katesar Nala Section	112 MA / 126 MA	Early Aptian	Concordant
Glauconite	Basal Member of Umia Formation	Amiu Village Section	138 MA / 142 MA	Early Berriasian	Concordant

Fig. 1.13 Fission track radiometric ages of Kachchh Mesozoic glauconites (modified after Krishna 2002)

and gradually pouring in mostly from exposed successions, and can be only referred to as preliminary. There is large quiet normal polarity interval in the middle part of Cretaceous, so also similar quiet zone is being increasingly invoked in the Middle Jurassic, and even later. In the Indian context, there is realized some data in the terminal Cretaceous part from the Deccan traps, while data generation in the Bajocian–Tithonian interval in Kachchh basin has been initiated in recent years (Fig. 1.10—Venkateshwarlu et al. 2008, 2014, 2016). Most Kachchh samples so far have indicated normal polarity in the Middle Jurassic, which is rather disappointing. May be, there is a long quiet interval of normal polarity. Apart from the polarity data, paleopole estimations in the Indian Mesozoic should also throw light on the latitudinal shift of India as part of Gondwana, for example on India's latitudinal framework prior and after the terminal Triassic to Early and Middle Jurassic and even thereafter during the interval of the oceanic isolation of India from

Cenomanian–Turonian onward. However, on the basis of ammonoids, a reverse interval has been picked up near the Bajocian/Bathonian transition in basal Bathonian Macrescens Subzone of Zigzag Zone (Fig. 1.11) (Venkateshwarlu in unpublished project progress report 2014). Another reversal is suspected in late Middle Oxfordian within the Orientalis Zone at the Schilli Subzone/Rotoides Subzone boundary (Fig. 1.12—Krishna et al. 2011). These reversals accord well with data from ETM as also with the recent geologic timescales (2012–2015). Both normal and reverse polarities have emerged in the terminal Maestrichtian intrusives. The international M-scale, available earlier only down to Callovian stage of the Middle Jurassic, has been extended farther below to include even Triassic, however, mostly from the exposed European successions. The integrated studies realized in the present work have facilitated correlation of the magnetic anomalies, igneous events, sequence stratigraphic surfaces, and the excellently resolved ammonoid-based ages.

The paleomagnetic studies on the Kachchh Cretaceous igneous rocks and Jurassic sediments have shown promise since recent initiation (Venkateshwarlu et al. 2008). Conspicuous clustering of directions both in initial NRM and ChRM is observed, which indicates that the primary magnetization rests in detritial magnetization DRM (Fig. 1.10 from Venkateshwarlu et al. 2008). The studies when completed under an extremely high chronologic resolution are expected to provide meaningful results from GTM. It is desirable to examine major magnetic polarity changes on the two margins of the Neotethys.

1.6 Radiometric Scenario

There is substantive data available on Meghalaya, Myanmar, Bengal, Jharkhand, High Himalaya, Madagascar, Pakistan, Kachchh, Deccan, Karnataka, Kerala, Tamil Nadu Mesozoic igneous rocks. The pre-Deccan St Marys and coeval volcanism is now well established. A few Cretaceous glauconites have also been dated, mostly by fission track technique (Srivastava et al. 1994; Krishna 2002, Fig. 1.13), however, there is strong desirability of additional data. The FT dates match well with actual ages of the glauconites indirectly constrained from the overlying and underlying ammonoids, so also constrained by ages of associated sequence surfaces. Glauconite samples from more than one stratigraphic level have been resampled for radiometric dating.

1.7 Igneous Activities

Several important igneous activities occurred during the span of the intra-Permian –intra-Paleogene mega-sequence (Panjal of ~289–243 ma from Early Permian to Middle Triassic; ~183 ma igneous event in Pakistan, also suspected in Kachchh, and Saurashtra; Rajmahal ~126–107 ma episode along with the preceding Late Jurassic to Neocomian volcanism of ~159–136 ma; St. Marys/Mannar/Kerala/Kachchh igneous events of ~94 to ~84 ma, and Deccan 71–61 ma), respectively, in India's north, east and west sectors, the oldest one at or near the start, the youngest somewhat terminus with the close of the sequence, while the middle ones in the Jurassic–Cretaceous interval. The tectonic and igneous events provided the primary triggers, and are considered to have chartered the course of geodynamic evolution during the span of the mega-sequence. The sequence stratigraphic study of the resultant sedimento-stratigraphic facets has facilitated the precise age constraining of the igneous episodes along with links to major sequence surfaces. The stratigraphically constrained igneous rocks encountered in the Kachchh and Saurashtra wells require priority dating for ascertaining their suspected intra-Jurassic age.

1.8 Oceanic Anoxic Events

The oceanic anoxic events (OAEs) are characterized by widely present interregional marked positive shift in the stable carbon (C) isotope in marine systems across the globe. OAEs are said to be triggered by volcanic activity that in turn leads to quick succession of or somewhat simultaneous happenings. There is increase in atmospheric CO_2, temperature, nutrient availability, productivity and rate of preservation of organic matter, etc. In the Late Permian–Early Paleogene mega sequence, volcanism as stated above has been rather frequent, and so also the OAEs, particularly in the Cretaceous. OAEs are of short duration of an ammonoid zone/zone of less than one my. Presence of black shales in the marine sediments facilitates the identification of the OAEs. The OAEs often even affected shallow environments. The anoxic events seem related to hydrocarbon resources, and hence increasing emphasis on them. The anoxism on the GTM has been interpreted as genetically linked to most of the above-stated igneous events and sequence surfaces, and has been appropriately discussed in the later chapters.

1.9 Eustatic Fluctuations

The Neotethyan span of time is known to have the highest Mesozoic sea-level in Late Cretaceous, and the second all time high of the Phanerozoic. The Haq et al. sea-level charts (1987) attracted special attention of the geosciences community to sea-level fluctuations with graded rise since the dawn of the Triassic to its Late Cretaceous Mesozoic peak to trigger research on this important aspect. Based on the Indian geological data sea-level, curves are developed in the three Mesozoic systems and compared with a few of the best-known curves proposed in recent decades.

References

Ayyasami (2011) Recognition of the Cretaceous stage boundaries in Southern Indian Cretaceous based on Invertebrate Fauna, 78–87. In: Jaitly AK, Singh AD, Pandey B, Nath S (eds) Paleontology and stratigraphy: basics to applications. Banaras Hindu University, Varanasi

Bhandari LL (1983) Introduction. In: Bhandari LL, Venkatachala BS, Kumar R, Swami SN, Garg P, Srivastava DC (eds) (1983) Petroliferous Basins of India

Bhargava ON, Krystin L, Balini H, Lein R, Nicora A (2004) Revised litho and biosequence stratigraphy of the Spiti Triassic. Albertiana supplement 30:21–38

Biswas SK (1971) Note on the geology of Kutch. Quart J Min Met Soc Ind 43(4):223–235

Biswas SK (1977) Mesozoic rock-stratigraphy of Kutch. Quart J Min Met Soc Ind 43(3&4):1–52

Biswas SK (1993) Geology of Kutch, 1, K D Malaviya Institute of Petroleum Exploration, Dehradun, p 450

Biswas SK (2012) Stratigraphy of Kutch—an overview, (abstract) DST contact programme-cum-field workshop, Structure, tectonics and Mesozoic stratigraphy of Kachchh, Bhuj (Kachchh), Gujarat, pp 11–13

Blanford WT (1869) On the geology of a portion of cutch. Mem Geol Sur Ind 6(1):17–39

Cariou E, Hantzpergue P (1997) Biostratigraphie du Jurassique ouest-europeen et mediterraneen. Bull Centr Recher Elf Explor Prod Mem 17:1–440

Cox LR (1940) The Jurassic lamellibranch fauna of Cutch (Kachh). Pal Ind 9.3(3):1–157

Dave A (2011) Recent developments in sequence stratigraphy and depositional sequences of Indian petroliferous basins. In: Jaitly AK, Singh AD, Pandey B, Nath S (eds) Palaeontology and stratigraphy: basics to applications. Banaras Hindu University, Varanasi, pp 88–112

Dave A (2012) Depositional sequence stratigraphy and petroleum systems of Kachchh basin—Status Talk. In: Thakkar MG (ed) Geology of Kachchh basin, Western India, Bhuj–Kachchh, pp 94–97

Dubey N (1992) Mesozoic sedimentary sequences and their depositional environments in Northwestern part of Kachchh Mainland, Gujarat (unpublished Ph.D thesis), BHU, Varanasi

Dubey N, Chatterjee BK (1997) Sandstone of Mesozoic Kachchh basin: their provenance and basinal evolution. Ind J Petrl Geol 6(1):55–68

Gradstein FM, Ogg JG, Smith AG (2005) A geological time scale 2004

Grant CW (1837) Memoir to illustrate a geological map of Cutch. Trans Geol Soc Lond 2(5):289–326

Gregory JW (1893) The Jurassic fauna of Cutch–I. The Echinoidea. Pal Ind 9.2(1):1–11

Gregory JW (1900) The Jurassic fauna of Cutch–II The Corals. Pal Ind 9.2(2):12–196

Haq BU, Hardenbol J, Vail PR (1987) Chronology of fluctuating sea-level since the Triassic. Science 235:1156–1167

Kitchin FL (1893–1900) The brachiopods of Kutch. Pal Ind 9.3(1)

Krishna J (1983a) Callovian–Albian ammonoid stratigraphy and paleobiogeography in the Indian subcontinent with special reference to the Tethys Himalaya. Him Geol 11:43–72

Krishna J (1983b) Reappraisal of the marine and/or "mixed" Lower Cretaceous sedimentary sequences of India; paleogeography and time boundaries, Cretaceous of India, pp 94–119

Krishna J (1986) Bearing of paleontological record on plate tectonics with reference to Indian plate. Short Term Course in Paleontology and Stratigraphy. COSIP-ULP in Geol, pp 257–269

Krishna J (1987a) Biological evidence for better appreciation of the Indian Gondwana. Palaeobotanist 36:268–284

Krishna J (1987b) The Jurassic-Cretaceous ammonoid geography in the Indian subcontinent vis-à-vis marine seaways and plate tectonics in the Indian ocean region, In 3 Decades of Development in Paleontology and Stratigraphy in India. Spl Publ Geol Surv Ind 11(1):453–482

Krishna J (1987c) An overview of the Mesozoic stratigraphy of Kachchha and Jaisalmer. J Pal Soc Ind 32:136–145

Krishna J (1991) Discovery of Lower Berriasian (Lower Cretaceous) ammonoid genus Argentiniceras from Kachchh (India) and its relevance to the Jurassic/Cretaceous boundary. Newsl Stratigr 23(3):141–150

Krishna J (1994) Environmental reconstruction studies of the Mesozoic sedimentary sequences in Western Kachchh (ONGC sponsored project), (unpublished technical report), pp 1–143

Krishna J (1996) Origin of the gondic (trans-Gondwana) corridor: the ammonoid evidence vol 2, Gondwana Nine, Oxford and IBH Publishing Company, Delhi, pp 1091–1099

Krishna J (2002) Mesozoic Microstratigraphy. DST contact programme-cum-field workshop, Structure, Tectonics and Mesozoic Stratigraphy of Kachchh, Bhuj (Kachchh), Gujarat, vol 6, pp 96–121

Krishna J (2004a) Outcrop based geological refinement and sequence stratigraphic studies in India: Example of Kachchh Mesozoic, Compendium on the National Workshop on Sequence Stratigraphy, Bangalore, 1–10, Figs. 1–10 (Keynote Address)

Krishna J (2004b) Refinement of time in the Indian geological record: Case sheet of the Jurassic in Kachchh. J Geol Soc Ind 63:120–123

Krishna J (2004c) Stratigraphic refinement in the Indian shallow marine exposed sedimentary successions as modeled in the Kachchh Jurassic, Compendium on the National Workshop on Sequence Stratigraphy, Bangalore, 20–40, Figs. 1–10 (Lecture Note)

Krishna J (2005) Update on Kachchh Jurassic ammonoid zonation: summarised intra-basinal to global perspectives. Spl. Publ.-1, Association of Petroleum Geologists, pp 63–68

Krishna J (2009) Genetic relationships among extensional tectonics, regional geology, sequence stratigraphy and biotic evolution: example of the Gondwanian Tethyan Mesozoic contextual to hydrocarbon exploration as modeled for the Kachchh sector, India. J Ind Geol Cong 1(1): 30–34

Krishna J (2014) Ammonoid-sequento-magnetostratigraphic integration in the Kachchh Jurassic (unpublished project report)

Krishna J, Cariou E (1990) Ammonoid faunal exchanges during lower Callovian between the Indo – East-African and Submediterranean Province: Implications for the Long Distance East–West Correlations. Newsl Stratigr 23(2):109–122

Krishna J, Ojha JR (1996) The Callovian ammonoid chronology in Kachchh (India). Geo Research Forum 1–2:151–166

Krishna J, Ojha JR (2000) The intra-basinal correlation in the Middle Jurassic Callovian stage of Kachchh (Gujarat) and ammonoid-microfossil integration. Geophytology 28(1&2):101–120

Krishna J, Pathak DB (1991) Ammonoid biochronology of the Kimmeridgian stage in Kachchh. India J Paleont Soc India 36:1–13

Krishna J, Pathak DB (1993) Late Lower Kimmeridgian—Lower Tithonian virgatosphinctins of India: evolutionary succession and biogeographic implications. Geobios 15:217–226

Krishna J and Pathak DB (1989) Kimmeridgian in Ler–Katrol area of Kachchh Western India: ammonoid systematic and biochronology, (abstr). In: 28th International Geological Congress, Washington, pp 228–229

Krishna J, Pathak DB (1994) Stratigraphic, biogeographic and environmental signatures in the amonoid bearing Jurassic–Cretaceous in the Himalaya on the south margin of the Tethys. J Him Geol 4(2):151–167

Krishna J, Pathak DB, Pandey B (1994a) Discovery of Middle and Late Oxfordian in Kachchh (India) with first ever approximation of the Oxfordian/Kimmeridgian boundary in the Indian subcontinent (abstr) Oxfordian–Kimmeridgian Joint Working Groups Meet, Lyon, p 10

Krishna J, Melendez G, Pandey B, Pathak DB (1994b) Middle Oxfordian Perisphinctinae and biostratigraphy from Kachchh (India). Jornadas de Paleontologia 106–108

Krishna J, Pathak DB, Pandey B (1994c) New ammonoid evidence on the Jurassic/Cretaceous boundary in Kachchh (India). Geobios, MS 17:327–335

Krishna J, Melendez G, Pandey B, Pathak DB (1995a) Characterisation of the ammonoid genus Larcheria (Middle Oxfordian) in Kachchh (India): paleontology, biostratigraphic and paleobiographic evaluation in the context of North-Tethyan occurences. C R Acad Sci Paris, 321, lla, 1187–1193

Krishna J, Pathak DB, Pandey B (1995b) Kimmeridgian–Tithonian (Upper Jurassic) ammonoid zones in Kachchh (Western India) and their correlations. Proc Rec Res Geol W Ind Badodara, pp 323–347

Krishna J, Melendez G, Pandey B, Pathak DB (1996a) Middle Oxfordian Ammonites (Perisphinctinae) from Kachchh (India): biostratigraphic and paleobiogeographic Implications, Revista Espanola De Paleontologia, Numero Extraordinario, pp 140–147

Krishna J, Pandey B, Pathak DB (1996b) Ammonoid chronology in the Tithonian of Kachchh (India). Geo Research Forum, Transtec Publications, Switzerland 1–2:205–214

Krishna J, Pathak DB, Pandey B (1996c) Quantum refinement in the Kimmeridgian ammonoid chronology in Kachchh (India) vol 1–2. Geo Research Forum, Transtec Publications, Switzerland, pp 195–204

Krishna J, Pathak DB, Pandey B (1998) Development of Oxfordian (Early Upper Jurassic) in the most proximal exposed part of Kachchh basin at Wagad outside the Mainland Kachchh. J Geol Soc Ind 52:513–522

Krishna J, Pathak DB Pandey B, Ojha JR (2000) Transgressive sediment intervals in the Late Jurassic of Kachchh, India. Geo. Res Forum, 6:321–332

Krishna J, Pandey B, Pathak DB (2009a) Characterization of Dichotomoceras in the Oxfordian of Kachchh (India). J Geol Soc Ind 74:469–479

Krishna J, Pandey B, Ojha JR, Pathak DB (2009b) Reappraisal of the age framework, correlation, environment and nomenclature of Kachchh Mesozoic lithostratigraphic units in Wagad. J Sci Res 53:1–20 Banaras Hindu University, Varanasi

Krishna J, Pandey B, Pathak DB (2010) Update on the Jurassic stratigraphy of the Indian subcontinent with focus on the tectonically controlled regional transgressive—regressive couplets. Earth Sci Front 17(Sp Iss):138–140

Krishna J, Pandey B, Pathak DB (2011) Current status of the Jurassic ammonoid stratigraphic framework in the Indian subcontinent with focus on the tectonically controlled regional transgressive—regressive couplets. Mem Geol Soc Ind 78:140–176

Krystyn L, Balini M, Nicora A (2004) Lower and Middle Triassic stage and substage boundaries in Spiti. Albertiana 30(supplement):39–52

Lyell C (1853) Principles of Geology. John Murray, London, pp 459–464

Ogg JG, Hinnov LA (2012) Jurassic. In: Gradstein FM, Ogg JG, Schmitz MD, Ogg GM (eds) The geological timescale. Elsevier Publications, pp 731–792

Ojha JR (1996) The Callovian at Keera, Kachchh, India: ammonoid systematics, chronology and geography. J Sci Res BHU 46:233–244

ONGC (1993) Lithostratigraphy of the Indian petroliferous basins documents, KDMIPE–ONGC, Dehradun

ONGC (2008) Petroleum systems sequence stratigraphy, KDMIPE–ONGC, Dehradun

Pandey B (1993) The Kachchh Jurassic Virgatosphinctinae, high resolution evolutionary succession and chronology. Unpublished Ph.D. thesis, Banaras Hindu University, Varanasi, pp 1–473

Pandey J, Dave A (1993) Studies in Mesozoic in Foraminifera and chronostratigraphy of Western Kutch Gujarat. Pal Ind KDMIPE–ONGC 1:1–221

Pandey B, Pathak DB (2015) Record of Early Bathonian ammonoids from Kachch, India: biostratigraphic and paleobiogeographic implications. J Pal Soc Ind 60(1):33–44

Pandey B, Krishna J, Pathak DB, Kumar A (2013a) Ammonoid biostratigraphy of Bathonian succession at Jumara, Kachchh, Western India. J Ind Geol Cong 4(2):7–18

Pandey B, Pathak DB, Krishna J (2013b) Calliphylloceras heterophylloides (Oppel, 1856) from basal most Jurassic succession of Sadhara dome, Kachchh, India. J Pal Soc Ind 58(1):61–65

Pascoe EH (1959) A manual of geology of India and Burma, 2. Govt. of India press, Calcutta

Pathak DB (1989) The Kimmeridgian-Tithonian at Ler-Katrol, Kachchh, India, Ammonoid Systematics, Biochromology and Biogeography. Unpublished Ph.D. thesis, Banaras Hindu University, Varanasi, pp 1–496

Prasad B, Kumar A, Raju DSN (1996) Chronostratigraphic subdivisions of Pre-Aptian (Pre-Uttaturaian) marine sediments in the Krishna-Godavari basin, India. Mem Geol Soc Ind 37:193–208

Rai J, Jain S (2012) Early Jurassic Gondwana break-up—a nanno-fossil story. In: Geology of Kachchh basin, western India (abstract), pp 31–32

Rai J, Jain S (2013) Pliensbachian nanno-fossils from Kachchh: implications on the earliest Jurassic transgressive event on the western Indian margin. Zitteliana A 53:105–120

Rajnath (1932) A contribution to the stratigraphy Of Kutch. Quart J Min Met Geol Soc Ind 4 (4):161–174

Raju DSN (2008) Stratigraphy of India: mega chart–I major hiatuses, stratigraphic sequences, lithology, thickness and rates of sedimentation of litho units of the Proterozoic and Phanerozoic of India. ONGC Bulletin 43(1) Sp. Iss:1–44

Raju DSN (2011) Cretaceous stratigraphic record on Indian plate: an overview. Mem Geol Soc Ind 78:177–226

Ramkrishnan M, Vaidyanathan R (2008) Geology of India, 1-2. Spl Publ Geol Soc, Ind 994p

Reeves CV (2014) The position of Madagascar within Gondwana and its movement during Gondwana dispersal. J Afr Earth Sci 94:45–57

Singh NP (2006) Mesozoic lithostratigraphy of the Jaisalmer basin, Rajasthan. J Pal Soc Ind 51 (2):1–25

Sowerby J (1840) A notice respecting some forms collected in Kutch. Trans Geol Soc Lond 21 (5):715–719

Spath LF (1924) On the Blake collection of ammonites from Kutch, India. Pal Ind 9(1):1–29

Spath LF (1927–33) Revision of the Jurassic Cephalopod fauna of Kach (Kutch). Pal Ind 9(2): 1–945

Srivastava AP, Rajagopalan MK, Krishna J, Pathak DB, Ojha JR (1994) The first ever absolute age determination from the Jurassic of Kachchh (India). Geobios MS 17:529–533

Valdiya KS (2010) The making of India; geodynamic evolution, 1st edn. McMillan, India, p 816

Valdiya KS (2016) The making of India; geodynamic evolution (2nd edn), Springer, p 924

Venkateshwarlu M, Pandey B, Pathak DB, Krishna J (2016) Magnetostratigraphy of the Middle Jurassic sediments from Kachchh basin, Western India. Int J Geo Sci 7:301–310

Venkateshwarlu M, Ojha JR, Pandey B, Krishna J (2008) Preliminary remanent magnetism results from the Jurassic sediments of Kachchh, India. Curr Sci 725–727

Venkateshwarlu M, Pandey B, Papanna G, Pathak DB, Krishna J (2014) New palaeomagnetic evidences about Deccan Trap volcanic activity from the magmatic bodies of Kachchh basin, Northwest India. J Ind Geophys Uni 18(1):99–107

Waagen W (1871) Abstracts of results of examination of the ammonite fauna of Cutch with remarks on their distribution among the beds and probable age. Rec Geol Surv Ind 4(4):89–101

Waagen W (1873–75) Jurassic fauna of Cutch, Cephalopoda. Pal Ind 9(1):1–247

Wynne AB (1872) Memoir on the geology of Kutch, to accompany the map. Mem Geol Surv Ind 9:1–289, compiled by Wynne, AB Fedden, F., during the session 1867–1869

Chapter 2
Mesozoic Stratigraphic Framework in India with Focus on the Jurassic Geological Record in the Kachchh Basin

Abstract It updates the biostratigraphic refinement in the Indian Mesozoics with focus on the Triassic in Spiti, Jurassic in Kachchh, and Cretaceous in Kachchh and Cauvery. As an example, the ammonoid stratigraphic zonation in Kachchh basin inclusive of the new Oxfordian zones, as also the origin of the basin are developed at length. Also are touched upon the salient pre-sequence Paleozoic features as back drop, along with the start and termination of the sequence. State of the art is presented on the integrated high resolution scales for the three periods. The Early-amd Middle Triassic update includes integration of ~ 35 ammonoid zones to ~ 15 conodont zones in the lithostratigraphically and splendidly resolved Spiti sections. The Indian Late Triassic to Early Jurassic is largely ammonoid devoid but for scanty Early Jurassic presence at Lamayuru in Ladakh. The Jurassic includes ~ 35 ammonoid zones, over 50 subzones, and yet more number of Horizons in the Middle and Late in well-resolved Kachchh sections. The Cretaceous update is compositely presented of 18 informal ammonoid zones with ammonoid presence almost in all the 12 stages in some or other Indian basin yet lacking the desired level of lithostratigraphic differentiation.

Keywords Lithostratigraphy · Ammonoids · Lineages · Heterochrony · Range charts · Zones/Subzones/Horizons · Biostratigraphy · Correlation · IEAP · Kachchh · Spiti · Cauvery · Bivalves · Forams · Conodonts · Palynomorphs · Mesozoic · Triassic · Jurassic · Cretaceous

2.1 Origin of the Kachchh Basin

Kachchh is a pericratonic rift basin in the extreme west of India that was proximal and united to NE Madagascar in the erstwhile Gondwana during most of the Mesozoic (Fig. 1.7). Earlier, the intra-Permian origin of the Neotethys witnessed

© Springer Science+Business Media Singapore 2017

J. Krishna, *The Indian Mesozoic Chronicle*, Springer Geology,
DOI 10.1007/978-981-10-2477-1_2

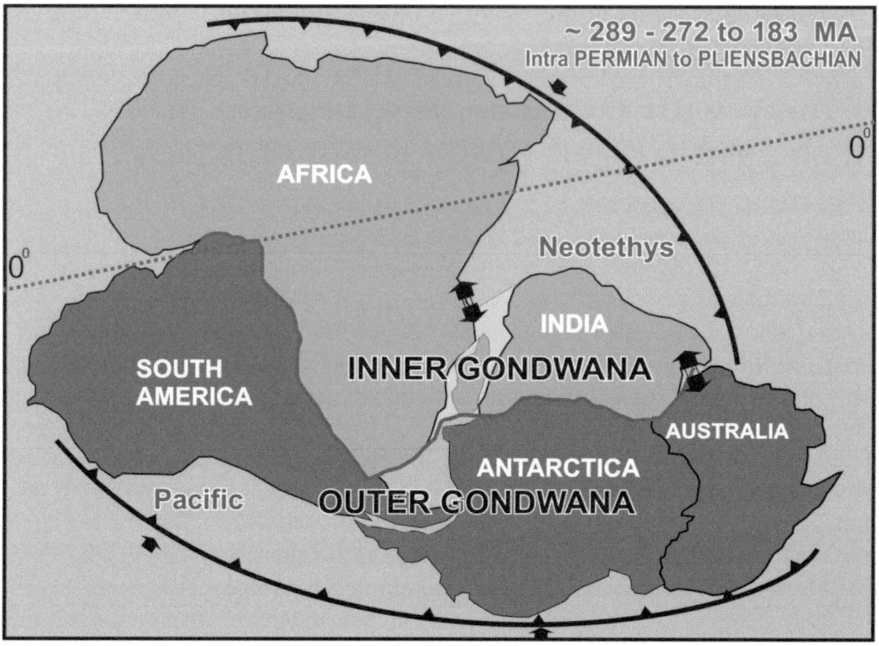

Fig. 2.1 Reorganization of plates and associated rifting on either side of India (inclusive of Madagascar) in India–Africa and India–Australo-Antarctica sectors as possible consequence of compression extented inward from the Neotethyan and Pacific margins of the Gondwana (modified after Geiger and Schweigert 2006)

tilting/subsidence of the just carved new north and northwest margins of India towards the new ocean. The crustal sagging/attenuation/extension/infra-rifting-related progressive subsidence during the Late Permian – Mid-Triassic interval gave rise to number of pericratonic rift basins on India's west, north and east margins inclusive of Kachchh due to reactivation of the faults of the Precambrian times. The mentioned extensional tectonics resulted presumably due to the build-up of threshold level inward radially directed stresses all round the Gondwana (Fig. 2.1).

2.1.1 Reactivation of the Precambrian Weak Zones

The ancient Precambrian weak zones, parallel to the east–west striking Narmada fracture, reactivated in the Kachchh region, which was located north of the mentioned fracture between the Nagarparkar, high in the north and the Saurashtra–Kathiyawar high in the south (Fig. 1.6). The rifting proceeded from north to south in sympathetic consequence of the reactivation of the India–Africa north–south-directed rifting, almost parallel to the future fracture zones of 40°E and

50°E. The mentioned fractures east and west of Madagascar later formed India's west margins in succession, first inclusive of Madagascar and subsequently exclusive of Madagascar.

2.1.2 *Infra-rift Sagging*

The rifting in the Kachchh region took place in steps, with presumable initiation of the infra-rift sag stage at the start of the Late Permian. The infra-rifting is considered linked to the origin of the Neotethys in the north sector. This interpretation is based on the record of reworked Late Permian palynomorphs in Late Jurassic and Miocene sediments in different parts of Kachchh (Venkatachala 1970). In this context, the record of the reworked Permian palynomorphs in Salt Range north of Kachchh, and in Kerala south of Kachchh, provides credence to the above interpretation in a regional frame. The record of the reworked Late Permian palynomorphs is followed up by the oldest insitu sediments in the Kachchh basin of the Anisian–Pliensbachian age bracket in multiple wells (Koshal 1984).

2.1.3 *Intra-Triassic Rift Initiation*

The actual/effective origin of the rift basin is interpreted in the MiddleTriassic. In the progression of rifting, each due south step went deeper than the previous one. The Kachchh region sloped due west at the start of the rifting. Thus was developed the resultant southwest sloping basin. The initiation of the mentioned intra-Triassic rifting in Kachchh has been précised near the close of the Mid-Triassic Anisian stage, based on the application of sequence stratigraphy to the Triassic succession of Spiti (Krishna in Bajpai et al. 2012) (discussed in detail in Chap. 3) above the Late Anisian Hollandites Zone–Kellnerites Zone in bed 34 (Krystyn et al. 2004). Small shallow east–west directd non-marine basin/s may have originated with progressive rifting. The presence of non-marine sediments in the Banni and Nirona wells close to the KMU (Kachchh Mainland Uplift) suggests existence either of a ~ 100 km long east–west striking basin or of two or more basins of smaller sizes. The exercise may have even involved migration of the rift locii in the early phase of the rift basin.

2.1.4 *Early Non-marine Phase Prior to the Neotethyan Transgression*

The basin in its initial phase escaped inundation by the north–south-directed arm of the Neotethys, and instead received only localized shallow non-marine sediments as evidenced in the Banni and Nirona wells (Fig. 1.5 locality map). It could be transgressed by the Neotethys only in early Early Toarcian as a result of the cumulative effect of the start of sea level rise and tectonically caused submergence near the Pliensbachian–Toarcian boundary. The reworked Late Permian non-marine sediments with possible extension into Early Triassic, subsequently may have been eroded away. The non-marine Late Permian–Early Jurassic phase preceded the initiation of the marine regime in the basin in basal Toarcian. The initial transgression was sudden as evidenced by the widespread presence of the Toarcian–Aalenian coccoliths (Rai and Jain 2012, 2013) almost in the entire exposed sites of the basin, not just in the relatively distal Mainland, but also in the proximal Island belt in the north and Wagad in the east of the exposed margin.

2.2 Salient Features in Brief Prior to the Mesozoic

Significant multifaceted geological transformation occurred immediately prior to the Late Permian–Early Paleogene mega-sequence at the Early/Late Permian boundary on the Paleotethyan East Gondwana margin. Emphasis has been given to the drastic geological changes associated with or without gap, not just in the Indian Mesozoic geological record, but also below the Mesozoic. These are summarized here to provide a backdrop to the present chapter on the current status of the stratigraphic refinement in the Indian Mesozoic geological record.

2.2.1 Intra-Devonian Origin of the Paleotethys

In recent publications (Metcalfe 2011 and others), a widespread interpretation has gained ground of the origin of the Paleotethys out of Paleopacific/Prototethys (Fig. 2.2) as late as intra-Devonian along with the spreading away of Tarim, North China, South China, Indo-China and a few other microcontinents from the north margin of the Indo-Australian plate.

2.2.2 Major, Widespread Gaps in India and Neighbourhood Prior as also During the Late Permian–Early Paleogene Mega-Sequence

Large gaps are known prior to, as also after the commencement of the mega-sequence which provide significant clues to the integrated geodynamic

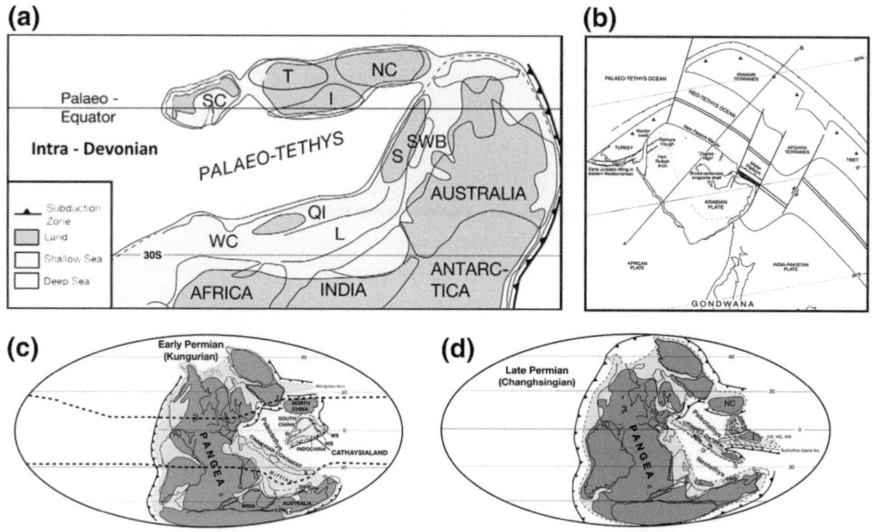

Fig. 2.2 Origin of Paleotethys and Neotethys (**a**, **c**, **d**) (modified after Metcalfe 2011 and **b** modified after Sharland et al. 2001)

evolution. These are invariably subaerial, and are evidenced in the geological record in several Indian basins including the non-marine Gondwana basins. A few such gaps are present even in the GTM basins. The gaps up to the Early Permian are on the Paleotethyan margin, and later ones on the Neotethyan margin. Among such gaps, the most widely spread and one of the largest is the almost continuous one on the proximal Paleotethyan margin from the terminal Precambrian to Carboniferous of more than 240 my in several belts and basins. The Early Permian marked a widespread rifting event in High Himalaya, Lesser Himalaya, Narmada–Son–Damodar belt, and elswhwere. In the Upper Indus basin of Pakistan, the gap is from Cambrian to Carboniferous. In Lesser Himalaya, Assam, and West Rajasthan Bikaner and Nagaur basins, the gap ranges at least from the late Middle Cambrian to the close of Carboniferous. In Ganga basin, it includes gap of most of the Ordovician (Prasad 2011), later part of Devonian and entire Carboniferous. In High Himalaya are present relatively much shorter gaps from late Cambrian to Early Ordovician, from Wenock to Middle Devonian, Middle Permian, Changhsingian to Induan (Bhargava 2008). Smaller gaps are also widely present in Indian basins during the Middle Triassic, Late Jurassic and Late Cretaceous.

In the Indian northeast, the gap extends from the Cambrian up to the close of Mesozoic but for a short transient Early Permian in many basins. In the K-G basin, near continuous Permian to Quaternary succession is present, while the gap is from the Precambrian to Carboniferous. In Cauvery, the gap, but for a short Permian intervention, extends up to late Middle Oxfordian. In West Bengal, the gap is from the Late Precambrian to Carboniferous. In the Gondwana basins, the gap extends from Precambrian to Carboniferous and then again from Toarcian to late Middle Oxfordian.

The mentioned gaps of different durations are probably caused by the tilting down/up of the distal/proximal locations of the basins episodically at important tectonic events, for example, in the distal High Himalaya the gaps are causally associated with tilting down of the plate. Similarly, tilting up of the proximal locations of the submerged part of the Indian plate prompted the gaps in Lesser Himalaya, Vindhyan and Ganga basins. The gaps included in the mega-sequence are discussed and differentiated with regard to their subaerial or submarine character in appropriate context in later chapters.

2.2.3 *Widespread Basal Permian Extensional Tectonics, and Glaciation, Followed up Further by Major Transgressive Event*

The widespread basal Permian extensional event supposedly reactivated the pre-existing fracture zones. Linear graben/half-graben basins formed along High Himalaya, Lesser Himalaya, Gondwana and a few other localities. One of the all-time major Gondwana events was of widespread glaciation that ranged from the terminal Carboniferous to basal Permian. The glaciation in India occurred in basal Permian, and is evidenced through the basal Talchir boulder and coeval units of explicit glacial impress. As the climate improved and temperatures picked up immediately after the glaciation, the melting of ice prompted the rise of sea level. It in turn resulted in an important and widespread transgression (Fig. 2.3) not only on the Paleotethyan margin but also in West Rajasthan, Bihar, Madhya Pradesh, Orissa, Tamilnadu, Uttarakhand, northeast states and elsewhere, also suspected in Kachchh.

2.2.4 *Intra-Permian Origin of the Neotethys*

The most important geological event on the GTM has been the intra-Permian origin of the Neotethys. The basal Permian rift developed into spreading at the start of Late Permian, locally even earlier at the Sakmarian/Artinskian boundary as suggested by the start of volcanism around this time (Shellnut et al. 2009). The start of the spreading marks the basal SB of the mega-sequence. A few microplates (Cimmeria continent) broke and spread away due north to subsequently unite with the then Asian congregation, and thus in size grew the Neotethys.

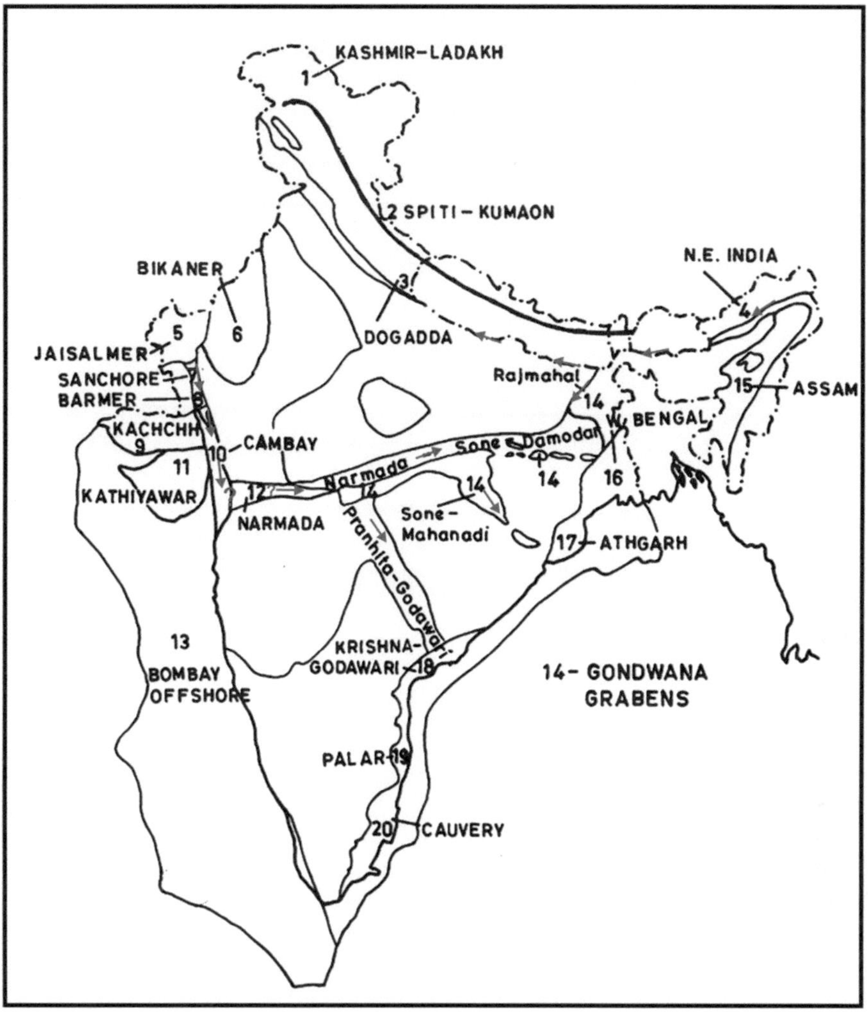

Fig. 2.3 early Early Permian transgression along the Gondwana basins from the east with suspected connection to the west sector through Barmer–Sanchore–Cambay–Narmada (modified after Krishna 1987a)

2.2.5 Wide Spread Early/Late Permian Change in Inland Non-marine Gondwana Basins at the Start of the Mega-Sequence

Widespread lithological and faunal/floral change and stratigraphic break are known between the Barakar and the coeval units below and the Raniganj and the coeval

Period	LH	K	TH	HH	Kumaon	Bhutan
Late Permian		Mamal Fn.	Shyoke Volcanics	Kuling/ Gungri Fn.	Kringkrong Fn.	
Stratigraphic Gap / Diverse Geological Changes						
Early Permian	Jogira Member	Panjal Volcanics	Khalsar Fn.	Phe Volcanics	Girthi Fn.	Shodug Fn.

North Sector

Period	Madagascar	Indus	Jaisalmer	Bikaner-Nagaur	Kachchh	Saurashtra	Kerala Konkan
Late Permian	Sakamena Gp.	Zaluch Fn.			Reworked non-marine Palynomophs	non-marine (Suggestive)	Reworked non-marine Palynomophs
Stratigraphic Gap / Diverse Geological Changes							
Early Permian	Sakoa Gp.	Nilawan Fn.	Bhuwan Fn.	Bhadaura Fn.	?	?	?

West Sector

Period	K-G	Cauvery	Reva	S-M	Damodar	Assam	Rajmahal
Late Permian	Chintalpalli Fn.		Pali-Tiki Fn.	Raniganj Fn. Barren Measures	Raniganj Fn. Barren Measures		Barakar Fn.
Stratigraphic Gap / Diverse Geological Changes							
Early Permian	Komugudden Fn.	Pundi Shale	Barakar Fn.	Barakar Fn.	Barakar Fn.	Marine	Barakar Fn.

East Sector

Fig. 2.4 Widespread geological changes with or without stratigraphic gap between Early and Late Permian at the base of the mega-sequence

units above within the intervening Barren Measures. The post Barakar gap is at least of the Middle Permian Wordian stage of the broad D. indicus palynozone in almost all the inland Gondwana (Satpura, S. Rewa, Damodar, Rajmahal and Pranhita–Godavari) basins (Fig. 2.4). The lithostratigraphic succession of the Talchir–Karharbari–Barakar units is nearly identical in all the inland non-marine Gondwana basins. However, the mentioned uniformity is not maintained further up due to a widespread post Barakar event. The Permian is otherwise featured with widespread Gondwanian *Glossopteris* flora which allows easy differentiation of the GTM Permian from the neighbouring Cathyasian and Angaran land assemblies and their diagnostic floral elements.

2.2.6 *Early/Late Permian Change in Marine Basins*

The disconformity between the Early Permian Pundi Shale and the Early Cretaceous sediments in the Cauvery basin well, between the Komuguddem Formation of Early Permian age and Late Permian Chintalpudi Formation in K-G basin in outcrops as also in wells, is well-documented. Similar disconformities are

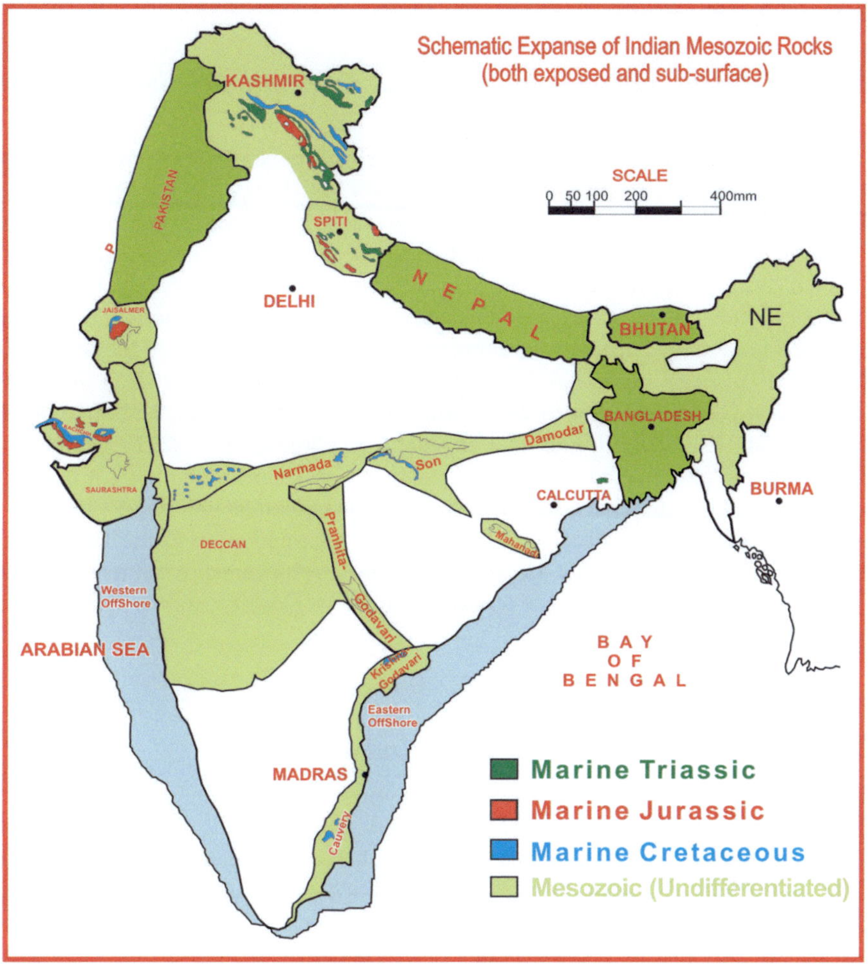

Fig. 2.5 Schematic expanse of Indian Mesozoic sediments (both exposed and subsurface)

recorded in the Palar basin, in the Lesser Himalaya between the marine Early Permian and the younger sediments, also in Shilong (Assam) above the Early Permian, in West Rajasthan above the Early Permiam Bap and Badhaura units. These disconformities are manifestations of a widespread mid/late Permian Pan-Indian event or even Pan IEAP event (Fig. 2.4).

The detailed current status of the Indian Mesozoic stratigraphic distribution (Fig. 2.5) and refinement along with salient geological events.

2.3 Triassic

2.3.1 Triassic System—International and National Status

The Triassic system was so named in 1832 by Alberti in North Germany, in view of its tripartitite division into Bunter, Muschelkalk and Keuper in ascending order. Soon it was realized that the first and third divisions were mostly non-marine, and thus difficult to subdivide and refine further as also correlate to the far located basins. This in turn prompted discovery and description of the exclusively marine ammonoid-rich Triassic in Alps as replacement for the earlier divisions. The Alpine Triassic found ready acceptance among the Triassic stratigraphers of the time. The Triassic is currently organized into a succession of seven stages, two in the Early (Induan and Olenekian), two in the Middle (Anisian and Ladinian) and three in the Late Triassic (Carnian, Norian and Rhaetian). According to the 2012 Geological Time Scale (Gradstein et al. 2012), the Triassic ranges from 252.2 to 201.3 ma with a duration of 50.9 my. The system is defined by the base of its first stage—Induan—and more precisely by the base of its first conodont based Parvus Zone. The system includes a succession of about 45 ammonoid zones with average zonal resolution of 1 my. Conodont studies have also provided a moderately well-resolved parallel scale into ~25 zones.

2.3.2 Development of Triassic in India

Among the best developed and early studied Triassic in the world is held the Triassic of the Indian High Himalaya with best studies to date in the Spiti basin. Apart from High Himalaya, marine Triassic is not exposed any where else in India. On the contrary, non-marine Triassic is well-developed in several Gondwana graben basins. Subsurface Triassic is also encountered in part or full in a few East Coast basins, particularly in K-G and Bengal basins. Subsurface Triassic is also present in the Indian Jaisalmer and Pakistani Indus basins.

2.3.3 Triassic Succession in Spiti

The Spiti basin of High Himalaya includes one of the best developed ammonoid marine Triassic in the world with a gap over the Late Permian. Stoliczka (1866), Hayden (1904) were the earliest workers who investigated the Triassic stratigraphy of Spiti. Srikantia (1981) mapped the area and classified the succession into five formations. Bhargava et al. (2004) have organized the succession into four mappable groups and ten formations, and subdivided them further into member units (Figs. 2.7, 2.8, 2.9, 2.10, 2.11, 2.12, 2.13, 2.14, 2.15, 2.16 and 2.17), along with detailed description of the lithological and paleontological features of the formations and members. The ~1400 m thick Triassic sediments unconformably ovely the Wuchiapingian Gungri unit. In between is a thin ferruginous layer. Changhsingian makes a gap below the Triassic. Mikin Formation—the basal most of the Triassic—is extremely rich in the stratigraphically significant ammonoids, compared to the overlying succession of the Kaga to Kioto units. Other significant fossil groups present are bivalves and conodonts.

2.3.4 Intra-Triassic Interstage Boundaries

In Triassic, the interstage boundary demarcation is possible only in Spiti basin, and is based mostly on Bhargava et al. (2004) and Krystyn et al. (2004) (Figs. 2.6, 2.7, 2.8, 2.9, 2.10, 2.11, 2.12, 2.13, 2.14, 2.15, 2.16 and 2.17).

The base of the Induan stage or for that matter of the Triassic system is best approximated in the Spiti (High Himalaya) stratigraphic sections between the underlying Gungri Formation of the Kuling Group or its coeval units in spite of some gap exclusively on the Permian side of full or part of Changhsingian. The youngest brachiopods of the Gungri Formation are ranged up to Early Changhsingian (Bhargava 2008). The boundary lies in the early part of a normal magnetic polarity chron. The basal *Otoceras* ammonoid bed includes the basal Triassic *Hindeodus parvus* Zone (Krystyn et al. 2004). Immediately underlying is a marker ferruginous layer. In turn underneath the ferruginous layer, is picked up the Europium enrichment event in the youngest black shale (Bhargava 2008). In the non-marine Gondwana basins the P/T boundary is approximately delineated between the Damodar Group/Formation and the Panchet Formation on the basis of the succession of palynoassemblages/events. Vijaya (1997) demonstrated the similarity of the late Permian and Induan palynoassemblages of non-marine Gondwana basins and the Spiti and Niti basins of Higher Himalaya and termed them as the palyno-assemblage/event 4 of Late Permian age and palyno-assemblage/event 5 of the basal Triassic Induan age.

The demarcation of interstage intra-Triassic boundaries is reproduced here as determined by Bhargava et al. (2004) and Krystyn et al. (2004) (Figs. 2.7, 2.8, 2.9, 2.10, 2.11, 2.12, 2.13, 2.14, 2.15, 2.16 and 2.17).

The Induan/Olenekian Boundary is placed between the thick limestone bed 13 and the thinly bedded limestone bed 14 (Fig. 2.7b). It is based on the *Flemingites* bearing beds below and the *Parahedenstroemia* inclusive beds above in several sections of the Spiti basin. On the basis of conodonts, the boundary is placed at the FAD of *Neospathodus waageni* (Fig. 2.7a). An other good level is the FAD of *Rohilla* that marks the base of the beds with *Flemingites* sensu lato. Another possibility is suggested at the FAD of *Flemingites* sensu stricto.

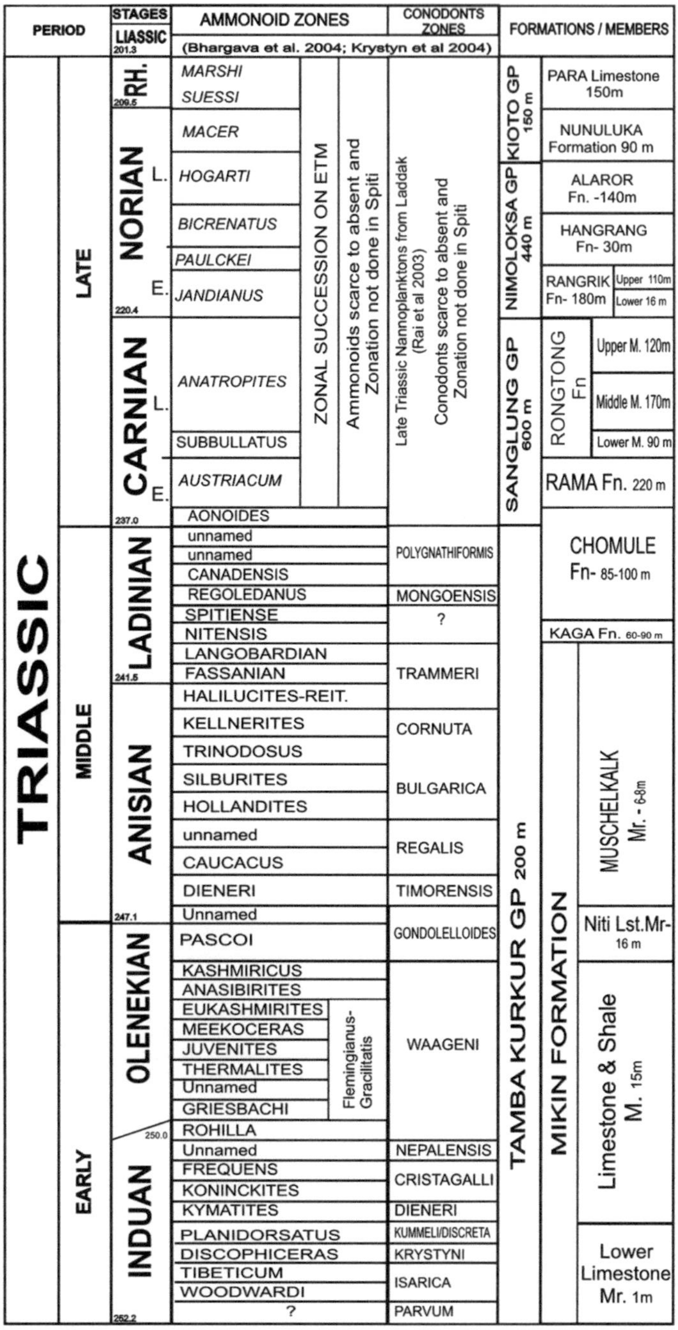

Fig. 2.6 Triassic lithostratigraphic units and ammonoid/conodont zones of Spiti (modified after Bhargava et al. 2004; Krystyn et al. 2004)

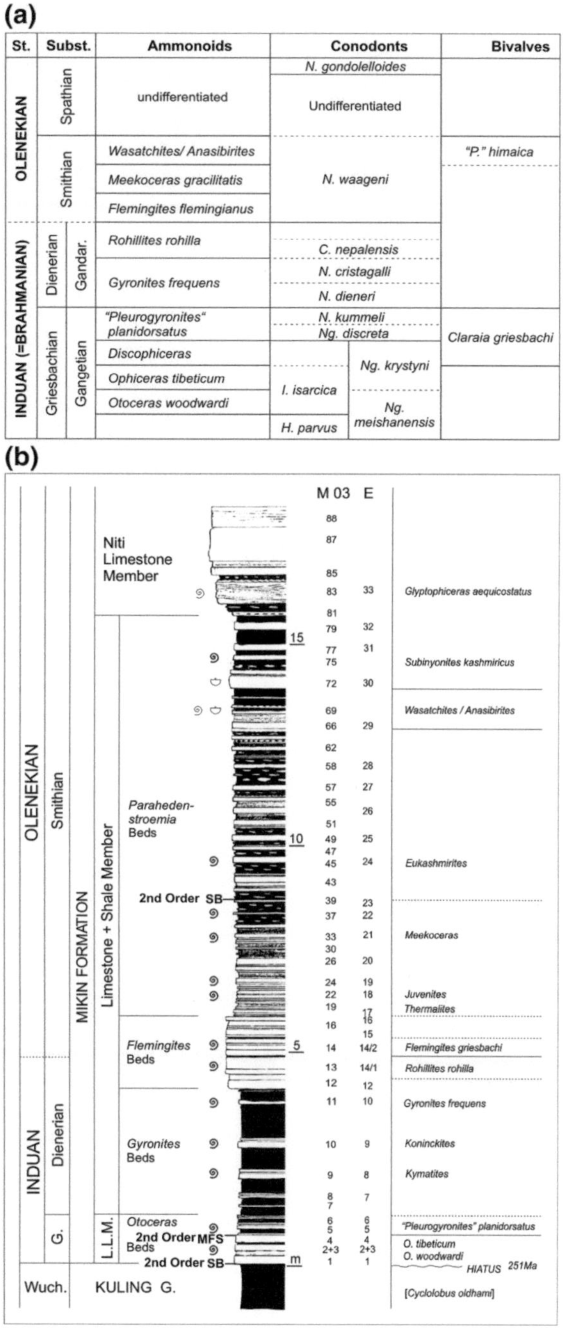

Fig. 2.7 **a** Induan/Olenekian boundry in the Spiti Triassic (after Bhargava et al. 2004; Krystyn et al. 2004). **b** Interstage boundaries along with SBs and MFSs in the Triassic ammonoid zonal framework of Spiti (after Bhargava et al. 2004; Krystyn et al. 2004)

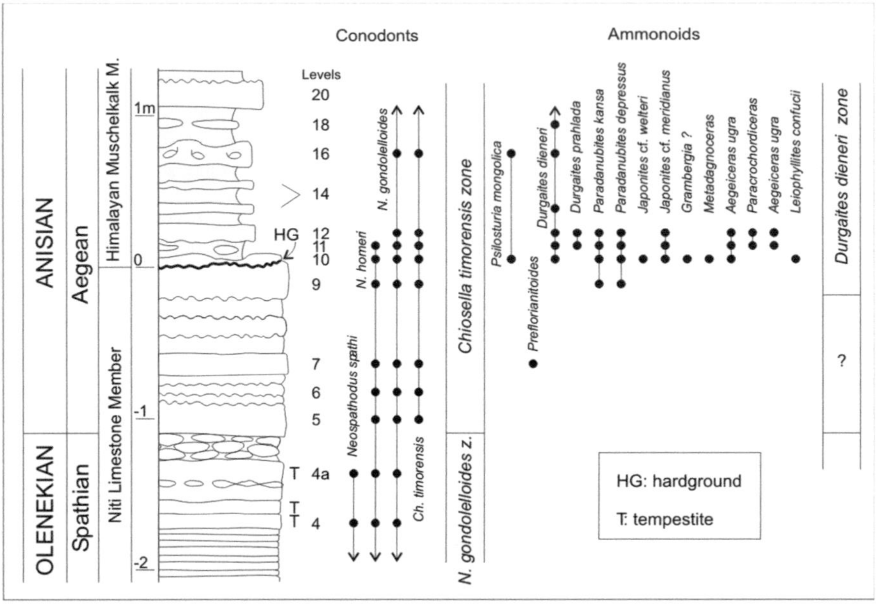

Fig. 2.8 Olenekian–Anisian Boundry at Muth in Spiti Triassic (after Bhargava et al. 2004; Krystyn et al. 2004)

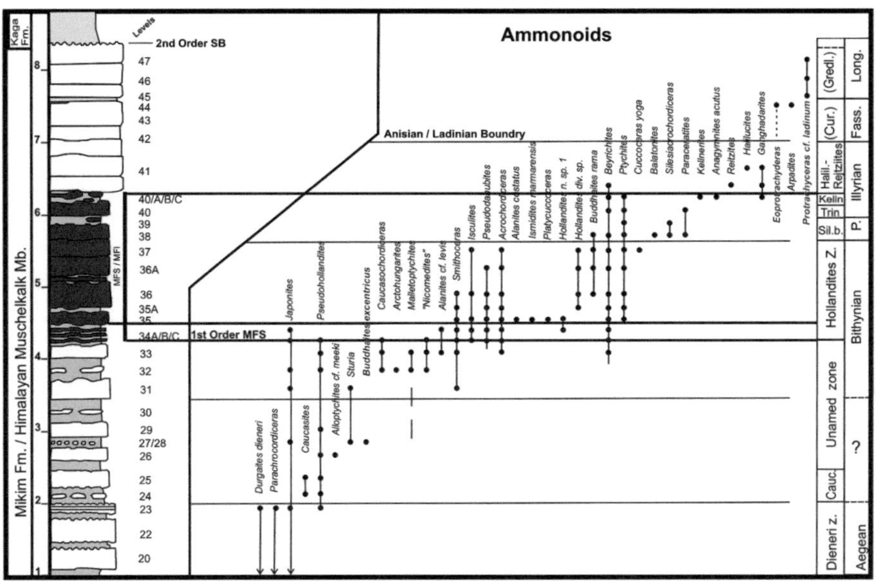

Fig. 2.9 Ammonoid stratigraphy of the Himalayan Muschelkalk with Anisian/Ladinian boundary. Note the strongly reduced thickness in the Late Anisian and the sequence boundary (SB) in the upper part of the section of the Ammonoid stratigraphy in Krystyn et al. (2004)

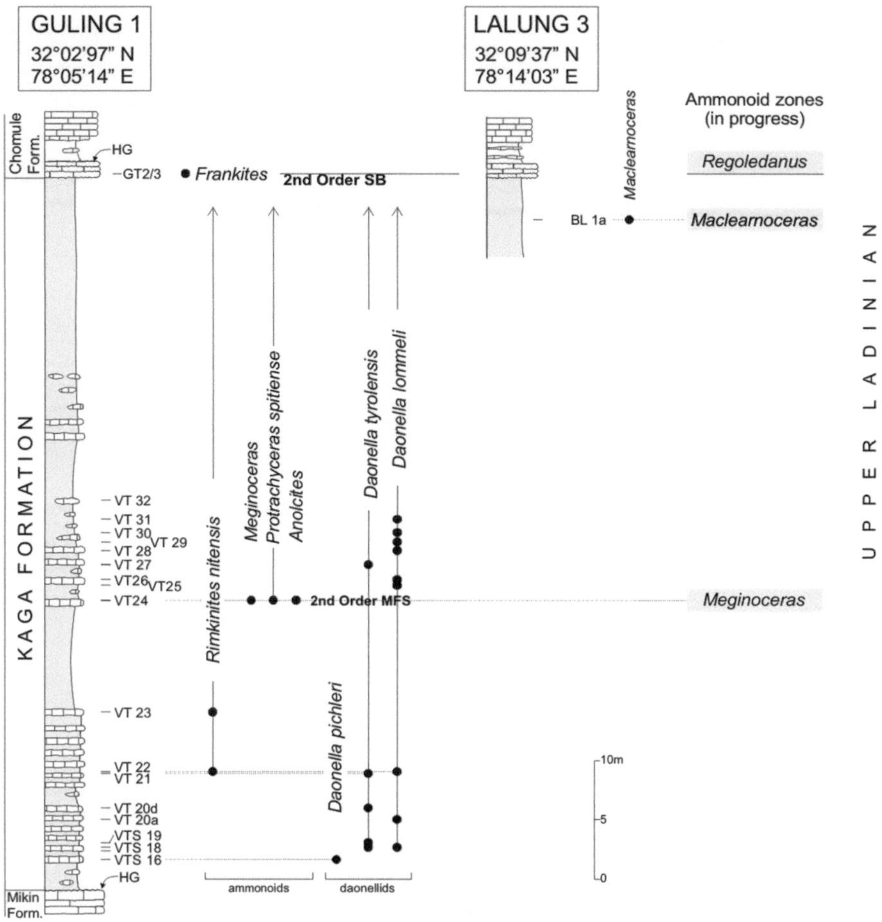

Fig. 2.10 Second-order MFS and SB determined in the stratigraphic log of the Kaga Formation at Guling (after Bhargava et al. 2004; Krystyn et al. 2004)

The Olenekian/Anisian Boundary (also the Early/Middle Triassic boundary) is précised between beds 4 and 5 at the FAD of the conodont *Chiosella timorensis* within the Niti Limestone one m below its top above a marker nodular bed (Fig. 2.8). The interval near the boundary is ammonoid devoid, and the oldest ammonoid found in Anisian is in bed 7–30 cm above the boundary.

The Anisian/Ladinian Boundary is placed between beds 41 and 42 (Fig. 2.9), however, a revised international consensus has favoured the boundary at the FAD of *Protrachyceras* between beds 44 and 45. Formal GSSP is yet to be decided.

The Ladinian/Carnian Boundary (also the Middle/Late Triassic boundary Fig. 2.11) is placed according to one view within one to three m of the base of Chomule Formation. Other views place the same at the LAD of *Frankites*, or at the

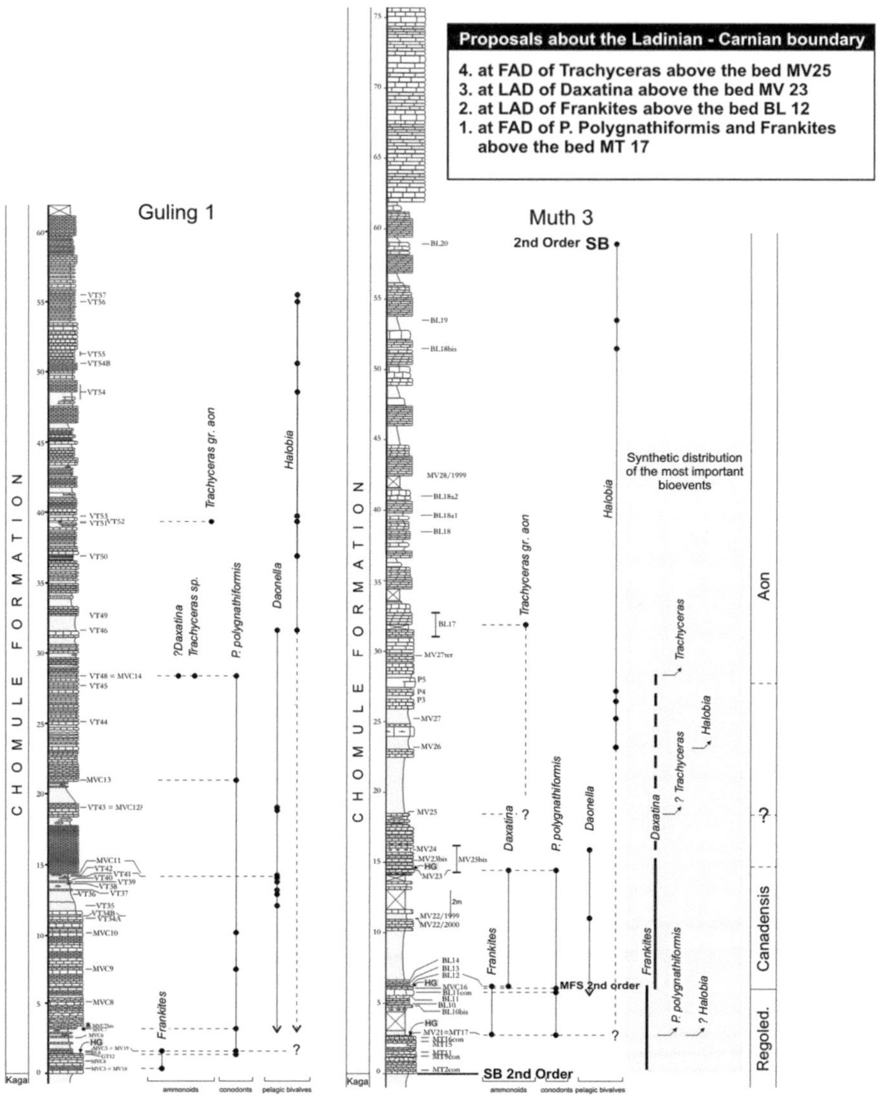

Fig. 2.11 Ladinian–Carnian boundary proposals at guling 1 and muth 3 along with second-order SB and MFS in the Spiti Triassic (after Bhargava et al. 2004; Krystyn et al. 2004)

LAD of *Daxanites*, or at the FAD of *Trachyceras*, or at the FAD of the conodont *P. polygnathiformis*. All these candidates are within the lower half of the Chomule Formation. As per bivalves, the boundary is placed between the LAD of *Daonella* and the FAD of *Halobia*.

The Carnian/Norian Boundary has been only tentatively suggested between the Rongtong Formation and the Rangkrik Formation. That is, between the Nimoloksa

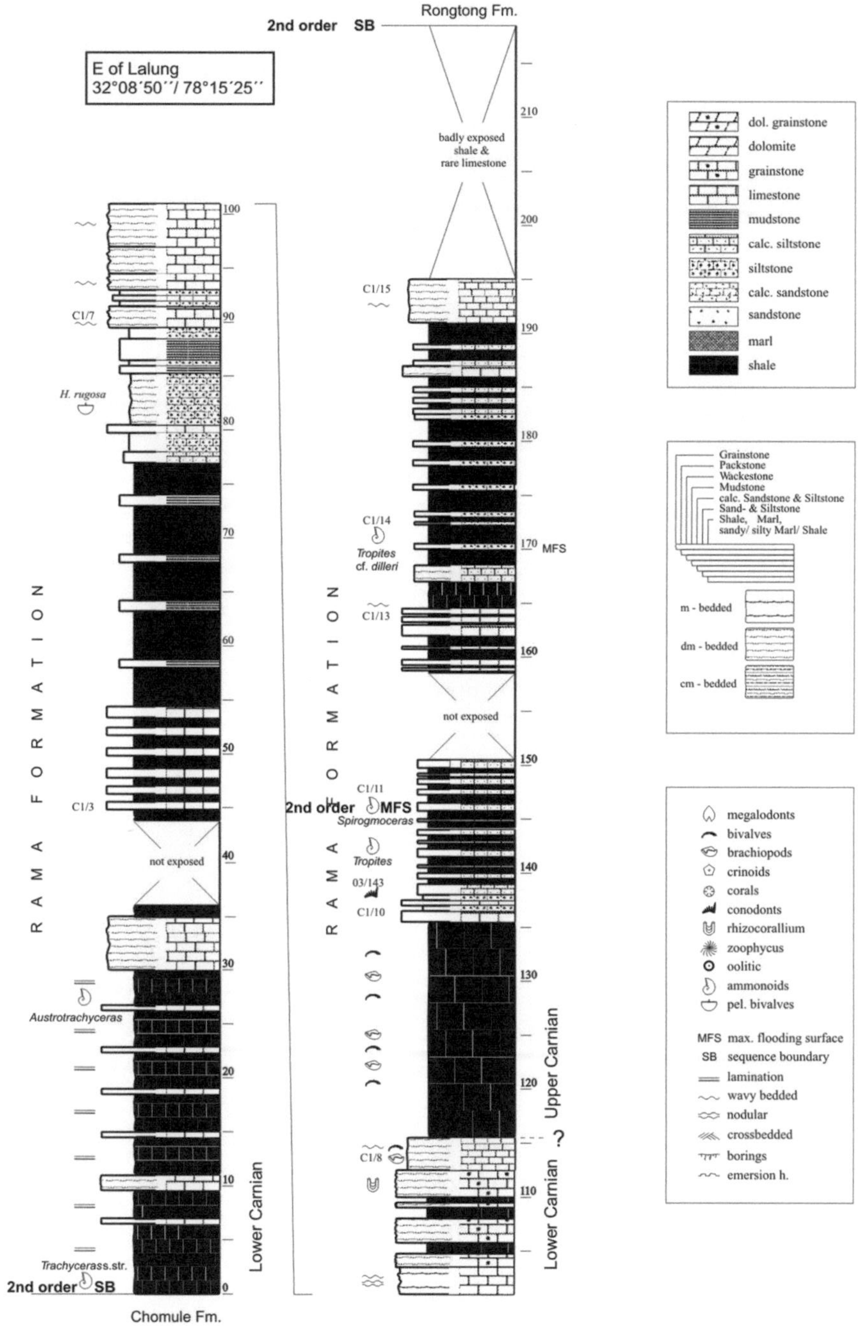

Fig. 2.12 Second-order SBs/MFSs interpreted in the Carnian column (after Bhargava et al. 2004)

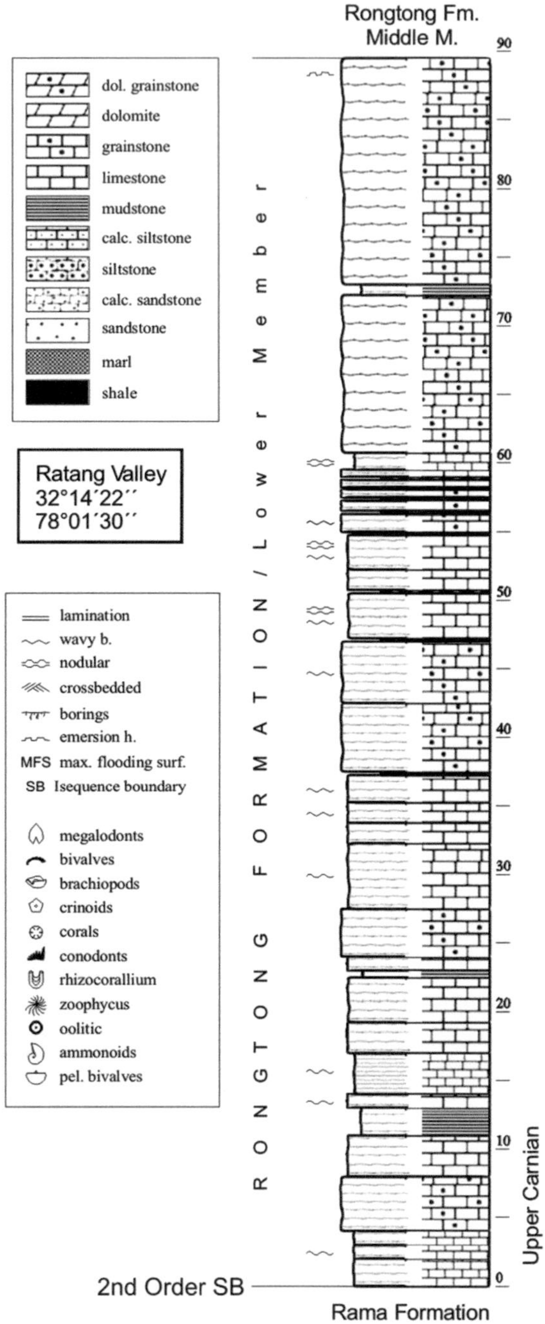

Fig. 2.13 Demarcation of second-order intra Carnian SB at the Rama/Rongtong Unit boundary in Spiti Triassic (after Bhargava et al. 2004)

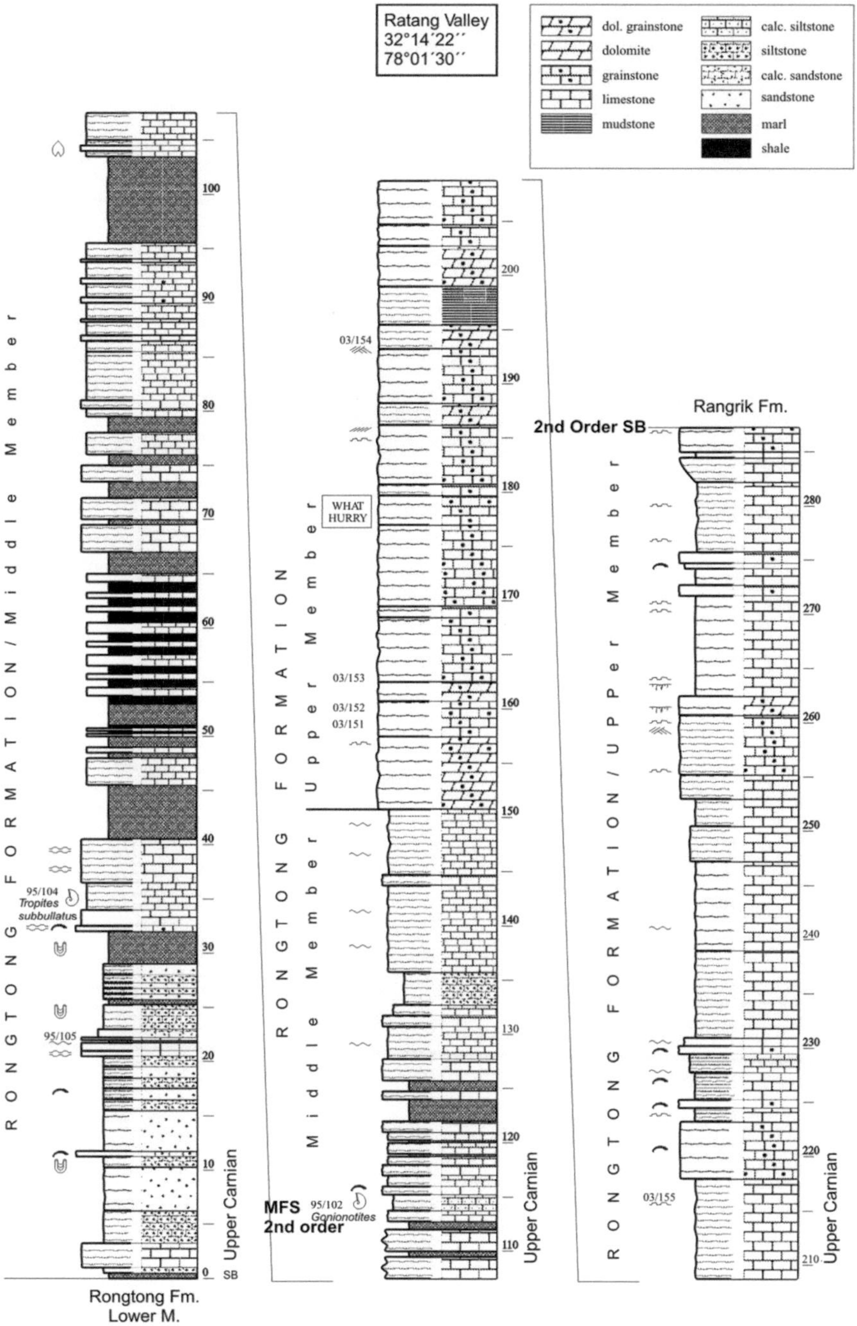

Fig. 2.14 Intra-Carnian second-order SB and MFS determined within the middle member of Rongtong Fn., and SB at the boundary with Rangrik Fn. in Spiti Triassic (modified after Bhargava et al. 2004)

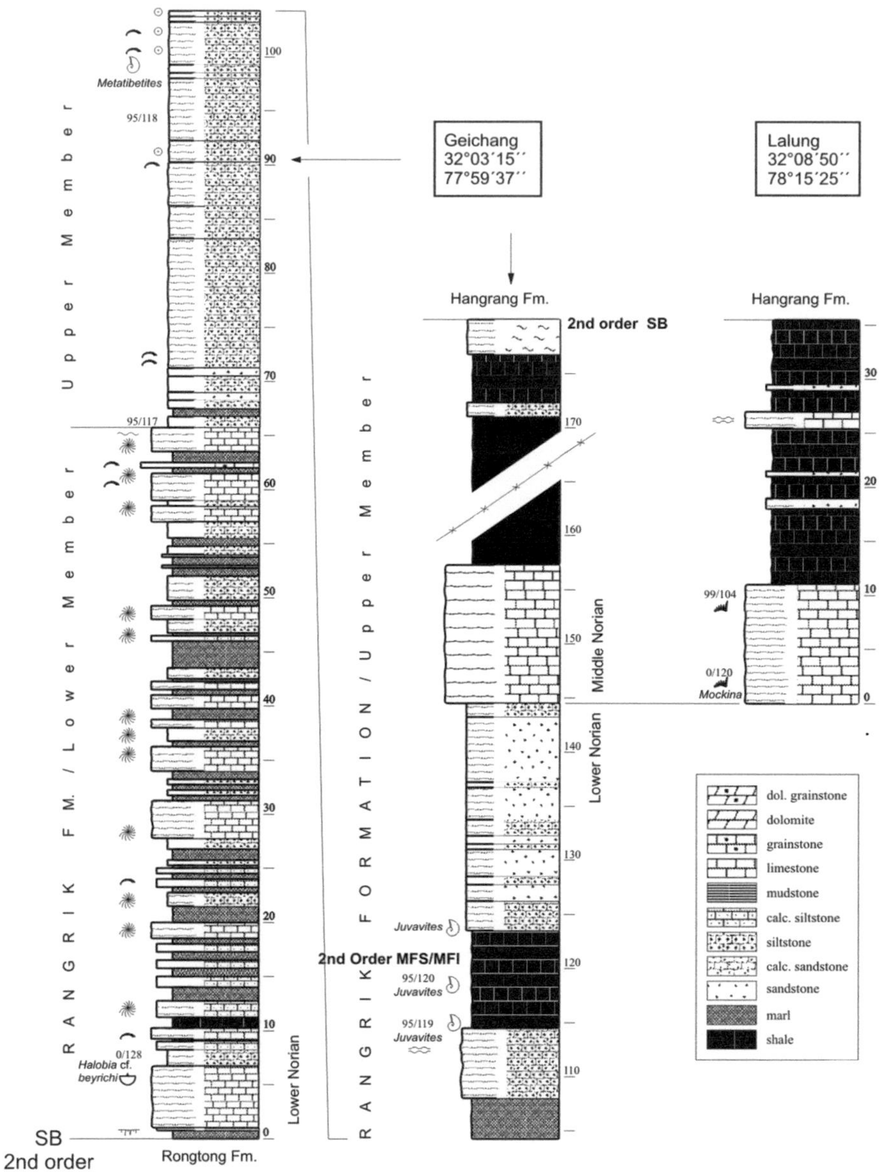

Fig. 2.15 Second-order SBs at the base and top of Rangrik Fn. and MFS/MFI in Juvavites bearing shales in Spiti Triassic (after Bhargava et al. 2004; Krystyn et al. 2004)

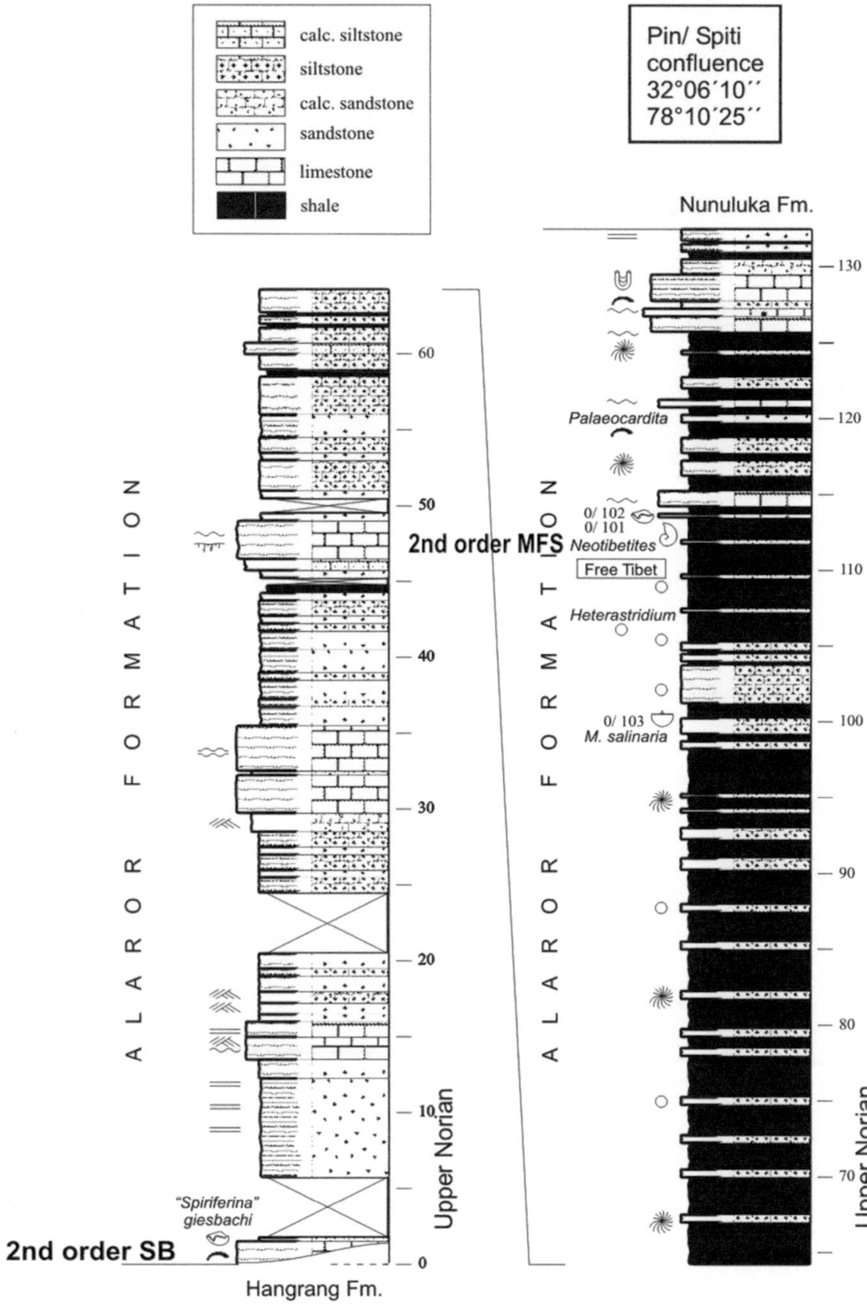

Fig. 2.16 Second-order SB at the base of Alaror Fn. and MFS in the *upper part* of Alaror Fn. above the Neotibetites above the bed 102 (after Bhargava et al. 2004; Krystyn et al. 2004)

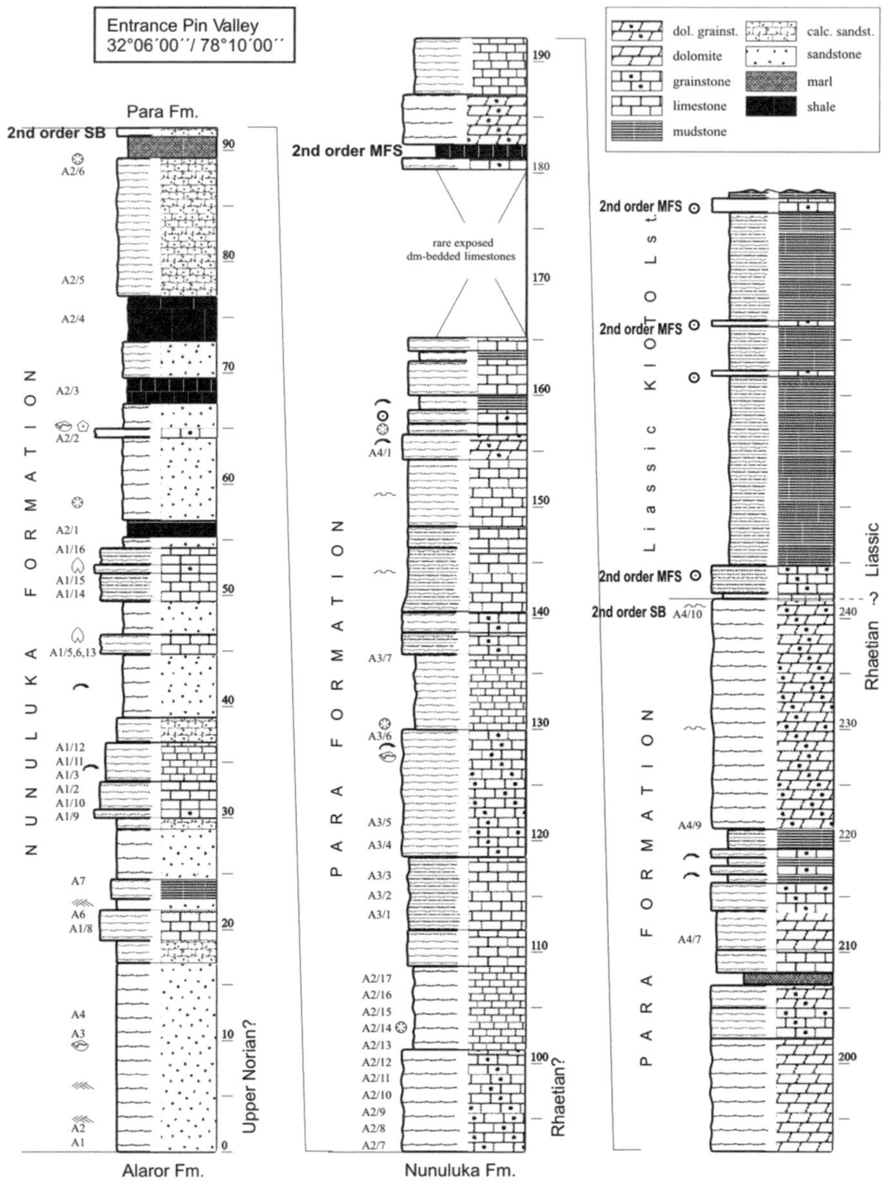

Fig. 2.17 Second-order SB at the base of Para Fn., MFS in the upper part of Para Fn. and Intra Liassic MFSs in the Kioto Limestone of Early Jurassic age in Spiti (after Bhargava et al. 2004; Krystyn et al. 2004)

Group and Alaror Group between *Tropites* beds and *Juvavites* beds at the start of *Zoophycus* rich limestones above the LAD of *Halobia* in the basal limestones of Rangkrik Formation (Fig. 2.12).

The Norian/Rhaetian Stage boundary is tentatively suggested between the Alaror Formation and the Nunuluka Formation. Another placement is between the sands of the early Nunuluka Formation and the overlying grainstones/limestones between beds A2/6 and A2/7 (Fig. 2.13).

The Rhaetian/Liassic boundary is only suggestively placed between the Para Formation and Tagling Formation (Kioto Group) (Fig. 2.17). The latter begins with the thick interval of thinly bedded platy mudstone beds interspersed with thin oolitic limestones.

2.3.5 Paleogeographic Framework at the Start of the Triassic Contextual to Kachchh

Kachchh region until the origin of the basin was a positive peneplained craton since the Precambrian. The Indian plate then occupied the axial/core/central position along with Africa within the Gondwanaland, and included Madagascar as its integral part. This axial/inner Gondwana (Fig. 1.1), comprising India (inclusive of Madagascar) and Africa, was bordered by the outer Gondwana, consisting of Australo-Antarctca in the east and South American continent in the west. There were significant weak divergent ancient tectonic fault zones, particularly on both east and west of the Indian plate. In Kachchh and neighbourhood, there existed east–west striking faults of varying depth parallel to the east–west striking Narmada graben, which were activated, and subsequently exercised control on the Mesozoic geological dynamics in the region. Stresses had been building up for a long time radially inside from the periphery in the region, which triggered the initiation of the rifting, between India and Somalia and in the vicinity of Kachchh and also between India and Australo-Antarctica.

The invasion of the Neoethys later in early Early Toarcian may be related to the deeper somewhat farther south basal Toarcian rift episode. The recently interpreted expanse of the Early Toarcian coccoliths throughout the Kachchh basin (Rai and Jain 2012, 2013) also suggests the near full growth of the Kachchh basin at the onset of the basal Toarcian transgression. The Toarcian ammonoids suggest communication links to Arabia. The Neotethyan invasion advanced from the north at the start of a new first-order sequence. The absence of the benthic ostracods suggests oxygen depleted basins unfavourable to ostracods (Mette 2004). The Gondic shallow marine corridor had yet not evolved and communication farther south to SE Pacific was presumably prevented by the Mozambique ridge (Fig. 1.7).

2.3.6 Suspected Presence of the Late Triassic in the Non-marine Nirona Formation as Encountered in Nirona and Banni Wells of Kachchh

Koshal (1984) recorded Rhaetian–Liassic pollens and spores. It implies an older origin of the basin during a long phase of restrictive, localized and shallow rifting and extension that prolonged for ~45 my. The rifting may have begun much earlier and farther north of the Banni well close to the Nagarparker ridge. The initial intra-Triassic origin of the Kachchh basin may be localized and related to the later Middle Triassic to mid Early Liassic Gondwana rift-framework. Additionally, the record of reworked non-marine Late Permian palynomorphs in younger reworked sediments may extend the localized non-marine sedimentation even as back as Late Permian.

The Late Triassic–Early Jurassic in the small shallow basin/s may have started receiving non-marine sedimentation as evidenced in the Banni and Nirona wells through the study of palynomorphs of Rhaetian–Liassic age in Zone 4 (Koshal 1973) between 1620 and 1760 m immediately above the basement in well Banni 2. In a 1993 0NGC publication, the mentioned Nirona Formation has been assigned to Late Triassic. It may be noted that the initiation of the Kachchh basin occurred immediately south of the Nagarparkar High ~50 km north of the Nirona and Banni wells. Obviously, the non-marine sediments close to the Nagarparker margin of the Kachchh basin are expected to be distinctly older of late Middle and Late Triassic (Ladinian to Norian) stages than the recorded Rhaetian (Koshal 1973) in the Banni and Nirona wells.

2.3.7 Broad Stratigraphic Correlation of the Non-marine Nirona Formation to Non-marine Gondwana Units as also to Spiti Marine Units

In a regional tectono-stratigraphic context, the non-marine Nirona Formation sediments of Banni and Nirona wells correspond broadly to the Mahadeva/Durgpur/Pachmahi/Tikki units of the different Gondwana graben basins, also to the combined nine unit marine succession of Kaga to Kioto units of Spiti, discussed later in Chap. 4.

2.3.8 Marine Triassic in Neighbourhood

In general, the marine Triassic in Pakistan is organized into a succession of three formations; Mianwali Formation, Tredian Fromation and Kingrali Formation, among which Mianwali Formation includes rich ammonoids while the younger two units are ammonoid scarce to devoid (Fatmi 1972). It is disconformable over Permian, and in turn is overlain unconformably by the Jurassic. Thus the ammonoid bearing Mianwali Formation can be best corresponded to the pre bed 40 part of the Mikin Formation of the Induan–Anisian age up to the ammonoid-rich hard grounded part of the Muschelkalk, while the ammonoid devoid Tredian and Kingrali units to the ammonoid scarce part of the Kaga to Kioto units.

2.4 Biostratigraphic Zonation in the Triassic

2.4.1 Ammonoid Zones in the Spiti Triassic

On the GTM, the Triassic is best refined into a succession of ammonoid zones in Spiti (Krystyn et al. 2004), however only Early and Middle Triassic is so zoned. The Late Triassic in Himalaya is mostly ammonoid devoid but for occasional record. There are ~35 zones in the ~15.2 my long Early and early Middle Triassic (Fig. 2.6), which averages to ~400 ky per ammonoid zone.

2.4.2 Conodont Zones in the Triassic of Spiti

Krystyn et al. (2004) provided conodont taxon range charts against detailed Spiti sections up to the close of the Chomule Formation, i.e. approximating to the Ladinian/Carnian or Middle/Late Triassic transition, however, zonation is done exclusively within the Early and Middle Triassic. A succession of 17 conodont zones is developed in the ~15.2 my long Early and early Middle Triassic, which approximates to an average conodont zonal reolution of ~800 ky which is almost half to that of ammonoid zones in the same interval.

2.4.3 Palynozones in the Non-marine Triassic

In the Gondwana non-marine exposed and subsurface basins is developed a succession of seven palynozones (Prasad 2011). Initially, a succession of 4 zones (Prasad et al. 1995) was developed in the Late Schythian to Norian interval. Later one more zone was added for the Early Schythian, while the Fossulatus Zone was presumably subdivided into two zones, and one zone added at the top, thus totaling a succession of seven zones. The first three broadly correspond to Schythian and at least Early Anisian which in turn corresponds to the Panchet Group. The younger four palynozones correspond to the Mahadeva Group of late Anisian to close of

Triassic. The correlation of marine and non-marine biostratigraphic zones has been mutually developed and integrated in Chaps. 3 and 4 with the help of second-order sequence framework.

2.5 Salient Mid-Triassic Features

2.5.1 Biotic and Abiotic Changes Across the Mid-Triassic Anisian/Ladinian Boundary in Spiti

Drastic changes are noticed across the Anisian/Ladinian transition like sudden marked decrease in (i) density, diversity and frequency of ammonoids, (ii) extent and frequency of hard grounds, (iii) carbonate content, (iv) number of concretion/pebble/nodular beds and increase in (i) sedimentation rate, (ii) size of structures, (iii) size of beds, increase in coarseness of texture etc. Quantitatively, there are 34 ammonoid levels in ~13 my long interval up to bed 40 of Muschelkalk Member that correspond to 34 ammonoid levels in in ~38.5 my long post bed 40 part of Spiti Triassic succession. The same amounts threefold decrease in frequency of ammonoids across the Anisian/Ladinian boundary. The pre-Ladinian thickness is ~40 m while post-Anisian thickness is ~1400 m. Thus in context of thickness, the decrease in the frequency of ammonoid is 35-fold. The maximum generic diversity is 10 in bed 34 of the Mikin Formation at or around the boundary (in beds 34–40) which in the post-Anisian Triassic reduces to 3 in the Ladinian Kaga and Chomule units, and to just one in Carnian–Norian interval, and finally absence of ammonoids in the terminal Rhaetian–Plienbachian interval of a major phase of eustatic fall. (detailed discussion on sequence stratigraphy in Chap. 3).

2.5.2 Mid-Triassic Unconformity and Associated Changes in Inland Gondwana Non-marine Basins as also in East and West Sector Coastal Basins

Disconformity with floral and lithological change is observed in K-G basin (Prasad 2011). Mid-Triassic unconformity along with a large stratigraphic gap up to the Pilensbachian or up to the Late Jurassic Kimmeridgian duration is interpreted in the Bengal basin (Vijaya 1997, 2009). Even in Kachchh, the disconformity is present along with a large gap from Late Precambrian to Mid-Triassic until the origin of the Kachchh basin.

2.5.3 *Intra-Triassic Localized Rifting/Igneous Event in India*

Casshyap and Khan (2000) attribute the profound change in the east–west striking Koel–Damodar, Son–Mahanadi, Rajmahal and Satpura basins to uplift of the source area in the southeast and the resultant non-deposition/deformation and, angular unconformity between the Late Permian–Early Triassic sediments below, and the superposed coarse to very coarse, sandstone, conglomerate, pebbly sandstone above. Evidence is known in West Kashmir of the intercalation of volcanics with Mid-Triassic Muschelkalk and associated sedimentary units (Wadia 1934). Calkins et al. (1975) have made similar observations in north Pakistan. Even the Siberian igneous event had its terminal pulse in the terminal Anisian. Thus, there could be either extended or prolonged Panjal igneous activity or even a separate mid-Triassic event in the Indian north sector as well. The Permian and Early Triassic sediments invariably in all the inland Gondwana basins feature uniform development which is distinctly disturbed by the mid-Triassic event (Krishna 1987a). Mid-Triassic crustal stretching or attenuation has been interpreted in the Spiti High Himalaya (Draganits et al. 2004), so also the origin of the pericratonic rift basins of Kachchh and Saurashtra. In the Spiti sedimentary succession, diverse drastic changes have been inferred within the Middle Triassic.

In Palamau (Bihar) close to the Koel–Auranga Gondwana basin are known Rhyodacite within the Precambrian of K-Ar date of $\sim$217 ma (Sarkar et al. 1996). Also is recorded a $\sim$265 ma intrusive hornblende peridotite dyke at Richguta (Ghose et al. 1973).

2.5.4 *Triassic Palyno-Record in the West Sector*

In Jaisalmer, in a stratigraphic table (ONGC 1993), the Bhuvan Formation in the Bhuvan well is shown against Triassic period on the basis of palynomorphs present therein. It appears that in the West Indian sector, both Early and Late Triassic sediments are recorded. In Pakistan, there is nearly complete presence of the marine Triassic with ammonoids being restricted to the Early Triassic Mianwali Formation (Shah 1977). The reevaluation of the Banni well palynomorphs can throw more light as to how far back they go in Triassic.

2.6 Jurassic

2.6.1 *Jurassic System—International and National Status*

The Jurassic system was named in 1829 by Brongniart on a succession in the Jura mountain range of Europe. It is presently organized into 11 stages (4 in Early

Jurassic—also known as Liassic, 4 in Middle Jurassic—also known as Dogger, and 3 in Late Jurassic—also known as Malm). The succession of Jurassic stages in ascending order is Hettangian, Sinemurian, Pliensbachian, Toarcian, Aalenian, Bajocian, Bathonian, Callovian, Oxfordian, Kimmerdgian and Tithonian. According to 2015 time scale (Cohen et al. 2015), it ranges from 201.3 to 145.0 ma with a duration of 56.3 my. Currently, on the basis of succession of the rapidly evolving ammonoids, it is organized into ~65 zones, ~150 subzones and more than 200 horizons. The system is defined by the base of its first stage—Hettangian, more precisely by the base of its first ammonoid zone—the Psiloceras planorbis Zone.

2.6.2 Development in the Indian Subcontinent

There is widespread development of marine Jurassic along the entire High Himalaya (India, Pakistan, Nepal, Bhutan and South Tibet), Kachchh, Jaisalmer, Axial belt and Indus basin of Pakistan with the best development in Kachchh (Fig. 2.18) with continuous late Middle Bathonian–Late Tithonian ammonoid succession. Early Bathonian ammonoids are recorded only recently (Krishna unpublished report 2014; Pandey and Pathak 2015a) However, the ammonoid record in Kachchh is from Bajocian onward of the younger six stages, whereas Hettangian, Sinemurian, Toarcian ammonoids are known from High Himalaya. The Bajocian ammonoid record in Kachchh is extremely poor, but rich in Madagascar, Pakistan, Nepal, and South Tibet. Toarcian ammonoid record is rich in Madagascar and Pakistan. In recent years, important marker coccoliths have also been discovered (Krishna et al. 1983; Rai and Jain 2012, 2013) of Toarcian to Kimmeridgian stages along with reworked taxa of Early Toarcian interval in Kachchh. The presence of terminal Jurassic is also evidenced through the record of the Late Oxfordian–Tithonian palynomorphs in the east coast basins, in particular, in the K-G basin (Prasad 2011).

2.6.3 Outline of the Ammonoid Biogrographic Framework

A distinct IEAP (Indo-East African Province) has been interpreted between Early Toarcian and Late Tithonian on the basis of the ammonoid distribution (Krishna 1983 onward Fig. 2.21). Ammonoid biogeographic reconstruction studies evidenced the availability of the shallow marine Gondic corridor only from late Late Tithonian (Krishna 1983 and in many later papers) from the Neotethyan margin of the Gondwana (GTM) to the SE Pacific margin of the Gondwana. Detailed ammonoid geographical treatment has been made in Chap. 6.

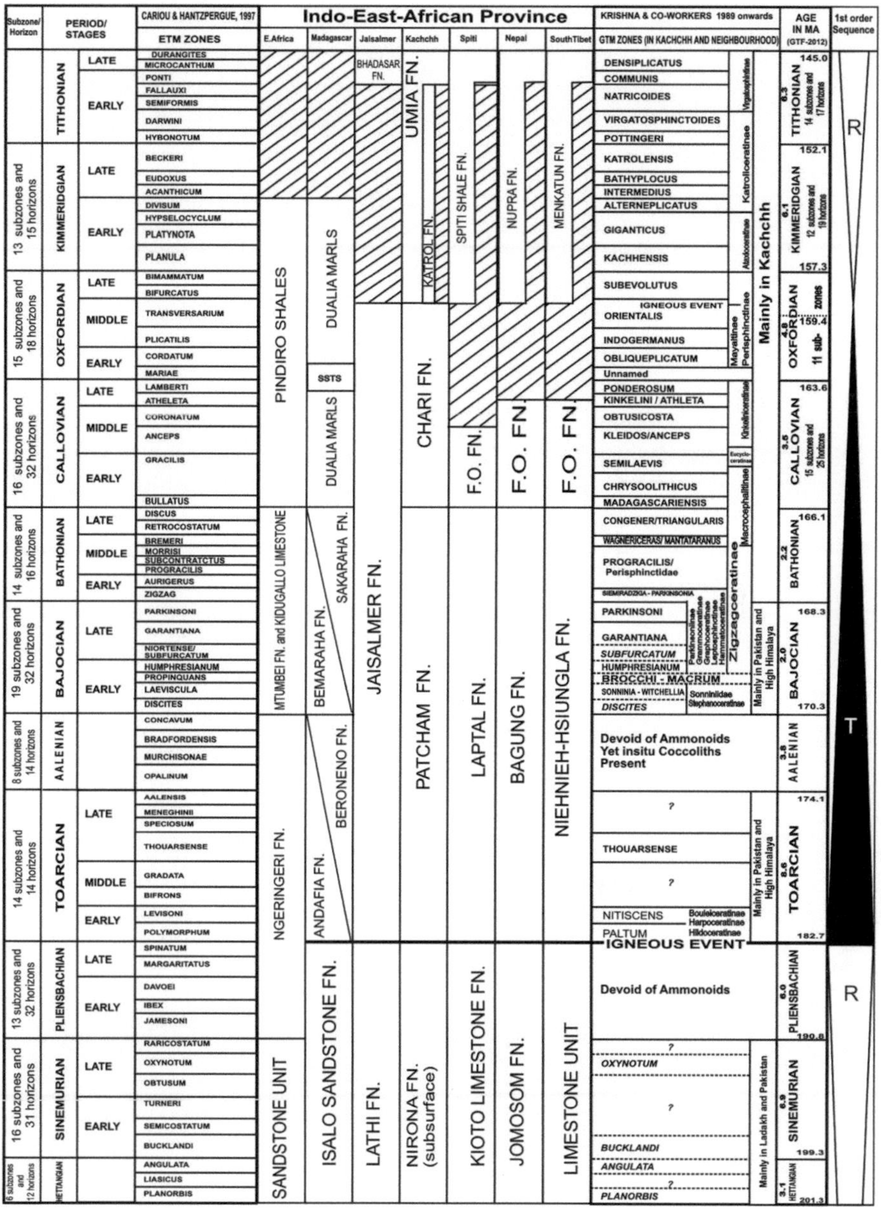

Fig. 2.18 Jurassic units in Indo-East-African Province (modified after Krishna 2002)

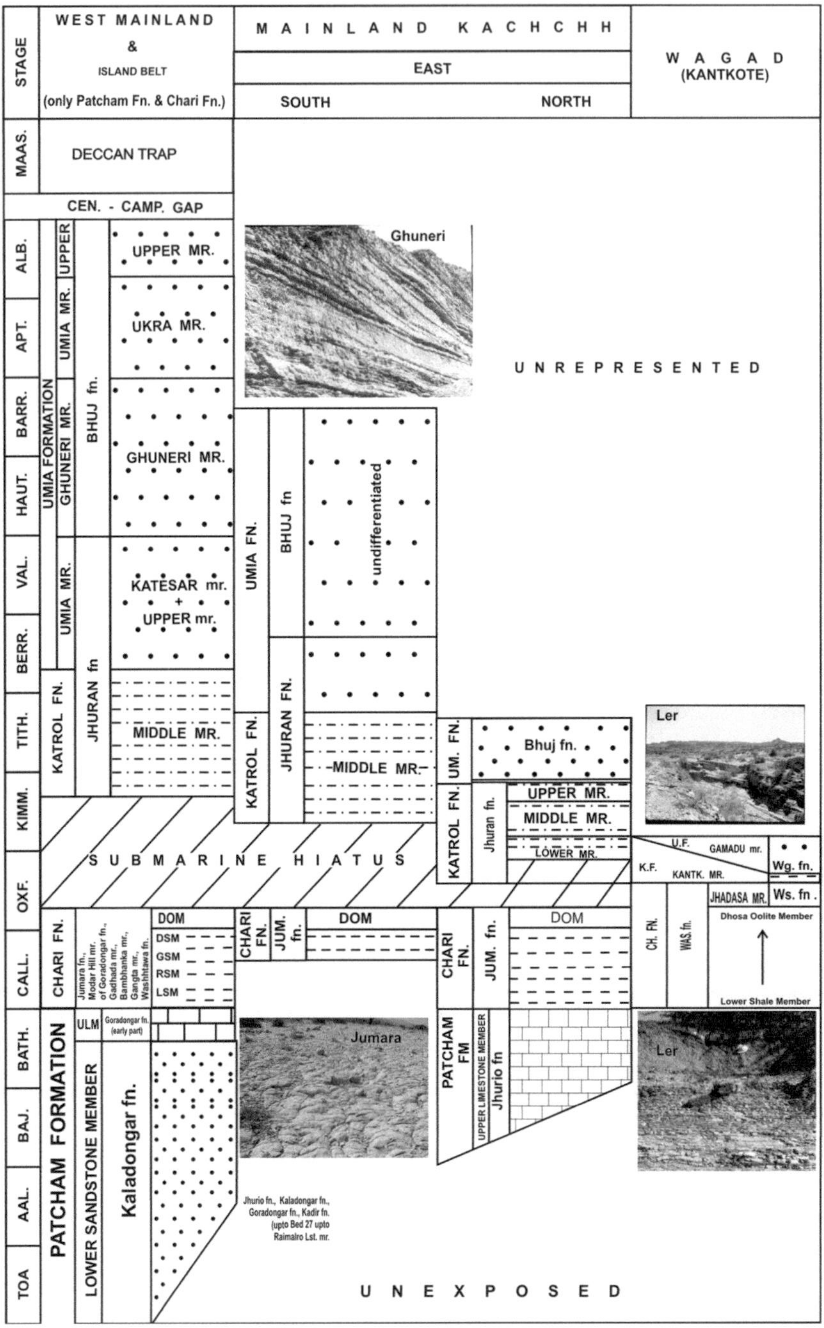

Fig. 2.19 Integration of Biswas (1993) lithostratigraphic units into Stoliczka's units in the Kachchh Mesozoic (modified after Krishna 2002)

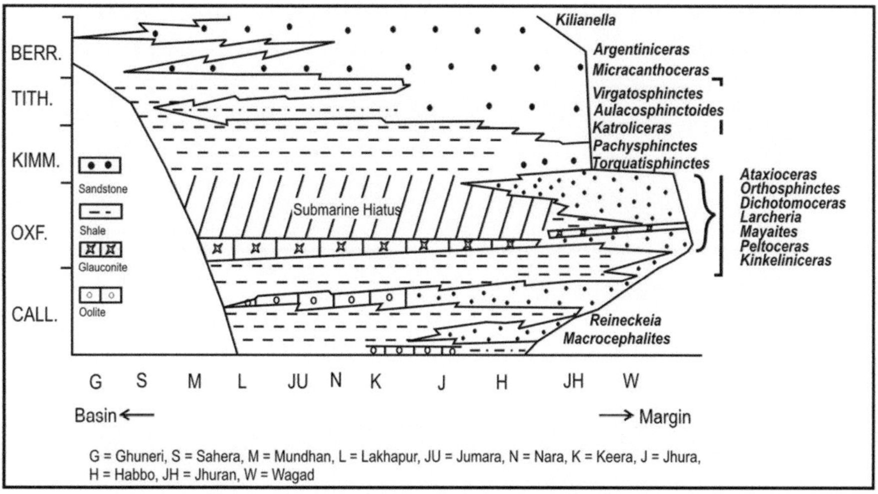

Fig. 2.20 Schemetic chronostratigrapic profile of the Callovian–Berriasian interval in Kachchh between Wagad and Mainland (modified after Krishna 2002)

2.6.4 Marine Jurassic in Jaisalmer

As in neigbouring Kachchh, based on the regional context of the Indian subcontinent and GTM, the deposition of marine sediments in Jaisalmer presumably, also, started in Early Toarcian over the non-marine Ladinian–Pliensbachian. The ammonoids are mainly of Callovian and Tithonian stages. A few ammonoids are also known of Late Bathonian, Oxfordian and Kimmeridgian. The Middle Oxfordian to early Early Tithonian makes a gap which has been interpreted as submarine as in Kachchh (Krishna et al. 2011).

2.6.5 Marine Jurassic in High Himalaya with Focus in Spiti

The marine Jurassic succession is known to rest conformably over the marine Triassic and in turn is conformably overlain by the marine Cretaceous in Spiti and elsewhere in High Himalaya. At least outer shelf maximum depths are realized in the Late Jurassic. The Middle Oxfordian to at least Early Kimmeridgian interval, if not more, is unrecorded in High Himalaya, and is interpreted as submarine gap. Additional smaller gaps are also present above the Early Bajocian in a few basins. The Hettangian–Pliensbachian record is in carbonate facies, the Toarcian–Bathonian in mixed fine carbonate and clastics, and the Callovian–Tithonian interval in black shales that is rich in nodular concretions. Ammonoids are known of Toarcian, Bajocian, Bathonian, Callovian, Early Oxfordian, Late Kimmeridgian

Photo 2.1 Golden oolitic limestone of Bajocian age from Jhura

Photo 2.2 Late Bathonian nodular limestone of Patcham Formation from Jumara

and Tithonian stages, however, abundant only in Tithonian as nucleii of concretions. Guide dinocysts, mostly of the Tithonian age have also been recovered (Garg et al. 2003) (Photos 2.1, 2.2, 2.3, 2.4, 2.5, 2.6, 2.7, 2.8, 2.9, 2.10, 2.11, 2.12, 2.13, 2.14, 2.15 and 2.16).

2.6.6 Brief Summary of the Marine Jurassic in the Neighbouring Pakistan

The Upper Indus basin of Pakistan has been a constituent basin of the large Indus shelf, which is located only few hundred km away from the Kachchh basin. Both

Photo 2.3 Shales of Chari Formation at Ler

Photo 2.4 Dhosa Oolite marker at Ler

Photo 2.5 Dhosa Oolite hard ground at Lakhapur

Photo 2.6 Sandstones of Katrol Formation at Ler

Photo 2.7 Sandstones of Katrol Formation at Ler

Photo 2.8 Sandstones of Umia Formation at Ler

Photo 2.9 Jaisalmer fort on the Fort Member

Photo 2.10 Panaromic view of the Spiti Jurassic–Cretaceous

these basins share similar geological history since the early Early Toarcian initiation of the marine sedimentation in Kachchh. The Upper Indus basin on the other hand includes as well the marine ammonoid scarce Hettangian–Pliensbachian. The Early Bajocian in the Axial belt of Pakistan includes the ammonoid-rich Early Toarcian and Early Bajocian. Thus, Pakistan provides an appropriate Hettangian–Early Bajocian compliment to the Kachchh Late Bajocian–Tithonian ammonoid record. It also makes the record of the marine Jurassic of the Indian subcontinent nearly complete. The ammonoid record however still excludes the Pliensbachian and

Photo 2.11 Giumal Formation at Giumal in Spiti

Photo 2.12 Nodular Limestone Member of Bagh Formation (*courtesy* B. Pandey)

Photo 2.13 Cretaceous in the Narmada basin (*courtesy* B. Pandey)

Photo 2.14 Development of asymmetric sequences in the Karai Formation (*courtesy* B. Pandey)

Photo 2.15 Marker sandstone bed in the Late Albian of the Cauvery Cretaceous (*courtesy* B. Pandey)

Photo 2.16 Transition from the Spiti shales to Guimal sands at Giumal

Aalenian stages, which, though present in High Himalaya marine facies, are neither ammonoid bearing nor even stratigraphically differentiated from the Sinemurian below and the Bajocian above.

2.7 Jurassic in Kachchh

2.7.1 Initiation of the Marine Transgression in Kachchh

The marine transgression that began regionally in early Early Toarcian, is also so interpreted in the Kachchh basin (Krishna and Pathak 1994; Krishna et al. 2010, 2011). At this point of time occurred a regional tectonic event on the entire GTM, with associated igneous expression in all the sectors. The geological signature of this regional event is well-recorded all over the GTM as the major intra-Jurassic first-order SB. In the regional perspective, it is expected that even the non-marine Ladinian to Norian sediments may also be locally present underneath the Rhaetian in the exposed Patcham sediments immediately after the origin of the basin.

2.7.2 Improbability of the Transgression in Early Pliensbachian During the Terminal Phase of Eustatic Fall

Minor transgressions in the terminal part of the first-order RST interval have been suspected in a few Arabo-African basins, e.g. Oman (Rabu 1993) in Early Pliensbachian Ibex Zone MFS. These are based on coccolith and/or other micro-fossil evidence. Any definite Pliensbachian ammonoids have neither been knwon nor any marine incursion otherwise confirmed over the last 100 years in East Africa or Madagascar, or even Indus basin of Pakistan and the Indian Jaisalmer basin, or even in wells. Hardly any one ever interpreted a significant eustatic rise but of a third order in an over all major fall since the Ladinian until the close of Pliensbachian. In the entire neighbourhood, thick unfossiliferous sands are known to make the terminal Sinemurian–Pliensbachian interval. The coccoliths ages are based on ammonoids and not otherwise. In Kachchh, the presence of the reworked Early Plienbachian coccolith mixed with reworked Early Toarcian elements (Rai and Jain 2012, 2013) is difficult to explain. For example in an insitu assemblage, the *C. pliensbachia*, with its LAD within Early Pliensbachian, can not co-occur with *B. finchii* having FAD in Late Pliensbachian. So, also, is impossible the cooccurrence of the *P. liasicus* having LAD within Early Toarcian and *T. tiziense* which appeared first only in Aalenian. These examples demonstrate that either the above cooccurrences are anomalous or the ranges in India are different than those in Europe. Even in Europe the ranges are diachronous from the Boreal to Mediterranean to Submediterranean regions.

At the time of intra-Triassic origin of the Kachchh basin, the Neotethys, in spite of being present in the vicinity through its due south progressing arm, witnessed a major eustatic fall phase (Haq et al. 1987 and others). On the other hand, the infant Kachchh basin most probably was extremely shallow and localized, and hence away from the reach of the Neotethyan arm. During the major eustatic fall phase, the Kachchh basin is interpreted as surrounded by emerged relatively high land conditions and several hundred km away from the Neotethys in spite of the over all due west basinal slope. The surrounding high-lands coupled with shallow localized basin/s did not permit any transgression. Subsequently, the start of eustatic rise combined with significant deepening of the rift in Early Toarcian facilitated the transgression. The flooding of the basin could not have occurred during relatively quieter intervening phase of eustatic fall between the successive rift events of the terminal Anisian and basal Toarcian. Prior to start of the Toarcian, any minor order rise is considered insufficient for the transgression in Kachchh during the pre-ceeding Pliensbachian interval as claimed by Rai and Jain (2013). The published coccolith data lacks decisive evidence. On the same grounds is also contested the pre-Toarcian initiation of transgression in any East African basin.

But for the above anomalies and discrepancies, the Early Plienbachian ingression cannot be totally ruled out in a basin which was undergoing rift-related deepening. It could have been transgressed if the rate of rift-related subsidence exceeded the rate of eustatic fall as an extreme remote possibility. It is also important to evaluate as to which of the two transgressions was stronger in the Kachchh–Early Toarcian or Early Pliensbachian. The answer emphatically is Early Toarcian in a regional context. Doubts also arise since Rai and Jain (2012) reported the assemblage without any mention of reworking, and interpreted the NJ5 to NJ7 Late Pliensbachian–Early Toarcian age. The NJ5 to NJ7 range owerwhelmingly favours Early Toarcian. The same assemblage from the same sample is claimed to be reworked and assigned as well to NJ4 by Rai and Jain (2013) without any explanations about reworking and rvised age assignments. It is here contextual and significant to note that the coccolith based ages have small overlaps/discrepancies with the undisputed ammonoid based ages in Callovian, and has been occasionally found distinctly older to the ammonoid based ages. In this context, it is emphasized that the coccolith ranges are standardized by integrating them with the ammonoid biostratigraphy in the Mesozoic for unreserved age assignment of the ammonoid devoid sediments. The Hettangian–Pliensbachian coccoliths need be explored in High Himalaya Spiti and Ladakh, Axial belt and Lower Indus basins of Pakistan. In spite of the above uncertainties, the date of rifting and concomitant transgression in the Indo-East African province at the start of Toarcian is here maintained.

2.7.3 Oldest Jurassic Ammonoid Evidence

The oldest ammonoid evidence in the exposed geological record since the early 1980s (Jaitly and Singh 1983) has been the single incomplete 'Late Bajocian *Leptosphinctes*' whose taxonomic status and age have been reviewed in the present work. The '*Leptosphinctes*' level is underlain by the ammonoid devoid plus 200 m sediments in Kala Doongar near the north margin of the basin. Another stratigraphic section in the same Patcham region at Gora Doongar, during sampling for magnetochronology yielded a single *Calliyphylloceras* of suggestive Early Bajocian age, if not older (Pandey et al. 2013a) from near the base of that section. This level of tentative correlatibility is underlain by the older sediments of the Dingy Member at Dingy hill and Kuar Bet in the Kala Doongar section.

2.7.4 Oldest Jurassic Nannoplankton Evidence

The older Aalenian and Toarcian stages were always suspected even in the exposed sediments of Kachchh (Krishna and Pathak (1994) and later). Reworked Toarcian nannoplanktons (Rai 2006; Rai and Jain 2012, 2013; Rai et al. 2005) in the Late Callovian non-gypseous shales (collected under guidance of the present author) of Lakhapur/Jara dome of Mainland Kachchh suggested presence of marine Toarcian sediments in Patcham region of the north margin of the basin which later worked as provenance for the Lakhapur Late Callovian sediments (Krishna 2012). The author's so held interpretation was confirmed through the discovery as well of insitu Toarcian–Aalenian coccoliths in the middle part of the Dingy beds at Kuar bet (Rai and Jain 2012) of Lower Sandstone Member (LSM) of Patcham Formation. It substantiated that the marine transgression commenced in Kachchh in Early Toarcian as elsewhere in several basins on the GTM.

2.7.5 Regional Evidence of the Early Toarcian Marine Transgression

There is evidenced major transgression in early Early Toarcian all over the Gondwanian Tethyan Margin (GTM) from Arabia to NW Australia over a regional subaerial gap at the basin margins (Krishna et al. 2011), particularly in Madagscar and Pakistan that were located close to Kachchh in the Jurassic. Even in Tethys Himalaya, is witnessed drowning of the carbonate platform near the Pliensbachian/Toarcian boundary, so dated by the record of Early Toarcian *Alocolytoceras* (Stoliczka 1866) in the marls immediately above the Kioto Limestone in the Spiti basin.

2.7.6 Jurassic Stratigraphic Record of Kachchh and Neighbourhood

Among the three Mesozoic systems in Kachchh, the best developed is Jurassic, also in comparison far better than the overlying Cretaceous, while the Triassic record is meagre as discussed above. Out of the 11 Jurassic stages, the younger six are definitively differentiated in the exposed record through ammonoids. Insitu Toarcian–Aalenian nannoplanktons have been recorded at Kuar bet in Dm, soon after the discovery of reworked Toarcian–Aalenian nannoplanktons in Late Callovian of West Mainland (Rai and Jain 2012). In a regional sratigraphic framework, these two stages had long been anticipated in the exposed basal part that is included in Dm of Patcham Formation at Patcham, which otherwise had been best assigned to Bajocian. The non-marine undifferentiated Liassic is known in the Banni well (Koshal 1973, 1984). The exposed Jurassic stratigraphic sections are of high quality with good over all presence of the stratigraphically significant ammonoids, and their potential has been successfully exploited by the author and associates in developing the high resolution ammonoid based zonal scales in Kachchh for the GTM in the relatively better endowed early Early Bathonian–late Late Tithonian interval. There is developed a well-defined succession of about 27 zones, 50 subzones and 80 plus horizons in the ~26.0 my long late Bajocian–Tithonian interval (Fig. 1.8). The Jurassic development elsewhere in India, even outside in Pakistan and Madagascar, which were part of the erstwhile Tethyan margin of the Indian subcontinent, even of GTM, allowed further recognition of the additional ~20 ETM ammonoid zones on the GTM. To summarize, out of the ~65 zone ETM Jurassic zonal scale ~45 zonal intervals have found differentiation in the Indian subcontinent among which are 25 indigenously developed ammonoid zones together with their subzones/horizons. Among the mentioned 20 ETM zonal intervals, five are differentiated in the Kachchh Bajocian–early Middle Bathonian interval, and rest 15 recognized in the other neighbourly located basins of the erstwhile Indian subcontinent. Excellent stratigraphic sections and the refined scale so developed further allowed us to formulate a highly refined first to third-order sequence framework for the GTM in the Indian subcontinent that is detailed later in Chap. 3. The sequences are mostly developed and differentiated in Kachchh, and then demonstrated elsewhere on GTM as best as possible in spite of severe impediment of the non-availability of the desired level of biostratigraphic refinement in most of the basins.

2.7.7 Lithostratigraphic Framework in the Jurassic of the Kachchh Basin

The succession of Patcham, Chari, Katrol and Umia lithostratigraphic units (2.17–20) developed by Stoliczka in his field diary in 1867 for the exposed Mesozoic

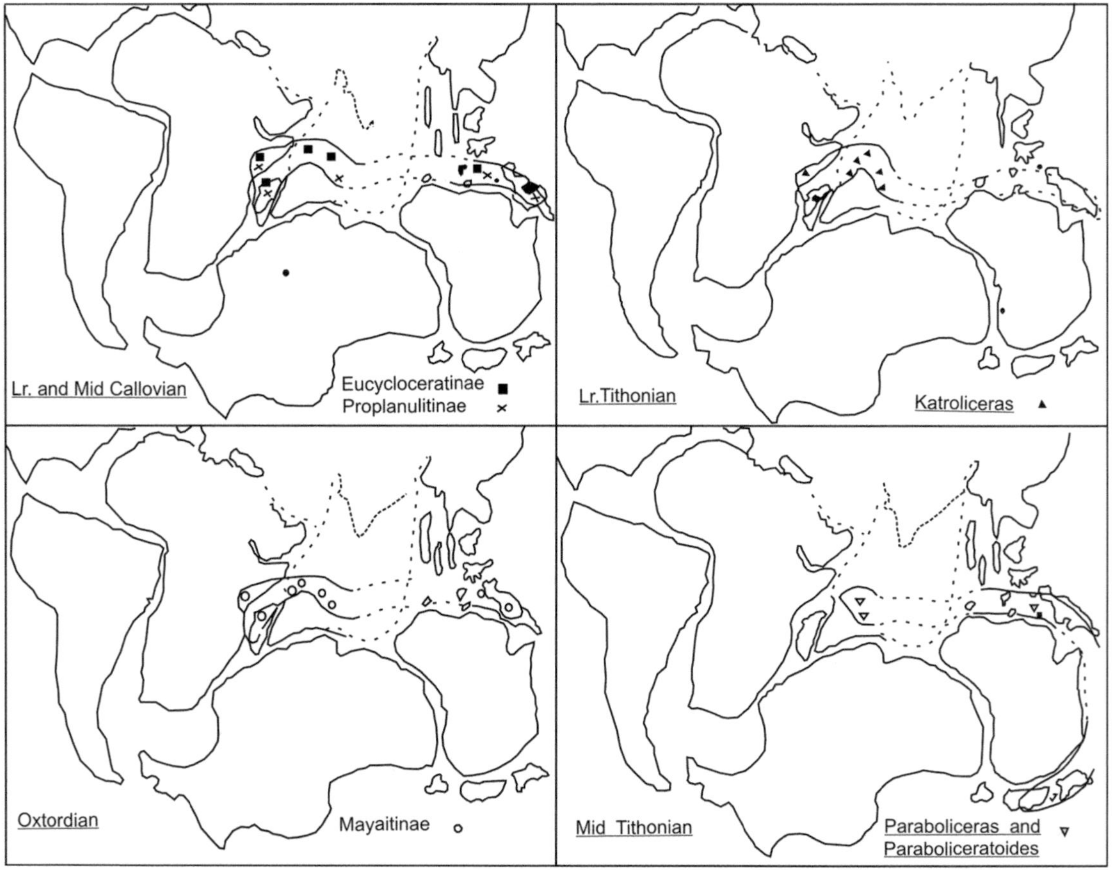

Fig. 2.21 Probable extent of the Indo-East-African Ammonoid Faunal Province (within the Tethyan Realm) for successive stage and substage time intervals (Lower Callovian to Tithonian) (after Krishna 1983)

sediments of the entire Kachchh basin was later published in Waagen (1871), who himself never visited Kachchh. The Stoliczka's scheme was based primarily on Stoliczka's lithostratigraphic considerations that he recorded in his field diary.

Biswas (1977) and a few others misunderstood Stoliczka's units as biostratigraphic and on that pretext coined new lithostratigraphic names. On the contrary, the present author recognized all the four units in the entire basin with emphasis on physical and lithological attributes.

Biswas' Several Unwarranted Objections to Stoliczka's Units Clarified Once Again

1. Stoliczka's units lacked original definitions and were considered biostratigraphic by Biswas.
   Stoliczka—a non-ammonoid field geologist—classified the sequence during his only stay in Kachchh in his field diary on the basis of his field observations. He did not have the opportunity of any taxonomic determinations of the collected fossils even after his return due to his untimely death within a year of his field visit to Kachchh. Such a classification cannot be biostratigraphic. The taxonomic determinations were made later by Waagen (1871) in the same publication in which he included Stoliczka's fourfold classification.
2. Stoliczka's units were not linked to any type sections.
   This was not considered obligatory in the nineteenth century. The very names of the units explicitly implied the type section/locality. Any code for stratigraphic nomenclature did not then exist. Nevertheless, it is clear that the Patcham Group/Formation was named and based on the exposed sedimentary succession in Patcham region up to the contact of the light coloured limestones and relatively darker grey shales, while the Chari Group/Formation was so based in Keera dome up to the top of the marker Dhosa Oolite in the neighbourhood of Chari–Fulae village settlement. It was neither obligatory then nor at all important even at present as to whether Chari village stood on Chari Group sediments or not. Even Biswas could not find a better or alternative differentiation of his Jumara Fromation which only later turned out as a junior synonym of Stoliczka's Chari Formation with the same content and boundaries in another locality.
   Similarly, Stoliczka's Katrol Group/Formation and Umia Group/Formation were based on their respective dominantly shaly and dominantly sandy character, respectively, in the Katrol hill and the Umia region. The base of Umia unit was defined by the base of the green glauconitic sands above the Katrol shale succession in Umia. Krishna could easily recognize the Katrol unit in Ler–Katrol above the Dhosa Oolite.
   To clarify the nomenclatural ambiguities and uncertainties Krishna (1987b, Table 4) published an elaborate framework with stratotype sections, essential physical/lithological features of these units, their lithostratigraphic boundaries, and facies variation from basin to margin.
3. The mappability of Stoliczka's units was never tested, yet lacked mappability.

This is not true. Rajnath mapped and published Stoliczka's units in west Mainland. The applicability of the mentioned units has been demonstrated in the entire basin from east to west and north to south.

4. Stolickzka's units were distinguished by fossil content. This is untrue, as the mentioned units have been primarily recognized and differentiated on gross lithology and marker beds.
5. Boundaries remained disputed and remained untraceable in poorly fossiliferous east Kachchh.
   This is again untrue; the boundaries of Stoliczka's units are tracable in the entire basin on gross lithology and lithological marker beds.
6. Stoliczka's classification and units are ambiguous and defy code of stratigraphic nomenclature.
   There is no such ambiguity found by the present author. No code existed in the nineteenth century. The units as originally implied by Stoliczka and later understood by the present author and numerous other workers are well in tune with all such codes that have come up from time to time.
7. Stoliczka's units had their time ranges, disputed and constantly under revision. This has neither any relevance nor obligation to the validity of the units. Time ranges of lithostratigraphic units are regularly revised as per new information.
8. Stoliczka's units are only applicable to Mainland while correlation problems prevent their extension to other parts of the basin not studied by Stoliczka. The applicability of Stoliczka's units in all parts of the basin has been amply well-demonstrated through new data and vastly improved correlations that have allowed reasonably precise age to every cubic centimetre of sediments through the length and breadth of the basin.

In light of the above, every new worker in any basin needs to honestly stand by the existing unit names irrespective of either lack of minimum or sufficient information or even faulty information. There is always scope of supplementing information towards making them more objective and understandable by fresh crop of workers without changing the names of the units, and thus avoiding possibility of addition to confusion. The science of geology and stratigraphy may soon reach a stable pedestal that it can even do away with plethora of names of lithostratigraphic units, and yet arrive at improved high resolution objective comprehension through event and sequence stratigraphic approach, events and resultant sequences that transcend intrabasinal and interbasinal limitations and instead have regional applicability.

2.7.8 Adoption of the Member Units of Biswas and Others

Stoliczka in his incomplete, rather preliminary, field studies did not attempt member units, and as such member units of Biswas and others in the stratotype localities of Stoliczka's units have been readily and easily incorporated and integrated in

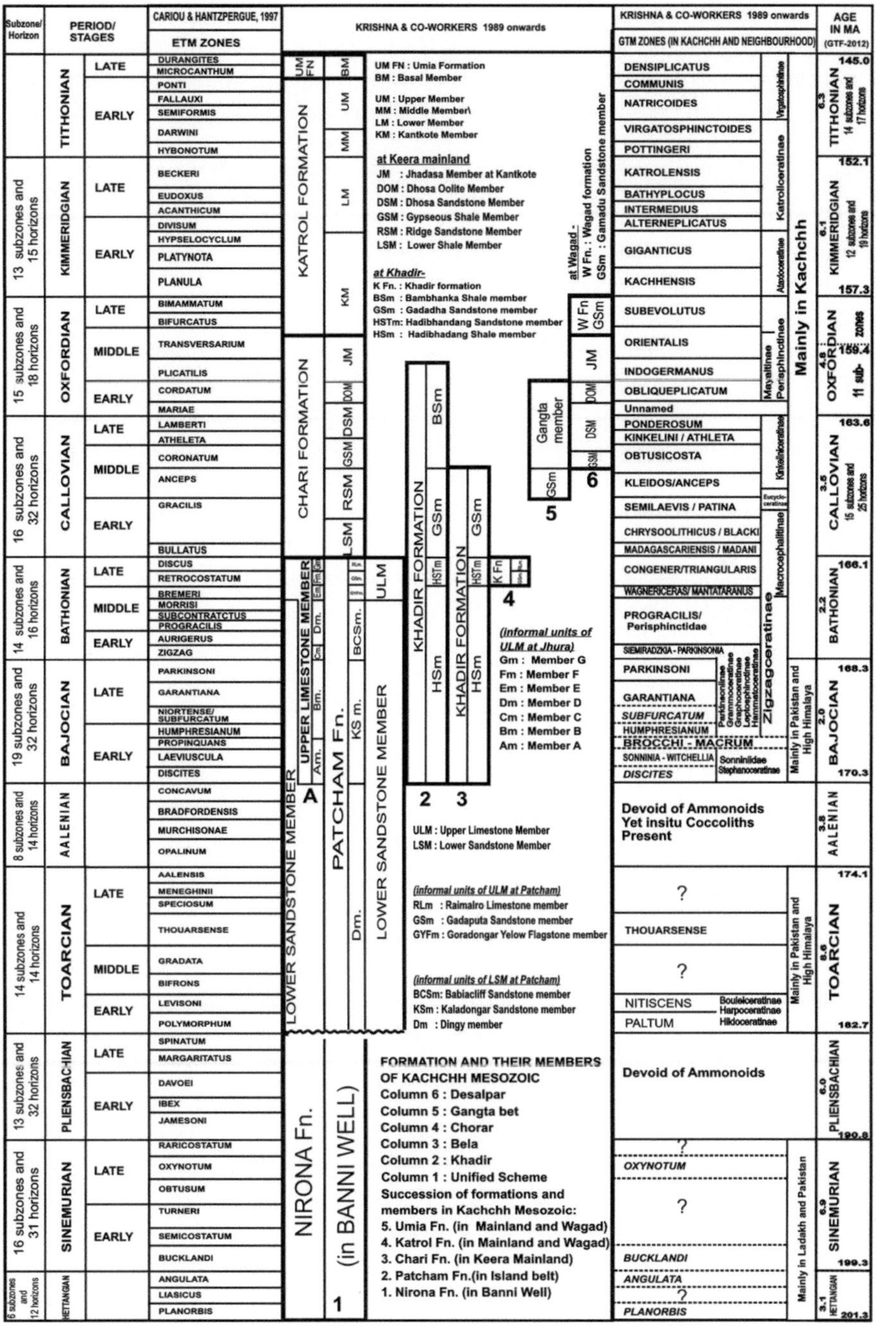

Fig. 2.22 Stoliczka's units demonstrated in the entire kachchh basin

Stoliczka's scheme while 'formations' created later have been abandoned and suppressed as junior synonyms (Krishna 2002, 2005 and later). A few new formal and informal units at member level have been necessitated, and so here introduced. The so integrated and unified lithostratigraphic framework is reproduced here (Fig. 2.22). In the present work, the abandoned members and formations, if and when, informally used are mentioned with suffixes m and fn.

Patcham Formation
The base of the Patcham Formation, i.e. contact with the basement is nowhere exposed in the basin. The marker Raimalro Limestone member and other homotaxial limestones make the top of the Patcham Formation in Mainland and in 'island belt'. The present author has traced the Patcham/Chari lithostratigraphic boundary meticulously. In its stratotype in Patcham region, the Patcham Formation includes two newly introduced members, viz., **Lower Sandstone Member (LSM)** with three informal members of Biswas and **Upper Limestone Member (ULM)** again with three informal members of Biswas in their stratotypes in Patcham. LSM is exclusive to the 'Island belt'. ULM is present both in Mainland (Habo, Jhura, Nara and Jumara localities) and 'Island belt' (Fig. 2.22). The complete ULM is present only in Jhura, yet its base is not exposed there. The contact with the underlying LSM is present all over the 'Island belt'.

Chari Formation
It is incomplete throughout the Mainland where the marker Dhosa Oolite represents its youngest part. Its terminal part is absent all over the Mainland in view of the overlying large significant submarine gap and the associated Chari/Katrol disconformity (Pathak 1989; Krishna and Pathak 1989, 1991 and since then). The terminal part of the Chari Formation is exclusively present in Wagad and in the neighbourhood of Khadir. The best development of the mentioned terminal part is in a $\sim$10 m thick ammonoid abundant hardgrounded succession in the Kantkote section (included in Jhadasa Member), that is well over two ammonoid zones ($\sim$2 my) younger to the Dhosa Oolite of the Mainland. But for near absence of oolites, the mentioned 10 m thick succession is physically closest to Dhosa Oolite in its hard grounded, nodular, bumpy, conglomeratic features as also richness in ammonoids (only as visually observable physical features in the field). The Chari Formation includes six members (Fig. 2.22), the first five in Mainland and the sixth member in Wagad. Among the Mainland members, the first four members are created in Biswas (1977), while the younger two members have been created later. The first three members of Biswas are Lower Shale Member (LSM), Ridge Sandstone Member (RSM) and Gypseous Shale Member (GSM). Biswas' 'Dhosa Oolite member' has been further differentiated by Fursich into Dhosa Sandstone Member (DSM) and Dhosa Oolite Member (DOM) inclusive of the youngest conglomerate bed in Mainland. The youngest member—Jhadasa Member (JM)—is created in (Krishna et al. 1998). JM is present exclusively in Wagad, Gangta bet, Khadir and neighbourhood. The older five are present in Mainland, while all the six in Wagad, and Khadir. The younger five are present in Gangta bet. The older two members only are present in Patcham and Bela, while just the oldest member in Chorar.

'Members' created elsewhere are held extensions of the formal members developed in the stratotype. Chari Formation is also incomplete in Patcham, Bela and Chorar uplifts. The base of the youngest member—the Jhadasa Member—is defined at Gangta Bet in GBm (=Gangta Bet Member) above the bed 13 of Biswas while the top is the top of the Chari Formation at Kantkote.

Katrol Formation

The Katrol Formation is also incomplete all over the Mainland. There is absence of its basal part in Mainland because of the inclusion of the corresponding time in the submarine gap at the Chari/Katrol disconformable boundary in Mainland. The basal part of the Katrol unit is present only at Wagad as Basal/Kantkote Member because of the absence of the mentioned submarine gap there at the Chari/Katrol unit boundary.

Katrol Formation includes four members; the basal member is present only in Wagad with Kantkote–Bharodia section as the type locality as Kantkote Member. The younger three members (Lower Member, Middle Member and Upper Member) are developed in Mainland with stratotypes in the Ler–Katrol section on the right flank of the Kukma–Hajipir Road exposed on a small hillock/mond (Photos 2.4, 2.6 and 2.7). The Lower Member includes shales/silts with interbeds of pebbly/nodular/concretionary conglomeratic sandstones in its early part that alternate with shales/clays/silt, while the later part of the laminated and flaggy sandstones up to bed 13 in East Ler. The Middle Member mainly includes sandstones with subordinate shales/siltstones from bed 14 to bed 24, while the Upper Member mainly includes shales/siltstones. The Lower Member and Middle Member combinely are broadly equivalent and lithologically similar to the 'Lower Member of Jhuran Formation' of Biswas at Jhuran as also to early part of the Katrol Formation at Ler–Katrol up to the marker belemnoid bearing marly sands. The Upper Member is lithologically similar and equivalent to the Middle member of the Jhuran Formation of Biswas.

Umia Formation

Krishna (1987b), in absence of the Umia Glauconite marker bed, placed the Katrol/Umia unit boundary in East Mainland and Wagad at the start of the first high ridge forming hard sands (start of the 'Upper Member' of 'Jhuran Formation of Biswas in Mainland and start of the 'Gamadu Sandstone' in Wagad) above the Basal or Kantkote Member of the Katrol Formation (Fig. 2.22). The Umia Formation is observed to rapidly prograde towards the basin. Umia Formation is considered physically undifferentiable in the east into member units that have been recognized and formalized only in the west. Four to Five member units are distinguished—Basal Member, Ghuneri Member, Ukra Member, Upper Member and the Amarsar Member. The youngest member—the rather thin green glauconite bed as an alternative—could be included in the Upper Member, subject to its conformity with the rest of the Upper Member that underlies it. In fact, the sandy facies that typifies the Umia Formation may have conceptually started immediately after the climax of the major transgression in the basin farther east at the margin near Radhanpur. The mentioned sands now may be lying buried under the cover of the post Mesozoic sediments.

2.7.9 Fundamentals of Stratigraphic Refinement and Ammonoid Zonation

The past two centuries witnessed refinement of the geological time to the welfare of the human society through the exploitation of diverse geological resources that are in substantive manner tied to specific geological intervals of the earth's history. Initially, the principle of order of superposition of strata was used. The Geological Time Scale (GTS) primarily evolved between 1820 and 1840 with the enunciation of the principle of faunal and floral successions by William Smith. It allowed correlation of fossiliferous beds of widely separated localities through guide/index fossils. Prior to this, the progress of stratigraphy was extremely slow and limited in geographical extent for a century and half. Smith's discovery made it fast, and organization of the rock successions could be rapidly spread through out Europe into erathems and systems.

The Mesozoic systems were named in Europe between 1822 and 1834. By 1860, the Jurasic and the Cretaceous guide fossil rich successions were already subdivided into stages and zones. The Jurassic in England as early as 1858 was differentiated into 28 ammonoid zones, even many decades earlier to the discovery of the radioactivity by Becqurel in 1894, and subsequent initiation of the radiometric age determinations.

At present, the average zonal resolution in the Phanerozoic systems is of the order of 1–2 my. In the twentieth century, of all the systems, the ~56 my long Jurassic received maximum refinement in Western Europe into 11 stages, 65 zones, 150 subzones (Figs. 2.24, 2.25 and 2.6) and well over 300 horizons with average horizonal resolution of ~200 ky. In recent years, in conjunction with the third and yet finer order sequence stratigraphic data, it has been possible to unearth depositional cycles of about 400, 200, 100, 40 and 20 ky durations in Europe and occasionally even elsewhere inclusive of Kachchh, that are supposedly governed uniformly across the globe by earth's orbital dynamics. Internationally, the refinement of geological time is a high priority research area, and is being actively persued under the aegis of the UNESCO/IUGS through the International Commission of Stratigraphy (ICS) and its sub-commissions for various systems, and stage boundary GSSP working groups.

2.7.10 Lithostratigraphic Differentiation and Guide Fossil Collection in the Field Followed by Systematic, Bio- and Chrono-Stratigraphy in the Laboratory

Detailed field investigations were undertaken only after selection of stratigraphic sections in the basin through literature survey and reconnaissance traverses. Parameters that carried paramount significance in the selection of sections were good nearly continuous frequency and density of guide fossils, fine textured, somewhat slower than moderately deposited sediments under subtidal low energy framework, and also having good potential for parallel microfossil, radiometric, geochemical and magnetochronologic studies.

The first and most essential step after the selection of the sections has been meticulous high resolution lithostratigraphic differentiation, measurement and record of the stratigraphic column, approaching near optimum possible differentiation (Krishna 2004) into say **nl** number of the smallest possible beds/bands/bedules as the smallest order lithostratigraphic units. Even this first step of high resolution lithostratigraphic differentiation has not been adequately accomplished in the Indian geological record in most of the Phanerozoic systems, although good progress has been registered in the last decade. For a Jurassic stage of 5 my average duration **nl** desirably need be of the order of 300 beds in the selected sections.

The next most important exercise has been of the guide fossil collection from as many precise lithostratigraphically differentiated positions as possible, say **np** positions. It is found that **nl** is always way greater than **np**,

$$\mathbf{nl} \gg \mathbf{np} \tag{2.1}$$

since all the beds in any stratigraphic section for environmental and preservational factors do not contain the guide fossils. The collected guide fossils were numbered, carefully packed and safely brought to the laboratory.

Laboratory studies included cleaning and preparation of the collected guide fossils, transfer of field numbers to laboratory numbers to make the material ready for taxonomic determination. Taxonomic studies were followed by heterochronic evolutionary lineage studies with macro and micro trends, preparation of biostratigraphic range charts, and finally establishment of zones, subzones and horizons. Let the number of the smallest order biostratigraphic units realized is **nb**. It is observed that **np** is much larger than **nb** which is expressed as under.

$$\mathbf{np} \gg \mathbf{nb} \tag{2.2}$$

This is so, since even the rapidmost evolutionary rates are slower compared to the durations of the finest individual depositional cycle components. From (2.1) and (2.2), it is easily deduced that

$$\mathbf{nb} \lll \mathbf{nl} \tag{2.3}$$

nb requires to be large for high resolution stratigraphic and chronologic refinement which in turn necessarily requires several fold larger **nl**. This obviously implies hard painstaking patienceful field input.

In this context, a few other parameters like chronologic resolution were also explained and defined. Also, conceptual presentation was included on the desired high quality of lithostratigraphic differentiation within a Jurassic or for that matter Phanerozoic stage of about 60–300 beds towards realizaion of high resolution bio/chrono-stratigrphic refinement. Based on the so-defined parameters, a crude comparison of relative stratigraphic refinement in the different periods of the earth

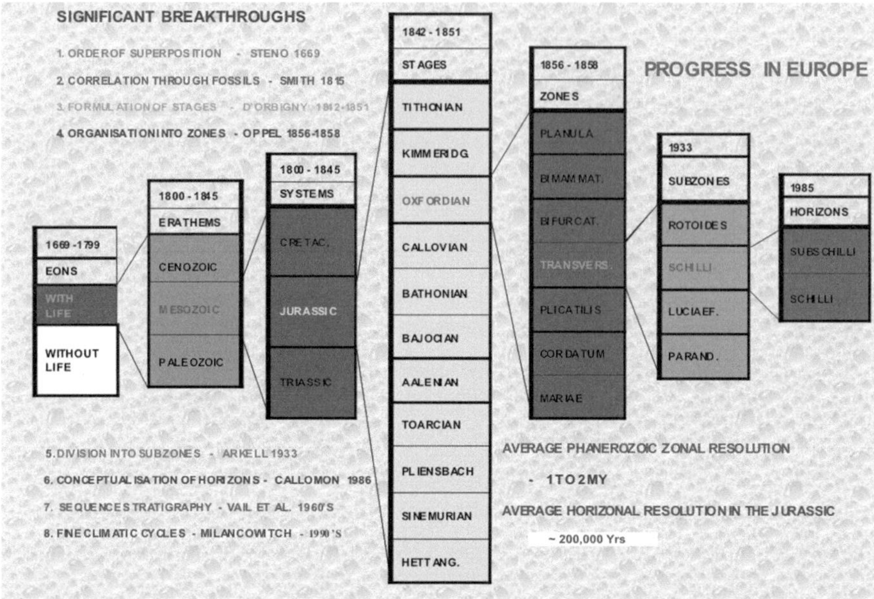

Fig. 2.23 Significant breakthroughs in ammonoid stratigraphic refinement since Smith 1813

AVERAGE DURATION OF A JURASSIC / MESOZOIC / PHANEROZOIC STAGE	55 / 11	OR	175 / 29	O R	540 / 90	=	5 to 6 MY
AVERAGE DURATION OF FINE ORDER CLIMATIC CYCLIC SEQUENCES	= CA 100,000 YRS TO CA 20,000 YRS						
AVERAGE NUMBER OF FINE ORDER CYCLICS IN A STAGE	= 50-60 TO 250-300						
MINIMUM NUMBER OF PHYSICALLY DIFFERENTIABLE UNITS IN EACH FINE SEQUENCE	= 2						
NUMBER OF OPTIMUM SMALLEST ORDER LITHOSTRATIGRAPHIC UNITS	= 100-120 TO 500-600						
MODERATELY GOOD TO VERY GOOD TO EXCELLENT TO OUTSTANDING LITHOSTRATIGRAPHIC DIFFERENTIATION	= 60% AND MORE OF OPTIMUM = 60-72 TO 300 - 360						

Fig. 2.24 Desired lithostratigraphic differentiation for high resolution ammonoid stratigraphy in the Mesozoic

history in international vis-à-vis Indian Mesozoic does not provide a comfortsome picture. Examples are presented of good quality lithostratigraphic differentiation in the Kachchh Mesozoic units. The history of the ammonoid stratigraphic refinement is also traced over the last two centuries from the differentiation of eonthems to the differentiation of horizons in the Jurassic (Figs. 2.23 and 2.24).

Indian Scenario

In contrast to the above progress in Euro-America, the slow progress in India has been frustrating in spite of foundation studies between 1840 and 1940. Nothing to speak of zonal differentiation, even systems and stages are not precisely differentiated in spite of collection of vast amount of geological data since the establishment of the Geological Survey of India way back in 1851. More depressing is the fact, that such studies in India are not allotted priorities. Currently, researchers seem more inclined to instrument based studies hardly involving any sustained painstaking field work.

Author's Strides and Failures of the Past 3 Decades

Such a dismal scenario prompted the author to persue this seemingly lack-lustre area in the Jurassic of Kachchh. The journey has been extremely slow and struggleful involving only occasional strides forward amidst innumerable failures and set-backs. However, in this venture, my long continued association as a member of ISJS and near continuous interaction with colleagues in Europe helped me in substantive manner to realize meaningful progress.

Kachchh Jurassic has been a relatively better and constantly studied area for nearly two centuries. However, the author had to begin almost from scratch. Numerous molluscs and other marine invertebrates were collected without the required level of lithostratigraphic differentiation. Thus only limited lithostratigraphy was possible. Ammonoid stratigraphic range charts could not be developed, for worthwhile objective ammonoid stratigraphy. It was in such backdrop, the author was baptized in the Kachchh Jurassic at the start of 1980s, with the first field excursion in the winter of 1980–1981.

2.8 Ammonoid Stratigraphic Refinement in Kachchh—Principal Results at a Glance

Ammonoid-rich six stages spanning through ~25.0 my are studied. In respective stratotypes, the study included ~900 beds vertically spread through a composite thickness of ~700 m. Ammonoids were collected from over 350 stratigraphically précised positions out of ~900 precisely differentiated beds, as a rare extreme high quality exercise. The high resolution scale so developed included ~27 zones, 50 subzones, and yet greater number of horizons with average horizonal resolution of ~200 ky. The stage wise description is undertaken as under.

2.8.1 Hettangian–Pliensbachian Record (Non-marine)

The early three of the ammonoid devoid five older Jurassic stages are differentiated in Kachchh on the basis of palynomorphs, and have already been discussed earlier. A review of the Nirona and Banni well microflora may in near future allow broad if not precise differentiation of the mentioned three stages from one another as also from the older Late Triassic part from the younger early to mid Early Jurassic part.

2.8.2 Toarcian–Aalenian Record (Largely Marine, yet Ammonoid Devoid)

The presence of Toarcian in the exposed Jurassic sediments of Kachchh based on record of nannoplanktons has been discussed earlier. The ~300 m thick early part of LSM (Dingy member beds 1–36 at Dingy in Patcham of Patcham Formation Figs. 2.25 and 2.26) exposed only in Dingy and Kuar bet localities of Kaladoongar is considered Toarcian–Aalenian in Kachchh. Lithologically, the duration is made up of conglomerates, sandstones and sandy limestones with shale/silt/clay interbeds. The thin calcareous sands/silts are mostly fossiliferous except in the basal part. The mega-invertebrates are present only in a few levels, significantly in the middle part of the member. These are mostly bivalves (corbulids, gervillids, astartids, trigonids, etc.) and corals. In sequence stratigraphic context, detailed in Chap. 3, possibilities have come forth to distinguish the Aalenian from Toarcian below and Bajocian above. It should also be possible to differentiate the Aalenian in High Himalaya. The author's interpretation (Krishna 1987b; Krishna and Pathak 1994) on the initiation of the marine succession in Kachchh in Early Toarcian is validated by insitu record of the Toarcian coccoliths in the Dm at Kuar bet, and regional framework, and detailed in Chap. 3. Beds 37–38 of Dingy Member (Biswas 1977, 1993) are included at the base of the overlying KSm, while Dm is held to comprise beds 1–36 on an unknown unexposed basement (Figs. 2.25 and 2.26). The basement most likely is Precambrian, though there is also a possibility of its being the non-marine Nirona Formation.

Pandey and Dave (1993) speculated without any evidence of the thick marine Aalenian in the Banni well, and created the stage—'Bannian'. In a regional framework, in all probability, the marine Aalenian should be present in the Kachchh wells, as already reported, although undifferentiated, in the exposed sediments in 'Island belt' as Toarcian–Aalenian from the early part of the LSM (Dm) in the Kaladongar section of the Patcham hill at Kuar Bet and Dingy. Biotically, the differentiation of Aalenian has yet not been possible even from any region of the Indian subcontinent, although there is continuous record of the marine Early and Middle Jurassic in neighbouring Pakistan, Madagascar, and High Himalaya. However, the application of sequence stratigraphy has allowed tentative differentiation of the Toarcian from Aalenian in the Dm as a second-order RST of coarse cross-bedded sands near the

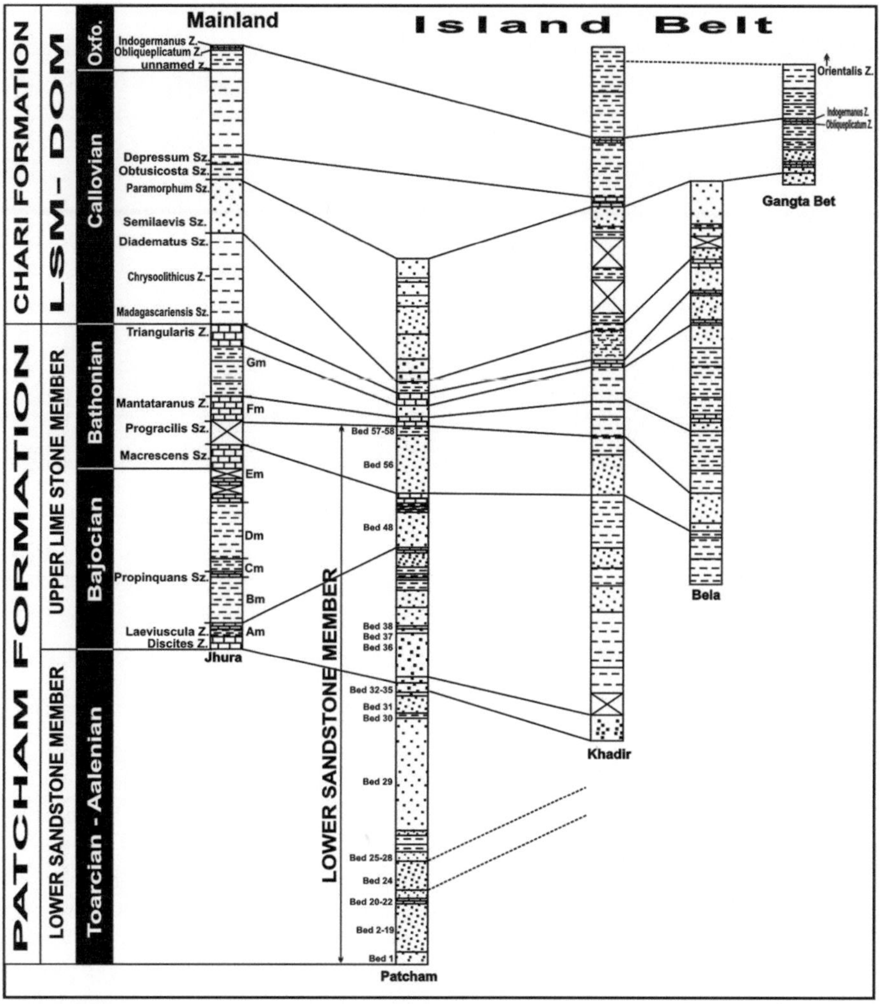

Fig. 2.25 Lower Sandstone Member at Patcham and Upper Limestone Member at Jhura alongwith intrabasinal correlation among Mainland and Island belt in the Kachchh Jurassic in the Toarcian–Middle Oxfordian interval with bed nos. and simplified columns after Biswas (1993)

margin within the Dm, suggestively, from the bed 25 to bed 36 (Figs. 2.25 and 2.26), presumably so also on the margin of other GTM basins.

2.8.3 *Bajocian*

The exposed Bajocian sediments, ~100 m or less thick, are here suggested to occur widely in 'Island belt' (Kaladongar and Goradongar sections) as the informal middle part of LSM and Mainland (Jhura and Nara) as the early part of the ULM. In 'Island belt', in view of the near margin relatively proximal locations, the Bajocian is expressed almost exclusively in coarse clastics with thin interbeds of fine textured limestones/marls/shales (Figs. 2.25 and 2.26), while in Jhura the Bajocian includes alternations of golden ooltic limestones and shales as Am (mostly golden ooltic limestomes, occasionally also conglomeratic) and Bm (mostly shales). Am consists mostly of the golden oolitic limestones that are formed below normal wave base conditions. In Nara, only, late Bajocian may be present, if so, near the base immediately above the underlying igneous body (Fig. 1.11) in view of the recent discovery of the early Early Bathonian ammonoids in the overlying limestones of the bed 9. The Nara sedimentary succession may be most probably conformably underlain by unknown and unexposed Bajocian–Toarcian sediments over the igneous plug of presumable ~183 ma widely expressed igneous activity already recorded in the Kachchh and Saurashtra wells.

The Late Bajocian sediments included in the BCSm of LSM were claimed from the Kala Doongar section of the Patcham region near Kuran. The claim was based on the discovery of a single incomplete entirely septate ammonoid specimen determined as '*Leptosphinctes*' and assigned to Late Bajocian (Jaitly and Singh 1983). The so called *Leptosphinctes* level is underlain by ammonoid devoid ~150 to 200 m thick sediments. In a few other sections of Patcham hill, both in Kaladoongar and Goradoongar, nearly equally thick sediments underly the levels otherwise considered correlative to the so called *Leptosphinctes* level. In Kaladoongar, the nannoplankton evidence and regional stratigraphic framework extended the succession as far back as Toarcian. In view of the overlying well-dated late Middle to early Late Bathonian ammonoid bearing Goradoongar Yellow Flagstone member (GYFm) in Kaladoongar, the immediately underlying sand dominating Kaladoongar Sandstone member (KSm) and Babia cliff Sandstone member (BCSm) of LSM are here considered as belonging to the Bajocian–mid. Middle Bathonian interval. Similar thick sands in the Goradongar sections are also here assigned to the Bajocian–mid. Middle Bathonian interval below the ULM.

Pandey and Dave (1993) determined the start of Bajocian from the base of GYFm and included the same in the early part of their newly created 'Patchamiam' stage. The base of the GYFm based on enclosed ammonoids is otherwise firmly placed in late Middle Bathonian. Recently, a single septate *Calliphylloceras* has been discovered in Goradoongar in the basal limestones of the middle part of the LSM. The mentioned limestones in sequence stratigraphic context are later in Chap. 3 tentatively assigned best to Early Bajocian. It now becomes the oldest Jurassic as also the sole Bajocian ammonoid of Kachchh. However, it is hoped to find in near future Late Bajocian ammonoids in the Nara section below the newly discovered early Early Bathonian (Pandey and Pathak 2015a).

Ammonoid Zones in the Kachchh Jurassic Stages

The presentation of the recently developed ammonoid zonation in the Kachchh Jurassic stages is preceded by general remarks and summaries of lithostratigraphic framework of each stage. The individual zones are characterized by their

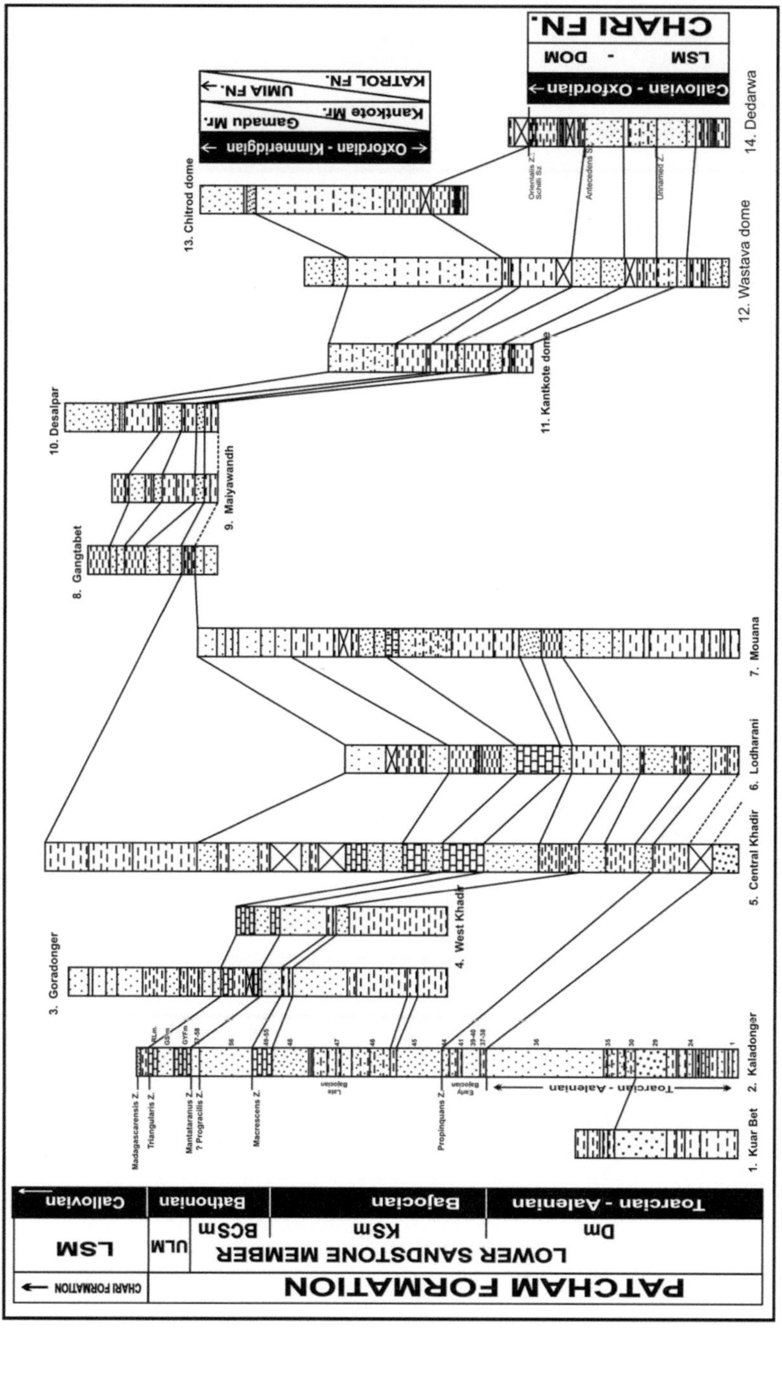

Fig. 2.26 Ammonoid based precise intrabasinal correlations among lithocolumns of Island belt and Wagad region of Kachchh basin during the early Toarcian–early Kimmeridgian Interval with bed nos. and simplified columns after Biswas (1993)

assemblages as a whole. The zones have been named after a characteristic and commonly present species appearing at the base or otherwise significant in the zone. The ammonoid zones, subzones and horizons are based on bed by bed collection under the rigorest possible stratigraphic precision. The successive units are initially developed as biostratigraphic units. It is emphasized that once the ammonoid zonal boundaries are confirmed as time boundaries through intrabasinal and long distance correlations, the mentioned ammonoid based biostratigraphic units, since fulfilling the requirements and definition, were also formalized as chronostratigraphic units. Many taxa of the Kachchh subzones and even horizons are present elsewhere in the Indo-East African province cf. Madagascar, Pakistan and East African countries on GTM, and even in ETM basins.

2.8.4 Bathonian

Recent discovery of the Early Bathonian ammonoid bearing section in Nara
The author's research group recently discovered an ammonoid bearing Early to mid Middle Bathonian section at Nara. The preliminary evaluation yielded a moderately ammonoid-rich ~1.5 m limestone bed ~3.5 m above the exposed base of early Early Bathonian Zigzag Zone Macrescens Subzone (Fig. 1.11—Krishna unpublished report 2014; Pandey and Pathak 2015a). Improved bed-by-bed collection were, subsequently, made in the ~23 m thick column from beds 1 to 15 which yielded ~100 fragmented ammonoid specimens, most from bed 9. These belong to seven families and 13 genera (Pandey and Pathak 2015a). The bed 9 includes *Parkinsonia*, *Siemiradzkia*, *Prohecticoceras*, *Cadomites*, *Berbericeras*, *Zigzagiceras*, *Telermoceras*, *etc*. (Pandey and Pathak 2015a). The presence of *Micromphalites* in bed 15 in context of earlier records from Jhura, Sadhara, and Jumara (Jain 2014) is considered significant for intrabasinal correlation in the exposed Bathonian in the Kachchh basin. Over all, the discovery is very important as the only ammonoid bearing Early and early Middle Bathonian in IEAP, and one of the rare ones on the entire GTM of pre-Macrocephalitinae pre-Morissi Zone age. The ammonoid devoid beds 16–18 above the ammonoid bearing beds 1–15 are considered pre-Morissi Zone in age. Distinct overlapping is evident between the oldest exposed at Jumara and the youngest at Nara of the early Middle Bathonian prior to the Macrocephalitinae bearing succession.

The complete Bathonian in Kachchh is exposed in Patcham, Khadir and Bela in the 'island belt', Nara and Jhura section of East Mainland, where it is underlain by the Bajocian and older sediments. The Jumara section of the West Mainland includes mid/late Middle to latest Bathonian, while the Habo section also includes part of the terminal Bathonian. The Nara section (Fig. 1.11) of West Mainland has only in the last decade yielded Bathonian ammonoids, while the presence of the Early to early Middle Bathonian ammonoids in the Nara section is for the first time determined. As expected, the dominant lithology in the 'Island belt' is of sandstones. Even in Jhura thick sandy limestones are present that are suspected to be of the early part of Bathonian, while the relatively distal Nara and Jumara sections of West Mainland exhibit fine textured shales/marls and limestones. There is noticed large facies variation among the exposed proximal and distal sections. In addition, is observed a marked temporal facies contrast between the Toarcian to mid Middle Bathonian LSM at Island belt, and Bajocian–Bathonian ULM at Goradongar, Jumara, Nara and Jhura. The LSM comprises ammonoid scarce to ammonoid devoid coarse sand dominating sediments, while the ULM includes poorly to richly ammonoid bearing golden oolitic sediments, nodular limestones and shale interbeds. The maximum age range of the ULM is suggested as Bajocian – Bathonian at Jhura and minimum at Habo of only the terminal part of the Late Bathonian. Thus, Bathonian (Figs. 2.25 and 2.26) includes the sand dominating early part (BCSm of LSM) and the limestone/shale dominating later part in the 'Island belt' as ULM. The Bathonian in Habo, Jhura, Nara and Jumara exclusively includes the limestone/shale succession as ULM of the Patcham Formation.

2.8.4.1 Early and Early Middle Bathonian

In 2012, under sponsored research, is discovered a limestone/marl/shale succession close to the igneous plug in Nara which included as many as 19 beds (Krishna unpublished project report 2014 Pandey and Pathak 2015a). Ammonoids are present all through, particularly in the middle part of the $\sim$23 m dominantly limestone succession (Fig. 1.11). In total, 10 levels of ammonoids have been differentiated with good density and diversity. The ammonoids collected do not include any Macrocephalitinae that are present in younger beds in another section. Thus, the succession in consideration appears to be of pre-Macrocephalitinae age, i.e. older than mid Middle Bathonian, and includes Early to mid Middle Bathonian, since at least in distal Nara; even subaerial break can not be thought of.

Initially, in absence of the confirmatory evidence, the principles of sequence stratigraphy were utilized. In thick sand dominating sediments at relatively proximally located 'Island belt', Washtawa and even East Mainland (close to KMF), the presence of ammonoids is invariably observed in the terminal part of the successive second-order TSTs close to the MFS, e.g. early to mid Early Callovian in Habo, Washtawa, Bela, Mouwana, Khadir and Patcham, late Early Oxfordian in Kakindia limestone, late Middle Oxfordian at Kantkote, in early Late Callovian in Washtawa, Gangta bet, Khadir, in basal Kimmeridgian in Bharodia, early to mid Late Kimmeridgin in Jhuran etc. The mentioned almost invariable observation strengthens the probable anticipation that the bed 9, the richest of all the 10 ammonoid levels devoid of the macrocephalitins at Nara below the late Middle Bathonian Macrocephalitinae bearing sediments in the newly discovered section, should correspond to the early Early Bathonian Zigzag Zone Macrescens Subzone. The sequence stratigraphic interpretations were subsequently confirmed (Pandey and Pathak 2015a) through the record of number of ammonoid genera (*Zigzagiceras*,

Procerozigzag, *Siemiradzkia*, *Berbericeras*, *Parkinsonia*, *Ebrayiceras*, *Telermoceras*, *Micromphalites*, *Prohecticoceras*, *Procerites*, *Cadomites*, etc.).

The above framework necessitated the reevaluation of the so called *Leptosphinctes* in the terminal part of the BCSm as of early Middle Bathonian Progracilis Zone Progracilis Subzone Perisphinctinae. The taxonomic status of the claimed *Leptosphinctes* may require review since Leptosphinctinae already extincted by the close of the Macrescens Subzone.

The above discussion demonstrates that the Bathonian in Nara and Jumara is ammonoid bearing. It includes a rich succession of relatively less shallow water Macrocephalitinae only from mid/late Middle Bathonian onwards. This faunal assemblage and fine textured carbonate/shale lithology suggests increase in bathymetry from Early to Late Bathonian, and farther up until at least the mid/late Early Callovian boundary.

The ammonoid record from the Bathonian sediments of Kachchh is poor except for Jumara and the newly discovered Nara where the sediments are highly fossiliferous as also rich in ammonoids. Available reports mentioned only a few ammonoid genera/species over the last many decades from a maximum of six ammonoid levels in the ~300 m thick Bathonian sediments exposed in Kachchh. Only occasional stray attempts have been made in recent past (e.g. Roy et al. 2007) towards enlargement of the Bathonian ammonoid stratigraphic record. The lithostratigraphic differentiation in the above work is inadequate and the ranges of the determined genera/species are vague. The attempted zonation does not meet the rigours of precision zonation. The new '*Prohecticoceras majalense*' Horizon has been assigned to the Middle Bathonian based on a new species with no taxonomic details. Unfortunately, the act renders the new species invalid, so also the horizon created after it.

A part of the biostratigraphic study of the Bathonian sediments of Kachchh, is presented as a preliminary biostratigraphic scheme for the sediments exposed at Nara in Mainland that are included here in a new informal member NLm within the ULM. Also included in the discussion here is the '*Leptosphinctes*' bed of Kuran in Kaladongar part of Patcham. In the present preliminary effort, only two zones are proposed as under:

Nara Limestone member (NLm)

1. **Parkinsonia–Siemirdzkia Zone**
   It is based on the co-occurrence of *Parkinsonia* and *Siemiradzkia* at the top of the ~1.5 m thick bed 9 of the newly discovered Late Bajocian–early to mid Middle Bathonian ammonoid inclusive section (Fig. 1.11). The other ammonoid genera have already been listed above (Pandey and Pathak 2015a). The zone is tentatively extended from bed 1 to bed 9, and assigned to Zigzag Zone Macrascens Subzone, however, in view of the presence of *Parkinsonia*, its extension down into late Late Bajocian is not ruled out.
2. **Perisphinctinae Zone**
   It is created tentatively on the revised determination of the earlier claimed *Leptosphinctes* as a Perisphinctinae instead of Leptosphinctinae below the

ammonoid bearing Middle Bathonian above, and correlation of the basal part of the BCSm with ammonoid bearing Early Bathonian at Nara. It may be noted that the claimed *Leptosphinctes* came from the terminal part of BCSm at Kuran ~10.0 m below GYFm. The zone is speculatively assigned to the early Middle Bathonian Progracilis Zone–Subcontractus Zone interval.

2.8.4.2 Late Middle to late Late Bathonian

Good and near continuous presence of ammonoids is noted only in the Jumara section, and hence the same is included in a new informal member JLm of the ULM. The ~51 m thick Bathonian sedimentary succession exposed at Jumara constitutes the later part of the Patcham Formation (Fig. 2.27). Its base is not exposed. The contact with the overlying Chari Formation is conformable. It has been lithostratigraphically differentiated into two distinct sediment intervals I and II having a sharp erosional contact. These two major lithostratigraphic intervals have been further subdivided, respectively, into 24 and 63 beds/bands (Pandey et al. 2013a). The Patcham Formation yielded over 130 ammonoid specimens from 36 levels. The entire Bathonian succession at Jumara exhibits the dominance of the relatively fine textured well-bedded limestone and shale/silt/marl. The sediment interval I of the older ~18 m thick succession largely comprises alternations of relatively thicker yellowish/greenish white, occasionally gypseous shale/marl and yellowish grey to earthy limestone (packstone) packed prolificly with corals (both in variety and number) in multiple relatively harder levels that project out, as also enclosing other diverse invertebrate fauna including ammonites, nautiloids, gastropods, bivalves, brachiopods etc, which correspond (Fig. 2.27) to beds 23–26 of Rajnath (1932). The younger ~33 m thick succession consists of alternations of greyish white/whitish marly limestone and earthy shale/silt which is here corresponded to bed 22 of Rajnath. The characteristic whitish grey nodular limestone makes easy recognition of bed the 22 from distance. Almost all the limestone beds contain ammonoids and other invertebrate fossils. Crinoid stems in lower part and sponges in the upper part are particularly observed at many levels in the sediment interval II. In general, the ammonoid density increases from the older to younger succession and is maximum in bed II-30, subsequently followed up by slow and gradual decrease in ammonoid density between bed II-31 to 53. The overlying succession from the beds II-54 to 63 is the poorest in ammonoids. The limestone/shale Bathonian succession as described above is included in the ULM.

Also, at Jumara Dome based on ranges of seven subfamilies, 10 genera, their many species in the limestone succession of JLm, two ammonoid zones have been created (Pandey et al. 2013a). Among the subfamilies, Macrocephalitinae has had its origin in SW Pacific while the rest arrived on the GTM from ETM. The Macrocephalitinae is thus the indigenous and the dominant GTM element with about 33 % in terms of individuals.

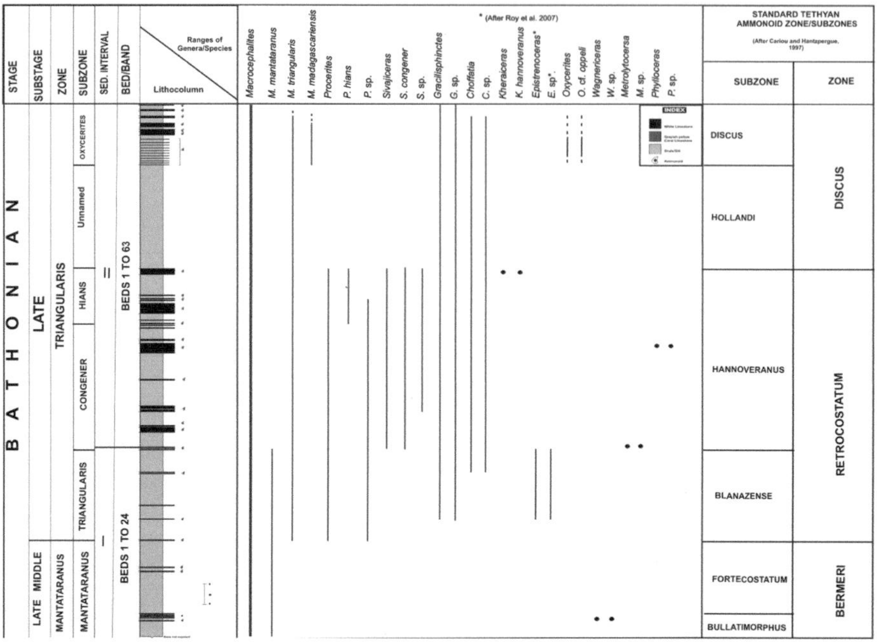

Fig. 2.27 Ammonoid ranges and zones in the later part of Bathonian in Jumara, Kachchh after Pandey et al. (2013a)

The two ammonoid zones are as under:

Jumara Limestone member (JLm)

The JLm, presumably, is confirmably underlain by the unexposed limestone succession of the Early and early Middle Bathonian age.

1. **Mantataranus Zone**
2. **Triangularis Zone** (Spath 1927–33, later emended in Krishna and Westermann 1985, 1987).

These two zones have been further subdivided into six subzones. All the proposed zones/subzones have also been correlated with the Standard Tethyan Scheme developed in Europe (Cariou and Hantzpergue 1997).

Mantataranus Zone/Subzone Pandey, Pathak and Krishna: beds I-01 to I-15 (Fig. 2.27)

This Bathonian ammonoid zone is marked by the first appearance and also dominance of *M.* cf. *mantataranus* Boehm. *Wagnericeras* is also present (Krishna 1994 unpublished report) here. The genus *Wagnericeras* is widely known from Middle Bathonian of NW Germany, Sula Island, New Guinea, (Westermann and Callomon 1988) and India (Krishna and Pathak 1994). Thus the Mantataranus Zone can be safely correlated with Bremeri Zone of the Standard Tethyan scheme developed on the ETM (Cariou and Hantzpergue 1997). The correlation is also strengthened by record of *Epistrenoceras* (Roy et al. 2007) of early Late Bathonian Retrocostatum Zone in the immediately overlying sediments at Jumara. In context of an exclusive Perisphinctidae based Late Bajocian–Late Tithonian ammonoid zonal succession, the zone can be alternatively named *Wagnericeras* Zone.

Triangularis Zone Spath beds I-16 to II-59 (Fig. 2.27)

This Zone is marked by the first appearance of *Macrocephalites triangularis* Spath in bed I-16, and is further differentiated into five subzones as under

Triangularis Subzone beds I-16 to I-23

The base of this subzone is marked by the first appearance of *M. triangularis* Spath with *Procerites* sp. in bed I-16. The genera *Gracilisphinctes* and *Choffatia* also appear in this subzone. The genera *Phylloceras* is also present. *M.* cf. *mantataranus* continued from below. *M. triangularis* Spath has been since long used as the zonal index for the Late Bathonian on GTM (Gondwanian Tethyan Margin) (Krishna and Westermann 1985, 1987; Krishna and Cariou 1993). In context of the Perisphinctidae, this zone alternatively can also be named as Congener Zone.

Congener Subzone beds I-24 to II-16

The base is marked by the first appearance of *Sivajiceras congener* (Waagen). *M.* cf. *mantataranus* Boehm marks its last appearance at the base of the subzone, however, *M. triangularis* Spath, *Gracilisphinctes* and *Choffatia* continue from below. The genus *Phylloceras* is also present.

Hians Subzone The base is marked by the first appearance of the subzonal index in bed II-17. The subzonal index is also restricted within this subzone. *Kheraiceras hannoveranus* (Roemer) is also present. *Gracilisphinctes* and *Choffatia* continue from below. *S. congener* (Waagen) makes its last appearance.

Unnamed Subzone Bed II-31

Oxycerites Subzone beds II-32 to II-63. The base is marked by the first **O.** cf. *arkelli* (Elmi) and *M. madagascariensis* Lemoine at the base in bed II-32. M. *triangularis* Spath, and *Gracilisphinctes*, mark their last appearance in this subzone.

M. triangularis Spath is the zonal index for the Late Bathonian on GTM (Krishna and Westermann 1985, 1987; Krishna and Cariou 1993). Roy et al. (2007, Fig. 2) have recorded *Epistrenoceras* having nearly world wide distribution from the early Late Bathonian Retrecostatum Zone from sediments of Triangularis Subzone (Dietl 1981). The Triangularis Subzone is correlated with the Retrocostatum Zone of the Standard Tethyan scheme (Cariou and Hantzpergue 1997). *K. hannoveranus* (Roemer) is recorded from Hians Subzone here proposed. Thus the Triangularis Subzone–Hians Subzone interval of the proposed scheme can be safely correlated with the ETM Retrecostatum Zone. The superjacent Oxycerites Subzone is easily assigned to late Late Bathonian Discus Zone. The zone can also be named Epistrenoceras Zone.

The age assignments of the GYFm by the Kachchh Jurassic workers in their publications from time to time are reviewed. The GYFm is present only in Patcham and Jhura. Biswas (1977, 1993) placed the unit in Callovian. Fursich et al. (2013, Table 4, p. 33, and also in several other earlier publications) have repetitively over the last two decades in their tables indicated the unit in Early to Middle Bathonian while in the accompanying texts assigned it to Middle Bathonian. Similarly, coral rich limestone of Jumara in the table is indicated in Middle Bathonian, in the text on p. 127 ranged from Middle to Late Bathonian, and on the same page again assigned to Progracilis Zone and Triangularis Zone. Progracilis Zone is an early Middle Bathonian zone while Triangularis Zone is exclusively Late Bathonian, and the two are separated by a large interval of three ammonoid zones—Subcontractus Zone, Morrisi Zone and Bremeri Zone—all of Middle Bathonian. Thus, the coral rich limestone of Jumara can be assigned if at all either to the Progracilis Zone of the early Middle Bathonian or to the Triangularis Zone of Late Bathonian, and not in any way to both the mentioned zones as done by Pandey and Callomon (1995). Several such multidimensional age discrepancies are exampled in Chap. 4. The unit in Patcham and Jhura is known to exclusively include the late Middle to early Late Bathonian ammonoid bearing limestones, and Krishna and coworkers have consistently ranged it from late Middle to early Late Bathonian. The recent discovery of Early Bathonian ammonoids (Krishna 2014 unpublished; Pandey and Pathak 2015a) in Nara below the first Macrocephalitnae in Nara or for that matter even elsewhere in Kachchh clinches the age of the GYFm as late Middle to early Late Bathonian. There is also the issue of additional inclusion of $\sim$25 m of immediately underlying sands in the GYFm by Fursich et al. (2013) which instead may be included in the underlying BCSm of the LSM. The coeval upper BCSm sands below the GYFm in Kaladongar are assigned to Late Bajocian (Jaitley and Singh 1983 to Fursich et al. 2013). Having no possibility of a subaerial gap, the mentioned sands are indirectly considered early and mid Middle Bathonian, without inclusion of any part of Early Bathonian, nothing to speak of Late Bajocian.

2.8.5 *Callovian*

Development of Callovian in Kachchh

The Callovian span of time is included in the Chari Formation, and is widely exposed through out the basin, in Mainland, 'island belt' and Wagad. The sediments are mostly fine textured shale/silt/clay with not so thick sand interbeds in the relatively distally located Mainland sections west of the median high. In the proximal sections east of the median high, 'island belt' and Wagad, much thicker sand intervals are present in addition of the shales. The early part in the Keera section locally includes golden oolitic limestones cyclically alternating with shales. The distally located Callovian sedimentary columns west of the median high profusely enclose ammonoids and other marine mega-invertebrates in pebble/concretion/nodular beds. The Keera sedimentary succession overlies the unexposed basement probably of ULM as in Jumara.

The proposition of a high resolution ammonoid chronologic scheme of 7 zones, 13 subzones and 25 horizons for the Callovian stage in Keera section of Kachchh (Ojha 1996a, b; Krishna and Ojha 1996, 2000; Krishna et al. 1993) is noteworthy. One of the significant features of the study is the initial lithostratigraphic differentiation of the Callovian at Keera into ~400 beds. The Keera section is the stratotype of the Chari Formation, so chosen by Stoliczka, as also the Indian stratotype for the Callovian stage. It is one of the several isolated block-faulted and domally uplifted Jurassic localities in the Kachchh Mainland, about 70 km northeast of Bhuj, bounded by longitudes 69° 14′ 15″ to 69° 14′ 33″ E and 23° 35′ 13″ to 23° 35′ 05″ N. The Chari Formation is well-exposed in an elliptical outcrop with Dhosa Oolite as its youngest member there.

Lithostratigraphic Framework (Figs. 2.28, 2.29 and 2.30)

A summary of the Callovian sedimentary succession (from base to top) belonging to the Chari Formation is given below, and the vertical ranges of the ammonoid taxa are indicated in Figs. 2.28, 2.29 and 2.30. The succession is easily differentiable into its members; LSM (sediment intervals 0 and I), RSM (sediment intervals II to VI), GSM (sediment intervals VII, VIII and IX), and DSM (early part of sediment interval X up to bed 392).

Sediment Interval 0 (Beds 1–32, 15 m): Alternation of shelly limestones, ferruginous calcareous sandstones and shales, scarcely fossiliferous.
Sediment Interval I (Beds 33–92, 55 m): Golden oolitic limestones intercalated with shales, highly fossiliferous.
Sediment Interval II (Beds 93–118, 25.59 m): Alternation of variegated limestones and gypseous shales, fossiliferous.
Sediment Interval III (Beds 119–130, 06.60 m): Alternation of variegated soft muddy, somewhat oolitic sandy limestones and shales/siltstones, highly fossiliferous.
Sediment Interval IV (Beds 131–160, 05 m): Alternation of white marly/pebbly bands and shales or maroon fractured marly bands and shales, highly fossiliferous.
Sediment Interval V (Beds 161–192, 25 m): Variegated, hard, jointed, pebbly calcareous sandstones intercalated with shales/siltstones, highly fossiliferous.
Sediment Interval VI (Beds 193–218 as here revised of reduced thickness of 25.5 m): Variegated, fractured, marly to pebbly hard bands capped by fossiliferous sandy limestones, highly fossiliferous.
Sediment Interval VII (Beds 219–334 as here revised with increased thickness of 35.4 m): Maroon to white pebbly fractured bands intercalated with gypseous shales, highly fossiliferous.
Sediment Interval VIII (Beds 335–336, 0.37 m): Whitish maroon limestone intercalated with yellowish brown shale, highly fossiliferous.
Sediment Interval IX (Beds 337–372, 11.5 m): White to maroon, marly to pebbly, somewhat fractured shelly bands alternating with yellowish brown gypseous shale, highly fossiliferous.
Sediment interval X (Beds 373–392, 4.0 m)
Zones, subzones and horizons and correlations with ETM

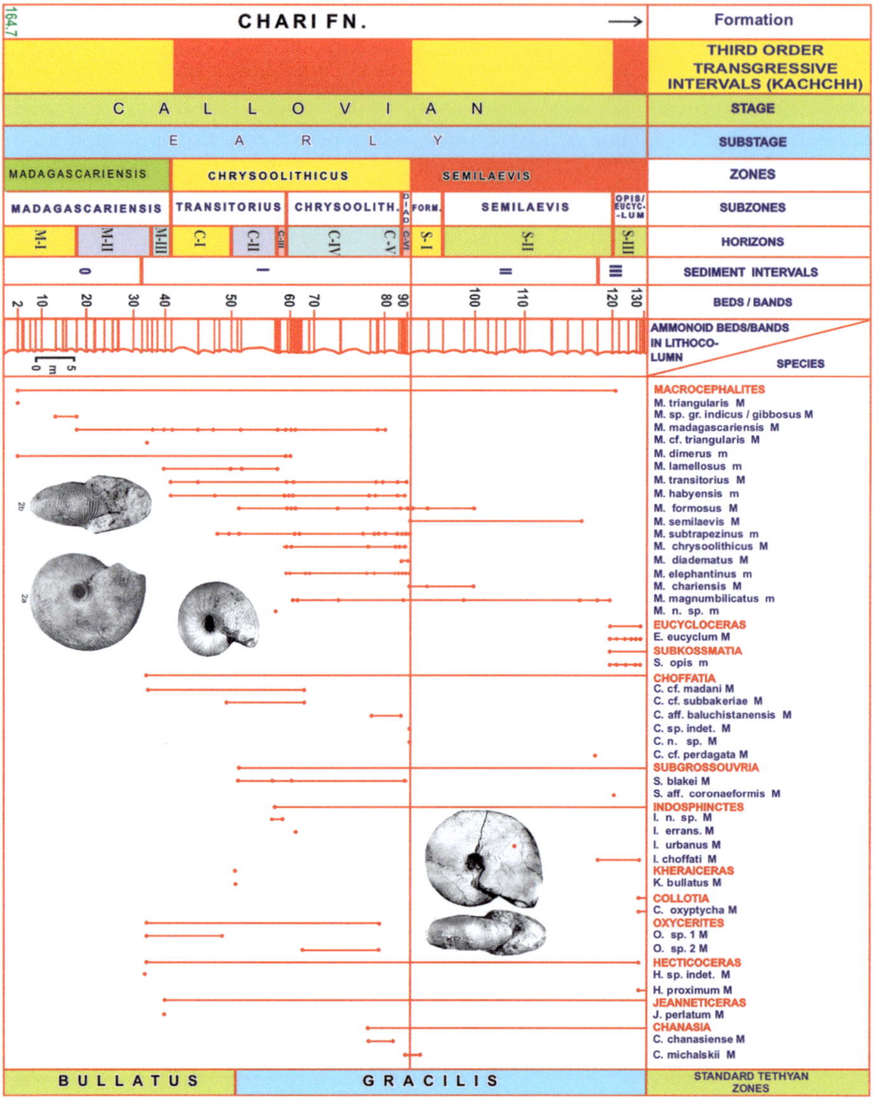

Fig. 2.28 The stratigraphic occurrence, ranges of the ammonoid species and third-order sequences in the Early Callovian of Kachchh (modified after Ojha 1996a, b)

Early Callovian

This interval includes 3 zones, 6 subzones and 12 horizons as under:

Lower Shale Member (LSM)

Madagascariensis Zone Krishna and Cariou (1990): Used first as assemblage by Krishna and Westermann (1987) and formalized later as a zone in Krishna and

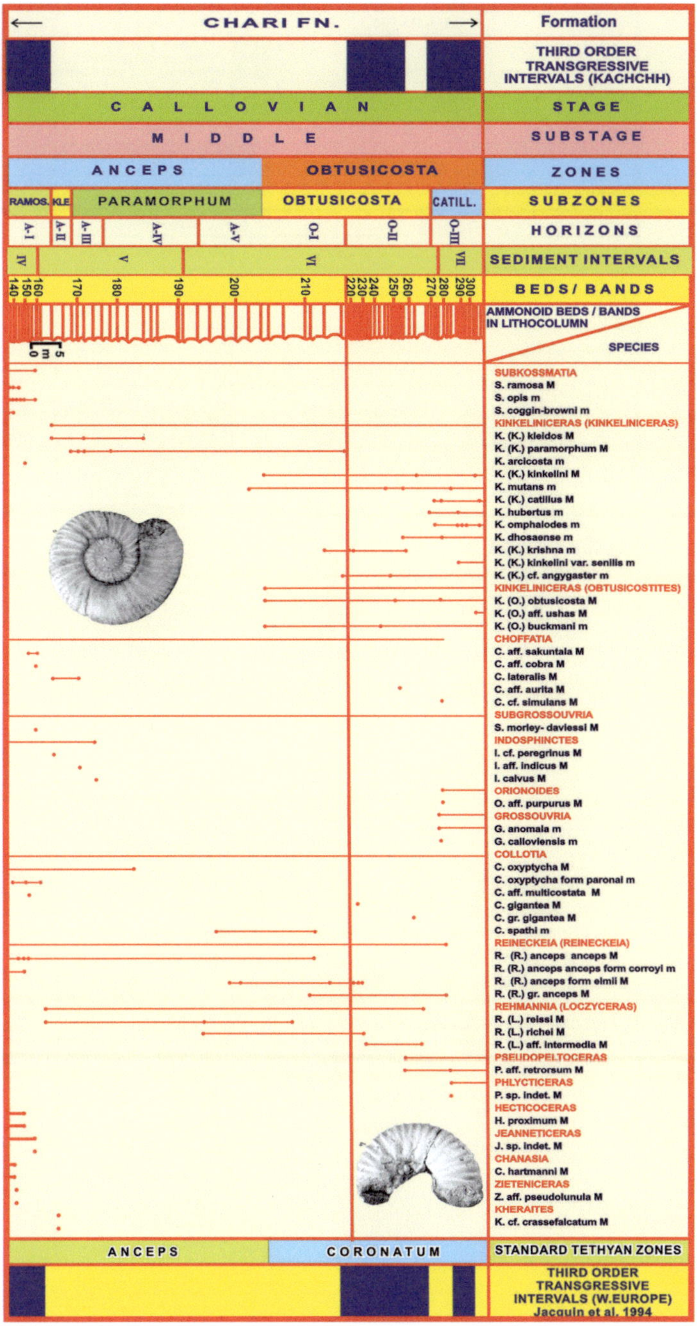

Fig. 2.29 The stratigraphic occurrence and ranges of the ammonoid species vis-a-vis third-order sequence in the Middle Callovian of Kachchh (modified after Ojha 1996a, b)

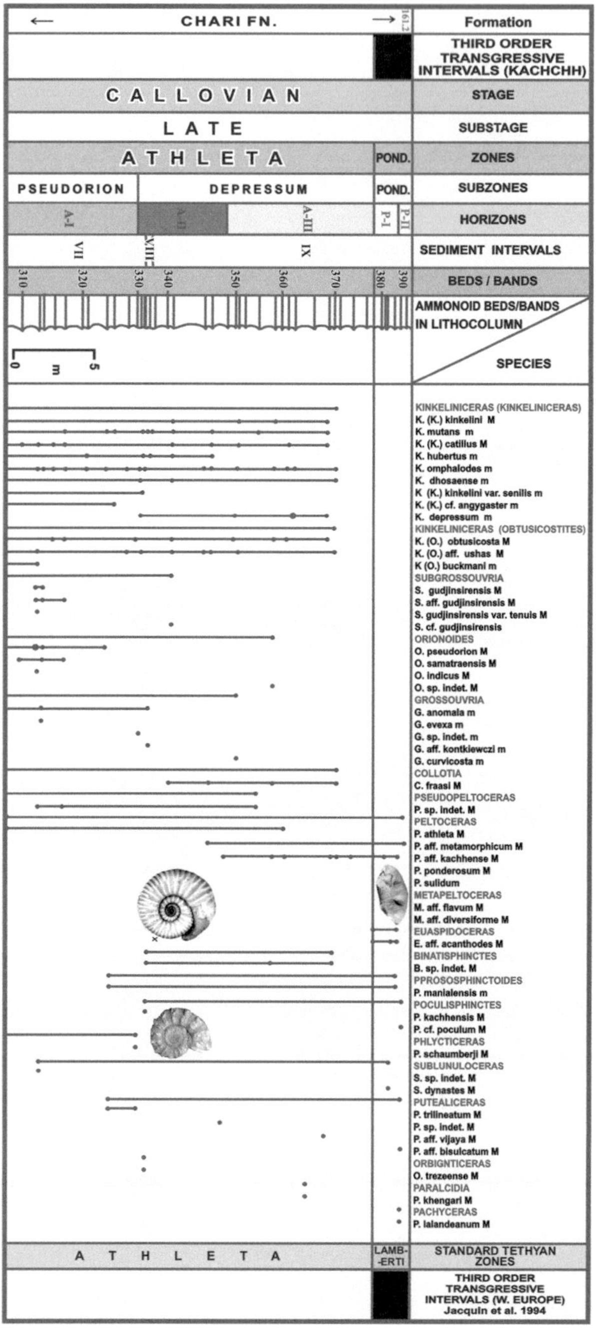

Fig. 2.30 The stratigraphic occurrence and ranges of the ammonoid species vis-a-vis third-order sequence in the Late Callovian of Kachchh (modified after Ojha 1996a, b)

Cariou (1990). It is characterized by the dominance of the zonal index *M. madagascariensis* Lemoine M with exclusion of *M. transitorius* (Spath) M.

M-I Horizon Krishna and Ojha (1996) (beds 1–17): It is indicated by scarce yet near exclusive *Macrocephalites* in Keera, e.g. *M.* cf. *madagascariensis* Lemoine M, *M.* cf. *triangularis* Spath M and *M.* gr. *gibbosus/indicus* (Spath) M, the last species with thick, blunt and sparser primaries.

M-II Horizon Krishna and Ojha (1996) (beds 18–35): It is indicated by the first *M. madagascariensis* Lemoine M morph I exclusive of the morph II of the species. Rare compressed nuclei of *M.* cf. *triangularis* Spath M continue from below.

M-III Horizon Krishna and Ojha (1996) (beds 36–41): It is indicated by the first *Choffatia* gr. *madani* Spath M and *Oxycerites* sp. indet. M in association with *M. madagascariensis* Lemoine M continuing from below. *M. madagascariensis* Lemoine M, the nominal and the most significant species of the Madagascariensis Zone is considered morphologically very close to the European *M. verus* Buckman of the Keppleri Subzone, basal Herveyi Zone (Callomon et al. 1989) and could be a coeval provincial subspecies. Thus is allowed precise subzonal correlation.

Chrysoolithicus Zone Krishna and Cariou (1990): Used first as an assemblage by Krishna and Westermann (1987) and later formalized as a zone in Krishna and Cariou (1990), considerably emended by Krishna and Ojha (1996). The base is defined (Ojha 1996a, b, Krishna and Ojha 1996) by the first *M. transitorius* (Spath) M and its possible dimorph *M. habyensis* (Spath) m. *M. chrysoolithicus* (Waagen) M is the dominant species in this zone along with its dimorph *M. elephantinus* (Sowerby) m.

Transitorius Subzone Krishna and Ojha (1996):
Its base is defined as for the zone. It includes two horizons

C-I Horizon Krishna and Ojha (1996) (beds 42–49): It is indicated by the first transient morph I of *M. transitorius* (Spath) M while *M. madagascariensis* Lemoine M continues.

C-II Horizon Krishna and Ojha (1996) (beds 50–55): It is characterized by *Kheraiceras bullatus* (d' Orbigny) M and its successor smaller variant with the first *Subgrossouvria* (gr. *blakei* Spath) M. *Choffatia* aff. *subbackeriae* (d' Orbigny) M is also present. The *Kheraiceras* provides good basis of correlation with Bullatus (uppermost part) and Prahecquense Horizons of France (Cariou 1984).

C-III Horizon Krishna and Ojha (1996) (beds 56–57): It is characterized by the first *Indosphinctes*. It is significant to note that the first *Indosphinctes* in France occurs in Laugieri Horizon, permitting correlation of the respective horizons of Kachchh and France.

Chrysoolithicus Subzone Krishna and Cariou (1990), emended (Krishna and Ojha 1996): Its base is defined by the first *M. chrysoolithicus* (Waagen) M along with its dimorph *M. elephantinus* (Sowerby) m. It includes two horizons

C-IV Horizon Krishna and Ojha (1996) (beds 58–75): It is marked by the presence of *Indosphinctes errans* Spath M.

C-V Horizon Krishna and Ojha (1996) (beds 76–87): It is marked by the presence of *Hecticoceras chanaziense* Parona and Bonarelli M common in the

Michalskii Horizon of France for correlation. *Choffatia* cf. *baluchistanensis* (Noetling) M is also present.

Diadematus Subzone Krishna and Cariou (1990): It is defined by significant *M. diadematus* (Waagen) and includes only one horizon.

C-VI Horizon Krishna and Ojha (1996) (beds 88–91): It is indicated by the last extremely inflated *M. diadematus* (Waagen) M. This horizon corresponds to the Subboreal Galillaeii Horizon near the top of the Koenigi Zone, which also includes similarly inflated *Macrocephalites*.

Ridge Sandstone Member (RSM)

Semilaevis Zone Krishna and Cariou (1990): Its base is defined by the first *M. semilaevis* (Waagen) M and acme of *M. formosus* (Sowerby) M.

Formosus Subzone Krishna and Cariou (1990) emended (Krishna and Ojha 1996): Its base is defined as for the Semilaevis Zone. It includes just one horizon.

S-I Horizon Krishna and Ojha (1996) (beds 92–95): It is marked by the acme of *M. formosus* (Sowerby) M and its dimorph *M. subtrapezinus* (Waagen) m together with *M. chariensis* (Waagen) M, the latter being exclusive to this horizon. *H. michalskii* Lew provides correlation with Michalskii Horizon of France.

S-II Horizon Krishna and Ojha (1996) (beds 96–119): It is marked by near exclusive *M. semilaevis* (Waagen) M, *Indosphinctes urbanus* Spath M and *I. choffati* Spath M, suggesting correlation with Michalskii and Proximum Horizons of France.

Eucyclum Subzone Krishna and Ojha (1996): It is characterized by the fist *Eucycloceras eucyclum* (Waagen) M and *Subkossmatia opis* (Sowerby) m, and includes only one horizon. Eucycloceratinae is a geographically restricted subfamily, and the associated ETM elements are indecisive about the firm age of its oldest subzone—the Eucyclum Subzone. The Eucyclum Subzone thus alternatively can also be included at the base of Middle Callovian, thereby including the entire subfamily at the base of Middle Callovian. Nevertheless, it is a mere suggestion, and for the present, the Eucyclum Subzone is maintained as the terminal subzone of the Early Callovian.

S-III Horizon Krishna and Ojha (1996) (beds 120–131): It is characterized by the abundance of *Eucycloceras eucyclum* (Waagen) M and *S. opis* (Sowerby) m. *H. proximum* Elmi M and*Collotia oxyptycha* (Neumayr) M in common between Kachchh and France.

Middle Callovian

This interval includes 2 zones, 5 subzones and 8 horizons

Anceps Zone Oppel: It is indicated by significant *Reineckeia anceps* (Reinecke) M, while the endemic *Subkossmatia ramosa* Spath M, *Kinkeliniceras kleidos* (Spath) M and *K. paramorphum* (Waagen) M are also significant in the early part.

Ramosa Subzone Krishna and Ojha (1996): It is characterized by *S. ramosa* Spath M with *R. anceps* (Reinecke) M, and includes one horizon.

A-I Horizon Krishna and Ojha (1996) (beds 132–163): It is defined by significant *Reineckeia anceps* (Reinecke) M, *S. ramosa* Spath M and single *Chanasia hartmanni* Zeiss M, the latter providing correlation with Bannense Horizon of France.

Kleidos Subzone Krishna and Ojha (1996): It is defined by the first *Kinkeliniceras kleidos* (Spath) M at the base, and includes one horizon.

A-II Horizon Krishna and Ojha (1996) (beds 164–167): It is characterized by the first *Kinkeliniceras kleidos* (Spath) M and *Choffatia lateralis* Spath M. *Indosphinctes* aff. *peregrinus* Spath M is exclusive to this horizon. The common *I.* cf. *peregrinus* Spath M suggests correlation of Kleidos Subzone with the Bannense and Medea Horizons of France.

Paramorphum Subzone Krishna and Ojha (1996): It is defined by the first *Kinkeliniceras paramorphum* (Waagen) M at the base, and includes three horizons.

A-III Horizon Krishna and Ojha (1996) (beds 168–175): It is characterized by the first *K. paramorphum* (Waagen) M at the base. *Indosphinctes* aff. *indicus* (Siemeradzki) M. *I. calvus* (Sowerby) M is exclusive to this horizon.

A-IV Horizon Krishna and Ojha (1996) (beds 176–193): It is characterized by the last *K. kleidos* (Spath) M with the significant exclusion of *Indosphinctes*.

A-V Horizon Krishna and Ojha (1996) (beds 194–203): It is characterized by the first *Reineckeia (Loczyceras) richei* (Roman) M and *Collotia spathi*, providing correlation with Jason Horizon of France.

The Paramorphum Subzone, on the strength of common *R. (L.) richei* (Roman), M and *C. spathi,* correlates well with the Jason Subzone of France.

Obtusicosta Zone Krishna and Ojha (1996): It is defined by the FAD and significant presence of the zonal index *Kinkeliniceras (Obtusicostites) obtusicosta* (Waagen) M. *K. paramprphum* (Waagen) continues. It includes two subzones

Obtusicosta Subzone Krishna and Ojha (1996): The base is defined as for the Obtusicosta Zone. *R. (L.) richei* (Roman) M is still present while *R. (L.)* aff. *intermedia* m is ecxclusive to the subzone. *Collotia gigantea* Bourquin M is also present. *Pseudopeltoceras* marks its FAD in this subzone, so also *K. dhosaeinse* m. It includes two horizons.

O-I Horizon Krishna and Ojha (1996): (beds 204–217): It is indicated by the first *K. (O.) obtusicosta* (Waagen) M. *R. (L.) richei* (Roman) M continues. *R (L) reissi* becomes extinct in this horizon.

Gypseous Shale Member (GSM)

O-II Horizon Krishna and Ojha (1996) (beds 218–269): It is indicated by *C. gigantea* Bourquin M and *R. (L.) intermedia* (Bourquin) M, the latter being exclusive to this horizon.

Catillus Subzone Krishna and Ojha (1996): It is marked by first *Kinkeliniceras catillus* (Waagen) M and *Orionoides*. It includes just one horizon

O-III Horizon Krishna and Ojha (1996) (beds 270–307): It is marked by the first *K. catillus* (Waagen) M, *K. hubertus* m and *K. omphalodes* m, also first *Orionoides*, providing correlation with Rota Horizon of France.

Late Callovian

This interval includes 2 zones, 2 subzones and 5 horizons

Athleta Zone It is indicated by the first scarce *Peltoceras* through *Peltoceras athleta* (Phillips) M and first *Orionoides pseudorion* (Waagen) M at the base. *Binatisphinctes* is exclusive to this zone, while *Poculisphinctes* registers its first occurrence in this zone. The last *Subgrossouvria* occurs in this zone. It includes two subzones

Pseudorion Subzone Krishna and Ojha (1996): Its base is defined as for the Athleta Zone. It includes one horizon

A-I Horizon Krishna and Ojha (1996) (beds 308–331): It is marked by the first *P. athleta* (Phillips) M and *O. pseudorion* (Waagen) M. *Subgrossouvria* gr. *gudjinsirensis* Spath M is also significant.

Depressum Subzone Krishna and Ojha (1996): Its base is defined by the first *Kinkeliniceras depressum* Krishna and Ojha m. It includes two horizons:

A-II Horizon Krishna and Ojha (1996) (beds 332–347): It is indicated by the first *K. depressum* Krishna and Ojha m, *Binatisphinctes* and *Poculisphinctes* and the last *Subgrossouvria.*

A-III Horizon Krishna and Ojha (1996) (beds 348–377): It is marked by the first *Peltoceras* aff. *kachhensis* Spath M, the last kinkeliniceratins, the last *Binatisphinctes*, the last *Orionoides* as also the last *P. athleta* (Phillips) M along with solitary *Paralcidia* cf. *khengari* Spath M.

The Depressum Subzone is correlated with the French Collotiformis Subzone in view of common *Collotia fraasi* (Oppel) M, *P. metamorphicum* Spath M, *P. vijaya* Spath M, and morphologically close *Binatisphinctes* and *Paralcidia.*

Dhosa Sandstone Member (DSM)

It in fact begins in the terminal part of Depressum Zone from bed 373.

Ponderosum Zone Krishna and Ojha (1996): It is defined by the first *Peltoceras ponderosum* (Waagen) M. It includes two horizons

P-I Horizon Krishna and Ojha (1996) (beds 378–385): It is indicated by the first *P. ponderosum* (Waagen) M, the first *Euaspidoceras*, the first *Metapeltoceras* and the first *P. solidum* Spath M, suggesting good correlation with the Subtense Horizon of France.

P-II Horizon Krishna and Ojha (1996) (beds 386–390): It is indicated by the first *Pachyceras llandeanum* (d' Orbigny) M, *Prososphinctoides manialensis* Spath m, and single *Poculisphinctes* cf. *poculum* (Lackenby) M which suggest correlation with the Athletoides Horizon of France.

The Ponderosum Zone neither includes any Early Oxfordian pandemic Perisphinctinae, nor the Mayaitinae, nor the Kinkeliniceratinae that are geographically restricted to GTM. It is instead featured by the diverse ETM elements of Mediterranean origin, e.g. Peltoceratinae, Aspidoceratinae and Pachyceratinae. It is interpreted that the Kinkeliniceratins underwent near extinction to drastic reduction near the second-order MFS at the close of the ammonoid-rich gypsiferous clays. The change from the ammonoid-rich gypsiferous clays to the ammonoid devoid/scarce non-gypseous silt/sand is also easily picked up in the well logs, and is interpreted to mark the start of the second-order RST. The ammonoid assemblage in the Horizon P-II in general suggests a small third-order transgression which is in

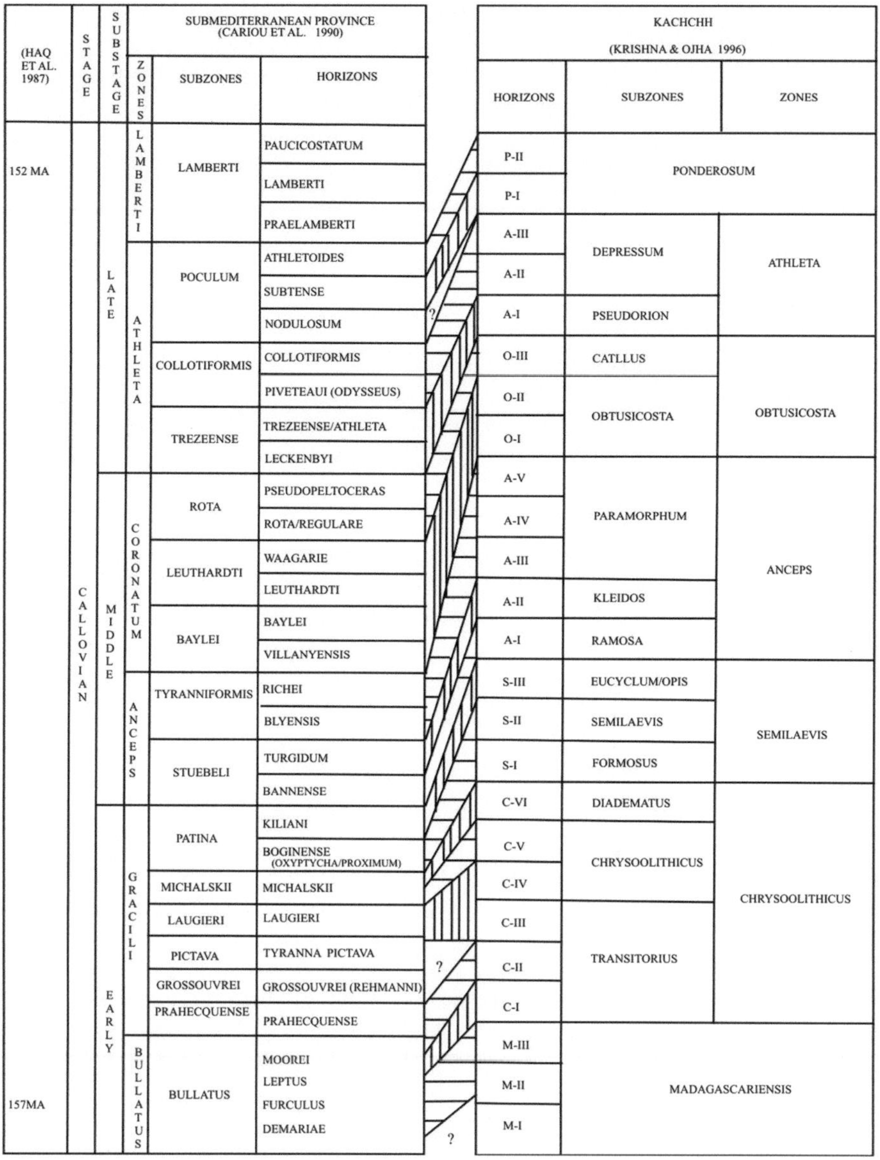

Fig. 2.31 The correlation of the Kachchh Callovian Horizons with the ETM standard (modified after Krishna and Cariou 1990)

turn followed up by the early Early Oxfordian ammonoid devoid thick sand/silt present throughout the basin from Mainland Kachchh to Wagad and Khadir. Such sands are even found in different basal Oxfordian sections in Madagascar of probable Mariae Zone age that are deposited in the regressive phase. The possibility of a small subzonal subaerial gap above the ammonoid devoid non-gypseous sands is discounted in view of the location of Keera a good distance away from the margin, however, the presence of such a gap is not ruled out in proximal sections in East Mainland, Khadir, Gangta bet and Wagad. High resolution precise correlation with the Submediterranean province of ETM is indicated in Fig. 2.31. The entire DSM could also be included in the terminal most Late Callovian.

2.8.6 *Oxfordian*

Development of the Oxfordian in Kachchh
Zonal correspondence with ETM

The development of the Oxfordian (s. l.) on the Gondwanian Tethyan Margin (GTM) between Arabia and Australia is invariably incomplete. Also, almost all over the GTM, it is expressed in the moderately to highly condensed oolitic facies, particularly, in the Early and early Middle Oxfordian in proximal locations. Such characteristic features are on account of the inclusion of the first-order Jurassic MFS on the GTM in the Middle Oxfordian along with the accompanying submarine stratigraphic gap at sedimentation sites away from the margin. The GTM Oxfordian interval is more interesting and significant than on the ETM (European Tethyan Margin) since the intra-Jurassic first-order MFS on the ETM instead of falling in the Oxfordian is included in the Kimmeridgian stage. In spite of the features of incompleteness and condensation, it has been all the more desirable to refine the ammonoid stratigraphy and zonation in the Oxfordian of the GTM. The prospects of progress in the Oxfordian ammonoid zonation improved in context of the recent discovery of uninterrupted, moderately to rapidly sedimented, complete Oxfordian in the proximal part of the Kachchh basin at Gangta Bet and Kantkote in Wagad (Krishna et al. 1994a, c, 1995a, 1996a, b, 1998, 2000 and later). The ammonoid-rich Early Oxfordian has been already known since the classic monographs of Waagen (1873–75) and Spath (1927–33). Almost all the ETM Oxfordian zonal intervals are found recognizable in Kachchh (except the doubtful nearly ammonoid devoid equivalent of the early Early Oxfordian Mariae Zone) those of Early to early Middle Oxfordian throughout the basin, while the Middle and Late Oxfordian exclusively in proximal Kantkote and Gangta Bet. Here, the author proposes a preliminary tentative succession of ammonoid zones in the Oxfordian of Kachchh as the first such effort anywhere on the Gondwanian Tethyan Margin. The Oxfordian sediments invariably are conformably underlain by the exposed Callovian sediments.

The Oxfordian duration includes two highly contrasting stratigraphic intervals of unequal duration in Kachchh (Krishna et al. 1998, 2000). The older is present all over the exposed basin from the distal-most Lakhapur in the west to proximal-most Wagad in the east, also in Khadir of 'Island belt' while the younger is exclusive to Wagad (Figs. 2.32 and 2.33) and Khadir. The older interval is ~2.6 my long up to the early Middle Oxfordian Vertebrale Subzone of the Plicatilis Zone of the ETM inclusive of the terminal part of the DSM, and the DOM. It is ~12–18 m thick (thickness increasing towards the margin), mostly fine textured, slow to extremely slow sedimented, ammonoid and other body fossil rich, irregularly based, nodular, boxworked, hardened to hard grounded, pebbly/nodular/bumpy/conglomeratic, densely oolitic limestone (in the distal relatively deeper sites to much less oolitic and less calcareous beds in the proximal sites). The dominantly oolitic succession although present in the ~300 km wide exposed shelf, is best characterized in the western half of the Mainland Kachchh, west of the MH (Median High). The coeval Early and basal Middle Oxfordian part is also included in Wagad and Khadir, which is overlain by the younger mid Middle to latest Late Oxfordian. The mid Middle to Late Oxfordian is exclusive to Wagad and Khadir on the east and north margins outside the Mainland Kachchh. It is ~140 m thick of ~3.6 my duration, and is compositely best developed in Khadir, Gangta Bet, Desalpar and Kantkote–Bharodia stratigraphic sections. It includes the terminal part of the DSM and the full JM of the Chari Formation up to the late Middle Oxfordian. The Chari Formation is conformably superposed by Katrol Formation in Wagad.

Lithostratigraphic Framework

Among the several Oxfordian sections studied, only three are detailed here (1. Lakhapur section as representative of the Mainland Kachchh, coeval sediments of Early and early Middle ages are also present particularly in Khadir, Gangta bet, and Kakindia, however, mid Middle to Latest Oxfordian sediments are absent on the Mainland, 2. Gangta bet section and 3. Kantkote–Bharodia, i.e. the K–B section of Wagad). These account for the first tentative, yet, independent ammonoid zonal formulation for the Kachchh Oxfordian in Particular, and for the GTM in general.

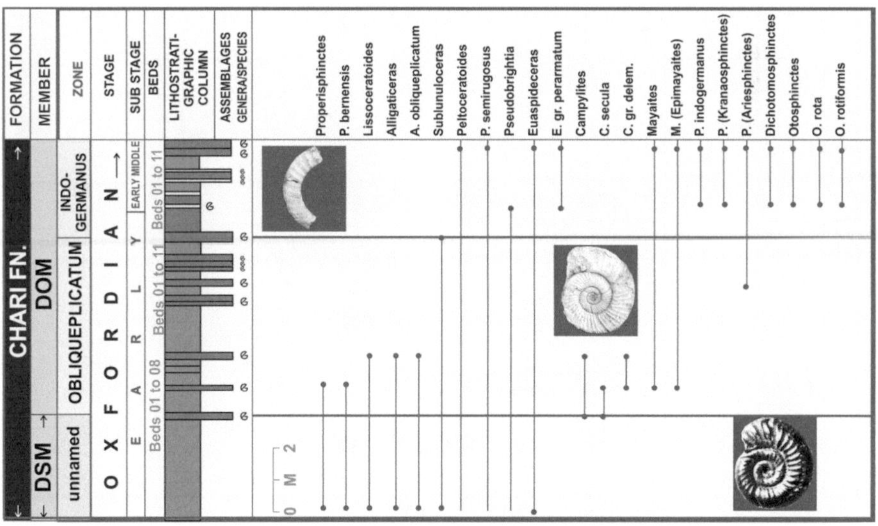

Fig. 2.32 Ammonoid zones in the Early and early Middle Oxfordian of Kachchh in Lakhapur section

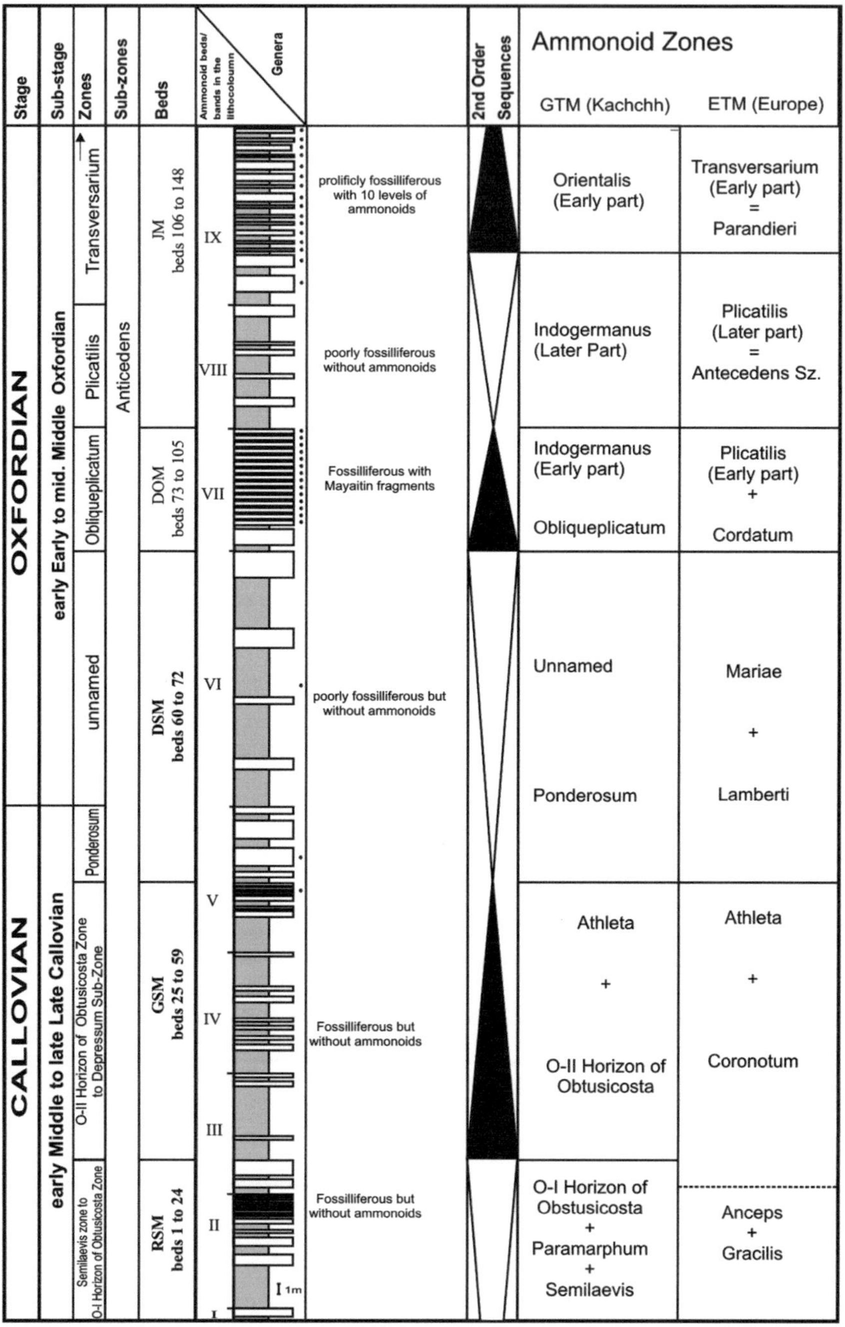

Fig. 2.33 Preliminary zonation in the newly discovered Callovian–Oxfordian stratigraphic column at Gangtabet vis-a-vis second-order sequences

The Oxfordian sediments all over the basin are conformably underlain by the exposed Callovian sediments.

These sections are

Lakhapur Stratigraphic Section

The stratigraphic section is organized into three sediment intervals (LI to LIII) which are further differentiated into a succession of 30 beds (LI-beds 01–08, LII-beds 01–11 and LIII-beds 01–11) with 15 ammonoid levels at the distal most exposed part of the basin (Early and early Middle part of Oxfordian representing the younger part of Chari Formation which is followed by a large submarine stratigraphic gap up to the early Early Kimmeridgian Hypselocyclum Zone (Krishna et al. 1998).

Sediment Interval LI (Beds 01–08): 5.30 m thick, earthy non-gypseous shale with maroon to greenish maroon, medium to fine grained, pebbly, marly, fractured, calcareous, fossiliferous, sandstone bands, yielding ammonoids from 3 different levels.

Sediment Interval LII (Beds 01–11): 3.80 m thick, earthy shales intercalated with yellow to greenish maroon, pebbly, nodular, fractured, very hard, occasionally oolitic, sandy limestone bands, highly fossiliferous, yielding ammonoids from six different levels.

Sediment Interval LIII (Beds 01–11): 2.90 m thick, earthy to yellowish green shale/siltstone, occasionally enclosing nodules, alternating with medium to fine grained, hard, compact, blocky, jointed, pebbly, nodular, yellowish to greenish yellow, ammonoid-rich, highly fossiliferous, sandy limestone bands, yielding ammonoids from six different levels.

It may be noted that the coeval sediments are also studied at Gangta Bet, Kakindia bet, Khadir and Wagad outside the Mainland Kachchh where the thickness is distinctly greater, ooilte presence minimal and ammonoid density and diversity substantively reduced in view of their proximal locations.

Gangta Bet Section

It is organized in nine sediment intervals which are further differentiated into 64 beds with at least 25 ammonoid levels in the proximal-most exposed part of the basin of Early to late Middle Oxfordian age. The first five sediment intervals belong to the Callovian while the younger four to thc Oxfordian (Fig. 2.33). The mentioned Oxfordian up to the early Middle Oxfordian is underlain by at least Middle to Late Callovian. The entire late Middle Callovian to Middle Oxfordian included earier in the column of the abandoned ‘Gangta bet member’ is unambiguous extension of the Chari Formation in Gangta bet.

Sediment interval I—0.80 m hard, medium to fine grained, maroon to brown, highly jointed pebbly sandstone.

Sediment interval II—10.80 m earthy yellow shale alternating with thick (50 cm to 1.00 m thick) hard medium to fine grained, brown concretionary sandstone bands in the lower part and with thin (15–20 cm thick) ash-maroon to maroon, box worked nodular concretionary sandstone bands.

The above two sediment intervals correspond to the later part of RSM.

Sediment interval III—12 m ochre yellow to earthy shale alternating with hard medium to coarse grained cross-bedded nodular fractured box worked reddish to yellowish maroon thick (40–1.5 m thick) sandstone/sandy limestone bands, highly fossiliferous in younger levels with numerous shell fragments, yet no ammonoids.

Sediment interval IV—12 m earthy shale alternating with hard medium to fine grained greenish yellow to ash-maroon and even reddish grey (40–60 cm thick) sandy limestone bands with numerous shell fragments, yet without ammonoids.

Sediment interval V—14.80 m earthy yellow shale alternating with hard medium to fine and even coarse grained, occasionally flaky and jointed, greenish grey, yellowish maroon to even reddish sandstone/sandy limestone bands, fossiliferous with occasional or rich in shell fragments, and two levels of rare ammonoids.

Sediment intervals III–V are broadly correlated to GSM with two levels of Kinkeliniceratinae ammonoids in sediment interval V of Depressum Subzone.

Sediment interval VI—24.60 m grey to earthy yellow shale alternating with hard and compact, yellowish to greenish maroon, medium to fine, even coarse grained, box worked, pointed to blocky with occasional concretions and nodules sandstone, sandy limestone bands, mostly poorly fossiliferous while a 2 m thick medium grained friable variegated sandstone at the top with a marker conglomerate in the middle. Ammonoids as yet not collected in this interval.

Sediment interval VII—11.40 m earthy shale alternating with hard fractured medium to fine maroon 10–20 cm thick sandstone bands with indeterminate perisphinctin and mayaitin fragments in upper part while a 1.6 m thick very hard compact coarse grained grey coloured jointed blocky sandstone bed in the lower part. This is perhaps the correlative of the marker Kakindia Limestone of Kakindia bet (beds 8–12 of Gangta bet and bed 36 of Khadir).

Sediment interval VIII—11.20 m earthy shale alternating with hard grey to maroon medium to fine sandstone bands (20–60 cm thick) capped by 1.60 m thick medium grained very hard and compact poorly fossiliferous sandstone.

Sediment interval IX—17.00 m thick earthy shale with hard coarse even gritty, occasionally friable pebbly nodular fractured jointed blocky yellowish to yellowish maroon or even grey calcareous 20–80 cm thick sandstone bands, highly fossiliferous yielding ammonoids from at least 16 ammonoid-rich levels. This is perhaps the famous ammonoid-rich Gangta ammonoid bed.

Kantkote–Bharodia (K–B) Stratigraphic Section (Figs. 2.34 and 2.35): It is located at the proximal-most part of the basin from Kantkote to Bharodia in Wagad (late Middle Oxfordian to early Early Kimmeridgian in age representing the youngest part of the Chari Formation and the Basal (Kantkote) Member of the Katrol Formation exclusively present in Wagad). It may be noted that the sediment intervals III to VII of the Fig. 2.5 after Krishna et al. (2009) correspond to the sediment intervals IV to VIII in the here updated Fig. 2.34.

Sediment Interval K–B I (Beds 01–05 of K–B I & II): 24.50 m thick, earthy shale alternating with medium to fine sandstone bands in the lower part capped with hard, medium to coarse grained, rippled and nodular sandstones, yielding ammonoids from two levels.

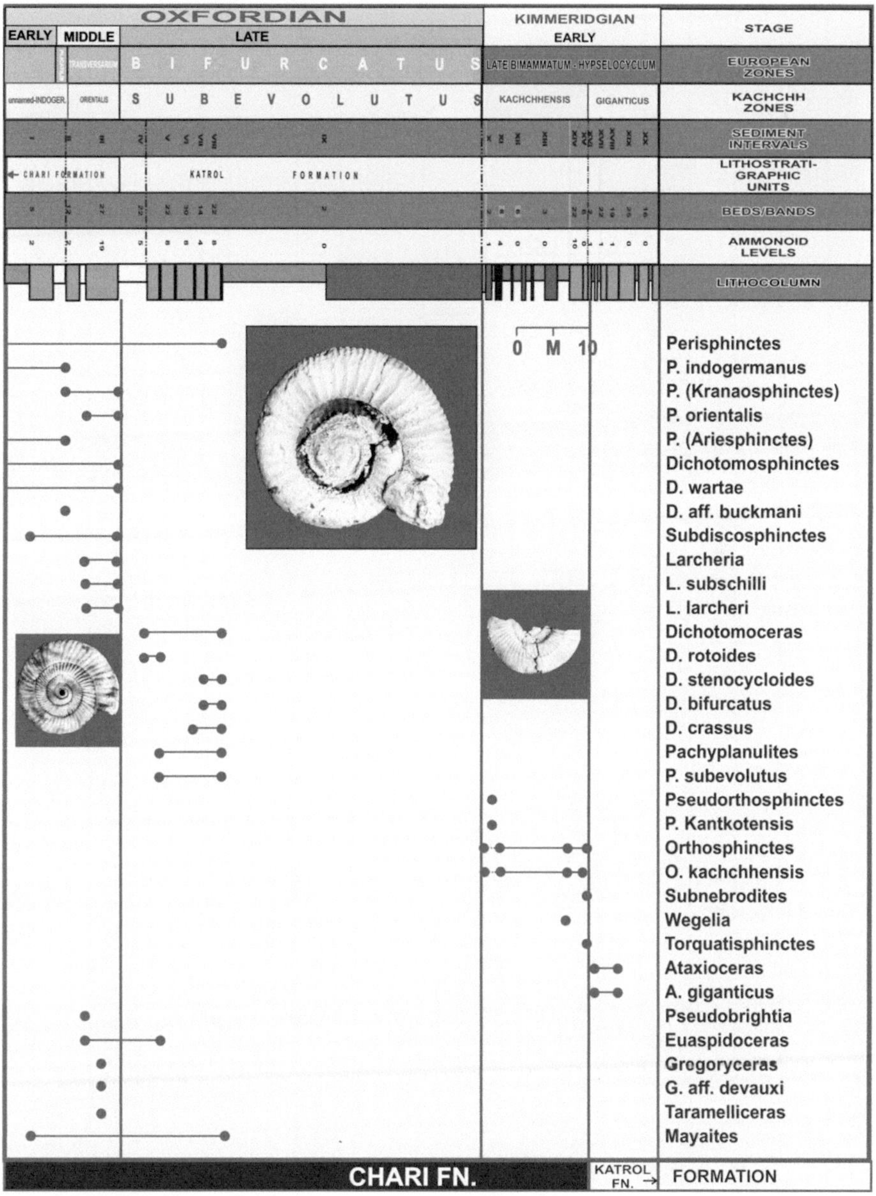

Fig. 2.34 The stratigraphic occurrence and ranges of the ammonoid species in the Middle and Late Oxfordian–Early Kimmeridgian of Kachchh in Kantkote–Bharodia section

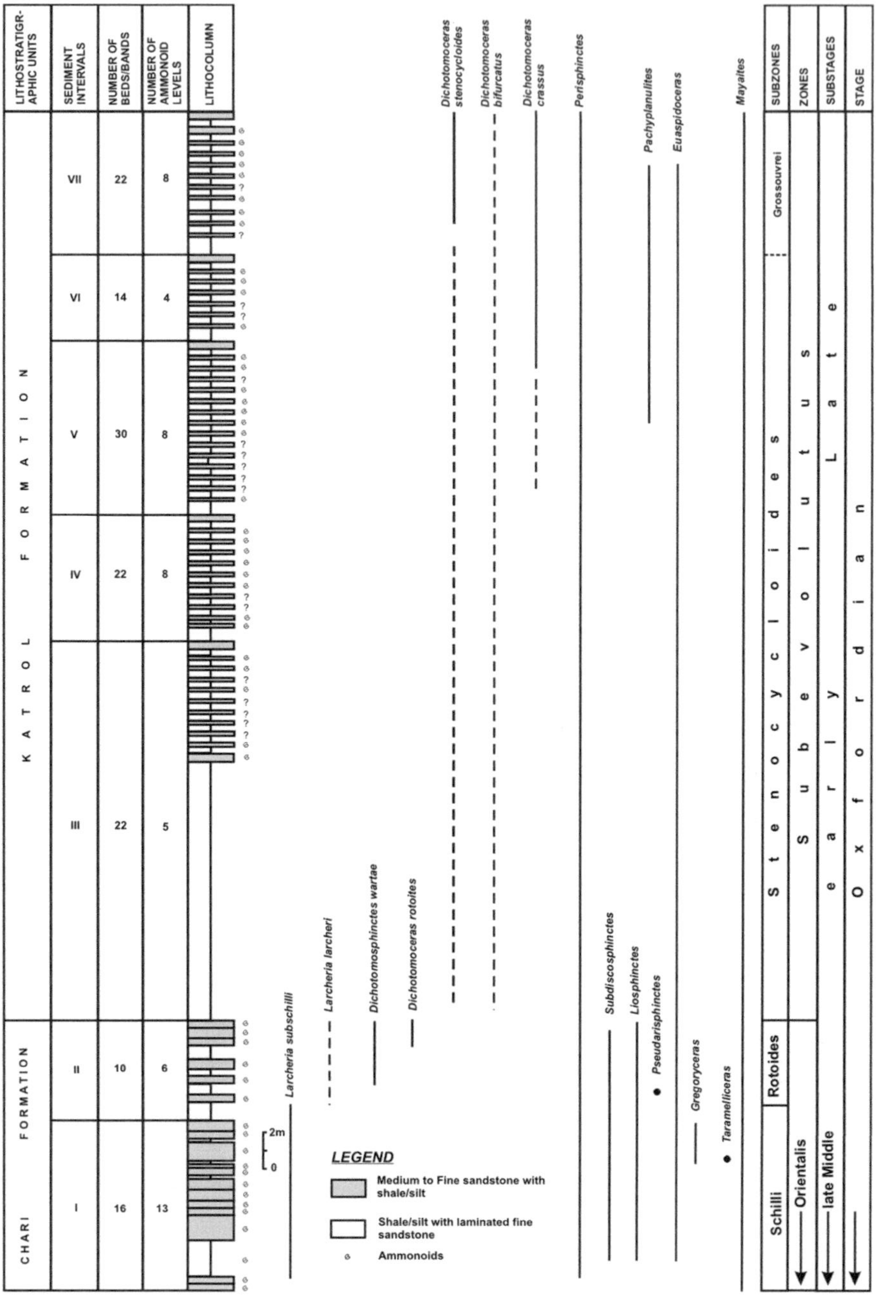

Fig. 2.35 Ammonoid zones in the late Middle Oxfordian–Early Kimmeridgian column of Kantkote–Bharodia section after Krishna et al. (2009)

Sediment Interval K–B II (Beds 01–12 of K–B I & II): 13.50 m thick, earthy gypseous shale alternated with nodular, sometimes bioturbated sandstone bands, poor in ammonoids (only two levels).

Sediment Interval K–B III (Beds 01–27) m thick, shale with maroon to brick red, medium to fine grained, hard, nodular sandstone, highly fossiliferous with plenty of ammonoids (19 different levels), frequent wood and belemnites.

Sediment Interval K–B IV (Beds 01–22): 22.0 m thick, earthy-grey gypseous silty shale with several thin, variegated, nodular, flaky sandstone bands in the lower part and shale alternating with maroon to yellowish maroon, medium to fine grained, pebbly, fractured bands in the middle part, capped with (50–70 cm thick) massive, jointed, blocky, hard and compact, grey, medium to fine grained sandstones, poorly fossiliferous, yielding ammonoids only from five different levels.

Sediment Interval K–B V (Beds 01–22): 9.0 m thick, earthy-grey silty shales intercalated with several thin bands of sandstone and topped with the massive, jointed and blocky sandstone giving similar physical and lithological appearance as that of Sediment Interval IV as above, yielding ammonoids from eight different levels.

Sediment Interval K–B VI (Beds 01–30): 10.0 m thick, earthy-grey silty shale and thin sandstone bands having similar lithological appearance as that of Sediment Intervals IV and V and yielding ammonoids from eight different levels. (lower portion) to hard and compact, yellowish to maroon coloured, occasionally bioturbated, variegated calcareous sandstone with belemnites, ammonoids, woods and other fossil shell fragments.

Sediment Interval K–B VII (Beds 01–14): 6.0 m thick, repetition of similar lithology as that of Sediment Intervals IV, V and VI, yielding ammonoids from four ammonoid levels.

Sediment Interval K–B VIII (Beds 01–22): 8.0 m thick, sedimentary succession of the same physical and lithological appearance as that of Sediment Intervals IV, V, VI and VII, yielding ammonoids from eight different levels.

Sediment Interval K–B IX (Beds 01–02): 34 m thick, variegated shale alternating with thin sandstone bands (lower 54 m part), capped with massive, cross-bedded, medium to coarse grained, friable to hard, variegated sandstone (80 m upper part), unfossiliferous.

Dhosa Sandstone Member (DSM)
The terminal part of DSM may include the unnamed zone of early Early Oxfordian.

Ammonoid zones and Correlation

Early and early Middle Oxfordian (Mainland Kachchh at Lakhapur Section)

early Early Oxfordian (unnamed zone = Mariae Zone of the ETM): There is determined a third-order TST within the sediment interval VI near the top in a marker conglomerate bed. Applying isochroneity of the third-order sequence surfaces across the world, this marker conglomerate bed may be assigned to the early part of the early Early Oxfordian Mariae Zone. This marker conglomerate is otherwise age-constrained between the *Kinkeliniceras* bearing sediment interval V of the early part of early Late Callovian Athleta Zone below and *Mayaites* bearing sediment interval VII of late Early Oxfordian Obliqueplicatum Zone above. Thus, the major part of the Sediment interval VI underlying the transgressive marker Conglomerate bed is suggestively corresponded to the Dhosa Sandstone Member, while the Callovian/Oxfordian boundary in Kachchh may be suggested within the sediment interval VI below its youngest marker sandstone bed. Alternatively, the sediment interval VI could also exclusively belong to terminal Late Callovian, which is farther up disconformably superposed by late Early Oxfordian ammonoid bearing sediment interval VII. Any ammonoids as such have not been discovered in the mentioned conglomerate or sandstone bed, which could be due to shallowing of the basin followed up by possible transient emergence. The absence of ammonoids does not permit any formal name to this terminal Callovian/basal Oxfordian zone. The mentioned marker conglomerate presumably is the bed 6 of Biswas.

Dhosa Oolite Member (DOM)

Obliqueplicatum Zone (new) (Sediment Interval LI, Beds 02 to LII bed 11, Fig. 2.32) (equal to Cordatum Zone): The zone is indicated by the first Mayaitinae, first *Alligaticeras* through *A. obliqueplicatum* (Waagen) and other species, first *Properisphinctes* through *P. bernensis* (de Lor.) and associated other *Properisphinctes* in the sediment interval LI, beds 02–08 of the Lakhapur section in association with Hecticoceratinae and Peltoceratinae. It is tentatively correlated to the Claromontanus Zone of Europe on the basis of the common *Properisphinctes bernensis* (de Lor.) in Claromontanus Zone (Cariou and Hantzpergue 1997) and Kachchh, which in turn, is correlated with the Bukowskii Subzone of the Cordatum Zone. The morphologically similar *Peltoceratoides* and *Alligaticeras* suggest the duration of the zone up to the Cordatum Subzone of the Cordatum Zone of the ETM. There is possibility of even further differentiation into at least two subzones based on the first *Properisphinctes* and the first *Alligaticeras* in the early part of the bed 01, and the first *Mayaites* in the bed 04.

The early Early Oxfordian interval of the Mariae Zone is unknown. Any clue to that is not picked up even in the sequence stratigraphic context. There could even be a subaerial break particularly in the proximal sections of Gangta bet, Habo and Jhura at the second-order SB between the Dhosa sand/silt and Dhosa Oolite. Whatever, be the reason, neither the latest Late Callovian nor the earliest Early Oxfordian subzones are yet differentiated through ammonoids in Kachchh. However, in the Nara section Kanjilal and Singh (2012) suggested the presence of the latest Late Callovian Lamberti Zone through a rare not easily explainable record of the cold water boreal *Quenstedtoceras* cf. *lamberti* in the middle part of the DSM. This in turn could relate to the widespread transient cooling phase near the Middle/Late Jurassic transition as in Europe of Dromart et al. (2003), also discussed later elsewhere. At present, in the absence of the early Early Oxfordian ammonoids, using isochroneity of the third-order TSTs between the GTM and the ETM, there is determined a thin oolitic, conglomeratic slow sedimented bivalve bearing bed (bed 6 of Biswas) as a third-order TST, and assigned to early Early Oxfordian in the Gangta bet section less than ~10 m below the Mayaitinae bearing Kakindia Limestone marker of definite late Early Oxfordian Obliqueplicatum Zone age.

Biswas (1993) correlated the mentioned conglomeratic sandstone to an ammonoid bearing bed of the Khadir section, although the present author did not find any. May be later searches yield in the mentioned bed ammonoids of the early part of early Early Oxfordian. Which may allow naming of this zone as also confirm its being of the early part of early Early Oxfordian.

Indogermanus Zone (new) (Sediment Interval LIII, Beds 02 to 11—upper limit not exposed in Lakhapur section but determined in the Gangta Bet section, Fig. 2.33) (time equivalent of the ETM Plicatilis Zone): The base is indicated by the first appearance of *Perisphinctes indogermanus* (Waagen) in sediment interval LIII, bed 02 in Lakhapur stratigraphic section in association with Peltoceratinae, Mayaitinae and Haploceratinae. It is correlatable to the Plicatilis Zone of Europe (Cariou and Hantzpergue 1997) on the basis of morphologically close Perisphinctinae and Peltoceratinae. Only the early part of this zone, that is suggestively equivalent to Vertebrale Subzone, is present in the Mainland Kachchh on account of a large submarine stratigraphic gap above. The later part of this zone is recognized recently in the Gangta Bet section up to the top of sediment interval G—VIII, and beds 11–13 of Biswas with ammonoids of the Antecedens Subzone.

Jhadasa Member (JM)

Mid. Middle to Late Oxfordian at Kantkote–Bharodia section of Wagad

Orientalis Zone (new) (at Kantkote–Bharodia Section at Wagad) (Figs. 2.34 and 2.35) (time equivalent of the late Middle Oxfordian Transversarium Zone of ETM): The Orientalis Zone is exclusively developed at Wagad in the Kantkote, Khadir and Gangtabet sections. The precise base is determined in the Gangta Bet section at the base of the sediment interval G—IX through rich *Dichotomosphinctes* cf. *buckmani*. The single and septate example of *Dichotomosphinctes* aff. *buckmani* in sediment interval K–B II, bed 04 (Fig. 2.34) at Kantkote, also, suggests the base of the Orientalis Zone. In Biswas, the base is indicated at the base of bed 13 at Gangta bet and also includes bed 15. Bed 13 includes *D.* gr. *buckmani* of Parandieri Subzone of the ETM. It is the most abundantly ammonoid bearing zone in the entire Kachchh Jurassic and named after the most common Perisphinctinae species found in this interval. It is correlatable with the Transversarium Zone of Europe (Cariou and Hantzpergue 1997) on the basis of the presence of rich *Larcheria*, particularly the species *Larcheria subschilli* (Lee) and a few *Gregoryceras* aff. *devauxi* (Bert and Enay), and *Dichotomoceras rotoides* (Ronch.) in later part of the zone in sediment interval K–B III, beds 05–27 (Figs. 2.34 and 2.35). It is further differentiable in several subzones as on ETM based on the ranges of *D.* cf./aff. *buckmani*, *D. lucaeformis* (Krishna et al. unpublished), *Larcheria schilli* and, *Dichotomoceras rotoides* (Krishna et al. 2000, at the Gangta Bet and Kantkote–Bharodia sections, which shall be formalized at a later stage.

Informal Patasar Shale Member of Kantkote Member (KM)

It starts at the lithostratigraphic contact of the JM of the Chari Formation below and the KM of the Katrol Formation at the ammonoid stratigraphic Schilli Subzone/Rotoides Subzone boundary within the Orientalis Zone. It makes the basal part of the KM in the Kantkote section.

Subevolutus Zone (new) (K–B IV, bed 02–K–B IX, bed 02, Fig. 2.35): The base is indicated by *Pachyplanulites pagri* Spath and *P. subevolutus* (Waagen) in sediment interval K–B IV, bed 02 in association with the significant presence of species of *Dichotomoceras* like *D. bifurcatus*, *D. crassus*, *D. stenocycloides* which otherwise characterize the Bifurcatus Zone in Europe (Cariou and Hantzpergue 1997) and hence, the correlation with the European Bifurcatus Zone. This zone is quite thick and ammonoid scarce with large indeterminate varicostate Perisphinctinae fragments. Though the terminal few metres are completely devoid of ammonoids, yet are tentatively included in the terminal Oxfordian Subevolutus Zone, prior to the onset of Kimmeridgian through the first Ataxioceratinae *Orthosphinctes*. The revised Oxfordian/Kimmeridgian boundary suggests tentative correspondence of the upper boundary of the mentioned zone to the early part of Bimammatum Zone.

2.8.7 *Kimmeridgian*

Develoment of the Kimmeridgian in Kachchh

The Kimmeridgian ammonoid bearing succession is best represented in the Kachchh Mainland (Fig. 2.1). Six different sections have been studied (Figs. 2.34, 2.36 and 2.37), which from the margin to the basin are: (i) Bharodia that includes early and mid. Kimmeridgian. (ii) Jawahar Nagar (iii) Southeast Ler (iv) Southwest Ler (v) Walakhawas and (vi) Lakhapur. Sections (ii) to (vi) include late Early and Late Kimmeridgian. The Kimmeridgian all through the Mainland is unconformably underlain by the exposed Early and early Middle Oxfordian sediments.

The Kimmeridgian is not present in the island belt on account of the complete withdrawl of the sea from that region even prior to the close of the Middle Oxfordian. Also, only early and mid Early Kimmeridgian is known through ammonoid bearing sediments belonging to the genera *Orthosphinctes* and *Ataxioceras* presumably of Planula–Hypselocyclum suprazonal interval at Bharodia in the informal Adhoi member of the Kantkote Member of Katrol Formation in Wagad. It may incidently be the youngest Jurassic ammonoid bearing geological record of the Kachchh basin outside the Kachchh Mainland. The immediately superjacent ammonoid devoid, rapidly deposited sands could also be included in the Hypselocyclum Zone. Thus the Wagad succession is not younger than the Early Kimmeridgian. The Early Kimmerdgian in Wagad is underlain conformably by the Late Oxfordian sediments.

Lithostratigraphic Framework
At Bharodia

The Bharodia section of early and middle Early Kimmeridgian age is described below:

Sediment Interval K–B X (Beds 01–02): 5.0 m thick, shale with very hard and compact, jointed, yellowish, medium to fine grained, calcareous sandstone band with shell fragments, rare ammonoids.

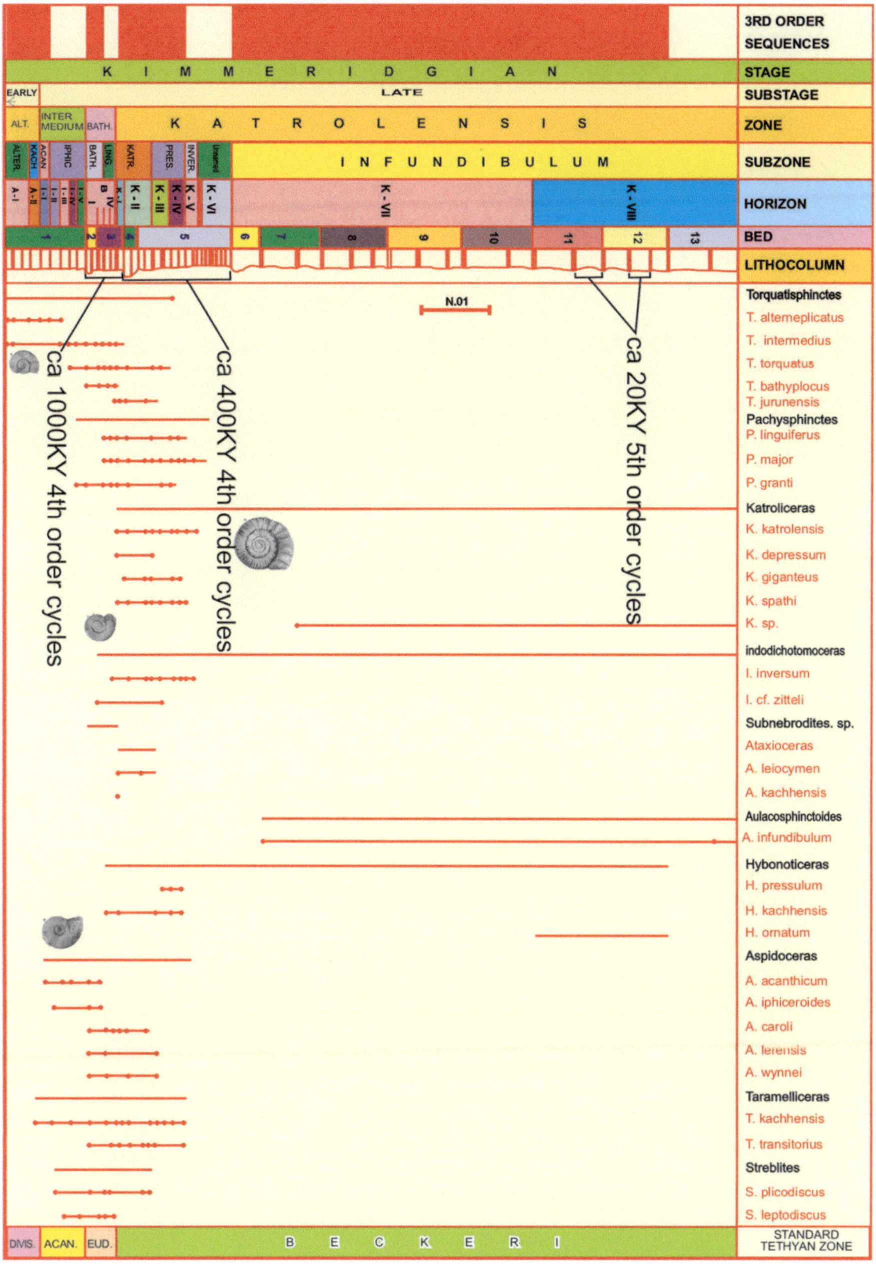

Fig. 2.36 The stratigraphic occurrence, ranges of the ammonoid species and third-order sequences in the Kimmeridgian of Southwest Ler, Kachchh after Krishna et al. (1995b)

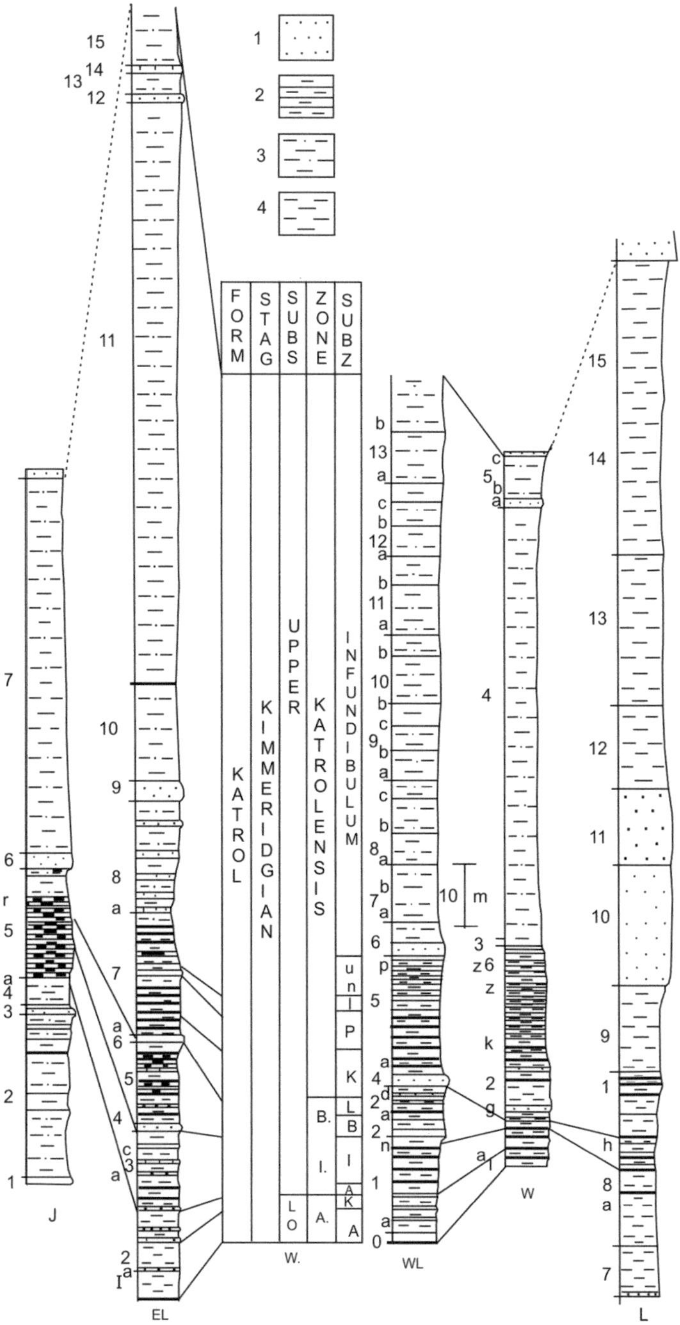

Fig. 2.37 Correlation in the Mainland Kachchh during the Kimmeridgian interval; 1–4 stand, respectively, for sandstone, shale/sandstone alternation, siltstone/shale and shale; J,WL, W and L abbreviate, respectively, for Jawahar Nagar, East Ler, West Ler, Walakhawas and Lakhapur localities after Krishna et al. (1995b)

Sediment Interval K–B XII (Beds 01–02 of K–B XI and XII): About 36.0 m thick sedimentary succession of yellowish shale intercalated with coarse grained, gritty, hard, maroon colored vertically burrowed sandstone with rare wood, belemnites and very rare indeterminate ammonoids, frequent shell fragments. Sediment Interval K–B XIII (Beds 01–03): 15 m thick shale with variegated, massive, occasionally cross-bedded, coarse grained, friable sandstone, capped by highly ferruginous, bioturbated, maroon colored, unfossiliferous sandstone.

Sediment Interval K–B XIV (Beds 01–22 of K–B XIV & XVIII): 14.0 m thick, shale alternating with hard, nodular, fractured, maroon, ash-maroon to yellowish maroon sandstone bands, fossiliferous, yielding ammonoids from 10 different levels.

Sediment Interval K–B XV (Beds 01–06 of K–B XIV & XVIII): 2.30 m thick, gypseous shale intercalated with medium grained, flaky sandstone bands.

Sediment Interval K–B XVI (Beds 01–02 of K–B XIV & XVIII): 1.80 m thick, shale, capped with yellowish maroon, highly fossiliferous, medium grained, densely packed with trigonids and rare ammonoids.

Sediment–B XVII (Beds 01–22 of K–B XIV & XVIII): 8.30 m thick, gypseous silty shale with variegated sandstone bands, poorly fossiliferous, ammonoids scarce.

Sediment Interval K–B XVIII (Beds 01–19 of K–B XIV & XVIII): 6.80 m thick, shale with several hard and compact, coarse grained, calcareous, fossiliferous, trigonid bands, ammonoids rare.

Sediment Interval K–B XIX (Beds 01–25): 8.30 m thick, shale with variegated sandstone bands, poorly fossiliferous.

Sediment Interval XX (Beds 01–16): 11.80 m thick unfossiliferous shale with thin, occasionally massive, variegated sandstone bands.

At Ler

The basal 100–150 m part of the Katrol Formation represents the Kimmeridgian in the Mainland Kachchh. It disconformably overlies the Chari Formation. The contact is distinctively marked by a hard persistent oolitic bed (Dhosa Oolite Member) representing the top of the Chari Formation (Callovian–early Middle Oxfordian) of early Indogermanus Zone age.

The late Early and Late Kimmeridgian ammonoid successiosn is best developed in Southeast and Southwest Ler. In the latter section, ammonoids have been collected from 37 levels within the thickness of ca 110 m (subdivided into 13 beds), while in Southeast Ler the Kimmeridgian ammonoids come from 41 levels in ca 150 m thick succession (split up into 15 beds).

Adhoi member of Kantkote Member (KM)

Ammonoid Zones

The interval includes two zones Kachchensis Zone and Gignaticus Zone based on new species which are here described prior to the introduction of the mentioned new zones.

Genus: ***Orthosphinctes* Schindewolf 1925**

***Ammonites tiziani* Oppel 1863**

Species: ***Orthosphinctes kachchhensis* n. sp.**

(Figures 1–3, plate 8)

Material: Three specimens, fragmented, moderately preserved.
tirely septate.

Locality: Kantkote–Bharodia section, Wagad, Kachchh, India.

Measurement:

S. No.	Types	D	H	W	U	H/W %	P/2	S/2	Remarks about body chamber
1.	Holotype	124.0	44.0	36.0	51.0	122	26	–	Septate
		92.0	32.0	28.0	38.0	114	–	–	
		70.0	23.0	21.0	30.0	110	23	–	
2.	Paratype	106.0	32.0	29.0	50.0	110	27	–	Septate
		80.0	26.0	24.0	36.0	108	–	–	
		60.0	20.0	19.0	27.0	105	26	–	
3.	Paratype	110.0	38.0	37.0	49.0	102	27		Septate
		84.0	29.0	28.0	38.0	103	–		
		65.0	21.0	23.0	30.0	91	26		

Diagnosis: Size moderate to large, entirely septate, compressed, whorl section quadrate to subquadrate, evolute, umbilicus wide, umbilical wall almost vertical, umbilical shoulder curved, laterals slightly convex to nearly flat, venter rounded, primaries from near the base of umbilical wall, slightly concave forward around umbilical shoulder, prorsiradiate on laterals, sharp, distinct, well-spaced, coarseness and spacing gradually increasing with diameter, sescondaries from near the middle of the whorl height or slightly above, mostly biplicate with occasional single ribs up to 105–110 mm D, triplication appearing somewhere between 90–110 mm D, first triplicate rib at 93 mm D, thereafter 2–3 biplicate ribs between two successive triplicate ribs until the preserved end of 124 mm D, slightly projected forward and uninterrupted on venter, suture distinct but undecipherable.

Remarks: It is comparable to *Orthosphinctes* (*Orthosphinctes*) *polygyratus* (Quenstedt) in its prorsiradiate ribbing pattern. *Orthosphinctes* (*Praeataxioceras*) sp. gr. *laufenensis* (Siemiradzki) differs from the present Kachchh species in its distinctly smaller size, more compressed whorl section and almost radiating pattern of ribbing.

Stratigraphic Horizon: Kachchhensis Zone (Sediment Interval K–B X, bed 01 to Sediment Interval K–B XV, bed 06).

Age: Kachchhensis Zone.

Genus–*Ataxioceras* Fontannes 1879

Perisphinctes hypselocyclum Schindewolf 1925

***Ataxioceras giganticus* n sp**

(Figure 4, plate 8)

Material: One specimen made up of several small fragments.

Locality: Bharodia in Wagad, Kachchh.

Size and Remarks: The specimen has an unusually large size of 500 mm among the known species of *Ataxioceras*, and hence the creation and naming of the new species as *A. giganticus*. The specimen is compressed, moderately evolute, and featured by typical ataxioceratid twice bifurcated ribs, first at the mid lateral and later near the ventrolateral shoulder. The preserved bodywhorl is characterized by strong sharp cuniform primaries.

Stratigraphic Horizon: KB XVI to KB XVIII

Age: **Giganticus Zone**

Ammonoid zones, subzones and horizons

(Bharodia)

Kachchensis Zone (new) Krishna (KB-XIV beds 01–22 and KB-XV beds 01–06) (time equivalent of Planula Zone of ETM): The base of the zone is indicated by the first Ataxioceratinae through moderate *Orthosphinctes kachchensis* n. sp. In bed 02. Single fragment of *Wegelea* aff. *gredingensis* (Wegele) and again single nucleus of *Subnebrodites* cf. *minutum* (Dietrich) are present in bed 22 which have prompted correspondence of this zone to collective interval of Haufianum Subzone of Bimammatum Zone and Planula Zone of ETM. Fursich et al. (2013) have placed the early red/maroon part of this zone for the terminal ammonoid-rich part of Chari Formation in the Kantkote river section, and assigned to the late Middle to mid Late Oxfordian instead of early Early Kimmeridgian, which is contrary to our findings. Their indentification of *Orthosphinctes as Larcheria* and *Dichotomoceras* and farther in the section, the determination of *Orthosphinctes* and *Ataxioceras* as species of *Torquatisphinctes* occurring within the younger *Astarte–Gryphaea* interval as Late Kimmeidgian is negated. The gap of the entire Late Oxfordian–Early Kimmeridgian ~5 my between the underlying early Early Kimmeridgian red beds of Kachchhensis Zone and the conformably overlying younger *Astarte–Gryphaea* bearing sands of mid Early Kimmeridgian Giganticus Zone proposed by Fursich et al. (2013) is non-existent.

Giganticus Zone (new) (KB-XVI to XVIII) (time equivalent of the collective interval of Platynota Zone and Hypselocyclum Zone of ETM): The zone is featured by rare *Ataxioceras* at three levels. The youngest level includes a new species of relatively large size (*A. giganticus* n. sp.). Further up to the top of sediment interval KB-XVIII, only, the lag bivalve astartid–trigonid assemblage is observed. This zone on the basis of *Ataxioceras* is tentatively corresponded to the collective interval of Platynota Zone and Hypselocyclum Zone of ETM.

Lower Katrol Member (LKM)

Ammonoid zones/subzones/horizins

All the undermentioned zones and most subzones were first defined by Krishna and Pathak (1991, 1993) while 3 subzones and 19 horizons were introduced for the first time in Pandey (1993 unpublished) and Krishna et al. (1995b, 1996c) with minor revision in a few zones.

The ammonoid zonation at Ler is based on a sedimentary succession, which unconformably overlies the early Middle Oxfordian Indogermanus Zone (Fig. 2.36). In absence of common ammonoids, it is not possible to establish any correspondence between the oldest late Early Kimmeridgian Alterneplicatus Zone of Ler to the youngest Kimmeridgian Giganticus Zone developed in Wagad. However, in view of the suggested correlations with the ETM zonation scheme, it is surmised that the Alterneplicatus Zone follows the Giganticus Zone without any significant gap, not exceeding a subzone or so.

Alterneplicatus Zone Krishna and Pathak (1991) (beds 0–1c)

Alterneplicatus Subzone Krishna and Pathak (1991)

A-I Horizons Krishna et al. (1995b) (beds 0–1b): Its base is marked by the first appearance of the index *Torquatisphinctes alterneplicatus* M.

Kachhensis Subzone Krishna and Pathak (1991)

A-II Horizons Krishna et al. (1995b) (bed 1c): It is marked by the first appearance of *Taramelliceras kachhensis* (Waagen) M.

Intermedius Zone Krishna and Pathak (1991) (beds 1d–1h)

Acanthicum Subzone Krishna and Pathak (1991)

I-I Horizon Krishna et al. (1995b) (bed 1d): This horizon is marked by first *Aspidoceras acanthicum* (Oppel) M.

Iphiceroides Subzone Krishna and Pathak (1991)

I-II Horizon Krishna et al. (1995b) (bed 1e): This horizon is marked by the first appearance of *Aspidoceras iphiceroides* (Waagen) m and *Streblites plicodiscus* (Waagen) M.

I-III Horizon Krishna et al. (1995b) (bed 1f): It is indicated by the first appearance of *Streblites leptodiscus* Spath M.

I-IV Horizon Krishna et al. (1995b) (bed 1g): This horizon is marked by first appearance of *Torquatisphinctes torquatus* (Sowerby) m.

I-V Horizon Krishna et al. (1995b) (bed 1h): It is marked by first appearance of *Pachysphinctes granti* Spath M.

Bathyplocus Zone Krishna and Pathak (1991) (beds 2–3c): In the present work this zone has been divided into two subzones, namely Bathyplocus Subzone and Linguiferus Subzone. It includes four horizons (B-I to B-IV Horizons). B-I and B-II Horizons correspond to the Bathyplocus Subzone, while B-III and B-IV Horizons correspond to the Linguiferus Subzone. These horizons are based on the four successive transient morphs (morph I, morph II, morph III and morph IV) within the stratigraphic range of *T. bathyplocus* (Waagen) m. These informal/quasiformal stratigraphically differentiated morphs of *T. bathyplocus* (Waagen) m are based on the clearly expressed distinct microevolutionary unidirectional centripetal extension (hypermorphic peramorphosis) with time in the position of the initiation of triplicate ribbing on the body chamber from the last quarter to near the start of the body chamber.

Bathyplocus Subzone Krishna et al. (1995b) (beds 2–3a): The base of this subzone is marked by the first appearance of the zonal index *Torquatisphinctes bathyplocus* (Waagen) m (genetic assignment here transferred from *Pachysphinctes* to *Torquatisphinctes*) and *Aspidoceras caroli* Spath *M*. *Torquatisphinctes intermedius* Spath m, *Torquatisphinctes torquatus* (Sowerby) m, *Pachysphinctes granti* Spath M, *Taramelliceras kachchhensis* (Waagen) M, *Streblites plicodiscus* (Waagen) M, *Streblites leptodiscus* Spath M, *Holcophylloceras mesolcum* (Dietrich) M, *Aspidoceras acanthicum* (Oppel) M and *Aspidoceras iphiceroides* (Waagen) m continue from bellow. *Indodichotomoceras* cf. *zitteli* Spath m and *A. lerensis* Spath m & M mark their first appearance within this subzone.

B-I Horizon Krishna et al. (1995b) (bed 2): It is marked by the first appearance of morph I of *T. bathyplocus* (Waagen) m in which the first triplicate rib appears in the last quarter of the body chamber.

B-II Horizon Krishna et al. (1995b) (bed 3a): It is marked by the first appearance of morph II of *T. bathyplocus* (Waagen) m in which the first triplicate rib appears at the end of the second quarter of the body chamber. *Aspidoceras lerensis* Spath m & M also starts in this horizon.

Linguiferus Subzone Krishna et al. (1995b) (beds 3b–3c): The base of this subzone is marked by the first appearance of *Pachysphinctes linguiferus* Spath m as also of the first appearance of morph III of *T. bathyplocus* (Waagen) m. *Torquatisphinctes intermedius* Spath m, *Torquatisphinctes toquatus* (Sowerby) m, *Torquatisphinctes bathyplocus* (Waagen) m, *Pachysphinctes granti* Spath M, *Indodichotomoceras* cf. *zitteli* Spath m, *Taramelliceras kachhensis* (Waagen) M, *Streblites plicodiscus* Waagen m, *Aspidoceras caroli* Spath M, *Aspidoceras lerensis* Spath m & M, *Holcophylloceras mesolcum* (Dietrich) M, *Streblites leptodiscus* Spath M continue from below. *Torquatisphinctes jurunensis* Spath M, *Pachysphinctes major* Spath M, *Indodichotomoceras inversum* (Spath) M, *Indodichotomoceras* sp. m, *Taramelliceras transitorius* Spath M and *Aspidoceras wynne*i Waagen M also mark their first appearance in this subzone.

B-III Horizon Krishna et al. (1995b) (bed 3b): It is indicated by the first *P. linguiferus* Spath m as also of morph III of *T. bathyplocus* (Waagen) m in which the first triplicate rib appears in the beginning of the second quarter of the body chamber.

B-IV Horizon Krishna et al. (1995b) (bed 3c): Its base is marked by the first appearance of *T. transitorius* Spath M as also of morph IV of *T. bathyplocus* Waagen m in which the first triplicate rib appears in the first quarter of the body chamber.

Katrolensis Zone Krishna and Pathak (1991) (beds 3d–13b): The Katrolensis Zone has been revised to include Infundibulum Zone, earlier placed at the base of the Hybonotum Zone. This zone includes five subzones, namely Katrolensis Subzone, Pressulum Subzone, Inversum Subzone, Unnamed Subzone and Infundibulum Subzone. These have been further subdivided into eight horizons (K-I to K-VIII Horizons). Among these, K-I and K-II Horizons correspond to the Katrolensis Subzone, K-III to K-IV Horizons correspond to the Pressulum Subzone.

K-V and K-VI Horizons belong to the Inversum Subzone and the unnamed subzone and K-VII to K-VIII to Infundibulum Subzone, respectively. Horizons of the Katrolensis Subzone and Pressulum Subzone are developed/strengthened with the first appearance of the successive four transient morphs (morphs I–IV) of *Katroliceras katrolensis* (Waagen) m within a part of its stratigraphic range. These informal/quasiformal stratigraphically differentiated morphs of *K. katrolensis* (Waagen) m are based on the clearly expressed distinctly unidirectional microevolutionary centripetal extension with time in the position of the first appearance of the abruptly modified (cuniform) primary ribs from the last quarter to the start of the body chamber in the successively younging morphs (hypermorphic peramorphosis). There is also observed size increase in successively younging morphs with stratigraphic age upto the level of morph IV. On the other hand, K-V horizon is based on the sudden reduction in size of *K. katrolensis* (Waagen) m. K-VI Horizonal interval is devoid of determinable ammonoids although fossil wood and belemnoids are present. Horizons K-VII and K-III are based, respectively, on the first *Aulacosphinctoides* and *H. ornatum* Spath m (Fig. 2.36).

Katrolensis Subzone Krishna and Pathak (1991) (beds 3d–5b)

K-I Horizon Krishna et al. (1995b) (bed 3d): It is marked by the first appearance of morph I of *K. katrolensis* (Waagen) m in which the first abruptly modified (cuniform) primary rib appears in the last quarter of the body chamber.

K-II Horizon Krishna et al. (1995b) (beds 4–5b): Its base is marked by the first appearance of the morph II of *K. katrolensis* (Waagen) m, in which the first abruptly modified (cuniform) primary rib appears in the middle of the second quarter of the body chamber.

Pressulum Subzone Krishna and Pathak (1991) (beds 5c–5f): The base of this subzone marked by the first appearance of the zonal index *Hybonoticeras pressulum* (Neumayr) m and is additionally strengthened by the first morph III of *K. katrolensis* (Waagen) m. It is also correlatable to Early Beckeri Zone Subeumela Subzone Horizon II of ETM (Schweigert et al 1996).

K-III Horizons Krishna et al. (1995b) (beds 5e–5d): Its base is marked by the first appearance of the morph III of *K. katrolensis* (Waagen) m in which the first abruptly modified rib appears in the beginning of the second or at the end of the first quarter of the body chamber, in addition of the first *H. pressulum* (Neumayr) m.

K-IV Horizon Krishna et al. (1995b) (beds 5e–5f): Its base is marked by the first appearance of the morph IV of *K. katrolensis* (Waagen) m, in which the first abruptly modified rib appears between the start and the middle of the first quarter of the body chamber. It is suggestively correlatable to Subeumela Subzone Horizon III.

Inversum Subzone Krishna and Pathak (1991) (beds 5g–5h)

K-V Horizon Krishna et al. (1995b) (beds 5g–5h): In addition to the significant presence of *I. inversum* (Spath) m, this horizon is marked by the sudden decrease in size of *K. katrolensis* (Waagen) m.

Unnamed Subzone Krishna et al. (1995b) (beds 5i–5p): It includes only K-VI Horizone which is marked by the presence of scarce indeterminate virgatosphinctin

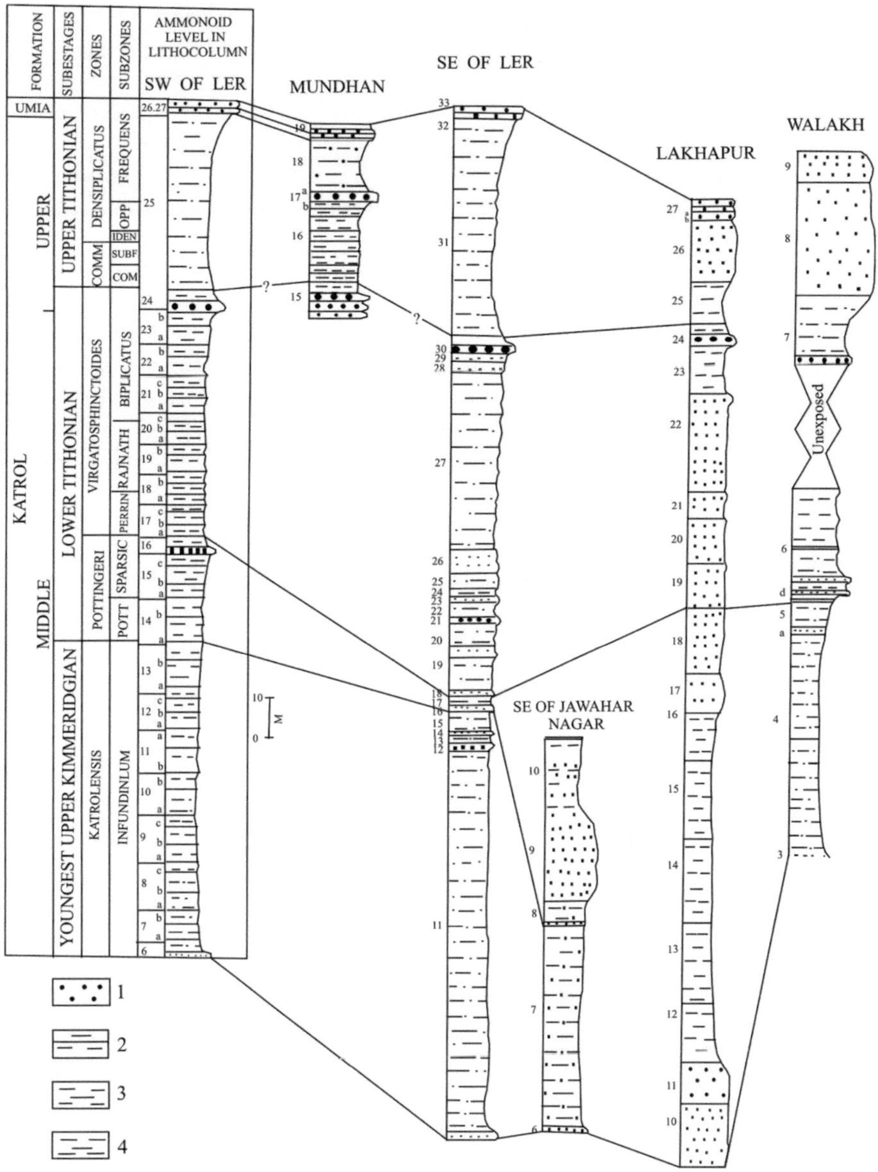

Fig. 2.38 Correlation in the Mainland Kachchh during the youngest Upper Kimmeridgian–Tithonian interval; 1–4 stand, respectively, for sandstone, shale/sandstone alternation, siltstone/shale and shale after Krishna et al. (1995b)

ammonoids. *K. katrolensis* (Waagen) m seems to have become extinct. Fossil wood and belemnoids are present.

Infundibulum Subzone Krishna and Pathak (1991) (beds 6–13b)

K-VII Horizon Krishna et al. (1995b) (beds 6–10 b): Its base is marked by the first *Aulacosphinctoides* with *A. infundibulum* (Uhlig) m.

K-VIII Horizon (beds 11a–13b): Its base is marked by the first *H. ornatum* Spath m which is also restricted in this horizon. Here, it is correlated to mid Setatum Subzone of Beckeri Zone.

In the above scheme, the Kimmeridgian/Tithonian boundary is revised and raised from the earlier (within bed 7) to the top of bed 13 in Southeast Ler to additionally include the Inversum Subzone, Unnamed Subzone and Infundibulum Subzone. This became necessary in view of the revised taxonomic affiliation of *H. ornatum* Spath m to *H.* gr. *beckeri* (Neum.) m instead with *H.* gr. *hybonotum* (Oppel) m as thought earlier (Krishna and Pathak 1993). Thus, in all the sections the occurrence of *H. ornatum* Spath m are now included in the Kimmeridgian.

The Kimmeridgian/Tithonian boundary is revised to the base of Pottingeri Zone at the level of the first *K. pottingeri* (Sowerby) m between the youngest Kimmeridgian Infundibulum Subzone and the oldest Tithonian Pottingeri zone. However, alternatively, bed 13 could still represent either the youngest part of Setatum Subzone exclusively or even be inclusive of the Ulmense Subzone. The Ulmense Subzone is either absent or represented by the early part of the Pottingeri Zone. In that case, the Kimmeridgian/Tithonian may be included within the Pottingeri Zone. In sequence stratigraphic context, the K/T boundary marks the start of a third-order TST which imparts better candidature of bed13/14 boundary as K/T boundary, so also in context of the heterochronic evolutionary cut between *K. katrolensis* and *K. pottingeri*.

2.8.8 Tithonian

Development of Tithonian in Kachchh

The ammonoid bearing Tithonian stage is fairly well-developed in the Mainland Kachchh (Figs. 2.39 and 2.40). In spite of some good sections, the Tithonian biochronology has remained much less known in comparison to the pre-Tithonian part. In India, after the early works (Waagen 1871, 1873–75; Uhlig 1903–10; Spath 1927–33). Krishna and Pathak (1993) established a scheme of four zones and five subzones in the Early Tithonian at Southeast Ler in the eastern part of the Mainland Kachchh. The Tithonian in the East Mainland is invariably underlain conformably by the Late Kimmeridgian sediments, while in the West Mainland it is either conformable over the Late Kimmeridgian or uncoformable over the Early and early Middle Oxfordian sediments.

Six different sections were studied. These are: Jawahar Nagar, Southeast Ler, Southwest Ler, Walakhawas, Lakhapur and North Mundhan (Figs. 2.38, 2.39 and 2.40). Among these, the Early Tithonian ammonoid succession is best developed in Ler, while the Late Tithonian at Lakhapur and Mundhan. The collection includes 27–48 successive levels of ammonoids in the Tithonian within the ~130 m thick succession, which has been organized into 14 beds. All the Early Tithonian zones and subzones of Krishna and Pathak (1993) were later confirmed in the principal section at Southwest Ler (Pandey 1993 unpublished, Krishna et al. 1995b, 1996b) with slight reorganization and further refinement in to horizons. However, the first ever zonation was realized yet later (Krishna et al. 1995b) in the Late Tithonian at Mundhan. An integrated zonation scheme with 5 zones, 14 subzones and 17 horizons in Tithonian is presented here.

Lithostratigraphic Remarks The Tithonian stage broadly corresponds to the major part of Katrol Formation (Middle Member and Upper Member) and basal part of Umia Formation. The Katrol Formation is further organized in to four members namely, Kantkote Member, Lower Member, Middle Member and Upper Member. The ~36–51 m thick Lower Member (mostly Kimmeridgian) characteristically comprises alternation of hard, highly fractured, deep brown to maroon coloured, highly fossiliferous, fine to medium grained sandstone bands and gypsiferous to sandy shale with occasional concretions. The overlying ~166–222 m thick Middle Member (mostly Early Tithonian) is separated from the Lower Member by a massive, hard and compact, dark brown, fine to medium grained, persistant sandstone at the base. The Middle Member mainly comprises alternation of maroon, calcareous sandstone bands and siltstone/shale. The lower part of this Member (~70–90 m thick) is relatively very poorly fossiliferous in comparison to the ~40–50 m thick upper part. The ~45–62 m thick Upper Member mainly consists of sandy shale/siltstone intercalated with occasional thin, maroon, fractured sandstone bands and concretions. Its contact with Umia Formation appears structurally disturbed.

Middle Katrol Member (MKM)
Ammonoid zones
The Kimmeridgian/Tithonian boundary placement in Kachchh (Pathak 1989 unpublished, Pandey 1993 unpublished, Krishna and Pathak 1993) is revised (Krishna et al. 1995b, 1996b). The Infundibulum Zone characterized by the presence of *A. infundibulum* (Uhlig) m and *H. ornatum* Spath m is now included as the youngest Kimmeridgian. In comparative studies undertaken with the European type material (Schweigert et al. 1996), *H. ornatum* Spath m was found closest to *H.* gr. *beckeri* (Neumayr) m instead of *H. hybonotum* (Oppel) m as previously considered (Kishna and Pathak 1993). The revised base of the Tithonian stage in Kachchh is now marked at the base of the Pottengeri Zone with the first *K. pottingeri* (Sowerby) m. The proposed scheme (Krishna et al. 1995b, 1996b) now includes

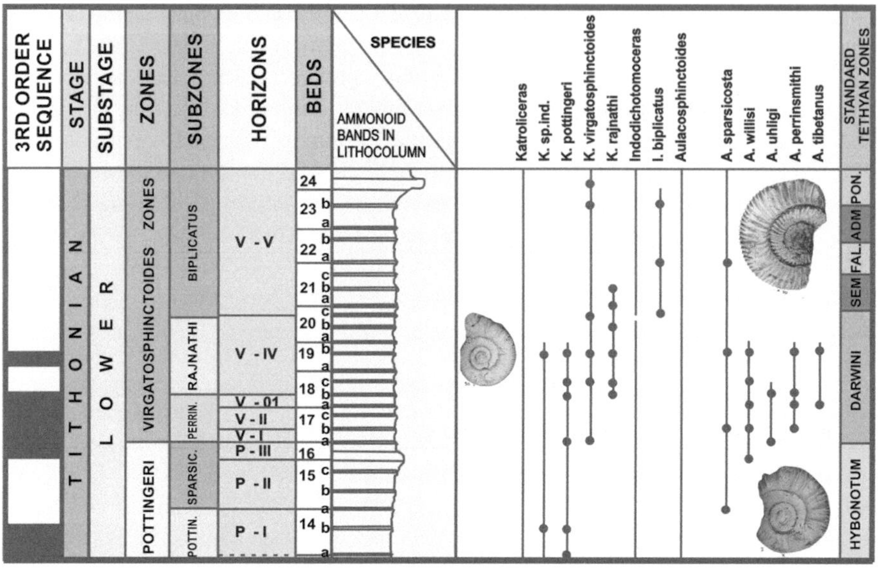

Fig. 2.39 The stratigraphic occurrence, ranges of the ammonoid species and third-order sequences in the Early Tithonian of Southwest Ler, Kachchh (modified after Krishna et al. 1995b)

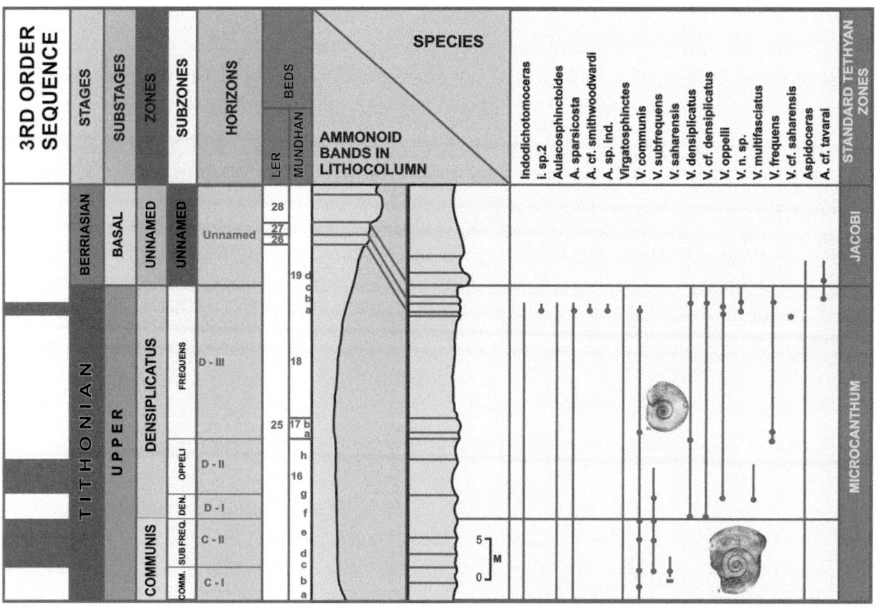

Fig. 2.40 The stratigraphic occurrence, ranges of the ammonoid species and third-order sequences in the Late Tithonian of Southwest Ler, Kachchh (modified after Krishna et al. 1995b)

two zones, five subzones and eight horizons in the Early Tithonian and two zones, five subzones and five horizons in the Late Tithonian. The resolution realized is of the order of 500,000 years. The horizons are either marked on the basis of the first appearance of a species or distinct evolutionary morphological change in the species in successive stratigraphic levels.

Zones, Subzones and Horizons

Early Tithonian (At Southwest Ler)

Pottingeri Zone Krishna and Pathak (1993)

Pottingeri Subzone Krishna and Pathak (1993)

P-I Horizon Krishna et al. (1995b) (beds 14a–14b): The base of this horizon is indicated by the first *K. pottingeri* (Sowerby) m as for the zone.

Sparsicosta Subzone Krishna et al. (1995b) (beds 15a–16): The base of this subzone is marked by the first *A. sparsicosta* (Uhlig) m. *K. pottingeri* (Sowerby) m and *Katroliceras* sp. indet. M continue from below, while *A. willisi* (Uhlig) m appears first within this subzone.

P-II Horizon Krishna et al. (1995b) (beds 15a–15c): Its base is marked by the first appearance of *Aulacosphinctoides sparsicosta* (Uhlig) m.

P-III Horizon Krishna et al. (1995b) (bed 16): Its base is marked by the first appearance of *Aulacosphinctoides willisi* (Uhlig) m.

Virgatosphinctoides Zone Krishna and Pathak (1993) (beds 17a–24): In its revised definition, it also includes the Rajnathi Zone of Krishna and Pathak (1993). Its base is marked by the first appearance of the zonal index *Katroliceras virgatosphinctoides* Krishna and Pathak m and *Aulacosphinctoides uhligi* Spath m which are also restricted within this zone. *Indodichotomoceras* and *Aulacosphinctoides* continue from below while *Katroliceras* marks its last occurrence within this zone.

Perrinsmithi Subzone Krishna et al. (1995b) (beds 17a–18a): Its base is marked by the first appearance of *Aulacosphinctoides uhligi* Spath m and *K. virgatosphinctoides* Krishna and Pathak m. *Aulacosphinctoides perrinsmithi* (Uhlig) m is the most significant/abundant species within this subzone. *Katroliceras* sp. indet. M, *Katroliceras pottingeri* (Sowerby) m, *Aulacosphinctoides willisi* (Uhlig) m and *Aulacosphinctoides sparsicosta* (Uhlig) m continue from below. *Aulacosphinctoides perrinsmithi* (Uhlig) m and *Aulacosphinctoides tibetanus* (Uhlig) m mark their first appearance within this subzone.

V-I Horizon Krishna et al. (1995b) (bed 17a): Its base is marked by the first appearance of *Aulacosphinctoides uhligi* Spath m and *Katroliceras virgatosphinctoides* Krishna and Pathak m.

V-II Horizon Krishna et al. (1995b) (beds 17b–17c): Its base is marked by the first appearance of *Aulacosphinctoides perrinsmithi* (Uhlig) m.

V-III Horizon Krishna et al. (1995b) (bed 18a): Its base is marked by the first appearance of the nominal species *Aulacosphinctoides tibetanus* (Uhlig) m.

Rajnathi Subzone Krishna and Pathak (1993): In view of additional data, this zone of Krishna and Pathak (1993) is here emended as a subzone. In its emended definition, the base is indicated by the first appearance of the subzonal index *Katroliceras rajnathi* Krishna and Pathak m. *Katroliceras virgatosphinctoides* Krishna and Pathak m and *Aulacosphinctoides sparsicosta* (Uhlig) m continue from below. *Katroliceras* sp. indet. M, *Katroliceras pottingeri* (Sowerby) m, *Aulacosphinctoides uhligi* Spath m, *Aulacosphinctoides perrinsmithi* (Uhlig) m and *Aulacosphinctoides tibetanus* (Uhlig) m mark their last occurrence within this subzone.

V-IV Horizon Krishna et al. (1995b) (beds 18b–20b): The base of this horizon is defined as that of the Rajnathi Subzone.

Biplicatus Subzone Krishna et al. (1995b).

V-V Horizon Krishna et al. (1995b) (beds 20c–24): Its base, as also of the Biplicatus Subzone, is marked by the first *appearance of Indodichotomoceras biplicatus (Uhlig) m.* Katroliceras virgatosphinctoides Krishna and Pathak m and *Aulacosphinctoides sparsicosta* (Uhlig) m continue from below. *Katroliceras rajnathi* Krishna and Pathak m marks its last appearance.

Natricoides Zone Pandey and Krishna (2002) (beds 2d–21 at Jaisamer) (Fig. 3.00): This zone was introduced in the Indian Tithonian by Pandey and Krishna (2002) in Jaisalmer between Virgatosphinctoides Zone below and Communis Zone above in Early Tithonian, thus increasing the ammonoid zones in Early Tithonian from 3 to 4. Lithologically, the c 6.0 m thick strata include alternation of gypseous shale/silt and ferruginous maroon to deep brown medium to fine grained, concretionary, nodular, pebbly, boxworked and bioturbated sands. The interval is rather the richest of the entire c 56.0 m succession. Belemnoids and fossil wood also show significant presence. The base is marked by the first *Aulacosphinctoides natricoides* (Uhlig). *Hildoglochiceras latistrigatum* (Uhlig), *H. kobelli* (Oppel), *A. doghlaensis* (Fatmi), *A. linoptychus* (Uhlig), *A. hyderi* (Fatmi), *Virgatosphinctes pumpeckji* (Uhlig), *V. krsafti* (Uhlig), *Haploceras* cf *elimatum* (Oppel) and *Holcophylloceras mesolcum* (Dietrich) also have their FAD in this zone. Pandey and Krishna (2002) have suggested best possible correlation of Natricoides Zone to ETM Semiforme Zone.

It is further organized into four subzones/horizons as under:

Natricoides Subzone/N-I Horizon (beds 2d–2f): The base is indicated by the zonal index along with the first *H. latistrigatum* (Uhlig).

Doghlaensis Subzone/N-II Horizon (beds 2g–2h): The base is marked by the FAD of the subzonal index. *A. natricoides* (Uhlig) and *H. latistrigatum* (Oppel) continue from below.

Pumpeckji Subzone/N-III Horizon (beds 2i–2j): The base is indicated by the first subzonal index. *A. natricoides* (Uhlig), *A. doghlaensis* (Fatmi), *H. latistrigatum* (Oppel) continue from below.

Krafti Subzone/N-IV Horizon (beds 2k–2l): The base is marked by the first appearance of the subzonal index, first *A. linoptychus* (Uhlig) and *H. mesolcum* (Dietrich). *H. kobelli* (Oppel), *H.* cf *elimatum* (Oppel) and *A. hyderi* Fatmi also make their first presence in this subzone. *A. natricoides* (Uhlig), *A. doghlaenis* Fatmi and *V. pumpeckji* (Uhlig) continue from below.

Late Tithonian (At Mundhan Fig. 2.40)
The succession at Sahera, Mundhan and Katesar is rather quite similar. The long known Umia green glauconitic ammonoid beds in quick succession comprise ammonoid-rich maroon to brown ferruginous pebbly/nodular/concretionary glauconitic somewhat conglomeratic, at places even bioturbated sands and muds. Bivalves, brachiopods, belemnoids are also present. More than one Gryphaea bands have also been noticed. The J/C boundary is determined even in Ler section, where the pre-boundary strata are characteristically medium to coarse grained and cross-bedded

Communis Zone Krishna et al. (1995b): The base of this zone is marked by the first appearance of the zonal index *Virgatosphinctes communis* Spath M. It also indicates the first appearance of the genus *Virgatosphinctes*. The genera *Indodichotomoceras* and *Aulacosphinctoides* continue from below. This zone includes two subzones.

Upper Katrol Member (UKM)
Communis Subzone Krishna et al. (1995b)
C-I Horizon Krishna et al. (1995b) (beds 16a–16b): Its base as also of the Communis Subzone is marked by the first appearance of *Virgatosphinctes communis* Spath M.
Subfrequens Subzone Krishna et al. (1995b)
C-II Horizon Krishna et al. (1995b) (beds 16c–16e): Its base as also of the Subfrequens Subzone is marked by the first appearance of *Virgatosphinctes subfrequens* Uhlig m and *Virgatosphinctes saharensis* Spath m. The latter is also restricted within this subzone. *V. communis* Spath M continues from below.
Densiplicatus Zone Krishna et al. (1995b): Its base is marked by the first appearance of the zonal index *Virgatosphinctes densiplicatus* (Waagen) m and *V.* cf. *densiplicatus* (Waagen) m which are also restricted within this zone. The genera *Indodichotomoceras* and *Aulacosphinctoides* continue from below. This zone includes three subzones. Ammonoids have been collected from 6–8 close levels (beds 17a–19 d). Besides Virgatosphinctinae found as the principal elements, *Aspidoceras*, *Argentiniceras*, *Micracanthoceras*, *Blanfordiceras*, *Corongoceras*, *Aulacosphinctes*, *Himalayites* have also been found in the youngest Tithonian level in the last but one green bed.
Densiplicatus Subzone Krishna et al. (1995b)
D-I Horizon Krishna et al. (1995b) (bed 16f): Its base as also of the Densiplicatus Subzone is marked by the first appearance of *Virgatosphinctes densiplicatus* (Waagen) m and *V.* cf. *densiplicatus* (Waagen) m.
Oppeli Subzone Krishna et al. (1995b) D-II Horizon Krishna et al. (1996a, b, c) (beds 16g–16h): Its base as also of the Oppeli Subzone is marked by the first appearance of *Virgatosphinctes oppeli* Spath M and *V. multifasciatus* (Uhlig) m. The latter species is also restricted within this subzone. *V. communis* Spath M, *V. densiplicatus* (Waagen) m and *V.* cf. *densiplicatus* (Waagen) m continue from below.

Basal Member of Umia Formation

Frequens Subzone Krishna et al. (1995b)

D-III Horizon Krishna et al. (1995b) (beds 17a–19c): Its base as also of the Frequens Subzone is marked by the first appearance of *Virgatosphinctes frequens* (Oppel) M which is also restricted within this unit. *V. communis* Spath M, *V. densiplicatus* (Waagen) m, *V.* cf. ***densiplicatus*** (Waagen) m and *V. oppeli* Spath M mark their last occurrence while *Aulacosphinctoides sparsicosta* (Uhlig) m, *Aulacosphinctoides* sp. indet. m, *Aulacosphinctoides* cf. *smithwoodwardi* (Uhlig) m, *Virgatosphinctes* n. sp. m and *Virgatosphinctes* cf. *saharensis* Spath m are restricted within this unit.

The comments made by Enay (2009), and Enay and Cariou (1999) on the Tithonian ammonoid zonal succession in Kachchh of Krishna et al. (1996b), Pathak (1989 unpublished thesis) and Pandey (1993 unpublished thesis) are subjetive and unsubstantiated, hence unwarranted, and uncalled for. The Tithonian ammonoid stratigraphic studies in Nepal and Tibet Himalaya by Enay are not based on precise bed by bed collection and lack adequate lithostratigraphic differentiation and ammonoid stratigraphic range charts in comparison to the studies in Kachchh, Jaisalmer and Spiti Himalaya. Enay and others have ammonoid collections from just seven imprecise levels from a highly generalized lithostratigraphic section without any bed numbers (Fig. 5 on p. 673) compared to 42 stratigraphically précised ammonoid levels in Kachchh from a section differentiated in great detail into 70 beds or their subdivisions along with precise range charts of a large number of ammonoid taxa. The High Himalaya species have been determined by us in Kachchh after actual comparison with their types stored in GSI Museum at Calcutta. Enay's surmise that these forms are either microconchs of large Katroliceratinae or inner whorls or juvenile individuals is far fetched. The mentioned species of *Aualcosphinctoides* from Kachchh are mostly from stratigraphically higher levels than those of large Katrloiceratinae (Krishna and Pathak 1991). The large Katroliceratinae are invariably restricted to pre-Beckeri Zone part of Kimmeridgian that either became extinct or drastically reduced in density, frequency and diversity. *A. infundibulum* is found in the terminal Kimmeridgian and based on cooccurrence with *Hybonoticeras ornatum* common between ETM and GTM is firmly assigned to late Late Kimmeridgian Setatum Subzone of ETM (Schweigert et al. 1996). Other determined species are *A. sparsicosta*, *A. willsi*, *A. uhligi and Tibetanus* again do not occur together with large Katroliceratinae (Fig. 2.36). These taxa are stratigraphically well-constrained in the Virgatosphinctoides Zone of early to mid Early Tithonian after the extinction of *A. infundibulum*. *A. sparsicosta* ranges long from base to close of Tithonian. There emerges an explicit succession of *Aulacosphinctoides* from late Late Kimmeridgian to late Late Ttihonian of 1. *A. infundibulum* (Katrolensis Zone = Beckeri Zone), 2. *A. sparsicosta* of early Early Pottingeri Zone (=to ETM Hybonotum Zone), 3. *A. perrinsmithi*. *A. uhligi*, *A. willisi*, *A. tibetanus* within the early part of Virgatosphinctoides Zone which in spite of absence of any common elements with ETM has been convincingly dated as of early to mid early Tithonian, thus indirectly age-constrained with well-correlated terminal Beckeri Zone below and mid to late

Early Natricoides Zone above, and 4. *A. natricoides*, *A. doghlensis*, *A. linoptychus and A. hyderi* of GTM Natricoides Zone in turn tentatively correlated to ETM Semiforme Zone of mid Early Tithonian below the relatively better dated and well-defined terminal or late Early Tithonian Commuis Zone. The correlation of GTM Natricoides Zone to ETM Semiforme Zone is based on common *Haplophylloceras strigale*. The above succession of *Aulacosphinctoides* in a limted measure is present in Tibet through *A. infundibulum and A. tibetanus* and only tentatively placed in early to mid Ealy Tithonian. Their co-occurence instead of presence in succession is on account of condensed section owing to slow sedimentation after a submarine gap in which the authors failed to differentiate between the underlying terminal Kimmeridgian from Early Tithonian. Enay has similarly found *V. densiplicatus* and *V. frequens* together and placed them in the late Early Tithonian instead of early Late Tithonian. Thus there is unanimity regarding the succession of species of *Aulacosphinctoides and Virgatosphinctes* arrived at by Krishna and coworkers and Enay (2009), though the ages assigned to these species by these authors are different. In Kachchh and Jaisalmer *V. frequens* begins later than *V. densiplicatus*, yet occur together in the Frequens Subzone of the Densiplicatus Zone. Enay and associates have the last *V. densiplicatus* in early Late Tithonian, while Krishna and associates extend the same up to close of Tithonian in view of its last occurrence together with *M. micracanthum*, *Corongoceras* and *Blanfordiceras* in Kachchh, Jaisalmer and Spiti which again does not conflict with the ages given to *Corngoceras* and *M. micracanthum* by Enay and associates. The dispute is only with regard to presence of *V. densiplicatus* in terminal Tithonian Micracanthum Zone (inclusive of undifferentiated *Durangite*s fauna) at the top of Tithonian.

For reasons stated above, the top of the Densiplicatus Zone with *Corongoceras*, *M. micracanthum* and *Blanfordiceras* is inclusive of the undifferentiated *Durangites* fauna, and at present makes the best candidate for the top of Tithonian in India. Citation of Madagascar zones and ammonoid occurrences is hardly of any biostratigraphic significance, since not a single ammonoid there was collected with any precision by Collignon's troops. *Aspidoceras tavarae* in its Spanish stratotype ranges from terminal Tithonian to basal Berriasian, and accordingly its occurrence together with *V. densiplicatus* marks terminal Tithonian and its younger presence without *V. densiplicatus* suggests basal Berriasian. There is absolutely no contradiction about the ranges of *Aspidoceras tavarae* in Spain and Kachchh. Thus the succession of species of *Aulacosphinctoides*, *Virgatosphinctes Aspidoceras*, *Corongoceras*, *Micracanthoceras*, *Blanfordiceras*, *Himalayites*, *Spiticeras* and others have as expected same ranges on GTM in Kachchh, Jaisalmer, Tibet, Pakistan and Madagascar and ETM in France, Spain etc. The apparent differences and discrepancies are on account of condensation, starvation and difficulties of precision collection in High Himalaya. The studies in Kachchh and Jaisalmer have provided an incotrovertible ammonoid succession of GTM not just in Tithonian but almost in the entire Jurassic on the whole of GTM. Fursich et al. (2013) also comment that the J/C boundary can not be fixed needs reconsideration. The youngest Tithonian late Densiplicatus Zone Frequens Subzone ammonoids in

Kachchh, Jaisalmer, Spiti and Malla-Johar, as discussed above are identical and at present make the best parameter for the J/C boundary in IEAP.

Uniqueness of the Kachchh Jurassic Ammonoid Zonal Scale
Unlike the scale developed on ETM which is based on genera/species of several ammonoid families, the ~22 m IEAP scale is near exclusively based on a single superfamily—the Perisphincticeae (Fig. 1.8)—except for the basal Macrocephalitinae based late Middle to Late Bathonian part. Even the succession of zones based on Macrocephalitinae species, also, have alternative Perisphinctinae taxa as names of the mentioned zones. Also, the scale is uniquely developed in a single basin with continuos and splendid outcrops/sections. The IEAP scale has its zonal/subzonal/horizontal boundaries precisely determined in Kachchh, and most of the ammonoid zones well-correlated to those of ETM. The Kachchh ammonoid zones are also found applicable on GTM in spite of invariable absence of the development of ammonoid zonal successions in most basins.

2.8.9 Intra-Jurassic Interstage Boundaries

Triassic/Jurassic system boundary The only Indian locality where basal Hettangian ammonoids (*Psiloceras* and others) are found is Lamayuru in Ladakh close to the Indus suture on the Indian side (Krishna et al. 1997). In High Himalaya in Spiti only a broad surmise is possible above with Triassic. In subsurface Rhaetian/Liassic palynomorphs are known from Banni well, though no clue to demarcate the boundary has as yet come forth. Same situation exists in Jaisalmer wells. The author in sequence stratigraphic context suggests possibility of the tentative boundary in Gondwana non-marine successions.

Hettangian/Sinemurian boundary can be tried in the exposed Lathi Formation of Jaisalmer, also in Kachchh subsurface. In Spiti, it could be explored through microfossil studies in Kioto Limestone.

Sinemurian/Pliensbachian boundary may also be explored in the exposed Jaisalmer sediments and Kachchh subsurface besides Spiti in Himalaya through study of microfossils, particularly coccoliths.

Pliensbachian/Toarcian boundary may be broadly understood at the base of LSM in Dingy and Kuar bet sections of Kaladongar of 'Island belt' in Kachchh. In Spiti and Nepal and other places in High Himalaya, the mentioned boundary is placed at the contact of Kioto Limestone and coeval units and Bagung/Laptal and coeval units.

Toarcian/Aalenian boundary may again be explored in LSM and its coeval fine textured sediments in distally located well sections. The first broad sequence stratigraphy based differentiation in Kachchh is discussed in Chap. 3.

Aalenian/Bajocian boundary is for present suggested at the base of Am in Jhura and at the base of basal limestones at Sadhara, and near the revised KSm/BCSm transition in Kaladongar.

Bajocian/Bathonian boundary is for the present best suggested at the base of bed 3d of Nara immediately underlying the Early Bathonian Macrescens Subzone second-order MFS. In Jhura, it can be placed between Bm below and the massive golden oolitic limestone bearing Cm above. In Kaladongar and Goradongar section, it is placed, respectively, between KSm and BCSm, and between RSm below and LFm above.

Bathonian/Callovian boundary is much better understood in several 'Island belt' and Mainland sections between the youngest light coloured limestone bed of RLm of ULM of Patcham Formation and basal dark-coloured grey shales of LSM of Chari Formation.

Callovian/Oxfordian boundary requires refinement within the topmost sand/silt beds of DSM of Chari Formation in West Mainland sections, above the midLate Ponderosum Zone, however, the obstacle to this so far has been absence of Mariae Zone ammonoid fauna. The boundary has been recently also suggested in sequence stratigraphic context in Khadir and Gangta bet at the top of a fossiliferous yet ammonoid devoid conglomeratic sand bed which is interpreted as a third-order MFS. It is suggestively assigned basal Oxfordian basal part of Mariae Zone age in between the immediately overlying Obliqueplicatum Zone of late Early Oxfordian and the underlying early Late Callovian Athleta Zone which are firmly based on enclosed ammonoids. Thus, the boundary can be placed within or immediately below the mentioned conglomerate bed in Khadir and Gangta bet.

Oxfordian/Kimmeridgian boundary is best demarcated at the base of Am in Bharodia section based on first *Orthosphnctes*.

Kimmeridgian/Tithonian boundary is best placed in Ler section at the base of *K. pottingeri* bearing beds above the largely unfossiliferous laminated sand between beds 13 and 14. Bed 12 includes the last *H. ornatum* and bed 14 starts with first *K. pottingeri*.

Jurassic/Cretaceous boundary is best placed in West Mainland and Ler sections above the beds with late Late Tithonian Densiplicatus Zone ammonoids. Possibilities also exist in east coast well sections between Late Tithonian and Early Berriasian palynoassemblages.

Applicability of the Kachchh scale in the Indo-East-African faunal province and GTM The so developed Jurassic ammonoid zonal scale is the first such effort on GTM, and until this independent indigenous formulation the ETM zones were tentatively extended to GTM. The Kachchh scale is the best and fully applicable to IEAP region in view of the unrestricted geographical spread of IEAP ammonoid subfamilies in the region. Good correlations, mostly at zonal level have been demonstrated with ETM. In Arabia and SW Pacific part of GTM, workable correlation has been managed partly based on common ETM elements, and partly based on common IEAP ammonoid elements.

2.9 Cretaceous

2.9.1 International Status

The system was established in 1822 by a Belgian geologist d'Halloy on a chalk bearing succession in the Vocontian basin of France, and was so named after the dominant chalky lithology there. It is the longest of the three Mesozoic systems, distinctly longer than the other two. Also, it is globally among the most eventful not only of Mesozoic but of the entire Phanerozoic. Its eventfulness in Indian context is noteworthy. It is defined by its basal stage Berriasian, more precisely, by the basal Berriasian ammonoid zone Berriasella Jacobi Zone. It includes 12 stages, 6 each in its Early and Late divisions. The stages are of highly varying duration, Santonian being the shortest of ~2.7 my and Aptian being the longest of ~13.3 my duration. It is classically organized into ammonoid zones, yet providing, the highest stratigraphic resolution, however, micro-planktons (foraminifers, dinocysts and coccoliths) have come up fast with closely comparable resolutions, particularly during the Late Cretaceous.

2.9.2 Development in India

The Cretaceous system has had much larger development in India than either the Jurassic or Triassic with marine, non-marine and igneous expressions all round the Indian plate as also in inland Gondwana basins. Over the past three decades, maximum hydrocarbon reserves globally have been located in the Jurassic–Cretaceous sequences, and the Indian Cretaceous is being increasingly investigated. Yet the stratigraphic refinement both in surface and subsurface record inclusive of offshore wells leaves much to be desired in terms of zonal differentiation and correlation, stages, and international stage boundaries.

2.9.3 On the Local/Regional Stages in the Indian Cretaceous

In a recent review (Raju 2011), a succession of 8 Cretaceous regional stages is included. These stages have been created by ONGC scientists in recent decades in Indian east coast exposed and subsurface well sections. Among these eight stages, the younger five stages are based in the exposed Cauvery basin sections. These holostratotype sections have hardly ever been adequately lithostratigraphically diffrerentiated and described, in spite of such description of stratotypes being an essential prerequisite to the proposition of the stages. Stages or for that matter any chronostratigraphic unit essentially comprise and refer to the column of sediments which requires to be measured and described in detail with respect at least to its field observable lithological features, physical and biogenic structures, as also body fossil content. It may also be emphasized that guide fossil taxa included in the column of sediments are only used to define the basal time plane of the mentioned stage. It is the column of sediments which makes the stage not the guide fossils included in the column of sediments which merely serve the important purpose of defining the basal time plane. Any taxon range chart of the planktonic foraminifera used as guide fossils either have not been published either in adequately differentiated exposed lithostratigraphic columns or have escaped the present author's notice. ~30 microfossil taxa are indicated in a table against the succession of 12 international stages, 12 planktic foraminifera based biochrons have been introduced in the ~32.5 my Late Cretaceous Cenomanian–Maestrichtian interval in comparison to ~70 international planktic foraminiferal datums in the entire Cretaceous, most of which have been picked up in some other basin with deeper neritic to bathyal signatures. The eight regional stages based on relatively deeper water planktic foraminfera are hardly usful in the study of the shallow marine exposed Indian Cretaceous successions. The best studied are the Cauvery and K-G subsurface well Cretaceous sections. Subsurface Cretaceous has also been studied for dinocysts, coccoliths, pollen and spores also mostly by the ONGC and the BSIP scientists. Composite presence of as many as 18 international or internationally correlatable ammonoid zones is realized in Kachchh (seven zones) and Cauvery (thirteen zones).

Much unlike the older Triassic and Jurassic systems, high resolution ammonoid based zonation could not be developed in the Cretaceous system of India. Even the international inter/intrastage boundaries are not precisely differentiated. Instead, a succession of local/regional stages has been created by the ONGC scientists. Parallel zonations have been published based on planktonic foraminifera/ dinoflagellates/coccoliths and pollen/spores without integration.

2.9.4 Integrated Indian Cretaceous Ammonoid Zonal Succession

In Kachchh, only three Early Cretaceous stages are expressed through ammonoids which are Berriasian, Aptian, and Albian. Maximum of eight stages are expressed through ammonoids in Cauvery basin which are the younger ones from Aptian to Maestrichtian. The integration of the two principal Indian Cretaceous basins of Cauvery and Kachchh indicates presence of ammonoids in as many as nine of the twelve stages. Ammonoids of Turonian and Coniacian stages are also known from Narmada basin. In High Himalaya, e.g. in Spiti, almost complete succession of marine Cretaceous is present, however, ammonoids are present in Berriasian and Valanginian at the start, and then again in Aptian and Albian, thus found restricted to the mentioned four stages of Early Cretaceous. Taking the entire Indian Cretaceous, the stages which so far have not yielded any ammonoids are Hauterivian and Barremian, although unconfirmed suggestions have been made in the east coast basins. The ammonoid presence in the Indian Cretaceous is for the first time organized into an integrated succession of at least ~18 ammonoid assemblages without precise differentiation of stage and zonal boundaries. In this context Ayyasami (2011), on the basis of ammonoids has provided tentative demarcation of a few interstage boundaries in the Cauvery basin. The integrated succession of the ammonoid assemblages is produced here (Fig. 2.42) as under

1. Basal Berriasian Aspidoceras tavarae/Argentiniceras loncochensis Zone, 2. Late Berriasian Kilianella Zone, 3. Early Late Valanginian Olcostephanus verrucosum Zone, 4. Early Aptian Deshayesites deshayesi Zone, 5. Late Aptian Australiceras jacki Zone, 6. Early Albian Douvilleiceras mammilatus Zone, 7. Early Middle Albian Hamites rotundus Zone, 8. Late Albian Mortoniceras rostratum Zone, 9. Early Cenomanian Mantelliceras vicinale Zone, 10. Middle Cenomanian Calycoceras asiaticum Zone, 11. Late Cenomanian Eucalycoceras pentagonum Zone, 12. Early Turonian Pseudoaspidoceras footeanum Zone, 13. Early Middle Turonian Romaniceras ornatissimum Zone, 14. Late Middle Turonian Romaniceras deverianum Zone, 15. Coniacian Kossmaticeras theobaldianum Zone, 16. Latest Santonian Texanites roemeri Zone, 17. Campanian Karapadites karapadense Zone, and 18. Maestrichtian Pachydicus otacodensis Zone. In addition, ammonoid presence is also suspected in the Hauterivian–Barremian sediments in Spiti (Pandey and Pathak 2015b; Pandey et al. 2013c; Cauvery and K-G; Spath 1927–33)

Among the above 18 zones, 1 and 2 are in Kachchh; 3 in Spiti; 4, 5, 6 and 7 in Kachchh/Jaisalmer; and 8–18 in Cauvery basin.

2.9.5 Exposed Cretaceous record in Kachchh and Jaisalmer

There is evidenced in Kachchh and Jaisalmer substantive record of the exposed Cretaceous sediments at least west of the median high of Early and early Late Cretaceous age recorded in detail under ONGC sponsored project (Fig. 2.41 Krishna 1987b; Krishna 1994; Dubey 1992; Singh 2006; Singh et al. 2000). The Kachchh Cretaceous includes extremely sparsed and rare presence of ammonoids of Berriasian through *Argentiniceras*, *Negreliceras*, *Groebericeras*, *Kilianella*, *Ducaliceras*, *Aspidoceras*, etc. Aptian through *Deshayesites*, *Dufreynia*, *Epicheloniceras*, *Australiceras*, *Tropaeum* etc., and Albian through *Lemuroceras* (Krishna 1980,

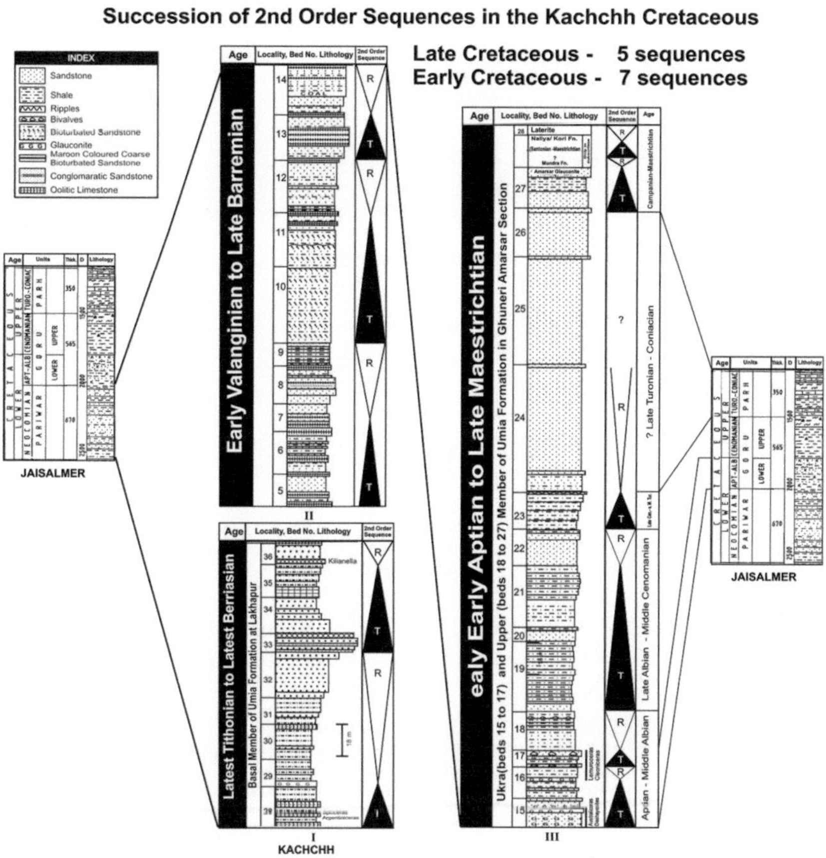

Fig. 2.41 Stratigraphic correlation in the Cretaceous between Kachchh (modified after Krishna 1994 unpublished report, Dubey 1992) and Jaisalmer (modified after Singh 2006) columns vis-a-vis second-order sequences

1983, 2002, 2012 and unpublished). The Kachchh Cretaceous also includes the youngest Cretaceous exposed marine sedimentary unit through relatively thin glauconitic green shale bed underlying the Traps in the Ghuneri–Amarsar section of probable Middle Turonian or Santonian/Campanian age. The non-marine intertrappeans in East Kachchh are assigned Maestrichtian age (ONGC 1993), so also the Deccan Trap themselves. Thus, almost the entire Cretaceous seems present and well-exposed in Kachchh.

2.9.6 *Subsurface and Offshore Record*

In subsurface and offshore wells (sp-1-1 and K1-1A), respectively, located aout 80 and 150 km in the dip direction from the thin glauconite bed locality, presence of inner to outer shelf sediments (Naliya Fn. in SP-1-1 about 80 km and Kori Fn. About 150 km away) has been interpreted on the basis of planktonic foraminifera and coccoliths of Coniacian–Maestrichtian interval (ONGC 1993).

2.9.7 *Early Vis-à-Vis Late Cretaceous in Kachchh*

The exposed Early Cretaceous record in Kachchh basin is marine, continuous and confirmed only up to Albian, while the marine Cenomainan to Santonian/Campanian is suggested on broad foraminifer, radiometry, and sequence stratigraphic interpretations. As such, the Albian/Cenomanian or Cenomanian/Turonian stage boundaries can not be differentiated. In context of the equivalent Early Cretaceous record in other Indian basins as also in the Indian subcontinent, the Kachchh Early Cretaceous is better measured and recorded, as also stratigraphically evaluated. In comparison, the Kachchh Late Cretaceous record inclusive of onshore/offshore wells is much less known, and is somewhat discontinuous, scattered and fragmentary. Reliable datable elements merely include microfossils, mostly foraminifers and palynomorphs, which have to some degree facilitated the progress of the Indian Cretaceous stratigraphy.

2.9.8 *Interstage Cretaceous Boundaries*

Tithonian/Berriasian Boundary This boundary can be placed in Kachchh between the last *Virgatosphinctes* in bed 27 below and bed 28 above with basal Berriasian ammonoids in Ler–Katrol transect as also in West Mainland sections of Umia, Lakhapur, Sahera and Mundhan (Krishna et al. 1994b, 1996b; Dubey and Chatterjee 1997). In the K-G basin, the Bapatala Sandstone and its facies variant Nellore Claystone include most of the Berriasian, and late Tithonian, if present, should be included in the lower ~200 m. Thus, the best tentative suggestion to this boundary is between the shale dominated lower part and sand dominated middle part of the Bapatla Sandstone.

Berriasian/Valanginian Boundary The best suggestion to this boundary in east coast basins is the lithostratigraphic contact between Bapatala Sandstone and Penner Shale in the KS-A-4 well of the K-G basin. In Kachchh, it may be best placed above the *Kilianella* bearing beds in the West Mainland Sahera and Mundhan sections. In Spiti, the best possibilities exist above the Late Berriasian and below the Early Valanginian ammonoid bearing black shales. This boundary may also be explored in east coast well sections. Ammonoid studies under much better stratigraphic control are in progress, which in near future should provide precise determination of the Tithonian/Berriasian, and the Berriasian/Valanginian boundary, as also ammonoid zonation in the mentioned stages.

Valanginian/Hautervian Boundary In the K-G subsurface wells, this boundary is suggested above the Penner Shale or at the contact of Penner Shale and the overlying Golapille Formation. In Kachchh, the best possibilities exist in the early part of the Ghuneri Member sands. In Spiti, this boundary may be best expored between the early Late Verrucosum Zone beds below and basal Aptian ammonoid bearing beds above.

Hauterivian/Barremian Boundary It is the most difficult to approximate in any basin of India due to complete absence of ammonoids or other guide fossils. The best places are east coast well sections. The thin wedge shaped occurrence of the lagoonal coal in the Ghuneri section of Kachchh on composite indirect evidence is best assigned to Late Barremian.

Barremian/Aptian Boundary It is approximated widely in the Indian subcontinent between Pariwar Formation and Abur Frmation in view of start of the Abur Formation little above the Barremian/Aptian boundary in the second oldest Aptian Deshayesites Zone. The above approximated Barremian/Aptian boundary can be extended to the Ghuneri Member/Ukra Member boundary in Kachchh, between Sivaganga Formation and Dalmiapuram Formation in Cauvery basin, within the Andimadam Formation below the level of *Australiceras jacki*, below the level of *Dufreynia/Australiceras* between the Golapille Formation and the Raghavapuram Formation, within the Guimal Formation above the youngest ammonoid devoid sands and *Deshayesites* bearing marl (Pandey et al. 2013b, c). In regional stratigraphic context the boundary in the Narmada basin may be explored between the principal non-calcareous early part of the sandstones below and the overlying bivalve bearing calcareous sands above of the Nimar Formation. However, there are no clues to the actual ages of the older non-calcareous and younger calcareous sands. Jafar (1982) has recorded Turonian coccoliths from the thin fine textured intercalations of the younger calcareous unit of the Nimar Formation. There could even be a large subaerial gap between the lower and upper units of the Nimar Formation allowing the calcareous sands of the upper part to be as young as Turonian. In the neighbouring Kachchh basin, also elsewhere in the Indian

Cretaceous successions, the thick non-calcareous pre-Turonian sands are mostly of the Hauterivian–Barremian age of Ghuneri Member of Umia Fn in Kachchh and its coeval units in many other Indian basins. However, similar sands are also of the younger Late Albian–Early Cenomanian age of the Upper Member of Umia Formation in Kachchh with the marine slow sedimented ammonoid bearing Ukra Member in between the Ghuneri and Upper units. The younger bivalve bearing calcareous sandstone unit in the Narmada basin could even be in part shallower near coast facies variant of the overlying Nodular Limestone Member of Bagh Formation/Group. This boundary can also be strengthened with FAD of charlottae.

Aptian/Albian Boundary It is tentatively placed between the Late Aptian *Epicheloniceras* and Early to early Middle Albian *Douvilleiceras* in the Kuchri–Habur traverse in Jaisalmer, while in subsurface below the FAD of Roberti. The mentioned boundary is included in the early part of Karai Formation in view of the recent determination of the Mammilatum Zone (Gautam et al. 2015).

Albian/Cenomanian Boundary Ayyasami (2011) placed the Albian/Cenomanian boundary in a rather rare but good section east of the Uttatur village in the otherwise badland country within the Karai Shale Formation suggestively in the middle of the unit between the levels repectively of late Late Albian *Mortoniceras rostratum* and early Early Cenomanian *Mantelliceras mantelli* (fig/map, GSITI, Course Material pp. 4–5).

Cenomanian/Turonian Boundary Ayyasami and Bannerjee (1984) and Ayyasami (2011) determined this boundary east of the Odium village within the Karai Shale around the middle of its Kunnam Member below the $\sim$30 m interval of shales with calcareous nodules between the levels, respectively, of late Late Cenomanian *Eucalycoceras pentagonum* below and early Early Turonian *Pseudoaspidoceras footeanum* above (fig./map GSITI, Course Material, pp. 6–7). However, in the section near Kullakalnuttam village, Early Turonian ammonoids seem absent, and the Karai Formation succession not extending into Turonian.

Turonian/Coniacian Boundary Ayyasami (2011) placed this boundary in the gully north of the Andur and Varagur villages above the $\sim$50 m shales with calcareous nodules near the top of the Alundalippar Formation of Trichinopoly Group. It is near the contact of Alundalippar Formation below and Varagur Formation above between the levels repectively of the early Middle Turonian *Lewesiceras anapadense* and early Early Coniacian *Proplacenticeras tamulicum*. The contact of the units is disconformable, and a gap is evident of late Middle to Late Turonian.

Coniacian/Santonian Boundary This boundary can only be tentatively suggested immediately above the Anaipadi Formation of Coniacian age, since the Early and Middle Santonian mark a stratigraphic gap.

Santonian/Campanian Boundary Ayyasami (2011) have suggested this boundary north of the Melmattur village on the basis of the levels, respectively, of Late

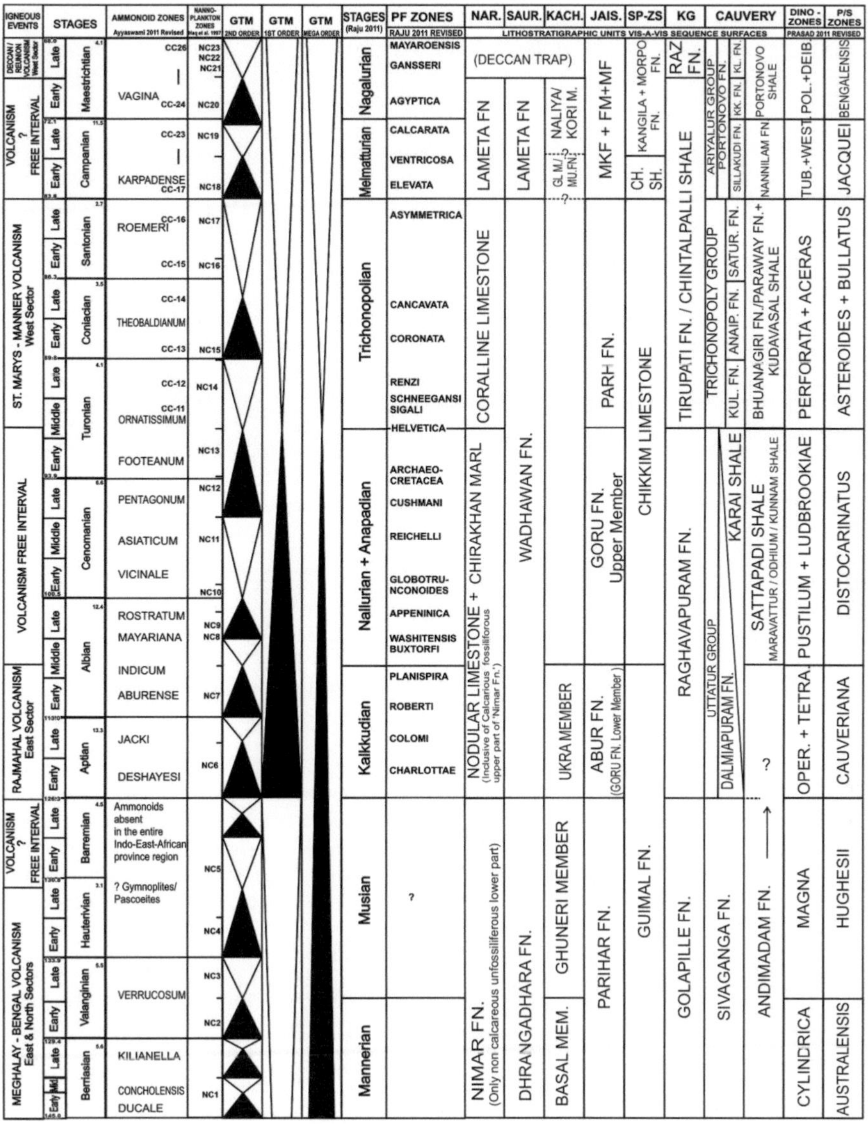

Fig. 2.42 Indian cretaceous lithostratigraphic units vis-a-vis second-order sequences, planktonic, dinoflagellate, pollen–spore zones

Santonian *Inoceramus balitus* below and of Early Campanain forams above. The boundary can also be explored in the east coast well sections.

Campanian/Maestrichtian Boundary Ayyasami (2011) placed this boundary in a gully section north of the Ariyalur–Vilangudi road near Periyanagalur below the first occurrence of *Eubaculites vagina* and the other Early Maestichtian heteromorph ammonoids. The boundary again may be explored in the east coast well sections (Fig. 2.42).

References

Ayyasami K (2011) Recognition of the Cretaceous stage boundaries in Southern Indian Cretaceous based on invertebrate fauna, 78–87. In: Jaitly AK, Singh AD, Pandey B, Nath S (eds) Palaentology and stratigraphy: basics to applications. Banaras Hindu University, Varanasi

Ayyasami K, Bannerjee RK (1984) Cenomanian—Turonian transition in the Cretaceous of Southern India. Bull Geol Soc Denmark 33:21–30

Bajpai S, Prasad GVR, Prasad V, Krishna J, Sarkar A (2012) Recent advances on Phanerozoic biodiversity, bioevents and climates in India. PINSA 78(Sp. Iss.):445–456

Bhargava ON (2008) Paleozoic succession of the Indian plate. Mem Geol Soc Ind 74:209–244

Bhargava ON, Krystyn L, Balini H, Lein R, Nicora A (2004) Revised litho and biosequence stratigraphy of the Spiti Triassic. Albertiana Suppl 30:21–38

Biswas SK (1977) Mesozoic rock-stratigraphy of Kutch. Quart J Min Met Soc Ind 43(3 and 4):1–52

Biswas SK (1993) Geology of Kutch, 1, K D Malaviya Institute of Petroleum Exploration, Dehradun, 450

Calkins JA, Offield TW, Abdullah KM, Ali ST (1975) Geology of the Southern Himalaya in Hazara Pakistan and adjacent area. U.S.G.S. Prof Pap 716 C:29

Callomon J H, Dietl G, Niederhofer (1989) The ammoniten faunenhorizonte im grenzbereich Bathonium/Callovium des Schwabischen Jura und deren correlation mit W. Frankreich und England. Stuttg Beitr Nat B-148:1–13

Cariou E (1984) Les Reineckeiidae (Ammonita Callovien) de la Tethys occidentale; dimorphism et evolution, Etude a partir des gisments du centre-ouest de la France. Doc Lab Geol Lyon 8:559

Cariou E, Hantzpergue P (1997) Biostratigraphie du Jurassique ouest-europeen et mediterraneen. Bulletin des Centres de Recherches. Elf Explor Prod Mem 17:1–440

Casshyap SM, Khan A (2000) Tectono-sedimentary evolution of the Gondwanian Satpura basin of Central India: evidence of pre-trap doming, rifting, and paleoslope reversal. J Afr Earth Sci 31 (1):65–76

Cohen KM, Finney SC, Gibbard PL, Fan JX (2015) The ICS International Chronostratigraphic Chart. Episodes 36:199–204

Dietl G (1981) Uber macrocephalites ammonoidea aus dem aspidoides oolith und die bathonium/callovium grenz schisten der zollernalb (S W Deutschland). Stuttg Beitr Naturk B-68:1–15

Draganits E, Grasemann B, Janda C, Wiesmayr G (2004) Brief outline of the tectonic history of the NW Himalayas with emphasis on Spiti. Albertiana 30(Supplement):11–20

Dromart G, Garcia JP, Picard S, Atrops F, Lecuyer C, Sheppard SMF (2003) Ice age at the middle/late Jurassic transition: geological and geochemical evidence. Am J Sci 303:667–707

Dubey N (1992) Mesozoic sedimentary sequences and their depositional environments in Northwestern part of Kachchh Mainland, Gujarat (unpublished Ph.D thesis), BHU, Varanasi

Dubey N, Chatterjee BK (1997) Physical and biological changes across the Jurassic/Cretaceous boundary in Northwestern Kutch basin. Palaeobotanist 46(1–2):141–148

Enay R (2009) Les faunes d'ammonites de l'Oxfordien au Tithonien et la biostratigraphie des Spiti-Shales (Callovien Superieur–Tithonien) de Thakkhola, Nepal Centrale. Docum Lab Geol Lyon 351

Enay R, Cariou E (1999) Jurassic ammonite faunas from Nepal and their bearing on the palaeobiogeography of the Himalayan belt. J Asian Earth Sci 17:829–848

Fatmi AN (1972) Stratigraphy of the Jurassic and Lower Cretaceous rocks and Jurassic ammonites from the northern areas of Pakistan. Bull Brit Mus Nat Hist Geol 20(7):297–380

Fursich FT, Alberti M, Pandey DK (2013) Stratigraphy and paleoenvironments of the Jurassic rocks of Kachchh (field guide). Beringeria 7(Spl. Issue):1–174

Garg R, Ateequzzaman K, Krishna J, Jain KP (2003) Biostratigraphic potential of dinoflagellate cysts recovered from the late jurassic ammonites of the tethys himalaya. India J Pal Soc India 48:41–58

Gautam J, Pandey B, Pathak DB, Jaitly AK (2015) Recognition of the Early Albian Douvilleiceras mammillatum Zone in the Cauvery Basin. SE India. J Earth Sci Ind 8(4):100–111

Geiger M, Schweigert G (2006) Toarcian—Kimmeridgian depositional cycles of the southwestern Morondova basin along with rifted continental margin of Madagascar. Facies 52:85–112

Ghose NC, Shmakin BM, Smirnov BN (1973) Some geochronological observations on the Precambrians of Chotanagpur, Bihar. Geol Mag 110:481–484

Gradstein FM, Ogg JG (2012) A Phanerozoic time scale. Episodes 19(1–2)

Haq BU, Hardenbol J, Vail PR (1987) Chronology of fluctuating sea-level since the Triassic. Science 235:1156–1167

Hayden HH (1904) The geology of Spiti. Mem Geol Surv India 36(I):1–129

Jafar SA (1982) Nannoplankton evidence of the Turonian transgression during Narmada valley, India, and Turonian—Coniacian boundary problem. J Pal Soc Ind 27:17–30

Jain S (2014) The Arkelli chronozone: a new early Middle Bathonian standard ammonite zone for Kachchh, western India (southern Jurassic Tethys). Zitteliana A 54:91–135

Jaitly AK, Singh CSP (1983) Discovery of Late Bajocian Leptosphinctes Buckman (Jurassic Ammonitina) from Kachchh, Western India—Neues Jahrbuch fur Geologie und Paleontologie monatshefte, 91–96

Kanjilal S, Singh SR (2012) Arrival of Quenstedtoceras Hyatt: a Late Callovian ammonite in kachchh and its significance. Geology of Kachchh basin 27

Koshal VN (1973) Palynozonation of Mesozoic subsurface sediments of Banni Kutch Gujarat. Q J Min Met Soc Ind 47:79–81

Koshal VN (1984) Differentiation of Rhaetic sediments in the subsurface of Kutch based on palynofossils. Petr Asia J 7:102–105

Krishna J (1980) Uncoiled ammonites of Middle Albian (Lower Cretaceous) age from Habur Series, Jaisalmer, Rajasthan. J Pal Soc Ind 23 and 24:49–54

Krishna J (1983) Callovian–Albian ammonoid stratigraphy and paleobiogeography in the Indian subcontinent with special reference to the Tethys Himalayas. Him Geol 11:43–72

Krishna J (1987a) Biological evidence for better appreciation of Indian Gondwana. Paleobotanist 36:268–284

Krishna J (1987b) An overview of the Mesozoic stratigraphy of Kachchh and Jaisalmer. J Pal Soc Ind 32:136–149

Krishna J (1994) Environmental reconstruction studies of the Mesozoic sedimentary sequences in Western Kachchh (ONGC Sponsored Project), (Technical Report), 1–143, pts, 1–10

Krishna J (2002) Mesozoic microstratigraphy. DST contact programme-cum-field workshop, structure, tectonics and Mesozoic stratigraphy of Kachchh, Bhuj (Kachchh), Gujarat, 6:96–121

Krishna J (2004) Refinement of time in the Indian geological record: Case sheet of the Jurassic in Kachchh. J Geol Soc Ind 63:120–123

Krishna J (2005) Update on Kachchh Jurassic ammonoid zonation: summarised intra-basinal to global perspectives. In: Raju DSN et al (eds) Spl. Publ.-1, Association of Petroleum Geologists, pp 63–68

Krishna J (2012) Litho-bio-chrono-magneto-igno-sequento-stratigraphic integration in the Kachchh Mesozoic, Status Talk—Mesozoic Biostratigraphy, In: Thakkar MG (ed) Abstract volume, Geology of Kachchh basin, Western India, pp 21–25

Krishna J (2014) Ammonoid—sequence-magneto stratigraphic integration in the Kachchh Jurassic. Unpublished DST project report, 60 p

Krishna J, Cariou E (1990) Ammonoid faunal exchanges during Lower Callovian between the Indo–East-African and Submediterranean province: implications for the long distance East-West Correlations. Newsl Stratigr 23(2):109–122
Krishna J, Cariou E (1993) The Tethyan Macrocephalitinae: evolutionary, environmental and dispersal strategies. Geobios 15:227–236
Krishna J, Ojha JR (1996) The Callovian ammonoid chronology in Kachchh (India). GeoResearch Forum 1–2:151–166
Krishna J, Ojha JR (2000) The Intra-basinal correlation in the Middle Jurassic Callovian stage of Kachchh (Gujarat) and ammonoid-microfossil integration. Geophytology 28(1 and 2):101–120
Krishna J, Pathak DB (1989) Kimmeridgian in Ler–Katrol area of Kachchh, Western India: ammonoid systematics and biochronology. In: Abstracts 28th international geological congress, Washington, pp 228–229
Krishna J, Pathak DB (1991) Ammonoid biochronology of the Kimmeridgian Stage in Kachchh, India. J Pal Soc Ind 36:1–13
Krishna J, Pathak DB (1993) Late Lower Kimmeridgian – Lower Tithonian virgatosphinctins of India: evolutionary succession and biogeographic implications. Geobios MS 15:217–226
Krishna J, Pathak DB (1994) Stratigraphic, biogeographic and environmental signatures in the amonoid bearing Jurassic-Cretaceous in the Himalaya on the south margin of the Tethys. J Him Geol 4(2):151–167
Krishna J, Westermann GEG (1985) Progress report on the Middle Jurassic ammonoid zones of Kachchh, Western India. Newsl Stratigr 14(1):1–11
Krishna J, Westermann GEG (1987) Faunal associations of the Middle Jurassic ammonite genus Macrocephalites in Kachchh, Western India. Can J Earth Sci 24:1570–1582
Krishna J, Singh IB, Howard JD, Jafar SA (1983) Implications of new data on the Mesozoic rocks of Kachchh, India. Nature 305:790–792
Krishna J, Ojha JR, Pandey B (1993) Comment on research note—a note on the Middle Jurassic stratigraphic succession of Keera dome, Kachchh District, Gujarat by S. Prasad 1993. J Geol Soc Ind 42(1):102
Krishna J, Melendez G, Pandey B, Pathak DB (1994a) Middle Oxfordian Perisphinctinae and biostratigraphy from Kachchh (India). Jornadas de Paleontol, pp 106–108
Krishna J, Pathak DB, Pandey B (1994b) New ammonoid evidence on the Jurassic/Cretaceous boundary in Kachchh (India). Geobios 17:327–335
Krishna J, Pathak DB, Pandey B (1994c) Discovery of Middle and Late Oxfordian in Kachchh (India) with first ever approximation of the Oxfordian/Kimmeridgian boundary in the Indian subcontinent (abstr) Oxfordian–Kimmeridgian Joint Working Groups Meet, Lyon, p 10
Krishna J, Melendez G, Pandey B, Pathak DB (1995a) Characterization of the ammonoid genus Larcheria (Middle Oxfordian) in Kachchh (India): paleontology, biostratigraphic and paleobiographic evaluation in the context of North-Tethyan occurrences. C R Acad Sci Paris 321, lla, 1187–1193
Krishna J, Pathak DB, Pandey B (1995b) Kimmeridgian–Tithonian (Upper Jurassic) ammonoid zones in Kachchh (Western India) and their correlations. In: Proceedings of recent researches in geology in Western India, Badodara, pp 323–347
Krishna J, Melendez G, Pandey B, Pathak DB (1996a) Middle Oxfordian Ammonites (Perisphinctinae) from Kachchh (India): biostratigraphic and paleobiogeographic implications. Revista Espanola De Paleontologia, Numero Extraordinario, pp 140–147
Krishna J, Pandey B, Pathak DB (1996b) Ammonoid chronology developed in the Tithonian of Kachchh (India). GeoResearch Forum 1–2:205–214
Krishna J, Pathak DB, Pandey B (1996c) Quantum refinement in the Kimmeridgian ammonoid chronology in Kachchh (India). GeoResearch Forum, Transtec Publications, Switzerland 1–2:195–204
Krishna J, Sinha AK, Upadhyay R (1997) The first South Tethyan Hettangian Psiloceras from the Indus Suture, Ladakh Himalaya: diverse implications. J Him Geol 18:45–51

Krishna J, Pathak DB, Pandey B (1998) Development of Oxfordian (Early Upper Jurassic) in the most proximal exposed part of Kachchh basin at Wagad outside the Mainland Kachchh. J Geol Soc Ind 52:513–522

Krishna J, Pandey B, Ojha JR, Pathak DB (2000) Transgressive Sediment Intervals in the Late Jurassic of Kachchh. India, GeoResearch Forum 6:321–332

Krishna J, Pandey B, Pathak DB (2009) Characterization of Dichotomoceras in the Oxfordian of Kachchh (India). J Geo Soc Ind 74:469–479

Krishna J, Pandey B, Pathak DB (2010) Update on the Jurassic stratigraphy of the Indian subcontinent with focus on the tectonically controlled regional transgressive—regressive couplets. Earth Sci Frontiers China 17:138–140

Krishna J, Pandey B, Pathak DB (2011) Current status of the ammonoid stratigraphic framework in the Indian subcontinent with focus on the tectonically controlled regional transgressive—regressive couplets. Mem Geol Soc Ind 78:140–176

Krystyn L, Balini M, Nicora A (2004) Lower and Middle Triassic stage and substage boundaries in Spiti. Albertiana 30 supplement, pp 39–52

Metcalfe I (2011) Tectonic framework and Phanerozoic evolution of Sundaland. Gond Res 19:3–21

Mette W (2004) Deciphering and interpreting Jurassic diversity and distribution trends for marine ostracods in Gondwana—the role of biologic, endogenous and exogenous processes. Beringeria 34:3–35

Ojha JR (1996a) The Callovian at keera, Kachchh, India: ammonoid systematics, chronology and geography. J Sci Res BHU 46:233–244

Ojha JR (1996b) The Callovian at Keera, Kachchh, India; Ammonoid systematic, Chronology and Geography. Unpublished Ph.D thesis, Banaras Hindu University, Varanasi

ONGC (1993) Lithostratigraphy of Indian petroliferous basins documents. KDMIPE, Dehradun

Pandey B (1993) The Kachchh Jurassic Virgatosphinctinae, high resolution evolutionary succession and chronology. Unpublished Ph.D. thesis, Banaras Hindu University, Varanasi, pp 1–473

Pandey DK, Callomon JH (1995) Contributions to Jurassic of Kachchh, Western India III, the Middle Bathonian ammonoid families Clydoniceratidae and Perisphinctidae from Pachchham Island. Berringeria 16:125–145

Pandey J, Dave A (1993) Studies in Mesozoic foraminifera and chronostratigraphy of Western Kutch, Gujarat. Pal Ind 1:1–221

Pandey B, Krishna J (2002) Ammonoid biostratigraphy in the Tithonian (Late Jurassic) of Jaisalmer, Western India. Geophytology 30(1 and 2):17–25

Pandey B, Pathak DB (2015a) Record of Early Bathonian ammonoids from Kachch, India: biostratigraphic and paleobiogeographic implications. J Pal Soc Ind 60(1):33–44

Pandey B, Pathak DB (2015b) Status of the Indian Early Cretaceous ammonoid record in light of recent observations in the Spiti Valley, Himachal Himalaya. Himalayan Geol 36(1):1–8

Pandey B, Krishna J, Pathak DB, Kumar A (2013a) Ammonoid biostratigraphy of Bathonian succession at Jumara, Kachchh, western India. J Ind Geo Cong 4(2):7–18

Pandey B, Pathak DB, Krishna J (2013b) Calliphylloceras heterophylloides (Oppel, 1856) from basal most Jurassic succession of Sadhara dome, Kachchh, India. J Pal Soc Ind 58(1):61–65

Pandey B, Pathak DB, Krishna J (2013c) Preliminary remarks on new ammonoid collection from freshly exposed succession of the Spiti Formation between Lidang and Giumal, Spiti valley, Himachal Himalaya, India. Himalayan Geol 34(2):124–134

Pathak DB (1989) The Kimmeridgian-Tithonian at Ler-Katrol, Kachchh, India, Ammonoid Systematics, Biochromology and Biogeography. Unpublished Ph.D. thesis, Banaras Hindu University, Varanasi, pp 1–496

Prasad B (2011) Palynology: principles, methodology and applications in biostratigraphy and oil exploration, In: Jaitly AK, Singh AD, Pandey B, Nath S (eds) Palaeontology and stratigraphy: basics to applications, pp 16–35

Prasad B, Jain AK, Mathur YK (1995) A standard palynozonation scheme for Cretaceous and pre Cretaceous subsurface sediments of Krishna-Godavari basin, India. Geo Sci J 16(2):151–232

Rabu B (1993) Stratigraphy and structure of the Oman mountains. Bur Recher Geol Min Paris 221:262

Rai J (2006) Reworked Pliensbachian—Aalenian nannofossils from Jara dome Kachchh; early Jurassic palaeobiogeography of western India. In: XIth International Nannplankton Association, Nebraska, pp 84–85

Rai J, Jain S (2012) Early Jurassic Gondwana break-up- a nanno-fossil story, in Geology of Kachchh basin, western India (abstract), pp 31–32

Rai J, Jain S (2013) Pliensbachian nanno-fossils from Kachchh: implications on the earliest Jurassic transgressive event on the western Indian margin. Zitteliana A 53:105–120

Rai J, Krishna J, Ojha JR (2005) Late Callovian nannofossils from Jara dome Kachchh, western India; biochronology and calibration by ammonites. In: XXth CMS, Visakhapatnam, p 40

Rajnath (1932) A contribution to the stratigraphy Of Kutch. Q J Min Met Geol Soc Ind 4(4): 161–174

Raju DSN (2011) Cretaceous stratigraphic record on Indian plate: an overview. Mem Geol Soc Ind 78:177–226

Roy P, Bardhan S, Mitra A, Jana SK (2007) New Bathonian (Middle Jurassic) ammonite from Kutch, India. J Earth Sci 30:629–651

Sarkar A, Datta AK, Poddar BC, Kolhapur VK, Bhattacharya BK, Sanval R (1996) Geochronological studies of Mesozoic igneous rocks from Eastern India. J SE Asian Earth Sci 21:188–193

Schweigert G, Krishna J, Pandey B, Pathak DB (1996) A new approach to correlation of the Upper Kimmeridgian Beckeri Zone across the Tethys. N Jb Geol Paleont 202(3):345–373

Shah SM (1977) Stratigraphy of Pakistan. Mem Geol Surv Pakistan 5(12):250

Sharland PR, Archer R, Casey DM et al (2001) Arabian plate sequence stratigraphy. Geoarabia 9(2):371

Shellnut JG, Bonit GM, Brookfield ME, Jaba BM (2009) No link between the Panjal traps (Kashmir) and Permian mass extinction. Geophys Res Lett 38:219308. doi:10.1829/2011GL059052

Singh NP (2006) Mesozoic lithostratigraphy of the Jaisalmer basin, Rajasthan. J Pal Soc Ind 51(2):1–25

Singh NP, Singh B, Bharktya DK, Srivastava S (2000) Subsurface stratigraphy, sedimentation model and hydrocarbon prospects of Saurashtra Basin, India. Bull ONGC 37(1):167–179

Spath LF (1927–33) Revision of the Jurassic Cephalopod fauna of Kach (Kutch). *Paleont. Indica*, Calcutta 9(2):1–945

Srikantia SV (1981) The lithostratigraphy, sedimentation and structure of the Protorozoic—Phanerozoic of the Spiti basin in higher Himalaya in Himachal Pradesh, India. In: Sinha AK (ed), Contemporary Geoscientific Research in the Himalaya, Bishen Singh Mahinder Pal, Dehradun 1, pp 31–48

Stoliczka F (1866) Geological section across the Himalayan Mountains in Spiti. Mem Geol Sur Ind 5(1)

Uhlig V (1903–10), The fauna of Spiti Shales. Pal Ind Ser 15(4):511

Venkatachala BS (1970) Palynology of the Mesozoic sediments of Kachchh, West India; reworked Permian Pollens from the upper Jurassic sediments—a discussion. Paleobotanist 18(1):45–49

Vijaya (1997) Benchmark Permian—Triassic palynoevents in peninsular India and northwest Himalaya. J Pal Soc Ind 42:117–126

Vijaya (2009) Palynofloral changes in the Upper Paleozoic and Mesozoic of the deocha-pachami area, Birbhum Coalfield, West Bengal. Sci China Ser D-Earth Sci 52(12):1932–1952

Waagen W (1871) Abstracts of results of examination of the ammonite fauna of Cutch with remarks on their distribution among the beds and probable age. Rec Geol Surv Ind 4(4):89–101

Waagen W (1873–75) Jurassic fauna of Cutch, Cephalopoda. Paleont Indica, Calcutta 9(1):1–247

Wadia DN (1934) The Cambrian–Triassic sequences of N W Kashmir. Rec Geol Surv Ind 68:121–167

Westermann GEG, Callomon JH (1988) The Macrocephalitinae and associated Bathonian and Early Callovian (Jurassic) ammonoids of the Sula Island and New Guinea. Paleontographica A 203:1–90

Chapter 3
Outcrop-Based Sequence Stratigraphic Studies on GTM with Focus on the Kachchh Mesozoic

Abstract The history of the relatively new interdisciplinary and highly applicative discipline of sequence stratigraphy is summarized. Basics, terminology used in the present work, as also author's approach to outcrop-based studies in the Kachchh basin are briefly outlined. The five mega-sequence framework in the Indian Phanerozoic is introduced with focus on the fourth one of intra-Permian–intra-Paleogene span from the origin to the closure of the Neotethys. The said mega-sequence is differentiated into 3 first-order sequences in the Indian record, and 37 second-order sequences in Spiti, Kachchh and Cauvery. The Kachchh record is further exampled with third-, fourth- and fifth-order differentiation. The SBs and MFs are précised at ~400/200 ky subzonal/horizonal resolution or its multiples. The chapter concludes with crude investigation of the sequence periodicity at different orders in which the second-order sequences are found to correspond with cyclicity of ~400 ky or its multiples (Tables 3.1, 3.2 and 3.3).

Keywords Sequence stratigraphy · Depositional sequence · Sequence boundary · Maximum flooding surface · Unconformities · Correlative conformities · Transgressive/regressive systems tract · Subaerial/submarine gaps · Condensed/ starved · Biotic/abiotic · Ammonoid · Other body fossils · Parasequences · Stacking · Bathymetry · Spatiotemporal migration · Proximal/distal · GTM · IEAP · Kachchh · Concordant/discordant · Tectonics · Eustatics · Submergence/ uplift · Pro/a/retro-gradation · Frequency · Magnitude · Density · Diversity · Extinction/origin · First/Second/Third order · Earths orbital dynamics · Cyclicity · Periodicity · 400 my cycles · Homogeneous tectonostratigraphic regions

3.1 Sequence Stratigraphic Studies

3.1.1 Brief History of Sequence Stratigraphic Studies

Depositional sequences and sequence stratigraphy have a history of mere five decades since the organization of the North American geological record in a

© Springer Science+Business Media Singapore 2017

J. Krishna, *The Indian Mesozoic Chronicle*, Springer Geology,
DOI 10.1007/978-981-10-2477-1_3

succession of six super/mega-sequences by Sloss (1963) as Sauk, Tippecanoe, Kaskaskia, Absaroka, Zuni and Tejas, so named after the North American aboriginal tribes. The Late Permian–terminal Cretaceous mega-sequence comprised the Absaroka and Zuni sequences with further differentiation as under

1. UPPER ABSAROKA (UAA 1-4)
2. UPPER ABSAROKA (UAB 1-4)
3. LOWER ZUNI (LZB 1-4)
4. UPPER ZUNI (UZA 1-4)

The Exxon group of Vail and colleagues developed the subject in the 1970s and 1980s (Vail et al. 1987; Haq et al. 1987 and others) in context of the study of seismic profiles related to the exploration of hydrocarbons in the Euro-American world. In the early decades, the sequence stratigraphic studies, in context of the hydrocarbon exploration were mostly confined to Cenozoic well sections. Obviously, the developed sequence stratigraphic formulations were also, mostly, limited to the Cenozoic well geology. Yet, adequately refined bio/chronostratigraphy always remained an essential prerequisite to any worthwhile sequence stratigraphic investigation. The availability of the excellent ammonoid-based stratigraphy in the West European exposed Mesozoics subsequently allowed the rapid expansion of the sequence stratigraphic studies to shallow marine outcrops and exposed sections since the mid-1980s (Haq et al. 1987). Over the last few decades, the best-resolved sequence stratigraphy in Europe is being realized in the Jurassic exposed geological record (Graciansky et al. 1993; Jacquin et al. 1998a; Jacquin and de Graciansky 1998; Catuneanu 2006 and later). The subject in spite of the diverse conceptual discrepancies, mostly of academic nature, has emerged as of unparalleled influence and fundamental importance, apart from revolutionizing the hydrocarbon exploratory stratigraphic and geological studies.

3.1.2 Sequence Stratigraphic Studies on GTM

Efforts are known for scattered sequence stratigraphic studies in individual basins of the GTM, largely, restricted to selective time spans. Only a few have endeavoured sequence formulations on a regional basis. Very often, sea level curves developed in a basin have been compared to the Euro-American curves with a covert effort to confirm the applicability of the Euro-American curves in their respective individual GTM basins. Outcrop-based studies have been still fewer and limited in geographical extent. However, as an exception Sharland et al. (2001) made an extremely comprehensive publication in Arabia through composite expertise of specialists working for the hydrocarbon industry in the region. The said work integrated all the available published and unpublished, outcrop, subsurface, shallow and offshore well data.

3.1.3 Update on Sequence Stratigraphic Studies by the Author

The studies began in the mid-1980s. Eustatic curves and sequences were developed initially in the Kachchh basin for stages, extended later in steps to other Indian basins, then to IEAP basins, and yet later to GTM (Krishna and Cariou 1986; Krishna and Pathak 1989, 1991, 1993, 1994; Krishna et al. 1998–2010, 2011). Succession of Indian Jurassic and Cretaceous sequences was developed in an extended West India-High Himalayan region (Krishna 1994 unpublished project report; Krishna and Pathak 1994). The late Middle Oxfordian Schiili Subzone MFS for the entire GTM was précised in Krishna et al. (1998). Succession of ten second-order sequences on the GTM was floated in Krishna et al. (2011). The succession of second-order sequences in the GTM Triassic and Cretaceous has been worked out in the last few years. The composite first-order Mesozoic sequence stratigraphic framework is also realized in Krishna (2012) and Krishna (in Bajpai et al. 2012). Composite application of tectonics, magmatism, sedimentation framework and paleontological data facilitated the differentiation of the entire Phanerozoic into 5 mega and 13 first-order sequences. In the Mesozoic, the first-order sequences are further differentiated into second-order sequences. Many second-order sequences have been further organized into third- and fourth-order ones. The present Mesozoic framework includes 3 first-order and 35 second-order sequences (Figs. 3.1, 3.2, 3.3 and 3.4).

3.2 Elements of the Depositional Sequences

A depositional sequence (DS), since the advent of the subdiscipline, is understood to include a complete sea level/base level cycle. It embodies the genetically related stratal package/s that formed during such a cycle for example the first-order Toarcian–Barremian sequence of the GTM determined in Kachchh and subsequently demonstrated throughout the GTM. The said stratal package corresponds to a cycle of transgression and regression. It comprises an early transgressive component (transgressive systems tract—TST) as the ammonoid rich late Early–mid-Late Kimmeridgian second-order TST of the Ler–Katrol region is in Mainland Kachchh (Fig. 2.36). The said TST is followed up later by a regressive component (regressive systems tract—RST) of the ammonoid scarce late Late Kimmeridgian RST. The sequence is bounded both at start and close by regionally to globally spread subaerial unconformities or their correlative conformities in relatively shallow proximal depositional sites of the individual basins for example the regional subaerial unconformities at the base and top of the Toarcian–Barremian first-order sequence in proximal sites of several GTM basins. Their correlative conformities are featured in relatively deeper distal sites for example in the Spiti basin above the Kioto Limestone and below the ammonoid rich terminal part of the

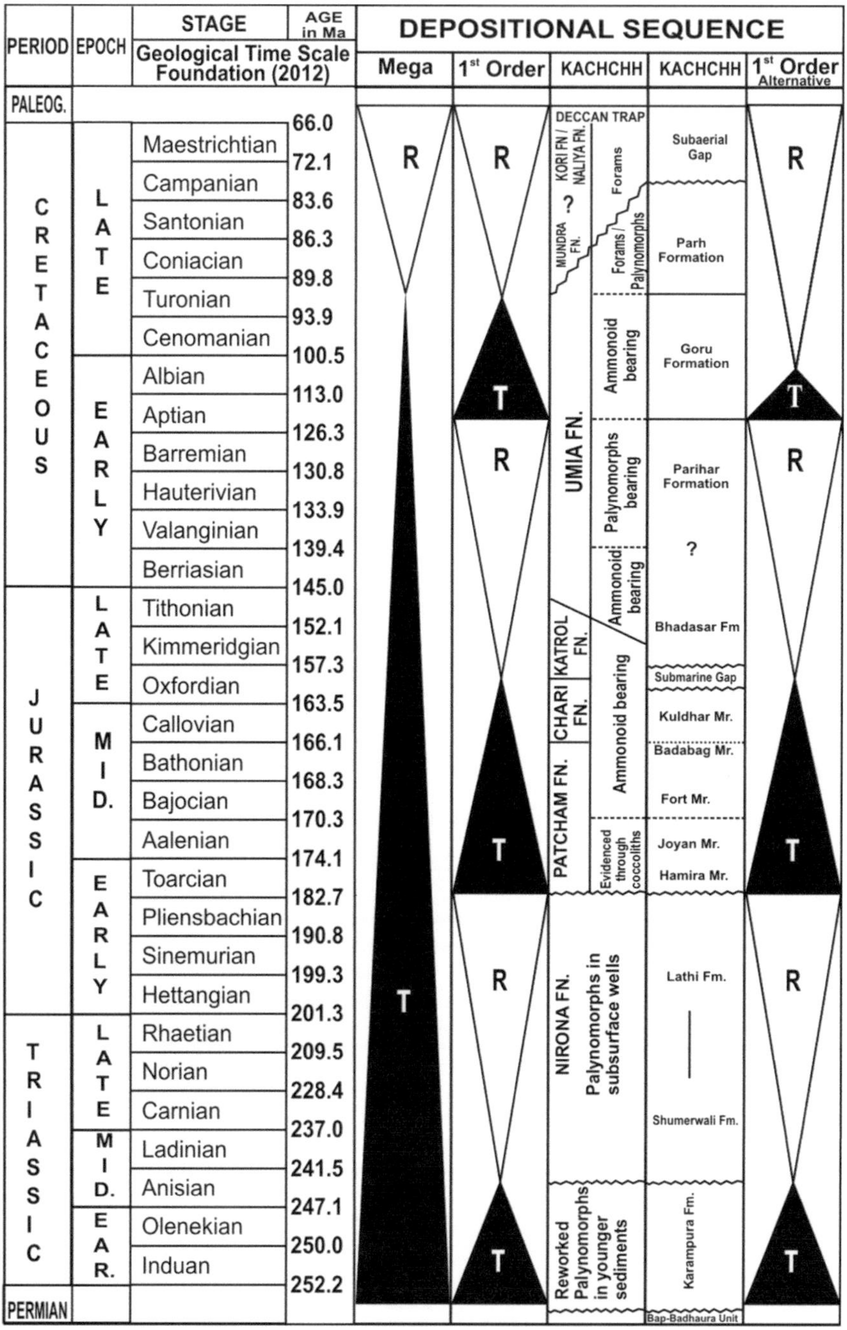

Fig. 3.1 First-order sequences on GTM determined in Indian subcontinent with focus on the Kachchh and Jaisalmer successions

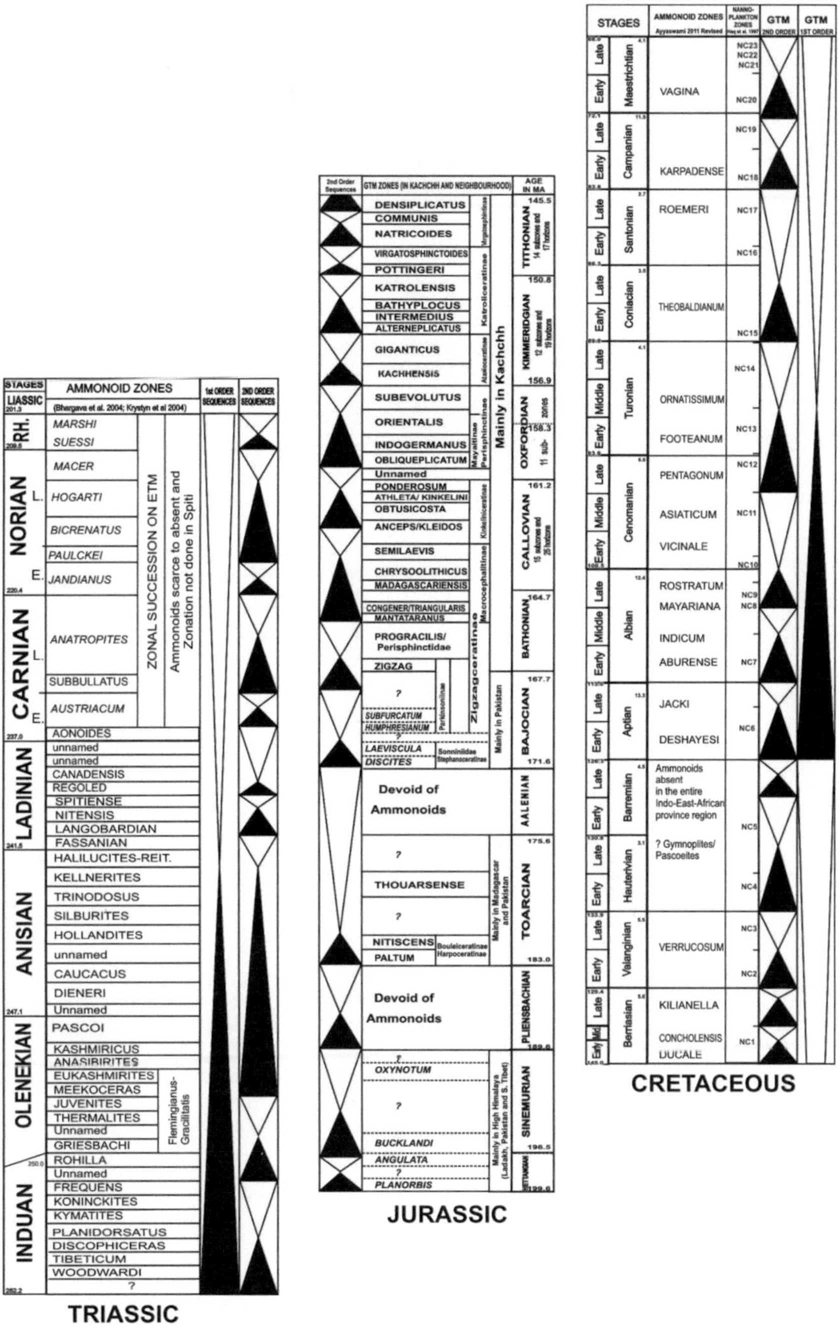

Fig. 3.2 Succession of second-order sequences in the Indian Mesozoic

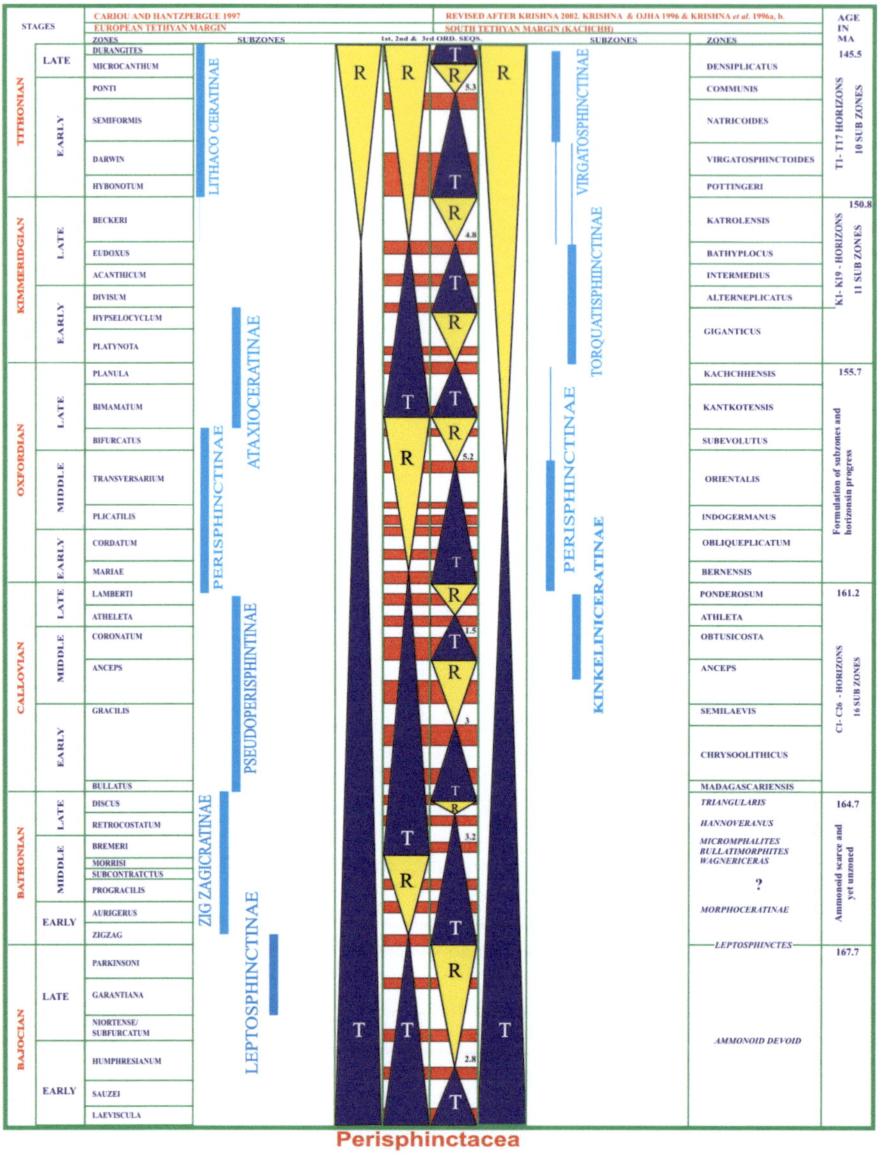

Fig. 3.3 Diachronous first- and second-order and isochronous third-order sequence stratigraphic framework between GTM and ETM (after Krishna 2005)

Giumal Formation. The bounding surfaces of a depositional sequence are termed as sequence boundaries (SBs). Besides the bounding SBs, each depositional sequence includes one additional important surface within the sequence called maximum flooding surface (MFS) at which the transgression and even bathymetry are practically maximum at any depositional site with farthest encroachment on the land of the depositional sequence in question as the first-order late Middle Oxfordian MFS determined for the entire GTM in the Kantkote section of Kachchh (Fig. 1.6).

3.2.1 Sequence Stratigraphic Surfaces, and Types of Unconformities at Sequence Boundaries

The unconformities included near the margin at SB's are of two types. In type 1, the regression moves the shoreline to the shelf edge or beyond exposing the entire shelf to subaerial processes involving significant erosion and subsequent development of the LST. In type 2, only a part of the shelf is exposed through regression of a nearly flat shelf. It does neither include significant erosion nor the development of LST. Instead parasequences continue to develop one atop the other, however changing from a progradational to agradational stacking pattern. Most of the unconformities discussed in the present work are of type 2, for example the second-order SBs in the Bajocian–Tithonian interval in Kachchh.

3.2.2 Shoreline Migration Processes

The mechanism responsible for the migration of shoreline involves important phenomena (Fig. 3.4); (1) Transgression, i.e. landward migration of the shoreline on account of net increase of bathymetry in the basin, (2) Regression, i.e. basinward migration of the shoreline due to net decrease of bathymetry in the basin, (3) retrogradation, i.e. landward migration of the shoreline due to decline in sediment supply leading to the retreat of land from the sea by slow to non-deposition, erosion, compaction or subsidence, (4) progradation, i.e. basinward migration of the shoreline due to increase in sediment supply leading to the advance of the land towards the sea through rapid sedimentation. The landward migration of the shoreline is splendidly displayed by the shift of shoreline to the extreme margin of the basin at the first-order intra-Oxfordian MFS in Kachchh, and basinward retreat of the shoreline through fast prograding sands during the first-order terminal Middle Oxfordian–Late Barremian RST.

3.2.3 Terminology and Definitions

Most of the terminology related to the depositional sequences has originated in context of the studies of seismic profiles, and is readily available in numerous

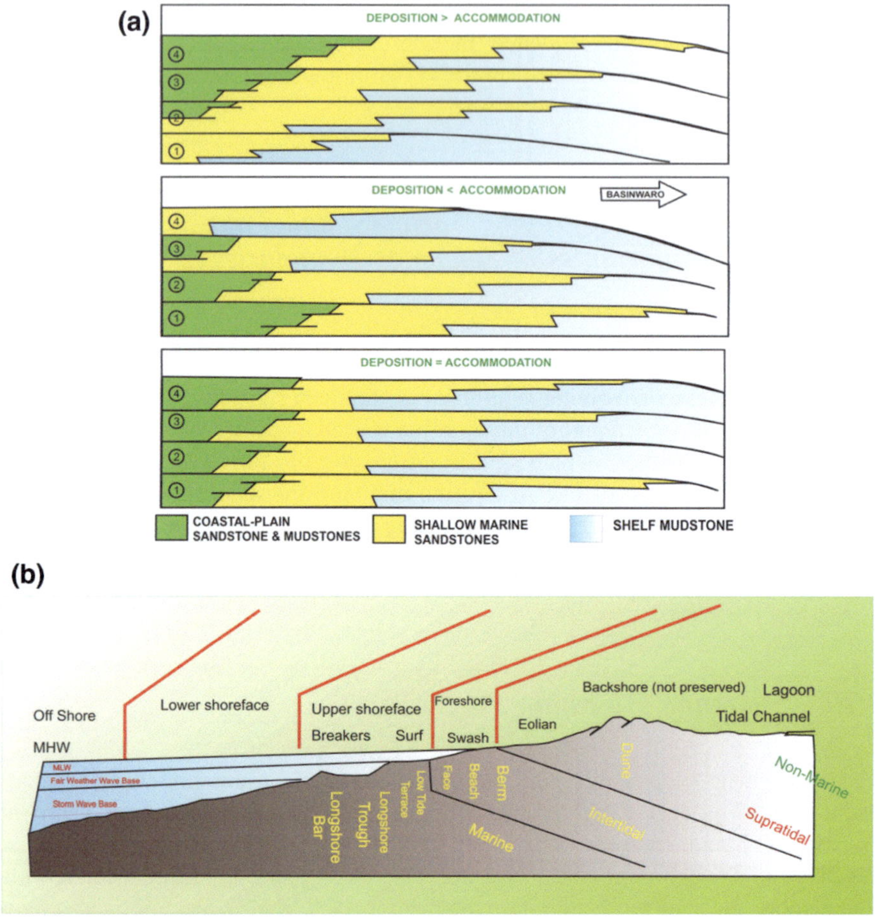

Fig. 3.4 **a** Stacking patterns of parasequences **b** shoreline depositional processes (modified after Van Vagoner et al. 1988)

publications. But for the widely applied basic terms, the detailed terminology is not that desperately essential for outcrop-based sequence stratigraphic studies, yet the same is of substantive practical advantage at least to the beginners in the trade. To me a depositional sequence includes sediments that are laid during successive phases of a single sea level/base level cycle irrespective of size and span of the sequence. More appropriate references to base level changes do involve conceptual progress (Catuneanu 2006 and many references there in), however, there is hardly any significant difference to the practical application and over all stratigraphic and geological comprehension.

3.2.4 *Parasequences as Building Blocks of the Sequences*

The parasequences are the shoaling upward stratal units as the fundamental building blocks of the sequence and the same within a sequence are included as parasequence sets (Fig. 3.4). The siliciclastic parasequences are invariably progradational with an upward coarsening facies association. The progradational pattern ends with increase in the net bathymetry and the resulting transgression. The parasequence is then transgressed by the MFS, and is readily recognizable in outcrops and well sections.

3.2.5 *Parasequence Stacking Patterns*

A parasequence set is made up of several parasequences having similar genetic relationship and pattern. Three types of vertical organizations of parasequence sets are included: (1) backstepping or retrogradational when accommodation is greater than sediment input, (2) vertical or agradational when accommodation is less than sediment input. The above three stacking patterns of parasequence sets collectively constitute the systems tracts. The increase in net bathymetry causes transgression and the preceding prograding parasequence sets move towards the margin. The retrogradationaly stacked parasequence set is called the TST. As the rise in sea level slows subsequently, it reaches a high stand. The shoreline transgression is maximized, and henceforth the parasequence stacking pattern first becomes agradational and later progradational which is termed as the RST. Thus systems tracts are parasequence sets with a common stacking pattern.

3.2.6 *Subaerial and Submarine Stratigraphic Gaps*

The SB near the margin includes subareial stratigraphic gaps related to decrease of net bathymetry and resulting shallowing and emergence. The subaerial stratigraphic gaps underlie and overlie shallow water, coarse textured, high energy, above wave base, rapidly sedimented facies rich in relatively large size physical structures, large size thickening and coarsening upward sequences. These gaps in general also lack nectic and planktic fossils. If at all present, the said fossils are poor in density, diversity and frequency, for example the subaerial gap on the GTM (Fig. 3.8) near the Pliensbachian/Toarcian boundary (Fig. 3.5a), and in the Indian east and west coast basins around the first-order Barremian/Aptian boundary SB (Fig. 3.6). In contrast the less known and less understood submarine stratigraphic gap is included close to the MFS and is present in the basinal part away from the margin for example in the

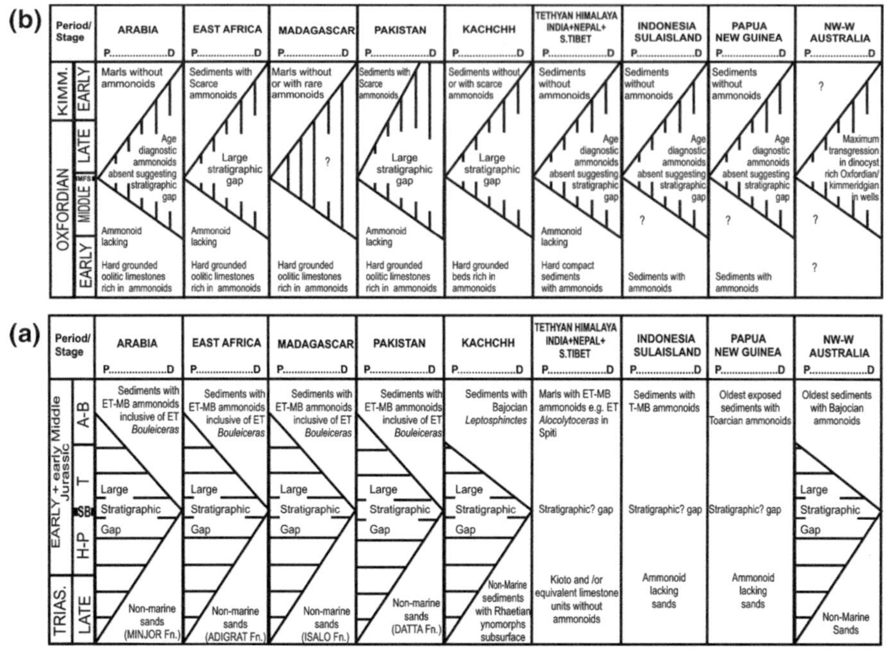

Fig. 3.5 Schematic depiction of widespread first-order subaerial **a** and submarine **b** gaps on GTM between Arabia and Australia after Krishna et al. (2011)

different regions of Pakistan, Arabia, East Africa and Madagascar (Figs. 3.5, 3.7 and 3.8). It results from net increase in bathymetry and rapid deepening. The submarine gaps underlie and overlie sediments that are relatively much less shallower, fine textured, low physical energy, biologically well reworked, slow sedimented, structure less, often nodular/pebbly/concretionary/conglomeratic, thinning and fining upwards, rich in nectic and planktic fossils with high density, diversity and frequency for example the large submarine gap below and above the first-order MFS on GTM, beautifully exampled in Kachchh and other IEAP regions (Figs. 3.5b, 3.7 and 3.9).

3.2.7 *Bathymetric Changes in Shallow Marine Near Coast Systems*

The net bathymetric changes in the shallow shelf marine near coast framework are the outcome of the combined interaction of sea level or base level changes and are global in character, with some basinal to regional parameters as vertical tectonics inclusive of sediment weight related submergence, and sediment input comes from the neighbouring hinterland. The above three parameters can unidirectionally combine to make rapid bathymetric changes, particularly increase in bathymetry for

Fig. 3.6 Regionally widespread subaerial gap on GTM at the Barremian–Aptian first-order SB

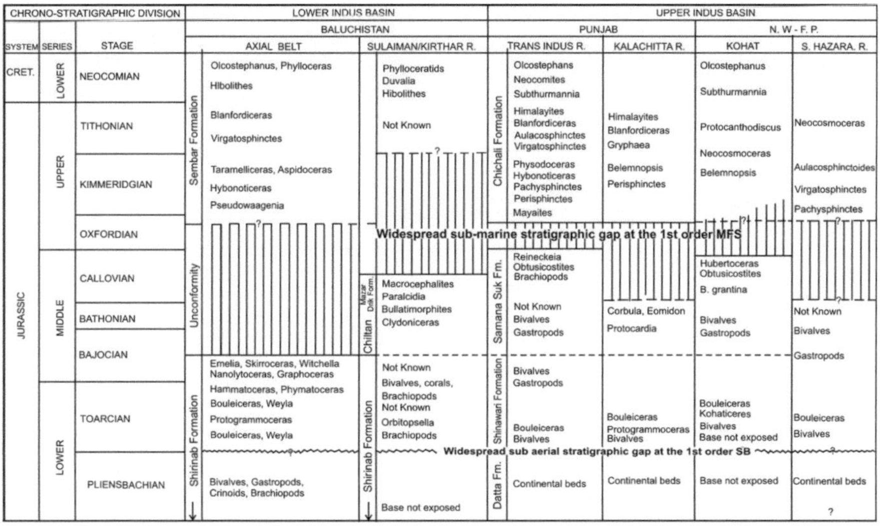

Fig. 3.7 Widespread subaerial (at the first-order SB) and submarine (at the first-order MFS) stratigraphic gaps in Pakistan determined in the table after Fatmi (1984)

example in the late Early to late Middle Oxfordian on GTM by several tens of meters. Increase in bathymetry is equal to net subsidence if sea level remains static. Increase in bathymetry equals the eustatic rise in absence of any subsidence, tectonic or non-tectonic. Bathymetry can increase, even in case of eustatic fall and can decrease even in case of eustatic rise as during terminal Middle Oxfordian to Late Kimmeridgian on GTM, and in turn bathymetric changes in shallow marine framework lead to transgressions and regressions of shoreline.

The distribution and geometry of sediments, over and above the foresaid parameters, is also crucially influenced by the nature and relief of the basin floor besides the influence rendered by prevailing climate.

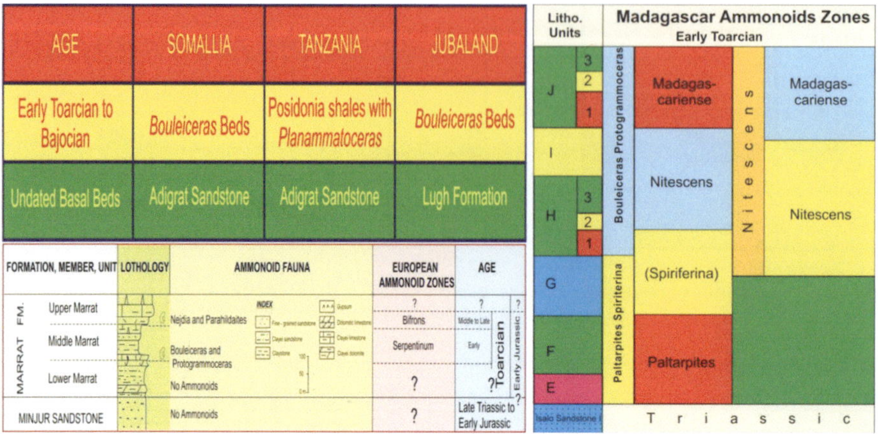

Fig. 3.8 Widespread sub-aerial gap at the intra-Jurassic first-order SB in Madagascar, East Africa and Arabia (modified after Blaison 1967; Arkell 1956; Sharland et al. 2001)

3.2.8 Sequence Surfaces Versus Spatiotemporal Distribution of Ammonoids

In sequence stratigraphic context, it is observed that thin slow sedimented low energy below wave base subtidal fine textured ammonoid and other body fossil rich nodular/pebbly/concretionary intervals constitute the TSTs, as exampled here by the upper part of Muschelkalk Member in Spiti (Fig. 2.9), and terminal part of the Jhadasa Member at Kantkote in Kachchh (Figs. 2.34 and 2.35), while the following ammonoid scarce/devoid high energy above wave base coarse textured rapidly sedimented structure rich thick sediment intervals make the RSTs as exampled by the RSM of Chari Formation in Mainland Kachchh (Fig. 2.28). The presence of long lineages allow the development of better scales, for example the Bathyplocus lineage of Katroliceratinae in the early and mid-Late Kimmeridgian in the Ler–Katrol region of Kachchh (Figs. 6.9 and 6.10).

3.2.9 Asymmetric Sequences in Context of Thickness

In extreme marginal shallow conditions, the sequences, in context of thickness are highly asymmetric, and it is relatively easier to determine and differentiate the thin fossiliferous, even ammonoid-devoid TSTs, as fine as finer fourth- to sixth-order, for example in the terminal Barremian in West Mainland during the terminal phase of a first-order RST (Fig. 2.41, Krishna 2000).

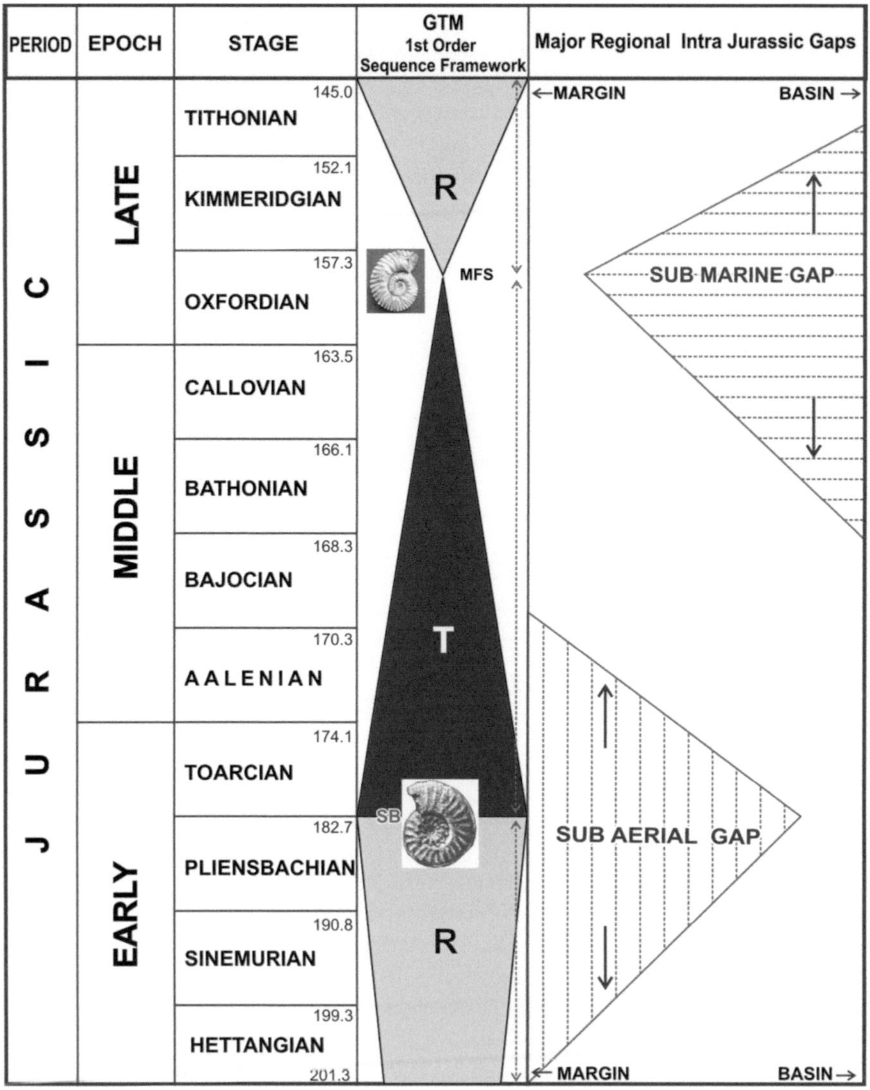

Fig. 3.9 Major regional intra-Jurassic gaps on GTM after Krishna et al. (2011)

3.2.10 *Bathymetric Changes and Ammonoid Heterochronic Evolution*

The morphological evolutionary changes in the dominant unidirectional lineages result in unidirectional changes in morphological parameters like the ribbing, whorl section, coiling during the second-order TSTs in phase of graded yet unidirectional bathymetric rise. The study of heterochronic evolutional changes also thus directly relate to bathymetric change which in sequence stratigraphic context take place either at the MFSs or SBs, facilitating in turn the determination of MFSs and SBs. Sudden change in the heterochronic framework, for example from peramorphism to paedomorphism or vice versa takes place either at MFSs or SBs. The origin, evolution and extinction or drastic reduction of the ammonoid lineages are found to take place at MFSs and SBs. The changes in density, diversity and frequency of ammonoids like lithology, sedimentology, physical and biogenic structures follow predictable patterns both temporally and spatially across the MFS and SB or between margin and basin within TST or RST for determination of MFSs/SBs and differentiating TSTs and RSTs. On the contrary, in relatively low latitude subtropical to tropical conditions, the ammonoid faunas are made up of several stocks and their responses to environmental changes are not similarly revealing enough to decipher any such direct responses and relationships. The influence of regional extensional tectonics and the resultant bathymetric changes on ammonoids in particular and organic evolution in general has been profusely exampled in Chaps. 6 and 7.

3.2.11 Sediment Accumulation Rates, Facies and Their Depositional Environemnts

The significance of sequence stratigraphy stems from its crucial influence on the distribution of sediments in space and time, and their relationship to geological phenomena and processes. The distribution of sediments is a function of changes in shelfal accommodation and sediment supply. The TST's are invariably marked by slow sediment accumulation rate, that is evidenced in condensed to starved facies particularly near the MFS as exampled by the members DOM and JM of Chari Formation in the terminal part of the first-order TST of late Early to late Middle Oxfordian duration (Fig. 2.34). The RST's have moderate to rapid sedimentation rates as in the Ghuneri Member in Kachchh during the terminal Hauterivian–Barremian of the first-order RST (Fig. 2.41).

The recognition of the depositional environments and subenvironments is based on the study of the spatiotemporal litho/bio facies relationships through interpretation of physical structures, biogenic features, body fossil distribution, sedimentology, etc., mostly in the field. The marine facies include offshore muds and shallow marine sands. Slow sedimented starved to condensed sediment intervals rich in planktic and nectobenthic micro/macro fossils are significant and very revealing, so also are the rapidly sedimented intervals.

3.2.12 Biotic and Abiotic Parameters

The parameters (Figs. 3.10, 3.11 and 3.12) applied mostly are field observable ones like density, diversity and frequency of fossils particularly ammonoids; ranges of index taxa and lineages; benthic/planktic/nektic, dino/pollen-spore, epi/infauna, fauna/flora and ichno/biofacies (Fig. 3.10); heterochronic evolution, differentiation

ABIOTIC PARAMETERS OBSERVED

DISTRIBUTION AND SPATIO-TEMPORAL MIGRATION OF

- GROSS LITHOLOGY, BED CONTACT, BED GEOMETRY, PETROGRAPHY, PHYSICAL STRUCTURES, TEXTURE,
- PEBBLES / NODULES / CONCRETIONS,
- CARBONATES / NON-CARBONATES, OOLITES / GLAUCONITES,
- THINNING/THICKENING, SHALLOWING/DEEPENING, FINING/COARSENNING
- PHYSICAL ENEGRY FRAMEWORK, BATHYMETRY,
- SEDIMENT ACCUMULTION RATE, SLOW / CONDENSED / STARVED TO MODERATE TO RAPID SEDIMENTATION,
- SHALE/SAND
- SITE AND ENVIRONMENT

Fig. 3.10 Abiotic parameters observed in the field for the determination of sequence surfaces

BIOTIC PARAMETERS OBSERVED

DISTRIBUTION AND SPATIO-TEMPORAL MIGRATION OF

- MACROINVERTEBRATES, PLANTS, VERTEBRATES; MICOFAUNA AND FLORA,
- NECTOBENTHIC, NECTIC AND PLANCTIC FOSSILS
- BENTHIC CALCAREOUS/SILICEOUS,
- AMMONOIDS, NAUTILOIDS AND BELEMNOIDS.
- AMMONITINA, PHYLLOCERATINA LYTOCERATINA, AND ANCYCLOCERATINA
- AMMONOID MORPHOLOGY, DENSITY, DIVERSITY, AND FREQUENCY.
- GEOGRAPHICALLY RESTRICTED VIS-VIS-VIS WIDELY DISTRIBUTED
- SIZE, PRESERVATION, TAPHONOMY.
- INVOLUTE/ EVOLUTE ; INFLATED/COMPRESSED ; SMOOTH /ORNAMENTED
- HETEROCHRONIC EVOLUTIONARY PROCESS IN THE LINEAGE PERAMORPHOSIS- PAEDOMORPHOSIS

Fig. 3.11 Biotic parameters observed in the field in sequence stratigraphic studies

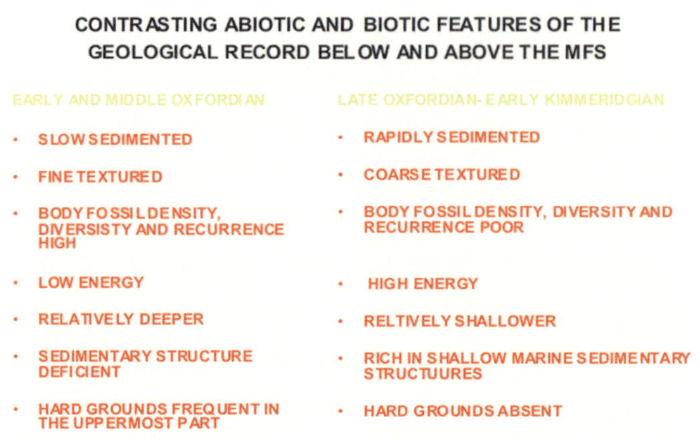

Fig. 3.12 Contrasting abiotic and biotic features of the geological record below and above the MFS

between submarine and subarial stratigraphic gaps and their varying durations in different parts of the basins, also field observable abiotic features (Fig. 3.10) like lithology, texture, bed geometry, contact and size, structures, presence of pebbles/nodules/concretions, hard grounds, etc. Applicability of the sequence formulations heavily depends upon meeting the specific crucial test/requirement of regional (up to second order) to global (third and finer order) recognition and correlatibility of the sequence surfaces.

3.2.13 Refined Chronostratigraphy—An Essential Requisite for Sequence Stratigraphic Studies

The depositional sequence apart from including the guide fossil-based stage/zonal/subzonal/horizonal boundaries, also provide a few near isochronous surfaces, i.e. the SB, MFS, flooding surface and transgressive surface which are correlatable for large even intercontinental distances. It may be emphasized that the SBs and MFSs are natural chronostratigraphic boundaries having greater practical utility in geological interpretations compared to the not that natural arbitrarily chosen system/stage/zonal boundaries. The sequence SBs and MFSs together with guide fossil based chronostratigraphic boundaries strengthen and further improve the stratigraphic refinement and precise correlations across large distances. Yet, high-resolution sequence stratigraphy cannot be realized without the high-order guide fossil-based chronostrtigraphic refinement, and in turn high-order chronostratigraphic refinement cannot be developed without detailed centimetre level lithostratigraphic differentiation in the field. It does also improve stratigraphic refinement and correlations. Thus, these two types of stratigraphic studies in a large measure complement each other.

3.2.14 Sequence Stratigraphy as Here Understood

Outlining the author's understanding of sequence stratigraphy, it is not much different from a highly refined and micro-chronostratigraphically controlled realistic integrated comprehension of the rock relationships of genetically related cyclic stratal units that are based on inputs from tectonics, igneous events, eustatic fluctuations, sedimentology, paleontology and stratigraphy, climate and others.

Such sequences are classified into mega, first to third order or even finer fourth to sixth order varying in duration between a few hundred my to $\sim$20 ky, however the fundamental unit is a third-order sequence which is invariably of less than $\sim$1.5 to 2 my duration.

3.2.15 Rationale of Sequence Stratigraphic Studies

Stratigraphy is a synonym of integrated geology to which maximum possible realism and objectivity can only be brought about through sequence stratigraphy on intrabasinal to regional to global scale. It has been demonstrated earlier (Krishna et al. 2000a onwards) that the geological record (shallow marine or deep, exposed or unexposed, clastics or carbonate) includes succession of depositional sequences of different orders, magnitudes and durations that are either correlatable regionally (longer than 2 my duration sequences controlled by regional tectonics) or globally (smaller than 2 my duration sequences governed by manifestations of Earth's orbital dynamics like the global sea level changes). Being the considered products of regional tectonics, the ages of the sequence surfaces of the longer duration sequences provide precise ages of the tectonic events that supposedly produced them. Sequence surfaces in basins having high potential of geological record can be dated to precision of $\sim$200 ky which is much more precise to ages of tectonic events arrived at any time earlier through any other technique. Moreover, the age is much more authentically linked to a regionally spread event with resultant litho-bio-chrono-radio-magneto-sequento expressions and associated mutually interlinked responses.

3.2.16 Uniformity of Sequences in Broadly Homogeneous Tectonostratigraphic Regions

Another important premise of the sequence formulations is their essential uniformity throughout a broadly homogenous tectonostratigraphic region like GTM and ETM. This in turn allows to date a sequence surface in one section or basin of high stratigraphic potential and its subsequent application in some other distant basin of the broadly uniform tectonostratigraphic region, e.g. the interpretation of marine transgression having begun in Kachchh in early Early Toarcian at a first-order SB that is already determined in Madagascar (Krishna et al. 2011), or the drowning of carbonate platform in High Himalaya at the same first-order intra-Jurassic SB.

3.2.17 Temporally Changing Geographical Extent of the Broadly Uniform Tectonostratigraphic Regions

The extent of the broadly uniform tectonostratigraphic region/s may change in successive first/second-order sequences. The extent, until near Barremian–Aptian major sequence boundary, includes entire GTM from North Africa to NW Australia. There after until the Middle Turonian intra-Cretaceous first-order MFS, the extent is reduced merely to a loosely held Axial Gondwana of India, Madagascar and Africa, while for the rest of the Cretaceous, the extent is further reduced to oceanically isolated India inclusive and exclusive of Madagascar.

3.2.18 Sequence Stratigraphic Studies in the Indian Geological Record

The first detailed effort in this context in recent decades has been by Shanker et al. (1994) who differentiated 15 major super sequences in the post-Archean geological record of Indian Himalaya. Yet earlier, Krishnaswami and Swaminath (1965), Srikantia (1977), Bhargava (1987) and Bhargava and Bassi (1996) have advanced the cause of the differentiation of the High Himalaya geological record into sequences based on regionally widespread marker beds, stratigraphic gaps, marked change of facies, and depositional environments. Also propositions are made in this regard in specific sections and basins studied by them. Shanker et al. (1994) demonstrated the succession of their sequences (sequences VI–IX) spanning through Early Permian to Early Cretaceous also in the Indian Gondwanian geological record. In comparison, the mega, first- and second-order sequences developed are well defined regionally. Among the five mega-sequences in the

Phanerozoic, the one studied is geologically constrained with the intra-Permian origin of the Neotethys and closure with disappearance of the said ocean in Paleogene. It includes three first-order sequences and ~35 second-order sequences (Figs. 3.1 and 3.2).

3.2.19 *Applicative Attributes of the Sequence Stratigraphic Studies*

Sequence stratigraphic studies have excellent potential not only for fundamental progress in geological comprehension at substantively improved resolution, but also for diverse economic geology exploration; hydrocarbon, coal, atomic minerals and others. It is most applied to hydrocarbon exploration for (1) age determination in the seismic profiles to systems level resolution, (2) improvement in facies identification in unknown stratal columns, (3) differentiation of source and reservoir rocks, (4) sedimentation modelling and basin analysis, (5) facies correlation and interpretation of environments, (6) strengthening chronostratigraphy and many others. These applications in turn lead to not only new discoveries of hydrocarbons but also enhanced recoveries from long exploited fields. It is an excellent process and response-based predictive tool as to where to find what kind of facies and sediments for selection of well sites. Apart from intrabasinal comprehension, sequence stratigraphic formulations are also applied to plate-tectonic interpretations.

3.2.20 *Isochronous Surfaces of the T/R Sequences*

The present study is largely restricted to shallow marine systems hardly ever going beyond mid-shelf. The study also is invariably outcrop-based. Thus maximum transgression involves maximum bathymetry, and maximum regression implies minimum bathymetry, there by implyiing practical geological isochroniety of SBs and MFSs. A T/R sequence in its basic simple organization is made up of a transgressive component, i.e. the trangressive systems tract (TST) corresponding to a phase of net increase in bathymetry in the basin which is followed up by the regressive component, i.e. the regressive or high-stand systems tract (RST/HST) related to the phase of net decrease in bathymetry. Minor conceptual differences are made between RST and HST, which appear to me more academic. The upper surface of the TST representing the maximum bathymetry and also maximum transgression is termed MFS, while the surface related to maximum regression is known MRS or SB. The individual sequence is understood as a genetically related stratal package bounded by subaerial unconformities near the margin or their correlative conformities in the basin. The seismic/well-log specialists, however, subdivide the TST into an early low-stand systems tract (LST) and later TST at a surface termed as the transgressive surface (TS). However, the author is personally more comfortable with the much simpler undifferentiated TST in his outcrop-centric studies, and sequences which conform to T/R couplets. The differentiation within the TST is particularly relevant in the hydrocarbon exploration based on well-log/seismic studies. In the present step by step approach, the target is translation of the first- and second-order sequence surfaces MFSs/SBs into well logs and seismic profiles. Once that is realized, it may be easier to constrain additional surfaces within the TSTs in conjunction with well log and seismic profile specialists. Moreover, LST is formed only in rare cases when the entire shelf gets exposed prior to a fresh transgression.

3.3 The Mega Mesozoic Sequence and Its Three First-Order Sequences

The Mesozoic mega-sequence conceived and formulated earlier (Krishna 2005; Krishna et al. 2011) (Fig. 3.1) begins with intra-Permian origin of the Neotethys to its intra-Paleogene closure. It includes three first-order sequences that are punctuated by two SBs and three MFSs in the Mesozoic. The SBs, respectively, are near the base of Toarcian determined in Pakistan and Madagascar in the immediate erstwhile neighbourhood of Kachchh, and near base of Aptian determined in Kachchh itself while the MFSs are intra-Anisian determined in Spiti, intra-Oxfordian in Kachchh, and either intra-Albian developed better in the west and north or intra-Turonian in east sector basins. The Albian–Turonian duration appears to have maintained the highest Mesozoic sea levels with only slightly inferior levels during Cenomanian, so interpreted from the corresponding geological record in the Indian basins.

There is further differentiation of the first-order sequences into second-order (Fig. 3.2) and third-order sequences (Fig. 3.3), and occasionally even into fourth and fifth-order sequences (Fig. 2.36) as discussed in detail later in this chapter. The differentiation in the Triassic is done exclusively in Spiti Himalaya. The most detailed formulation is developed in the Kachchh Jurassic. The Cretaceous is studied compositely in different basins. The guide fossil-devoid Kachchh Late Albian–Turonian is only homotaxially age-constrained in context of the guide fossil-based differentiation elsewhere in Cauvery, Spiti and Jaisalmer.

The varied gaps within the mega-sequence in Kachchh are interpreted both in outcrops and subsurface wells. In Kachchh, among the three first-order sequences, the first one is expressed only in bits in the wells as also on reworked palynomorphs in younger exposed sediments. The second one of the Toarcian–Barremian duration is almost fully and splendidly exposed. The third is compositely expressed in exposures (TST and perhaps early part of the RST) and wells (later part of the RST).

In the Kachchh Mesozoic sequence framework, the five first-order SBs near the margin are included in subaerial gaps (Figs. 3.1 and 3.2). The first first-order SB is between the Precambrian basement and the interpreted Late Permian so suggested through reworked palynomorphs. It is linked to the rifting/spreading in the north sector that marked the start of the mega-sequence with origin of the Neotethys. The Kachchh region prior to its rift-genic origin of the basin may have sagged as suggested for the Gondwana basins during the Late Permian for the sporadic deposition of the evidenced non-marine Late Permian. The sags may not have been soon followed by rifting of consequence. The true rifting may have occurred only later at the Middle Triassic first-order MFS as evidenced by reworked Late Permian sediments in wide apart locations of the basin. Reworked Late Permian non-marine sediments are recorded in Pakistan, Kerala–Konkan, Cauvery and elsewhere. The Early/Late Permian boundary event is widely present in the Indian basins all round the Indian plate in east, west, north and south through diverse geological changes with or without unconformity (Fig. 2.4). The igneous activity is largely restricted to the north sector in High Himalaya. The second first-order SB is between the subsurface non-marine Nirona Formation of the Ladinian–Pliensbachian age-bracket and the unconformably overlying marine Patcham Formation of Toarcian–Bathonian age, and is linked to the initial rifting/spreading between the East and West Gondwana both at the north and south extremes of the sector. It is supposedly triggered by Karoo and coeval volcanism in the west sector with localized expression in other sectors as well. The third first-order SB is interpreted between the Ghuneri and Ukra units in West Mainland, which marked the initiation of the Rajmahal igneous event in the east sector with coeval igneous expression in other sectors inclusive of Kachchh. The terminal first-order SB is placed near the close of the Madh Formation of early Paleocene age, while the first significant impingement of the rapidly drifting Indian plate on Asia with major shrinking of the Neotethys is determined in Late Ypressian. Among the three first-order MFSs, the first coincides with the origin of the rift basin, and is interpreted between the non-existent unknown pre-rift source sediments of the reworked Late Permian palynomorphs below and subsurface non-marine Late Triassic–Liassic Nirona Formation above. It is linked to the origin of Kachchh, and Saurashtra basins. The second first-order MFS is the most important of all the mega-sequence MFSs/SBs, and is also the best and precisely, and, also, exclusively determined in Kachchh above the late Middle Oxfordian Orientalis Zone Schilli Subzone Subschilli Horizon Bed I/III-08 or the revised KB III-17 in Fig. 3.29 in the Kantkote section of ~159 ma age.

The early Late Jurassic Oxfordian stage according to the GTS 2015 (Cohen et al. 2015) ranges from 163.5 to 157.3 ma, and has duration of 6.2 my. In the Neotethyan realm, the Oxfordian is currently organized into five ammonoid zones with each zone averaging to ~1.24 my. The Orientalis Zone thus ranges from ~158.44 to ~159.68 ma. The late Middle Oxfordian Orientalis Zone is further differentiated into four subzones, and thus the top of the Schilli Subzone is best assigned to ~159.06 ma. The Schilli Subzone includes a succession of 13 hardgrounds, and among these the eleventh from the base or third from the top makes the MFS. Each of the 13 hardgrounds constitutes a sedimentary cycle of ~200 to 400 ky and in this case of ~200 ky. Further differentiation approximates the MFS between 158.9 and 159.0 ma. Thus, the best approximation is $\sim$159 ma for the MFS. However, there does not exist any approximation in the late Middle Oxfordian Orientalis Zone Schilli Subzone Subschilli Horizon Bed I-08 subzonal/horizonal precision, while the radiometric dates vary with technique, sample and margin of error of at least plus/minus one my.

The geological event dated through the intra-Oxfordian MFS is manifested through its innumerable facets that are discussed later in Chap. 7. The intra-Jurassic submarine gap is present in all the exposed distally located Mainland sections, and the MFS is determined extreme precisely to horizonal ($\sim$200 ky) resolution (Photo 7.2). It is a real physically touchable and visually observable undulating bumpy nodular classical marker hardground bedding surface.

Third, the youngest as also the last first-order MFS has been difficult to single out and two alternatives are suggested of near equal significance. The older one is placed between the Ukra and Upper member units of mid-Middle Albian age. The MFS is widely recognized in Madagascar, Pakistan, Jaisalmer, Kachchh, Cauvery and Spiti, even in the K-G basin. The age is précised in Jaisalmer above the Abur Formation in outcrops, and between Lower and Upper members of Goru Formation in wells. The younger and somewhat more evident MFS across the globe is the early Middle Turonian one a little below the top of the Karai Formation in Cauvery wells, at the top of the Raghavapuram Formation in K-G, near the top of the Nodular Limestone in Narmada, and also elsewhere in Spiti and Jaisalmer. In Kachchh, the said intra-Turonian MFS as per fission track radiometric date could be at the top of the Upper Member of Umia Formation above the thin glauconitic bed localized in the Ghuneri–Amarsar section. An alternative is also suggested in the Middle of the Upper Member above the bed 23.

In view of the absence of the exposed marine Triassic in Kachchh as also elsewhere except High Himalaya, the sequence stratigraphy in the Triassic is only explored in the Spiti basin of High Himalaya (Fig. 3.13). Earlier Garzanti et al. (1995) have suggested 20 odd sequences, mostly in the relatively well expanded Late Triassic. Bhargava et al. (2004) accepted a few of the SBs of Garzanti et al., however, even in their preliminary synthesis, they spelled out many of their lithostratigraphic formation/member boundaries, about 11 in number, as SBs. Krishna et al. (2010) determined the major first-order intra-Triassic MFS of the GTM in Spiti in the Late Anisian Hollandites Zone–Kellnerites Zone interval between beds 34–40 of the Lalung section (Fig. 2.9) (Bhargava et al. 2004; Krystyn et al. 2004). The determination was based on the ammonoid frequency, diversity and density, proportion of relatively deeper water leiostraca ammonoids, extremely slow sedimentation, vertical concentration of condensed/starved intervals, hard grounds and even possible small submarine gaps, and other characteristic features. The MFI, however, has been ranged between beds 34–40, in which the bathymetric maximum remained somewhat stable. This late Anisian first-order intra-Jurassic MFS on GTM determined in Spiti is distinctly older in age by few my to the corresponding first-order intra-Ladinian MFS on the ETM of Haq et al. (1987). The basal and terminal SBs of the concerned first-order sequence were determined outside the Triassic, respectively, in the Permian and Jurassic.

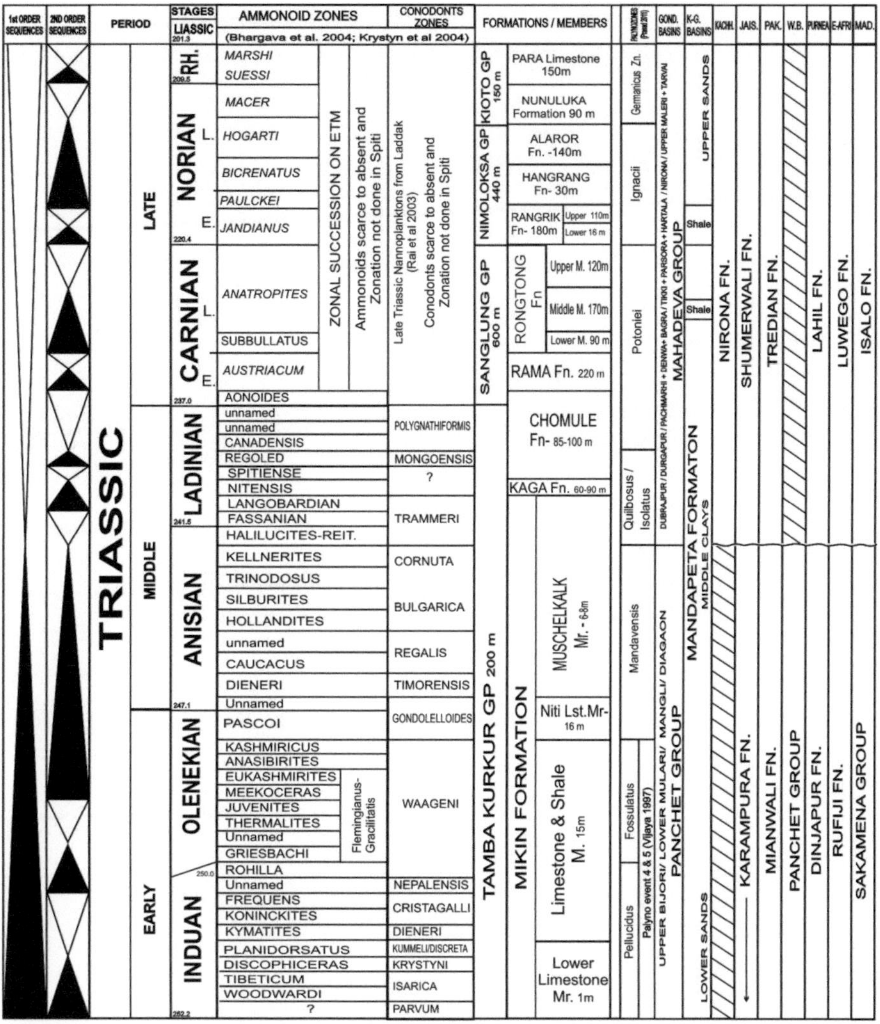

Fig. 3.13 Triassic lithostratigraphic units, ammonoid zones of Spiti and GTM second-order sequences vis-a-vis Gondwana and other units/palynozones

Apart from defining the first-order sequence from the origin of the Neotethys to the drowning of the carbonate platform that included the entire Triassic development in High Himalaya, Krishna, also, for the first time any where on the GTM, differentiated the Triassic succession of GTM into a succession of ten semiformal second-order sequences in the Spiti basin of High Himalaya in India, which is skeltalised as under (Figs. 2.7–2.22). The number of these second-order sequences broadly corresponds to the number of SBs listed in Bhargava et al. (2004), and half of the second-order SBs appear to accord with their suggestions. The formal ordering of the first- and second-order sequence surfaces is introduced for the first time, so also the ammonoid based precise dating and demarcation of the intra-Triassic MFSs.

3.3.1 T-I. Basal Triassic to Intra-induan Frequens Zone (Limestone + Shale Member) Second-Order Sequence

It includes beds 1–11 of the Muth section (Fig. 2.7) with MFS either within or at the top of Discophiceras Zone (below bed GU 4/4).

3.3.2 T-II. Basal Olenekian Rohilla Zone to Intra-olenekian Flemingites-Euflemingites Zone Second-Order Sequence

The sequence comprises beds 12–39 of the Muth section of Limestone + Shale Formation—Fig. 11 of Krystyn et al. 2007 (Fig. 2.7a) with MFS within the bed 15 (Intra-Olenekian Flemingites griesbachi Zone.

3.3.3 T-III. Late Olenekian Meekoceras Zone–Early Ladinian Fassinian Curionii Zone Second-Order Sequence

It is made up of the terminal part of Limestone + Shale Formation, entire Niti Limestone, and up to bed 47 of Himalayan Muschelkalk with MFS/MFI in Hollandites Zone to Kellnerites Zone (Figs. 2.7a and 2.9).

3.3.4 T-IV. Intra-Ladinian Langobardian Gredlerensis Zone–late Early Ladinian Maclerni Zone) Second-Order Sequence

The sequence includes entire Kaga Formation with MFS in or above bed VT 24 of early Late Ladinian Meginae Zone age inclusive of the *Daonella*-bearing (Figs. 2.9 and 2.10) interval.

3.3.5 T-V. late Early Ladinian Sutherlandi Zone–Early Carnian Aon Zone (up to the Level BL 20) Second-Order Sequence

The sequence includes the entire Chomule Formation with MFS in or above the hard ground level BL 12 at Muth 3 at the boundary of Regoled Zone and Canadensis Zone in the terminal part of Ladinian (Fig. 2.11). The demarcation of

Ladinian/Carnian boundary although yet not formalized, differ substantively in context of ammonoid and conodont markers within the Chomule Formation or near its boundary with the overlying Rama Formation.

3.3.6 T-VI. Mid to late Early Carnian Second-Order Sequence

The sequence comprises the 220 m thick Rama Formation with MFI in the siltstone/shale cyclic interval of Rama Formation enclosing *Tropites* and *Spirogmoceras* in late Early Carnian (Fig. 2.12). This is in contrast with Early/Late Carnian boundary as suggested by Bhargava et al. (2004) within Rama Formation. The said boundary is here suggested between Rama and Rongtong units. The interpreted second-order MFS late Early Carnian MFS may also mark the Pan-Tethyan (Hornung et al. 2007) crisis in carbonate production with concomitant increase in the influx of siliciclastics. The crisisful event appears somewhat protracted over couple of million years. Similar crisis in carbonate production has occurred in Jurassic as well at the close of the first first-order sequence above the Kioto Group in High Himalaya, and near the Bathonian/Callovian boundary in Kachchh, Jaisalmer and Madagascar. It is also related to latitudinal positioning of the GTM within specific climatic zones. The GTM sequence framework suggests their occurrence only occasionally at the first- or second-order sequence surfaces often involving regional to global level reorganization of plates.

3.3.7 T-VII. Latest Early to Late Carnian Second-Order Sequence

It includes the entire Rongtong Formation with MFS in *Gonionotites*-bearing calcareous siltstone in the Middle Rongtong Member of mid-Late Carnian age (Figs. 2.13 and 2.14).

3.3.8 T-VIII. Early Norian Second-Order Sequence

The sequence is made up of the entire Rangrik Formation with MFS in ~8 m thick *Juvavites*-bearing shales of late Early Norian age in Upper Member (Fig. 2.15). The Early/Norian boundary is also précised above the Juvavites beds between thick sandstone/siltsone below and thick limestones above, both lying above the

Juvavites-bearing thick shale bed, instead of the same boundary marking the Lower/Middle Norian boundary in Bhargava et al. (2004).

3.3.9 T-IX. Middle to Late Norian Second-Order Sequence

It includes the entire 28 m thick Hangrang Formation plus entire 130 m Alaror Formation and entire Nunuluka Formation with MFS in the *Neotibetites*-bearing shale dominated upper part of Alaror Fn of Late Norian age (Fig. 2.16). The Nunuluka/Para unit boundary is interpreted as a second-order SB in view of marked lithological change from sand dominating to limestone/grainstone dominating units, i.e. from an explicit regressive to transgressive unit. Bhargava et al. (2004) have only tentatively, in absence of any guide fossils, suspected this boundary with question marks which is here supported. In this context, may be referred the two terminal SBs in the Dampier basin wells of NW Australia namely SB2 and SB3 (Westphal and Aignar 1997, Fig. 5.7), respectively, of the Norian/Rhaetian and the Rhaetian/Hettangian boundaries, which further strengthen the ages of the said boundaries.

3.3.10 T-X. Rhaetian Second-Order Sequence

The sequence comprises Para Formation with MFS in the lone shale bed of the unit (Fig. 2.17). Unlike the record of ammonoids of Late Rhaetian in South Tibet (Yin et al. 2007), the Para Formation sequence is only suggestively assigned to Rhaetian. It could be merely a third-order sequence, since in both Spiti and South Tibet, there is post Nunuluka (Norian) graded increase in bathymetry that extends into Early Hettangian. Bhargava et al. (2004) also interpreted disconformity at the T/J boundary (Para/Tagling lithostratigraphic boundary), along with anoxic event between lower and upper parts of the Para Formation, and related distinct deepening.

In view of the focus of the present work, particularly, on the second-, third- and fourth-order sequence description in the Kachchh Jurassic, the detailed description of the Triassic and Cretaceous second-order sequences is deferred for a later occasion through a separate work.

Also, the detailed demonstration of the validity of the Triassic second-order sequences east and west of High Himalaya on GTM is also deferred. However, a summarized comparison of the first-order surfaces and a few significant second-order ones is undertaken in Chap. 5 along with Jurassic and Cretaceous sequence framework.

3.4 Sequence Stratigraphy in the Kachchh Jurassic

The sequence stratigraphic framework in the Jurassic is developed in Kachchh, and there after its applicability is demonstrated in India, in IEAP and GTM. The discussion begins with a brief review of the ammonoid and other guide fossil evidences from Kachchh and neighbourhood.

The oldest ammonoid record in Kachchh is the recently discovered *Calliphylloceras* (Pandey et al. 2013b) of Early Bajocian age from the basal most beds in the Sadhara section of LSM of the Patcham Formation which has been otherwise held younger than Dm. The age relationship between the basal most beds of the two sections of Patcham Formation, respectively, located on northwest and southeast ends of the Patcham hill has also found favour of the recently developed sequence stratigraphic framework (Krishna et al. 2010, 2011 and unpublished).

The record of the first-order sequences is outlined earlier in this chapter. In view of the non-marine subsurface guide fossil-devoid character of the Hettangian–Pliensbachian interval in Kachchh, the said interval of the first-sequence is here organized outside Kachchh in High Himalaya into three second-order sequences. The differentiation is based on the ammonoid records of Hettangian in Ladakh and S. Tibet, and of Sinemurian in Pakistan. Krishna (1987b onwards) based on regional tectonostratigraphic framework has been consistently suggesting the start of the Kachchh marine Jurassic sedimentary succession in Early Toarcian. The three second-order GTM sequences developed in different basins of IEAP in view of poor and disjunct record of guide fossils have only their MFSs determined with some confidence. These are detailed below.

The third sequence is also partially recorded both in outcrops and distally located wells (Raju 2011). Subaerial gap is present between the first two first-order sequences, and submarine gaps are interpreted in the second and third first-order sequences. The intra-Jurassic submarine gap is in the relatively distal Mainland, while the first-order MFS is precised in the proximal Kantkote in Wagad. The intra-Cretcaceous gap is better discerned in the distally located subsurface wells (SP 1-1, K 1-1A and GK 23-1). In the exposed part, the MFS is manifested in thin glauconites in the Ghuneri–Amarsar section where gap is interpreted at least between the glauconites and the Deccan traps (Fig. 2.41). Foraminifers and palynomorphs (Mathur et al. 1970) suggest marine Cretaceous sedimentation through the entire Late Cretaceous duration. If so, these Late Cretaceous sediments are here interpreted to have been eroded away after the Mesozoic along with leaking away of microfossils into older sediments. Any definitive conclusions in such circumstances are difficult to draw. However, evidences from neighbouring Pakistan, Madagasacar, Narmada and even from distant Cauvery support the interpretations made here of bathymetric increase during Aptian to early Middle Turonian duration of the major TST.

The focus in the present work is exclusively on the exposed shallow marine sedimentary succession of the Kachchh Jurassic. The exposition is initiated with history of sequence stratigraphic studies along with outlining of the fundamentals and terminology. Then follow-up the approaches and results in the Kachchh Mesozoic. The sequence stratigraphic formulations are developed through the litho/bio/chronostratigraphy, and concluded with emphasis on salient intrabasinal to global applicative perspectives of the formulations realized.

There have been published several other papers on differentiation of sequences in the Indian Cretaceous geological record mostly from well data, however without ordering and precision determination of the sequence surfaces. The recognition of five cycles in Aptian–Albian and 7–9 MFSs during Late Albian–Late Maestrichtian interval is mentioned in Cauvery basin wells, though not detailed in text (Raju 2008), also any ordering of either the sequences or MFSs is not provided. Comparison therefore of such differentiation with the first to third-order Cretaceous sequences developed in the present work is rather difficult. Govindan (2000), Govindan et al. (1996, 1998, 2000), Nagendra et al. (2002a, b), Nagendra et al. (2011) have registered significant progress in sequence stratigraphic studies in the Cauvery basin. Four major transgressions are recognized by Govindan and Ravindran (1996) in Albian, Coniacian, Campanian and Maestrichtian in Cauvery basin. These efforts however have not been extended to other Indian basins.

3.4.1 Advantage of Relative Subtemperate High Latitude Location of Kachchh for Greater Part of the First-Order Toarcian–Barremian (~56.40 my Long) Sequence

Kachchh is interpreted to have occupied subtemperate latitudes for most of its exposed Jurassic Toarcian–Tithonian ~37.70 my interval. The Bajocian–Bathonian ~4.20 my is represented in carbonate facies; Toarcian to early Middle Bathonian mostly by clastic facies on the margin in island belt. In the Mainland, Bajocian–Bathonian ~4.20 my duration is in mixed carbonate-siliciclastic facies. Basal Callovian onwards up to the end of the Jurassic, there is near complete absence of any carbonates. Thus, Kachchh maintained marginal subtropical to marginal subtemperate latitudes and humid climate except for part or full of Bajocian–Bathonian. The subtemperate ammonoid faunas are known to have dominance of single lineages within a sequence, and invariably just one species of the lineage at any particular time. At any point of time this single species supposedly interacted directly with the prevailing environment, and underwent adaptive evolutionary morphological modifications with change in environment. The relationship is largely direct. Interpretation of the heterochronic morphological changes between the mother and daughter species, thus, provide reliable clues to changes in bathymetry, and associated parameters like energy framework, rate of sediment accumulation, etc. The subtemperate latitudinal setting of Kachchh for most of the Jurassic allowed interpretation of environment, particularly, bathymetry, through heterochronic evolutionary studies of the ammonoid lineages as an additional tool, besides application of field and laboratory sedimentological and paleontological features for reconstruction of environments, and in turn determination of sequence stratigraphic surfaces. In contrast, the subtropical setting and resultant high diversity fauna did not allow the same kind of direct process–response relationship between the ammonoid fauna and environment.

3.4.2 Domination of the Kachchh Ammonoid Fauna by a Single Subfamily in Each Second-Order Sequence

The Kachchh ammonoid faunas are invariably dominated by geographically restricted regional ammonoid GTM subfamilies with only minor/auxiliary/secondary ETM share. The ammonoids are heavily skewed in favour of the regional elements.

3.4.3 Auxiliary/Minor/Secondary Extraneous Elements Crucial for GTM/ETM Correlation

On the other hand, the minor/auxiliary/secondary elements, that are more significant during the terminal TSTs/MFSs close to the eustatic maxima, facilitate precise correlations with ETM as exampled by the ammonoids of the subfamilies Streblitinae, Aspidoceratinae, Hybonotoceratinae in the Late Kimmeridgian in spite of these subfamilies constituting only a small percentage of the over all near exclusive IEAP fauna.

3.4.4 Strategic Significance of the Kachchh Mesozoic

Kachchh is a pericratonic rift basin which originated in the East/West Gondwana divergent fracture region in Middle Triassic as part of the early rifting history of the Gondwana dismemberment. Located in the vicinity of the divergent action theatre, it witnessed the happenings and thus bore imprints and evidences of important geological events. In view of its substantively thick 3000 m plus Meso-Cenozoic sediment pile coupled with its position along the subcontinent's prime hydrocarbon belt in the west sector, it has been for long considered a significant hydrocarbon prospective basin. Fortunately, the basin has in recent years proved commercially exploitable petroliferous basin, to soon start producing hydrocarbons.

3.4.5 Ammonoid Heterochrony also Used in Relative Chronometry

The ammonoid heterochronic evolution and its main processes of paedomorphosis and peramorphosis are also applied for stratigraphic refinement (Figs. 6.6, 6.7, 6.8, 6.9 and 6.10). The increasingly refined ammonoid stratigraphic data on one hand has been used to comprehend high resolution intrabasinal geological dynamics, while on the other hand to age constraint the plate-tectonic history of the Gondwana. The T/R sequence stratigraphic framework summarized earlier (Krishna et al. 2008, 2009, 2010, 2011) is a reflection of intrabasinal geology through precise ammonoid dates.

3.4.6 Geological Dynamics Through Outcrop Based Sequence Stratigraphic Studies in the Indian Mesozoic

The geological dynamics is developed through the succession of thirteen plus outcrop based second-order depositional sequences that are age precised and differentiated in Kachchh and/or neighbourhood. The developed sequence stratigraphic framework, otherwise, has wide applicability on the entire GTM.

3.4.7 Geographically Restricted GTM Ammonoid Subfamilies as Representative of Second-Order Sequences

Apart from other significant credentials, each of the successive second-order sequences is invariably characterized by a single dominant geographically restricted ammonoid subfamily/lineage which is exclusive to the particular sequence with either origin/extinction or sudden marked reduction at either of the sequence stratigraphic surfaces of the particular sequence, at either the MFS or SB (Figs. 6.4 and 6.5). The subfamily/lineage is replaced by environmentally/bathymetrically better adapted subfamily/lineage in the successively younger depositional sequence.

3.4.8 Second-Order and Bigger Sequences Controlled by Regional Tectonics

As earlier interpreted (Krishna et al. 2000a, b, 2009, 2010, 2011), the low-frequency high-magnitude second-order and yet longer sequences (bigger than ~1.5 to 2 my) are consequences of the regional tensional divergent tectonics, and also are net controlled by tectonics. In such sequences the global eustatic influence in terms of magnitude is overweighed by the regional tectonic influence. The bigger/longer sequences have their sequence surfaces only of regional extent. The individual regions of the like of GTM and ETM have sequence stratigraphic surfaces independent of each other within their regional extent. Such regional surfaces have discordant relationship corresponding to those of other regions.

3.4.9 Third-Order and Smaller Sequences Governed by Earth's Orbital Dynamics and so Global

High frequency and low magnitude sequences, finer and smaller than ~1.5 to 2 my, are eustatically guided sea level fluctuations. These sequences have globally (interregionally) extended isochronous sequence stratigraphic surfaces. Thus, within a broadly uniform tectonostratigraphic region, the entire sequence stratigraphic framework from the mega order to third-, fourth- and even fifth-order is same.

In this context, the author suggests relook and review at a few frameworks (Sharland et al. 2001 in Arabia—a constituent of GTM, Alberti et al. (2011, 2012a, b, 2013), Fursich et al. (2013) in Kachchh—a constituent of GTM, Li and Grant-Mackie (1993)—in case of South Tibet—also a constituent of GTM, Geiger (2004) in case of Madagascar—a constituent of GTM, and several other efforts in basins between Arabia and NW Australia), as these often did not lend the desired significance in their sequence stratigraphic studies to the regionally restricted character of the GTM tectonostratigraphic framework.

3.4.10 Origin of the Kachchh Basin and Principal Sequence Surfaces Between the Precambrian and the Time of Origin of the Kachchh Basin

Small shallow East–West directed non-marine basin or basins may have originated as true rifting gradually progressed southward in Kachchh. These enfant basin/s may soon have started receiving non-marine sedimentation as evidenced in the Banni and Nirona wells through the study of palynomorphs of Rhaetian–Liassic age

(Fig. 3.1). The first explicit major sequence surface in Kachchh is thus interpreted between the Precambrian and the subsurface non-marine Nirona Formation sediments of the Banni and other wells of Late Triassic–Early Jurassic age.

3.4.11 SB of the Mega Order Sequence

The mega basal SB necessarily has to be regionally older than the origin of the Kachchh basin, which is otherwise interpreted at the intra-Triassic first-order MFS. It is impossible to date the mega basal post Precambrian and pre- intra-Triassic SB in the Kachchh basin in view of its origin at a later date. It can be dated in the neighbouring Indus basin of Pakistan, Madagascar and even in Spiti where the presence of older marine sediments is known.

The said major SB should be the resultant local expression of a major regional tectonic event. There are two such major events to note that are well evidenced in the geological record. The first is the long spectacular regional gap in all the inland Gondwana basins, also in Jaisalmer, even suspected in Kachchh between Precambrian and Permian and the onset of widespread rifting of presumably basal Permian age. The other event is intra-Permian origin of the Neothethys. The intra-Permian event of the origin of the Neotethys is preferred, since, the mega-sequence is earlier defined from the origin to the closer of Neothethys (Fig. 3.1).

Mention in this context is made of the suspected reworked Permian palynomorphs in the palynomprph yield of the younger Jurassic Katrol Formation sediments by Venkatachala (1970), and Venkatachala and Tiwari (1987) which perhaps escaped notice of researchers, so also is noted the isolated loose collection of the Early Permian brachiopod *Streptorhyncus* in Kachchh by Rajnath (1934) in his bed 22 of the Jumara section. These apparently small records are considered significant in regional tectono-sequento-stratigraphic scenario. If the pollens are held as contaminated, where did the *Streptorhyncus* came from in bed 22 at Jumara ? Reworked Late Permian is also interpreted in Pakistan, Kerala–Konkan (Potonie and Sah 1960), Cauvery and elsewhere, so is Early Permian marine incursion on the Indian plate in all the sectors, west, north, southeast and southwest, even in the inland Gondwana grabens as the widest of all the Phanerozoic marine incursions since the Late Precambrian–Early Cambrian one. The Early Permian marine incursion is present in the Indus basin and West Rajasthan, and Narmada, respectively, north and south of Kachchh which is in complete accord with the evidence in Kachchh. Inquiry is desirable into the Kachchh wells for the confirmation or the rejection of the marine Early Permian evidence. If true, the oldest, but not the most important SB, would be at the base of the Early Permian above the Precambrian. The basal mega SB between the Precambrian and reworked Late Permian has wider extent compared to the same SB between the suspected marine

Permian and the non-marine Permian. The more significant interpretation made is the possible tectonostratigraphic similarity of the Kachchh region with other Indian plate margin basins in the west and east sectors. These basins inclusive of Kachchh with the initial infra-rift sagging of pre-existing weak zones went into sedimentation after a long gap up to the Carboniferous/Permian boundary.

3.5 First- to Third-Order Sequence Stratigraphy in the Kachchh Jurassic

The Kachchh sequence stratigraphic framework demonstrates near perfect accord at the third order with the global Vailean model (Fig. 3.3), yet exhibiting unambiguous disagreements in the first- and second-order sequence structure (Fig. 3.3). The same is explained by the low-frequency high-magnitude character of the basinal/regional tectonics to interfere and distort the first- and second-order formulations. Such low-frequency and high-magnitude intraregional tectonics, however, is unable to interfere with the high-frequency low-magnitude third-order sea level cyclicity, and hence the accord at the third-order. The third-order and yet finer/smaller sequences are otherwise governed by globally pervading Earth's orbital dynamics.

As many as ten plus second-order sequences have been differentiated into the Jurassic of Kachchh and neighbourhood (Krishna et al. 2011). There are differentiated three additional second-order sequences immediately below the said ten plus second-order sequences in the High Himalaya basins of Spiti and Nepal as under, which although present in Kachchh wells, could not be differentiated in the coeval non-marine record in the Banni and Nirona wells of Kachchh. Thus the total number of sequences in the GTM Jurassic in the present work stands at 13 plus with possibilities of further addition through anticipated split of the present Toarcian–Aalenian second-order sequence of ~12.4 my duration.

The Hettangian–Pliensbachian regressive regime of ~16.4 my duration includes three second-order sequences in Spiti. This first-order regressive regime interval began in early Middle Triassic latest Anisian. The regime is neither exposed nor present even in the subsurface in marine facies in Kachchh, since the early sediments in the Kachchh basin corresponding to this interval are non-marine. However, the regime is expressed well in Ladakh, Spiti, Nepal and South Tibet, and axial belt of Pakistan. In India the basal Hettangian is differentiated only in Lamayuru locality of Ladakh through the record there of the basal Hettangian Planorbis Zone through *Psiloceras gr planorbis* in deeper water succession of calcareous and argillaceous turbidites (Krishna et al. 1997). Rhaetian to early Middle Hettangian (Yin and Enay 2000; Yin et al. 2007) ammonoids are known in

South Tibet in a continuos succession without any evidence of any break at the T/J boundary. The ammonoid record improved in density, diversity and frequency, also geographically first in Early Hettangian and later in Sinemurian Bucklandi Zone and Oxynotum Zone in Pakistan and South Tibet, presumably on account of improved sea levels after the Triassic/Jurassic boundary drop of sea level and associated major extinctions. It is here interpreted that the Liasicus Zone (Tibeticum Zone to Pluronotum Zone) and Oxynotum Zone represent the successive second-order MFSs (Fig. 3.14). Incidentally, these two zones also make second-order MFSs on ETM (Haq et al. 1987).

In absence of ammonoids, the Pliensbachian stage is not substantiated any where either on the GTM or the Indian subcontinent. This is so, possibly, on account of the drop in sea levels for most of the Pliensbachian and associated wide spread regression during the terminal part of the first-order RST. Only in a few basins on GTM, there have surfaced, suggestive, not decisive, evidences of a small intra-Pliensbachian transgression (Somalia in Bosselini 1992; Oman in Rabu 1993). The temporal restriction of the ammonoids in the Hettangian to late Late Sinemurian is tentatively interpreted here as successive second-order MFSs. On the other hand, farther up there is an additionally interpreted second-order MFS in Ibex Zone of Pliensbachian although without ammonoids in Somalia and Oman. The evidence is based on unconfirmed extension of the coccolith ranges from ETM to GTM.

In this context, in the Spiti section of Bhargava et al. (2004), the tentative Hettangian–Pliensbachian interval is shown to include three second-order sequences with intervening oolitic limestones. These are speculated as the three second-order sequences with oolitic limestones as TSTs-cum-MFSs of the mid-Middle Hettangian Liasicus Zone, late Late Sinemurian Oxynotum Zone and Early Pliensbachian Ibex Zone. The first second-order sequence is evidenced in Lamayuru (Ladakh) in ammonoids, and without ammonoids also speculated in Spiti, the second in Ladakh, Pakistan and South Tibet in ammonoids, and also speculated in Spiti. The speculative differentiation is done with a view to looking in near future for evidences particularly ammonoids in support or otherwise on GTM. Ammonoids of the said interval have not been found west of the Indian subcontinent, while in the east, the only localities with ammonoids are Sula Island and Timor of Indonesia. These include the genera discovered at Lamayuru with faunal similarity. The three sequences are included in the terminal part of the Latest Anisian–Pliensbachian first-order RST of the previous first-order sequence, which are summarized as under (Figs. 3.14 and 3.15).

The second-order sequences J-I to J-III are developed in Spiti and J-IV to J-XII in Kachchh and/or Jaisalmer as under:

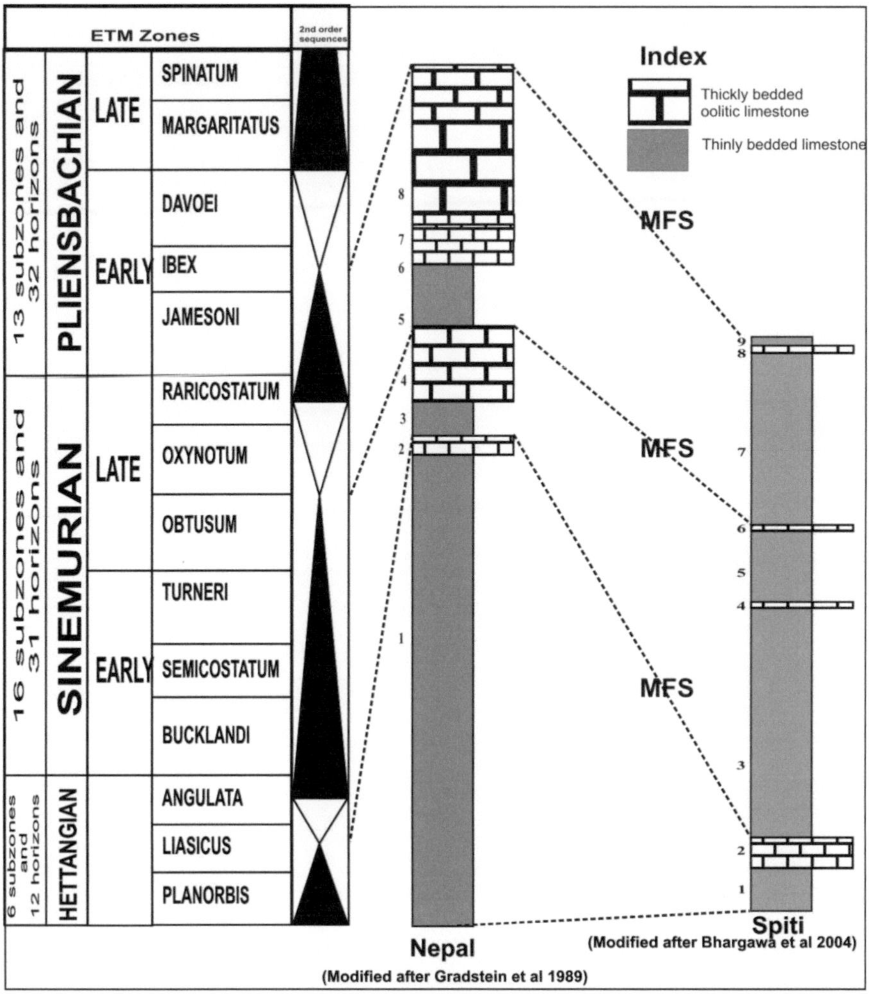

Fig. 3.14 Second-order sequences in the Hettangian–Pliensbachian successionof Spiti and Nepal

3.5.1 J-I. Hettangian Second-Order ~2.0 my Long Sequence with Middle Hettangian Liasicus Zone MFS

In Spiti, in the Hettangian–Pliensbachian interval, the first iron oolite bed immediately above the Tagling Formation is here speculated as of the Middle Hettangain Liasicus Zone with its top as the MFS. The following clays denote the RST of presumable late Middle and Late Hettangian. Hettangian ammonoids in Ladakh and South Tibet evidence the presence of this sequence there as well, i.e. widely in High Himalaya. The basal to Middle Hettangian ammonoids have been known through

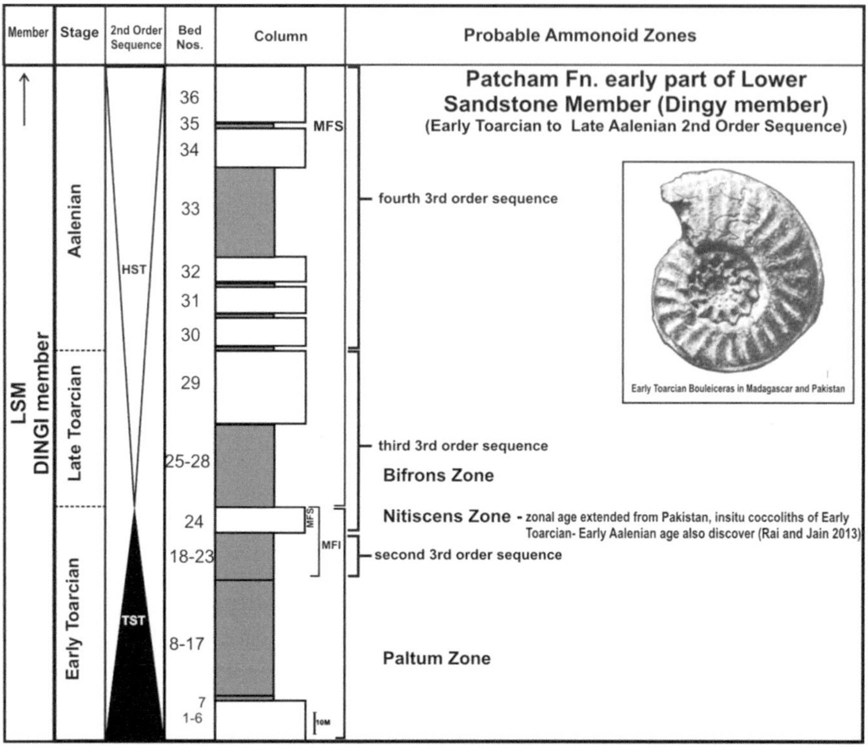

Fig. 3.15 Second- and third-order sequences in the Dingy member of Lower Sandstone Member with bed numbers after Biswas (1993)

recent and old records of *Psiloceras*, *Lytotropites* and *Fuciniceras* (Krishna et al. 1997), and *Euphyllites*, *Pleuracanthites* and *Rhacophyllites* (Diener 1908), respectively, from Ladakh, Spiti and Pakistan. In addition *Schistophyloceras*, *Juraphyllites*, *Analytoceras*, *Euphyllites* and *Schlotheimea* have been recorded from olistoliths at Kiogad (Diener 1908).

3.5.2 J-II. Hettangian–Sinemurian Boundary to Rauricostatum Zone ~8.5 my Second-Order Sequence with Oxynotum Zone MFS

The second iron oolite bed is interpreted as TST and its top as MFS. The thick shales that follow make the RST. This sequence is best evidenced in Pakistan, where ammonoids of Oxynotum Zone have long been known.

3.5.3 *J-III. Latest Sinemurian to End Pliensbachian ~8 my Long Second-Order Sequence with Ibex Zone MFS*

In Spiti the third oolitic limestone bed is speculated as TST with its top as the MFS. The Pliensbachian without any guide fossils is also definitely included in the Axial belt of Pakistan, where from Sinemurian and Early Toarcian ammonoids are known. The absence of Pliensbachian ammonoids could be due to decreased bathymetry. Reworked Pliensbachian coccoliths (Von Salis and Immenhauser 1997) at Masriha island of Oman support the interpretation of Rabu (1993). In the Somalia and Mandawa basins is also interpreted the intra-Pliensbachian transgression. Several basins on ETM have attested a second order relatively minor transgression in or close to Ibex Zone. On the contrary, the Kachchh basin compared to the said Arabo-African basins was located somewhat proximal on the Indus shelf and so away from the reach of the Neotethys. The Early Toarcian evidence is interpreted as the first second-order TST of the second Neotethyan first-order sequence as discussed below.

The early Early Toarcian – late Middle Oxfordian intra-Jurassic transgressive regime of ~24.7 my duration includes plus four second-order sequences (Fig. 3.2). The first second-order sequence is differentiated in Kachchh, and in absence of the ammonoid guide fossils, the ages of the sequence surfaces are extended from the neighbouring GTM basins of Pakistan and Madagascar. The younger three plus sequences are differentiated for the first time in Kachchh, and only subsequently extended to other basins of the GTM.

3.5.3.1 Geological Dynamics in the Kachchh Basin Since Toarcian

Ammonoid density, diversity, frequency, intrabasinal to interbasinal geographical extent suddenly exhibit marked improvement in early Early Toarcian. However, Toarcian ammonoids are yet not known from Kachchh, Jaisalmer, Kumaon and South Tibet. The ammonoids also began to acquire their regional GTM character for example the subfamily Bouleiceratinae. It is somewhat geographically restricted to the West GTM relatively subtropical localities between North Africa and Pakistan with occasional spread to Portugal and Spain of ETM.

3.5.3.2 Transform Boundary Tectonics and Origin of the Transgression in Kachchh Basin

As rifting advanced southward in steps and grew relatively deeper in Kachchh, the transform tectonics between India and Africa produced longer and deeper north to south inroads for the Neotethys to transgress the relatively young Kachchh basin. The transgression in Kachchh is considered genetically linked to the tectonically controlled first-order transgressive event on the entire GTM shelf from North Africa to North–West Australia. However, better evidences are recorded between North Africa and Pakistan (Krishna et al. 2010, 2011) by way of regionally spread Early Toarcian *Bouleiceras*-bearing (Fig. 3.15) fine textured silts and limestones unconformably overlying the thick unfossiliferous crossbedded sands.

3.5.3.3 Absence of Ammonoids in Kachchh During Toarcian–Aalenian

The recently interpreted Toarcian age to the early part of the Dm of LSM of Patcham Formation has so far yielded only benthic bivalves, corals, gastropods as mega-marine invertebrates, rich frog beds, and nannoplanktons. However, confirmatory ammonoid evidence is still lacking. One sample from the Kuar bet section has suggested late Early Aalenian, implying that the underlying sediments may be of Toarcian age.

3.5.3.4 Poor Bajocian Ammonoid Record in the Kachchh Basin

Just one incomplete ammonoid specimen (*Leptosphinctes* by Jaitly and Singh 1983) was claimed as of Bajocian age over the last 170 years. *Calliphylloceras* of Bajocian age is reported from a level older than that of *Leptosphinctes*, while the level with '*Leptosphinctes*' is relegated to Bathonian age. Also, presence of the terminal Late Bajocian ammonoids has been suspected in the distal Mainland Nara section. In contrast, in the neighbouring relatively distal Indus basin axial region, there are present multiple levels with good density and diversity of Bajocian ammonoids in carbonate facies in subtidal below wave base depths.

3.5.3.5 Ever Increasing Density, Diversity and Frequency of Ammonoids in the Early Bathonian–Late Middle Oxfordian Later Part of the First-Order TST

As one could expect, sufficient bathymetry developed prior to Bathonian at least in the distal exposed Mainland part of the Kachchh basin, for ammonoids to flourish in the ecosystem, and registered graded increase up to the late Middle Oxfordian MFS. In the distal exposed Mainland, the bathymetry maximized at the first-order MFS of at least mid-shelf kind. In spite of the start of decrease of bathymetry in early part of the following first-order RST, there continued sufficient depths to let ammonoids enhabit the ecosystem, and get later fossilized.

The second-order sequences in the Toarcian–late Middle Oxfordian first-order TST are

3.5.4 J-IV. Second-Order Early Toarcian–End Aalenian Second-Order Sequence of ~12.5 my Duration with MFS in the Nitiscens Zone

3.5.4.1 Unique Among the Kachchh Jurassic Second-Order Sequences

This is the first second-order sequence of the Kachchh Jurassic exposed succession. It is unique in being almost devoid of guide mega-fossils, also largely bereft, of other tools of age determination except for a single coccolith based late Early Toarcian age from Kuar bet section of Dm of LSM. It includes large thickness in the exposed sections, and also is the longest in duration among the Kachchh Jurassic second-order sequences (Fig. 3.15). These peculiarities make it difficult to either stratigraphically refine or develop in it any kind of yet finer sequence stratigraphic differentiation.

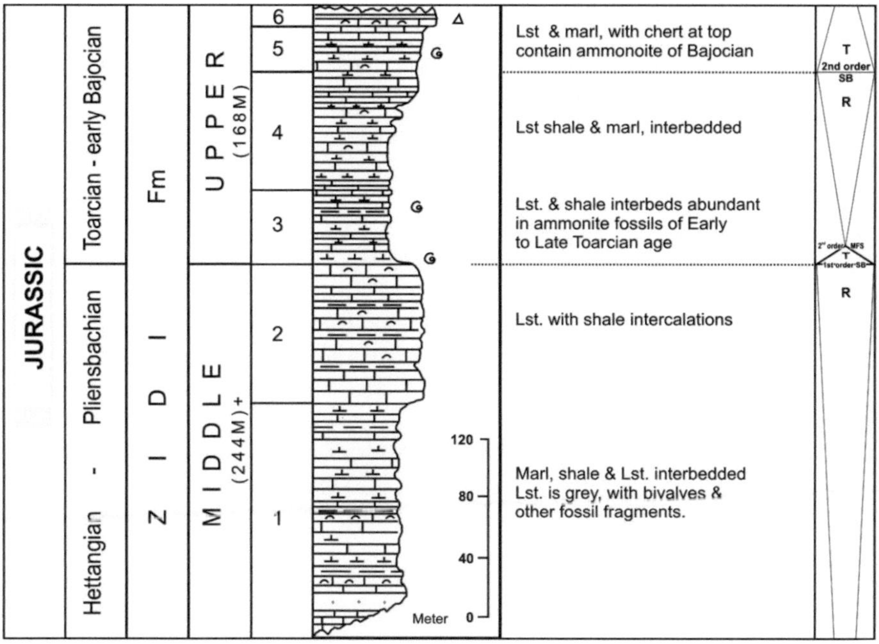

Fig. 3.16 First- and second-order sequences with SBs and MFSs indicated in the column of Zidi Formation east of Khuzdar (Goru section) (modified after Fatmi et al. 1986)

3.5.4.2 Differentiation of the Sequence on GTM

The sequence has been first interpreted (Krishna et al. 2010, 2011) between North Africa and Pakistan in the neighbourhood of Kachchh on GTM. It was differentiated by the sudden appearance of Early Toarcian *Bouleiceras*, along with marked increase of ammonoid frequency, density and diversity and their geographic extent, unconformably over the non-marine crossbedded sands in proximal parts of several basins. In India, Early Toarcian *Harpoceras* is recorded from Lamayuru locality of Ladakh Himalaya from the Indian side of the Indus-Tsangpo suture zone (Krishna et al. 1997) as also Early Toarcian *Alocolytoceras* from Spiti (Stoliczka 1866), however, in these localities the differentiation of sequence surfaces is not possible due to lack of lithostratigraphic details.

3.5.4.3 The Geographically Nearest Early Toarcian Record from Pakistan

The nearest definite Toarcian is known from the Goru section of Pakistan in the carbonate dominating ~50 m thick bed 3 of the Upper Member of Zidi Formation (Fatmi et al. 1986; Krishna et al. 2011) (Fig. 3.16) The Lasbella district localities (South Baluchistan) of the Axial belt (250–300 km from Kachchh) provide a rich record of Early Toarcian ammonoids. All the Jurassic Pakistani basins, together with the Indian basins of Jaisalmer and Kachchh were neighbouring parts of the Indus shelf, along with similar sequence stratigraphic frameworks. The exposed Kachchh shelf itself is ~300 km wide between Radhanpur in the east and Ghuneri in the west. The 30 m thick Early Toarcian – Early Bajocian ammonoid rich sediment interval in Lasbella localities is underlain by a thick ammonoid-devoid marine carbonate succession similar to the Kioto Limestone of the Indian High Himalaya. The Upper Indus basin Early Toarcian ammonoid-bearing succession is underlain by non-marine clastics as in Jaisalmer and Kachchh. It is exposed in Jaisalmer while encountered only in wells in Kachchh.

3.5.4.4 Qualitative and Quantitative Account of the Pakistan Early Toarcian Ammonoids

The ammonoid fauna includes about 40 described specimens, 5 genera and 13 species from at least 5 levels in the total Early Toarcian sediment thickness of ~25 m (Fatmi 1986). As early as 1909 Holland recorded the Early Toarcian ammonoids from Lasbella district which included 6 genera. Total number of Early Toarcian ammonod genera from Lasbella district is 8, while the total number of specimens should far exceed 40. Early Toarcian ammonoids are also known from the Upper Indus basin of Pakistan. Fatmi (1972) interpreted the rich Pakistan record as a major marine transgression in Pakistan, in the Tenuicostatum Zone–Falciferum Zone interval.

3.5.4.5 Speculative Extrapolation of Sequence Surfaces from Madagascar and Pakistan

The LAD of *Bouleiceras* at the top of Nitiscens Zone has been also interpreted as the second-order MFS (Krishna 1987a; Krishna et al. 2010, 2011). The sequence is suggested in Kachchh, exclusively in Kuar bet and Dingy localities of Patcham in consequence of the discovery of the Toarcian–Aalenian nannoplanktons (Rai and Jain 2013). The entire Kuar bet succession is broadly equated to the beds 1–24 of the Dm of LSM of Patcham Formation at Dingy (Biswas 1993). The entire Dm of beds 1–36 of Dm is included in this second-order sequence with MFS/MFI in the thin fossiliferous oolitic shale/marl beds 18–24 (Figs. 3.17 and 3.18). It is difficult to classify the relatively thicker marker bed 24 with fossil wood, bivalves, corals, etc. in context of the TST/MFI. Bed 24 is a bioclastic conglomerate with evidences of reworking only ~10 m above the beds 18–21. Beds 3–17 are also relatively fine textured, without any body fossils. The MFS is here speculatively assigned to Nitiscens Zone. The second-order RST of beds 25–36 is assigned Late Toarcian–Intra- or end Aalenian age. There is well known Early Bajocian widespread revival of rifting on GTM both east and west of the Indian subcontinent, along with increased bathymetry, and rich record of Early Bajocian ammonoids in several GTM constituents as detailed in Chap. 7 (Arabia, East Africa, Madagascar, Pakistan, Nepal and South Tibet, Sula, Iriyan Jaya, Timor). The same has prompted the author to mark the upper limit of this sequence only tentatively as Late Aalenian, though it could be anywhere in Aalenian. On ETM, the second-order SB has been placed either in Early Aalenian Opalinum Zone or Late Aalenian Concavum Zone. Moreover, as yet no Aalenian ammonoids are known in the above mentioned basins/regions except for doubtful passing mention of Aalenian ammonoids without any names in Majunga wells of Madagascar (De Jongh 1990 unpublished data of SHELL in Geiger 2004, Geiger and Schweigert 2006). The near absence of Aalenian ammonoids on GTM particularly in IEAP is also indirectly suggestive of reduced mostly above wave base bathymetry and associated regressive regime during the RST while the MFS suggests intertidal depths.

In view of the widespread record of Early Bajocian ammonoids, the tectonostratigraphic homotaxiality of the next second-order sequence beginning in Early Bajocian is without any doubt. The Bajocian sequence with or without ammonoids invariably on GTM is followed up by the late Middle and Late Bathonian ammonoids, which are present in Jhura, Nara, Jumara, Sadhara and even at Kaladongar as its youngest sediments. The lithostratigraphic columns going down from the said late Middle Bathonian at the mentioned localities exhibit clear tectonostratigraphic homotaxiality. In Kaladongar (Biswas 1993), the next older significantly fossiliferous relatively fine textured calcareous or argillo-calcareous unit without ammonoids is of the beds 49–55, in Sadhara again without ammonoids, is LYFm (Fursich et al. 1994). In Jhura it is Cm again without ammonoids (Biswas 1993), and in Nara with ammonoids of Early Bathonian age (Pandey and Pathak 2015a). The intervening several fold thicker sediments are unfossiliferous sands in Kaladongar at the base of BCSm, unfossiliferous sands of MSm in

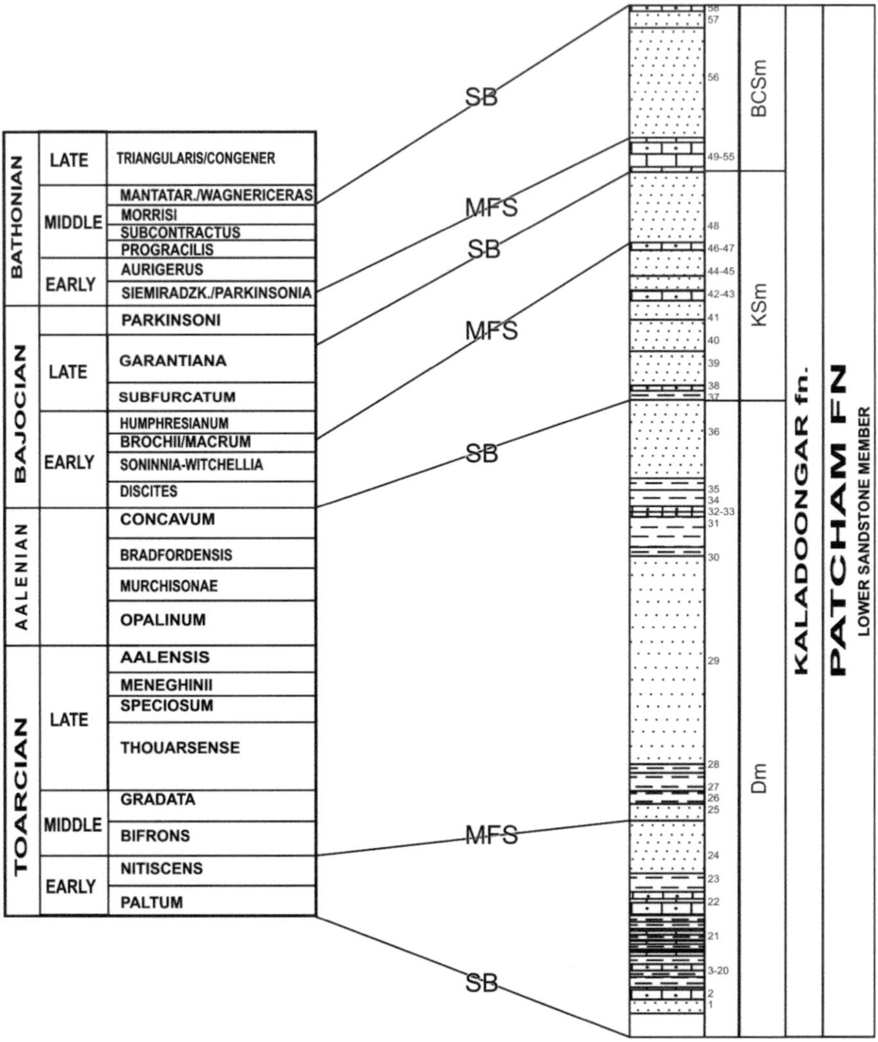

Fig. 3.17 Second-order sequences as interpreted in the Lower Sandstone Member of Patcham Fn. at Kaladoongar

Sadhara, ammonoid-devoid sandy limestones in Jhura, and ammonoid scarce argillo-calacareous sediments in Nara as a reflection of the relative bathymetry and distance from margin of the respective sedimentation sites. In context of the tectonostratigraphic homotaxiality of the said columns, the relatively much thinner fossiliferous ammonoid-devoid calcareous to argillocalcareous or arenocalcareous beds of basal BCSm in Kaladongar, LYFm of Sadhara, and in Em of Jhura should also be of Early Bathonian age. Yet farther down in the said columns, after intervention of unfossiliferous, several fold thick dominantly sandy or

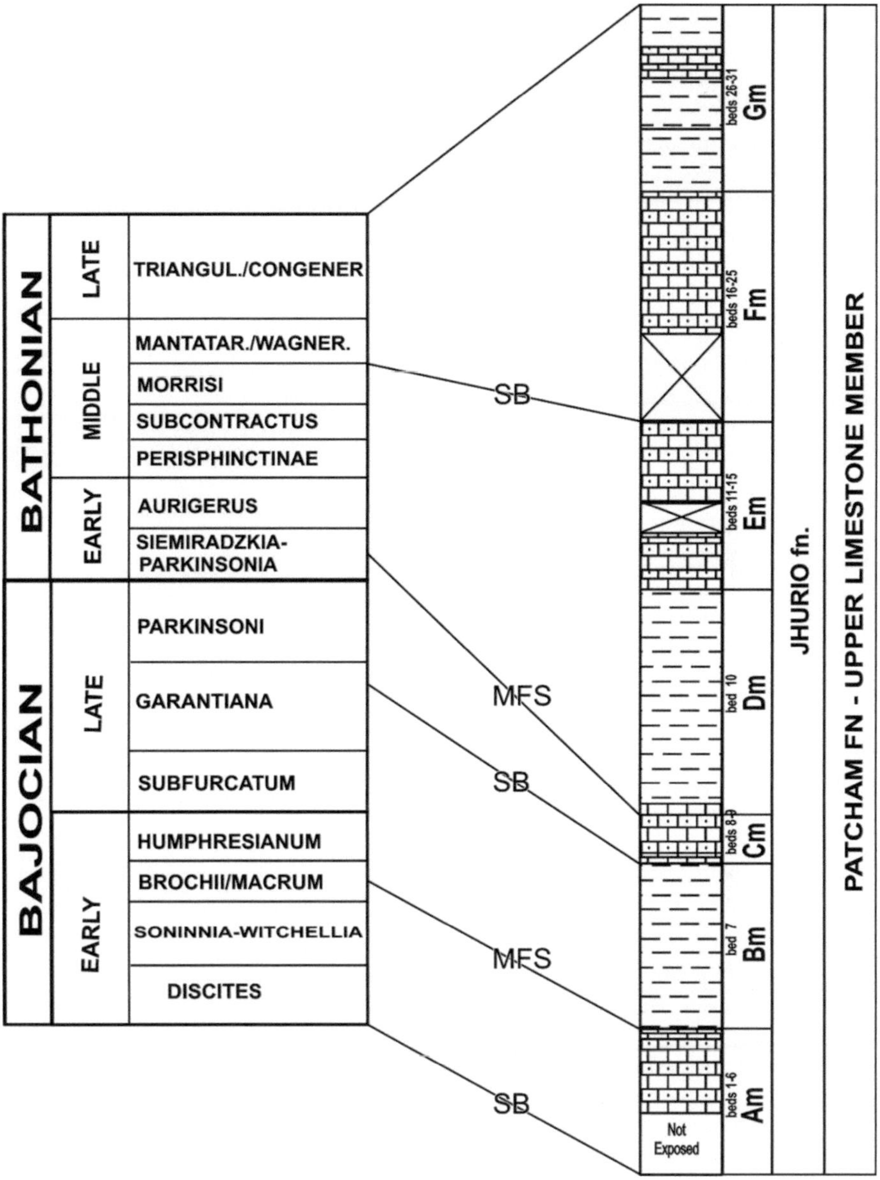

Fig. 3.18 Second-order sequences in the Jhura Bajocian–Bathonian

argillo-calcareous/areno-calcareous unit, occur the thin fossiliferous marker interval (beds 42–47 of the early to Middle part of KSm in Kaladongar, SCLm in Sadhara, and Am at Jhura). Among these only the SCLm at Sadhara has yielded *Calliphylloceras heterophylloides* possibly of Early Bajocian age. Even otherwise the said fossiliferous fine textured calcareous unit at Kaladongar, Sadhara, and Jhura should reflect either the terminal part of TST or MFS/MFI of Eary Bajocian age. If that be correct, as it should, the immediately underlying similar thickness wise asymmetric sequence becomes Toarcian–Aalenian in age. This interpretation is further strengthened by undisputed Early Toarcian base of the Jurassic transgression in several basins of GTM. The recent discovery of the insitu Toarcian coccoliths (Rai and Jain 2013) from the early part of Dm of Toarcian–Aalenian duration of the basal second-order sequence in the Kaldongar column, and its exclusive differentiation as yet in Kachchh, is considered very important.

3.5.4.6 Thin Fossiliferous Intervals in an Otherwise Extremely Thick Fossil-Devoid Massive Sand Build-up

There are about five thin significant fossiliferous intervals in the Dm of LSM inclusive of beds 18–21, three below the MFI and two above (Figs. 2.25 and 2.26). The combined thickness of these thin fossiliferous beds is about 30–35 m out of a total thickness of ~ 300 m. These thin beds/intervals are interpreted as thin fossiliferous reworked fine textured third-order TSTs mostly as transgressive lags, particularly those that are followed up by thick unfossiliferous coarse crossbedded sands as typical RSTs inclusive of the short non-marine interludes within a largely marine yet fluctuating coastal environments.

3.5.4.7 Suggestive Correlation Between Kuar Bet and Dingy Hill Sections

One thin characteristic bed, a little underneath bed 24, is characterized by *Corbula* bivalve assemblage. Biswas (1993) correlated this bivalve assemblage with the terminal beds at Kuar bet that included a similar assemblage. The Kuar bet part of the Dingy Member also includes two fossiliferous levels, respectively, ~ 2 and ~ 22 m below the *Corbula* bed that are included in the five such levels mentioned above. Based on the above broad correlation, the Kuar bet section may be accepted as the early part of Dingy Member, so also as the early part of the TST. The basal ~ 65 m sediments at the 90 m thick Kuar bet section are unfossiliferous, so also the early 28–30 m of the succession at Dingy hill.

3.5.4.8 Highly Stressed High Energy Extremely Shallow Near Shore Strand Line Environments in the Toarcian–Aalenian Interval and the Asymmetric Third-Order Sequences

But for rare low diversity brackish water bivalves, the early 80–90 m sequence is almost bereft of body fossils. The presence of several successive levels of rootlets is suggestive of near coast reduced salinity and frequently fluctuating environments.

The later fossiliferous fine textured sediments are very thin compared to the thick unfossiliferous sands that alternate with them within the Dm. These thin intervals also mostly include low-diversity bivalves and other benthics which suggest highly stressed environment and near coast regime, some what away from the normal open sea conditions, with salinity and shoreline fluctuations, and even inclusive of short but noticeable episodic emergences. The transgressive encroachment began in the strandline conditions that were not only highly variable but also additionally included several micro-environments. These thick sands exhibit varied mixed coastal environments inclusive of non-marine interludes of localized conglomerates, and occasional channel cut and fills in early part of the Dm, while easily recognizable coastal marine environments include lagoons and sand bars.

Granite–pebble conglomerates interspersed in Dingy/Kuar bet localities in the basal beds of the Patcham Formation (the oldest exposed of Kachchh marine Jurassic) suggest proximity of the granitic basement that witnessed rapid intermittent uplift and erosion. Such restrictive highly stressed coastal regimes in rapidly fluctuating salinity conditions persisted perhaps at least until the close of Aalenian or at least well into Aalenian, even up to Bajocian in small bits at least in the ‘Island belt’. The Toarcian–Aalenian interval is evidenced only through nanno-planktons. Ammonoids are yet not discoveded in this sequence in Kachchh. Sands with almost regular intercalations of granite–pebble conglomerates suggest periodic uplift of the provenance and associated sinking of the basin while the initial invasion by the sea graduated in bathymetry. The sequence although exposed only in Patcham is suggested to extend down the basin beneath the Mainland if not through out the basin. In the West Mainland, west of Jhura, the basin was relatively less shallow, where rare small fragments of ammonoids may be searched for in cores/cuttings.

The rifting and associated tilting of the basin facilitated the invasion by the southward advancing arm of the Neotethys in a second-order eustatic rise. Once through the second-order MFS, perhaps, the sea regressed, and the whole peneplained shelf emerged. Number of such third- or fourth- or fifth-order short but wide transgressions and regressions facilitated the reworking and redistribution of sediments during the RST interval of the first second-order sequence. As it is the case in marginal sites, extreme shallow invariably above normal wave base conditions are expected to have prevailed in the basal second-order sequence. Repetitive emergence and associated erosion particularly after the second-order MFS provided reworking and mixing of the sediments and included biota.

Further deepening and associated tilting down in the basin and graded eustatic rise farther up in the second second-order sequence of Early Bajocian – mid-Middle Bathonian facilitated the development of the carbonate ramp. The ramp reached Patcham by the mid/late Middle Bathonian under rather stable marine conditions in the basin. The development of the carbonate ramp in Mainland and Island belt in turn enabled ammonoids to gradually become part of the depositional environment. The ammonoid favouring environment started from relatively distal and deeper exposed parts of the basin from Jumara and Nara in West Mainland to Jhura in the East Mainland and proximally farther to Goradongar and Kaladongar in Patcham.

The sediments of the Dm inclusive of the Kuar bet section are thinly bedded and fine textured until the second-order MFS/MFI (beds 18–21) for the first ~50 m. The said sediment interval is interpreted as the second-order TST, and suddenly there after, the remainder Dm became thickly bedded and coarsened upward for the rest of the second-order sequence as RST of ~250 m thickness up to the top of Dm as here revised at the top of bed 36 instead of bed 38.

3.5.4.9 Origin/Extinction of the Ammonoid Subfamily/ies

The representative GTM ammonoid subfamily of the sequence is the short lived Early Toarcian Bouleiceratinae which climaxed and terminated at the close of the Early Toarcian Nitiscens Zone. A few other Pan-Tethyan subfamilies (Harpoceratinae and Hildoceratinae) also terminated at this point of time on the GTM (Figs. 6.4 and 6.5).

Several third-order highly asymmetric sequences are tentatively differentiated as under of unknown ranges and durations:

J-IV. (1) Third-Order Sequence
It is differentiated exclusively in Kuar bet in the basal ~70 m sediments. The first third-order MFS is a level of *Corbula*, *Astarte*, and gastropods ~22 m below the top of the Kuar bet section which is underlain by 65 m of unfossiliferous sediments, and overlain by ~4 to 5 m unfossiliferous sediments. The sediments below the interpreted MFS are relatively finer textured than those above the MFS, and are here interpreted mostly as transgressive shallow marine. The relatively coarser silt and fine sand above the determined third-order MFS are interpreted as channel fill to shore face micro-environments. It is likely that the basal ~65 m thick unfossilierous sediments may in future reveal one or more fossiliferous beds for further differentiation of the first third-order TST. The sequence based on globally isochronous sequence surfaces suggestively includes most of the Early Toarcian.

IV. (2) Third-Order Sequence
The sequence is compositely differentiated in both Kuar bet and Dingy hill sections. The second third order is expressed in ~25 m sediments, and includes two thin green fossiliferous clay beds with bivalvs *Corbula*, *Gervilla and Trigonia/Astarte* ~4 m above the first third-order MFS, and the prolificly fossiliferous sandy limestone of presumable intertidal depths near the top of the Kuar bet section. In Dingy hill the MFS is determined in a similar bivalve rich bed ~60 m above the base. The basal 35 m sediments are by and large unfossiliferous. The MFS level is the same as the second-order MFS of beds 18–21 followed by the RST in beds 22–23 (Fig. 3.16). The MFS is suggestively placed at the top of late Early Toarcian Nitiscens Zone.

J-IV. (3) Third-Order Sequence
The beds 24–29 make the sequence, with the rich fossil wood, broken bivalve, coral and gastropod-bearing-bed 24 as TST of suggestively terminal Thouarsense Zone.

The thick crossbedded unfossiliferous coarse sands constitute the RST of likely late LateToarcian age. It can be further differentiated into two fourth-order sequences with the bivalve-bearing conglomeratic sand bed 30 as the fourth-order TST, and unfossiliferous much thicker siltstone as the fourth-order RST of the first fourth-order sequence. Further up in the column follow beds 32–34 as the second fourth-order TST having MFS of the rich broken bivalve-bearing conglomerate. Unfossiliferous thick sands of beds 35–36 make the RST.

J-IV. (4) Third-Order Sequence

The fourth third-order TST is interpreted in the conglomeratic beds 30–34 with broken reworked shells of bivalves, corals and fossil wood. These are followed up by the ~40 m thick crossbedded sandstones of beds 35–36 as RST. If present in wells Manera-1 and Ambatolahy-1 of Majunga basin in Madagascar, the ammonoids should be of Early Aalenian age, corresponding to the beds 30–34 of Dm. The basal SB of this sequence, if extended from ETM, comes close to the Toarcian/Aalenian boundary, which in turn suggests broad Aalenian age of the fourth third-order sequence.

Here are at present provisionally conceived at least two second-order sequences as I (a) and I (b) with informal second-order MFS above the bed 21 and SB above the bed 23 of I (a). These could be given formal status in future. The first two third-order sequences may be informally included in the first such second-order sequence as 1(a), while the younger two third-order sequences in I (b). On ETM as well, two second-order sequences have been differentiated in about the same interval. Under such informal division, the respective durations may be of more or less equal of ~6.2 my each with the informal second-order SB speculated somewhere near the close of Late Toarcian, however, thicknesses are highly unequal and textures different. The older one is one-third thick compared to the younger one, and also relatively fine textured. Thus, the entire Kuar bet and basal one-third thickness of the Dingy member (Dm) is tentatively assigned to Toarcian and the younger sand dominated part to Aalenian, with two third-order sequences each in Toarcian and Aalenian. In this context, the informal second-order MFS of the informal second second-order sequence would correspond to the top of the Late Toarcian Thourascens Zone. Also, the terminal Late Toarcian would have constituted the shallowest Toarcian inclusive of non-marine coastal environments.

The Late Toarcian ammonoids are exclusively recorded from another Axial belt locality of Khuzder in Pakistan. As in External Alps in France and some other areas of ETM, are interpreted condensed to starved relatively deeper water conditions in Axial belt of Pakistan along with possible submarine gap in the vertical proximity of the second-order MFS. It is possible, that the less deep Khuzder included the Late Toarcian ammonoid record although not as rich as that of Early Toarcian in Lasbella district. For the same reason, the Khuzder Early Bajocian ammonoid record is also not as rich as that of Lasbella.

Development of Carbonate Ramp during Early Bajocian to Late Bathonian

The carbonate ramp development is interpreted to have initiated with the start of Bajocian as elsewhere on GTM with the earliest indication at Jhura in Mainland at the Aalenian/Bajocian boundary SB. The ramp existed for the entire ~4.2 my Bajocian–Bathonian duration for plus two second-order sequences; first up to mid-Late Bajocian ?Garantiana Zone while the younger one from Garantiana Zone to mid-Middle Bathonian Morrisi Zone (Figs. 3.17, 3.18, 3.19, 3.20 and 3.21) The dominantly carbonate development in the proximal exposed margin in Island belt at Patcham was, however, possible only from late Middle Bathonian Bremeri Zone. During the first sequence of ramp formation in Bajocian, ammonoids are absent in the entire basin but for the single *Calliphylloceras* (Pandey et al. 2013b) at Sadhara. However, ammonoids could be present in the subsurface distal West Mainland Nara and Jumara sections, since the exposed Early Bathonian at Nara and Late Bathonian at Jumara collectively include near continuous succession of ammonoids (Pandey et al. 2013a; Pandey and Pathak 2015a; Krishna unpublished report 2014). Even *Calliphylloceras* should have found its way to Sadhara from the neighbouring Pakistan and Madagascar.

Kaladongar and Sadhara sections: The basal red beds with caliche nodules (Fursich et al. 1994) suggest short time exposures and non-marine sedimentation during the terminal phases of the second- and third-order RST durations mostly in Aalenian, also less so even in Bajocian, prior to the fresh transgressive pulses. This transgressive pulse close to the MFS is tentatively correlated to the beds 37–38 at Kaladongar (Biswas 1993) here transferred to the base of KSm of LSM. A few tens of meter thick dominantly large crossbedded sandy RSm at Sadhara with occasional rare bivalve levels is invariably coastal marine in lagoonal and sand bar framework

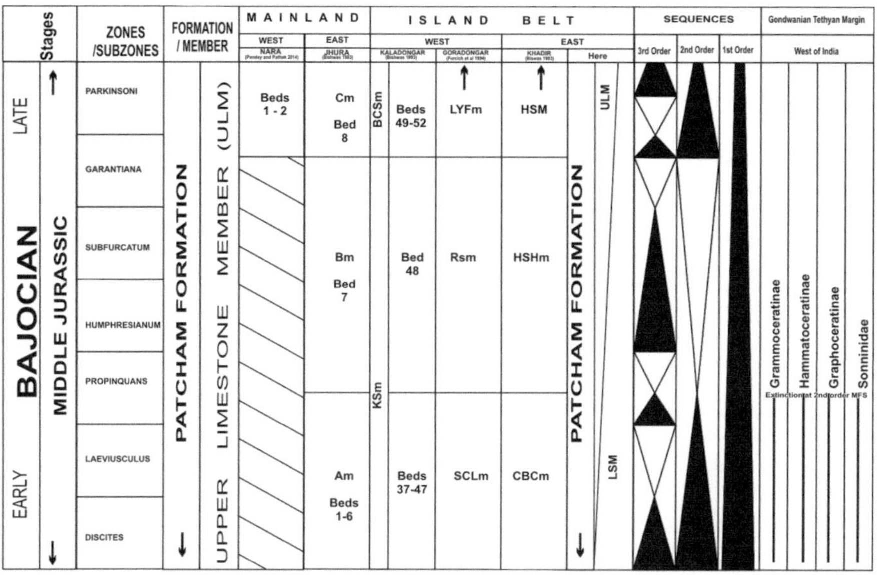

Fig. 3.19 Intrabasinal correlations vis-à-vis first- to third-order sequences during Bajocian interval in Kachchh basin

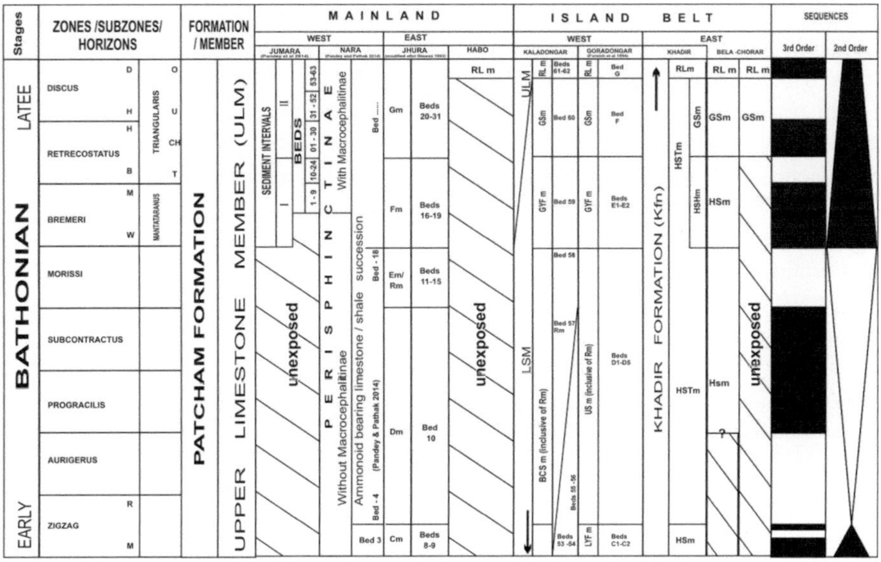

Fig. 3.20 Intrabasinal correlations vis-à-vis firsst to third-order sequences during Bathonian interval in Kachchh basin

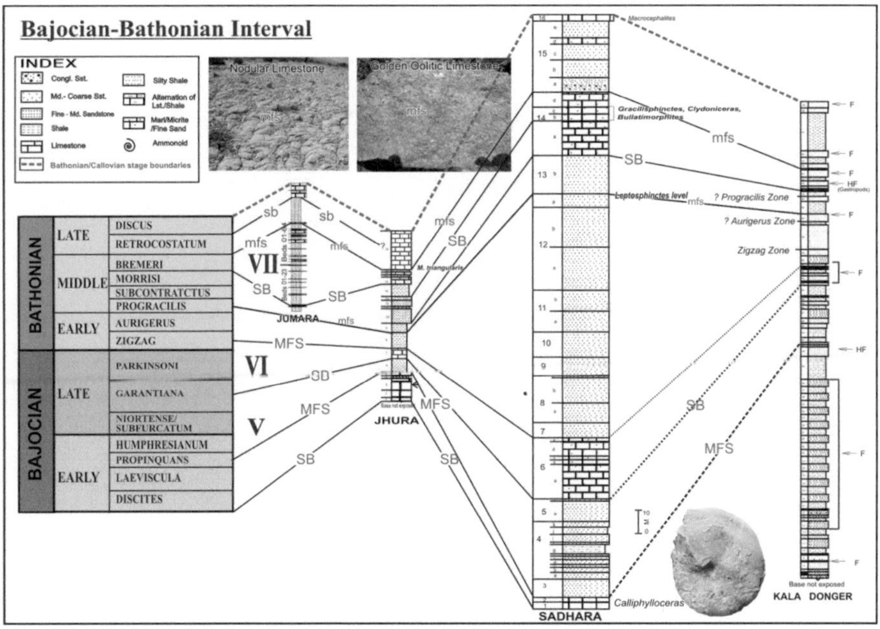

Fig. 3.21 Second- and third-order depositional sequences in the Jumara, Jhura, Sadhara and Kala Donger sections of Kachchh

with interludes of emergences and associated non-marine coastal plain sediments. The third-order differentiation, outlined later, provides possible non-marine interludes with subaerial emergences during the terminal parts of the RSTs. Similar non-marine interludes are also expected and interpreted at Kaladongar within KSm in the beds 39–40, 44–45 and 47–48 as third-order RSTs and their non-marine equivalents. The corresponding beds 37–38, 42–43, and 46 make the fine textured shell-bearing third-order TSTs.

Spatiotemporal distribution of carbonate facies in the exposed sections: For the duration of the first two sequences until the mid/late Middle Bathonian SB, the proximal/shallower ramp in the Island belt is expressed in the coarse rippled crossbedded thick sands as the 2nd and 3rd, and yet inferior order RSTs. These are interspersed with thin fossiliferous limestone/shale transgressive lags as inferior order TSTs at Kaladongar and Sadhara depositional sites. The middle to deeper/distal ramp is made up predominantly of the diverse thick massive limestones inclusive of the long known golden oolitic limestones of Jhura. The exposed Nara section depositional site represents the farthest distal ramp in continuos ammonoid-bearing terminal Bajocian to mid-Middle Bathonian for nearly the entire duration of the second second-order sequence with basal Bathonian Macrescens Subzone MFS. So would have been presumably the framework as well in the Jumara subsurface.

Improvement in bathymetry and spread of ramp to the Island belt: Improved bathymetry resulted in the spread of the golden oolitic limestone facies as GYFm to Patcham during the Triangularis Subzone–Congener Subzone–Hians Subzone interval i.e. Bremeri Zone to Retrecostatum Zone interval of the late Middle to early Late Bathonian, and of the non-oolitic limestones as RLm of the terminal Late Bathonian Discus Zone in the entire Island belt. The ramp in essence reached the Island belt only at the mid/late Middle Bathonian boundary in Patcham and Khadir in predominant carbonate facies.

Differentiation within the ramp during the Bajocian sequence: The relatively distal/basinal/deeper/outer to proximal/shallower facies developments on the ramp during the Bajocian sequence are expressed between Jhura and Island belt, respectively, (Patcham and Khadir). The typically golden oolitic Am–Bm informal units of ULM at Jhura make the middle to relatively deeper part of the ramp in subtidal low-energy framework between normal and storm wave base of oolitic and non-oolitic limestones and shales during the Early Bajocian Propinquans Zone terminal TST and MFS to mid-Late Bajocian Garantiana Zone SB. The coeval proximal thick silty/sandy facies of KSm (beds 37–48) are known in the above normal wave base high-energy coastal environmental framework at Patcham and of HSHm at Khadir.

For the first time since the start of Bathonian, the deeper basinal ammonoid-bearing thickly bedded nodular limestones and shales are expressed at Nara and Jumara in West Mainland of Early to Late Bathonian in slow sedimented low-energy beyond storm wave base even reaching near mid-shelf depths. At Jumara the limestones include rich corals and sponges. At Jhura in East Mainland the middle ramp golden oolite dominated limestone facies is exposed. Golden oolitic facies continue even to proximal Patcham as GYFm at the start of a new second-order sequence during the late Middle to early Late Bathonian, while the ramp terminated in the entire Island belt with the majestically looking non-oolitic cliffs of RLm in varying/fluctuating subtidal to below storm wave depths and environments of mixed high and depleted energy frameworks along with corresponding physical and biogenic features.

The ramp development is more prominent in the west part of the GTM and less so in the east part in Spiti, Nepal and South Tibet. The near exclusive to mixed silici-carbonates of Bemaraha and Sakaraha units of Madagascar, Mtumbei and Kidugallo units of Tanzania, Hamanlei unit of Somalia, Laptal unit of Spiti, Bagung of Nepal and others) are manifested by thick massive limestones, carbonate mud, varying oolitic and pelletoidal limestones. The India–Madagascar correlation and comparison is discussed in Chap. 6.

3.5.5 J-V. Basal Bajocian–Mid-Late Bajocian ~1.7 my Long Second-Order Sequence with MFS in Early Bajocian Propinquans Zone

The basal SB of this sequence or the terminal SB of the previous second-order sequence is placed between the relatively thick unfossiliferous crossbedded sands of bed 36 of Biswas (1993) below and thin fossiliferous marl/lime of bed 37 above in the Kaladongar section. In context of thickness are again observed asymmetric inferior order sequences (with ~tenfold thicker RSTs) (Figs. 3.17, 3.18, 3.19, 3.20 and 3.21). The second-order MFS is determined at the top of the bed 43. The thick coarse unfossiliferous crossbedded sands of the beds 44–48 constitute the RST. Beds 37–38 are here transferred from the top of the Dm to the base of KSm.

It may be re-stressed that exclusive Early Toarcian nannoplankton taxa are also present in the discovered assemblage, and it is not justified to restrict the assemblage to Early Aalenian. Aalenian ammonoids are absent from the entire GTM but for doubtful mention from Majunga wells (De Jongh 1990—unpublished data recorded by Shell in Geiger 2004). This sequence is almost ammonoid devoid like the earlier one in Kachchh except for the recently discovered (Pandey et al. 2013b) *Calliphylloceras* in Sadhara section. The Early Bajocian Propinquans Zone MFS obviously is extended from the neighbouring Axial belt region of Pakistan. The basal SB may be of intra-Aalenian age or of Aalenian/Bajocian boundary age while the terminal SB is here suggested without evidence as of Late Bajocian Garantiana Zone as on ETM. The Aalenian/Bajocian boundary basal SB is favoured by the author in regional IEAP and GTM context.

3.5.5.1 Age of the MFS Extended from Pakistan

The single level, rich ammonoid fauna includes rich lytoceratins and phylloceratins in addition of ammonoids of Early Bajocian Propinquans Zone in Pakistan (Fatmi 1986), that without any hesitation allowed the determination of the level as second-order MFS. There are three sections (two in ‘Island belt’ and one at Jhura in Mainland) in Kachchh which supposedly include the Early Bajocian MFS. Prevalence at least at Jhura of intertidal to subtidal depths and quite energy conditions is interpreted. The age of the MFS is reliable, since extended from neighbouring Pakistan only $\sim$275 km away, also a constituent of IEAP and ETM.

3.5.5.2 Early Bajocian Ammonoid Record in IEAP Particularly in Pakistan and Madagascar

The Early Bajocian ammonoid record in Africa, Madagascar, Pakistan, Nepal and South Tibet is much richer and widespread than that in the previous sequence, and also in carbonate facies suggesting widespread rise of sea level and increased bathymetry on GTM. The interpreted deepening event is due to the cumulative effect of eustatic rise and rift related submergence.

3.5.5.3 Lithology and Sedimentation Features in Kachchh

The sediments in the proximal Kaladoongar and Goradongar localities are still dominated by the marine-fossil-devoid coarse clastics with extremely thin fine textured fossiliferous intervals as transgressive lag deposits with ample evidences of reworking. Thus, in spite of general improvement of depth, not only the entire sequence remains ammonoid devoid in all the exposed sections at Jhura and Patcham, but for the MFS inclusive limestone at Sadhara which has recently yielded an incomplete individual of *Calliphylloceras heterophylloides*. At least, in the entire Island belt, varying subtidal conditions from storm wave base or deeper to below wave base prevailed during the TST, while straddling the shoreline with multiple emergences for the RST. Yet, there is good supply of coarse clastics in a near continuously rejuvenating provenance with occasional emergences as in the preceding Early Toarcian – Aalenian duration. In the Kaladongar section, it includes the beds 37–48 of KSm of LSM of Patcham Formation with MFS above the bed 44 (Figs. 3.17, 3.18, 3.19, 3.20 and 3.21). At Sadhara, the basal limestone is interpreted as later part of the TST, its top above the bed 2 as the MFS, while the following thick red strandline sands of beds 3–5 make the RST. If not complete, the major part of this sequence is here suggested to be also present in Mainland at Jhura in alternating golden oolitic limestones included in the Am (beds 1–5) as TST and the following non-golden oolitic limestones/shales included in the Bm (bed 6) as RST of the early part of ULM of the Patcham Formation. The entire sequence at Jhura suggests subtidal to storm wave base depths.

3.5.5.4 Speculative Ages of the SBs

The basal SB as determined in Kaladongar is at the base of the thin fossiliferous limestone/shale interval included here as the basal part of the KSm (beds 37–38) without any clue to its age either within or at the top of Aalenian. In such circumstances, the age of this second-order SB is provisionally adopted from ETM at the top of Opalinum Zone (Graciansky et al. 1993; Jacquin et al. 1998a; Jacquin and de Graciansky 1998). The argument applied is that if the MFS is nearly isochronous on ETM and GTM, the SBs should similarly be the same on ETM and GTM. Another alternative is placement of this SB at the base of the Late Aalenian Concavum Zone. It may even be at the base of Bajocian, since rifting related deepening is regionally in IEAP interpreted from the start of Bajocian, c.g. in the Somalia basin on account of sudden appearance of abundant deep water ammonoids. The terminal SB is also similarly extended from the ETM within the Late Bajocian Garantiana Zone. The 3rd and fourth-order RST intervals are interpreted to involve short emergences.

A few third-order sequences are differentiated at Kaladongar within the above second-order sequence as under (Fig. 3.21).

J-V. (1) First Third-Order Sequence
It includes beds 37–38 as the TST and the superjacent beds 39–40 as the RST with the MFS above the thin benthic gastropod-bearing fossiliferous limestone at Kaladongar in the early part of KSm of Patcham Formation. There could be short emergence interlude during the RST. Beds 39–40 represent the much thicker rapidly deposited crossbedded mostly unfossiliferous sands as the regressive or non-marine component. It is rather difficult at this stage to differentiate this third-order sequence outside Patcham. The differentiation in Patcham has been possible on account of the thick, rapid, expanded expression in the mixed coastal environmental framework. Using the concept of globality of isochronous third-order surfaces, the third-order sequence is corresponded here to the early Early Bajocian Discites Zone–Laeviuscula Zone interval with MFS at the top of the Discites Zone. In Narewari, the most proximal exposed section, the sequence begins above a ~ 10 m thick marker bed featured with caliche nodules that marks the emergence related closure of the preceding almost mega-marinc invertebrate-devoid Toarcian–Aalenian second-order sequence. The TST includes only a few bivalve-bearing thin fine textured siltstone beds. The RST is made up of the thick unfossiliferous laminated/rippled crossbedded sands with occasional thin clay/silt intervals. In Sadhara, the differentiation is rather difficult. The relatively more fossiliferous (*Calliphylloceras* bearing) SCLm at the base of the section along with early cross-bedded part of RSm make the first third-order sequence.

J-V. (2) Third-Order Sequence
It includes beds 41–45 of the informal early middle part of the KSm with the MFS above the fossiliferous interval of the beds 41–44 as that of the second-order TST. The thick rapidly deposited crossbedded coarse sands of the bed 45 represent the corresponding RST with possible terminal emergence. The MFS as in the previous

third-order sequence is speculated in early part of Propinquans Zone which also is the second-order MFS. The terminal SB of this third-order sequence is at the top of the Propinquans Zone. In Narewari, the third-order TST is readily picked up with multiple levels of bivalve-bearing thin clay/silt beds with occasional sands of the fourth order. The RST is mostly made up of the thick unfossiliferous crossbedded sands. In Sadhara, the middle part of RSm includes multiple thin levels of broken and reworked shells in clay/silt/sand beds that suggest good near or just above normal wave base depths, the best of RSm. These suggest third-order TST which is superposed by more than one relatively thicker sand beds as RST. At Jhura, the golden oolitic Cm makes the second third-order TST, while the superposed marls constitute the RST.

J-V. (3) Third-Order Sequence It includes the beds 46–48 with thin fine textured fossiliferous beds 46–47 as TST and thick unfossiliferous crossbedded coarse sands of bed 48 as the regressive/non-marine RST interval. The top of the corbulid abundant bed 47, the most fossiliferous bed of the second-order sequence is interpreted as the MFS of the third as also second-order sequence. Using synchroniety of the third-order sequence surfaces, the second-order MFS is assigned to the top of Subfurcatum Zone while the terminal SB is suggested within the Late Bajocian Garantiana Zone. At Jhura, the composite Em includes this sequence, with lower golden oolitic limestones as Em makes the TST of suggestively Humphresianum Zone–Subfurcatum Zone interval, while the superposed clay/silt constitute the RST of early part of Garantiana Zone. At Sadhara, the upper part of the RSm includes rare thin silt/sand broken reworked shell beds as third-order TST that are overlain by the thick rippled and cross-bedded sands as the RST immediately below the LYFm.

3.5.5.5 Intra-Bajocian Ages and Correlations in Kachchh

The exposed Bajocian in Kachchh is better than the earlier Toarcian–Aalenian interval, yet restricted in the proximal exposed part to the Island belt in Patcham and Khadir, to Jhura in East Mainland about 50 km from Patcham. The terminal part of the Bajocian may also be present in the basal part of the Nara section, where studies have only recently begun. Ammonoids are almost absent except for *Calliphylloceras* from near the base of Sadhara dome. The long held Late Bajocian *Leptosphinctes* is now revised as a Bathonian Perisphinctidae. The succession in Island belt is dominated by the unfossiliferous medium and coarse clastics with a few intervening thin fine textured fossiliferous limestone/shale/marl intervals, while at Jhura there is alternation of limestones, mostly golden oolitic, and shales/marls. In absence of the guide fossils, the correlation (Figs. 3.17, 3.18, 3.19, 3.20 and 3.21) is based on sequence stratigraphy with age of the second-order sequence surfaces extended from other constituents of IEAP, and those of the third-order speculatively extended from the ETM. Based on experience of the younger guide fossil-bearing intervals in Kachchh, it is observed that in the proximal sand dominated rapidly deposited unfossiliferous sediments, the intervening thin fossiliferous intervals invariably constitute the second or third-order TSTs. Among the precisely age constrained units of possible help on the upper side is the ammonoid-bearing GYFm, with base as the second-order SB of late Middle Bathonian Bremeri Zone Bullatus Subzone, i.e. at the ETM Morrisi Zone/Bremeri Zone boundary and at the base of the GTM Mantataranus Zone, so marked by the first Macrocephalitinae in Kachchh. On the lower side, the Aalenian/Bajocian boundary, due to absence of ammonoids in Kachchh, is extended from Pakistan as a second-order SB. The precisely age constrained units in Jhura section include three relatively thin oolitic/non-oolitic fossiliferous limestone intervals, alternating with unfossiliferous to scarcely fossiliferous thicker shales/marls. In Kaladongar, also, between the Toarcian–Aalenian second-order sequence below and GYFm above, there are three similar relatively thin fossiliferous limestone/marl intervals that alternate with thick unfossiliferous sands. In Sadhara, only two such intervals are present. These slow sedimented fossiliferous limestone intervals are here interpreted as the second-order TSTs, however, could also include even third-order TSTs. On the other hand, an Early Bajocian second-order TST is widely known in Pakistan, South Tibet, Nepal and Madagascar (Krishna et al. 2011). Therefore, the basal SCLm at Sadhara, Am at Jhura, and beds 37–44 at Kaladongar are interpreted as the second-order TSTs of the Early Bajocian age with MFS at the top of the basal limestone unit at Sadhara, the top of Am at Jhura, and the top of bed 44 at Kaladongar. The MFS is here assigned to the Early Bajocian Propinquans Zone. So, the top of the Red Sandstones (RSm), top of Bm at Jhura (same as top of Badi White Limestone member, i.e. BWLm) are tentatively suggested within the mid-Late Bajocian Garantiana Zone. Thus Am–Bm succession, i.e. BWLm suggestively ranges from the basal Bajocian Discites Zone to the mid-Late Bajocian Garantiana Zone. The Am/SCLm, also beds 37–44 of Kaladongar suggestively range from the Discites Subzone to the early part of Propinquans Subzone. The Bm, RSm and the upper part of KSm (beds 45–48) range from the late Propinquans Zone to Early Garantiana Zone. Major parts of Cm, LYFm, and BCSm range from the late Garantiana Zone to the close of Bajocian, i.e. Parkinsoni Zone.

Unlike the correlation of the Cm of Jhura with the terminal part of BCSm and basal part of GYFm by Fursich et al. (2013), Cm as above is suggestively correlated to the basal fossiliferous fine textured part of BCSm as another second-order TST of the mid-Late Bajocian late Garantiana Zone to the basal Bathonian Macrescens Subzone interval. The Macrescens Subzone is otherwise firmly determined in the newly discovered ammonoid rich Nara section. Thus beds 1–3 of Nara are easily correlated with the basal part of BCSm (Beds 49–54). The top of LYFm of Sadhara is also assigned to Macrescens Subzone. The bases of these TSTs are tentatively dated at the Aalenian/Bajocian boundary in context of the rift relating deepening event in Somalia and Axial belt of Pakistan at the start of Bajocian as the second-order SB in IEAP.

The third-order sequence surfaces and their extrapolated ages are only introduced as speculations in the Kaladongar and Sadhara sections. The

Aalenian/Bajocian boundary is placed at the LSM/ULM contact at Jhura and at the base of the Sadhara section. In Kaladongar, the boundary is tentatively placed between KSm and BCSm, between beds 48 and 49. The Bajocian/Bathonian boundary is interpreted as included in Nara between beds 2 and 3, and within beds 49–54 at Kaladongar. In Jhura, the said boundary is included within Cm, and at Sadhara within LYFm. The third-order systems tracts also allow vague differentiation of the ammonoid zones. Discites Zone makes the first third-order TST, Laeviuscula Zone the first third-order RST. The basal SB of the second third-order sequence defines the Laeviuscula Zone/Propinquans Zone boundary. While the top of the Propinquans Zone marks the terminal SB of the second third-order sequence. The Humphresianum Zone – Subfurcatum Zone interval makes the third third-order TST, while the terminal SBs of the 3rd as also second-order sequences fall in the middle of the Garantiana Zone.

3.5.6 J-VI. Late Bajocian Garantiana Zone–Mid. Middle Bathonian Morrisi Zone ~1.5 my Long Second-Order Sequence with Early Bathonian Macrescens Subzone MFS

This sequence is present at Kaladongar, Goradongar, Jhura, and Nara with MFS precisely determined in the early Early Bathonian Macrescens Subzone (Figs. 3.17, 3.18, 3.19, 3.20 and 3.21). In Kaladongar, it includes the entire BCSm along with the basal sands of GYFm. In Jhura it includes Cm–Dm units, and in Sadhara, it includes LYFm and USm. In the relatively less proximal Sadhara and rather distal exposed Jhura, there is little doubt about the second-order MFS above the LYFm and Cm respectively, i.e. above Rm. In the most proximal exposed, the relatively some what expanded fine textured clastics dominated succession, there have emerged as discussed later four third-order MFSs, one among them being the second-order MFS of Macrascens Subzone. It could either be the third or fourth third-order MFS, however, of the basal Bathonian Macrescens Subzone. Thus, either the basal ~ 15 m range up to the Macrascens Subzone or the entire BCSm inclusive of the marker Rm ranges up to the Macrescens Subzone. Macrescens Subzone age of the Rm is favoured, however, even younger age within the Progracilis Zone–Subcontractus Zone interval is not ruled out. In both the possibilities, the Rm is assigned younger basal Early Bathonian or early to mid-Middle Bathonian age compared to it's consistently given earlier Late Bajocian age. In case of the older of the two—the Macrescens Subzone age of the Rm, the solitary *Leptosphinctes* included in the Rm remains under Leptosphinctinae. However, in case of the yet younger Middle Bathonian age, the so determined *Leptosphinctes* may have to be taxonomically reassigned as an indeterminate Perisphinctinae.

3.5.6.1 Discovery of the New Ammonoid-Bearing Nara Section of late Late Bajocian–Early/Mid. Middle Bathonian Duration

The only ammonoid evidence from this sequence is the recently discovered ammonoid inclusive terminal Late Bajocian–Early Bathonian limestone succession NLm at Nara (Pandey and Pathak 2015a; Krishna 2014 unpublished report). The oldest ammonoids thus far known from Nara were mainly mid/late Middle Bathonian Macrocephalitinae. The preliminary studies reveal MFS of this second-order sequence above the basal Early Bathonian (Zigzag Zone, Macrescens Subzone). The determination of the Macrescens Subzone MFS is based on the maximum density and diversity of ammonoids in the particular MFS level. In the Nara section, the nearly complete sequence is interpreted in the early ~16 m thick ammonoid inclusive greyish white limestone and sandy limestone with shale/silt/marl interbeds without golden oolites of the tentatively Late Bajocian–early and mid-Middle Bathonian age. The MFS is best determined in the ~1.50 m thick highly fossiliferous ammonoid-bearing limestone that lies ~3.50 m above the unexposed base. The ~1.50 m bed has yielded two stratigraphically significant genera *Parkinsonia* and *Siemiradzkia* which constrain the age of the single exposed bedding surface precisely as basal Early Bathonian Macrescens Subzone of Zigzag Zone. The LAD of *Parkinsonia* is in Macrescens Subzone while the FAD of *Siemiradzkia* is in Macrescens Subzone. The richest presence of ammonoids is noted in the bed 9. Hence, the age of the MFS is Macrescens Subzone. There are a few relatively long ranging genera present such as *Prohecticoceras*, *Procerites*, *Wagnericeras*, *Telermoceras*, *Ebrayiceras*, *Berbericeras*, *Procerites*, *Procerozigzag*, *Zigzagiceras*, *Cadomites*, etc., as well in this assemblage (Pandey and Pathak 2015a). Most of the ammonoid specimens are incomplete whorl fragments. Large part of the ammonoid collections comes from the basal part of the bed 9. The particular Nara section adjacent to a relatively large igneous body is ~16.00 m thick and is almost entirely ammonoid inclusive. The Nara section may in future provide clues to the Bajocian/Bathonian stage boundary in the continuously ammonoid-bearing succession, which is tentatively suggested immediately below the ammonoid rich interval of beds 1–8 as the second-order TST/MFS level/bed. Never the less, this is the first second-order sequence of the continuously ammonoid-bearing Bathonian–Tithonian Kachchh Jurassic. Possibly, had Early Bajocian been exposed at Nara, it would have been ammonoid bearing. The terminal SB of this sequence in Jhura is placed above the Dm of thinly bedded limestones. The igneous body did not affect the overlying sediments by way of baking/charring or related alteration, and suggestively could be older of the first-order intra-Jurassic ~183 ma SB age. It is strongly desirable to date the said igneous body in context of the tectonostratigraphically similar igneous events in wells of the west sector in Kachchh, Saurashtra, Pakistan, etc.

3.5.6.2 Large ~12 my or Longer Intra-jurassic Gap in the Pakistan Axial Belt Jurassic Record

There is no record of Middle and Late Toarcian–Aalenian ammonoids in the ~5 m sediment interval between the ammonoid-bearing Early Toarcian below and Early Bajocian above. On the ETM in many basins, the ~8.0 my late Middle to Late Toarcian interval has been interpreted as a submarine gap on account of starved conditions. In Axial belt of Pakistan too the Middle Toarcian to Late Aalenian internal may be a submarine gap. In this context, it is interesting to discuss the more pronounced Late Bajocian–Oxfordian ~12 m long gap in the Axial belt of Pakistan which in some other areas is reduced to Middle Callovian–Oxfordian ~10.4 my gap.

3.5.6.3 Revised Interpretation from the Subaerial to Submarine Gap

These gaps have been explained earlier by one and all in Pakistan through tectonic uplift of the Axial Belt. Not just in Pakistan but the understanding on the entire GTM about the said gap has been subaerial. The tectonic uplift, if so, need be complimented by shallow water high energy coarse textured, preferably ammonoid-devoid sediments both below and above the gap. On the contrary, both the underlying and overlying sediments are of low energy, fine textured, relatively deeper water ammonoid-rich condensed sediments. The author earlier revised (Krishna et al. 1998, and all later papers) the similarly interpreted uplift in Kachchh to tectonic submergence and associated deepening. The deepening in turn first caused condensed sediments, and yet further up starved basin conditions. Such scenario leads to a submarine gap. The so interpreted intra- Jurassic gap is present in the vertical vicinity of the first-order intra-Jurassic MFS in late Middle Oxfordian all over the GTM from North Africa to NW Australia in relatively basinal sites away from the margin (Krishna et al. 2008, 2011). A part of the submarine gap is below the first-order MFS and part above the MFS. The earlier Middle Toarcian–Late Aalenian ~9.6 my gap is also similarly interpreted as submarine gap. As the depositional site moves towards the margin, the submarine gap gets reduced in duration.

3.5.6.4 Jhura Section

This sequence is included in the fine textured golden oolitic limestones alternating with the limestone/shale/marl succession (informal members Cm to Em) (Figs. 3.18 and 3.19). Fm and Gm include Macrocephalitinae, and are interpreted to make the early part of the second-order TST of the next younger second-order sequence beginning at mid/late Middle Bathonian (Krishna et al. 2011) boundary. The Em includes an early golden oolitic part followed up by an upper shaly part. Above the Em is the terminal SB of the present sequence of mid/late Middle Bathonian. Based on the homotaxial relationship of the Jhura and Nara sections with similarly determined terminal SB as above discussed, the second-order Macrescens Subzone MFS in Jhura may be placed above the Cm. The tops of Am and Cm are thus interpreted as the second-order MFSs, however, the same could also be of third order. Further, using the regionally spread synchroneity of the second-order sequence surfaces, the same, respectively, are here corresponded to the tops of the Early Bajocian Propinquans Zone and the early Early Bathonian Macrescens Subzone. In the regional context, there is widespread ammonoid rich Early Bajocian in several basins on the GTM particularly Madagascar, Pakistan, Nepal and South Tibet, which suggests the second-order MFS in the Propinquans Zone above the Am, however, there can not be any certainty. The Jhura Bajocian is only relatively proximal, and, it is desirable to search for ammonoids in both Am and Cm and even Em. Both Am and Cm are relatively slow sedimented compared to Dm and Bm. Thus, the Jhura section may extend down up to the base of Bajocian, if not earlier. The Em is interpreted to make one third-order sequence, which is here correlated to the last or fourth third-order sequence differentiated at Kaladongar, of suggestive Middle Bathonian Progracilis Zone–Morrisi Zone interval.

3.5.6.5 Sadhara Section

The tops, respectively, of the relatively thin carbonate intervals in the otherwise sand-dominated Sadhara section—the lower flagstone and basal limestone (Figs. 3.19 and 3.20)—are similarly interpreted as the second- or third-order MFSs, preferably second order. If that be so, the same are here suggestively corresponded to the basal Early Bathonian Macrescens Subzone and Early Bajocian Propinquans Zone. The basal limestone, if constituting a third-order TST, could also belong to the Late Bajocian Garantiana Zone or Subfurcatum Zone. However, more chances are in favour of the Early Bajocian Propinquans Zone in regional context as already discussed above. As such Sadhara too may include Early Bajocian.

3.5.6.6 Comparison of Jhura, Sadhara and Kaladongar Sections

The correlation suggested here is of the tops of the carbonate intervals, respectively of Jhura (Am beds 1–5), basal limestones at Sadhara and beds 37–43 of Kaladongar of KSm well as MFS between Mainland and Island belt. The same in turn allows tentative correlation of the intervening shaly intervals—Bm bed 6 of Jhura with the intervening relatively much thicker sandy intervals of Sadhara and Kaladongar, i.e. Bm correlated to red sands of Sadhara, and sand rich beds 44–48 of KSm as corresponding RSTs. Similarly further up are correlated the Cm–Em interval of Jhura to LFm–USm of Sadhara and beds 49–58 of BCSm as constituents of the next second-order sequence.

3.5.6.7 ETM Context

The ETM as well includes a second-order sequence starting in Early Bajocian with MFS only one subzone younger and terminal SB above the early Middle Bathonian. On ETM, there is significant presence of lytoceratins in the Recinctus Subzone, which is immediately younger to the Macrescens Subzone. The lytoceratins continue from below. This in turn suggests near similar maximum depth in the Recinctus Subzone as well, that may have prompted the placement of the MFS on ETM above the Recinctus Subzone. However, in Kachchh the ammonoid data is for MFS above the Macrescens Subzone. On the ETM, there could be considered Macrescens Subzone–Recinctus Subzone interval as MFI with MFS above the Macrescens Subzone. Also the GTM ammonoid fauna of this sequence has near exclusive ETM impress while the GTM as yet has not developed a distinctive fauna of its own which it does later in the next second-order sequence.

The second-order sequence is found readily divisible at Kaladongar into four third-order sequences, three in the second-order TST and one in the second-order RST.

J-VI (1–4) Third-Order Sequences
The thin multiple fragmented reworked bivalve-bearing clay/silt/lime beds alternate with the unfossiliferous silt/sand interbeds within the otherwise ~15 m thick interval of beds 49–55 of Biswas (1977, 1993) of the second-order TST. There are differentiated three third-order sequences of beds 49–50, 51–53, and 54–55. The third order includes ~50 m thick sands. The fourth third-order sequence is differentiated in the terminal part of RST in ~10 m thick beds 57–58 along with early sands of the otherwise limestone dominating bed 59. The top of the bed 58 at the top of BCSm makes the third-order MFS. The first third-order sequence is ranged up to the close of early Parkinsoni Zone with tentative MFS at the top of the mid-Late Bajocian Garantiana Zone. The second third-order sequence is ranged through the late Parkinsoni Zone. The third third-order MFS is the same as the second-order MFS at the top of the basal Bathonian Zigzag Zone Macrescens Subzone, while the sequence extends up to the close of the Aurigerus Zone. The last or the fourth third-order TST is included within the Progracilis Zone–Subcontractus Zone interval, and the terminal third-order RST of Morrisi Zone. It requires to be remembered that the above ages of the second- and third-order sequence surfaces are merely suggestive and tentative in absence of ammonoids. Among the above four third-order TSTs, the last or the fourth TST is the famous more than a meter thick coarse grained calcareous slabby sand bed Rm which is rich in broken reworked bivalve/gastropod/coral shells along with pebbles inclusive of igneous pebbles. This marker bed (included as the terminal 1 in Biswas 1993) immediately underlies the dominantly carbonate sediments of GYFm. The foresaid features of the marker Rm suggests a fresh significant transgression under slow, above to near wave base sedimentation.

It is interpreted as the second third-order TST with MFS as that of the second order as one of the alternatives. The other alternative is that the second-order MFS is of the first third-order being as well the second-order MFS. The RST is not recognizable, may be the same is either eroded away at the margin or present in the early sand dominated part of the GYFm. Thus the MFS depending upon its second- or third-order status is assigned either basal Bathonian Zigzag Zone Macrescens Subzone age or late Aurigerus Zone or early Progracilis Zone. The next younger overlying second-order MFS is in Late Bathonian. Thus, the so-called *Leptosphinctes* level of the Rm is between the Early Bathonian second-order MFS to the late Middle Bathonian third-order MFS.

A few interpretations that have emerged are; (1) the specimen described as *Leptosphinctes* may or may not be a Leptosphinctinae, (2) its earlier assigned Late Bajocian age is improbable since being incompatible with the ammonoid based late Middle Bathonian base of the immediately overlying GYFm, (3) It is best ranged between the basal Early Bathonian Macrescens Subzone and Middle Bathonian Progracilis Zone–Subcontractus Zone as a third-order MFS, (4) *Leptosphinctes* is not the oldest Jurassic ammonoid of the Kachchh basin, (5) the oldest Jurassic ammonoid of Kachchh basin is *Calliphylloceras heterophylloides* of Bajocian age from the basalmost limestone beds of Sadhara, perhaps Early Bajocian, which as per correlations detailed above belongs to a distinctly older level.

3.5.6.8 Origin/Extinction of the Ammonoid Subfamily/ies

The interpreted MFS above the Macrescens Subzone marks the extinction of the Pan-Tethyan Parkinsoniinae, and Leptosphinctinae, along with the origin of Zigzagiceratinae in Early Bathonian, and the origin of Macrocephalitinae at mid/late Middle Bathonian. The appearance of relatively deeper water Macrocephalitinae and its spread in the entire basin suggests substantive improvement in bathymetry even in the proximal-most ramp development in Island belt that signals below wave base subtidal depths.

3.5.6.9 Jhura Section

This sequence is included in the fine textured golden oolitic limestones alternating with limestone/shale/marl succession (informal members Cm to Em). The Fm and Gm include Macrocephalitinae, and are interpreted to make the early part of the second-order TST of the next younger second-order sequence beginning in mid./late Middle Bathonian (Krishna et al. 2011). The Em includes an early golden oolitic part followed up by an upper shaly part. Above the Em is the terminal SB of the present sequence of early/mid. Middle Bathonian. Based on the homotaxial relationship of the Jhura and Nara sections with similarly determined terminal SB as above discussed, the second-order Macrescens Subzone MFS in Jhura may be placed above the Cm. The tops of Am and Cm are thus interpreted as second-order

MFSs, however, the same could also be third order. Further, using the regionally spread synchroneity of the second-order sequence surfaces, the same respectively are here corresponded to the tops of the Early Bajocian Propinquans Zone and early Early Bathonian Macrescens Subzone. In the regional context, there is widespread ammonoid rich Early Bajocian in several basins on the GTM particularly Madagascar, Pakistan, Nepal and South Tibet, which suggests a possible second-order MFS in Propinquans Zone above the Am, however, there can not be any certainty. The Jhura locality is relatively proximal, yet in this context, it is required to search for ammonoids in both Am and Cm and even Em. Recently, a gastropod has been described from Am (Pandey et al. 2012a, b). Both the Am and Cm are quite slow sedimented compared to the Dm and Bm. Thus the Jhura section may extend down up to the Early Bajocian Propinquans Zone.

3.5.6.10 First Ever Firm Intra-Bathonian Ages and Correlations in Kachchh

Thus for the first time, on the basis of new ammonoid data from the Bathonian of Jumara, Nara and Jhura coupled with the comprehensive first- to fourth-order sequence stratigraphic differentiation, it has been possible to firmly date a few second- to third-order sequence surfaces, while a few others are suggestively dated on regional homotaxis in IEAP. In all the Bathonian sections, all the informal members of Biswas of ULM from Am to Gm in Jhura, different unnamed units of Jumara and Nara Bathonian, and GYFm, GSm and RLm of the Island bel are here precisely age constrained. The top of the Am is correlated to the Early Bajocian Propinquans Zone, the top of the Bm to the base of Garantiana Zone, the top of the Cm to the Early Bathonian Macrescens Subzone, and the top of the Em to the base of the late Middle Bathonian early part of Bremeri Zone, the top of the Fm to the top of the mid-Late Bathonian Hians Subzone, i.e. between the Retrecostatum Zone below and the Discus Zone above, and the top of the Gm to the Bathonian Callovian boundary between the Late Bathonian Triangularis Zone and the early Early Callovian Madagascariensis Zone. Zonation is proposed in the Bathonian at Jumara and Nara along with the extension of ages to other sections particularly in the Island belt. The top of the GSm in Goradongar is here tentatively correlated to the top of the late ammonoid scarce bed 31 as a third-order SB, and assigned to the top of the ETM Hollandi Subzone of late Late Bathonian Discus Zone. The bed 31 as such is interpreted as a third-order RST which is equivalent to most of the GSm except its marker sandstone at the base. The marker sandstone correlates with the TST in Jumara of beds 1–30 of mid-Late Bathonian Congener Subzone plus Hians Subzone interval of the Triangularis Zone (Pandey et al. 2013a). The following RLm is assigned to the terminal late Late Bathonian Orbis Subzone of Triangularis Zone. The GYFm is for the first time extreme precisely ranged from the Wagnericeras Subzone of Mantataranus Zone age, i.e. from the base of the Bremeri Zone to the Triangularis Subzone of the Triangularis Zone, i.e. the Blanazens Subzone of the basal early Late Retrecostatum Subzone of ETM which in turn implies the terminal Middle to basalmost early Late Bathonian. In Nara alone in the entire Kachchh basin, where studies are in progress, the complete Bathonian is exposed in the continuously ammonoid-bearing limestone dominating succession which has its boundaries seemingly conformable with the underlying terminal ammonoid-bearing Bajocian and the overlying ammonoid rich Callovian.

3.5.7 J-VII. Second-Order Late Middle Bathonian Mantataranus Zone / Bremeri Zone–O-I Horizon of Late Middle Callovian Obtusicosta Zone Obtusicosta Subzone Sequence of ~2.40 my Duration with MFS Above the Diadematus Subzone of Chrysoolithicus Zone

It is one of the most fossiliferous sequences of the Kachchh Jurassic as also lithostratigraphically the best differentiated. In the Jumara and Keera sections, it has been differentiated in about 270 beds of which about 80 levels have yielded ammonoids, the most anywhere in the world, the highest in the 13 plus sequences in the GTM Jurassic. The early part of the second-order TST of late Middle to Late Bathonian is in dominant carbonate facies of limestones and marl (Figs. 2.27, 3.22, 3.23, 3.24, 3.25, 3.26, 3.27 and 3.28), the limestones being often golden oolitic. The later part of the TST and entire RST is in siliclastic facies (Fig. 2.28) except for the early part of the Early Callovian in golden oolitic limestones alternating with clays, however, only locally at Keera. The contact between the Goradoongar Yellow Flagstone Member and the underlying Babia Cliff Sandstone is interpreted here as the second-order basal SB in the Kaladoongar section, also between the Em and Fm of Jhura (Figs. 3.22, 3.23, 3.24, 3.25, 3.26, 3.27 and 3.28). Thus, much unlike the placement of the base of GYFm older to the base of the Fm by Fursich et al. (2014), the two are here mutually correlated as the second-order SB while the BCSm is correlated to the combined interval of Cm, Dm and Em.

3.5.7.1 Drowning of the Carbonate Platform

The drastic lithological change from carbonates to siliciclastics within this sequence marks the drowning of the carbonate ramp near the Bathonian/Callovian boundary. This is also the lithostratigraphic boundary between the Patcham and Chari units. The probable cause is the change of climate from relatively warm and dry to less warm and wet. The other factor could be the transform movement of India inclusive of Madagascar towards south by a few degrees along the north–south striking Davies fracture zone. For a long time, the relative temperatures during the presence of the carbonate ramp have been interpreted higher than in the Toarcian–Aalenian before or after during the Callovian–Oxfordian. The temperature estimates during the

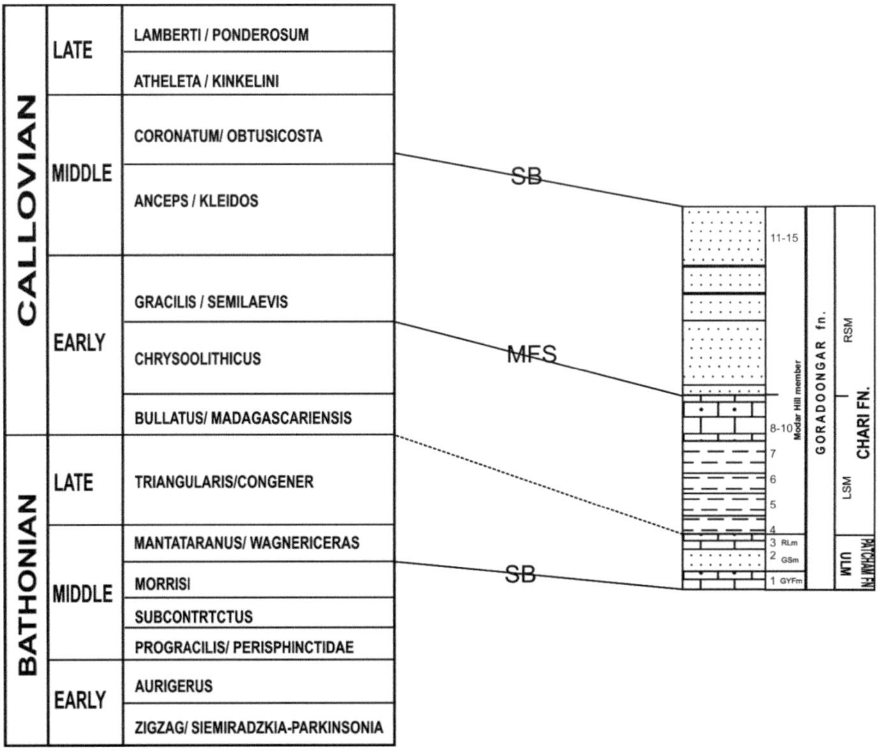

Fig. 3.22 Second-order sequence surfaces in ULM-LSM-RSM in Goradongar and precise ages of the members

Bajocian–Bathonian at the time of the carbonate ramp are of the order of 19°–24° centigrade (Fursich et al. 2005; Alberti et al. 2012a, b). No estimates are available for the Toarcian–Aalenian interval. During the Callovian–Oxfordian, the seawater temperatures are found reduced by several degrees to the range of 14°–16°. The interpreted transform movement let the Indian north and northeast margin (as a constituent of axial/core Gondwana) acquire subtemperate latitudes from the subtropical latitudes. These temperature estimates and differences seem to support well the repetitively interpreted north–south movement of India inclusive of Madagascar by a few degrees.

3.5.7.2 Correlation Among the Different Sections During Late Middle to Latest Late Bathonian

The sudden drowning of the carbonate platform allows correlation of the late Middle to Latest Bathonian part of the sequence among Jumara, Nara, Jhura, Habo and Island belt of the undivided ULM at Jumara beds 1–53 topped by the nodular

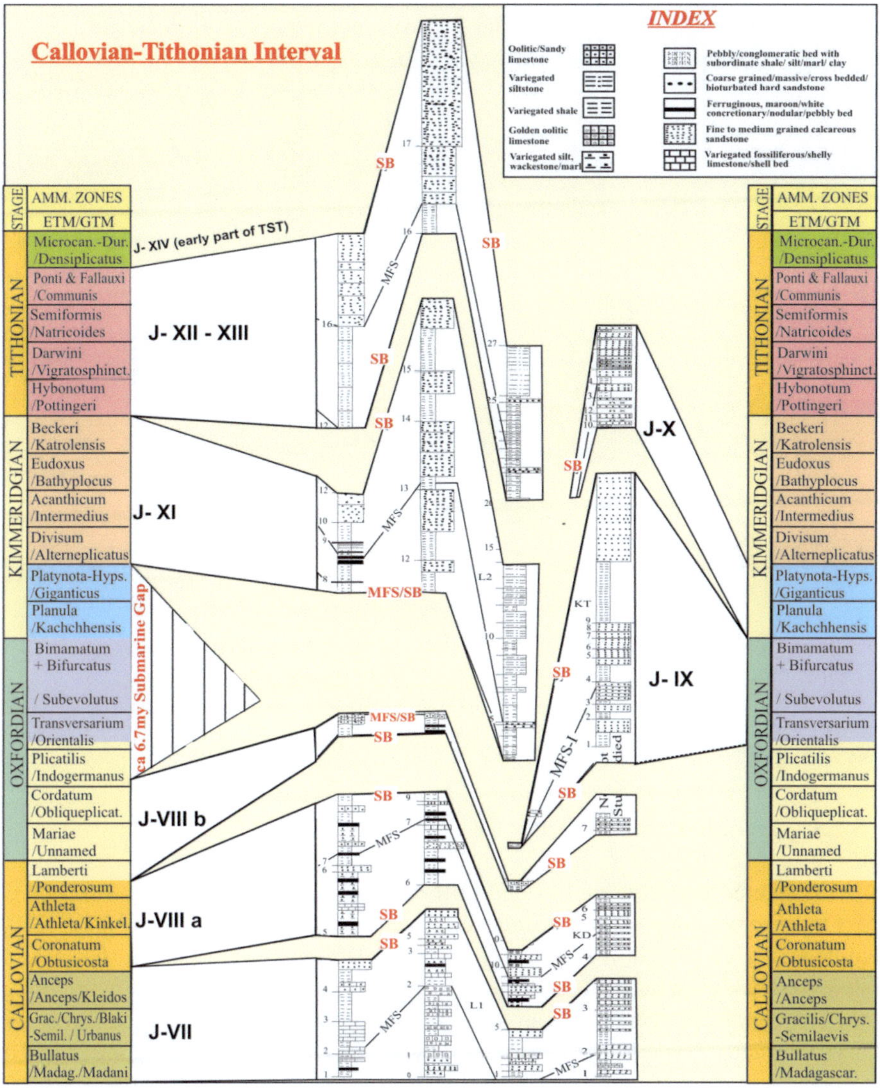

Fig. 3.23 Second-order depositional sequences in Kachchh Jurassic

limestones, Fm and Gm of Jhura also topped by nodular limestones, and Goradongar from the GYFm to RLm (3.22–28). More precise and finer correlations are discussed with the differentiation of the third-order sequences.

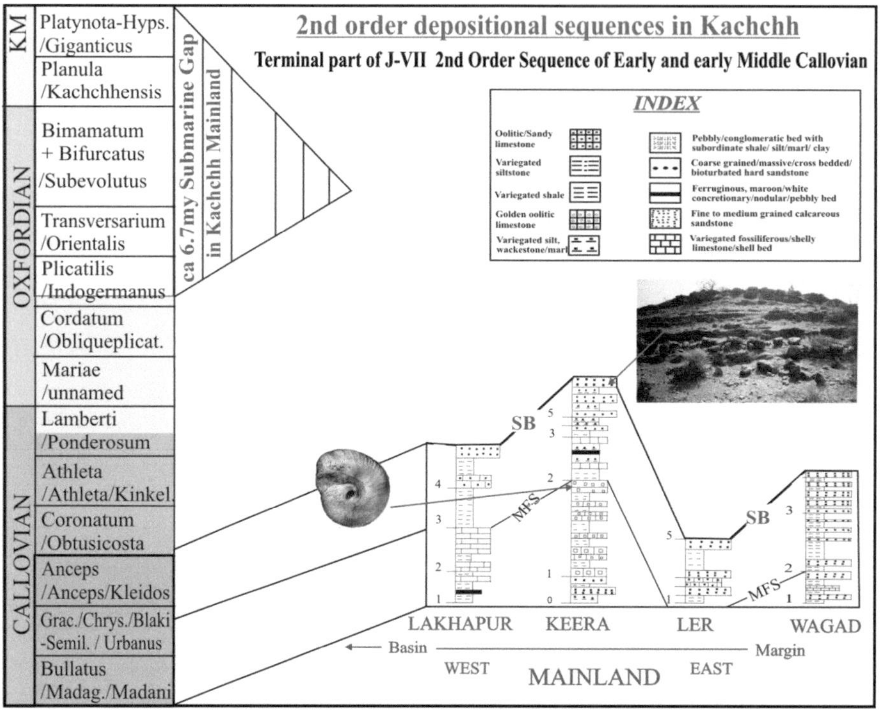

Fig. 3.24 Terminal part of J-VII second-order sequence of Early and early Middle Callovian

3.5.7.3 Terminal Second-Order SB

It is interpreted between the O-I and O-II horizons of the Obtusicosta Subzone of the Obtusicosta Zone of late Middle Callovian between the sediment interval VI and here revised base of the sediment interval VII of the Keera section (Fig. 2.29). Paleontologically, within this sequence, the maximum ammonoid density, diversity and frequency is observed at the MFS in the Diadematus Subzone of Chrysoolithicus Zone. There after there is noticed sudden and drastic reduction in the said parameters in the following RST. From a maximum of ~20 species at the MFS, the number of species in the taxon stratigraphic range chart is reduced to ~7, i.e. almost one-third while the chrysoolithicus lineage is terminated at the MFS in the Diadematus Subzone itself. Also, the MFS marks the start of the Semilaevis lineage. There are observed marked changes of abiotic factors as well across the MFS from the relatively fine textured thin bedded structure less to the thick coarse crossbedded hard resistant sands of the RSM.

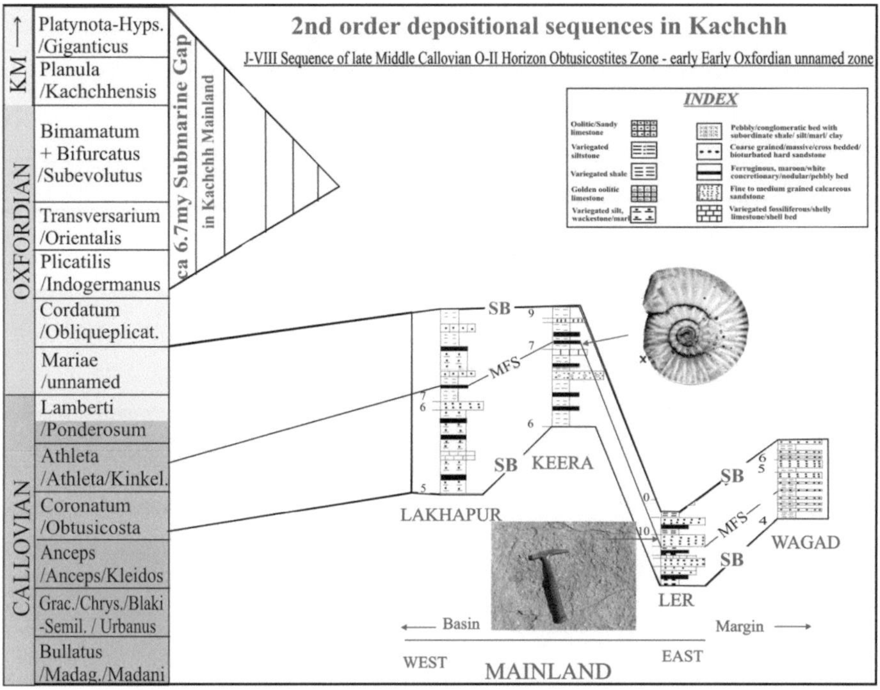

Fig. 3.25 J-VIII sequence of late Middle Callovian O-II Horizon Obtusicostites Zone–early Early Oxfordian unnamed zone

3.5.7.4 Constituent Units of the Second-Order Sequence

In terms of lithostratigraphic units, the above second-order sequence includes three younger members of Patcham Formation (GYFm, GSm and RLm) in Patcham, and two older members of the conformably overlying Chari Formation (LSM and RSM) in Keera. The early part of the sequence is also expressed in Habo, Jhura, Nara and Jumara in Mainland where later part of the Patcham Formation is exposed. Later part of the TST and RST are also exhibited in part or full in the entire basin, 'island belt', Mainland and Wagad.

3.5.7.5 Oolite Factory at a Shoal Near Keera

Chari Formation begins with shaly LSM which is full of varied marine invertebrate fossils, particularly ammonoids. The shale at the margin transforms into sandstones in the Island belt as early part of MHSm at Patcham or GSm elsewhere. Another feature of the LSM is presence of oolites which are in small unnoticeable manner as in Island belt and large densely packed way as at Keera—depending upon the relative distance from the oolite factory. Presence of an oolite factory at a shoal is

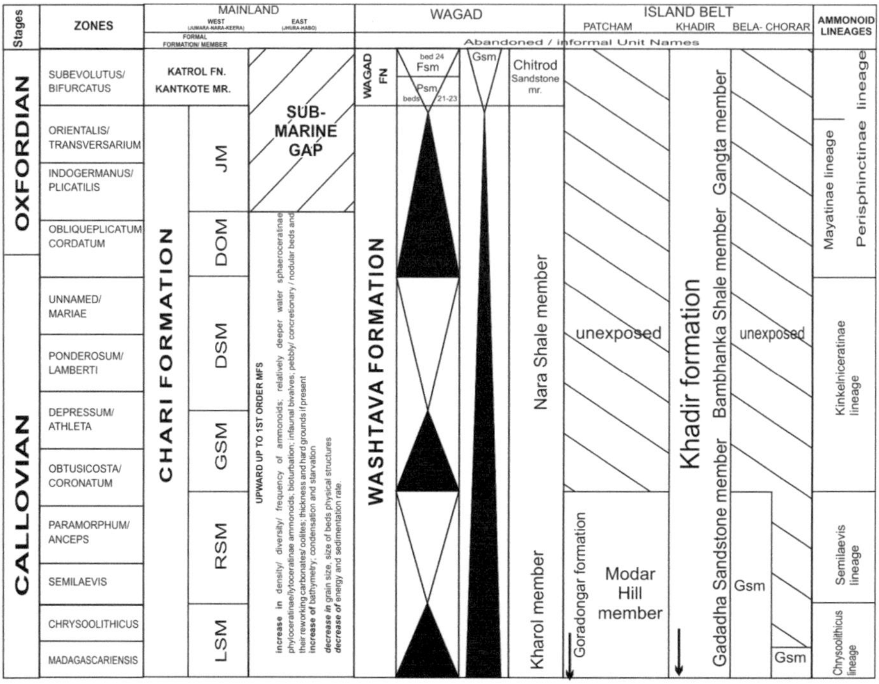

Fig. 3.26 Intrabasinal correlations vis-à-vis first- to third-order sequences during Callovian–Oxfordian interval in Kachchh basin

interpreted near Keera, from where the oolites are distributed by waves and currents. Bathymetry at any site in the basin is more than in the terminal part of the Patcham Formation. There is near absence of oolites in LSM at Washtawa dome section which has yielded a few *Macrocephalites*.

3.5.7.6 Development of Ridge Sandstone Member

The most striking lithological development of this second-order sequence is the Ridge Sandstone of the Chari Formation. It is a ridge forming hard resistant sandstone (also aptly named as China wall in Jhura and Habo) in East Mainland, Island belt and Wagad. Its development is explained due to tectonic reactivation of the provenance as also of the Median high at the MFS with slight tilting up along the Jhura–Patcham strike, and relative uplift of the 'Island belt' that caused the final emergence of Patcham, Chorar and Bela. This interpretation is supported by the presence of reworked Toarcian–Aalenian nanno-planktons that have been recovered by Rai and Jain (2013) in the sediments of the succeeding second-order sequence in the Lakhapur section. The emerged Patcham hill through weathering and erosion carried these Toarcian–Aalenian sediments to Lakhapur section of the Mainland.

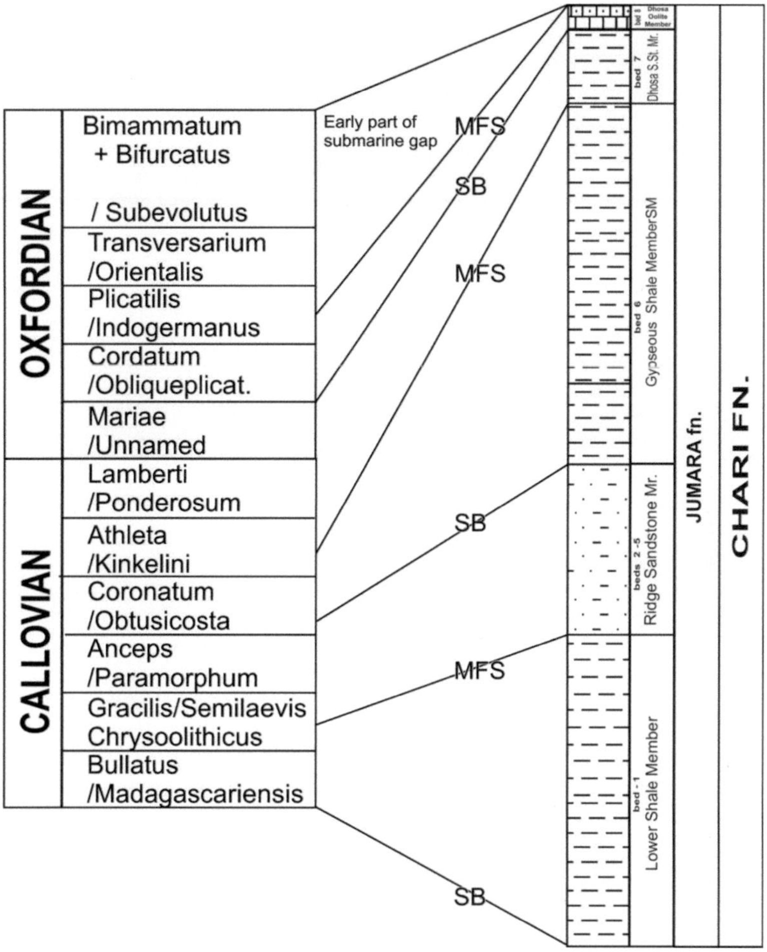

Fig. 3.27 Second-order sequences in the Callovian–Oxfordian at Jumara

RSM is found traceable and correlatable among Mainland, Island belt and Wagad, i.e. through out the basin with expected facies variation from the margin to basin. Thicker sands are developed in Patcham (upper part of MHm), at Khadir (upper part of GSm) and in Wagad and in East Mainland at Habo and Jhura. The sands near the margin are medium to coarse grained, coarsening, thickening and shallowing upward, rippled and crossbedded. In West Mainland sections, there is gradation from margin through decrease in size of structures, fining of texture, increase of bioturbation and increase of body fossils. Sheet like stacking of sand bodies is particularly featured as thick sand columns in West Mainland. Ammonoids are present mostly in West Mainland that too of geographically restricted IEAP elements, particularly during the third-order TSTs/MFSs as discussed later during the

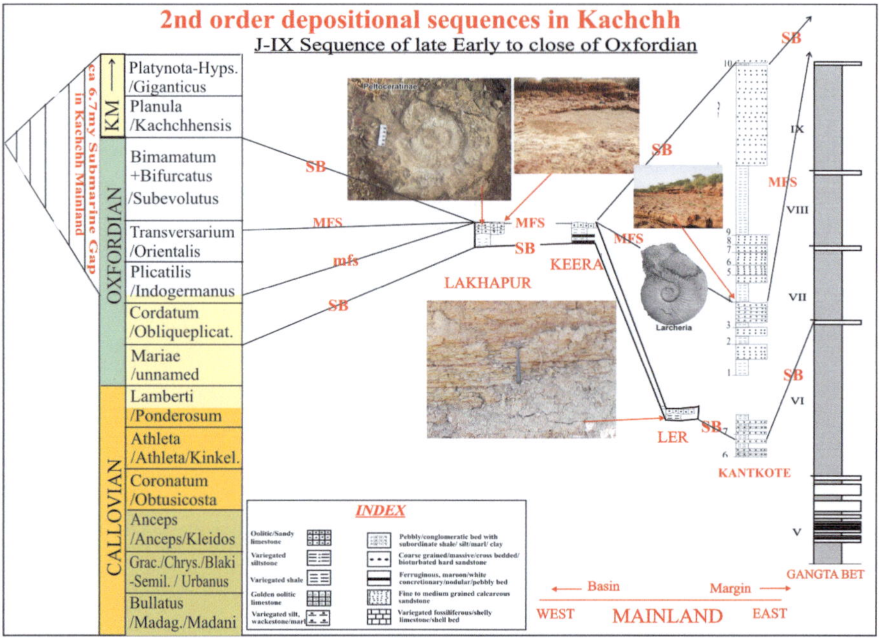

Fig. 3.28 J-IX sequence of late Early to close of Oxfordian

third-order differentiation. RSM is ammonoid age constrained from late Early Callovian Semilaevis Zone to O-I Horizon of the late Middle Callovian Obtusicosta Zone Obtusicosta Subzone (Fig. 2.29) instead of the exclusively Middle Callovian assigned by Fursich et al. consistently for two decades. At Keera, thick sands are not developed, significant sands occur intercalated between successive thin hard fossiliferous beds. Gadadha sands precisely correlate with the combined sediment interval of LSM and RSM and the top of Gadadha with the top of RSM as the second-order SB (Fig. 3.26). The top of Gadadha sands, top of RSM, or the top of Kharol Sandstone member of Wagad are practically isochronous. In most of the sections, two significant units are observed within the RSM which are intervened by relatively finer textured clay/silt part. Sand dominating lower part of RSM ranges from S-I Horizon of Formosus subzone to S-II Horizon of Semilaevis subzone while the upper one ranges from A-II Horizon of Kleidos Subzone to O-I Horizon of Obtusicosta Subzone. The relatively less thick intervening fine textured part ranges from S-III Horizon of Eucyclum Subzone to A-I Horizon of Ramosa Subzone. Even in most proximal Patcham, Khadir, Bela and Muouna on Island belt, the said two units of RSM are recognized, while in Wagad, the differentiation though present is not as explicit as elsewhere because of lack of continuos exposures.

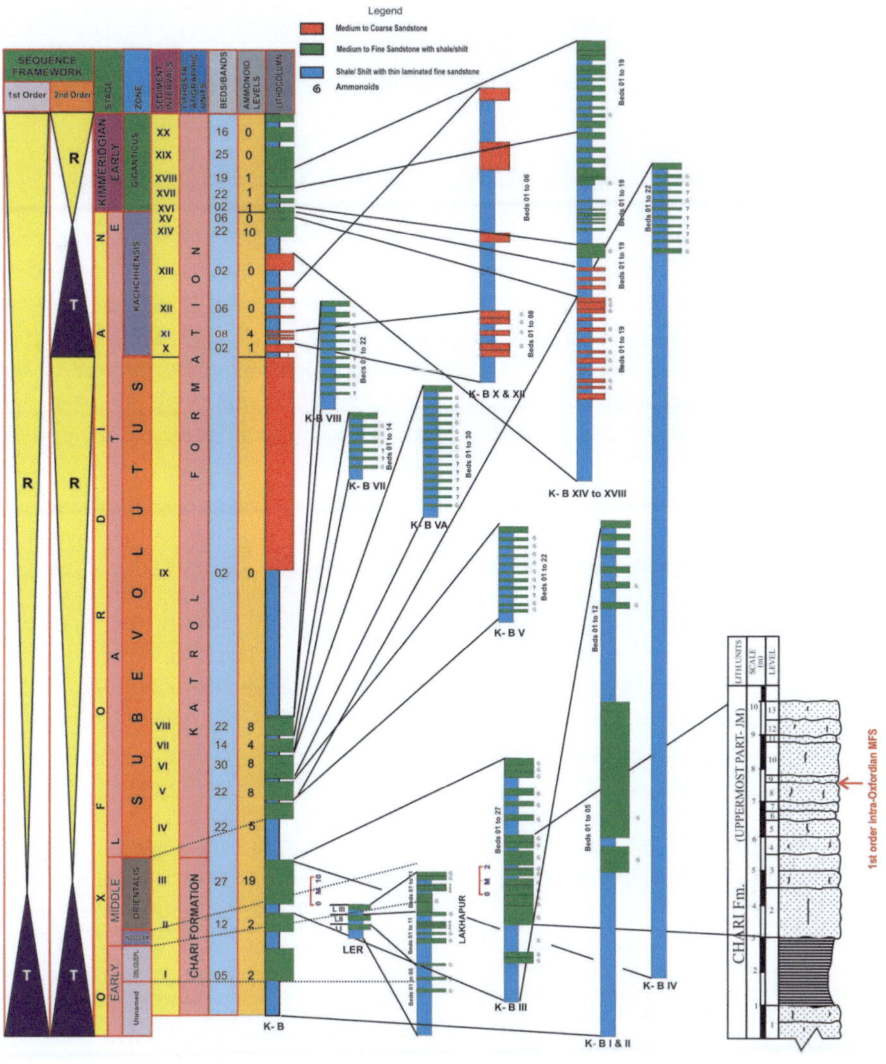

Fig. 3.29 Ammonoid zonal stratigraphy in the Oxfordian Early Kimmeridgian of Kachchh vis-a-vis first/second-order sequences after Krishna et al. (2009), also the succession of hard grounds is shown separately

3.5.7.7 Gradation of Biotic and Abiotic Features

There are several inferior order (third and fourth) thin interludes within the second-order RST of RSM. RSM in Kachchh basin is exposed for a length of well over 200 km, and obviously are present varied environments from storm wave base in West Mainland to above normal wave base in Island belt and Washtawa. There is noticed graded decrease of grain size and physical structures, energy framework and increase of bioturbation and body fossils, especially ammonoids from shallower margin in Island belt and Wagad to basinal sites of West Mainland.

3.5.7.8 Continuous Ammonoid Presence in West Mainland from Basal Bathonian to Close of Jurassic

The moderate to rich presence of ammonoids around mid-shelf in relatively distal/basinal exposed sections have allowed highly precise and refined zonal to horizonal level age refinement from the start of Bathonian onwards up to the end of the Jurassic. Pre-Callovian ages in the Kachchh basin are mostly supported by the ages of the second-order sequence surfaces in the neighbouring Pakistan and Madagascar constituents of GTM due to greater ammonoid presence there compared to Kachchh.

3.5.7.9 Differentiation of the Second-Order Sequence Surfaces

The second-order MFS is represented by a distinct thin hard visually observable top of the LSM bed 92 at the close of the sediment interval I. It marks the close of the Diadematus Subzone Chrysoolithicus Zone. Apart from the multiple abiotic parameters inclusive of the drastic change of litholgy, the MFS also characterizes the maximum frequency, density and diversity of ammonoids. There is a sharp drop in ammonoid density, diversity and frequency along with termination of the principal Macrocephalitinae lineage–the Chrysoolithicus lineage at the top of the chrysoolithicus Zone that had earlier begun at the base of the late Middle Bathonian Mantataranus Zone. The other lineage of the Macrocephalitinae continued up to the end of the second-order sequence, i.e. by the close of Ridge Sandstone Member. It is here surmised that the bathymetric decrease in the basin may have forced the Macrocephalitinae to retreat to deeper SW Pacific areas from where they had invaded the West Indian open sea shelf niches. The Macrocephalitinae may have continued to flourish in relatively deeper SW Pacific niches which have remained either not exposed or not age constrained in absence so far of good stratigraphic studies, since nearly identical Macrocephalitinae morphology did reappear in the form of Mayaitinae in Early Oxfordian after a gap of only late Middle to Late Callovian of over 2 my.

3.5.7.10 Suggestive Intrabasinal Tectonics

A regional tectonic event is suggested at the MFS of the late Middle Bathonian–early Middle Callovian second-order sequence of regional uplift and shallowing which is intrabasinally expressed in the tilting up towards north–northeast of the pre-existing NNE-SSW striking Median High, as also uplift and shallowing of the basin east of the Median High relative to west. The so interpreted tilting up of the Median High resulted in the uplift and emergence of the Patcham region since the Median High passed through Patcham, where the youngest Jurassic sediments are of late Early to early Middle Callovian age. In Mainland, particularly, east of Jhura (also located on the Median High) the Ridge Sandstone Member has a strong majestic thick monotonous ridge forming hard resistant sandy manifestation.

There is emergence as well of the Bela and Chorar regions along with Patcham. However, sedimentation continued in the Khadir region, since its being relatively less shallow than Patcham located on the Median High as evidenced by absence of thick sands of KSm and BCSm and instead development of a thick dominantly shaly/marly/limy succession in the immediately underlying Bajocian–early Middle Bathonian second-order sequence as the net resultant effect of the tectonic event which tilted up the Median High. The marginward tilting up of the Median High may also have caused uplift of Patcham and commensurate/complementary deepening of the immediately neighbouring Khadir.

ETM ammonoid elements are also present as auxiliaries/secondaries in the ammonoid fauna, particularly, so through out the TST. The present sequence includes three plus zones and seven subzones.

3.5.7.11 Ammonoid Lineage Evolution Linked to Intra-basinal / Regional Tectonics

The proliferation to decline and ultimate termination of the dominant ammonoid content of the two successive second-order sequences (the present and the previous one), was governed by regional tectonics, as exampled here on the GTM. Such unambiguous age constrained genetic link between regional tectonics and biological evolution has been for the first time demonstrated. The ammonoid lineages provide additional evidence of the regionality and net tectonic control of the 2nd and longer/bigger depositional sequences. The origin and extinction of the corresponding dominant ammonoid lineages on GTM and ETM have occurred at different times in the Jurassic in view of different ages of the tectonics on GTM and ETM, so also diachronous are their respective sequence stratigraphic surfaces.

3.5.7.12 Ammonoid Distribution and Compositional Changes in Mainland, Especially at Keera Across the Second-Order MFS

It is represented by the observable top of the LSM bed 92 at the close of sediment interval I which marks the close of Diadematus Subzone Chrysoolithicus Zone. Apart from the multiple abiotic parameters inclusive of the drastic change of lithology, the MFS also characterizes the maximum frequency, density and diversity of ammonoids. There is registered a sharp drop in ammonoid density, diversity and frequency along with termination of the principal Macrocephalitinae lineage–the

Chrysoolithicus lineage at the top of the chrysoolithicus Zone that had earlier begun at the base of the late Middle Bathonian Mantataranus Zone. The other lineage of the Macrocephalitinae–the Semilaevis lineage continued only up to the early part of the second-order sequence, i.e. by the close of RSM.

Several third-order sequences are differentiated as under (Figs. 2.28, 2.29, 2.30, 3.22, 3.23, 3.24, 3.25, 3.26, 3.27 and 3.28).

J-VII. (1) mid-Middle Bathonian–early Early Callovian third-order sequence
The differentiation of the first third-order sequence of the fore-discussed second-order sequence is approached through determination first of its MFS (Fig. 3.26).

Third-Order MFS Candidates
In the Patcham region in particular and 'island belt' in general three ammonoid-bearing beds in Gora Doongar (in the vicinity of Khawra and Sadhara) are present. The first two are vertically close together within a span of 2–5 m. The latter of these two successive close levels is relatively somewhat richer not only in ammonoids (*Gracilisphinctes*, *Micromphalites*, etc. of late Middle Bathonian) but in benthic elements as well, and is considered a third-order MFS candidate. The third ammonoid level, relatively less rich is in RLm with extremely scarce *M. triangularis* Spath together with *M. madagascariensis* Lemoine of late Late Bathonian age, is here interpreted as another MFS candidate. In the corresponding succession at Jumara, the bed 30 of sediment interval II, which, compared to Patcham, is much richer in ammonoids in density, diversity and frequency, obviously so, being located far down the basin under significantly greater bathymetry. The over all number of ammonoid levels is around 35–40 in Jumara, is much higher, and above all density is tens of fold higher. There is also sharp change of lithology and drastic reduction in ammonoid content across bed 30. In fact, beds 1–30 of sediment interval II at Jumara, physically, lithologically and paleontologically as already described in some detail in Chap. 2 differ with the underlying sediment interval I and overlying beds 31–63 of sediment interval II. Thus, the bed 30 is also interpreted as MFS candidate. In Jumara, another level (bed 4) near the base, the lone level within an otherwise ammonoid-devoid thickness of about 10 m, has also yielded a few ammonoid fragments of *Macrocephalites* and *Wagnericeras* and is here best assigned to late Middle Bathonian. It is also considered MFS candidate.

Ascending Order of MFS Candidates and Selection of the MFS
The correlation between the above two ammonoid-bearing MFS candidates at Patcham and two ammonoid-bearing MFS candidates of Jumara has been difficult in spite of all the four falling within the late Middle to late Late Bathonian bracket. The said beds 1–30 make a cohesive distinct unit in the Jumara column. At first, the four third-order MFS candidates are arranged in ascending order; (1) *Gracilisphinctes* level of Patcham, (2) *Wagnericeras* level near base of Jumara, (3) Bed 30 of Jumara, and (4) *M. madagascariensis* level of Patcham in RLm.

Bed 30 MFS of Jumara is the best, however, its correlation with any ammonoid-bearing coeval bed in Patcham and Jhura is difficult. The other choice could have been level 1. Both are part of the second-order TST, however, preference is given to the younger of the two in context of the conceptual increase in bathymetry as one goes up in the TST from basal SB to MFS. The older levels 1 and 2 could be close or of same age, while bed 30 in age falls between 1 and 2 below and 4 above. Once bed 30 of Jumara is selected as the third-order MFS, the other MFS candidates are here interpreted as fourth-order MFSs.

Tentative Correlation of the Selected Third-Order MFS

The bed 30 MFS is tentatively interpreted to fall at the boundary of Fm and Gm of Jhura, while in Patcham it should fall at or near the base of GSm between intra-GSm fourth-order MFS above and intra-GYFm fourth-order MFS below. In this context, it is here interpreted that bed 30 at Jumara makes the top of a marker brownish hard pebbly fossiliferous slow sedimented conglomeratic sandy limestone (bed 23 of Rajnath) which is also present at Jhura, Patcham, and perhaps also at Nara. Its thickness varies between less than a meter to ~ 15 m. Its biotic/abiotic features also vary substantively in a predictable graded manner between Jumara in the basin and Patcham at the margin. It is here firmly assigned to the top of Hians Subzone of the Triangularis Zone which in turn has been correlated to the top of early Late Bathonian Retrecostatum Zone Hannoveranus Subzone. Ammonoids remain unknown in Patcham at the intra-GSm level correlative marker sandstone of the said third-order MFS in the basal part of GSm. In Jhura ammonoids are present at the third-order MFS determined at the top of Fm. The ammonoid-devoid GSm is here best placed in the Hollandi Subzone of late Late Bathonian Discus Zone, which is broadly correlative of the unnamed subzone of Late Bathonian Triangularis Zone.

Most thickness of the first two sequences differentiated in Jaisalmer by Pandey et al. (2010) are included in the first third-order TST with MFS in the mid-Late Bathonian Triangularis Zone. The first third-order TST in the early part of the second-order TST unlike Kachchh is quite condensed and hard grounded with the Triangularis Zone ammonoid assemblage (Pandey et al. 2012a, b) near the GSM/RLM transition.

Determination of the SBs

The terminal SB in Keera is placed above the bed 42 at the boundary of Madagascariensis Zone and Chrysoolithicus Zone of early/mid-Early Callovian age. This is above the ammonoid scarce Madagascariensis Zone. Thus the composite RST is made up of beds 31–63 of the terminal part of Patcham Formation at Jumara and the beds 1–42 of the Chari Formation, at Keera. The entire third-order sequence contains scarce *Macrocephalites* spp., except at MFS where the density, diversity and frequency is moderate along with presence of secondary ETM elements.

The basal SB is between LSM and ULM, i.e. between BCSm and GYFm at Patcham. The oldest exposed beds at Jumara are probably younger to it, and so the above determined basal SB is not exposed. The basal SB at Jhura is placed between Em and Fm. The terminal SB in Jumara and other mainland sections fall within LSM of Chari Formation, although it is yet to be precised. The MFS in Jhura is determined at the top of Fm in the form of a marker pebbly fossiliferous hard calcareous sandstone. As per approximate durations of late Middle Bathonian to close of Bathonian of $\sim$ 1.1 my and of the superposed Transitorius Zone of $\sim$0.5 ma, the duration of the sequence comes out approximately 1600 ky.

A few fourth-order sequences are also here differentiated in the GYFm—intra-LSM third-order Sequence.

J-VII. (1a). The First Fourth-Order Sequence
It includes at least the entire GYFm. The ammonoid (*Gracilisphinctes* and others) bearing-golden oolitic limestones only a few meters below the top make the MFS of the late Middle Bathonian Bremeri Zone. The basal SB is marked at the base of the unit between BCSm and GYFm of suggestive Morrisi Zone. The terminal SB may either be the top of the unit or the terminal part of RST removed by weathering and erosion. In Mainland at Jumara the terminal sequence boundary is determined between the sediment interval I and II. The MFS is suggested in the *Wagnericeras*-bearing bed 04 of sediment interval I at Jumara, which is here correlated to the MFS at Patcham, and assigned to Bremeri Zone. The basal SB unfortunately is not exposed at Jumara. At Jhura the MFS is suggested between Fm and Gm. MFS should be exposed in the Nara section which, however, remains unstudied yet for sequence formulation.

J-VII. (1b). The Second Fourth-Order Sequence It includes $\sim$7–11 m thick marker hard fossiliferous pebbly brownish Sandstone at the base of GSm as the TST without presence of any ammonoids. The top of the marker sandstone makes the MFS at Patcham. There is noticed marked litho and bio contrast across the MFS. In the Jumara section down the basin, the MFS is observed in bed 30 of the sediment interval II on the basis of maximum ammonoid density and diversity in this bed. In view of the presence of *M. cf. triangularis* Spath together with *Procerites hians* Waagen, *Perisphinctes congener* Waagen and *Kheraiceras hannoveranus* (Roem) in the middle of Triangularis Zone at the top of the Hians Subzone, the top of the mid-Late Bathonian Hians Subzone is here equated to the top of the ETM Hannoveranus Subzone. The terminal SB at Jumara is determined at the top of the relatively thick ammonoid devoid to ammonoid scarce Nodular Limestone beds 31–52 of sediment interval II. Bed 53 thus becomes a good choice for the base of the youngest Subzone of the Late Bathonian Triangularis Zone. Bed 52 is here suggestively equated in age to the top of GSm. At Jhura also, the basal SB is determined by the base of the hard brownish pebbly fossiliferous marker sandstone which is followed up by the nodular limestone as RST between Fm and Gm and MFS at the top of Fm. As at Jumara, the top of the fossiliferous pebbly sandy limestone is determined as the MFS. Habo and Nara sections remain unstudied for sequence stratigraphy.

J-VII (1c). The third Fourth-Order Sequence
It begins with Raimalro Limestone as the TST in carbonate facies superposed by early part of Shales of the LSM of the Chari Formation of Madagascariensis Zone beds 1–11 as the RST of basal Callovian age. The topmost $\sim$4 m of marls/limestones at

Jumara are considered equivalent to the Raimalro Limestone and so also making the early part of the TST of this fourth-order sequence. This interval at Jumara is featured with overlapping presence both of *M. triangularis* Spath and *M. madagascariensis* Lemoine and is here assigned to the youngest subzone of the Late Bathonian Triangularis Zone equivalent to the ETM late Late Bathonian Discus Subzone of Discus Zone. The fourth-order sequence ranges from GTM terminal Bathonian Oxycerites Subzone to basalmost Callovian Horizon of basal Callovian Madagascariensis one.

However, the fourth-order sequence is neither that thick nor explicit in Jumara as in East Mainland Jhura and Habo and on the margin in Island belt in view of obvious non-development of GSm in relatively less shallow conditions. The typical crossbedded sandy character of GSm is obvious in East Mainland and Island belt sections between the two carbonate units of GYFm below and RLm above. It is replaced there by fine textured lime/marl/clay. The Yellow Flagstone *Gracilisphinctes* ammonoid levels of GYFm at Goradoongar and Sadhara may be correlated to the *Wagnericeras* ammonoid level of Jumara. Thus the GYFm ammonoid levels are here tentatively assigned to late Middle Bathonian, i.e. pre-Mantataranus Zone age. GSm is assigned to early part of the Discus Zone of late Late Bathonian which is equivalent to the Hollandi Subzone of Discus Zone, and Raimaro Limestone to later part of the Late Bathonian Discus Subzone. The important thing that has emerged from third- and fourth-order differentiation is the age equivalence of coral rich limestone of Jumara and GYFm of Patcham, along with the possibility of subaerial gap between GYFm below and GSm above. Fursich et al. (2013) have correlated GSm with the crinoid-bearing limestone in between the coral rich limestone below and sponge rich limestone above. This in turn implies a possible gap even between the coral rich limestone, i.e. sediment interval I beds 1–24 and crinoid-bearing limestone above of sediment interval II beds 1–30 near the late Middle/early Late Bathonian transition.

J-VII. (1d). Fourth Fourth-Order Sequence

Beds 01–18 make the TST and beds 19–32 as RST in the Keera section. The top of Horizon M-I makes the MFS while Horizon M-II the following RST.

J-VII. (2). Third-Order Mid-early Callovian latest Madagascariensis Zone M-III Horizon–late Early Callovian Semilaevis Subzone of Semilaevis Zone with MFS above Diadematus Subzone Chrysoolithicus Zone of Mid/late Early Callovian Age

The third-order MFS is the same as second-order MFS above bed 92, while the terminal SB is placed at the top of the Semilaevis Subzone of Semilaevis Zone above bed 120. The RST comprises thick ammonoid scarce fine crossbedded sandy limestones of early part of RSM.

J-VII. (3). Third-Order Basal Latest Early Callovian Eucyclum Subzone–O-I Horizon of late Middle Callovian Obtusicosta Subzone Sequence with MFS at the top of the Eucyclum Zone of early Middle Callovian age

Sequence Surfaces
The third-order TST is made up of thin ~10 m slow sedimented low energy thin bed pebbly/concretionary ammonoid-rich conglomeratic silt/sand at the terminus of late Early Callovian beds 120–160 of sediment intervals III and IV with MFS at the top of Ramosa Subzone of Eucyclum Zone. The RST comprises the majestic looking ~80 m thick hard resistant ridge/wall forming crossbedded rippled medium to coarse ammonoid scarce rapidly deposited sands with shale/silt interbeds from bed 161–bed 218.

The third-order differentiation allows division of the RSM into two distinct rather large parts of late Early Callovian and mid- to late Middle to early part of late Middle Callovian, which could be further divisible into 2–3 divisions even in West Mainland Keera, nothing to speak of the shallower East Mainland sections of Habo and Jhura. In West Mainland, the divisions are age constrained through ammonoids, while the divisions in East Mainland by way of fourth-order sequence differentiation.

Ammonoid Characteristics
There is marked origin and extinction of the Eucylum–Ramosa lineage respectively at the SB and MFS. The RST includes the main thicker upper part of Ridge Sandstone. The maximum density, diversity and frequency are observed at the top of the sediment interval IV.

Relatively Easier Differentiation of the Third-Order Sequences in Distal Sections
The third third-order sequence is differentiated relatively much easier in the distal relatively deeper sites. However, in the proximal sites, it is not impossible to miss the relatively thin ammonoid scarce third-order TSTs which are sandwiched in between the ammonoid poor thick crossbedded sands.

Differentiation at Keera
At Keera, the TST of this sequence includes 11–12 m thick fine textured pebbly maroon to purple slow sedimented ammonoid rich sediment intervals III and IV at the terminus of the late Early Callovian of Opis/Eucyclum Subzone to Ramosa Subzone age of early part of Anceps Zone. The RST comprises sediment intervals V and VI of thick ammonoid scarce sandy limestones to calcareous sandstones. In this sequence there is again drastic reduction of ammonoids to about one-third across the MFS between the TST and the RST. The TST is characterized by Eucycloceratinae, which started at or near the basal SB and became extinct at or near the MFS. The Eucycloceratinae may even have begun at the base of the Middle Callovian. If that be so, the Semilaevis Subzone may signal the top of Early Callovian. Equally significant is the origin of Kinkeliniceratinae at or near the MFS, however, Kinkeliniceratins continue temporally beyond the terminal SB of this fourth and the last third-order sequence of the second-order sequence under discussion.

Bathymetry
Normal fully marine subtidal and greater depths are witnessed for the first time in the exposed sediments of Kachchh in late Middle Bathonian in the limestone–

golden oolite association throughout the basin in Mainland Jhura and Jumara and almost all over the Island belt in Patcham, Khadir, Bela and Chorar. For the first time, moderate presence of ammonoids is observed in the relatively deeper Jumara fine textured fully carbonate development. Few ammonoids have also been found in Patcham, Jhura and Nara to facilitate the precise age assignment of late Middle to late Late Bathonian range to this most widespread carbonate development of the entire Kachchh Jurassic.

Arrival of Macrocephalitinae from SW Pacific

Widespread carbonate ramp like conditions developed on the GTM in most basins at least between North Africa to West India in Bajocian–Bathonian interval. The ammonoids are dominated by the relatively deeper water sphaeroceratid Macrocephalitinae which arrived from the SW Pacific. The sphaeroceratins having travelled and expanded from SE Pacific to SW Pacific, were presumably waiting for some time to advance further and invade the West Indian shelf in appropriate favourable environment (Krishna and Cariou 1993) that became available only in late Middle Bathonian. One significant change had been increase in bathymetry to allow the low to moderate order expansion of relatively cold water Macrocephalitinae in yet not sufficiently deep conditions in the Kachchh basin. Bathymetry for the first time is found just enough for ammonoids in a few exposed depositional sites on the Mainland. Conditions in the 'island belt' in spite of improvement than in Bajocian are still not favoured for habitation of ammonoids. It may be noted that the Macrocephalitinae could not enter the Tethys through the direct and nearest Central Atlantic seaway/corridor due to unavailability of appropriate bathymetric conditions, and instead had to travel very long distance from Central Atlantic to SE Pacific to SW Pacific, and still wait there until late Middle Bathonian Mantataranus Zone. The auxiliary/secondary ammonoid elements are perisphinctins, hecticoceratins and oxyceratins which arrived east and southward from ETM during the third-order TSTs particularly around the MFSs/MFIs. Thus so far during each second-order transgressive pulse, the sea level rise combined with the tectonic submergence to produce distinctive step wise increase in bathymetry. It so continued since the start of the transgression from early Early Toarcian until the close of Bathonian over a period of ~18 my, of which the ~15 my Toarcian–Bajocian interval went almost without ammonoids in the exposed Kachchh, in spite of good ammonoid presence almost all over the GTM. Moreover, further up in time, the proportion of Macrocephalitinae in the ammonoid fauna increased markedly during Callovian presumably on account of yet greater bathymetry, and relatively cooler temperatures. As also earlier stated, India inclusive of Madagascar as a constituent of the core/axial Gondwana underwent transform movement of a few degrees towards south latitudinally along the north–south striking Davies fracture vis-à-vis East Africa.

Increase of Sand in the System

In the distal Keera, Jumara and Jara sections, the said RST is expressed as poorly fossiliferous ammonoid scarce sandy limestone or calcareous sandstone. At Keera the terminal SB is marked by bed 118 at the boundary of sediment intervals II and

III. Marked litho and bio contrast across the MFS is observed. Ratio of sand in the depositional system increases rapidly while that of ammonoids decreases drastically. From basin to margin, somewhat similar contrast through rapid increase of sand towards the margin is observed.

Ammonoid Biotic Changes as Consequence of Regional Extensional Tectonics
There is as well drastic reduction in ammonoid content. In the most proximal Habo section on the Mainland, the *Macrocephalites Semilaevies* (Waagen) is almost absent while in the relatively deeper distal sections the RST is found characterized by this species with FAD at the start of the RST and extinction at the end of the sequence at the terminal SB. Thus in this sequence dominated by the Macrocephalitinae, the TST is characterized by the Chrysoolithicus lineage with extinction at the MFS while the RST is distinguished by the Semilaevis lineage with termination at the terminal SB. In context of ammonoids, the relatively shallower RST in the distal West Mainland sections is featured with dominance of the morphologically compressed shallow water Formossus–Semilaevis lineage of the Macrocephalitinae while the underlying relatively deeper TST by the dominance of the morphologically inflated relatively deeper water Chrysoolithicus lineage. The sequence demonstrates well the bathymetric control on ammonoid morphology, so also bathymetric changes in sequence stratigraphic context, and finally sequence stratigraphic surfaces governed by low frequency high magnitude regional to intrabasinal vertical tectonics. If that be so, the origin, extinction and evolution of higher order superfamilies to subfamilies is genetically linked to regional vertical tectonics, as elaborated in Chap. 6.

Macrocephalitinae Replaced by Kinkeliniceratinae at the Terminal SB
There is replacement of Macrocephalitinae by Kinkeliniceratinae. Compared to the relatively deeper water strongly involute and relatively smooth or less ornamented Macrocephalitinae and Eucycloceratinae subfamilies of the superfamily Stephanoceratacea, the shallower water strongly evolute and moderately ornamented Kinkeliniceratinae belong to the shallower water superfamily Perisphinctacea.

Interpreted Tectonically Induced Uplift/Shallowing
It appears that in the first-order TST, there is return of shallowing at the same depositional sites. Such shallowing in the basin is explained in terms of uplift and tilting up northward of the MH at the second-order MFS. There is rejuvenation of the source/provenance with resultant rapid increase in the supply of sands in the system. The RST of this sequence at Keera is almost twice as thick to that of the last third-order sequence. These two thick successive RSTs together with the sandwiched much thinner TST in the proximal sections constitute the RSM of Chari Formation which is so majestically expressed in Jhura and Habo sections. The paleoenvironment ranged from high energy above normal wave base at the margin to low energy deep subtidal and high energy storm depths in the distal sites, along with tremendous litho and biofacies variation along the margin to basin transact as described above.

3.5.8 J-VIII. Late Middle Callovian O-II Horizon Obtusicosta Subzone Obtusicosta Zone–early Early Oxfordian Second-Order Sequence of ~2.1 my duration with MFS in A-II Horizon Depressum Subzone Athleta Zone of early Late Callovian Age

3.5.8.1 Interpretation of Sequence Surfaces

This sequence is developed over the RSM in Mainland. The TST includes fine textured maroon to purple fractured gypsiferous shales of sediment interval VII bed 218 to sediment interval VIII bed 372 (Fig. 2.29). The contact of sediment interval VI and VII is here revised as between beds 217 and 218. The bed 372 marks maximum density, diversity and frequency of ammonoids and is interpreted as the MFS. The MFS also marks drastic reduction to near extinction of Kinkeliniceratinae. The TST is made up of beds 28–31 at Khadir, beds 18–19 at Wagad, and beds constituting sediment intervals IV and V at Gangta bet where Kinkeliniceratinae ammonoids have been discovered in later part of the TST. The TST is followed up by the Dhosa Sandstone or its equivalents as RST up to the top of sediment interval VI up to the base of Kakindia Limestone. As usual there is increase of sand in the systems tracts towards the margin with corresponding decrease of ammonoid density, diversity and frequency. Unlike the interpretation of Fursich et al. (2013), the GSM is considered having somewhat lesser or similar bathymetry than during LSM. This relatively somewhat shallower sequence is characterized by the geographically restricted (within the GTM) and also relatively shallower water perisphinctid Kinkeliniceratinae as ecological replacement of relatively deeper water Macrocephalitinae. There were two other good candidates for the MFS beds 312–314 and 334 where density and diversity was equally high, however, the latter two candidates did not indicate origin or extinction of over a dozen species, and more so near extinction of Kinkeliniceratinae which in fact characterizes the sequence. The MFS is precisely dated through Horizon A-III of the later part of Depressum Zone so also later part of Athleta Zone of almost early/late Late Callovian age. Within the Gypseous shales, the early part is distinctly thinly bedded while the latter part has relatively thicker beds. The terminal SB as indicated above is determined between sediment intervals VI and VII at the contact of DSM and DOM which in the Gangta Bet and Kakindia sections begins with the start of the Kakindia limestone. Much unlike the demarcation of Callovian–Oxfordian boundary at the contact of GSM and DSM, the contact is here placed decisively within the early Late Callovian Depressum Zone, i.e. within the early Late Callovian GTM Athleta Zone.

The third and fourth sequences in Jaisalmer of Pandey et al. (2010) correspond to the above TST of the second-order sequence J-VIII. The significant difference is condensation and lack of typical RST thick sands in context of deeper starved conditions compared to the corresponding interval of Kachchh. The said sequences in Jaisalmer too are characterized by the Kinkeliniceratinae. The Eucycloceratinae have totally disappeared.

3.5.8.2 Dhosa Sandstone RST at Distal Keera, Jumara, etc.

In West Mainland at Keera, Jumara, etc. because of their relative distal location, the Dhosa Sandstone mostly includes shale/silt instead of sand. It may be even interpreted that the GSM facies grades up to DOM. However, the RST is easily differentiable. The early part of the ~3 m RST is completely ammonoid devoid, followed up by ~3 m moderately ammonoid-bearing part, and then again the rest of Dhosa Sandstone is found ammonoid devoid. The middle ammonoid bearing part is included in the post Athleta Zone, post early Late Callovian Ponderosum Subzone of late Late Callovian. The terminal SB is tentatively suggested near the Mariae/Cordatum zonal boundary that also marks the origin of Mayaitinae. Thus in West Mainland, the GSM makes the combine of GSM + DSM somewhat in the same manner as Gadadha Sandstone member is combine of LSM + RSM in proximal Khadir (Figs. 3.28, 3.29, 3.30, 3.31, 3.32 and 3.33). There is interpreted tectonically induced regional sudden rapid deepening of the basin at the start of the next sequence.

3.5.8.3 Constituent Zones/Subzones

ETM ammonoid elements are also present as auxiliaries/secondaries in the ammonoid fauna, particularly, so through out the TST. The present sequence includes three plus zones and seven subzones.

3.5.8.4 Contents of the Sequence in Different Sections

This sequence is spread in Mainland, Khadir, Gangta Bet and Wagad. Rich presence of kinkeliniceratins is observed only west of the Jhura in the West Mainland west of the Median High, however, rare kinkeliniceratins have been found even in Washtawa, Gangta bet and Khadir. The terminal part of the Dhosa Sandstone even in West Mainland rarely includes any ammonoids, thus making difficult the precise dating of its upper SB. There could even be a transitory subaerial gap near the margin in Wagad, Khadir and even East Mainland, while the relatively less shallow West Mainland sections may some time in near future provide ammonoid evidence of the precise age of the second-order SB between the Dhosa Sandstone and Dhosa Oolite proper. The second-order MFS is interpreted above a thin coquina like ammonoid rich bed with maximum density, frequency and diversity of ammonoids in the Depressum Subzone.

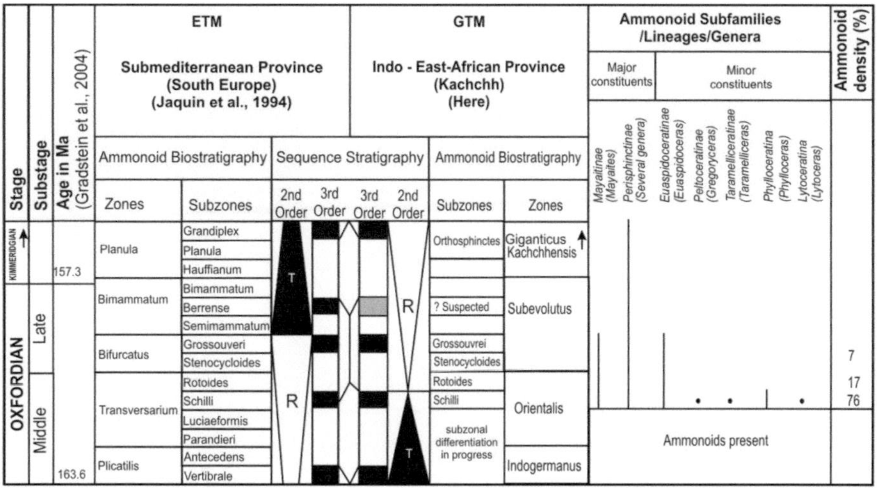

Fig. 3.30 Drastic reduction in diversity and density of ammonoids across the Kachchh first-order Intra-Oxfordian MFS after Krishna et al. (2000a)

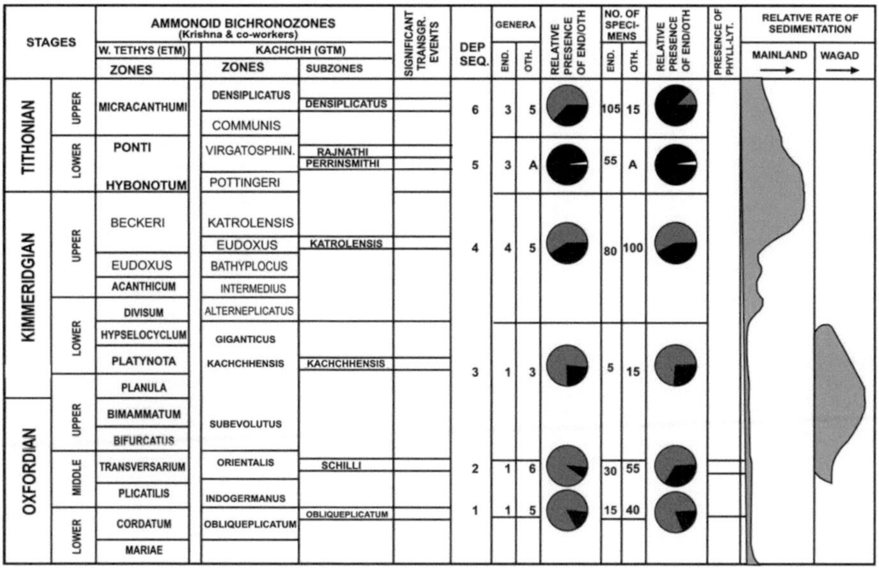

Fig. 3.31 Significant transgressive events through quantitative and compositional variation of ammonoids and relative rate of sedimentation in the Late Jurassic of Kachchh after Krishna et al. (2000a)

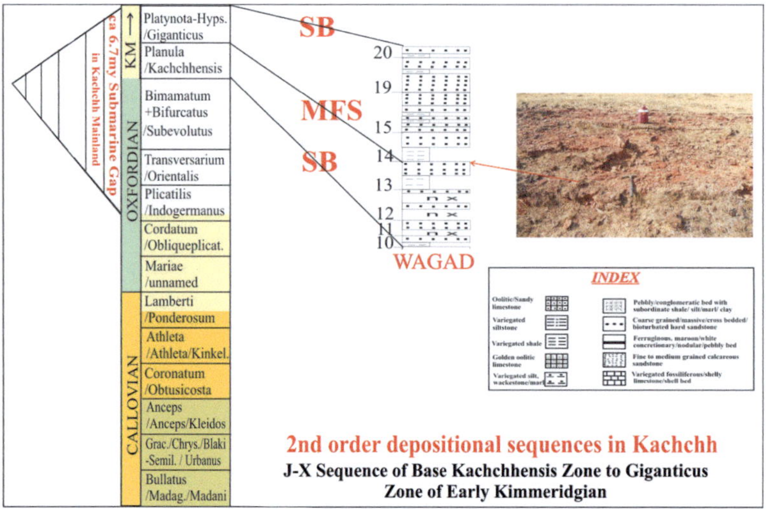

Fig. 3.32 J-X sequence of base Kachchhensis Zone to Giganticus Zone of Early Kimmeridgian

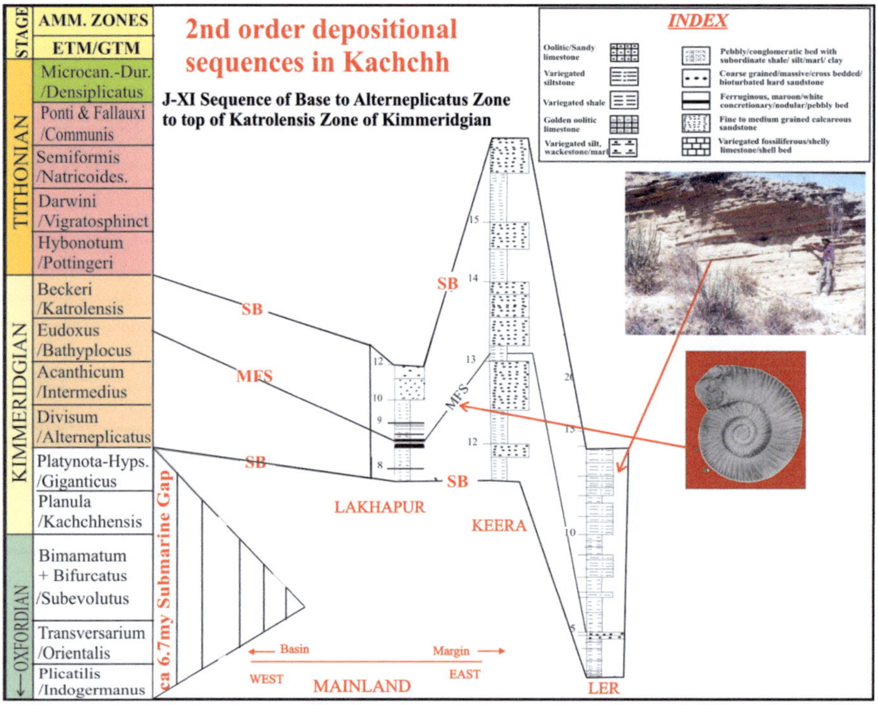

Fig. 3.33 J-XI sequence of base Alterneplicatus Zone to top Katrolensis Zone of Kimmeridgian

3.5.8.5 Influence of Bathymetry on Stephanoceratid/Perisphinctid Proportion in Successive Sequences

During the previous relatively deeper second-order sequence, the perisphinctids existed together with the stephanoceratids while in the present second-order sequence relatively deeper water stephanoceratids are completely absent. Thus, this observation also supports significant eustatic fall, with possible further support of the regional tectonically caused decrease of bathymetry, shallowing and emergence.

Five third-order sequences seem differentiable as given below.

J-VIII. (1). O-II Horizon of Late Middle Callovian Obtusicosta Subzone–Pseudorion Subzone of early Late Callovian Athleta Zone third-Order Sequence with MFS at the base of the terminal Middle Callovian Obtusicosta Zone Catillus Subzone

The basal SB is the same as that of the second-order sequence at the lithostratigraphic boundary of RSM below and GSM at the base of O-II horizon of Obtusicosta Subzone at the base of bed 218 at the base of sediment interval VIb. The MFS is placed above sediment interval VI at the base of Catillus Subzone. There is marked change of ammonoid fauna of the extinction of *K. peramorphum* along with the FAD of *K. catillus*, *K. mutans*, *K. Hubertus* of the Kinkeliniceratinae at the MFS. The terminal SB is placed at the top of sediment interval VII above bed 333 at the top of the Peudorion Subzone. There is marked change of lithology across the terminal SB from shales to sandy limestone. The sequence includes three subzones of ~800 ky duration.

J-VIII. (2). Early Late Callovian base of Depressum Subzone late Athleta Zone–bed 367 of Depressum Subzone top of early Late Callovian Athleta Zone third-Order Sequence with MFS/MFI in or Above the Sediment Interval VIII

This is a small third-order sequence of nearly one subzone duration with a thin less than a meter thick marker hard ammonoid rich sandy limestone or calcareous sandstone MFS as sediment interval VIII, beds 334–338 at Keera. Besides being ammonoid rich, it also marks a significant ammonoid faunal change with FAD/LAD of nine species. The TST is a slow sedimented low energy fine textured marker bed within GSM. The RST comprising beds 339–367 also includes good number of ammonoids. The sequence approximates to one subzone of ~400 ky.

J-VIII. (3). Intra A-III Horizon intra Depressum Subzone Athleta Zone–close of Athleta Zone third-order sequence with third-order MFS same as the Second-Order MFS

The beds 368–372, less than 2 m in thickness, rich in ammonoids constitute the TST. The MFS is placed above bed 372 at the lithostratigraphic boundary of GSM and the overlying silt/sand of DSM. There is near complete absence of ammonoids in the RST (beds 373–378) that is included in the early part of DSM. The sequence again is of ~400 ky duration.

J-VIII. (4). Ponderosum Zone–early Early Oxfordian Third-Order Sequence with MFS at the top of the P-II Horizon (bed 390)
Peltoceras ponderosum Waagen is a significant species in the TST. Among the ammonoids that are exclusive to this sequence are *P. solidum*, *E. aff. Acanthodes*, *Poculisphinctes cf. poculum*, *Sublunoloceras dynastes*, *Putealiceras aff. bisulcatum and Pachyceras llandeanum* which suggest placement of the TST in early part of latest Late Callovian Lamberti Subzone of ETM. The age assignment of the TST agrees well with the age of the global third-order TST of early Lamberti Zone age. The sequence involves the entire late Late Callovian Lamberti Zone with its SB straddling the Callvian/Oxfordian boundary.

J-VIII. (5). early Early Oxfordian Third-Order Sequence with MFS in or at the top of sediment interval VI of Gangta bet section
This sequence is differentiated at Gangta bet between Khadir and Wagad. In recent studies, Gangta bet section has been observed to include the major part of the Chari Formation. It is organized into nine sedment intervals. A thin marker fossiliferous pebbly nodular bed of sediment interval VI makes theTST underneath the ~2 m thick sandstone. The sandstone bed constitutes the following RST. Sediment interval V has yielded *Kinkeliniceras* of early Late Callovian Athleta Zone while sediment interval VII includes the first mayaitins of early Early Oxfordian Obliqueplicatum Zone. The Marker bed is interpreted as a slow sedimented reworked lag deposit. The thin marker maroon pebbly fossiliferous conglomerate determined asTST makes the bed 6 in Biswas (1993). It is followed up by at least five time thicker fossiliferous silty RST (bed 7 of Biswas). Unfortunately, ammonoids are yet not known. The tentative age assignment to the TST is based indirectly on the regional ages of the ammonoid-bearing third-order MFSs immediately below (Ponderosum Zone) and above (early Obliqueplicatum Zone). The sequence is also easily picked up at Khadir as well of beds 30–31. In Lakhapur and Keera it has not been possible to recognize this sequence in the terminal or later part of DSM. This rather small sequence is again of ~400 ky duration. This could as well be interpreted as the basalmost third-order sequence of the succeeding second-order sequence, along with the second-order SB between the two straddling the Callovian/Oxfordian boundary.

3.5.8.6 On the Short Global Ice Age of Dromart et al. (2003)

Dromart et al. (2003) interpreted a short ~800 ky global ice age interlude with eustatic minimum near the Henerici/Lamberti subzonal boundary within the late Late Callovian Lamberti Zone. The said level in Kachchh also makes a third-order MFS as on GTM and Boreal realm, and precisely corresponds to the MFS of the fourth third-order sequence at the top of bed 390 in the Keera section. The following RST of the third-order sequence spans through the entire Lamberti Subzone of Lamberti Zone of the terminal late Late Callovian and may even include the basalmost early Early Oxfordian. In the Nara section of Kachchh, Kanjilal and Singh (2012) even recorded a single fragment of Boreal *Quenstedtoceras*, which if true decidedly marks the southernmost Boreal cold water expansion limit of *Quenstedtoceras* to ~30° S latitude. It may be noted that this together with rare doubtful kosmoceratids makes the only record of any Boreal ammonoids in the Kachchh Jurassic. Another characteristic feature of the terminal Callovian–basal Oxfordian second-order sequence is the inclusion of small stunted pyritised ammonoids which could also be related and influenced by the cold interlude of the last but one third-order sequence. After the distinct decrease in bathymetry in the Kachchh basin during the late Early to Late Callovian, rapid deepening is interpreted during late Early and Middle Oxfordian up to the late Middle Oxfordian Orientalis Zone first-order MFS with return in the system of the relatively deeper water Mayaitinae in the late Early Oxfordian Obliqueplicatum Zone. There are in fact two expansion events of cold and rather deeper water ammonoid faunas in Kachchh, one in early to mid-Early Callovain of Macrocephalitine, and the other late Early late Middle Oxfordian of Mayaitinae. The Macrocephalitinae expanded northward even up to th NW European Subboreal province, however, the latter one appears related more to the due south transform sliding of India during the first-order Toarciaen–late Middle Oxfordian TST from subtropical to subtemperate latitudes. Compared to the said two marked expansion events, the terminal Callovian event is much less significant. The inherent genetic ability of the individual stocks may also be a significant factor in the faunal expansion, however, the factor of temperature changes can not be denied. A third-order transient phase of relative cooling with marked decrease of temperature is not discounted.

3.5.9 J-IX. Second-Order Obliqueplicatum Zone–Subevolutus Sequence with MFS Above the Late Middle Oxfordian Schilli Subzone of Orientalis Zone of ~4.8 my Duration

3.5.9.1 Exclusive Presence in Wagad

This sequence in full is only present in Wagad (Fig. 3.28) and Khadir although basal/early part of the TST is also present in Mainland. In Mainland, major part of the sequence (later part of TST and entire RST) is absent in view of the presence immediately above of a long submarine gap.

3.5.9.2 Multifaceted Manifestation of the Most Significant Second-Order Sequence

The Kinkeliceratinae bearing second-order sequence is followed up by the Mayaintinae-bearing sequence which is the most significant sequence of the Kachchh Jurassic in particular and of the entire GTM in general. It makes the terminal, the deepest, the most condensed, and also having the maximum intra-Jurassic density, diversity and frequency of ammonoids. Apart from the GTM restricted Mayaitinae, divers ETM elements are also present. The proportion of ETM elements is maximum in this sequence of the entire Jurassic, nearly 45 % at the MFS in view of maximum communication during eustatic maxima.

It includes succession of hard grounds (Figs. 3.28 and 3.29), followed up by submarine gap of substantive duration in the basinal exposed part of the basin in Mainland. It is also the most controversial with regard to its bathymetry, environment, presence or absence of the terminal gap, its submarine or subaerial character. It includes the Chari/Katrol lithostratigraphic boundary, also the best lithostratigraphic marker unit present throughout the Mainland—the Dhosa Oolite. It marks the extinction of not only the Mayaitinae, but also with that of the whole ammonoid superfamily of Stephanoceratacea. Also it discovered and collected the biggest Jurassic ammonoid of the world close to its MFS above the late Middle Oxfordian Orientalis Zone (GTM equivalent of the ETM Transversarium Zone) Schilli Subzone, along with several giant size Phylloceratina and Lytoceratina ammonoids (Figs. 3.28, 3.29, 3.30, 3.31, 3.32 and 3.33). The most interesting coincidence is the discovery of this abundantly ammonoid-rich spectacular section and splendid outcrops of the late Middle and Late Oxfordian sediments by us as late as 1994 after nearly 175 years of intensive and extensive studies in the Kachchh Jurassic first by the GSI, and later, particularly, by the ONGC. However, the sheer density of ammonoids at the MFS in the MFI afforded ample opportunity for more detailed studies, which the present author's team undertook. It also marks the abortion of rifting in the Kachchh basin, also reversal of tilting of the basin from due west to due east, and along with reversal of paleoslope, paleodrainage and paleocurrent in the non-marine inland Gondwana basins. The widespread tectonic event that is expressed in such diverse ways also signals the advent of the 'proto-Indian ocean' in the east between Greater India and Australo-Antarctica west of the NW Australian margin. It also initiates rifting between the Greater Indian plate and Australo-Antarctica along with shallow marine sedimentation in the K-G basin. This sequence also includes in the vertical proximity of its MFS the richest all time hydrocarbon bearing-source sediments of the globe on the GTM as demonstrated in hydrocarbon rich Arabia for example in the richest and the biggest oil field of the world—the Ghawar oil field of Saudia. Even the author has reasons to suspect that the source sediments of the K-G and Cauvery basin hydrocarbons may also significantly include the sediments of this sequence. The most recent hydrocarbon discoveries in the Kachchh and Jaisalmer basins may also have their source sediments as well in this sequence. So also is my suspicion of good source sediments of this interval in the thick subsurface pile of Mesozoic sediments underneath the Deccan traps.

3.5.9.3 Later Part of the TST in a Succession of Ammonoid Rich Hard Grounds

The TST of this second-order sequence is almost exclusively expressed in a succession of ammonoid-rich conglomeratic hard grounds not only in the distal mainland but also in the proximal-most exposed part of the basin in Kantkote. The hard grounded beds are almost stacked and piled one over the other with extremely thin intervening sediments both in the distal Mainland and proximal Kantkote (Fig. 3.29). The distal part is extremely slow sedimented and condensed, and called Dhosa Oolite inclusive of Dhosa Conglomerate bed as its youngest bed in West Mainland of late Early and early Middle Oxfordian (GTM Obliqueplicatum Zone and early Indogermanus Zone equivalent to the ETM Cordatum Zone and early Plicatilis Zone). Dhosa Oolite Member (DOM) is followed up by submarine gap of substantive duration. The terminal part of the second-order TST is Jhadasa Member in the Kantkote section of mid- and late Middle Oxfordian (later part of the GTM Orientalis Zone equivalent to Transversarium Zone Schilli Subzone). Unlike the indication of the age of the base of KAb (Kantkote Ammonoid bed) of Wagad within the interval of Dhosa Oolite Member only younger to the base of the Dhosa conglomerate bed in Mainland in most of the earlier studies, the base of the KAb is considered distinctly younger to the top of the Dhosa conglomerate bed by at least three subzones. The base of the KAb is the base of the Schilli Subzone while the base of the Dhosa conglomerate bed is the base of Plicatilis Zone Vertebrale Subzone.

The sequences 5–7 of Pandey et al. (2010) in Jaisalmer correspond to the third-order TST of the hard grounded late Early Oxfordian Obliqueplicatum Zone to the early Middle Oxfordian Indogermanus Zone sediment interval of Kachchh.

3.5.9.4 Composition of the RST

The RST of this sequence is made up of moderately to rapidly sedimented thick ammonoid sparse sands (inclusive of basal shale beds) of late Middle to Late Oxfordian age (Rotoides Subzone of the Orientalis Zone and Subevolutus Zone of GTM equivalent to ETM Rotoides Subzone of Transversarium Zone, Bifurcatus Zone and Semimmatum Subzone). In this sequence the MFS marks the extinction of Mayaitinae on GTM, while near the close of RST is indicated the extinction of ETM Perisphinctinae.

3.5.9.5 Drastic Reduction of Ammonoid Density, Frequency and Diversity

The intra-Jurassic maximum ammonoid density, frequency and diversity are expressed in this sequence at its MFS through presence of more than 10 ammonoid subfamilies (Figs. 3.30 and 3.31). The same is drastically reduced immediately above in the RST to low density and low diversity single subfamily Perisphinctinae. The sequence also includes the youngest ammonoids of the late Middle Oxfordian age MFS in the Island belt as also in most of Wagad except its Bharodia section.

3.5.9.6 Continued Firm Adherence to Maximum Bathymetry and MFS in Late Middle Oxfordian

In spite of the recent work of Alberti et al. (2012a, b), the long held interpretation (since our discovery of the late Middle Oxfordian beds in Kantkote in 1994) of the maximum intra-Jurassic bathymetry and the first-order intra-Jurassic MFS in Kachchh basin in the late Middle Oxfordian Schilli Subzone is firmly maintained. The sudden onset above the Dhosa sand/silt (DSM) of highly condensed ammonoid rich slow sedimented hard grounded sediments all over the Mainland is interpreted as a combination of tectonically induced submergence and eustatic rise of sea level. The cumulative bathymetric increase continued during the early part of the major intra–Jurassic submarine gap in the Mainland, while the condensed ammonoid rich slow sedimented hard grounded facies migrated to the relatively shallower exposed margin in Wagad at Kantkote. The interpretation of marked rapid deepening is further strengthened by the reappearance in this second-order sequence of the sphaeroceratid morphology in the included ammonoids through Mayaitinae. These are also considered homeomorphic dimorphs of Macrocephalitinae. It is noteworthy, that about two substages earlier similarly inflated deeper water Macrocephalitinae had disappeared at the onset of RST. It is here conjecturally speculated that in such a scenario, the sphaeroceratid morphology may have had a continuous presence in SW Pacific from Bathonian to Oxfordian. This speculation can be proved or rejected through near future ammonoid biostratigraphic studies in the SW Pacific.

Six third-order sequences are differentiated as given below.

J-IX. (1). Obliqueplicatum Zone Sequence with MFS at the Top of Sediment Interval VII of Late Obliqueplicatum Zone Age

This sequence is again better developed in Gangta bet and Khadir. Sediment intervals VII (TST) and VIII (RST) constitute this sequence. The beds 8–12 of Gangta bet and Beds 36–37 of Khadir of Biswas (1993) are included in this sequence. Thin marker unit of beds 8–10 makes the TST while the relatively much thicker beds 11–12 are interpreted as the RST. The sedimentation history is much the same as that of the second-order sequence. However, unlike the previous sequence, it includes perisphinctins along with the first mayaitins at the MFS at the top of the bed 10. The bed 10 is the famous Kakindia Limestone bed of Biswas

(1993), and ammonoids are included in concretions. In Mainland sections, the sequence includes much reduced thickness due to extremely slow sedimented condensation inclusive of hard grounds. The thick hard ammonoid rich slab like oolitic limestone and following thin shales of DOM excluding the top most conglomeratic unit constitute the sequence. The oolitic limestone makes the TST while the superposed thin shales the RST with MFS at the top of the oolitic limestone. The sequence approximates 800 ky.

J-IX. (2). Indogermanus Zone Third-Order Sequence with MFS in Early Indogermanus Zone of Basal Middle Oxfordian Age
In Mainland, the topmost ammonoid rich oolitic conglomeratic nodular limestone unit of the second-order sequence is determined as the third-order TST. The RST is absent due to prevention of sediment supply in such relatively distal Mainland sites. The shoreline has advanced much towards proximal Wagad and Khadir. The RST interval is interpreted as included in the early part of the large several subzone long submarine gap. The emergence of Mainland for substantive duration as modelled by Alberti et al. (2012a, b) is here discounted. Earthquakes and tsunamis are only sudden/instant phenomena, which did make instant catastrophic changes, however, over all geological history remained the same, also discussed at length later.

Moreover, evidences of emergence/s of Mainland need be recorded in reworked fossils of the time through weathering, erosion and redeposition in the sediments of the neighbouring sedimentation sites which is not the case. The multi-stage and multi-cycle history of DOM (inclusive of Dhosa Conglomerate bed) published recently (Alberti et al. 2012a, b; Fursich et al. 2013) has been repetitively detailed in Europe for such reworked units in later part of the 20th century, and also specifically for Kachchh DOM time and again including the author and associates (Krishna et al. 1982).

At Gangta Bet, the later part of the sediment interval IX makes the TST while the top of the same interval the MFS. Beds 13–14 of Biswas (1993) at Gangta bet make this sequence. Top of the thin marker bed 13 is determined as MFS, while bed 14 makes the RST which is indirectly assigned to late Indogermanus Zone, i.e. equivalent to the Antecedens Subzone of Plicatilis Zone. In Khadir, the sequence remains undifferentiated. It is noteworthy to emphasize the mixing of more than one ammonoid subzone in this sequence because of maximum depth and the resultant extreme condensation and following starvation. Such mixing occasionally also lets unworked younger ammonoid indices apparently even below the reworked older ammonoids. Such discrepant or anomalous occurrences are easily picked by ammonoid stratigraphers, however other field geologists could be misled. The included ammonoids are reworked, and the youngest age is provided by the least reworked ammonoids. This sequence again approximates one zone i.e. ~800 ky with mixing of both its subzones in West Mainland Dhosa Conglomerate bed as the youngest part of the DOM, however, both the subzones are found unmixed and separable in the proximal Gangta bet. The Gangta bet section is extremely important, since including rather moderately to rapidly deposited ammonoid-bearing late Early to early Middle Oxfordian, which is not exposed in the Kantkote section.

The ammonoid rich conglomeratic nodular oolitic limestone in the Nara section has also yielded a single septate *Gregoryceras cf. riazi* (Gross.) (Singh unpublished Ph.D. thesis 1992). *Gregoryceras* is a relatively tropical to subtropical low latitude warm water Mediterranean ammonoid (Krishna et al. 2008), which obviously would have arrived in relatively high latitude subtemperate Kachchh during a second/third-order TST that too preferably close to MFS. *G. riazi* in Mediterranean is known to range mostly in Plicatilis Zone which includes the Vertebrale Subzone TST. Thus the present third-order TST of early Indogermanus Zone may broadly be assigned early Middle Oxfordian early Plicatilis Zone Vertebrale Subzone age.

J-IX. (3). Early Orientalis Zone Third-Order Sequence with MFS in mid. Orientalis Zone

The TST of this sequence includes the Gangta Ammonoid Bed of Biswas (bed 15, 1993), which is also the youngest ammonoid rich bed at Gangta Bet, since superposed by nearly ammonoid-devoid shales. The MFS is determined at the top of the Gangta bet. The ammonoids are frequent in ferruginous concretions. The composition is 45 % perisphinctins and 55 % mayaitins. The RST is interpreted in the just mentioned shales. The RST is also determined in the Jhadasa section at Kantkote. It may be noted that the Jhadasa Member is the youngest member of the Chari Formation and includes beds 14 and 15 of Biswas as its basal beds at Gangta. The terminal part comprises the succession in the river section near Jhadasa village. In a way, the Jhadasa Member makes the terminal part of the 'Gangta member' while the RSM is the basal part followed by GSM, DSM, and DOM. Gangta beds informally representing the Gangta bet and neighbouring sections includes the later five members of the Chari Formation, although the 'Gangta member' of Biswas does not include RSM, GSM and DSM which make the older part at Gangta bet. The RST of this sequence is interpreted in the sands that underly the youngest hard grounded condensed ammonoid abundant 10 m thick succession discovered recently (Krishna et al. 1994a and later). Also, the bed 16 of the Gangta Member is better exposed in Nagalpur, Desalpur, Gedi and Maiya Wandh localities instead at Gangta bet. In the ETM ammonoid zonal reference, this sequence broadly corresponds to the combined interval of Parandieri Subzone and Lucaeformis Subzone of the Early Transversarium Zone.

It is significant that the Gangta ammonoid bed fauna so far has not yielded either any *Larcheria* or *Gregoryceras* elements which otherwise characterize the Schilli Subzone at Kantkote, and so decidedly is older than Schilli Subzone. In this context is commented upon the record in south half of the Gangta dome south of the east–west fault of post Schilli Subzone ammonoids of Late Oxfordian, even basal Kimmeridgian age by Pandey et al. (2013a, b, c, d). Gangta bet is one of a few isolated domes in between Khadir and Wagad uplift, geologically tied to Khadir than Wagad. All earlier works have put the upper age limit of the succession at Gangta Bet older or at the most equivalent to the top of Chari Formation (=top of abandoned Washtawa/Khadir fns) of late Middle Oxfordian, and the youngest age in Wagad of mid-Early Kimmeridgian Kachchhensis Zone, equivalent to ETM Hypselocyclum Zone. The so-called younger Late Oxfordian-basal Kimmeridgian age of the youngest beds in the Gangta bet dome possibly prompted by misidentifications of the enclosed Transversarium Zone perisphinctid ammonoids of the Gangta ammonoid bed/interval with similar looking Late Oxfordian perisphintid microconch forms, perhaps on account of slow micronch evolution compared to that of corresponding macroconchs. Moreover, the age diagnostic and decisive *Gregoryceras* and *Larcheria* of late Middle Oxfordian are missing in the Gangta fauna, which is suggestive of yet older Antecedens Subzone to Parandieri Zone age of the Gangta ammonoids. Biswas (1993) has also suggested the absence of Kantkote ammonoid bed or its equivalent in Gangta bet, and presence instead in the neighbouring Desalpur section of the Kantkote ammonoid bed of late Middle Oxfordian Orientalis Zone Schilli Subzone age. The earlier interpretation with regard to the succession at Gangta bet belonging exclusively to Chari Formation and not any younger is firmly maintained, there by disagreeing with Pandey et al. (2013a, b, c, d).

J-IX. (4). Schilli Subzone–Stenocycloides Subzone Third-Order TST with MFS Near the Top of Jhadasa Member Above bed I-08 of the River Section
This is the last third-order TST of the first-order TST, and also the most spectacularly ammonoid rich interval of not only Wagad but the entire East Mainland and 'Island belt' discovered recently (Krishna et al. 1994a, and later). Prior to that, the section was not even visited by geologists during the preceding 175 years since the first illustration of any Kachchh Jurassic ammonoids—a *Mayaites* by Sowerby in (1840). In addition to a diverse and very rich ammonoid fauna of perisphinctins, mayaitins and euaspidoceratins, the third-order TST also features good number of large lytoceratids and phylloceratids inclusive of the largest Jurassic ammonoid of the world—a large *Lytoceras* (Krishna et al. 1994a–2011) in the terminal part of the Chari Formation. Besides the ammonoids, the terminal part of the TST also includes large wood stems and abundant belemnoids along with rare gastropods. However, stratigraphically, the most significant and age diagnostic are a good number of *Gregoryceras gr. devauxi* and *Larcheria subschilli* in bed I-08. These two genera were discovered for the first time in the Indian subcontinent (Krishna et al. 1994a, and later). The RST is made up of the Stenocycloides Subzone in the Kantkote section with drastically reduced ammonoid density, diversity and frequency. About 10 ammonoid subfamilies present in the TST are reduced to mere one or two subfamilies in the early part of the RST of Basal Kantkote Member of Katrol Formation. Lytoceratids and phylloceratids which had peaked in the TST have completely disappeared. There is as well marked a drastic lithological, sedimentological, paleontological and environmental change from relatively deeper water (few tens of meters), low energy, abundantly fossiliferous, slow sedimented, condensed hard grounded concretionary/pebbled, bioturbated, ironstone sands to poorly fossiliferous moderately deposited thinly bedded fine shale/silt and fine sand across the first/second/third-order MFS. The third-order sequence precisely ranges from late Middle Oxfordian Orientalis Zone Schilli Subzone to early Late

Oxfordian Subevolutus Zone Rotoides Subzone in stead of ranging from Middle to Late Oxfordian (Pandey et al. 2013a, b, c, d).

It is strange, also scientifically unethical for Fursich et al. (2013) to avoid reference to our numerous publications (1994–2012) on the succession (since discovery in 1994) included in this third-order sequence in their description on pp. 157–158.

Abrupt Biotic and Abiotic Changes/Contrast Across the First-Order MFS As already remarked more than once earlier, there are observed drastic biotic and abiotic change temporally across the first-order MFS that took place precisely in the terminal part of the Chari Formation between late Middle Oxfordian Orientalis Zone Schilli Subzone and Rotoides Zone. The rifting is aborted, gradual tectonically induced shallowing commences. Sediment supply and rate of sedimentation increase fast. Coarsening of texture, improvement in size of sedimentary cycles, size of physical structures, crossbeds, ripples, etc., also increase. There is sudden drastic drop in the density, diversity and frequency of ammonoids. Stephanocerataceae disappear, along with increased domination of Perisphinctaceae. Deeper water lytoceratins and phylloceratins almost disappear after maximization at the first-order MFS. Yet, asymmetric cycles continue unabated.

In the successive second-order RST sand dominated intervals of the proximal sites in the Kachchh Mesozoic {Propinquans Zone–Garantiana Zone Bajocian RSm, Aurigerus Zone–Morrisi Zone Bathonain BCSm, Semilaevis Zone–Paramorphum Zone Callovian RSm, late Katrolensis Zone Kimmeridgian sands, Tithonian sands, Berriasian, Valanginian and Hauterivian–Barremian sands, Late Albian - ? Early Cenomanian sands} have lot in common with regard to their biotic and abiotic environmental parameters.

J-IX. (5). Grossourei Subzone Third-Order Sequence with MFS in the Terminal Part of the Grossourei Subzone

It is the second third-order RST of the first-order RST after the first-order TST. It is expressed exclusively in the Kantkote–Bharodia section (Biswas 1993, beds 1–17). In Mainland, the interval of this sequence is included in the large submarine gap of several subzones. Ammonoid record is poor with drastic reduction of density, diversity and frequency besides extinctions above the first-order MFS. However, ammonoids have not disappeared completely, since in spite of the start of decrease in basin bathymetry, there still is just bare minimum depth for ammonoids, particularly, and possibly so, during the third-order TSTs or around the third-order MFSs. Ammonoids present, are invariably small fragments, however, enough to provide precise ages. The sequence includes one zone and is of $\sim$400 ky.

J-IX. (6). Sixth Third-Order Terminal Oxfordian Sequence

It includes the succession between the Fort sandstone and Kuda village on both sides of the road between the Fort and Kuda (sediment interval X–XII). Inextractable and indeterminate ammonoids, often large perisphinctins or euaspidoceratins are present in gritty sands almost at two-third of the road distance between Kantkote and Kuda. *Epipeltoceras* is not observed. The MFS is placed in the ammonoid scarce grits. The ammonoid-devoid succession from the ammonoid bearing-grits up to the base of 'Adhoi beds' in the Bharodia section makes the RST. It is here tentatively assigned to the terminal part of Bifurcatus Zone in view of the suspected indeterminate large perisphinctins. However, these perisphinctins could actually be euaspidoceratins of early part of Bimmamatum Zone, i.e. the terminal Oxfordian as per ISJS revised/emended definition of the Oxfordian/Kimmeridgian boundary within the Bimmamatum Zone either at the base or top of the Bimmamatum Subzone. According to the proposed revised definition of the Kimmeridgian, it decidedly starts at the base of the Adhoi member as evidenced by the presence of *Wegelia gredingensis* (Weg.). Even a subaerial gap above the grits or prior to the Adhoi Member is not ruled out, since it is the terminal regressive part of the third-order RST as also of the second-order RST. The sequence includes almost half of the Bimammatum Zone, and so is of ~400 ky duration.

3.5.10 J-X. Second-Order early Early Kimmeridgian Kachchhensis Zone–Mid. Early Kimmeridgian Giganticus Subzone Sequence of ~2.0 my Duration with MFS Above the Kachchhensis Zone

This sequence is exclusively exposed in Bharodia (Wagad) (Fig. 3.32) with ammonoids only in second- or third-order TSTs, that too near exclusively of the subfamily Ataxioceratinae. Having already withdrawn earlier from other regions of the Island belt, the sea also finally withdrew from Khadir, at the second-order MFS of the immediately preceding sequence. The time duration of this sequence in the Mainland is included in the submarine gap. Ataxioceratinae marks its origin at the base of this sequence, while Perisphinctinae have extincted prior to the start of the sequence. The Perisphinctinae ammonoids of the last sequence in fact are replaced by Ataxioceratinae in this sequence. Ataxioceratinae is also included in the shallow water Perisphinctidae. There is continued tectonically induced shallowing during this sequence with complete withdrawal of the sea from entire Wagad by the close of this sequence. After the present sequence, the presence of the sea in Kachchh basin becomes restricted exclusively to Mainland for the remainder of not merely the Jurassic but also of the Cretaceous. The first genus to appear at the base of the sequence is *Orthosphinctes* of early Early Kimmeridgian Planula Zone at the base of Adhoi member. As per recently emended Planula Zone, it is now inclusive of the remainder of Bimmamatum Zone above the revised Oxfordian/Kimmeridgian boundary as the basal most zone of the Kimmeridgian. The sequence also marks the youngest presence of ammonoids of the Kachchh basin outside the Mainland of early and mid-Early Kimmeridgian duration (Kachchhensis Zone and Giganticus

Zone of GTM equivalent to the interval of Planula Zone, Platynota Zone and Hypselocyclum Zone of ETM). The sequence includes 2 zones and 4 subzones.

Fursich et al. (2013) while describing the succession included in this sequence have assigned the basal *Orthosphinctes*-bearing red beds late Middle to early Late Oxfordian age, and also excluded them from the Adhoi member. In view of the first *Orthosphinctes* in these basal beds at Bharodia, the precise age is certainly early Early Kimmeridgian Kachchhensis Zone, i.e. equivalent of the ETM Planula Zone. There is no overlap of these basal beds of Kantkote Member of Katrol Formation included in the abandoned Adhoi member, whatsoever, with the terminal Jhadasa Member of the Chari Formation, exposed in the Kantkote river bed. Their being part of Chari Formation (abandoned Washtawa fn) as interpreted by Fursich et al. (2013) is completely ruled out. Further up, in the two Astarte–Gryphaea intervals separated by about 10 m thick sands have been found a few species of *Ataxioceras* inclusive of a giant/large *Ataxioceras* from the upper Astarte–Gryphaea bed of mid-Early Kimmeridgian Giganticus Zone. Giant *Ataxioceras* on ETM are known in Platynota Zone, thereby making Gaganticus Zone as time equivalent of the ETM Platynota Zone. *Dichotomoceras* and *Torquatisphinctes* (Fursich et al. 2013) appear misidentified and confused with *Orthosphinctes*. Their age assignment of the referred beds as late Early to Late Kimmeridgian is rejected. Their interpretation of a large gap in the succession at Bharodia of most of Early Kimmeridgian is also grossly unfounded. Much unlike the indication of the base of Adhoi member in Oxfordian by Fursich et al. (2013), the same marks the revised Oxfordian/Kimmeridgian boundary in the Kuda–Bharodia section. Also, the broad correspondence of the Gamadu Sandstone member to Katrol Formation by Fursich et al. (2013) is incorrect. The said two units (Katrol Formation in Mainland and Gamadu Sandstone member in Wagad) do not even have the slightest of overlap relationship. In age, when GSm disappears in Wagad by the close of mid-Early Kimmeridgian, begins the Katrol Formation in Mainland in late Early Kimmeridgian Alterneplicatus Zone. The assertion of Fursich et al. (2013) that UAb makes the Adhoi member is also wrong. Instead UAb makes only the upper part of Adhoi member, and on the basis of the presence of *Ataxioceras* is assigned to Giganticus Zone, equivalent in Kachchh of the ETM Platynota Zone. Similarly, LAb is assigned to early part of Subevolutus Zone, which in turn is age equivalent of ETM Bifurcatus Zone. The sedimentary succession at Bharodia is without any doubt continuous, but for extremely short less than subzone unmeasurable intervals. Moreover, the large several million year gap in Kachchh Mainland is largely submarine and in the vertical vicinity of the first-order MFS, and so in no way comparable to the unmeasurable short subaerial gaps if any at all.

Two third-order sequences are differentiated:

J-X (1). First Third-Order Sequence
Sediment intervals XIII, XIV and XV are included in this sequence with basal SB at the base of the Adhoi beds below the sediment interval XIII, the terminal SB above the bed XV, and the MFS above the bed XIV. The dominant lihtology is shale/silt/marl with nodular fractured pebbly maroon ammonoid-bearing thin

slowly deposited sandstones. The sediment interval XI has relatively moderate to good presence of ammonoids. These are mostly fragmented inner whorls. The dominant genus is *Orthosphinctes* together with rare *Wegelia* and *Nebrodites*. The sequence includes interval of the emended Planula Zone and part of the Platynota Zone, of Kimmeridgian of either ~800 or ~1200 ky.

J-X (2). The Second Third-Order Sequence
In general it has significant increase in sand/shale ratio from the previous sequence, and associated drastic decrease in presence of ammonoids. There is dominance and cyclic repetition of the benthic bivalve Astarte–Gryphaea assemblage as transgressive lags of fourth/fifth order. Sediment intervals XV to XX are included in this sequence. The MFS/MFI is tentatively suggested in the middle of the sequence with extremely rare presence of ammonoids, one ammonoid level each in sediment intervals XVI to XVIII in bed XVI-02, bed XVII-14, and XVIII-04, respectively. The ammonoids are of large size of the genus *Ataxioceras*, and on the premise of isochronous third-order sequence surfaces and systems tracts, the TST can be tentatively correlated to ETM mid. Platynota Zone. There is no evidence of the presence in this sequence of the ETM Hypselocyclum Zone, however, it could be included in the youngest Gamadu sands. This sequence of part of Platynota Zone and entire Hyselocyclum Zone is of either ~800 or ~1200 ky, since the total duration of the second-order sequence is ~2.0 my.

It is noteworthy that the low diversity, moderate frequency, and moderate to high density *Astarte/Gryphaea/Trigonia*-bearing beds have found recurrent expression in the Kachchh Mesozoic succession particularly in the second- and third-order RSTs that too in the proximal depositional sites which have migrated towards the basin with each prograding sand sequence. In fact in most instances, the concerned RSTs become marine body fossil devoid after the last such bivalve bed, also ammonoid devoid well prior to the continued bivalve levels. This pattern is observed almost throughout the Kachchh Mesozoic, particularly in the basal and Ukra Member units of the Umia Formation. It suggests the extreme tolerance and survival capacity of such bivalves in extremely shallow rapidly deteriorating environments.

3.5.11 J-XI. Second-Order late Early Kimmeridgian Alterneplicatus Zone–late Late Kimmeridgian Katrolensis Zone of ~4.0 my Duration Sequence with MFS Above the Bathyplocus Zone

The sequence is expressed only in East Mainland. On one hand the sea had retreated from Wagad at the close of the previous sequence. On the other hand, the duration of this sequence in the exposed West Mainland does not have any geological record, since sediment supply possibly could not resume at such distal sites in spite of considerable decrease of bathymetry and retreat of shoreline. The ammonoid presence is further restricted to East Mainland in Ler–Katrol, Jawahar nagar, Walakhwas localities, the richest ammonoid record being in Ler–Katrol (Figs. 2.36 and 3.33). In West Mainland, the interval of this sequence is yet interpreted as included in the submarine gap. The sequence includes early ~100 m sediments of the Katrol Formation at Ler–Katrol over the Chari/Katrol disconformity. These are exposed east and west of the north–south striking Kukma–Hajipir highway.

The Ataxioceratinae ammonoids are replaced by the Katroliceratinae ammonoids in this sequence. Several ETM subfamilies are also present particularly in the TST part of this sequence as auxialary or secondary elements. These auxiliary/secondary elements facilitated precise correlations of the Katroliceratinae-based GTM Kimmeridgian zonal scale to the ETM scale even on a subzonal to horizonal level. Its TST unlike the TST of the previous sequence is rich in ammonoid presence (Fig. 3.35), perhaps the richest of the second-order TSTs of the post first-order MFS intra-Jurassic RST in view of its location on the relatively deeper Ler–Katrol belt. The sequence includes 4 zones and 10 subzones. The ammonoid richness in density, diversity and frequency is here attributed to location of the concerned ammonoid rich sections in much less proximal location with good bathymetry. Also, the TST interval of this sequence witnessed eustatic rise which may have considerably negated the effect of the tectonically induced shallowing in the basin. The second-order MFS of this GTM sequence coincides with the first-order MFS of the ETM first-order sequence. This in turn would have also supported better communication, and allowed ETM elements to migrate to GTM under improved bathymetry at Ler–Katrol. This could happen in spite of the second-order GTM TST being part of the first-order RST. There is yet greater bathymetry interpreted in sedimentation sites west and south of Ler–Katrol to explain the absence of the sediment supply there from rather distant provenance. Precise subzonal to even horizonal correlations have been possible between ETM and GTM (Schweigert et al. 1996), so realized due to rich and diverse presence of ETM elements as auxiliaries or secondaries to Katroliceratinae that were exclusive to GTM. The sequence is further differentiated into four third-order sequences.

J-XI (1). Early Alterneplicatus Zone–Late Intermedium Zone Third-Order Sequence with MFS in Early Part of Intermedium Zone

It includes beds 1–0 to 1.9 made up of shales, i.e. horizons A-I to I-V with MFS above I-I. The MFS marks the extinction of *T. alterneplicatus* (Waagen). The sequence is characterized by the FAD of *Torquatisphinctes* and prolificness of *T. alterneplicatus* (Waagen) (Figs. 2.36 and 3.34). It is the first third-order sequence after the resumption of sedimentation at this site above the submarine gap. It includes nearly half the duration of the second-order sequence of ~2000 ky.

J-XI (2). Bathyplocus Zone Third-Order Sequence with MFS in Linguiferous Subzone

It comprises beds 2.0–3.3, i.e. horizons B-I to B-IV with MFS at the FAD of *P. linguiferous* Spath as that of the second-order sequence. The TST starts with a marker ammonoid rich relatively thick conglomeratic pebbly slow sedimented bed which also marks the FAD of *T. bathyplocus* (Waagen) and closes with FAD of

T. jurunensis Spath. It includes nearly one-fourth duration of the second-order sequence, i.e. ~800 ky.

XI (3). late Late Kimmeridgian Katrolensis Subzone–Unnamed subzone Third-Order Sequence with MFS in Pressulum Subzone
This sequence (beds 3.4 and entire 5) also includes, at or near its base, a similar ammonoid rich marker sandstone as that of the previous sequence, and also similar thinly bedded shales superposed over the marker sandstone. The TST starts with FAD of *Katroliceras* while the MFS/MFI is marked by the short ranged Mediterranean ETM *H. pressulum* which expanded to GTM from ETM at the third-order MFS only about one zone above the second-order MFS. There is also associated marked reduction in ammonoid density, diversity and frequency above the third order MFS. The sequence includes four out of five subzones of the Katrolensis Zone, i.e. of ~800 ky.

XI (4). Latest Kimmeridgian Third-Order Sequence with Beds 11–12 in the Ler Section as MFS with H. Beckeri
This terminal third-order sequence of the present second-order sequence (beds 6–13) is physically, lithologically and paleontologically much distinct from the underlying three third-order sequences. It otherwise includes major part or most of second-order RST. Lithology is silt and sheet like bodies of sand of normal to storm wave base environment. The beds are relatively thicker, body fossil content extremely poor and ammonoids almost disappeared. The MFS/MFI is interpreted in beds 11–12 with *H. beckeri* which arrived from Mediterranean ETM at a third-order MFS. The sequence constitutes part of the Katrolensis Zone as Infundibulum Subzone of not more than ~400 ky.

3.5.12 J-XII. Second-Order early Early Tithonian Pottingeri Zone–Virgatosphinctoides Zone Sequence of ~2.8 my Duration with MFS in Rajnathi Subzone of Virgatosphinctoides Zone

This second-order sequence is exclusively present in the East Mainland, however, with further geographical restriction to Ler–Katrol, also with further drastic reduction in diversity, density and frequency of ammonoids, probably on account of further decrease in basin bathymetry under cumulative effect of eustatic fall and tectonically produced shallowing in the Ler–Katrol locality. There is complete withdrawal of the sea from the North belt of the East Mainland, particulary Jawaharnagar section during or at the end of this sequence. The TST comprises beds 14–19 in Ler–Katrol (Krishna et al. 1995, 1996a). The MFS is determined at the top of bed 19 in Rajnathi Subzone. In fact both beds 18 and 19 exhibit the same ammonoid diversity and thus make the MFI. Beds 20–26 are interpreted as RST with substantively reduced density, diversity and frequency of ammonoids.

Two third-order sequences are differentiated as stated below.

J-XII (1). early Early Tithonian Pottingeri Zone Third-Order Sequence with MFS Above bed 14 b of the Ler–Katrol Section

This is a small sequence, however, exclusively present and differentiated in Ler–Katrol section. The basal SB is between beds 13 and 14 at the base of the Pottingeri Zone, terminal SB is above bed 15 in the terminal part of the Potingeri Zone, and MFS above the bed 14 b in the Pottingeri Subzone. The sequence is featured with a few species each of just two genera *Katroliceras* and *Aulacosphinctoides*. There are no ammonoids common to ETM, yet the correlation is facilitated by isochronous third-order MFSs on the two Tethyan margins. The ETM TST is in the Hybonotum Zone while GTM TST is in the Pottingeri Zone, and therefore the GTM Pottingeri Zone is correlated with the ETM Hybonotum Zone. It includes duration of $\sim$800 ky since inclusion of less than one Tithonian ammonoid zone.

J-XII (2). Terminal Pottingeri Zone P-III Horizon Third-Order Sequence with MFS Above bed 18 b of Perrin-smithi Subzone

This sequence like the previous one is exclusive to Ler–Katrol, since the interval of this sequence in West Mainland being still included in the large submarine gap. The sea has withdrawn from 'Island belt' and Wagad while it presumably has shallowed beyond ammonoid depths in the East Mainland north belt. Hence, the sequence, even if present, in East Mainland north belt, can not be differentiated. The basal SB is at the base of the bed 16 at the base of P-III Horizon of the Pottingeri Zone, terminal SB above bed 19a within the Rajnathi Zone with MFS/MFI in beds 18a–b. Again there is not a single species common with the corresponding third-order sequence of ETM. The ETM TST is in the Semiforme Zone while the GTM TST corresponds to the combined succession of Perrin-smithi Subzone and Rajnathi Subzone. Thus, these two subzones of the Virgatsphinctoides Zone can be tentatively correlated to the Darvini Zone of ETM. However, such correlations require additional confirmation. The sequence is smaller than the previous one of $\sim$400 ky.

J-XII (3). Late Rajnathi Subzone bed 19b–Terminal Biplicatus Subzone bed 24 Third-Order Sequence with MFS at the Top of the bed 19b

The sequence has a thin condensed slow sedimented ammonoid rich TST with basal SB at the base of the bed 19b and MFS at the top of the bed 19b, and terminal SB near the top of the bed 24. The MFS is the same as that of the second-order sequence. The sequence includes $\sim$800 ky.

Development as Ridge Forming Guide Fossil-Devoid Sands of Uncertain Age

In Jawaharnagar, the development of the ammonoid-bearing previous second-order sequence was followed up by thick ridge forming sand dominating marine body fossil-devoid build-ups of Umia Formation. These thick ridge forming sands in age could be broadly of the same time duration as the present second-order sequence, however, there could even be a disconformity of unknown gap. In want of the post Kimmeridgian ammonoids in the North belt of East Mainland nothing could be said with any certainty about either the basal age of these ridge forming sands or their age duration in different sections. The Umia Formation rapidly progrades from east to west. The earliest of such sands in Wagad is Late Oxfordian and the youngest in Ghuneri–Amarsar section at least Late Albian, if not yet younger. In Ler–Katrol section which is located on the South belt of the East Mainland on east side of the road, the youngest ammonoid evidence is of the latest Late Tithonian or basalmost Berriasian, while from Jawaharnagar located on the North belt of the East Mainland, the youngest ammonoid evidence is the latest Kimmeridgian. The youngest ammonoid ages in Kachchh from east to west are late Midle Oxfordian to Early Kimmeridgian in Wagad, Late Kimmeridgian in North belt of East Mainland, Early Berriasian in Lakhapur section of West Mainland, Late Berriasian in Sahara section, and Middle Albian in the Ghuneri section. Similarly from north to south in the basin, the youngest ammonoid ages in the Island belt range in age from Early Callovian to late Middle Oxfordian, Late Kimmeridgian in Jawaharnagar on the North belt of East Mainland, and Tithonian/Berriasian boundary in Ler–Katrol.

Sudden Increase in Sand/Shale Ratio
The sand/shale ratio also increased in this sequence compared to the previous sequence, the two sequences having been compared in Ler–Katrol. Unlike the presence in the earlier sequence, of good number of ETM ammonoid subfamilies along with the dominant GTM restricted Katroliceratinae, this sequence exclusively includes Katroliceratinae with much reduced density, diversity and frequency. This is the only second-order sequence interval which includes nearly 100 % indigenous ammonoid fauna and near absence of any ammonoids common to Kachchh and ETM. Marine communication and intermigration of ammonoids between ETM and GTM is minimized, and correlations based alone on ammonoids become very difficult.

3.5.13 *J-XIII. Second-Order Natricoides Zone–Densiplicatus Zone Sequence of ~3.6 my Duration with MFS Above the Natricoides Zone*

The basal SB is interpreted between Virgatosphinctoides Zone and Natricoides Zone, tentatively between the beds 24 and 25. In this sequence the Katroliceratinae ammonoids are replaced by Virgatosphinctinae ammonoids. It is manifested (Fig. 3.36) in Ler Katrol (South belt of East Mainland) as also in Umia–Sahera Mundhan belt (North belt of West Mainland), however, the complete sequence is not well developed in either of the belts. In both the belts, the Natricoides Zone fauna is either absent or yet not known for different reasons. In Ler–Katrol, the said absence may be because of suspected structural disturbance. Nevertheless, the thick shale/silt bed 25 of Ler–Katrol may be on homotaxial basis corresponded to the Natricoides Zone succession as established in Jaisalmer (Fig. 3.36) where the sequence is better expressed. In West Mainland, the early part of the sequence interval is still included in the submarine gap. However, the Natricoides Zone is well developed in the neighbouring Jaisalmer and Indus basin of Pakistan along with the Communis Zone ammonoid fauna. The second-order TST near exclusively includes Virgatosphinctinae ammonoids with rounded depressed whorls, although rare *Hildoglochiceras* and *Haploceras* have been collected in Jaisalmer and Pakistan which allow assignment of the Natricoides Zone to mid-Early Tithonian Semiforme Zone of ETM. The sedimentary column of the Communis Zone in West Mainland localities is made up of alternations of silt/shale and ammonoid-bearing some what calcareous sands. In Ler–Katrol there are more sands compared to West Mainland as expected, so also lesser density, diversity and frequency of ammonoids. Bathymetry is somewhat more in West Mainland. The MFS is placed at or near the top of the Natricoides Zone. In Kachchh, the presence of *Hildoglochiceras* has been known from Gajansar, Lakhapur and Jara as evidence of the Natricoides Zone.

A few third-order sequences are differentiated.

J-XIII (1). Natricoides Zone–Communis Zone Communis Subzone Third-Order Sequence with MFS in Krafti Subzone of Natricoides Zone

This sequence is differentiated in the Rupsi section of Jaisalmer ~400 km north of Kachchh (beds 2–5). The third-order MFS is determined in beds 2 k–l with maximum density, diversity and frequency of ammonoids inclusive of the large number of Mediterranean ETM ammonoids *Haploceras* and *Hildoglochiceras (Fig.).* Lytoceratin *Pterolytoceras* is also present. The RST comprises beds 2–4 m in which the ammonoids are drastically reduced to scarce *Virgatosphinctes* fragments. Lithologically, the TST is made up of shale/silt alternations with occasional ammonoid-bearing concretion levels, while the RST includes the thick almost unfossiliferous fine to coarse crossbedded sands. It includes nearly two and half zones of ~2000 ky.

J-XIII (2). Communis Zone Subfrequens Subzone–Densiplicatus Zone Densiplicatus Subzone Third-Order Sequence with MFS Above the Subfrequens Subzone

This sequence is differentiated in the Bhadasar section of Jaisalmer. The basal SB is placed at the base of the Subfrequens Subzone, while the TST comprises beds 6–7. The MFS is determined in bed 7 on the basis of relatively high ammonoid density and diversity. The RST includes the thick almost unfossiliferous bed 8 of the shale/silt alternations, and only features scarce *Virgatosphinctes* fragments. The terminal SB is précised at the top of the bed 8. It includes less than one Tithonian ammonoid zone of ~800 ky.

J-XIII (3). Densiplicatus Zone Oppeli Subzone Third-Order Sequence with MFS in Middle of Oppeli Subzone

This sequence like the previous one is expressed also at Bhadasar in beds 9–11. Bed 9 makes the TST with MFS above it. The TST includes marker cocquina type coarse, even gritty, hard, compact pebbly/nodular/concretionary calcareous highly fossiliferous sandstone beds which are rich in ammonoids. Ammonoid density and diversity is distinctly reduced to moderate (in bed 11) to poor (in bed 12) presence of exclusive *Virgatosphinctes* spp. in beds 10–11 which are interpreted as RST. It also includes less than one Tithonian ammonoid zone of ~800 ky.

3.5.14 J-XIV. Terminal Part of Densiplicatus Zone Third-Order TST Of ~400 ky with MFS/MFI in Bed 19 A-B

This TST is better developed in West Mainland west of Umia in the early ammonoid-bearing slow sedimented condensed green oolitic glauconitic marker interval of Umia Formation (Fig. 3.36).

This second-order early part of TST is also differentiated in Ler–Katrol belt, however, the development in comparison is not that good. It is significant to note that this TST is absent in the north belt of the East Mainland east of the Median High where from by this time the sea seems to have withdrawn, at least for sure any ammonoid evidence is missing. This TST much like the late Early Kimmeridgian second-order TST is again rich in ammonoids of the GTM restricted Virgatosphinctinae of the *Virgatosphinctes* gr. *densiplicatus* (Waagen) ammonoids along with guide Himalayatinae ammonoids in common with ETM. Also are present SE Pacific elements like *Corongoceras* and *Argentiniceras* as a consequence of the expansion of such elements from SE Pacific through the Gondic corridor. The Gondic corridor is conceived as a long narrow transient shallow marine corridor like passage/route between the Gondwanian margin of the Tethys to the Gondwanian margin of SE Pacific that is interpreted to have originated at the start of Late Tithonian during a eustatic high (Krishna 1996) and certainly not any time earlier in the Jurassic period. Sharland et al. 2001 have emphasized on the rifting and related opening of the Indian Ocean during Late Tithonian, however, it is here opined that episodic rifting and transform movement coupled with eustatic rise only consequented the origin of the Gondic corridor instead of the origin at this time of the Indian ocean. The Indian Ocean in fact originated much later in the east sector between India and Australo-Antarctica, accompanied by oceanic spreading and igneous event between the two. Thus in this second-order TST in Kachchh basin are present ETM and SE Pacific elements in addition of the indigenous *Virgatosphinctes* gr. *densiplicatus*. The common presence of *V.* gr. *densiplicatus* (Waagen) almost through out GTM, both east of West Gondwana and west of East Gondwana easily discounts the origin of the Indian ocean at this time in the India–Africa divergent sector. However, much more significant support is of common land dinosaurs between Africa and India which simply could not be, had Indian Ocean originated at this time. The author interpreted this significant second-order transgressive interval for the first time in 1991, and has since then consistently elaborated upon this important MFS in Kachchh, Jaisalmer, High Himalaya and other constituents of IEAP. Fursich et al. (2013) have in spite of referring to our ammonoid zonation in Late Tithonian, have as their usual practise conveniently and intentionally skipped our repeated reference to this MFS since the early 90s.

3.6 Cretaceous

Cretaceous sediments are found well developed in the Kachchh basin, particularly, west of the median high in Mainland, and the sections are measured in detail (Krishna 1994 unpublished report). Cyclicity is excellently exhibited particularly within the Ghuneri Member of the Umia Formation through cyclic sand build-ups, however, more revealing cycles are of, relatively, much smaller periodicity compared to the above described second to fourth-order cycles. These unfortunately are not that easily picked up anymore as in the Jurassic. Ammonoids are absent in the Ghuneri Member, so are also absent any other biotic or abiotic age determination elements. Ammonoids when present, e.g. in the Basal Member underlying the Ghuneri Member are extremely scarce while in the Ukra Member that overlies the Ghuneri Member, the ammonoid presence is somewhat moderate.

Apart from Kachchh, the marine Cretaceous sediments are present almost all round the Indian plate in all the sectors (Fig. 2.6). Among the exposed Cretaceous marine successions, more important ones are of Cauvery, K-G, Narmada, Jaisalmer, and High Himalaya. Rich presence of ammonoids is known only in the Cauvery basin, particularly in early and mid-Late Cretaceous, however, sections have deprived measurement at desired scale because of the location of exposures/sections in badland topography. The Narmada succession also includes significant ammonoid presence but only in a few levels. The focus for sequence stratigraphic studies has been given to Kachchh and Cauvery, besides available information from other important Indian marine Cretaceous developments as already enumerated above.

First- and Second-Order Sequence Framework as Developed in the Indian Cretaceouswith Focus on the Kachchh–Cauvery Combine

There are two first-order sequence surfaces involved in the Cretaceous. The first-order intra-Cretaceous SB is determined between the mega-guide fossil-devoid Ghuneri Member medium to coarse crossbedded ripple marked thick sands below and prolificly fossiliferous ammonoid bearing-slow sedimented limestone/marls above of the Ukra Member. The SB with or without subaerial gap is widely interpreted in all the sectors of India (Figs. 3.7, 3.8, 3.9, 3.10, 3.11, 3.12, 3.13, 3.14, 3.15, 3.16, 3.17, 3.18, 3.19, 3.20, 3.21, 3.22, 3.23, 3.24, 3.25, 3.26, 3.27, 3.28, 3.29, 3.30, 3.31, 3.32, 3.33, 3.34, 3.35, 3.36, 3.37, 3.38, 3.39 and 3.40). In Cauvery basin, the said SB is determined between the Sivaganga and allied units below and the disconformably overlying Dalmiapuram Formation or Uttatur Group/Formation above.

Krishna (2005, 2012) determined the first-order intra-Cretaceous MFS in the early Middle Turonian Turoniense Zone/Helvetica Zone/Gartneri Zone in the Cauvery and Narmada basins based on ammonoid rich concretionary fine textured hardgrounded shales/limestones. An alternative is also suggested at the mid-Middle Albian in view of its equal or greater regional significance. In Kachchh, the younger of the two MFSs has been interpreted either in the thin green sands at the top of bed 27 at the top of the Upper Member immediately underlying the Deccan Trap of $\sim$89/92 my fission track age in the Ghuneri–Amarsar section (Fig. 1.13) or

Several 3rd order sequences are differentiated.

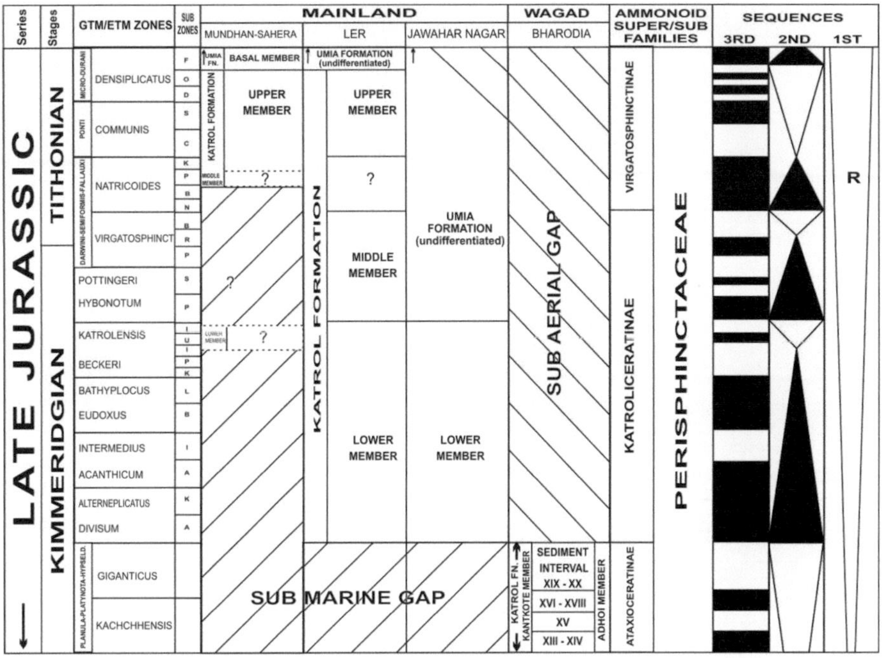

Fig. 3.34 Intrabasinal correlations vis-à-vis first to third-order sequences during Kimmeridgian–Tithonian interval in Kachchh basin

above the bed 23 within the Upper Member of the Umia Formation (Fig. 3.39). The radiometric age can be a few my either way and alone, it can only be suggestive requiring further research. Thus, the fission track age approximates well to the Turonian–Coniacian duration. Nevertheless, additional biostratigraphic data are invariably required for more precise and narrow age-constraints of 1–2 my or rejection of the radiometric date. The early Middle Turonian age of the MFS in many other basins, however, implies either of the two MFSs determined in Kachchh as the major MFS. On the other hand, based on the homotaxis of the sequence stratigraphic surfaces, the placement of the first-order MFS above the bed 23 also makes a good proposal for future research in this direction.

A succession of twelve second-order sequences is for present skeltalised here as under while a detailed account is deferred for a later work. The first two of the Cretaceous second-order sequences represent the terminal two second-order

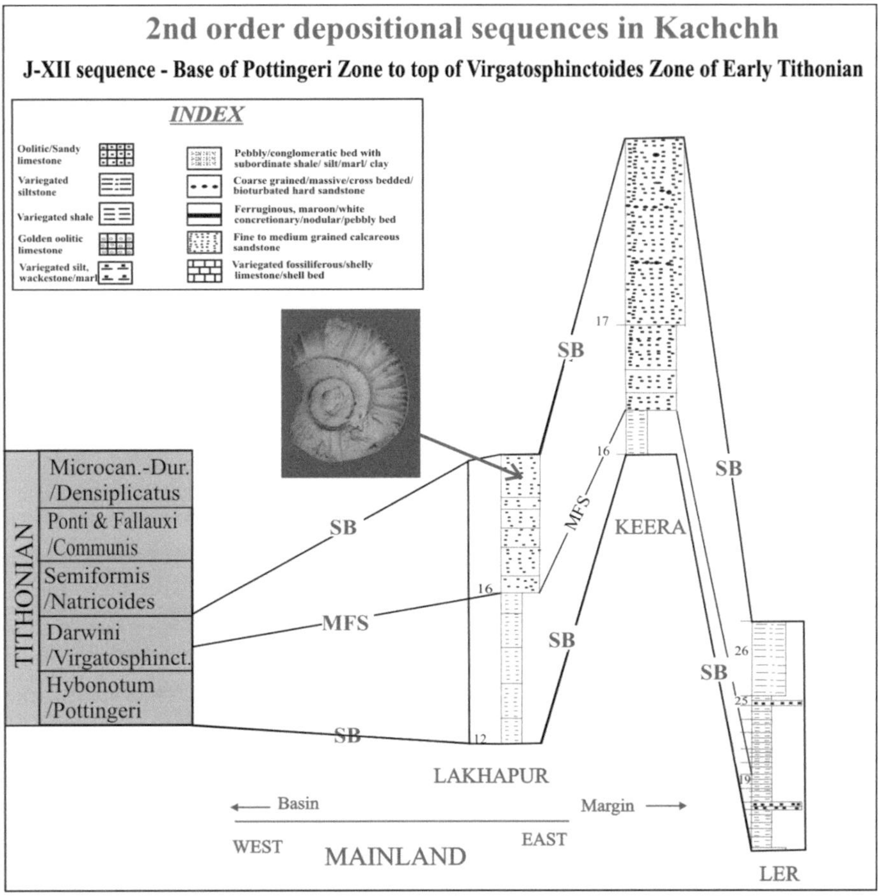

Fig. 3.35 J-XII sequence-base of Pottingeri Zone to top of Virgatosphinctoides Zone of Early Tithonian

sequences of the Toarcian–Barremian first-order sequence. The next five sequences, i.e. sequences C-III to C-VII make the Aptian–early Middle Turonian first-order TST (Table 3.3). The last three sequences make the early part the following first-order RST.

3.6.1 C-I. Early to Middle Berriasian Second-Order Sequence

The MFS is determined above the basal green glauconites with *Argentiniceras* and *Spiticeras* in the Lakhapur section at the Jacobi Zone / Occitanica Zone transition.

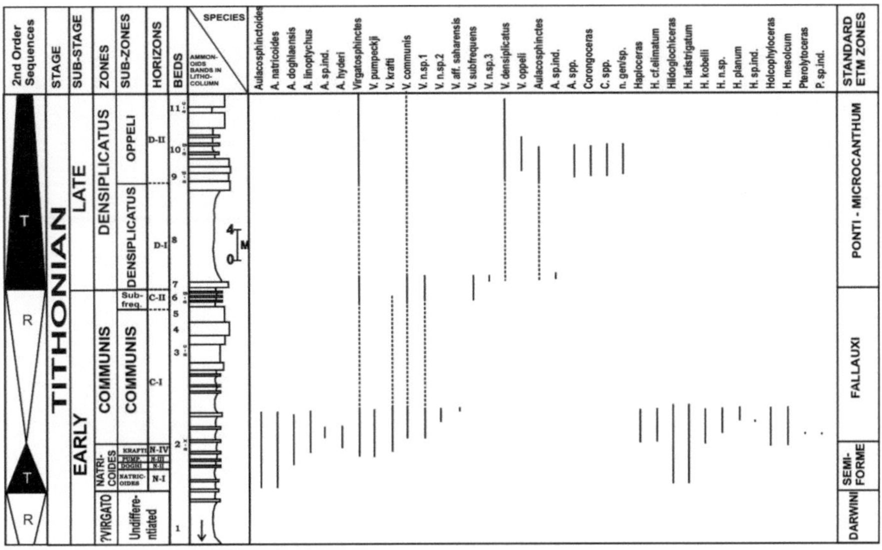

Fig. 3.36 Second-order sequences in the Tithonian of Jaisalmer (modified after Pandey and Krishna 2002)

3.6.2 C-II. Late Berriasian Second-Order Sequence

The MFS is determined above the *Kilianella*-bearing bed 35 in the Ghuneri section.

3.6.3 C-III. Valanginian Second-Order Sequence (beds 5–9 of the Ghuneri Member) in the Ghuneri–Amarsar Section

The MFS is placed within the bed 7 of early Late Valanginian Verrucosum Zone with *Olcostephanus* in the Spiti basin tentatively correlated to the top of the bed 7.

3.6.4 C-IV Hauterivian Second-Order Sequence (Beds 10–12 of the Ghuneri Member) in the Ghuneri–Amarsar Section

The MFS is determined at the top of bed 11 of terminal Hauterivian age.

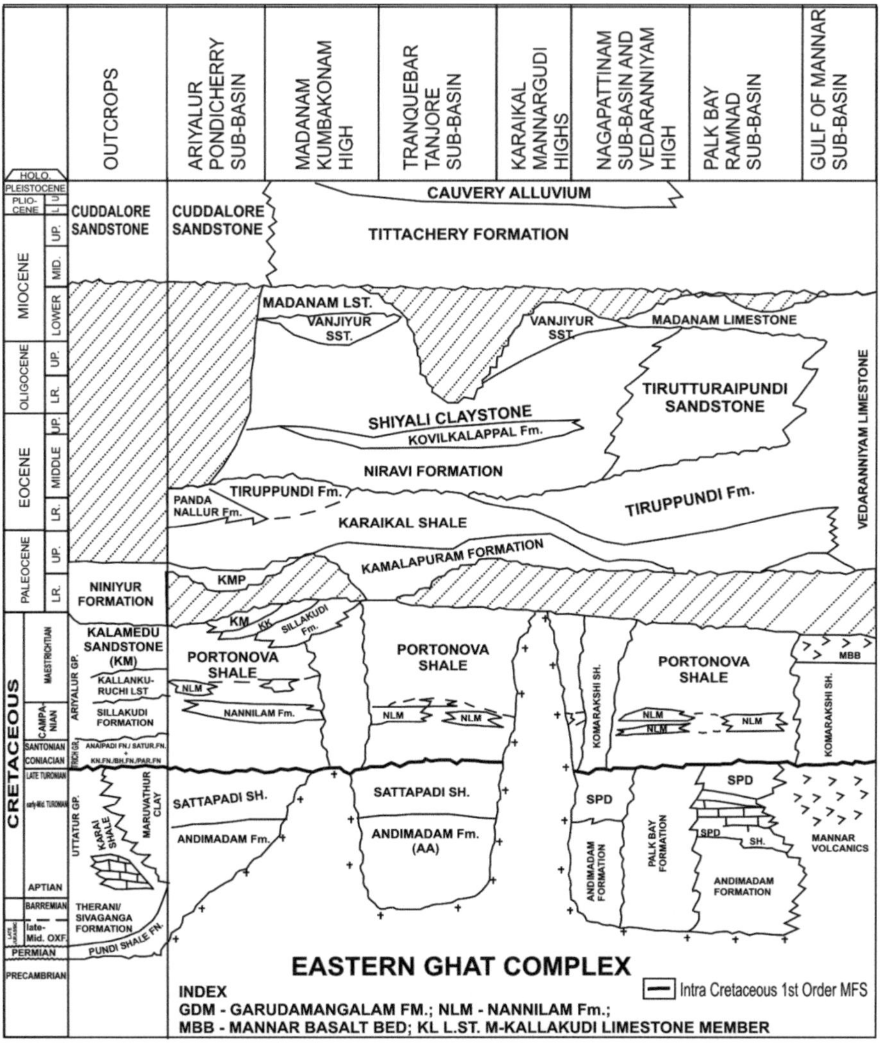

Fig. 3.37 Intra-Cretaceous MFS in the Cauvery wells (modified after Pandey and Dave 1998)

3.6.5 C-V Barremian Second-Order Sequence (Beds 13–14 of the Ghuneri Member) in the Ghuneri–Amarsar Section

The MFS is placed at the top of bed 13 of suggestively Middle Barremian age.

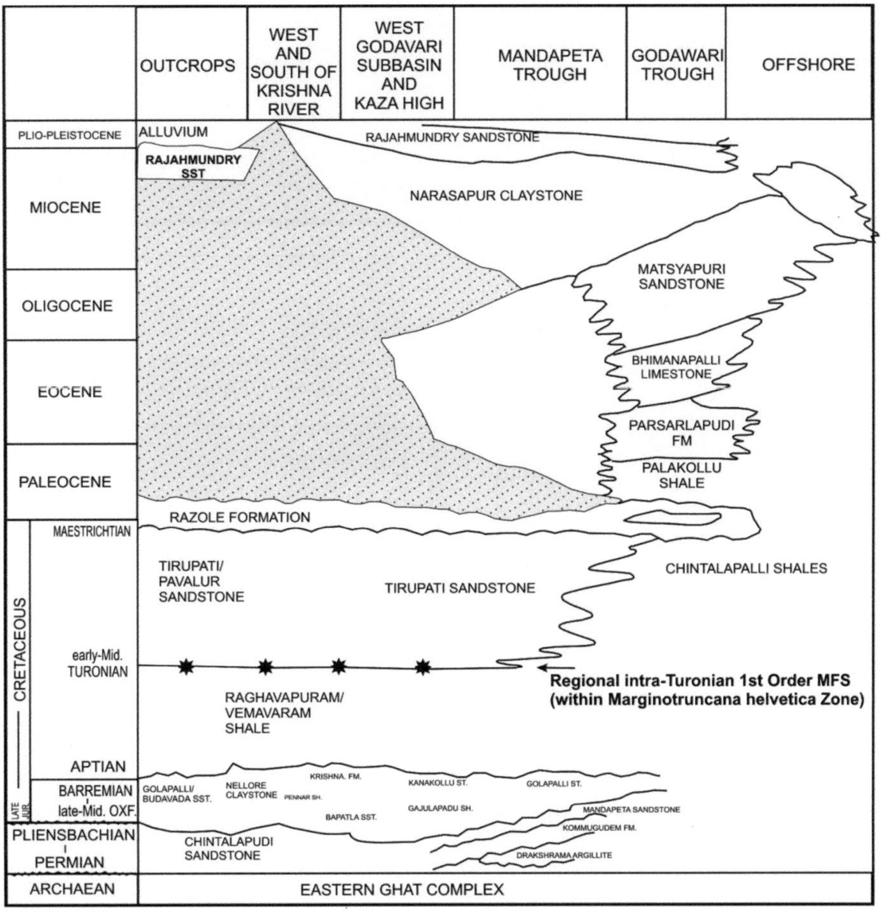

Fig. 3.38 MFS in the generalized lithostratigraphy of Krishna–Godavari basin (modified after Pandey and Dave 1998)

3.6.6 C-VI. Early Aptian Deshayesites Zone–late Aptian Furcata Zone Second-Order Sequence of the Early part of the Ukra Member (Beds 15 and 16)

The MFS is indicated within the bed 16 at the top of the green glaucontic shale/sands within the Ukra Member of Weissi Zone age.

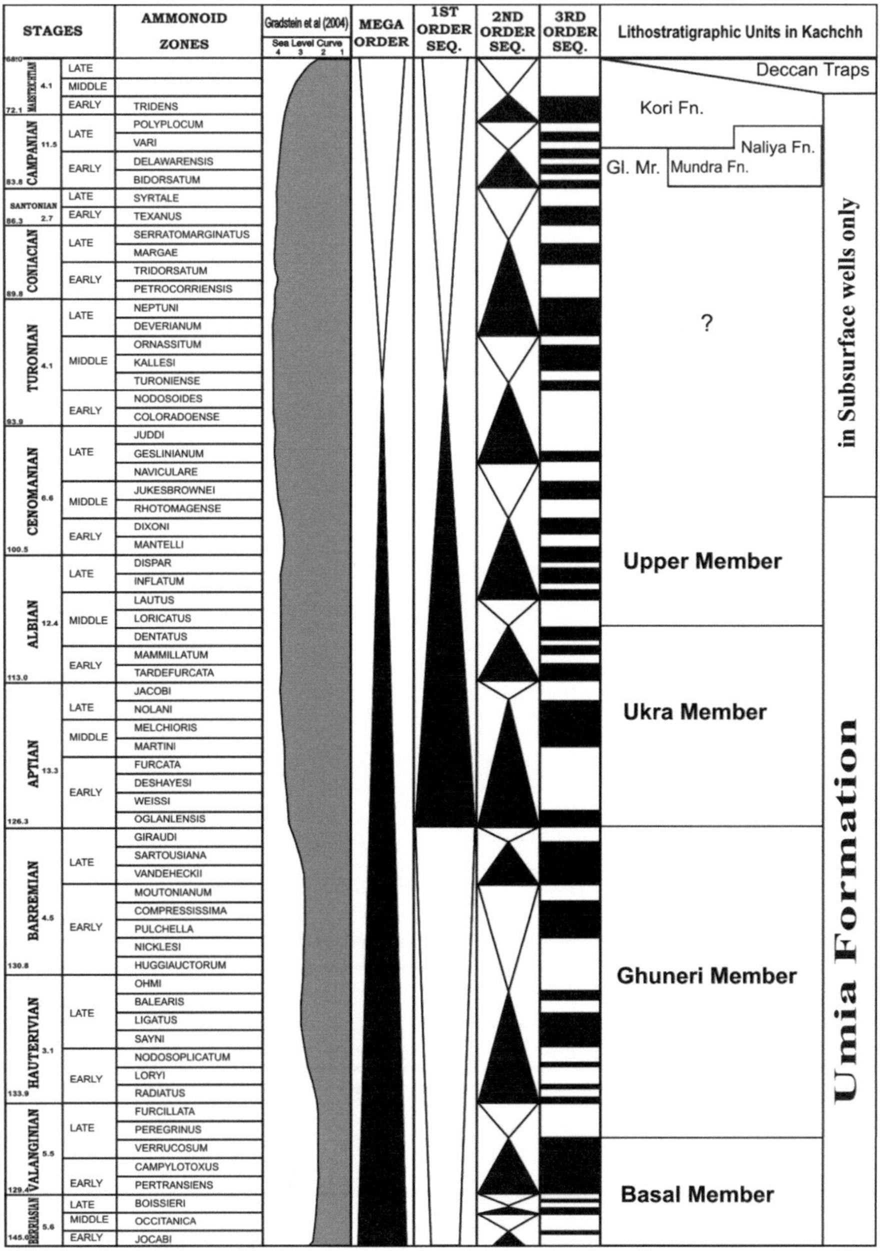

Fig. 3.39 First- to third-order sequences in Kachchh Cretaceous

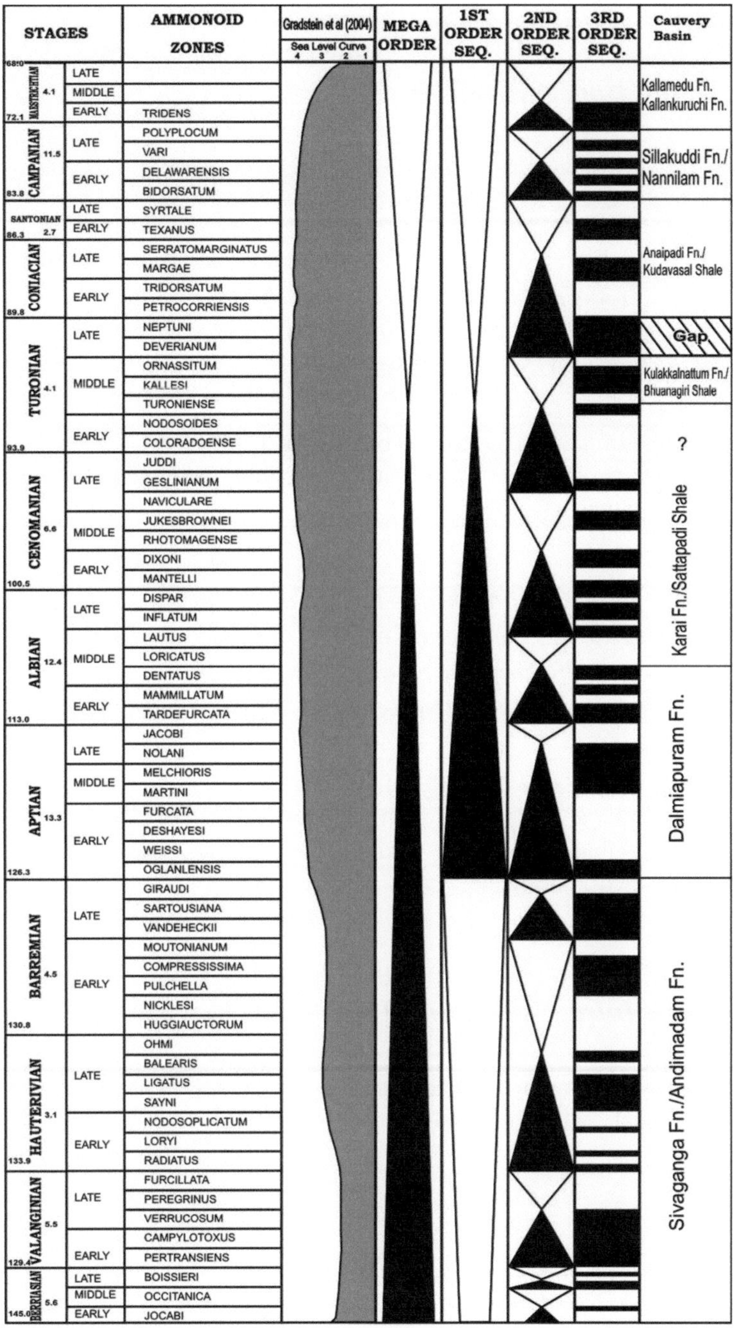

Fig. 3.40 First- to third-order sequences in Cauvery Cretaceous

3.6.7 C-VII. Early to Middle Albian Second-Order Sequence (Bed 17 of the Ukra Member and Bed 18 of the Upper Member)

The MFS is marked above the early Middle Albian Dentatus Zone at the top of the bed 17 i.e. at the top of the Ukra Member.

3.6.8 C-VIII. Late Albian–Middle Cenomanian Second-Order Sequence of Beds 19–22 of the Upper Member of Umia Formation

The MFS is determined at the top of bed 21 in the Ghuneri–Amarsar section of broad intra-Cenomanian age. The Albian/Cenomanian boundary at the top of the bed 21 in the Ghuneri–Amarsar section.

3.6.9 C-IX Late Cenomanian–Late Turonian Second-Order Sequence of the Upper Part of the Karai Formation/Sattapadi Shale in the Wells and the Overlying Bhuvanagiri Formation

The MFS is determined in the early Middle Turonian Turoniense Zone/Helvetica Zone–Gartneri Zone at or near the top of the Karai Formation. The sequence is also suggested in the Ghuneri–Amarsar section of Kachchh in beds 23–25 or 26 with MFS above the bed 23.

3.6.10 C-X. Coniacian–Santonian Second-Order Sequence with Intra-Coniacian MFS

The sequence includes the Anaipadi calcarenite as the TST at the early/late Late Coniacian boundary while the sandy Saturbhugam Formation makes the following RST.

3.6.11 C-XI. Late Campanian Second-Order Sequence with MFS Above the ?Early Campanian Delawarensis Zone

The sequence in the Cauvery basin includes the Sillakudi Formation.

3.6.12 C-XII. Maestrichtian duration Second-Order Sequence with MFS Above the Early Maestrichtian Tridens Zone

The sequence in the Cauvery basin includes the Kallankuruchhi Formation up to the top of the rich *Gryphaea*-bearing sediment interval as TST and the Ottakovil Formation/Kallamedu Formation as the corresponding terminal part of the RST.

The first seven second-order sequences are primarily developed in the exposed Cretaceous of Kachchh West Mainland. A few of these sequences are also interpreted in east coast basins. The younger five sequences are best recognized in the exposed Cauvery Cretaceous. The two second-order sequences in the Aptian–Middle Albian interval are also differentiated in the Cauvery basin. About half of these twelve second-order sequences are even easily picked up in all the Indian sectors, west, east or north.

3.7 Salient Features of the Indian Cretaceous Record

3.7.1 Intra- and Terminal Cretaceous Events Contextual to India

The significant geological events include origin of the transient Gondic corridor near the J/C boundary, the Dravidian corridor during Aptian on either side of India in the west and east, spreading across Australo-Antarctica and India, anticlock wise rotation of India, spreading between Madagascar and India, shaping of the Indian east and west coasts, Rajmahal, St Mary, Mannar and Deccan igneous activities, prior to the docking of India into Asia These events let the globe acquire geographical framework much closer to what it is at present. A detailed chronicle has been made later in Chap. 7.

3.7.2 Application of Sequence Stratigraphy

The limited and somewhat scattered biostratigraphic data lacking the desired refinement necessitated the application of the regional sequence stratigraphic formulations, if so developed and understood, in other basins of the region (detailed in Chaps. 4 and 7 later) towards integrated comprehension of the Indian Cretaceous. The major first-order intra-Cretaceous SB is determined between Ghuneri and Ukra member units of Kachchh in the Ghuneri, and Lakhapur sections in West Mainland Kachchh. It is placed between the marine body fossil devoid yet fully marine cyclic sandstone succession punctuated by thin slow sedimented concretionary/nodular to conglomeratic transgressive deep maroon densely bioturbated conspicuous markers of the Ghuneri Member and the deep green basal Ukra shale/sand of early Early Aptian age. The first major TST is differentiated in Ukra Member of Early Aptian–early Middle Albian age. Similarly, another significant intra-Cretaceous MFS is interpreted in the thin glauconitic shales immediately below the traps in the Ghuneri–Amarsar section. Subaerial stratigraphic gap is interpreted at the SB in the wells Banni 2 above the Bibat Formation. The neighbouring Narmada, Jaisalmer and Indus basins or even the distant Cauvery records also suggest somewhat equal bathymetry at the mid-Middle Albian and early Middle Turonian MFSs that makes the singling out of the first-order MFS among the said two MFSs rather difficult. In the exposed Cretaceous, ammonoids in view of their chronologic, bathymetric and paleogeographic significance, hold the key in conjunction with diverse microplaeontological information on forams, dino/pollen-spore, epi/infauna, fauna/flora and ichno/biofacies; heterochronic evolution, differentiation between submarine and subarial stratigraphic gaps and their varying durations in different parts of the basins, also field observable abiotic features like lithology, texture, bed geometry, contact and size, structures, presence of pebbles/nodules/concretions, hard grounds, etc.

3.7.3 Recent Independent Sequence Formulations in Different Indian Basins Vis-a-Vis Regional to Global Correlatibility of the Sequence Surfaces

Applicability of the sequence formulations heavily depends upon meeting the specific crucial test/requirement of regional (up to second order) to global (third and finer order) recognition and correlatibility of the sequence surfaces. Obviously, different formulations in different basins as skeltalised in some recent publications (ONGC 2008; Dave 2011, 2012), go against the fundamental concept of the large distance (at least interbasinal) correlatibility of the sequence surfaces for example among Kachchh, Spiti and Cauvery basins in three different sectors of India.

3.7.4 Regional Uniformity of the Sequences

As in the case of the Jurassic, it is here emphasized again that (1) sequence frameworks of any order of large or small duration through pre-Aptian Cretaceous are uniform throughout GTM in view of the broadly similar tectonostratigraphic homogeneity, while the post-Barremian Cretaceous framework is uniform only within the Indian subcontinent in view of the oceanic separation of India from Australo-Antarctica at the first-order intra-Cretaceous SB (break up unconformity) near the Barremiam–Aptian boundary, (2) only third- and finer-order sequences smaller in duration than $\sim$1.5 to 2.0 my exhibit global uniformity in view of their governance by Earth's orbital parameters (mostly eccentricity and precession periodicity based cyclicity of ~400, 100, and 20 ky).

3.7.5 Differences of Ages Among Recent Sequence Formulations

The intra-Cretaceous gaps and first-order MFS interpreted here differ significantly from those published in successive recent tables/charts (Raju 2011). Unfortunately, these charts also often differ substantially among themselves, e.g. the intra-Cretaceous gap in one chart is in Cenomanian and in another in Turonian. Ages given in text often differ from those indicated in the tables.

3.7.6 Sequence Framework in the Indian Cretaceous

In the above backdrop, the earlier outlined formulations principally on ammonoid based data in the exposed Indian Cretaceous geological record are further strengthened through integration of surface/subsurface/offshore micropaleontological and other geological data. Also in the process are integrated the major Cretaceous igneous episodes of the region in the sequence framework.

3.7.7 Foraminifer Evidence of yet Otherwise Unsubstantiated Presence of Late Cretaceous Stages in Kachchh

Mathur et al. (1970) recorded Neocomian to Coniacian palynoflora almost all over the Mainland in the exposed Umia Formation. In the same work, they also recorded planktonic and benthic foraminifera of Albian–Cenomanian age. However, these claims did never receive unreserved acceptance. The presence or absence of Cenomanian is extremely difficult to decide.

3.7.8 Speculation on Second-Order Late Albian–Early Cenomanian Sequence

The second-order Late Albian to Early Cenomanian sequence interval in Kachchh remains unsubstantiated, which may even be part of the subaerial gap between the Upper Member of Umia Formation and Deccan trap in most sections of the

Mainland Kachchh. However, in the neighbouring Jaisalmer subsurface this second-order sequence though undifferentiated is definitively included in the Goru–Parh units, and on similar lines is also, as an alternative, speculated in the Upper Member of Umia Formation in Kachchh.

3.7.9 Intra-Cretaceous First-Order SB with or Without Gap

The intra-Cretaceous SB is determined in late Early Cretaceous early Early Aptian Deshayesi Zone corresponding to the Coptospora cauveriana palynoevent). There is determined subarial (increasing towards the margin) gap in the vertical vicinity of the first-order SB in different basins inclusive of the Kachchh Cretaceous. The intra-Cretaceous subaerial gap near the first-order basal SB is also large in the exposed Indian East coast Cauvery and K-G basins near the Barremian/Aptian transition between either Precambrian or non-marine pre-Aptian Late Jurassic–early Cretaceous and marine post-Barremian sediments. This gap is also present in the Narmada basin between Precambrian and Nimar Sandstone. This SB, even without the gap has been widely determined in Kachchh, Jaisalmer, Spiti and elsewhere.

3.7.10 Significant Difference in Ages of the Said Gaps

The intra-Cretaceous gaps and first-order MFS/SB interpreted here differ significantly from those published in successive recent tables/charts (Raju 2011; Dave 2011).

The Cretaceous succession in Cauvery basin clearly exhibits the intra-Cretaceous first-order sequence boundary unconformity, and also evidences the first marine incursion in the East Coast basins between Precambrian and Uttatur Gp. or between Sivaganga and Uttatur Gp. of early Early Aptian age, so dated through *Australiceras* in Cauvery, Kachchh and Jaisalmer preceded by deep green glauconitic sand/silt/shale build-up in Kachchh. The Dalmiapuram Formation is further organized into three second-order sequences through second-order SB's of mid-Middle Aptian and late Middle Albian. The first-order intra-Cretaceous MFS is interpreted within the terminal part of Karai Formation.

3.7.11 Creation of Indian Stages in Jurassic and Cretaceous—An Uncalled for Exercise

The best Late Cretaceous exposed as also subsurface geological record is in the Cauvery basin, however, in spite of continuous presence of ammonoids, good hydrocarbon recovery, exposed sections have shockingly remained unmeasured.

Consequently, refinement potential of ammonoids and other visually observable features of the exposed record could not be fully exploited. Planktonic microfossil studies relate invariably to wells, as also not fully published by the petroleum companies. In a review Raju (2011) provides 14 planktonic foraminifer datum markers in the Indian Late Cretaceous as a whole and in turn organized them into so-called 5 Indian stages and 13 Indian zones exclusively in the Cauvery basin. The boundaries of the so-called local/regional Late Cretaceous stages and zones do not match with either boundaries of international stages or zones. The stratotype sedimentary columns of these Indian stages and zones have neither been lithostratigraphically differentiated, nor described, nor their correlation potential explored. Thus the primary requisite of minimum quality and potential, based upon multiple internationally well defined parameters of the stratotype sedimentary successions has hardly been evaluated/assessed prior to erecting the stages. The stages are violative of stratigraphic fundamentals, procedures and rules. Like the International Commission on Stratigraphy and its System Sub-commissions under the auspices of IUGS, we should develop appropriate bodies in India with elaborate procedure, mechanism and rules to streamline the Indian Phanerozoic geology.

3.7.12 Summary

See Tables 3.1, 3.2 and 3.3

Table 3.1 Summary of the Second-Order Sequence for the Triassic T-I to T-X

No. of second-order sequence	Age range of the sequence	MFS and comments
T-X	Rhaetian Para formation	MFS in the lone shale bed of the unit
T-IX	Middle to Late Norian Entire 28 m thick Hangrang Formation plus entire 130 m Alaror Formation and entire Nunuluka Fn	MFS in the *Neotibetites*-bearing shale dominated upper part of Alaror Fn of Late Norian age
T-VIII	Early Norian Entire Rangrik formation	MFS in 8 m thick *Juvavites*-bearing shales of late Early Norian age in upper member. The Early/Norian boundary is also précised above the *Juvavites* beds between thick sandstone/siltsone below and thick limestones above, both lying above the *Juvavites*-bearing thick shale bed

(continued)

Table 3.1 (continued)

No. of second-order sequence	Age range of the sequence	MFS and comments
T-VII	Latest Early to Late Carnian entire Rongtong Formation	MFS in *Gonionotites*-bearing calcareous siltstone in Middle Rongtong Member of mid-Late Carnian age
T-VI	Mid to late Early Carnian 220 m thick Rama Formation	MFI in the siltstone/shale cyclic interval of Rama Formation enclosing *Tropites* and *Spirogmoceras* in late Early Carnian. This is in contradiction with Early/Late Carnian boundary as suggested by Bhargava et al. (2004) within Rama Formation. The said boundary is here suggested between Rama and Rongtong units
T-V	late Early Ladinian Sutherlandi Zone–Early Carnian Aon Zone. Entire Chomule Formation	MFS in or above the hard ground level BL 12 at Muth 3 at the boundary of Regoled Zone and Canadensis Zone in the terminal part of Ladinian
T-IV	Intra-Ladinian Langobardian Gredlerensis Zone to late Early Ladinian Maclerni Zone). Terminal part of Niti Limestone and entire Kaga Formation	MFS in or above bed VT 24 of early Late Ladinian Meginae Zone age inclusive of *Daonella*-bearing interval
T-III	Late Olenekian Meekoceras Zone–Early Ladinian Fassinian Curionii Zone, Terminal part of Limestone +Shale Formation, entire Niti Limestone to bed 47 of Himalayan Muschelkalk	MFS/MFI in Hollandites Zone to Kellnerites Zone
T-II	Basal Olenekian Rohilla Zone-Intra-Olenekian Flemingites-Euflemingites Zone Beds 12–39 of the Muth section of Limestone + Shale Formation	MFS within the bed 15 (Intra-Olenekian Flemingites griesbachi Zone
T-I	Late Permian–intra-Induan Frequens Zone (Limestone +Shale Member) Beds 1–11 of the Muth section	MFS either within or at the top of Discophiceras Zone (below bed GU 4/4)

Table 3.2 Summary of the Second-Order Sequence for the Jurassic J-I to J-XIII

No. of second-order sequence	Age range of the sequence	MFS and comments
J-XIII	Natricoides Zone–Densiplicatus Zone Sequence of ~3.6 my Bed 25 at Ler to bed 18 at Mundhan	MFS above the Natricoides Zone.
J-XII	early Early Tithonian Pottingeri Zone–Virgatosphinctoides Zone Sequence of ~2.8 my Beds 14–24 at Ler	MFS above bed 19b in Rajnathi Subzone of Virgatosphinctoides Zone
J-XI	late Early Kimmeridgian Alterneplicatus Zone–late Late Kimmeridgian Katrolensis Zone of ~4.0 my Beds 1–13 at Ler	MFS above bed 3 in Ler of the Bathyplocus Zone
J-X	early Early Kimmeridgian Kachchhensis Zone–mid. Early Kimmeridgian Giganticus Subzone of ~2.0 my Sediment intervals X–XX at Bharodia	MFS above sediment interval XV of the Kachchhensis Zone
J-IX	Obliqueplicatum Zone–Subevolutus Zone Sequence of ~4.8 my. Beds 25–105 at Gangtabet.	MFS above bed I-08 of the late Middle Oxfordian Schilli Subzone of Orientalis Zone.
J-VIII	Late Middle Callovian O-II Horizon Obtusicosta Subzone Obtusicosta Zone–early Early Oxfordian of ~2.1 my Em to Fm of ULS in Jhura Section	MFS above bed 392 in Keera section in A-II Horizon Depressum Subzone Athleta Zone of early Late Callovian age
J-VII	late Middle Bathonian Mantataranus Zone/Bremeri Zone–O-I Horizon of late Middle Callovian Obtusicosta Zone Obtusicosta Subzone sequence of ~2.40 my Sediment intervals I to II at Jhumara to bed 217 at Keera	MFS above bed 92 in Keera section of the Diadematus Subzone of Chrysoolithicus Zone
J-VI	Late Bajocian Garantiana Zone–mid. Middle Bathonian Morrisi Zone of ~1.5 my Beds 49—58 of LSM at Kaladongar	MFS above bed 9 in Nara section of the Early Bathonian Macrescens Subzone
J-V	?Intra-Aalenian/Basal Bajocian–mid-Late Bajocian of ~3.6 to ~1.7 my Beds 37–48 of LSM at Kaladongar	MFS above bed 47 at Kaladongar in the Early Bajocian Propinquans Zone
J-IV	Early Toarcian–? Intra or end Aalenian second-order sequence of <~12.4 my Beds 1–36 at Kaladongar	MFS above bed 24 in Kaladongar section in Nitiscens Zone
J-III	Latest Sinemurian to end Pliensbachian of ~8 my Beds 6–9 in Spiti	MFS above bed 8 in Spiti of Ibex Zone

(continued)

Table 3.2 (continued)

No. of second-order sequence	Age range of the sequence	MFS and comments
J-II	Hettangian–Sinemurian boundary to Rauricostatum Zone of ~8.5 my Beds 4–5 in Spiti	MFS above bed 6 in Spiti of Oxynotum Zone.
J-I	Hettangian of ~2.0 my Beds 1–3 in Spiti	MFS at the top of bed 2 in Spiti of the Middle Hettangian Liasicus Zone

Table 3.3 Summary of the Second-Order Sequence for the Cretaceous C-I to C-XII

No. of second-order sequence	Age range of the sequence	MFS and comments
C-XII	Maestrichtian	MFS above the Early Maestrichtian Tridens Zone. The sequence in the Cauvery basin includes the Kallankuruchhi Formation up to the top of the rich *Gryphaea*-bearing sediment interval as TST and the Ottakovil Formation/Kallamedu Formation as the corresponding terminal part of the RST
C-XI	Late Campanian	MFS above the ? Early Campanian Delawarensis Zone. The sequence in the Cauvery basin includes the Sillakudi Formation
C-X	Coniacian–Santonian	intra-Coniacian MFS. The sequence includes the Kullakalnattum Formation/Anaipadi calcarenite as the TST at the early/late Late Coniacian boundary while the sandy Saturbhugam Formation makes the following RST
C-IX	Late Cenomanian–Late Turonian	MFS above the early Middle Turonian Zone/Helvetica Zone/Gartneri Zone. The sequence includes the Karai Formation with MFS in the terminal part of the Karai Formation
C-VIII	Late Middle Albian–Middle Cenomanian	MFS near the Albian/Cenomanian boundary
C-VII	Early to mid-Middle Albian Beds 17 and 18 of the Ukra Member	MFS above the early Middle Albian Dentatus Zone at the top of the bed 17
C-VI	Early Aptian Deshayesites Zone–late Aptian Furcata Zone Early part of the Ukra Member (beds 15 and 16)	MFS within the bed 16 at the top of the green glaucontic shale/sands within the Ukra Member of Weissi Zone age
C-V	Late Barremian Beds 13–14 of the Ghuneri Member) in the Ghuneri–Amarsar section	MFS in the terminal Berremian at the top of bed 13
C-IV	Hauterivian–Early Barremian	MFS in the terminal Hauterivian at the top of bed 11

(continued)

Table 3.3 (continued)

No. of second-order sequence	Age range of the sequence	MFS and comments
	Beds 10–12 of the Ghuneri Member) in the Ghuneri–Amarsar section	
C-III	Valanginian Beds 5–9 of the Ghuneri Member) in the Ghuneri–Amarsar section	MFS in the early Late Valanginian Zone with *Olcostephanus* in the Spiti basin tentatively correlated to the top of the bed 7
C-II	Late Berriasian Beds 33–36 at Lakhaper	MFS above the *Kilianella*-bearing bed 35 in the Ghuneri section
C-I	Middle Berriasian Beds 28–32 at Lakhaper	MFS at the transition of Jacobi Zone/Occitanica Zone with *Argentiniceras* and *Spiticeras* above the basal Green Glauconites of the bed 28 of the Basal Member of the Umia Formation in the Lakhapur section

3.8 Scales of Cyclicity in the Kachchh Jurassic Geological Record

3.8.1 Rationale of the Preliminary Exercise

The detailed text in the foregoing chapters in general and of Chap. 3 in particular demonstrates the presence of several orders of cyclicity in the Kachchh Jurassic. Also, asymmetries of highly variable rates of sediment accumulation with regard to space and time, and in the successive systems tracts of a sequence are obvious. Interference and apparent distortions are also evident in the resultant expression of different orders of cyclicity in each lithostratigraphic column. In regressive near margin geological record, the cyclicity seems often undistorted for a large thickness, however, without any clues to precise age determination. Thus this crude probing of scales of cyclicity as one undertaken below is like groping in a massive spread of dark matter around.

3.8.2 Asymmetric and Highly Uneven Duration of the Sequences and Stages

The first-order sequence involving major part of the Jurassic (~37.7 my out of a total of ~56.3 my) has a total duration of ~57 my inclusive of ~18.6 my of Cretaceous. The duration of the first-order sequence almost equals that of the Jurassic system. The youngest first-order sequence involving major part of the

Cretaceous has a duration of ~66 my. The oldest one besides involving the whole of ~50.9 long Triassic, also includes part of the underlying Permian as also part of the overlying Jurassic. It spans through the duration of ~89 my. The difference between the smallest and the largest is ~32 my, which is quite large. The average comes to ~71 my which is close to that of the youngest sequence. The average exceeds the smallest one by ~14 my, and is less than the longest by ~18 my.

Looking at the eleven stages of the Jurassic, the five smaller ones (Hettangian, Aalenian, Bajocian, Bathonian and Callovian) have a total duration of ~10.6 my with an average of ~2.0 my while the six bigger ones (Sinemuirian, Pliensbachian, Toarcian, Oxfordian, Kimmeridgian and Tithonian) total ~43.7 my with an average of more than ~7.0 my i.e. about three and half times that of smaller ones. Taking a different angle of the organization of the stages into Early, Middle and Late, it is found that the Middle Jurassic is more than four times less to Early and more than twice less to Late Jurassic. The ratio of their time spans in ascending order is 4:1:2 of Early, Middle and Late divisions of Jurassic.

3.8.3 Relatively Shorter Duration of the First-Order TSTs to Corresponding RSTs

The first-order TST of the first sequence is ~29 my which is ~31 my shorter to the ~60 my long RST. In the second sequence, the first-order TST from near the basal Toarcian to the late Middle Oxfordian is ~24.7 my while the following first-order RST is ~66 my, i.e. longer by ~41 my. The pattern is repeated in third sequence as well. In the third sequence, the ~24 my long TST is again distinctly shorter than the ~54 my long RST. Thus, at the first-order level, the RSTs are way larger in two sequences while in the third sequence the duration is nearly equal.

At the second-order level, the TSTs of the fifth to eleventh second-order sequences are either larger or some what equal in duration than their corresponding RSTs. At the third-order level, it is not possible to assert the durations of their TSTs and RSTs.

The differentiation of the fourth-order sequences has been realized only in selective third-order sequences. For example in the Bathonian–Callovian second-order sequence of ~2.4 my, four third-order sequences are differentiated. These are either of ~400 or ~800 ky in duration. One of the four third-order either of ~400 or ~800 ky has been in turn organized into four fourth-order sequences, which would thus either be of ~100 or ~200 ky each.

The fifth-order sequence durations can be crudely measured through precise counts of their alternating hard and soft bands, each couplet forming a fifth-order sequence. This was endeavoured in better measured Bathonian to Oxfordian sections, in which the fourth-order differentiation has been already made. There is any way some subjectivity involved. For example, in the mid-Midle to Late Bathonian column at Jumara, five relatively thick hard ammonoid-bearing beds are present in the second fourth-order TST, thereby implying the presence of five fifth-order sequences in the fourth-order TST. Such counting can not be made in the following fourth-order RST. The fourth-order sequence is, as above, either of ~100 or ~200 ky. Thus these five fifth-order sequences may be of ~10 or ~20 ky each. Similar counts are also made in Early and early Middle Callovian part where fourth-order differentiation has been realized, however, any consistent pattern did not emerge. In the Oxfordian of Kantkote–Bharodia section within the Subevolutus Zone (=to Bifurcatus Zone) the fourth-order sequences invariably include either 10, or 11 or 12 hard beds, which could be by rough estimates of ~20 ky each. Similarly, the late Late Kimmeridgian laminated sands in second-order RST of the first-order RST, exhibit ~20 ky sediment packages (Krishna et al. 2000a, b, c).

The first- to third-order sequence stratigraphic framework developed in Kachchh has precisely determined sequence surfaces. Such framework is restricted to the late Middle Bathonian–late Late Tithonian interval. Twenty-eight third-order sequences spread through nearly equal number of ammonoid zones, seven plus second-order sequences have been differentiated in ~21.6 my interval with an average second-order sequence duration of ~2.8 my, as also average ammonoid zonal duration of ~800 ky, and average subzonal duration of ~400 ky. Each third-order sequence may invariably include 2 to 3 ~400 ky units compared to 1 or 4 ~400 ky units. At the same time, it requires to be remembered that neither sequences nor ammonoid zones are of equal duration. There is also no reliable precise measure, yet, to evaluate the duration of the individual zones.

There are different scales of cyclicity in different exposed sections from margin to the basin. The cyclic patterns are more obvious in major transgressive phases in relatively deeper distal sites away from the margin, e.g. Callovian to early Middle Oxfordian in West Mainland Kachchh having relatively slow sedimented condensed ammonoid rich beds alternating with ammonoid devoid/scarce moderately sedimented beds. In contrast, the finer/smaller cycles are better expressed in major regressive phases in relatively shallower proximal sites near the margin, e.g. latest Middle Oxfordian to Early Kimmeridgian in Wagad.

The ammonoid subzones have been invariably observed to span through either one third-order TST or one RST of ~400 ky or its multiples. One good example could be cited of the Schilli Subzone third-order TST of late Middle Oxfordian of presumably of ~400 ky duration, which is in turn divisible into ten fourth-order TSTs of ~20 ky each. Thus a single third-order sequence can be considered to have absolute durations in multiples of ~400 ky units.

Apart from the ~400 ky cyclicity, are also noted ~100, ~40 and ~20 ky cycles.

There are three third-order TSTs in the mid-Middle to Late Bathonian of ~1.2 my duration with nearly three ~400 ky units, six in ~2.60 my Callovian with nearly 430 ky units, eight in ~6.2 my Oxfordian with average of ~775 ky units, i.e. two units each of ~388 ky, six in ~5.2 my Kimmeridgian with thirteen ~400 ky units, and six in ~7.1 my Tithonian with eighteen ~400 ky units.

There are three third-order sequences in the fifth second order of ~2.4 my duration with an average third-order resolution of ~800 ky, five third-order sequences in the sixth second order of ~2.1 my duration with an average third order of ~420 ky, six third-order sequences in seventh second order of ~4.8 my duration with an average of ~800 ky, two third-order sequences in ~2.0 my long eighth second-order sequence with perhaps one third order of ~800 ky and the other of ~1200 ky, four third-order sequences in the ninth ~3.6 my long second order with an average third-order duration of ~800 ky, three third-order sequences in the tenth ~2.8 my long second-order sequence with two third-order sequences of ~800 ky and the third of ~1200 ky, and again three third-order sequences in the eleventh ~2.8 my long second-order sequence with two third-order sequences of ~800 ky and the third of ~1200 ky.

The above speculative data suggest that the third-order sequences in the later part of the first-order TST (from mid-Middle Bathonian to late Middle Oxfordian with eleven third-order sequences in ~6.9 my) are possibly shorter than sequences of early part of the first-order RST (latest Middle Oxfordian to latest Tithonian with 15 sequences in ~14.7 my). Alternatively, may be it is relatively hard to differentiate the third-order sequences in the second-order RSTs, particularly in the body fossil scarce rapidly deposited thick sands. Alternatively, the longer third-order sequences in the second-order RST are invariably twice as large as the smaller ones.

The said eleven third-order sequences in the terminal part of the first-order TST involving 17 ~400 ky units imply that a maximum of eight sequences of one and three of two such units. On the other hand the fifteen third-order sequences in the early part of the first-order RST would be mostly having two to three ~400 ky units while none having one unit. The same perhaps is also true for the second-order sequences. The second-order sequences in the later part of the first-order TST have an average duration of a little over ~two my while those of early part of the first-order TST have an average of around c 3 my.

To conclude, the most significant outcome of the above exercise is its rudimentary testimony to the fundamental cycle unit of ~400 ky, i.e. of subzonal duration. In the late Middle Bathonian to latest Late Kimmeridgian interval, the third-order sequences have been further differentiated into fourth- and fifth-order sequences, and it has been possible to observe 200, 100 and 20 ky cycles. For example, in the first and second third-order mid-Middle Bathonian Mantataranus Zone sequences, four fourth-order sequences each have been differentiated, which provide average fourth-order duration of ~100 to ~200 ky. More such estimates need be undertaken in different basins of GTM, also in the Triassic and Jurassic of Kachchh basin to either strengthen or discard the crude scales of sequence cyclicity that have emerged forth in this exercise.

Even the duration of stages have substantively varied in recent decades, no appreciable credence can be expected as to the durations of different orders of sequences and orders of classical ammonoid based-chronostratigraphic units. Nevertheless, the present preliminary exercise, it is hoped, shall form edifice for future studies with increased precision.

3.8.4 Sequences, Periodicity and Durations

Order and No. of sequence	Geological range	Duration						
		Total			Absolute duration	TST	RST	Organization as of ~400 ky
		Basal SB (ma)	MFS (ma)	Terminal SB (ma)				
Mega order								
Third mega-sequence	Late Permian to Early Paleogene	~272	~92	55	~217 my	~180 my	~37 my	
First order								
First sequence	Late Permian to Pliensbachian	~272	~243	183	~89 my	~29 my	~60 my	
Second sequence	Toarcian to Barremian	~183	~159	126	~57 my	~24 my	~34 my	
Third sequence	Aptian to Paleocene	~126	~92	60	~66 my	~34 my	~36 my	
Second order								
J-I	Hettangian				~2 my			
J-II	Most of Sinemurian				~8.5 my			
J-III	Terminal Sinemurian to Plinsbachian				~8 my			
J-IV	Toarcian to Aalenian				~12.4 my			
J-V	Basal Bajocian to Mid-Late Bajocian				~3.6 my			3 of ~800 ky and 1 of ~1200 ky
J-VI	Late Bajocian to Mid-Middle Bathonian				~1.5 my			4 of ~375 ky
J-VII	late Middle Bathonina to late Middle Callovian				~2.4 my			3 of ~800 ky
J-VIII	late Middle Callovian to early Early Oxfordian				~2.1 my			5 of ~420 ky

(continued)

(continued)

Order and No. of sequence	Geological range	Duration						
		Total			Absolute duration	TST	RST	Organization as of ~400 ky
		Basal SB (ma)	MFS (ma)	Terminal SB (ma)				
J-IX	late Early to Close of Oxfordian				~4.8 my			6 of ~88 ky
J-X	early Early Kimmeridgian to mid-Early Kimmeridgian				~2.0 my			1 of ~800 ky and 1 of ~1200 ky
J-XI	late Early Kimmeridgian to Close of Kimmeridgian				~3.6 my			3 of ~800 ky and 1 of ~1200 ky
J-XII	early Early Tithonian to Mid-Early Tithonian				~2.8 my			2 of 800 ky and 1 of 1200 ky
J-XIII	late Early to late Late Tithonian				~3.6 my			3 of ~800 ky and 1 of ~1200ky
J-XIV	Terminal Late Tithonian				~400 ky			One of ~400 ky

References

Alberti M, Pandey DK, Fursich FT (2011) Ammonites of the genus Peltoceratoides Spath 1924 from the Oxfordian of Kachchh, western India. N Jb Geol Palaont Abh 262:1–18

Alberti M, Fursich FT, Pandey DK, Ramkumar M (2012a) Stable isotope analysis of belemnites from the Kachchh basin, western India: paleoclimatic implications for the Middle to Late Jurassic transition. Facies 58:261–268

Alberti M, Pandey DK, Fursich FT (2012b) The Oxfordian stable isotope record (d18O, d13C) of belemnoids, brachiopods and oysters from the Kachchh basin (western India) and its potential for paleoecologic, paleoclimatic and paleogeographic reconstructions. Palaeogeogr Palaeoclimateol Palaeoecol 344–345:49–68

Alberti M, Fursich FT, Pandey DK (2013) Deciphering condensed sequences: a case study from the Oxfordian (Upper Jurassic) Dhosa Oolite member of the kachchh basin, western India. Sed Geol 60:574–598

Arkell WJ (1956) Jurassic geology of the world. Oliver and Boyd, London, pp 1–806

Bajpai S, Prasad GVR, Prasad V, Krishna J, Sarkar A (2012) Recent advances on Phanorozoic biodiversity, bioevents and climates in India. PINSA 78(Special Issue):445–456

Bhargava ON (1987) Stratigraphy, microfacies, and paleoenvironment of the Lilang Group (Schythian-Dogger), Spiti valley, Himachal Himalaya, India. J Pal Soc Ind 32:92–107

Bhargava ON, Bassi UK (1996) The Spiti Kinnaur Tethyan basin, Himachal Himalaya, and regional sedimentological pauses. In: Proceedings of symposium on Himalaya and foredeep. Sp Publ Geol Surv Ind 21(1):19–27

Bhargava ON, Krystyn L, Balini H, Lein R, Nicora A (2004) Revised litho and biosequence stratigraphy of the Spiti Triassic. Albertiana Suppl 30:21–38

Biswas SK (1977) Mesozoic rock-stratigraphy of Kutch. Quart J Min Met Soc Ind 43(3 and 4): 1–52

Biswas SK (1993) Geology of Kutch, 1, K D Malaviya Institute of Peetroleum Exploration, Dehradun, 450 p

Blaison J (1967) Etude biometrique et revision de l'espece Protogrammoceras (Thevnin) 1908. Ann Sci Univ Besancon Geologie 3:59–115

Bosselini A (1992) The continental margins of Somalia. In: Watkins JS, Zhikiang F, McMillen KJ (eds) Geology and Geophysics of continental margins, 53:185–205

Catuneanu O (2006) Principles of sequence stratigraphy. Elsevier, Amsterdaum 375 p

Cohen KM, Finney SC, Gibbard PL, Fan JX (2015) The ICS international chronostratigraphic chart. Episodes 36:199–204

Dave A (2011) Recent developments in sequence stratigraphy and depositional sequences of Indian petroliferous basins, pp 88–112. In: Jaitly AK, Singh AD, Pandey B, Nath S (eds) Palaentology and stratigrahy: basics to applications. Banaras Hindu University, Varanasi

Dave A (2012) Depositional sequence stratigraphy and petroleum systems of Kachchh basin, Status Talk–Hydrocarbon Potential and offshore geology. In: Thakkar MG (ed) Abstract volume, Geology of Kachchh basin, Western India, pp 94–97

Diener C (1908) Ladinc, Carnic, and Noric faunae of spiti. Springer, Germany 15(5):157

Dromart G, Garcia JP, Picard S, Atrops F, Lecuyer C, Sheppard SMF (2003) Ice age at the Middle/Late Jurassic transition: geological and geochemical evidence. Am J Sci 303:667–707

Fatmi AN (1972) Stratigraphy of Jurassic and Lower Cretaceous rocks and Jurassic ammonites from northern areas of West Pakistan. Bull Brit Mus Nat Hist Geol 20:297–380

Fatmi AN (1984) Stratigraphy and stratigraphic problems of the Jurassic rocks of Pakistan. 2nd Int Symp Jur Stratigr 3:653–659

Fatmi AN (1986) Lower and Middle Jurassic ammonites from Windar Group of Phusi Jhal (Kanrach) and Sand (Windar Nai) sections of Las Bella, Baluchistan, Pakistan. Mem Geol Surv Pak Pal Pak 16:1–17

Fatmi AN et al (1986) Stratigraphy of Zidi Formation, Ferozabad Group and Parh Group, Khuzdar district, Baluchistan, Pakistan. Rec Geol Surv Pak 1:1–32

Fursich FT, Oshman W, Callomon JH, Jaitly AK (1994) Contribution to the Jurassic of Kachchh, Western India: II Bathonian stratigraphy and depositional environments of the Sadhara dome, Pachcham Island. Beringeria 12:95–125
Fursich FT, Singh IB, Joachimski M, Crumm S, Schlirf M, Schlirf S (2005) Paleoclimatic reconstructions of the Middle Jurassic of Kachchh (western India): an integrated approach based on paleontological, oxygen isotopic and clay mineralogical data. Paleogeography Paleoclimateology Paleoecology 217:289–309
Fursich FT, Alberti M, Pandey DK (2013) Stratigraphy and paleoenvironments of the Jurassic rocks of Kachchh (field guide). Berringeria 7(Spl. Issue):1–174
Garzanti E, Jadoul F, Nicora A, Berra F (1995) Triassic of Spiti (Tethys Himalaya, India). Riv Ital Paleontol Stratigr 101(3):267–300
Geiger M (2004) Sedimentary and stratal patterns in Jurassic successions of Western Madagascar —facies, stratigraphy, and architecture of Gondwana break up and drill sequences, Ph.D thesis, Uni. Bremen
Geiger M, Schweigert S (2006) Toarcian – Kimmeridgian depositional cycles of the southwestern Morondova basin along with rifted continental margin of Madagascar. Facies 52:85–112
Govindan A (ed) (2000) Cretaceous stratigraphy an update. Mem Geol Soc Ind 46:423
Govindan A, Ravindran CN (1996) Cretaceous biostratigraphy and sedimentation history of Cauvery basin, India. In: Proceedings of XVth Indian colloquium on micropaloentology and stratigraphy, Dehradun, pp 19–31
Govindan A, Ravindran CN, Rangaraju MK (1996) Cretaceous stratigraphy and planktonic foraminferal zonation in Cauvery basin, south India. Mem Geol Soc Ind 37:155–187
Govindan A, Yadagiri K, Ravindran CN, Kalyansundaram R (1998) The Trichy field guide. Int Sem Rec Adv Study Cret Sections, pp 1–53
Govindan A, Ananthanarayanan A, Vijayalakshmi KG (2000) Cretaceous petroleum system in Cauvery basin. India. Mem Geol Soc Ind 46:365–382
Graciansky PC, Dardeau G, Dumont T, Jacquin T, Marchand D, Mouterde R, Vail PR (1993) Depositional sequencecycles, transgressive-regressive facies cycles and extensional tectonics: example from the southern subalpine Jurassic basin, France. Bull Soc Geol France 164(5):709–718
Haq BU, Hardenbol J, Vail PR (1987) Chronology of fluctuating sea-levels since the Triassic. Science 235:1156–1167
Hornung T, Krystyn L, Brandner R (2007) A Tethys-wide mid-Carnian (Upper Triassic) carbonate productivity crisis: evidence for the alpine Reingraben event from Spiti (Indian Himalaya)? J Asian Earth Sci. doi:10.1016/j.jseaes.2006.10.001
Jacquin T, Dardeau G, Durlet C, de Graciansky PC, Hantzpergue P (1998a) The North Sea cycle; an overview of 2nd-order transgressive/regressive facies cycles in Western Europe. In: de Graciansky PC, Hardenbol J, Jacquin T, Vail PR (eds) Mesozoic and cenozoic sequence stratigraphy of European basins. Soc Sed Geol (SEPM), Tulsa (OK) 60:445–466
Jacquin T, de Graciansky PC (1998b) Major transgressive/regressive cycles; the stratigraphic signature of European basin development. In: de Graciansky PC, Hardenbol J, Jacquin T, Vail PR (eds) Mesozoic and cenozoic sequence stratigraphy of European basins. Soc Sed Geol (SEPM), Tulsa (OK) 60:15–29
Jaitly AK, Singh CSP (1983) Discovery of Late Bajocian Leptosphinctes Buckman (Jurassic Ammonitina) from Kachchh, Western India—Neues Jahrbuch Fur Geologie und Palaontlogie, Monatshefte, 2:91–96
Kanjilal S, Singh SR (2012) Arrival of Quenstedtoceras Hyatt: a late Callovian ammonite in Kachchh and its significance (abstract). In: Thakkar MG (ed) Geology of Kachchh basin, p 27
Krishna J (1987a) Biological evidence for better appreciation of Indian Gondwana. Palaeobotanist 36:268–284
Krishna J (1987b) An overview of the Mesozoic stratigraphy of Kachchh and Jaisalmer. J Paleontol Soc Ind 32:136–149
Krishna J (1994) Environmental reconstruction studies of the Mesozoic sedimentary sequences in Western Kachchh (ONGC Sponsored Project), (Technical Report), 1–143, pts, 1–10

Krishna J (1996) Origin of the Gondic (Trans-Gondwana) Corridor: the ammonoid evidence, vol 2. Gondwana Nine, Oxford and IBH Publishing Company, Delhi, pp 1091–1099
Krishna J (2000) An overview of the recent progress in the Cretaceous stratigraphy of Kachchh basin, India. Mem Geol Surv Ind 46:399–401
Krishna J (2005) Update on Kachchh Jurassic ammonoid zonation: summarised intra-basinal to global perspectives. Spl Publ -1, Ass of Petr Geol 63–68
Krishna J (2012) Litho-bio-chrono-magneto-igno-sequento-stratigraphic integration in the Kachchh Mesozoic, Status Talk–Mesozoic Biostratigraphy. In: Thakkar MG (ed) Abstract Volume, Geology of Kachchh basin, Western India, 21–25
Krishna J (2014) Ammonoid—sequence-magneto stratigraphic integration in the Kachchh Jurassic. Unpublished DST project report, 60p
Krishna J, Cariou E (1986) The Callovian of Western India: new data on stratigraphy, biogeography of the ammonoids and correlation with Western Tethys. Newsl Stratigr 17(1):1–8
Krishna J, Cariou E (1993) The Tethyan Macrocephalitinae: evolutionary, environmental and dispersal strategies. Geobios 15:227–236
Krishna J, Pathak DB (1989) Kimmeridgian in Ler–Katrol area of Kachchh, Western India: ammonoid systematics and biochronology (abstract). In: 28th International Geological Congress, Washington, 228–229
Krishna J, Pathak DB (1991) Ammonoid biochronology of the Kimmeridgian stage in Kachchh, India. J Pal Soc India 36:1–13
Krishna J, Pathak DB (1993) Late Lower Kimmeridgian – Lower Tithonian virgatosphinctins of India: evolutionary succession and biogeographic implications. Geobios 15:217–226
Krishna J, Pathak DB (1994) Stratigraphic, biogeographic and environmental signatures in the amonoid bearing Jurassic-Cretaceous in the Himalaya on the south margin of the Tethy. J Him Geol 4(2):151–167
Krishna J, Melendez G, Pandey B, Pathak DB (1994a) Middle Oxfordian Perisphinctinae and biostratigraphy from Kachchh (India). Jornadas de Paleontologia 106–108
Krishna J, Pathak DB, Pandey B (1995) Kimmeridgian – Tithonian (Upper Jurassic) ammonoid zones in Kachchh (Western India) and their correlations. In: Proceedings of recent researches in Geology in Western India, Badodara, 323–347
Krishna J, Pandey B, Pathak DB (1996a) Ammonoid chronology developed in the Tithonian of Kachchh (India). GeoResearch Forum 1–2:205–214
Krishna J, Pathak DB, Pandey B (1996b) Quantum refinement in the Kimmeridgian ammonoid chronology in Kachchh (India). GeoResearch Forum 1–2:195–204
Krishna J, Pathak DB, Pandey B (1998) Development of Oxfordian (Early Upper Jurassic) in the most proximal exposed part of Kachchh basin at Wagad outside the Mainland Kachchh. J Geol Soc Ind 52:513–522
Krishna J, Pandey B, Ojha JR, Pathak DB (2000a) Transgressive sediment intervals in the Late Jurassic of Kachchh, India. GeoResearch Forum 6:321–332
Krishna J, Pandey B, Ojha JR, Pathak DB (2000b) Ammonoid—foraminifer zonal integration in the Jurassic of Kachchh, India. ONGC Bulletin 37(1):9–21
Krishna J, Pathak DB Pandey B, Ojha JR (2000) Transgressive sediment intervals in the Late Jurassic of Kachchh, India. GeoResearch Forum 6:321–332
Krishna J, Pandey B, Ojha JR (2008) Gregoryceras in the Oxfordianof Kachchh (India): diverse eventful implications. Geobios 42:197–208
Krishna J, Pandey B, Pathak DB (2009) Characterization of Dichotomoceras in the Oxfordian of Kachchh (India). J Geol Soc Ind 74:469–479
Krishna J, Pandey B, Pathak DB (2010) Update on the Jurassic stratigraphy of the Indian subcontinent with focus on the tectonically controlled regional transgressive—regressive couplets. Earth Sci Frontiers China 17:138–140
Krishna J, Pandey B, Pathak DB (2011) Current status of the Jurassic ammonoid stratigraphic framework in the Indian subcontinent with focus on the tectonically controlled regional transgressive – regressive couplets. Geol Soc Ind Mem 78:140–176

Krishna J, Singh IB, Howard JD, Jafar SA (1982) New thoughts on the environment and paleogeography of the so called coastal Gondwana sediments of Kachchh, Western India. (abstract) Internat Sed Congr Hamilton, 229
Krishna J, Sinha AK, Upadhyay R (1997) The First South Tethyan Hettangian Psiloceras from the Indus Suture, Ladakh Himalaya: diverse implications. J Him Geol 18:45–51
Krishnaswami VS, Swaminath J (1965) Himalayan and Alpine geology: a review, Wadia Mem. Min Mettalurg Inst Ind 172–192
Krystyn L, Balini M, Nicora A (2004) Lower and Middle Triassic stage and substage boundaries in Spiti. Albertiana 30(supplement):39–52
Krystyn L, Bhargava ON, Richoz S (2007) A candidate GSSP for the base of the Olenekian Stage: Mud at Pin Valley; district lahul & Spiti, Himachal Pradesh (Western Himalaya), India. Albertiana 35:5–29
Li X, Grant-Mackie JA (1993) Jurassic cycles and and eustatic sea-level changes in southern Tibet. Paleogeography Paleoclimatology Paleoecology 101:27–48
Mathur YK, Soodan KS, Mathur K, Bhatia ML, Juyal NP, Pant J (1970) Microfossil evidences on the presence of Upper Cretaceous and Paleocene sediments in Kutch. ONGC Bulletin 7 (2):109–114
Nagendra R, Nagendran G, Narsimha K, Jaiprakash BC, Nallapa-Reddy A (2002a) Sequence stratigraphy of Dalmiapuram Formation, Kalakuddy quarry II, South India. J Geol Soc Ind 59:249–258
Nagendra R, Raja R, Nallapa-Reddy A, Jaiprakash BC, Bhavani R (2002b) Outcrop sequence stratigraphy of the Maestrichtian Kallankuruchchi Formation, Ariyalur group. Tamilnadu J Geol Soc Ind 59:243–248
Nagendra R, Kamalnarayanan BV, Gargi S, Gilbert H, Bakkalraj D, Nallapa-Reddy A, Jaiprakash B (2011) Sequence surfaces and paleo-bathymetric trends in Albian to Maestrichtian sediments of Ariyalur area, Cauvery basin. J Mar Petr Geol 28:895–905
ONGC (2008) Pteroleum systems sequence stratigraphy. KDMIPE, Dehradun
Pandey J, Dave A (1998) Stratigraphy of Indian petroliferous basins, 17th ICMS, 1–248
Pandey B, Krishna J (2002) Ammonoid biostratigraphy in the Tithonian (Late Jurassic) of Jaisalmer, Western India. Geophytology 30 (1 and 2):17–25
Pandey B, Pathak DB (2015a) Record of Early Bathonian ammonoids from Kachch, India: biostratigraphic and paleobiogeographic implications. J Pal Soc Ind 60(1):33–44
Pandey B, Pathak DB (2015b) Status of the Indian Early Cretaceous ammonoid record in light of recent observations in the Spiti Valley, Himachal Himalaya. Himalayan Geol 36(1):1–8
Pandey DK, Sha J, Choudhary S (2010) Sedimentary cycles in the Callovian–Oxfordian of Jaisalmer basin, Western India. Vol Jur 8:131–162
Pandey B, Pathak DB, Jaitly AK, Krishna J, Venkateshwarlu M (2012a) Record of Tethyan gastropod genus Astrohelix Szabo, 1984 from Late Bajocian (Middle Jursassic) of Kachchh, western India. Indian J Geosci 66(1):65–68
Pandey DK, Choudhary S, Bahadur T, Swami N, Poonia D, Sha J (2012) A review of the Lower–lowermost Upper Jurassic facies and stratigraphy of the Jaisalmer basin. Vol Jur 10:61–82
Pandey B, Krishna J, Pathak DB, Kumar A (2013a) Ammonoid biostratigraphy of Bathonian succession at Jumara, Kachchh, western India. J Ind Geol Cong 4(2):7–18
Pandey B, Pathak DB, Krishna J (2013b) Calliphylloceras heterophylloides (Oppel, 1856) from basal most Jurassic succession of Sadhara dome, Kachchh, India. J Pal Soc Ind 58(1):61–65
Pandey B, Pathak DB, Krishna J (2013c) Preliminary remarks on new ammonoid collection from freshly exposed succession of the Spiti Formation between Lidang and Giumal, Spiti valley, Himachal Himalaya, India. Him Geol 34(2):124–134
Pandey DK, Alberi M, Fursich FT, Glowniak E, Oloriz F (2013) Ammonites from the Oxfordian/Kimmeridgian boundary and the Lower-Upper Kimmeridgian of Kachchh, Western India. Vol Jur X:97–146
Potonie R, Sah SCD (1960) Sporae dispersae of the lignites from Cannanore beach on the Malabar coast of India. Paleobotanist 7(2):121–135

Rabu B (1993) Stratigraphy and structure of the Oman mountains, vol 221. Bur Recher Geol Min, Paris, p 262

Rai J, Jain S (2013) Pliensbachian nanno-fossils from Kachchh: implications on the earliest Jurassic transgressive event on the western Indian margin. Zitteliana A 53:105–120

Rajnath (1934) Detailed stratigraphy of the Jumara area, Cutch (abstr). In: Proceedings of 21st Indian Science Congress, p 346

Raju DSN (2008) Stratigraphy of India; mega—charts 1–2, Bull. ONGC, Spl. Issue, 43(1):1–44 and mega charts

Raju DSN (2011) Cretaceous stratigraphic record on Indian plate: an overview. Mem Geol Soc Ind 78:177–228

Schweigert G, Krishna J, Pandey B, Pathak DB (1996) A new approach to correlation of the Upper Kimmeridgian Beckeri Zone across the Tethys. N Jb Geol Paleont 202(3):345–373

Shanker R, Singh G, Kumar G (1994) Sequence stratigraphy in Himalaya and its reflections in Gondwana sedimentation. In: Proceedings of 9th Gondwana Symposium, Special Publications of Geological Survey of India, Hyderabad, India, pp 1159–1173

Sharland P, Archer R, Casey DM et al (2001) Arabian plate sequence stratigraphy, GeoArabia, Spl publ 2, p 371

Singh SR (1992) Ammonite biostratigraphy of the Callovian—Oxfordian (Jurassic) rocks of Nara dome, district Kachchh, Gujarat. Unpublished PhD thesis BHU, p 328

Sloss LL (1963) Sequences in the cratonic interior of North America. Bull Geol Soc Amer 74: 93–114

Sowerby J (1840) A notice respecting some forms collected in Kutch. Trans Geol Soc Lond 21 (5):715–719

Srikantia SV (1977) Sedimentary cycles in the Himalaya and their significance on the orogenic evolution of the mountain belt. Int Colloq Centre Nat Res Sci (CNRS) 268:395–407

Stoliczka F (1866) Geological section across the Himalayan Mountains in Spiti. Mem Geol Sur Ind 5(1):1–153

Vail PR, Colin JP, Chene RJ, Kuchly J, Mediavilla F, Trifilieef V (1987) La stratigraphie sequentielle et son application aux correlations chronostratigraphiques dans le Jurassique du bassien de Paris. Bull Soc Geol Fr 3(7):1301–1321

Van Vagoner JC, Posamentier HM, Mitchum RM (1988) An overview of the fundamentals of sequence stratigraphy and key definitions. In: Wilgus CK et al (eds) Sea-level changes: an integrated approach, *SEPM* Sp. Publ 42:39–45

Venkatachala BS (1970) Palynology of the Mesozoic sediments of Kachchh, West India; reworked Permian Pollens from the Upper Jurassic sediments- a discussion. Palaeobotanist 18(1):45–49

Venkatachala BS, Tiwari RS (1987) Lower Gondwana marine incursion: periods and pathways. Palaeobotanist 36:24–29

Von Salis AK, Immenhauser A (1997) Mesozoic calcareous nannofossil from Masirah island (Sultanat of Oman). J Nannoplank Res 19:95–101

Westphal H, Aignar T (1997) Seismic stratigraphy and subsidence analysis in the Barrow—Dampier basin, Northwest Australia. Am Ass Petrol Geol B 81:1721–1749

Yin, Enay R (2000) The earliest Jurassic psiloceratids in the eastern Tethyan Himalaya, South Tibet, Comptes Rendus de l'academie de Sciences, Paris. Sciences de la Terre et Planetes 331:601–608

Yin J, Smith PL, Palfy J, Enay R (2007) Ammonoids and the Triassic/Jurassic boundary in the Himalayas of Southern Tibet. Palaeontology 50(3):711–737

Chapter 4
Integration of the Micro/Macro Faunal/Floral Data into Ammonoid Stratigraphic Framework in the Indian Mesozoics

Abstract An integrative mutually correlative stratigraphic framework of the Mesozoic lithobiostratigraphic successions/units is developed in the Indian sedimentary basins. It is realized on the combined strength of the stratigraphic refinement achieved in Chap. 2, together with the succession of the first- and second-order sequence surfaces formulated in Chap. 3. The exercise obligatorily involved large-scale revision and refinement of the imprecise ages of the Indian Mesozoic lithostratigraphic units which in earlier studies were based on unsubstantiated and inadequate data on different macro/microfossil groups. Also, in the process are provided new alternatives to many inaccurate intra/interbasinal correlations and models. The integration of the entire available outcrop and subsurface guide fossil-based biostratigraphic data in conjunction with sequence formulations is applied to intra and interbasinal correlation of marine with non-marine, guide fossil bearing to guide fossil devoid, along with time correspondence among ammonoid, conodont, foraminfer, coccolith, dinocyst, pollen/spore-based zonal schemes across the Indian sedimentary basins in north, west and east sectors. The chapter concludes with precise ages to the Kachchh Mesozoic units (formal formations of Stoliczka, formal members of Biswas and others, also informal/abandoned members created outside the type sections yet otherwise considered useful).

Keywords Biostratigraphic integration · Mesozoic · Micro/mega · Faunal/floral · Marine/non-marine · Outcrop/well · Peninsular/extra-peninsular · Intra/interbasinal · Kachchh · Mainland · Wagad · Island belt · Spiti · Cauvery · Narmada · Jaisalmer · K-G · Bengal · Ammonoid · Forams · Conodonts · Coccoliths · Dinocysts · Pollen/spores · Benthic · Planktic

© Springer Science+Business Media Singapore 2017

J. Krishna, *The Indian Mesozoic Chronicle*, Springer Geology,
DOI 10.1007/978-981-10-2477-1_4

4.1 Indian Lithobiostratigraphic Refinement, Correlation and Integration

4.1.1 The Indian Sedimentary Basins with Geological Record of the Mega-Sequence

The Indian basins with geological record of the mega-sequence are located on the margin as also in the interior (Fig. 1.4). These are broadly organized into east (Cauvery, K-G, Mahanadi, Bengal, Purnea, Meghalaya, Assam, Assam–Arakan basins), north (High Himalaya basins of Pakistan, Kashmir, Spiti–Zanskar, Kumaon–Garhwal, Nepal, Bhutan and South Tibet) and west (Pakistan, Jaisalmer, Barmer, Saurashtra, Narmada, Cambay, Mumbai offshore, Kerala–Konkan basins) sectors. The inland basins are mainly the Gondwana graben basins. The Cauvery basin is spread between Palar and Pondicherry, the K-G basin from Ongole in the south to Visakhapatnam in the north, the Mahanadi basin from near Chilka lake to near Baitarani river, the Bengal basin stretches from the Baitarani river to the Surma river, Assam–Arakan basins are spread through Meghalaya, Assam and Arunachal, the Purnea basin is located partly in east Bihar and partly in north Bengal and the Andaman strikes north–south and somewhat south of Myanmar and west of Malayasia. Among the three Mesozoic systems, the Triassic is the best developed, also the best studied in Spiti; the Jurassic in Kachchh; and the Cretaceous in Cauvery.

4.1.2 On the Utilization of the Ammonoid Data

Inquiries on ages revealed that most of the earlier studies on the Indian Mesozoic, unfortunately, did not utilize all the available ammonoid data in conjunction with other biostratigraphic information. The importance of ammonoids in age determination of the Mesozoic marine sediments was only, seldom appreciated. It was side tracked that the good time frameworks in the Mesozoic are primarily based on the ammonoids, without which even huge non-ammonoid data cannot be meaningfully applied for various interpretations. Moreover, the scales based on the non-ammonoid mega/microfossil groups across the globe are near exclusively age constrained by ammonoids and in case of age-discrepanices, the ammonoid dates are to be relied upon.

4.1.3 The Indian Mesozoic Lithostratigraphy Scenario

It emerged that the Indian Mesozoic field geologists in the twentieth century could only seldom develope quality lithostratigraphic differentiation in spite of the same being the basic pre-requisite to bio/chronostratigraphic refinement. One such unfortunate example is the absence of good lithostratigraphic differentiation in the prolificly ammonoid bearing and hydrocarbon producing Cauvery Cretaceous even today after nearly two centuries of studies. However, an example of good lithostratigraphic differentiation existed since the 1970s in the Kachchh Mesozoic (Biswas 1977, 1981, 1982, 1987, 1993, 1999, 2005, 2012; Biswas et al. 1993). Foundation to this was laid much earlier by Rajnath (1932) in the specific Callovian–Early Oxfordian section at Jumara (Table 1.2). Even in Kachchh, the good lithostratigraphy could not be followed-up for a long time by good biostratigraphy mostly due to inadequate application of the available ammonoid-based ages in the synthesized accounts.

4.1.4 Biostratigraphic Range Charts

To date never could any ammonoid stratigraphic range charts be produced in the Indian Mesozoic other than those by the author's group in the Kachchh Jurassic and by Bhargava et al. (2004), Bhargava and Tangri (1996), Krystyn et al. (2004) in the Spiti Triassic. Unfortunately, almost none exist in the splendidly and extensively developed ammonoid rich Indian Cretaceous. Microfossil-based range charts have been published, however, mostly in well sections and only seldom in exposed sections that too, without adequate lithostratigraphic differentiation (e.g. Pandey and Dave 1993; Prasad et al. 1995; Kale et al. 2000; Mehrotra and Aswal 2003; Govindan et al. 1996; Chidambaram 2000; Mehrotra et al. 2002; Vijaya 2009 and others).

4.1.5 Integration of the Non-ammonoid Zonal Schemes to Ammonoid Schemes

In the Triassic, the ammonoid zones have been integrated to conodont zones, while the palynozones have so far not been adequately constrained either by ammonoids or conodonts. In the Jurassic, apart from the ammonoid-based high-resolution zonation schemes in the Kachchh basin, and its extension in Jaisalmer and High Himalaya, are developed palyno and foraminifer zonation in Kachchh, and generated substantive nannoplanktonic data. However, none are appropriately integrated with the primary and the better resolved ammonoid-based framework. In the Cretaceous, the planktonic foraminiferal, coccolith, dinocyst, pollen/spore and ammonoid zonations in different intervals have been developed in bits in different basins in the exposed surface or subsurface wells. These independent zonations lack mutual integration.

4.1.6 Enlargement of the Stage Level Stratigraphic Record in the Mega-Sequence

The present studies have resulted in considerable enhancement in the over all stratigraphic record of the mega-sequence in different basins. The Late Permian–Pliensbachian sequence interval in India is well recorded in Spiti with differentiation of all the involved Late Permian, Triassic and Liassic stages. This first-order sequence interval is also differentiated in the K-G basin wells in a succession of pollen-/spore-based palynozones. The sequence also seems present in bits in Jaisalmer, Kachchh, Saurashtra and Kerala–Konkan. The Toarcian–Barremian first-order sequence interval includes a total of 12 stages, eight of the Jurassic and four of the Cretaceous, which are individually précised in Kachchh, and to a lesser degree in other Indian basins. The Aptian–intra-Paleogene sequence interval is again reasonably well differentiated in Spiti, Cauvery, Kachchh and numerous other Indian basins.

4.1.7 Relative Degree of Stratigraphic Refinement Among the Three Mesozoic Systems in India

The ~40 odd stages (5 to 7 in Permian, 30 Mesozoic and 2 early Paleogene) included in the mega-sequence have been differentiated in the principal Indian basins. The globally famed and splendidly developed ammonoid rich Triassic succession of High Himalaya along its vast stretch after the early studies of Diener (1908, 1912) in Spiti could not be followed-up for zonation for over a century until Bhargava et al. (2004), Krystyn et al. (2004). They developed ~35 ammonoid zones in the ~15 my long Early to Middle Triassic of Spiti with average resolution of ~plus 400 ky per ammonoid zone, however, further studies are in progress with every possibility of improvement in zonal/subzonal resolution. In the globally famous and ammonoid rich Kachchh Jurassic, the ammonoid stratigraphic studies go back to Waagen (1873–75) and Spath (1927–33). The thread was picked up by Krishna and coworkers in the 1980s. The subzonal resolution at present is of ~400 ky. The subzones in Kachchh have been further resolved into horizons, thereby increasing the stratigraphic resolution to ~200 ky in the Callovian–Tithonian interval (Krishna and Ojha 1996; Krishna et al. 1995a, b, 1996a, b). In specific expanded sections, packages of even 20 ky are recognized (Krishna et al. 1996a). The ammonoid zonal resolution, thus, in the Spiti Triassic is nearly half to that of the Kachchh Jurassic. The stratigraphic studies in the well-developed ammonoid rich Cauvery basin exposed Cretaceous were, similarly, begun in the nineteenth century (Stocliczka 1861–63 and Kossmat 1895–1897), however, could not be equally well-exploited later for multiple reasons, the foremost being the inavailability of good sections, supposedly, on account of the bad land topography. As such, there could be developed only a broad composite informal Cretaceous ammonoid succession in Kachchh, Spiti and Cauvery, which includes a succession of ~18 crude zonal intervals in the ~60 my long Aptian–Maestrichtian interval with average zonal resolution of ~3.5 my that too, in absence of any true zonation exercise. Thus as far as ammonoids are concerned, the Jurassic is best resolved, while the Triassic comes next, and the Cretaceous scores the last place in ratio of 6:3:1, i.e. the Kachchh Jurassic is about six times better resolved than the Cretaceous in Cauvery basin. Efforts are currently in progress to improve upon the lithostratigraphic differentiation for improvement in the ammonoid stratigraphic refinement in the exposed Cauvery Cretaceous.

4.1.8 On the Benthic/Planktic Foram-Based Basinal/Regional Indian Stages

The benthic and planktic forams have been used to erect new stages (Jurassic and basal Cretaceous stages) separately in Kachchh by Pandey and Dave (1993), Jaisalmer by Kalia and Chowdhry (1983) and Cretaceous stages in Cauvery by Narayanan and Raju (1996), and Prasad et al. (1996). In context of the history of regional stages, example on the GTM could be cited of the temporarily introduced regional stages in New Zealand in the 1960s in absence of the guide fossils for correlation to the international stages, not just in the Jurassic or Mesozoic but in the Cenozoic as well. Apparently influenced by the New Zealand example, the practice was in a big way adopted in India in the 1970s in the deep marine planktonic rich Neogene succession of Andaman (Srinivasan 1978 and after). Subsequently, the trait was followed in the Indian Paleogene and Mesozoic as well, mostly by a handful of ONGC scientists.

The creation of stages and zonal schemes in the Mesozoic, however, is largely not backed by the essentially pre-requisited lithostratigraphic differentiation, and regionally correlatable time planes. Thus, even good close-interval sample positions in the column, and the taxon range charts developed, in absence of the detailed lithostratigraphic differentiation, remained unexploited to their optimum potential for developing the succession of zones, and, in-turn, well-resolved integrated intrabasinal to intraregional geological comprehension.

It need be stressed that even the regional stages, if at all temporarily required, in context of their wide and broad usefulness, require the essential basis of guide fossils having the best resolution that, if so present, are ammonoids. On the other hand, in any case, the regional stages need not be erected on fossil groups having poor stratigraphic resolution, and if at all created, require the earliest possible integration with the available ammonoid ages, prior to using them for correlation. The regional stage frameworks lack correspondence of their inter-stage boundaries to the international inter-stage boundaries, otherwise, unfortunately, artificially forced by Pandey and Dave (1993) in the Kachchh Jurassic stages. Thus, if the assignment of altogether unsubstantiated gross wrong ages, was a fundamental mistake, the creation of the benthic foraminifer-based stages in the Kachchh Jurassic amounted to compounding of error of sorts. The used benthic forams lack any inherent strength of their own in context of the different and longer ranges of the foraminifer taxa in Kachchh and elsewhere in India and outside. The creation of the seven Jurassic stages in Kachchh on such unuseful benthic foraminifer data, so also local/regional stages in the Cretaceous and Neogene is also held violative of the fundamentals of chronostratigraphy. Fortunately, the basinal/regional Meso-Cenozoic stages failed to evoke acceptance in India, even in ONGC.

The Late Cretaceous regional stages, however, are, invariably, based on the regionally to internationally correlatable planktonic forams, subsequently, also integrated to the interregional nannoplankton zones (Raju 2011). Their time planes invariably do not correlate to the international stages, originally defined on the guide-ammonoid-taxa. Institutionally, better would have been to strive for ammonoid zonation in the ammonoid prolific Cauvery basin exposed Cretaceous, with follow-up integration of the ammonoid and planktic groups in India as done internationally (Graciansky et al. integrated charts 1998) for presenting the splendidly developed Indian Cretaceous on a much more reinforced pedestal that it deserved.

4.2 Microfossil-Based Zonations in the Indian Mesozoic

The Indian Mesozoic palynozones, in the absence of any indigenous scheme over the last few decades, unfortunately, continue to be based on the Australian palynozonal schemes (Helby et al. 1987 and others) which in turn are claimed as linked in some measure to the associated dinocyst record, however, well short of good chronologic refinement. The ranges of the marker Australian pollen/spore/dinocyst taxa need neither conceptually nor necessarily, nor otherwise be the same in Australia and India inspite of their relative proximity until the Barremian/Aptian boundary first-order SB. Prasad and Phor (2009) have mentioned the diachronic ranges even of critical species. Even after the assumed yet untested extension of ranges from Australia to India, the individual palynoassemblage zones could only achieve broad supra-stage grade duration/resolution (Prasad 2011) that averages to ~10 my per Mesozoic palynozone; ~7 my in Triassic, ~14 my in the Jurassic, ~10 my in the Cretaceous. Similarly, fallacious are the benthic foram-based zonal schemes in view of the inherent long spatially diachronous ranges of the recovered species. The palynozones and benthic foram-based zones are mostly single basin or sectoral frameworks, e.g. the palynozones are the most restricted to the Indian east coast basins, while the foram-based efforts exclusive to Kachchh. The basic objective of any zonation is its application for dating and correlating the included sediments at least across the basin, if not intraregionally, and the above microfaunal schemes (based particularly on benthic forams, and pollen/spores) do not meet the normal expectations of their application to recognize and extend the microfossil-based zones intra and interbasinally. Some times, different zones have been developed without correlation even from one well to another (Mehrotra and Aswal 2003; Mehrotra et al. 2002), so in outcrops only few tens of km apart (Pandey and Dave 1993 at Jumara, Gaur and Sisodia 2000 at Keera), which themselves speak decisively on the extremely limited applicability of such efforts.

The impractical situation may be exampled through the analogy of time, e.g. the smaller subdivision of the hour between 00.00 and 1.00 h is included in the successive bigger subdivisions of the days, e.g. partly in Tuesday and partly in Wednesday. The pre-requisite, however, in any hierarchical succession of units, is the start of the first smaller subdivision at the start of the bigger subdivision of which it is part of, obligatory inclusion of all the smaller subdivisions within a single bigger subdivision without gaps and overlaps. It, unfortunately, is not met in the Indian Mesozoic palyno and benthic foram zonal successions. There does not seem any possibility of dividing any Mesozoic stage into palynozonal subdivisions in such a dismal state of the art. However, the regionally recognizable sequence surfaces have supplemented/complimented such units, with resultant improved age constraints in the stratigraphic refinement in the non-marine successions.

4.2.1 Integration of the Diverse Indian Biostratigraphic Record Vis-à-Vis Sequence Surfaces

In the above not very comfortsome backdrop, the developed succession of the first- and second-order sequence surfaces, is used for age constraining of the rather poorly resolved palynozones and benthic foram zones vis-à-vis sequence surfaces. Turner et al. (2009 and references therein) admitted that major climatic and microfloral changes within the Jurassic in Australia occurred in basal Toarcian and Late Oxfordian. They admitted that lithostratigraphic units, which were palynologically dated, differ by vast margins of several million years from their recent highly refined zircon dates. Coincidentally, the times of major intra-Jurassic microfloral changes are interpreted at or close to the determined Indian first-order intra-Jurassic surfaces. It reinforces the present age constraining of the Jurassic palynozones by the first/second-order sequence surfaces. This in-turn has facilitated the initiation of the integration of the palynozones to ammonoid/conodont/foraminifer zones, of the marine units to non-marine units, marine units to palynozones, and non-marine units to zones developed in the marine succession in the Indian Mesozoic geological record, so as the correlation of the guide fossil devoid units to guide fossil bearing units. The premise used is that the diverse surface and subsurface geological expressions inclusive of the marine and non-marine paleontological records are consequences of the same regional tectonic events. It is conceptually held that the third- and finer-order high-frequency extensional tectonics is invariably unable to distort the mega to second-order low frequency high magnitude regionally to globally isochronous sequence frameworks.

4.2.2 Naming of Bio/Chronostratigraphic Zones Exclusively Through Their Basal Boundaries

Another significant amendment introduced here is about the nomenclature and definition of the palyno and other microfossil-based zones exclusively on their basal boundaries. Conceptually, the upper boundary on its own gets defined through the basal boundary of the next younger zone. The bio/chronostratigraphic units inclusive of all kinds of zones, thus, need be defined exclusively through their basal boundaries as conventionally practiced for well over a century. Conceptually, in any case, one of the two boundaries may be used. Constraining both the boundaries does not serve any additional purpose of either clarity or strength. Instead, it allows possibilities of gaps or overlaps between two successive zones. The concept of zones was introduced in the 1850s by Oppel, and the practice of defining the zones through the basal boundaries and naming them exclusively on the specific part of the species name has been followed ever since. Accordingly, the palynozonal names are here emended. The ammonoid zonation in the Kachchh Jurassic in Chap. 2 has elaborately exemplified the practice. The amendment realized here is exampled through the Muderongea tetracantha–Cribroperidinium edwardsii Zone which is emended here as Tetracantha Zone as defined through the LAD of M. tetracantha at the terminal boundary of the zone. However, it would have been a better practice to define and name the palynozones through the FADs at the basal boundary.

4.3 On the Earlier Assigned Guide Fossil-Based Ages and Correlations in the Indian Mesozoic Successions

4.3.1 Bivalves

Apart from the fast evolving ammonoids, the only other mega-marine invertebrates with ubiquitous presence in the Mesozoic shallow marine successions have been the rather long ranging environmentally and facies controlled benthic bivalves. As exception, the pelagic bivalve records have been used for precise correlations, particularly in the absence of ammonoids, e.g. in the Triassic of High Himalaya. The notable presence of benthic bivalves in the extreme shallow marine sediments, that too in the absence of ammonoids, tempted workers to use them for tentative bivalve-based correlations. The bivalve-based correlations, particularly, in the Kachchh Mesozoic have been commented upon in the earlier chapters. Such correlations invariably later turned out incorrect except in the neighbouring closely located sections involving only a distance of a few km. The longer the distance between the correlated sections more diachronous are found the benthic bivalve-based correlations. It has been rather easy to note the frequently recurring bivalve assemblages in time and space as indicators of recurring specific plaeoenvironment/facies. Biswas (1977, 1993) often used the benthic bivalve assemblages for correlation as exampled here.

***Corbula–Gervilla* Assemblage-Based Correlations** Biswas used the *Corbula–Gervilla* assemblage for the correlation between the sediments of the Dingy Member at Dingy and neighbouring Kuar bet localities, both of probable Early Toarcian age, and between Bela and Chorar in the Late Bathonian–Early Callovian age range. In Chorar, the assemblage is of Late Bathonian age while in Bela of mid-Early Callovian Chrysoolithicus Zone, which only implies somewhat less shallow depth and associated environmental conditions at Bela in mid-Early Callovian compared to Chorar, where similar depths and environments prevailed only up to Late Bathonian. In the Island belt, there are present at least four levels of *Corbula–Gervilla* biofacies within the respective TSTs of the Early Toarcian, Early Bajocian, Late Bathonian and Early Callovian of similar bathymetric/environmental framework. The first three developments are within the Patcham Formation, while the youngest one is in the early part of the Chari Formation (2.25–26, intrabasinal correlation of columns).

Astartid–Trigonid Assemblage-Based Correlations The assemblage is found recurrent at numerous levels in different sections of different ages in the Kachchh Mesozoic during the Callovian–Middle Albian interval. Gross faulty correlation of the Early Kimmeridgian Bharodia astartid–trigonid assemblage of Wagad and the Early Cretaceous sediments of Umia unit in West Mainland has been discussed earlier at some length (Krishna et al. 2009a, b—Fig. 4.1).

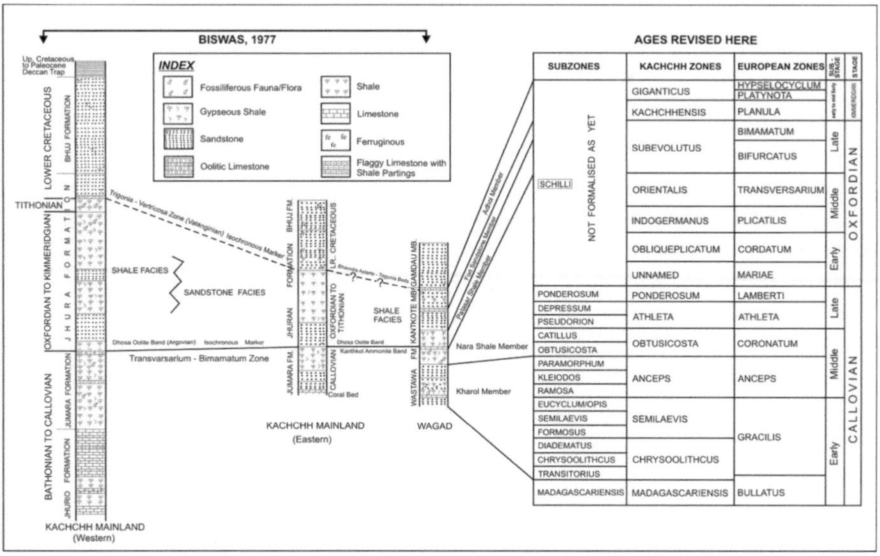

Fig. 4.1 Astartid/trigonid bivalve-based correlations of Biswas (1977) revised with ammonoid and sequence surface integration between mainland and island belt (modified after Krishna et al. 1998)

Revision of the Bivalve-Based Correlations Between the Mainland and Wagad Units The application of sequence surfaces permits revision of the broad time correlation of the Mainland succession of the abandoned 'Jumara', 'Jhuran' and 'Bhuj' units of Biswas to units in Wagad (in Biswas 1993, p. 165). The top of the Dhosa Oolite in the Mainland Kachchh is early Middle Oxfordian and not any younger while top of 'Khadir formation' and of 'Washtawa formation' is late Middle Oxfordian, i.e. younger by several subzones. Age equivalence of 'Jhuran formation' of Mainland to Kantkote Member of Wagad emphasized upon by Biswas is grossly incorrect. Kantkote Member ranges from Latest Middle Oxfordian to Early Kimmeridgian where as 'Jhuran formation' only begins in Late Kimmeridgian and goes up to Late Valanginian. Nothing to speak of their being time correlatives, there is absence of even any small overlap between the two units of Biswas. 'Gamadu Sandstone member' of Wagad similarly has been gross incorrectly suggested as time correlative of 'Lower Bhuj member'. The upper age limit of 'Gamadu member' should not be younger than Late Kimmeridgian. On the contrary 'Lower Bhuj member' in its stratotype in the neighbourhood of Bhuj town ship may not even start earlier than Late Ttithonian, and ranging farther up to Barremian. There is undisputed gross facies similarity between 'Washtawa formation' and Chari Formation, that is, why 'Washtawa formation' is abandoned and succession included as Chari Formation in Wagad. Kankote Member is included in Katrol Formation and 'Gamadu formation' included in Umia Formation in the integrated unified lithostratigraphic framework.

Once the ages of the formations and members turned out wrong, the intrabasinal correlations based on such ages obviously involved compounding of errors, as only gilempsed above with only a few examples.

4.3.2 Conodonts

Triassic The Early and Middle marine Triassic is organized into 15 conodont zones which are well integrated to the ammonoid zones (Figs. 2.6 and 3.13). The top of the Cornuta Zone is correlated to the top of the Kellnerites (ammonoid) Zone, and marks the first-order intra-Triassic MFS. Similarly, two intra-Triassic second-order MFSs are determined, respectively at the top of the intra-Induan Krystyni Zone coeval to the Discophiceras ammonoid Zone, and at the top of the intra-Ladinian Mongoensis Zone coeval to the Regoled ammonoid Zone. Bhatt et al. (1999), Chhabra and Sahni (1981), Sahni and Chhabra (1974) recorded late Early to early Middle Triassic conodonts from Kashmir and Kumaon Himalaya. Rare records also exist of the late Early Carnian Metapolygnathus carnicus from the terminal part of Kalapani Limestone in Kumaon (Chhabra and Kumar 1984), and early Middle Norian conodonts at the base of a second-/third-order TST in the terminal part of the Upper Member of Rangkrik Formation in Spiti (Bhargava et al. 2004). There is an old record of the Rhaetian *Neospathodus hernsteini* from the early Kioto Formation in NW Kumaon (coeval of Para Formation of Spiti) (Sahni and Prakash 1973) in the basal grey limstones of Early Rhaetian second-order TST. The rather rare Carnian, Norian, and Rhaetian conodont records suggest reduced bathymetry in the Carnian–Rhaetian interval of the first-order Ladinian–Pliensbachian RST.

4.3.3 Foraminifers

Triassic Singh (1979) recorded Late Triassic forams from the Kuti Shales of Kumaon High Himalaya, which are here corresponded to the first-order RST Ladinian–Pliensbachian age bracket.

Jurassic There are numerous benthic foram-based studies, invariably, without focus on zonation initiated in ONGC (Subbotina 1960), and later followed by others mostly in the West Indian basins except in the Banni well of Kachchh in the Middle Jurassic to Late Cretaceous span. In High Himalaya, the Spiti Shale unit in Nepal has been investigated for forams (Gradstein et al. 1991), and Middle Oxfordian to Tithonian age is assigned in conjunction with the included ammonoid fauna. Forams are also recorded in the Spiti Shales of Kumaon High Himalaya. Late Jurassic forams have also been recorded in Saurashtra wells. Late Jurassic forams, mostly long ranging, are also recovered from surface and subsurface sections in Jaisalmer.
The above, invariably, half baked single section or single basin studies lacked serious zonation effort. Even parallel zonations in the Kachchh exposed marine Jurassic–Cretaceous succession on any group of microfaunal guide fossils hardly have ever been endeavoured, and rarely developed, yet never integrated to the primary and distinctly better resolved ammonoid-based zonation.

The relatively detailed, yet chronostratigraphically not very useful, study is of Pandey and Dave (1993). They worked out a succession of ten foraminifer zones and 2 inter-biohorizons in the Late Bathonian–Late Tithonian interval. In view of the poor inherent resolution of the benthic forams, all their zones and inter-biohorizons, so zones of others enumerated above, are incorrectly/imprecisely dated. Their ages were subsequently revised on the basis of the known ammonoid record in the concerned sections (Krishna et al. 2000).

Regional even intrabasinal validation of the benthic foram-based zones could not be realized as exampled here. Gaur and Sisodia (2000) proposed four foraminifer zones along with a barren zone in the mid-Early Callovian–early Middle Oxfordian at Keera in Kachchh. The basal $\sim$60 m thick sediments are included in a barren zone. There are hardly any common elements between Pandey and Dave (1993), and Gaur and Sisodia (2000) even at genus level in the same interval in two different sections of Kachchh Mainland. Gaur and Sisodia did not even give ranges to their zones.

Heavy Bias of the Environmental Control on Benthic Taxa The benthic microfossil-based biostratigraphic differentiation in these works, is more in context of bathymetry and environment rather than any true biostratigraphic differentiation. For example, the change from dominance of calcareous to arenaceous foraminfera has taken place at the second-order MFS of the Bathonia–Callovian second-order sequence between mid- to late-Early Callovian from fine textured calcareous/argillaceous sediments (of biozone 2) of terminal second-order TST to arenaceous sediments (of biozone 3) of the following second-order RST. Again there is a change of bathymetry and environment from biozone 3 to biozone 4, i.e. from terminal RST to again TST. From biozone 4 to biozone 5 again, there is a drastic bathymetric change of sudden rapid increase in bathymetry. Here, it is interpreted that the successive assemblages thought to be biostratigraphic by their respective authors and are instead facies assemblages as consequences of changes in bathymetry and environment. These interpretations find further support from analysis of the range chart. Arenaceous foraminifera are exclusive to biozone 3 which otherwise comprises dominantly shallower water coarser textured sands. Similarly, maximum of 15 species are recorded from the ammonoid rich conglomeratic oolitic limestone as the youngest bed of Dhosa Oolite Member at the top of Chari Formation which otherwise has been interpreted having maximum bathymetry in the Jurassic of Kachchh Mainland. 12 species are recorded which otherwise represent the terminal part of the second-order TST.

However, in context of the succession of sequence surfaces, there is realized further refinement and credence to the foraminifer-based zones in the Kachchh Jurassic as under (Fig. 4.2).

The Lenticulina–Epistomina Zone corresponds to the RLm at Patcham with termination at the Patcham/Chari lithostratigraphic boundary, and is assigned to the terminal Late Bathonian Triangularis Zone.

Regularis–Ghoshi Zone was based by the authors compositely in two different sections. The fallacy found is that the lower part of the composite zone is found much younger than the upper part in two different far apart sections. The sediments representing the base and early part in the Patcham section begins in the early to mid-Early Callovian Chrysoolithicus Zone. The later/upper golden oolitic part of the composite stratotype at Jhura hill is included in the lower part of the ULM of the Bajocian age. The authors presumably misunderstood the FAD and LAD of the nominal species. The FAD of the nominal species is presumably in the golden oolitic limestone at Jhura in Early Bajocian, while the LAD is in the Patcham section in the early part of Chari Formation, just opposite to the opinion of Pandey and Dave. Moreover, both the FAD and LAD are lithology/facies controlled. The only possible reorganization in such unique and strange circumstances may be to restrict the early Bajocian golden oolitic limestone as a separate unnamed zone, with the status of a separate zone for the early Callovian interval at Patcham, since the separated zones include a large gap of about two stages of more than five million years.

Dilectiformis Zone is ranged through the second-order sequence from terminal Bajocian to early Middle Bathonian at Jhura in place of Middle Bathonian age by Pandey and Dave.

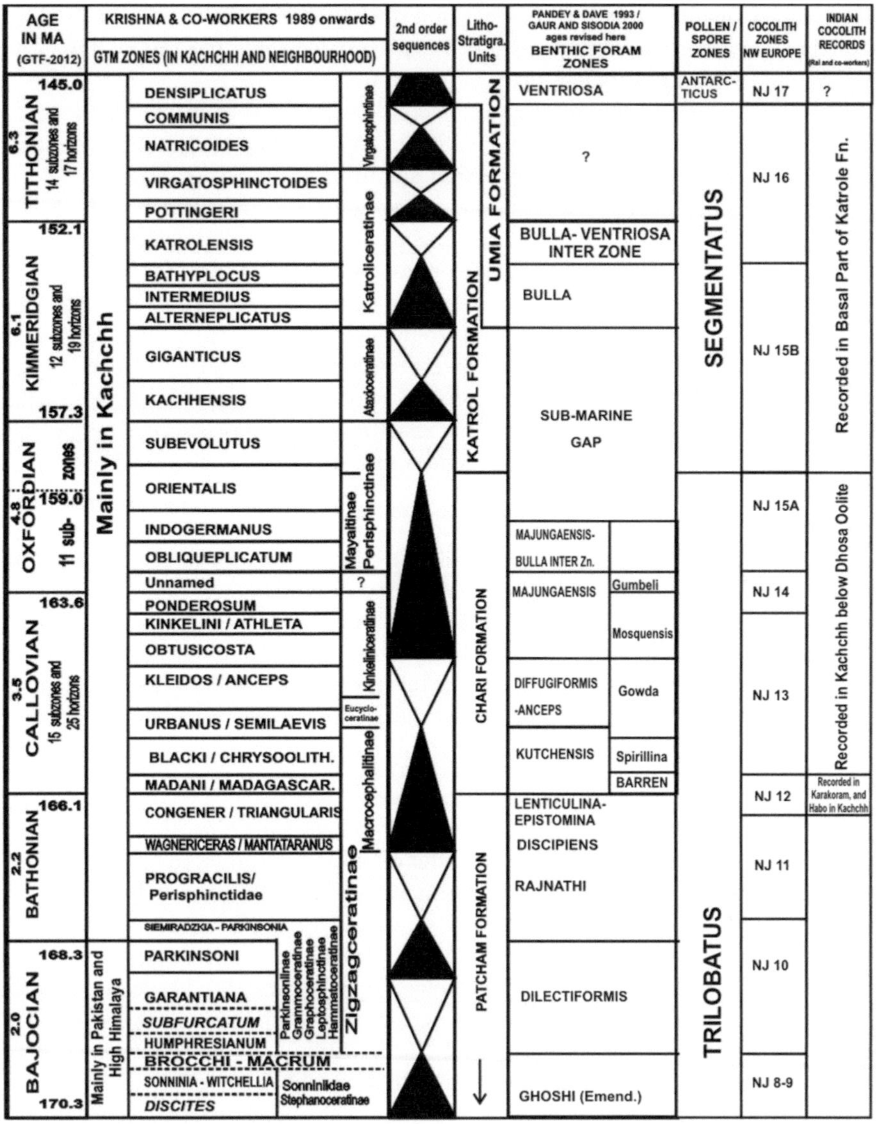

Fig. 4.2 Integration of ammonoid, benthic foram, pollen/spore, coccolith zones and second-order sequence surfaces in the Kachchh Jurassic

Rajnathi Zone begins at second-order SB, and is assigned to the later part of Middle and early part of Late Bathonian in place of early part of Late Bathonian by Pandey and Dave.

Discipiens Zone is ranged from the Late Bathonian to Early Callovian Madagascariensis Zone in the Jumara Dome in place of Early Callovian by Pandey and Dave. Thus, this zone, in its early limestone part at Jumara either includes or overlaps the earlier incorrectly dated E. regularis Zone (only the part at Jumara) of Pandey and Dave in Bajocian.

Kutchensis Zone is ranged up to a second-order MFS from the mid-Early to early Late Callovian up to the top of GSM in Keera, Jumara, and other Mainland sections. Pandey and Dave did not assign age to this zone.

Diffugiformis–Anceps Zone is ranged through a second-order TST, and is assigned to the early Late Callovian age in place of Late Callovian by Pandey.

Majungaenis Zone, ranging through a second-order RST, is assigned to the Late Callovian–Early Oxfordian in place of Middle Oxfordian age of Pandey and Dave.

Majungaensis–Bulla Zone is assigned to the late Early to early Middle Oxfordian interval, in place of imprecise reference to Oxfordian in Pandey and Dave.

Bulla Zone is ranged from late Early to mid-Late Kimmeridgian in place of Kimmeridgian of Pandey and Dave. This zone ranges through a second-order TST.

Bulla–Ventriosa I B H Zone is assigned terminal/latest Late Kimmeridgian in place of Late Kimmeridgian by Pandey and Dave. The zone terminates at a second-order SB.

Ventriosa Zone is assigned Late Tithonian age in place of Early Tithonian by Pandey and Dave. This zone begins at a second-order SB.

As per revised ages, there are five foraminifer zones in Bathonian, three in Callovian, one interbiohorizon in the Oxfordian, one zone, and one interbiohorizon in Kimmeridgian, and one in Tithonian of relevance that too exclusive to Kachchh. Surface and subsurface correlation charts included by Pandey and Dave are similarly full of incorrect ages, and stand revised here.

Cretaceous

The foraminfer studies worth stratigraphic significance have been realized principally in the Cauvery, Spiti, and Jaisalmer basins, unfortunately none of any stratigraphic relevance in the exposed Cretaceous of Kachchh, possibly, due to lack of favourable facies. Forams have also been studied in the Cretaceous of Kachchh, Narmada, Saurashtra, Mahanadi, Bengal and Assam.

Raju (2011) in a review included 16 planktonic foraminiferal bio/chrono zones in the ~36 my long mid-Late Albian to close of Maestrichtian with average resolution of ~2.25 my per zone. Chidambaram (2000), Govindan et al. (1996), Prasad et al. (1995), Narayanan and Raju (1996) earlier provided a succession of smaller foraminfera zones in the early part of Late Cretaceous. The FADs and LADs of the planktonic guide-microfossils, invariably, carry interregional chronostratigraphic significance and are well stabilized (Fig. 4.3). Bertle and Suttner (2005), Singh (2006) also have recorded similar assemblages of planktonic forams respectively in Spiti and Jaisalmer.

Foraminifer Zones

The Early Aptian–mid-Middle Albian is featured with the succession of Charlottae–Colomi–Roberti Zone in Jaisalmer wells. The top of the Appeninica Zone makes the top of Albian. The late Middle to Late Albian includes *Rotalipora ticinensis, Planomalina buxtorfi, and R. appeninica* in Jaisalmer, Spiti, and Cauvery (Fig. 4.3).

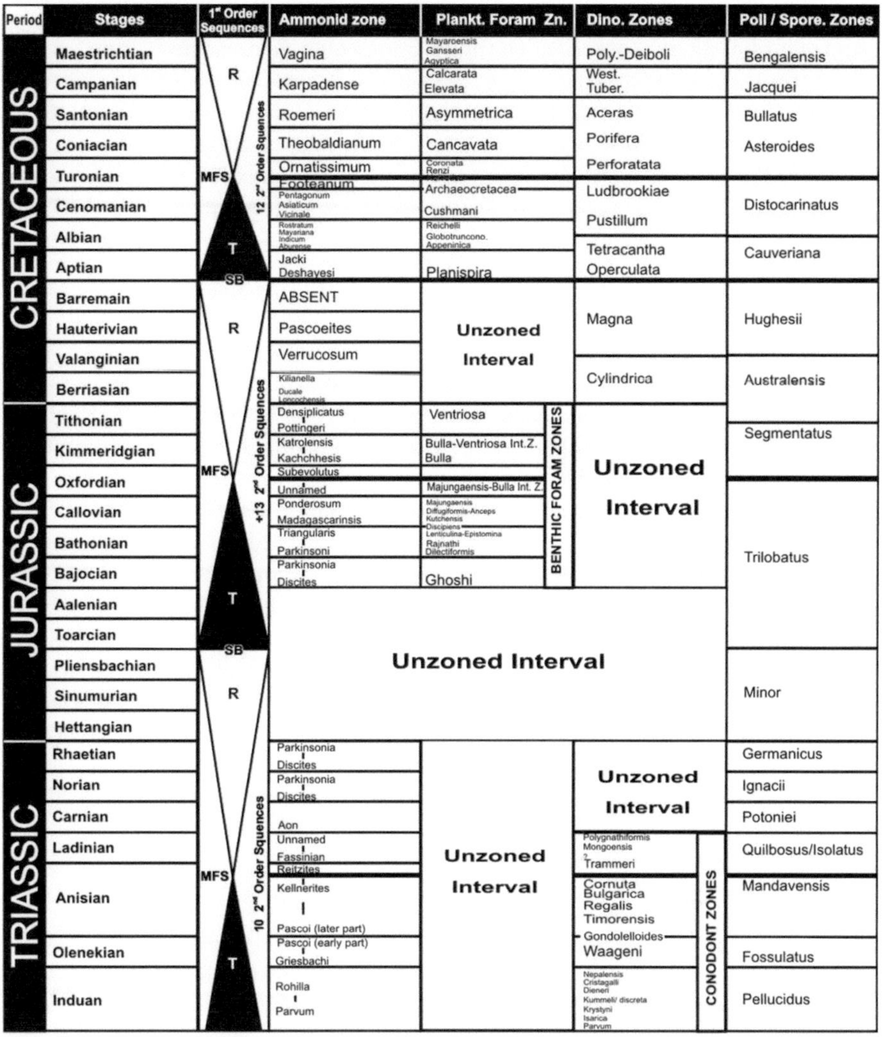

Fig. 4.3 Indian Mesozoic microfossil zones age constrained to ammonoid-based sequence surfaces

The combined interval of the Globotrunconoides Zone–Reichelli Zone–Cushmani Zone–Archaeocretacea Zone–Helvetica Zone (only early part) is constrained in the Cenomanian–early Middle Turonian interval, and its upper boundary placed at the major intra-Turonian Turoniense Zone MFS as determined in Cauvery. The succession of the Globotrunconoides Zone, Reichelli Zone, and Cusmani Zone is included within the Cenomanian stage. The Archaeocretacea Zone straddles the Cenomanian/Turonian boundary. The top of the Coronata Zone–Cancavata Zone interval is suggested at the terminal Late Coniacian second-order Theobaldinum

Zone MFS. The top of the Asymmetrica Zone makes the second-order Santonian/Campanian boundary Zone SB. The intra-Campanian Karapadense Zone second-order MFS is broadly corresponded to the Elevata Zone–Calcarata Zone boundary. The terminal Cretaceous second-order intra-Maestrichtian Vagina Zone MFS is broadly correlated to the Tricarinata Zone/Gansseri Zone Early/Late Maestrichtian boundary. The Cenomanian–Santonian succession of planktonic forams is present in Cauvery, and Jaisamer wells, and in the Spiti exposed columns. The integrated synthesis is included in Fig. 4.3.

Foraminfer-Based Stages Age Constrained in Sequence Stratigraphic Context

The 'Mannerian stage' is ranged up to the intra-Valanginan Verrucosum Zone second-order MFS as determined in Spiti (Pathak et al. 2011). The 'Musian' is ranged up to the first-order Barremian/Aptian boundary SB as précised in Kachchh, Jaisalmer and Spiti. The 'Kalakuddian stage' is corresponded to the second-order TST up to the mid-Middle Albian Dentatus Zone MFS as determined in Jaisalmer (Krishna 1980, 1983a, b). The 'Nallurian'–'Anaipadian' interval is broadly corresponded to the interval between the mid-Middle Albian second-order MFS and the early Middle Turonian Turoniense Zone first-order MFS (Krishna 2012) as determined in Cauvery basin. In fact the close of 'Anaipadian' is distinctly above the first-order MFS. The maximum bathymetry in Spiti is suggested in the Archaeocretacea Zone in the basalmost Turonian in Bertle and Suttner (2005). The 'Trichonopolian' stage is ranged from the first-order MFS to the Santonian/Campanian boundary second-order SB. The 'Melmatturian' stage is corresponded to the interval from the second-order Santonian/Campanian boundary SB to the Campanian/Maestrichtian boundary second-order SB. The 'Nagalurian stage' is ranged from the Campanian/Maestrichtian boundary second-order SB to the end Maestrichtian second-order SB.

Microflora

Among the significant works are Garg et al. (2003), Mehrotra et al. (2002), Mehrotra and Aswal (2003), Jain et al. (1980), Krishna et al. (1983), Jafar and Rai (1989), Upadhyaya et al. (2005), Rai (2006), Rai and Jain (2012, 2013), Rai et al. (2005, 2013a, b, 2015a, b), Kale et al. (2000), Kale and Phansalkar (1992), Prasad and Phor (2009), Prasad 2011, Vijaya (2009), references in the cited works, and others. The significant microfloral records are summarized separately as under.

4.3.4 Pollen–Spores and Dinocysts

The non-marine Mesozoic in the Gondwana basins and elsewhere are currently organized into 19 pollen-/spore-based palyno-assemblage zones (Prasad et al. 1995; Prasad 2011). These correspond broadly to the succession of 30 stages.

The pollen/spore and Dinocyst zones are here integrated to the succession of sequence surfaces.

4.3.4.1 Triassic

Correspondence of the Palynozones to the Triassic Ammonoid Zones and Lithostratigraphic Units of Spiti Himalaya

The sequence stratigraphic surfaces determined are exclusively in the marine Triassic of Spiti basin. Thus, the Tatarian–Pliensbachian non-marine succession in the K-G wells bounded by unconformities on both the sides is corresponded to the succession between the intra-Permian SB and intra-Jurassic SB in the exclusively marine Spiti succession. Unconformities involving SBs represent the subaerial gaps. The SBs and MFSs indicate time lines at which the regional tectonic events, invariably, take place. The Mandavensis Palynozone is carved out of the earlier Fossulatus Zone. Thus, the change above the Fossulatus Zone in fact implies the change above the Mandavensis Zone. The palynological change above the Mandavensis Zone, which otherwise corresponds to the lithological change across the Panchet/Mahadeva units, is found to occur at least above the Early Anisian. There is no more precision or constraint on the palynochange. The change is here interpreted at the first-order intra-Triassic MFS near the terminal part of the Muschelkalk member between beds 34 and 40 of the Gungri section in the terminal Anisian Hollandites Zone–Kellnerites Zone interval (Fig. 3.13). Thus the older three palynozones can be broadly corresponded to the Early and early Middle Triassic up to the Kellnerites Zone MFS, i.e. to most of the Mikin Formation, while the interval of the younger four zones is broadly equated to the collective sedimentary succession of nine formations from the Kaga to Kioto units (Fig. 2). It is altogether an innovative approach and requires further investigation and data in support or otherwise. The palynozonal changes were marked, however, ages of the palynozones remained vague and imprecise. The palynochanges are here linked to the diverse regional geological changes at the second-order sequence surfaces under the assumption of their being consequences of the same regional geological event. Another equally significant intra-Triassic palynochange is known above the Isolatus Zone in Ladinian in the Chomule Formation, which is here interpreted to occur at the second-order MFS in the Chomule Formation in the bed BL 12 assigned to the top of the mid- to late Ladinian Regoled Zone. Thus for the first time, the top of the palynozone Isolatus Zone is tied to the nearly 10 fold more precisely resolved and dated top of the Regoled ammonoid Zone of the mid- to late Ladinian age, while the palynozone itself is precisely ranged from the terminal Anisian Halilucites–Reitziites Zone to the mid/late Ladinian Regoled Zone interval. Alternatively, it could be placed at the Kaga/Chomule unit boundary at the base of the Regoled Zone. This in turn allows precision dating as well of most of the palynozones. The basal Pellucidus Zone is dated from the base of the Triassic to early Olenekian Flemingianus Zone. The Fossulatus Zone is précised from the Olenekian Gracilitatus Zone to the Late Olenekian Kashmiricus Zone. The Mandavensis Zone is ranged from the terminal Olenekian Pascoi Zone to the Late Anisian Kellnerites Zone interval. The Potoniei Zone is ranged from the Canadensis Zone to the uppermost part of Rongtong Formation of the terminal Carnian age. The Ignacii

Zone is ranged from the terminal Carnian/basal Norian to the mid/ Late Norian Hogarti/Macer zonal boundary, i.e. the Alaror/Nunuloka lithostratigraphic boundary, and the terminal Triassic Germanicus Zone from the terminal Norian to Rhaetian.

East Coast Subsurface Non-marine Triassic Vis-a-Vis Spiti Units
It may be noted that the top ~60 m of the Mandapeta Formation of K-G basin is devoid of palynofossils. This may range up to the first-order intra-Jurassic SB near the close of Pliensbachian. The rather thin red clay interval in the middle of the Mandapeta Formation is broadly correlated to the Muschelkalk Member of the Mikin Formation. The underlying Lower Sandstone Member to the pre-Muschelkalk part of the Mikin Formation and the overlying Upper Sandstone Member to the collective sedimentary succession of the Kaga to Kioto units.

Gondwana Palynoassemblages and Spiti Units
The lithostratigraphic boundary between the Panchet unit and the overlying Mahadeva/Pachmarhi/Durgpur units in different Gondwana basins is here suggestively corresponded to the intra-Anisian MFS. Vijaya (1997) has correlated her palynozones I and II (palynoevents 4 and 5) with abundance respectively of *Straitopodocarpites* and *Krempipollenites* to the basal most Panchet Formation and to the basal part of the Mikin Formation in Spiti, and assigned the same to the Induan stage of the Early Triassic (Figs. 3.13 and 4.3). The base of the Niti Limestone broadly corresponds to the base of the third Triassic palynozone, while the oldest two members to the first two Triassic palynozones. The entire Late Carnian shaly interval of the Maleri/Mandapeta Formation is broadly corresponded to the middle part of the Middle Member of Rongtong Formation, and farther up, the correlation of the Early Norian shaly interval of the same units to the Rangkrik Formation of the Spiti Triassic. The Parsora Formation is here in the sequence stratigraphic context assigned within the first-order RST Ladinian–Pliensbachian bracket.

Ages of the Gondwana Non-marine Units Against the Ammonoid Zones
Perhaps, for the first time, the Indian non-marine Gondwana units are being aged in terms of the ammonoid zones. The Panchet Group is ranged from the basal Triassic to the Anisian Kellnerites Zone, while the Mahadeva Group from the terminal Anisian Halilucites–Reitziites Zone to the end of Triassic. The Pachmarhi/Bheemaram/Tiki unis are precisely ranged from the terminal Anisian to mid/late Middle Norian. The Dharmaram/Bandhavgarh/Hartala units are suggestively ranged from the Late Norian to Rhaetian. The Durgapur unit ranges from the terminal Anisian Reitziites Zone to the close of Norian.

Jurassic
Prasad (2011) organized the ~56 my long Jurassic exclusively in Kachchh in a succession of four palynozones (Figs. 4.2 and 4.3), having a zonal resolution of ~14 my per zone. The alternate sequence surface approach allows the oldest Minor Zone to range up to the intra-Jurassic first-order SB at the Pliensbachian/Toarcian boundary, and corresponded to the Nirona Formation of the Kachchh wells, and the Lathi Formation (inclusive of the ‘Shumeriwali formation’). The second Trilobatus

Zone is here corresponded to the first-order Toarcian–late Middle Oxfordian first-order TST interval, as to the combined interval of the Patcham and Chari units. The third Segmentatus Zone is suggestively extended to the close of the second-order SB at the close of Kimmeridgian, and ranged up the close of Lower Member of Katrol Formation. The last Antarcticus Zone is corresponded to the entire Early Tithonian interval, as equated to the Middle and Upper units of the Katrol Formation. Based on palynoassemblages, palynostratigraphers (Prasad 2011 and others) invariably have interpreted most of the Jurassic as a long stratigraphic gap in the Indian east sector Gondwana successions (Fig. 4.4).

4.3.5 Nannoplanktons

The oldest Indian nannoplankton record is of rare Late Triassic nannoplanktons from the Ladakh region of High Himalaya. The origin of the nannoplanktons is here speculatively linked to the terminal Anisian first-order MFS extensional tectonic event.

The first Jurassic nannoplanktons from anywhere in India were recorded in the Banni well of Kachchh of Middle and Late Jurassic age in the 1970s. Krishna et al. (1983) recorded nannoplankton presence for the first time in several sections of the exposed Kachchh Jurassic, and listed the Middle Jurassic nannofossils recovered from the Chari Formation of Kachchh. They noted nannoplankton abundance in the Patcham Formation whose taxonomic determinations were prevented because of the recrystallized nature of the samples. Jafar (1996 and references there-in) in a review indicated the widespread presence of nannoplanktons of CC18 to CC26 in the Andaman flysch turbidites, Assam–Arakan, CC26 from South Shillong, CC26 from Lesser Himalaya, CC23 from Malla Johar High Himalaya, CC7 to CC26 in K-G wells, CC12 from Narmada, and CC7 from Jaisalmer ground water well. Rai and coworkers in recent years have described nannoplanktons in the Indian Mesozoic from many Jurassic and Cretaceous stages, mostly from the well-resolved sections, thereby providing further support to ammonoid and other guide fossil-based ages. The nannoplankton ages accord well with the earlier assigned ammonoid ages. Rai and coworkers in their studies spread through High Himalaya, Kachchh, Jaisalmer, Narmada, Cauvery from surface and well samples, assigned the studied samples to different Tethyan/international zones, which invariably matched well with the known ages and stratigraphic frameworks. Kale et al. (2000) identified the succession of international nanno-planakton zones (CC7 to CC12) in the Uttatur sediments of the Cauvery in the Aptian–Middle Turonian interval. As a welcome signal, the coccolith studies have grown in recent decades in the Indian Jurassic–Cretaceous successions in both east and west sectors as a powerful biostratigraphic tool, particularly, in the subsurface, and ocean drilling programmes. Zonation schemes have been developed in the Boreal and ETM. However, but for recognition of the international zones in bits of the Indian stratigraphic intervals, no concerted

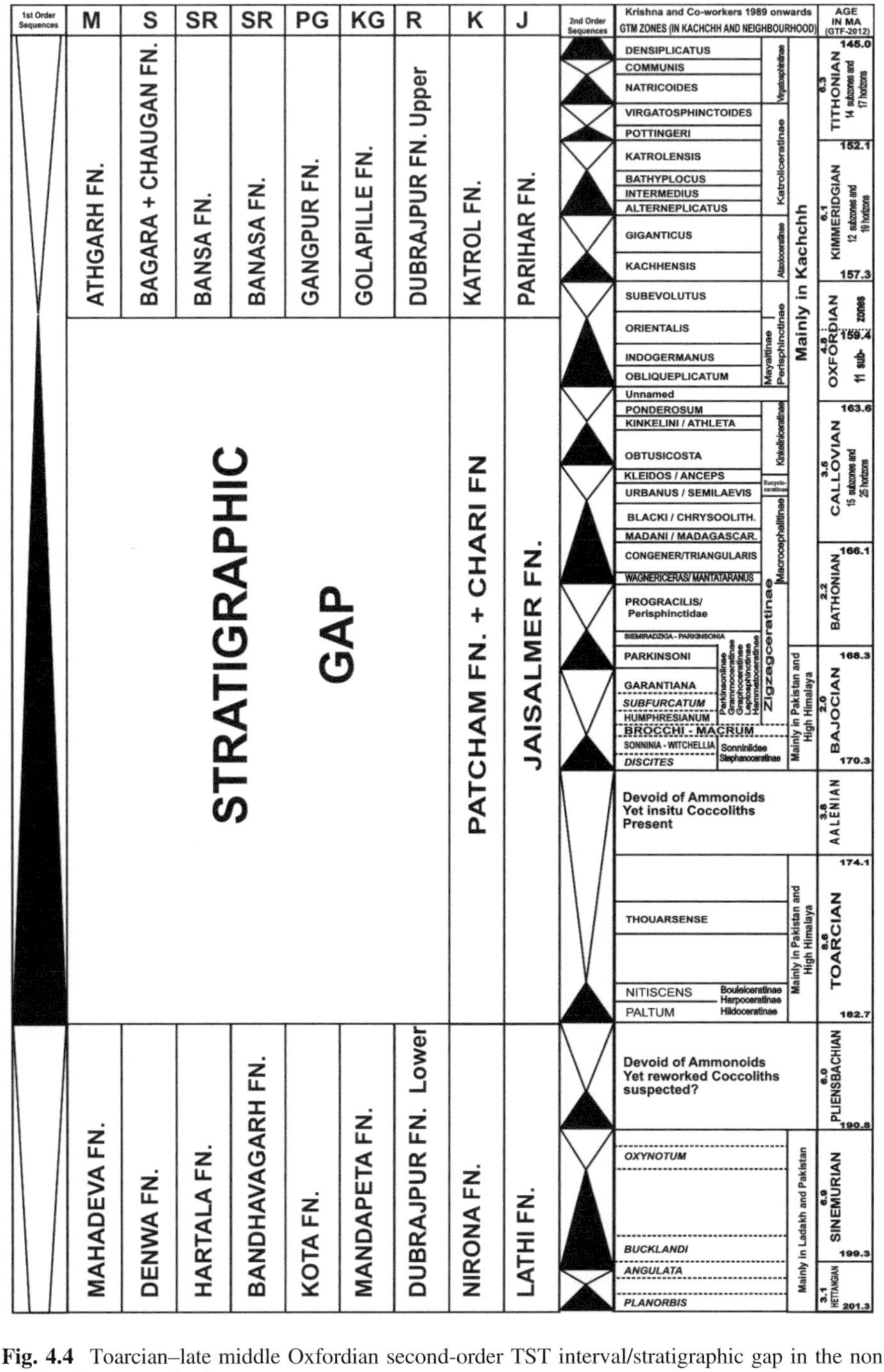

Fig. 4.4 Toarcian–late middle Oxfordian second-order TST interval/stratigraphic gap in the non marine Gondwana basins

efforts so far could be made for direly required comprehensive zonation in any of the three Indian Mesozoic systems. The nannofossil studies are considered significant, and require development of indigenous schemes along with integration to ammonoid based zones for their better utilization in the Indian Mesozoic geology. The limited studies are nevertheless interpreted along with other mega/microfossil groups in sequence stratigraphic context (Figs. 4.3, 4.4 and 4.5).

Triassic

The Late Triassic record is linked to the intra-Triassic terminal Anisian first-order MFS. The origin of the nannplanktons as such may be linked to the widespread Mid-Triassic extensional tectonic event which left diverse geological expressions across the globe, already discussed in earlier chapters, and again covered in Chap. 7.

Jurassic

In the NW European Boreal realm is developed a succession of 17 coccolith zones from NJ1 to NJ17. Many of these zones are extended to the Neotethyan margins ETM and GTM. The record from the Toarcian of Kachchh is good reinforcement of the widespread marine transgression on the GTM inclusive of Kachchh at the basal SB of the first-order Early Toarcian–Late Barremian sequence. It was so anticipated in a regional context in Krishna (1994), Krishna and Pathak (1994). The Late Pliensbachian record (Rai 2014 and earlier) as discussed earlier is reinterpreted as Toarcian.

Broad Imprecise Coccolith-Based Ages of the Kachchh Jurassic Sediments/ Units

In general, the coccolith tool is found quite promiseful in the Indian Mesozoics. However, the coccolith-based age assignments in the Indian ammonoid bearing sediments are considered broad and long ranging (Fig. 4.3) vis-à-vis firm precise ammonoid ages as exampled below.

Age of the Basal Beds in the Habo Section by Rai et al. (2014)

Rai et al. (2014) based on coccoliths assigned the Late Bathonian–Early Callovian age to the youngest part of the ULM of Patcham Formation in the Habo section, while the ammonoid-based age of the sediments is later part of the Triangularis Zone of Late Bathonian. The question of the Early Callovian age to the oldest sediments otherwise included in the terminal part of the Patcham Formation at Habo simply does not arise, however, integration with ammonoids would have ruled out Early Callovian.

Age of the Sediments Immediately Underlying the Dhosa Oolite

Similarly, the part of the Chari Formation underlying the Dhosa Oolite has been somewhat broadly assigned by Rai et al. (2014) to Late Callovian–Oxfordian. There is a complete absence of at least late Middle Oxfordian to Transversarium Zone to mid-Early Kimmeridgian Hypselocyclum Zone duration in Habo. The precise age of the sediments is from the Late Callovian Ponderosum Zone to the Early Oxfordian unnamed zone.

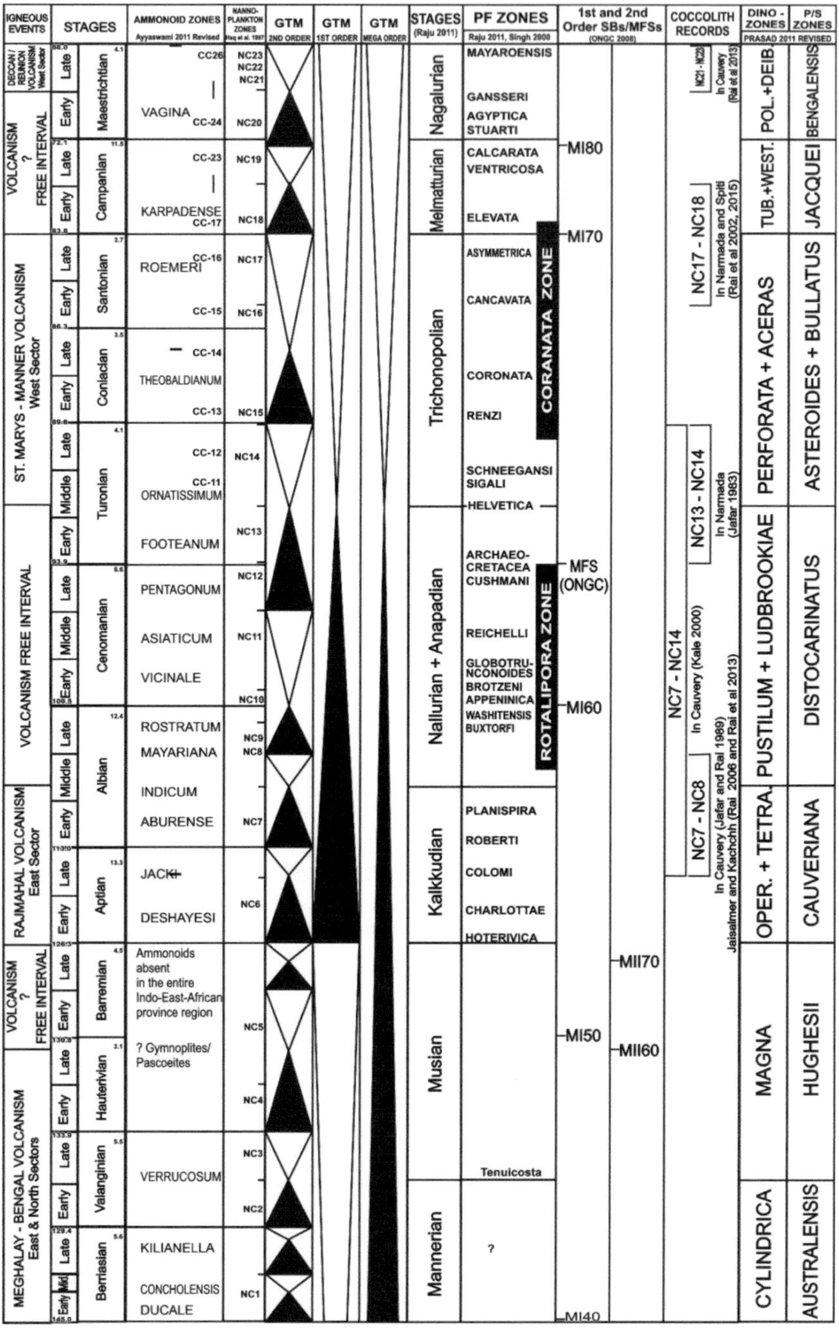

Fig. 4.5 Indian first- and second-order Cretaceous sequences vis-a-vis ammonoid, nanoplanktons, planktonic foram, dinocyst and pollen/spore zones

4.3.5.1 On Ages of Wagad Units

Age of Nara Shale Member
The shales in their lower part based on the presence of the Kinkeliniceratinae ammonoids range down to include at least Late Callovian, yet may extend farther down to the late Middle Callovian Obtusicosta Zone Obtusicosta Subzone Horizon O-II. On the upper side, the unit is maintained to extend up to the late Middle Oxfordian Schilli Subzone (Krishna et al. 2009a, b, and unpublished data). The dinocyst-/coccolith-based broad Early to Middle Oxfordian assignment (Rai et al. 2015a) is again several fold broad, and requires further convergence and improvement in context of known ammonoid ages. Moreover, in their Fig. 2, the unit is ranged from the base of the Oxfordian to the middle of Bifurcatus Zone, in Fig. 4, from mid-Cordatum Zone to Schilli Subzone, and Early to Middle Oxfordian in text which explicitly speaks of coccolith age discrepancies in the same publication.

Age of the Patasar Shale Member
The shaly unit, on the combined coccolith-dinocyst evidence has been recently assigned broad Middle Oxfordian–Early Kimmeridgian age (Rai et al. 2015a). The ammonoid-based several times more precise early Late Oxfordian Bifurcatus Zone age given earlier (Krishna et al. 2009a, b) is maintained (Fig. 4.1). The interval currently is included in the early part of the truncated/revised Late Oxfordian Subevolutus Zone of Kachchh. The unit does not at all include any part of the Kimmeridgian. Nothing to speak of the Patasar shale, even the superjacent Fort Sandstone member of the Kantkote Member of the Katrol Formation does not include the basalmost Kimmeridgian.

Fort Sandstone Member
The unit is ranged within the Late Oxfordian Bimammatum Zone.

Adhoi Member
The coccolith-based age of the Adhoi member from the Planula Zone to Hypselocyclum Zone is in accord with the ammonoid-based age to the unit (Krishna et al. 2009a, b).

Gamadu Sandstone Member The major part of the abandoned informal ammonoid devoid unit is considered facies variant of the Adhoi member in the proximal near margin East Wagad. Its base and top are diachronous in context of the rapidly receding shoreline immediately after the peak transgression in the late Middle Oxfordian Schilli Subzone. The unit represents the undifferentiated Umia Fotmation in Wagad. It is ranged only up to the top of the Early Kimmeridgian Giganticus Zone (=ETM Hypselocyclum Zone). Much unlike the depiction by Rai et al. (2015, Fig. 2), the unit does not have any age over lap with the Katrol Formation of Mainland. The sedimentation in the Mainland after the submarine gap resumed only after the complete withdrawal of the sea from Wagad by the close of the Giganticus Zone.

Age of the Basal Katrol Beds

The base of the Katrol Formation (same as the base of the abandoned 'Jhuran Formation') in Mainland is of late Early Kimmeridgian Alterneplicatus/Divisum Zone, however, on the basis of coccoliths, it has been vaguely assigned to an unusually long interval of Kimmerdgian–Tithonian.

In the above context of invariably broad imprecise coccolith-based ages of the Kachchh Jurassic units, the Late Bathonian–Early Callovian record from Karakorum directly from the ammonoid samples determined earlier by Krishna (in Upadhyaya et al. 2005) provides a new approach of multiple guide fossil evidence. In fact, more such studies are required either directly from the ammonoid samples or from ammonoid age constrained sediments.

In Jaisalmer, the Badabag Member / Kuldhar Member lithostratigraphic boundary between the hard grounded limestones of the terminal part of the Badbag Member below and the overlying shales of the the basal part of Kuldhar Member has been disputed with regard to its age (Jain 2008; Rai 2009) on ammonoid and nanno-plankton evidence. It makes the Bathonian/Callovian boundary as in Kachchh between the RLm of the ULM of Patcham Formation and the overlying shales of the Chari Formation at Goradongar, Jhura, Jumara and Nara. There is drowning of the carbonate platform on account of the deepening and associated latitudinally influenced climatic change at the Bathonian/Callovian boundary SB. The climatic change is influenced by slow due south transform sliding of the East Gondwana vis-à-vis West Gondwana along the north–south-oriented Davies fracture. The limestones are of the terminal Bathonian Triangularis Zone and the overlying shales are of the early to mid-Early Callovian Chrysoolithicus Zone, more precisely between the latest Late Bathonian Oxycerites Subzone and the basal Early Callovian Madagascariensis Subzone. As a third-order SB, it is at least isochronous through out the Neotethys, if not globally, which hardly permits any scope for the dispute on age.

The elaboration as above on the resolution of ammonoid and coccolith based ages in the Indian Mesozoic units in general and Kachchh Jurassic units in particular is primarily made to emphasize upon the scope of distinct further improvement of coccolith-based resolution through integration with already developed high-resolution ammonoid zonation over the last few decades in Kachchh. Coccoliths in context of resolution make the next best tool of stratigraphic refinement in the Jurassic. Parallel scales have been developed in the European Tethyan and Boreal regions, which in details differ among themselves, nothing to speak of the blanket extension of the European ranges to Kachchh, in spite of the ammonoid rich character of the studied Kachchh sediments. The fact is lost sight of that the Boreal and ETM coccolith zones are primarily based on ammonoid ages. The same also unfortunatley has prevented in the past and continues to prevent the development of indigenous Indian Jurassic coccolith-based scale. The same approach has all along the past many decades prevented the development of the parallel Indian Mesozoic dinocyst and other palynomorph-based indigenous scales in contrast of the development during the same duration of the independent Indian high-resolution ammonoid zonal scale (Fig. 1.8). Had blanket extension of the Australian palynotaxa was not done for short-term unreliable imprecise incorrect results, distinctly more refined parallel Indian schemes would have long ago emerged.

Cretaceous
Coccolith resolution, in spite of being over all fairly good, is distinctly inferior to that of ammonoids, yet holds certain edge over ammonoids in context of their presence even in extreme marginal environments, subsurface and deep sea, which are otherwise not suitable for ammonoid studies, e.g. the Albian coccoliths in Kachchh, Jaisalmer and Cauvery and many other such instances. In case of any discrepancy of ages as sometimes is the case, the coccolith data need to be evaluated in reference of the primary ammonoid record for appropriate correction and revision in the cause of over all multidisciplinary improved geological comprehension of the Indian geology.

The European coccolith record is organized into a succession of 23–26 zones (in Haq et al. 1987 and others) from NC1 to NC23 and CC01 to CC26. In the ammonoid devoid to scarce mid-Middle Albian to Maestrichtian part of the Cretaceous in the Indian west and north sectors, the microfossils particularly the planktonic forams and coccoliths assume particular significance. The indigenous planktonic foram zones have acquired stability while the coccoliths have so far only been age constrained against the international schemes in bits of stratigraphic columns. The coccolith records are here integrated with the Indian Cretaceous second-order sequence surfaces, since the same are considered invariably linked to the first- or second-order MFSs (Fig. 4.5).

4.4 Sequence Stratigraphic and Microfossil Zonal Integration in the West Indian Basins

The coccolith recoveries are invariably from near first- or second-order MFSs of the TSTs. The intra-Albian records of CC 7–CC 8 (Late Aptian to Early Albian) from the lower part of the grey shales of the Dalmiapuram Formation, from the calcareous sands of either the terminal part of the Parihar Formation or basal part of the Habur Formation, from the Tanot well in Jaisalmer, and from the Jakh temple section of Umia Formation of Kachchh are here age-constrianed during the major first/second-order TST from the basal Aptian start of the TST to the second-order mid-Middle Albian Dentatus Zone MFS. At the Jakh temple section in Kachchh, the coccolith record presumably marked a short transient reach of the shoreline only during the terminal part of the TST in the close vertical proximity of the MFS. The Neotethys had by the close of the preceding first-order sequence already withdrawn earlier from the East Mainland. The record in the exposed sediments of the Jaisalmer basin is considered as of CC 7 of Aptian age instead of the early part of Albian. In many Indian basins, e.g. Kachchh, Jaisalmer, Narmada in the west, Spiti in the north, and Cauvery in the south sectors, the aforementioned second-order TST in particular exhibits significant development of fossiliferous limestones or calcareous sands along with drastic change from non-calcareous to calcareous lithology across the Barremian/Aptian boundary first-order SB. In the then neighbouring Madagascar (Herngreen et al. 1982), the unfossiliferous sands of the terminal part of the first-order RST are followed-up by the second-order TST Aptian–mid-Middle Albian marls. The successively younger Late Albian–Early Cenomanian glauconites and marls up to the first-order MFS evidence graded increase in bathymetry.

Kale et al. (2000) based on coccoliths interpreted change in the position of the Cauvery basin from high to middle latitude between Late Albian to early Middle Turonian.

The intra-Turonian record (Jafar 1982) from the Narmada basin of near CC 13/CC 14 (Gartneri/Eximus zonal transition) is in tune with the first-order early Middle Turonian MFS, so it is the Early Campanian record (Rai et al. 2002) from the Lameta Formation at Chakrud in Narmada and the Upper Member of Chikkim Formation at Chichhim in Spiti (Rai et al. 2015b) at the second-order MFS near the CC 17–CC 18 (Cayern/Orbiculus) transition.

4.4.1 Full Second-Order TST Substantiated Only Through the Ammonoid Record in Jaisalmer

The full TST from the Early Aptian to mid-Middle Albian Dentatus ammonoid Zone is evidenced in the three basins in the west, north and east sectors of the Indian plate margin as the Ukra Formation in Kachchh, the Abur Formation in Jaisalmer and the Grey Shale part of the Lower Member of Dalmiapuram Formation. The Aptian to mid-Middle Albian ammonoids are known best from Jaisalmer (Krishna 1980). The full TST is not substantiated through the nannoplanktons in either of these basins, instead only the second-or third-order MFS is evidenced.

4.4.2 Interpretation of the Saurashtra and Narmada Cretaceous Successions in Sequence Stratigraphic Context

The rifting revived in the west sector between India and Madagascar, as in Jaisalmer, Kachchh, Saurashtra, and suggestively also in Narmada at the first-order Barremian/Aptian boundary SB, and episodically continued during the first-order TST at the second-order sequence surfaces. The event is also evidenced in the Cauvery, K-G, and High Himalaya Spiti and Nepal basins. Aptian–Albian ammonoids are recorded in High Himalaya, Cauvery, K-G, Pakistan, Kachchh and Jaisalmer. Ammonoid evidence is lacking in Saurashtra and Narmada in this interval, where ammonoids are known only in the relatively younger and deeper Turonian–Coniacian interval near the first-order MFS in the exposed Saurashtra and

Narmada successions. Occasional suspectful records are also made of Albian and Cenomanian ammonoids (Pal 1971) in Narmada. The early Middle Turonian first-order MFS is suggestively evidenced through ammonoids in Saurashtra and Narmada, and through volcanism in Madagascar.

In Saurashtra, even in the much more proximally placed Wadhwan Formation *Placenticeras stantoni* is known. *Placenticeras* is also known from Narmada and Cauvery. It is however, probably, not known from Australo-Antarctica and South America, and thus exclusive to Axial Gondwana constituents, thereby supporting the complete spreading away of the Outer Gondwana (South America, Antarctica, and Australia) from the Axial Gondwana (Africa, Madagascar and India). The constituents of the Axial Gondwana remained somewhat clung together, inspite of transform sliding of India vis-à-vis Africa in the west and Australo-Antarctica in the east on either divergent sectors, and associated limited spreading/distanting among themselves. The acute space constraints, forced the togetherness until the first-order MFS event of the India/Madagascar separation at early Middle Turonian.

4.4.3 Stratigraphic Framework in the Narmada Basin

As presented above, the Narmada Cretaceous is stratigraphically similar to that of Saurashtra only above the first-order intra-Turonian MFS with regard to presence of common Turonian ammonoids. Their pre-Turonian tectonostratigraphic history seems different in spite of physical proximity as adjacent basins in the west sector.

The first-order intra-Jurassic SB $\sim$183 ma rifting was widespread in the west sector with resultant shallow marine sedimentation in the Toarcian–late Middle Oxfordian age bracket up to the Saurashtra basin as already suggested earlier. The revival of rifting along the Narmada fracture presumably may have taken place later either at the first-order intra-Oxfordian MFS or during the first-order RST interval up to the intra-Cretaceous first-order Barremian/Aptian boundary SB. The oldest Mesozoic sedimentary unit in the Narmada basin is the unfossiliferous Nimar Sandstone sensu stricto as defined originally (Bose 1884). It is considered shallow marine in a regional tidal flat and associated marine depositional framework. The Nimar Sandstone sensu stricto is assigned to the first-order RST age bracket of Late Oxfordian–Barremian. The upper part of the Nimar Sandstone with 3–5 relatively thin bivalve rich but ammonoid devoid intervals, in absence of guide fossils, is interpreted within the early to middle part of the Aptian–early Middle Turonian first-order TST span. It could be a possible facies variant of the Nodular Limestone with a subaerial disconformity between the non-marine lower and marine upper parts. The post Nimar bivalve rich intercalations alternatively could also be interpreted as transgressive lags at the second-/third-order MFSs. There could be another view either of a rift reactivation at the early Middle Turonian first-order MFS after a subaerial gap of Aptian–Early Turonian, or the younger Nimar comprising calcareous sands being of Aptian–Early Turonian age. In absence of guide-ammonoids, the age framework of the Bagh Group continues to remain fluid.

The basin in a regional frame may have been transgressed by the Neotethys due to revival of rifting along the ancient Narmada fracture either at the Barremian/Aptian boundary first-order SB as in neighbouring Kachchh, Jaisalmer, Pakistan and even Spiti, or later at the second-order SB in Late Cenomanian, or latest by the mid-Turonian MFS. The maximum bathymetry is interpreted near the contact of the bioturbated Nodular Limestone (Chirakhan Member) and the overlying bryozoan rich Coralline Member in the Hardground II of Tripathi and Lahiri (2000) (Fig. 4.6). There is interpreted distinct decrease of bathymetry above the Hardground II MFS. The basin registered increase in bathymetry during the first-order TST, and the Late Turonian coccoliths, and even occasional ammonoids were recovered around the first-order intra-Turonian MFS in the terminal Albian–Coniacian interval (Jafar 1982). As anticipated, the Turonian interval is slow sedimented to condensed inclusive of a few hard grounds.

The entire area witnessed shallow marine sedimentation during either the first-order eustatic rise from basal Aptian to Early or early Middle Turonian or during the later part of the first-order TST. In such a framework, the Nodular Limestone along with its suggestive facies variant of fossiliferous marine upper 'Nimar sands' is assigned a maximum span of the Aptian–early Middle Turonian first-order TST span, although it may be mostly of Cenomanian–Turonian duration in the vertical proximity of the first-order MFS. This is disconformably followed-up by the phase of pre-Deccan volcanics in the Bagh area (Biswas 1990) of Turonian–

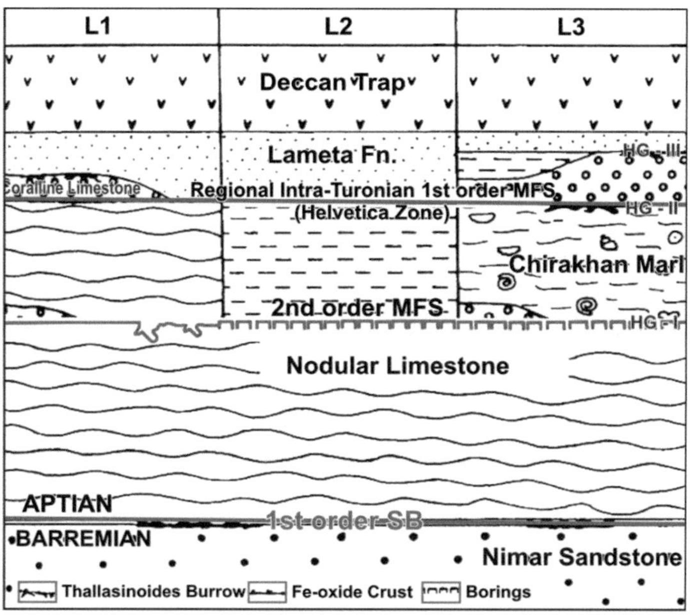

Fig. 4.6 Intra-cretaceous MFS/SB interpreted in the Narmada basin (modified after Tripathi and Lahiri 2000)

Santonian ~94 to ~84 ma age bracket with intra-Cretaceous spreading between India and Madagascar. The post Nodular Limestone part of the Bagh succession (Chirakhan Limestone/Marl) is considered to have been deposited during the LateTuronian second-order RST. The Coniacian Navgam Limestone, so age constrained through planktonic forams is interpreted as TST of the next terminal Turonian–Coniacian second-order sequence somewhat coeval to its better and fuller expression in Cauvery, Spiti and Jaisalmer. Yet later, the Bagh succession is discoformably superposed by the Lameta succession of Campanian–Early Maestrichtian. The Lameta succession in turn is terminally and disconformably overlain by the Late Maestrichtian–early Paleocene Deccan traps and associated intertrappeans.

The Bagh succession has been found overlain by the Turonian–Santonian volcanics underneath the overlying Lameta succession (Biswas 1990). The Lameta succession, in turn overlain by the Maestrichtian–early Paleocene Deccan Traps, is, thus, age constrained in the Campanian–Early Maestrichtian age bracket. Nodular Limestone, the principal stratigraphic unit of the Bagh succession in sequence stratigraphic context, is broadly correlated to the Uttatur Group / Karai Shale Formation, the Navgam Limestone to the Trichinopoly Group, and the Lameta Formation in the Ariyalur Group of the Sillakudi–Kallamedu succession.

4.4.4 West Indian Basins of Jaisalmer, Barmer, Cambay, Kachchh, Saurashtra, Narmada, Mumbai Offshore, and Kerala–Konkan

The Parihar Formation of Jaisalmer, Ghuneri Member of Kachchh, the Dhrangadhra/Dhanduka/Himmatnagar/Songir/Serau (only early part) units of Saurashtra, Narmada, Cambay basins and the unfossiliferous major part of the Nimar Formation are ranged up to the first-order intra-Cretaceous Barremian/Aptian boundary SB (Figs. 4.6 and 4.7). The succeeding Abur Formation of Jaisalmer, Ukra Member of Kachchh are ranged up to the mid-Middle Albian second-order MFS. The guide fossil devoid Wadhwan Formation of Saurashtra, Viramgam/Serau (only upper part) units of Cambay in homotaxial and sequence stratigraphic context are suggestively ranged up to the close of the Santonian second-order SB. The Bagh Formation of the Narmada basin is also ranged up to Santonian.

The Kerala–Konkan basin Mesozoic sedimentary succession is interpreted to start at the Pliensbachian/Toarcian boundary first-order SB. As in Narmada, it is interrupted by the rift-genic and plume-genic volcanics of the first-order early Middle Turonian MFS to the terminal Santonian second-order SB span, and later again by volcanics of the Late Maestrichtian age. As in the other West Indian basins, the second-order Early Campanian MFS is evidenced through the planktonic foram bearing transient mid-neritic sediments in the basin within the Coniacian–Maestrichtian sedimentary package.

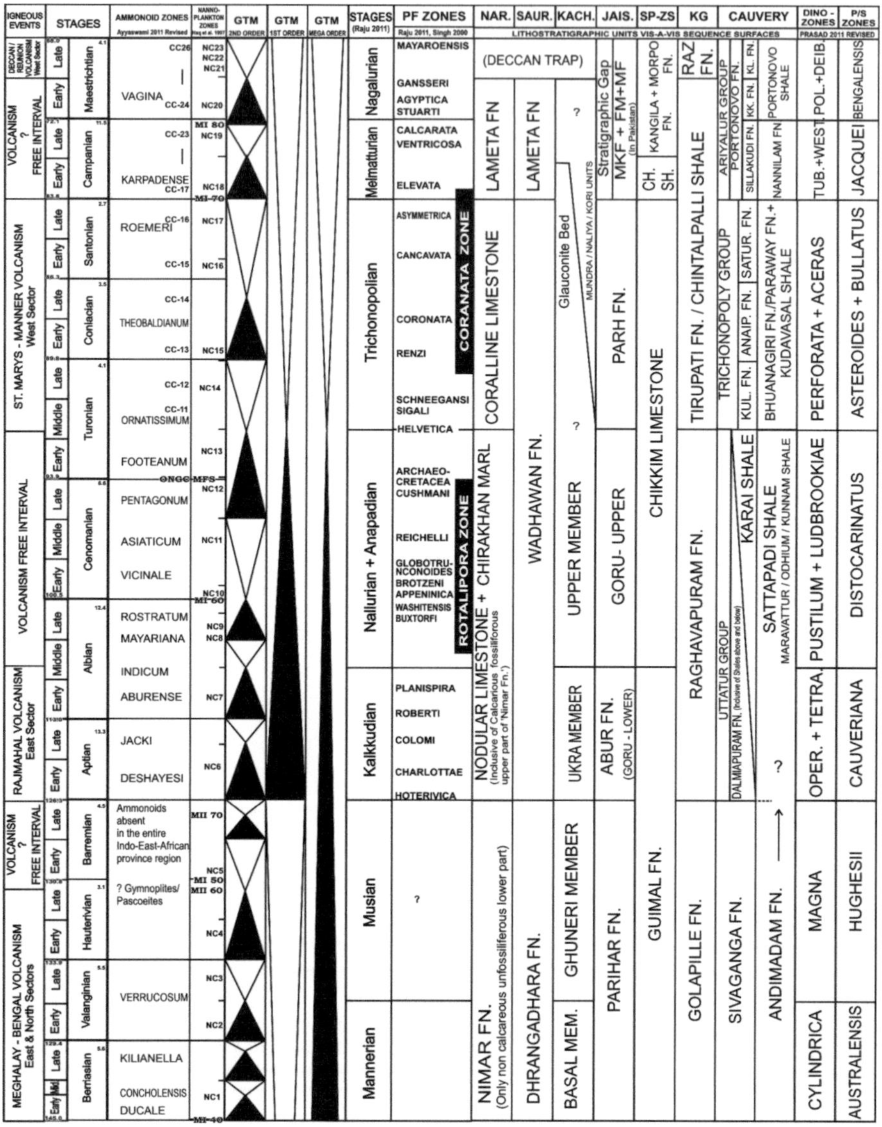

Fig. 4.7 Indian cretaceous lithostratigraphic units vis.-a.vis second-order sequences, planktonic, dinoflagellate, pollen–spore zones

The west and north sectors of the Indian plate margin also evidence the Rajmahal igneous event through the presence of volcanoclasts all along High Himalaya of ages U-Pb ages from ~162 to ~102 ma, i.e. from Oxfordian to Albian with a peak at ~132 ma (Hu et al. 2015). They interpreted younging of the ages in an east–west transect, and interpreted the same as westward progressive propagation of the rifting as of the drainage framework, instead of heterochronic progressively younging reach of the volcanics. The Bunbury–Bengal-Comei volcanism is here interpreted as a distinct rift-genic early to mid-Early Cretaceous igneous event independent of the younger Kerguelen influenced Rajmahal event. The older event for sure did neither reach Jharkhand, nor west of South Tibet nor south of Bunbury. The two distinct events are here related to the first- and second-order sequence surfaces, and discussed under Chap. 7. It is not true that volcanics and volcanoclasts are absent west of Nepal. The igneous event is widely evidenced in the Uttarakhand, Himachal, Spiti–Zanskar–Ladakh, and Pakistan High and Lesser Himalaya, even in Barmer, Jaisalmer, Kachchh and Saurashtra either in exposures or wells.

Exposed evidence of ~124 ma (Barremian/Aptian boundary) age from the Jhura igneous plug of Mainland Kachchh (Kalmalkar et al. 2014) has also emerged. The igneous history between the South Africa–India and Antarctica–South America being found increasingly similar to that of the India–Australo-Antarctica divergent sector.

4.4.5 Lumping of Several Morphologically Close Species of the Bagh Succession Under **Placenticeras mintoi**

Jaitley and Ajane (2013) have lumped a large number of morphologically close species that came from different levels of the Nodular Limestone. May be the morpho-variants represented an evolutionary lineage from the base to the top of the Nodular Limestone during a large chronospan. While lumping, it is essential to ensure that the intra-specific variants come from a small chronospan, i.e. from the same bed instead of being vertically spread through substantive thickness and significant time-span. The species lumped belonged to nine genera and three subfamilies spanning through ~27 my long Albian to Maestrichtian interval that included ~16 ammonoid zones, ~10 planktonic foraminiferal zones and ~12 nannoplankton zones. Among the nine genera, eight belong to Engonoceratidae and Placenticeratidae, the latter having evolved from the former, thus, probably constituting the principal lineage present there. More desirable in such circumstance is to work out the succession of species of the evolutionary lineage, which might in-turn lead to the succession of fine zones. The majority of workers over the last century or more have ranged the Nimar Sanstone–Nodular Limestone–Coralline Limestone succession from Neocomian to Maestrichtian, and only a few have favoured distinctly shorter duration. In sequence stratigraphic context, the calcareous upper part of the Nimar Sandstone and the superjacent Nodular Limestone combine is here bracketed between the Barremian/Aptian first-order SB on the lower side and the early Middle Turonian first-order MFS on the upper side. The overlying Coralline Limestone is ranged up to the first-order Santonian/Coniacian SB. The localized hard ground in the L3 section of Tripathi and Lahiri (2000) and associated sharp decrease in bathymetry at the hard ground in regional sequence stratigraphic context is here interpreted to mark the second-order Santonian/Campanian boundary SB which is unconformably followed-up by the Lameta Formation ranging from the second-order SB of the Santonian/Campanian boundary to the intra-Maestrichtian second-order MFS.

4.4.6 Post-Nimar Intra-Turonian Guide Fossil Records and Not Intra-Nimar

The intra-Turonian ammonoid and nannoplankton records from the Narmada basin are interpreted as from the fossiliferous calcareous sands that immediately overly the unfossiliferous Nimar sands sensu strict (Figs. 4.6 and 4.7). The younger fossiliferous calcareous sandy part ranges up to the early Middle Turonian first-order MFS, and may be the facies equivalent of the Nodular Limestone, and, preferably, included in the Bagh Formation instead of the Nimar Formation. Reactivation of the extensional tectonics along the Narmada fracture has taken place at the first-order intra-Cretaceous SB as at the first-order MFS. In fact, minor tilting down/up is interpreted even at the second-order surfaces within the first-order TST interval. The associated rift-genic igneous activity at the first-order MFS is widespread and genetically linked to the initiation of spreading between India and Madagascar.

4.4.7 Late Cenomanian–Early Turonian Condensed to Starved Interval at Site 258 Leg 26 of the Indian Ocean

Thierstein (1974) recorded a significant Late Cenomanian–Early Turonian condensed to starved slow sedimented interval at site 258 of the leg 26 which accords well with the first-order MFS at the top of the interval. The subsequent shallow nearshore early part of the first-order RST also is in tune with the anticipated eustatic fall up to the close of Santonian.

4.4.8 Evidence of Volcanism in the West Sector of the Indian Ocean

The volcanics are known even in the west sector of the Indian ocean leg 25 site 249 of Aptian age. Volcanics in siltstones are recorded at site 249 of leg 25 in the early part of the Aptian–early Middle Turonian first-order TST which may be coeval to the Rajmahal and associated volcanism in the east sector (Vallier DSDP report 1974). Coeval evidence of igneous activity is also met in the outcrops in the west and north sectors (Valdiya 2010, 2016).

4.4.9 Demarcation of Important Cretaceous Sequence Surfaces in Kachchh and Elsewhere

The first-order TST is also suggested in the guide fossil devoid interval in Kachchh through the correlation of the Kachchh sequence surfaces with the planktonic foraminifera bearing Spiti and Cauvery successions (Fig. 4.10). In fact, the Cretaceous is suggested as nearly complete all around the Indian plate inclusive of the west sector in Madagascar, Saurashtra, Kachchh, Narmada, Jaisalmer, Pakistan, S. Tibet, Spiti–Zanskar, Cauvery, K-G, and other basins of the IEAP.

Widespread Manifestation of the Intra-Campanian MFS in India The second-order intra-Campanian TST has its MFS in the mid-late Early Campanian which is widely evidenced in Cauvery, Bengal and Mahanadi, Andaman (ONGC 1993, 2008), Spiti (Bertle and Suttner 2005; Rai et al. 2015b), Narmada (Rai et al. 2002), and Kachchh offshore (Raju 2011) through planktonic forams, coccolith, and ammonoid evidence (*Karapadiceras karapadense, Gyptoxoceras indicum*, Ayyasami 2011). The Campanian transgression is rift-genic in most of the basins, and is here interpreted as a tectonically enhanced deepening. The widely expressed Campanian event is present in all the Indian sectors. There is either a fresh transgression after a non-marine interlude or marked change from a relatively shallower to deeper facies as exampled by the nannoplankton-based Early Campanian NC 18 (Rai et al. 2002) suspected short transient intra-Lameta incursion near Chakrud that in-turn implies older intra-Campanian start of the Lameta Formation vis-à-vis the presently opined intra-Maestrichtian base. For about a century, the unit based on dinosaurian and other vertebrates was considered Turonian to Santonian. Later, it was revised to Maestrichtian in context of the Lameta vertebrates being close to the overlying Maestrichtian intertrappean ones. The fresh recent evidence suggests the start of the Lameta Formation in Campanian.

4.5 Sequence Stratigraphy-Based Interpretations in the Bengal and Other East Sector Basins

4.5.1 The Important But Enigmatic Non-Marine Dubrajpur Formation in Bengal Basin Wells

The Dubrajpur Formation is a lithostratigraphic unit in the Rajmahal basin at Birbhum district of West Bengal. The unit as defined in the coal field boreholes lies unconformably over the Barakar Formation, and is unconformably overlain by the trap flows. The unit starts with a conglomerate and comprises coarse sands and silts/shales (Fig. 4.8). It is observed that the shale intercalations within the Dubrajpur Formation invariably have yielded richer but broad age diagnostic palynotaxa, invariably, in the middle part, less so in the upper part. This in turn allows better dating of the middle shaly part. The Dubrajpur Formation has been

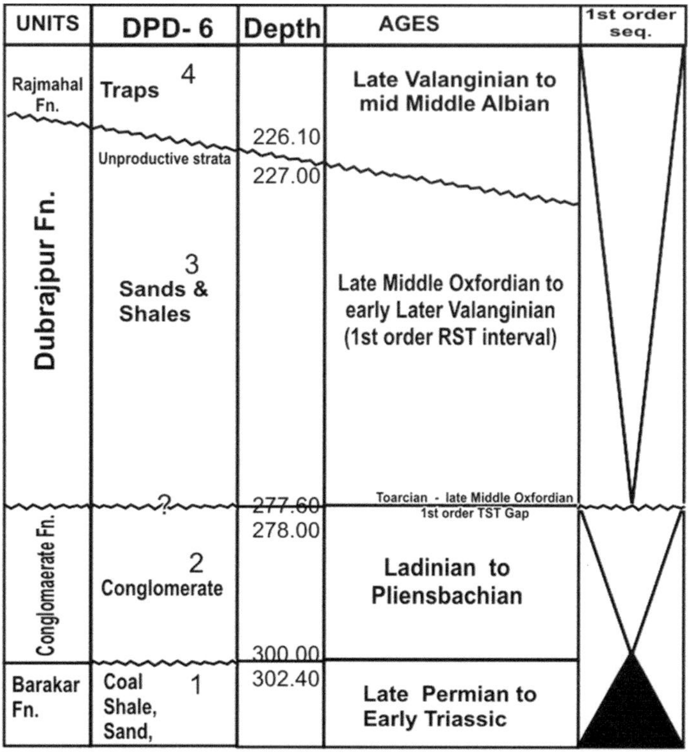

Fig. 4.8 First-order sequences interpreted in Rajmahal Basin (modified after Vijaya 2009)

assigned late Middle Jurassic to Late Jurassic age with suspected basal Berriasian, and the unit has been interpreted as disconformably underlain by the Early Permian to the basalmost Triassic sediments (Vijaya 2009). Vijaya in the same work has interpreted stratigraphic gaps below the Dubrajpur Formation as between the early (inclusive of the thick conglomerate at the base) and later parts of the Dubrajpur Formation, i.e. between the Cooksonae Zone below and the Florida Zone above. The palynozones are also not well age constrained and are only of temporary and limited usefulness in total absence of any other guide fossils. The gap, as suggested elsewhere may mark uplift, rejuvenation of the provenance and the resultant shifting of the depositional channels above the Cooksoniae Zone. The conglomerate in the early part in spite of poor record of palynotaxa is interpreted to mark the base of the Cooksonae Zone. Sen et al. in Vijaya (2009) assigned the unit to Early Triassic. Vijaya (2009) also gave variable ages to the unit in the same work (a) as Triassic–Jurassic, and (b) Bajocian–Berriasian in the same figure (Fig. 3), (c) Late Triassic/Early Jurassic in Table 1, (d) Middle to Late Jurassic in Fig. 10, and (e) Late Jurassic in Table 4, (f) Callovian–Tithonian in the abstract. The ages (a) to (e) put together involve a span of over 100 my. It is obvious that the used palynotaxa are long ranging, and so of little chronostratigraphic significance.

4.5.2 *Discrepancies About the Definition, Age and Geographical Expanse of the Dubrajpur Formation*

Apart from the large discrepancy in the age range of the Dubrajpur and Rajmahal units as above, serious questions arise about the precise lithostratigraphic characteristics and definitions of the, Barakar, Dubrajpur and Rajmahal units in Jharkhand and West Bengal. In case the volcanic interval in Jharkhand is of Early Aptian – mid-Middle Albian age, as probably is the case, would the immediately underlying infra-trappean sediments be included in Dubrajpur, or termed infra-trappeans, or included in a separate unit between the Dubrajpur Formation and the Rajmahal Formation or included in the Rajmahal Formation. Would it be advisable to subdivide the Dubrajpur unit into pre and post gap member units or even into two separate formations. May be the Dubrajpur unit in some other bore hole is without any palynological gap which suggests possible rejuvenation related break in the non-marine sedimentation. It is suggested to formally separate the conglomerate part at the base of the Dubrajpur Formation as Conglomerate Formation. The Conglmerate Member is here age bracketed in the Ladinian–Pliensbachian interval. The Conglomerate Formation and the disconformably overlying Dubrajpur Formation are here interpreted as tectonostratigraphically differentiated by the Toarcian–late Middle Oxfordian gap. The Dubrajpur Formation is considered having a diachronous upper boundary up to the first volcanic flow of Rajmahal Formation within the Rajmahal basin from Bengal to Jharkhand. The new unit has disconformable contacts at the base and top. The oldest volcanic interval/flow in the three wells is either of the same age, or the one in DPD 6 is older to the oldest of the other two wells. The ~40 m thickness of DPD 6 in between the oldest and the fifth flow is palynologically between the Cooksonae Zone below and the Australensis Zone above, and of suggestively Early Cretaceous age. The younger two flows are interpreted as post Barremian in view of the second youngest flow interpreted as the principal flow of Late Aptian age. The older four flows of DPD 6 are suggested in the Early Cretaceous Berriasian–Valanginian interval. It is also significant to converge on the lateral extent of the Rajmahal Formation exclusive to Jharkhand or inclusive of Bengal, Bangladesh and Meghalaya.

4.5.3 *Volcanism in the Bengal Basin*

Lower Stratigraphic Limit of Volcanics in the Bengal Basin Boreholes

Vijaya has strangely extended down the Rajmahal Formation in the borehole DPD 6 (1) only up to ~180 m, and (2) up to ~210 m in two different rows in the same table (Table 3), and (3) up to ~226 m depth in Fig. 3 and up to ~180 m in Fig. 4. Thus, three contradictory lower limits of the Rajmahal Formation are given by Vijaya in the same bore hole DPD 6. However, in DPD 6, the oldest volcanic interval is up to a depth of ~226 m, and as per definition, the Rajmahal Formation must extend down to the depth of ~226 m, as followed here.

4.5.4 Volcanic Intervals in the Bengal Basin Boreholes

Six volcanic intervals/flows are included in the Rajmahal Formation as above defined down to ~226 m in well DPD 6. From top to bottom of the unit, the first/youngest ~7 m thick volcanic interval has its base at ~33 m from the surface, the second ~78 m thick between ~40 m depth to ~118 m depth, the extremely thin third and fourth are the thinnest between ~149 m and ~145 m depth, the fifth, also equally thin as the third and fourth, has its base at ~180 m depth, and the oldest volcanic interval, at least ~10 m thick has its base at 226 m depth. Based on the relative thickness, the second ~78 m thick volcanic interval may be considered as the principal Rajmahal outpouring.

4.5.5 Bengal Basin Palynozones

Vijaya (2009) has developed a succession of five palynozones in the three Bengal bore holes DPD 3, DPD 6 and DPD 15. The younger three Australensis Zone, Wonthaggiensis Zone, and Asymmetricus Zone are included in the Rajmahal Formation as understood here, and ranged by her from Late Tithonian to Early Aptian on inadequate highly subjective considerations of the chronologically il-constrained FADs of the three species, respectively at ~161.20 m, ~160.80 m and at ~147 m. The Australensis Zone thus is only ~40 cm thick compared to the ~13.80 m thick Wonthaggiensis Zone. The yet older two zones Cooksoniae Zone and Florida Zone are included in the disconformably underlying Dubrajpur Formation. M. florida is not indicated in DPD 6, while the FAD of C. cooksoniae is marked between ~278 and ~300 m. The C. cooksoniae Zone is assigned a broad Bathonian–Callovian age, while the Florida Zone is ranged from Callovian to Kimmeridgian in Figs. 3 and 4. However, both these zones in Table 3 are included in Late Jurassic in contradiction to Figs. 3 and 4. Vijaya has also interpreted gap of unknown duration between the Cooksoniae Zone and Florida Zone. The contradiction of ages of zones, of the lower limit of Rajmahal Formation, the long suprazonal duration of the five zones, and imprecise unknown bounding surfaces of the zones undisputedly expose the poorly constrained durations of the zones. Scanning the palynozation in other east sector basins additionally brings out their highly subjective relevance. Vijaya has ranged the Asymmetricus Zone from Late Barremian to close of Aptian, while Mehrotra et al. (2002) have recorded the taxon in their Late Albian–Cenomanian Distocarinatus Zone in K-G basin instead of their Barremian-Aptian Hughesii Zone, which indirectly implies at least Late Barremian to Late Cenomanian ~30 my duration. Similarly the Cauveriana Zone is also differently dated by Vijaya (2009) and Mehrotra et al. (2002). The Barakar/Dubrajpur contact is also disconformable, and includes gap of most of the Late Triassic and Early Jurassic first-order RST interval.

4.5.6 Age Revision to the Bengal Basin Units and Palynozones Through Sequence Surfaces

The upper limit of the Barakar Formation in the Bengal basin is here ranged up to the intra-Triassic first-order MFS. The stratigraphic gap between the Barakar and the Dubrajpur units is age constrained within the first-order Ladinian–Pliensbachian RST interval. The Dubrajpur unit is age constrained in the first-order Toarcian–Barremian sequence interval. The intra-Dubrajpur disconformity between the Florida Zone and Cooksoniae Zone is corresponded to the marked multifaceted geological event at the first-order intra-Oxfordian MFS. The Cooksonae Zone is age constrained within the first-order Toarcian–late Middle Oxfordian duration. The four zone succession from the Florida Zone to Asymmetricus Zone is age constrained within the first-order terminal Middle Oxfordian–Barremian RST interval. The first-order intervals are isochronous and bounded by the isochronous sequence surfaces on the entire GTM between Arabia and NW Australia, which in-turn now allows correlation of the Bengal basin lithostratigraphic units to even the best resolved marine succession of the Kachchh basin. The premise used in the above correlation is the correlation of the tectonically produced unconformities or their correlative conformities between the excellently ammonoid rich marine succession and the guide fossil scarce non-marine sedimentary successions. The Bengal basin DPD 6 borehole succession inclusive of the older four volcanic intervals between ~118 and ~300 m of the Dubrajpur Formation in Jharkhand or the Dubrajpur Formation and the early part of the Rajmahal Formation in Bengal is age constrained in the Toarcian–Barremian first-order sequence interval.

4.5.7 Suggestive Ages to the Six Volcanic Intervals in the Bengal Basin Borehole DPD 6

The interest of the author is focussed exclusively on the betterment of the age constraints to the volcanic intervals of the Bengal basin bore holes. The application of sequence surface-based ages as above ranges the younger four volcanic intervals as per Fig. 11 of Vijaya above the Australensis Zone, while the older two volcanic intervals must be above the Cooksoniae Zone and below the Australensis Zone. The upper known limit of the volcanism is placed as mid-Middle Albian around the second-order MFS ~106 my, and the lower limit is determined between the late middle Oxfordian ~159 ma first-order MFS and early Late Valanginian. The Australensis Zone in all the palynozonal scheme begins in the Late Tithonian in both the east and west sectors of India, however, its upper limit in Vijaya (2009) is suggested up to the close of Berriasian while Mehrotra et al. (2002) extend it up to intra-Barremian. As per the shorter Late Tithonian–Berriasian range of the Australensis Zone, the younger four flows above the FAD of *C. australensis* need be Late Tithonian or younger age, however, if the upper limit of Berriasian is accepted, the younger four volcanic intervals may be ranged exclusively in Early Cretaceous. The ~78 m thick principal volcanic interval suggests Late Aptian interval of ~119 to ~112 ma, however, may also include Early Aptian. If that be so, the youngest flow may be ranged up to the known limit of east sector volcanism of mid-Middle Albian ~105 ma second-order MFS. The third and fourth volcanic intervals are thin and vertically close together in a small thickness of ~9 m between depths of ~140 and ~149 m, which include the FAD of *F. asymmetricus* at ~147 m. The FADs of *F. wonthaggiensis* and *C. australensis* are constrained in a small thickness of 40 cm in DPD 6 which in turn implies an age near the Berriasian/Valanginian boundary. Thus, the third and fourth flows are age constrained within the second-order TST up to the early Late Verrucosum Zone as per Vijaya's ranges of the Australensis Zone and Wonthaggiensis Zone. However, Vijaya has placed them in the Late Hauterivian–Barremian interval. The best age to the third and fourth volcanic episode could be Valanginian to Hauterivian. The fifth and sixth (also the oldest) volcanic episodes are below the Australensis Zone, and as per present interpretation are age constrained in the terminal Middle Oxforidan to Berriasian–Hauterivian, even up to Early Barremian as per Vijaya's ages to the younger third and fourth volcanic intervals. In context of the graded, due west spread of the volcanic intervals, the fifth and sixth intervals may be included in Berriasian. The older Late Jurassic age is much less likely, but not totally ruled out to the oldest flow. The Florida Zone being older to the oldest volcanic interval, thus, is here best suggested in the Late Oxfordian–Kimmeridgian interval between the first-order MFS and the second-order SB.

As a whole in the Indian Mesozoic, the microflora-based stratigraphic differentiation in the Cretaceous is much better than in the underlying Triassic and Jurassic. There are recorded seven pollen/spore zones in the ~80 my long Cretaceous of K-G basin subsurface and offshore wells (Mehrotra and Aswal 2003).

In these studies, it is rather forgotten that all kinds of stratigraphic units inclusive of the palynostratigraphic units need be differentiated only in the lithostratigraphically well-differentiated sedimentary columns, while the included palynofossils used as mere tools to characterize the palynostratigraphic unit in the column. Prior to developing biostratigraphic units in a sedimentary column, it is obligatory to realize the maximum possible lithostratigraphic differentiation within the column and record the same in publications along with the biostratigraphic zones. If the column is lithostratigraphically undifferentiated, the palynological differentiation within it, if not futile, substantially reduces its chronostratigraphic usefulness. Even the interval zones of one well do not match with those in other wells neither in name, nor age nor dinocyst content (Mehrotra and Aswal 2003). Intra-well correlation is neither tabled nor discussed. The point must not be side-tracked that the foremost objective of the biostratigraphic zonation is to refine and precise the ages of the rock successions and sharpen them for intrabasinal to interregional correlation which so far could not be realized through microfossil-based biostratigraphy in the Indian Mesozoic, particularly, so in the Cretaceous in spite of being the most studied interval of the Mesozoic.

Prasad (2011) published a composite scheme for the East coast basins of 8 palynozones. Most of the zonal boundaries of the scheme do not match with the international stage boundaries (Fig. 2). Regional/basinal stages have also been created in the Cretaceous of the Indian East coast basins. In most cases each Cretaceous palynozone, as in Triassic and Jurassic, includes parts of at least two stages, thereby defeating the very purpose of their application to the bigger cause of hig-resolution geodynamics of the Indian plate. Example is here cited of the Australiensis Zone of the intra-Tithonian–intra-Barremian range. The stages involved number five, i.e. Tithonian, Berriasian, Valanginian, Hauterivian, and Barremian with a total duration of ~26 my while the average duration of the five stages is ~5 my. It implies that the presence of the taxa could be anywhere within the ~20 my duration. The basic objective of the stage subdivision into multiple zones is of realizing finer and smaller chrons within the stage. Much better would have been to use the ranges of the palynotaxa for differentiation at least of the stages, instead of falsifying the concept of the multiple zones within the individual stages. However, the enigma here is the inclusion of multiple stages in a single zone that too exclusively to a single well or basin. The limited and localized usefulness of such zonation is more than obvious. Efforts have also lacked in testing and extending the Indian east coast Cretaceous palynozones to the other Indian basins. Also, the zonal ranges have significantly differed from one publication to another, even from one figure to another figure of the same publication, or from text to tables that too without any explanations (Vijaya 2009).

4.5.8 On the Palynology-Based Sequence Stratigraphic Surfaces in the Indian East Coast Wells

Mehrotra and Aswal (2003), Mehrotra et al. (2002) have (op. cit) additionally organized the well sediments into succession of sequence surfaces based on relative proportion of the pollens/spores and dinocysts in the successive assemblages (Fig. 4.9) without indication of the order of the determined SBs, first, second, or third. Neither, the zones nor sequence surfaces match from one well to another even in the same basin, thus violating the basic premise of the concept of sequence stratigraphy. For example, most of the Barremian but for the basal and terminal part in the well END-A makes the Longicornutum Zone while in the well DKR-1 it represents the Ceratophora Zone. Similarly, the Late Turonian interval in well END-A includes the fifth of the six SBs, which in the DKR-A well is included in Early Campanian, having a time difference of more than ten my. May be the SB in one well is not determined in the other well and vice versa. Moreover, the sequence surfaces marked in the tables are also not precisely dated. As such their correlatives

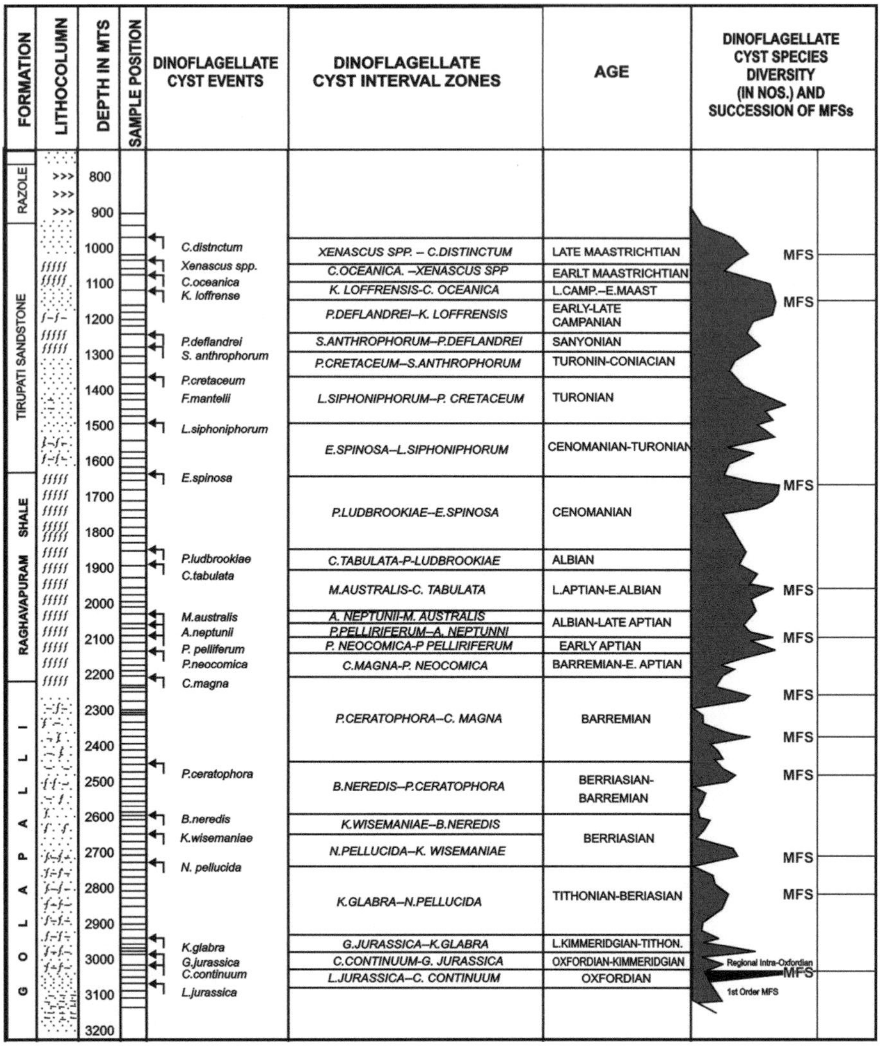

Fig. 4.9 Intra Oxfordian First-Order MFS in the K-G Well RCPM-A (modified after Mehrotra and Aswal 2003)

in other wells cannot be determined. Much contrary to the claim (Mehrotra et al. 2012) of one my time slices, and development of 9–10 SBs in the Indian Cretaceous compositely based on forams/dinocysts/pollen/spores, most of the zonal time planes, most of the SBs still remain imprecisely dated. The integrated ONGC sequence framework (2008) in the best studied Cauvery includes merely five microfloral/faunal biostratigraphic divisions, and just seven SBs (inclusive of just one first order) in the Cauvery basin in the ~80 my long Cretaceous. Similarly, in the K-G basin, the Cretaceous includes 11 biostratigraphic divisions, and only three

SBs inclusive of two first-order ones. Only a few of the biostratigraphically determined time planes are age précised. The ONGC framework includes a total of four first-order Mesozoic SBs (M 50–M 80). Conceptually and even as visually observable surfaces in the outcrops or lines in well logs and seismic profiles, the SBs are isochronous instants of geological time at least regionally, and as such need precise determination with zonal/subzonal resolution of one my or finer.

4.6 Ranges of the Palynozones Through Sequence Surfaces (Fig. 2.42)

4.6.1 Pollen–Spore-Based Palynozones

The Australensis Zone is assigned Late Tithonian to Neocomian age (Prasad et al. 1995; Prasad and Pundeer 1999). Vijaya (2009) ranged it from terminal Tithonian to Berriasian. Prasad (2011) in Fig. 19 assigned the zone from Late Tithonian to Early Hauterivian, while in Fig. 20 ranged it from Late Tithonian to Berriasian. The zonal time planes are arbitrarily placed. The basal time plane could be anywhere in Late Tithonian, so the upper any where in Berriasian to Barremian. The above age discrepancies speak unambiguously about the bounding time planes of the nominal palynotaxa of the microfloral zones, and their irrelevance to precise ages. The zonal ages also evidence large-scale discrepancies from one publication to another. Examples are here cited of Prasad et al. (1995), Prasad (2011), Mehrotra and Aswal (2003), Mehrotra et al. (2002). In such circumstances, the quantitative changes in palynotaxa of the Cretaceous microfloral zones are here interpreted at the first- and second-order sequence surfaces.

The zone is here ranged up to the early Late Valanginian Verrucosum Zone second-order MFS. Similar suggestive revision is also realized for the other zones. The top of the Cyclosporites hughesii–Ceratosporites equalis Zone is corresponded to the first-order Barremian/Aptian boundary SB. The Cauveriana Zone ranges through the second-order TST up to the mid-Middle Albian MFS. The Distocarinatus Zone is suggestively ranged up to the first-order intra-Turonian MFS. The Asterolectes Zone ranges through the second-order cycle from the second-order Late Turonian SB to Santonian/Campanian boundary SB to Late Coniacian MFS. The Jacquei Zone is ranged through the entire Campanian second-order sequence. The terminal Bengalensis Zone is similarly assigned to the terminal Cretaceous second-order sequence of Maestrichtian age.

4.6.2 Dinocysts

Dinocyst-based schemes are not developed in the Triassic–Jurassic interval. An updated succession of 8 dinocyst zones developed in ONGC in the Indian east coast

Cretaceous is provided in Prasad (2011). These are primarily based on the extension of the Australian marker taxa. Dinoflagellate studies in India are widespread spatially, as covering the entire Cretaceous chronospan with significant contribution to the Indian Jurassic–Cretaceous stratigraphy. Garg et al. (2003) in the terminal Jurassic–basal Cretaceous based their investigation on dinocysts directly extracted from the individual taxonomically determined and stratigraphically precised ammonoids. More such studies are desirable in the ammonoid bearing intervals, particularly in the excellently resolved Triassic of Spiti and Jurassic of Kachchh.

Prasad et al. (1995), Mehrotra et al. (2002) ranged the Wisemaniae Zone from intra-Tithonian to end-Valanginian. Prasad and Phor (2009) ranged the zone from intra-Tithonian to end-Berriasian, thus, excluding the Valanginian stage without giving any reasons. Prasad (2011) as a major departure restricted the zone exclusively to Late Tithonian. The discussion did not refer to the inclusion or exclusion of Valanginian from the range of K. wisemaniae. In the NW Australian Dampier basin is suggested the SB 7 at the Berriasian/Valanginian boundary between the Delta Sands of the Barrow Group and the overlying Mardie Green Sands. In Arabia is interpreted an ammonoid-based Early Valanginian MFS (Sharland et al. 2001). The second-order intra-Valanginian MFS is widespread across the Tethys of early Late Valanginian Verrucosum Zone, which is also demonstrated in Spiti (Pathak et al. 2011). There is sudden and drastic reduction of ammonoids above this MFS in Spiti. There is logically anticipated distinct change in the dinocyst scenario in the Indian east sector that was proximal to Australo-Antarctica prior to the oceanic separation during Aptian–Albian. As per range chart Fig. 7 of Prasad and Phor (op cit), there are better marked dinocyst changes at the Tithonian/Berriasian boundary and in mid-Valanginan than at the Berriasian/Valanginian boundary. Moreover, Prasad and Phor have clearly mentioned the diachrony in the ranges of the critical dinocyst species between Australia and India, in spite of the proximity of the two lands during the Neocomian. The studied lithocolumn indicates intra-Valanginian change in lithology, and absence of any lithological change either at the base or top of the Valanginian stage. Conceptual and fundamental requirements of zonation also demanded the differentiation of the Late Tithonian Jurassic part and the Berriasian–intra-Valanginian part as separate zones as done on rethinking by Prasad (2011), although he should have provided as well the revised definitions. Under the above additional regional sequence stratigraphic constraints, the Fromea cylindrica–Gardodinium lowii Zone is ranged up to the second-order intra-Valanginian Verrucosum Zone MFS. The Cassiculosphaeridia magna–Batiacasphaera asperata Zone is ranged up to the first-order Barremian/Aptian boundary SB. Again the studied column exhibits distinct lithological and palynological changes at the close of the Barremian and not at its base. The imprecise intra-Albian top of the Tetracantha Zone is here precised up to the first/second-order mid-Middle Dentatus Zone MFS. The Ludbrookiae Zone is ranged up to the first-order early Middle Turonian MFS. The Aceras Zone + Porifera

Zone + Pusillum Zone interval is here ranged up the second-order Santonian/Campanian boundary. The Tuberculata Zone + Westralinum Zone interval is ranged through the second-order Campanian sequence. The Polymorpha Zone + Diabelli Zone interval is suggestively corresponded to the terminal Mesozoic second-order sequence of the entire Maestrichtian duration.

4.7 Application of Sequence Surfaces in Precision Dating of the Cretaceous Units in the East Sector

The Indian Cretaceous in Chapter 3 is organized into 12 second-order sequences. The framework includes 23 second-order sequence surfaces, which is greater than any kind of macro/microfossil-based zones. The premise applied is the occurrence of sharp changes in micro/macro fossils invariably at sequence surfaces.

4.7.1 Cauvery Basin

The Sivaganga Formation is assigned broad Neocomian age in Raju (2011), the coeval Terani Formation given Aptian in Ayyasami (2011), Late Jurassic to earliest Cretaceous in (ONGC 1993), and Late Aptian in Kale et al. (2000). The marine body fossil devoid unit is here ranged up to the first-order SB at the Barremian/Aptian boundary. The Dalmiapuram Formation is variably ranged from Early Aptian to Early Cenomanian. Quoting only a few here, it is assigned Early Aptian–Early Albian in Bhatia and Jain (1969), Middle Albian in Kale et al. (2000), Aptian in Yadgiri and Govindan (1997), and Late Aptian–Early Albian in wells, Late Albian in Ayyasami (2011), Albian in Raju (2011), Late Albian–Early Cenomanian in course material. Early Albian nannoplanktons are recorded from the grey shales (Jafar and Rai 1989) of the early part of the Dalmiapuram Formation. Nagendra et al. (2011) has ranged it also differently; latest Albian to Cenomanian in the field guide-books, Albian–Middle Turonian in Fig. 5 and Albian–Cenomanian in Fig. 7 (Nagendra et al. 2011). The entire Uttatur Group inclusive of the Dalmiapuram Formation and its facies variant Karai Formation (in part) is ranged from the Barremian/Aptian boundary first-order SB to the first-order early Middle Turonian Turoniense Zone MFS. The Karai Shale Formation is assigned Middle Cenomanian–Early Turonian in Ayyasami (2011), Albian–Middle Turonian in Raju (2011), Albian–Cenomanian in Mehrotra et al. (2002), up to Middle Turonian in Chidambaram (2000), up to early Middle Turonian in Kale et al. (2000), Albian–Turonian in Govindan et al. (1998). Late Albian–Middle Turonian in Nagendra et al. (2011). The base of the Karai Formation is here ranged between the

Barremian/Aptian boundary first-order SB mid-Middle Albian second-order MFS with upper boundary up to the first-order MFS above the early Middle Turonian Turoniense Zone. The latter MFS is chronologically included in the planktonic foraminiferal Helvetica Zone interval. The ONGC (2008) has placed the MFS at the Cenomanian/Turonian boundary, whereas Nagendra et al. (2011) have not specified any MFS in the Turonian.

The early Grey Shale Member of the Dalmiapuram Formation is assigned to the second-order Early Aptian–mid-Middle Albian Dentatus Zone TST, the middle Limestone Member (algal limestones followed-up by bedded limestones) of late Middle Albian to intra-Cenomanian marked lithological change, while the upper marly part, otherwise held coeval to the upper part of the Karai Formation (also known as Marvattur/Odhium/Kunnam clay) is ranged through the later part of the Cenomanian up to early Middle Turonian. In a traverse/section east of the Odhium village, the transition from the concretion-devoid shale to the concretion rich shale ~50 m below the Uttatur/Tricinopoly contact marks the ammonoid-based Cenomanian/Turonian boundary. The first-order MFS is marked above the concretion bearing ammonoid rich shales ~20 m below the top of the Karai Formation. The unit is here precised from Early Aptian–mid-Middle Albian first/second-order MFS start determined in Jaisalmer to early Middle Turonian top determined in Cauvery. The top of the Uttatur in a few sections is also interpreted as a second-order SB on account of geological gap and disconformity.

Since the rift/drift transition at the first-order SB, the Indian plate inclusive of Madagascar spread mostly due west, and while doing so, it, presumably, met acute space constraints in the India–Africa sector. Thus began rifting and associated transform sliding in the India–Madagascar region at the first-order Barremian/Aptian boundary SB. The extensional event culminated with spreading at the south extreme in the vicinity of the Marion plume at the first-order MFS, and the basin thereafter registered graded uplift or tectonic inversion. Most of the Late Turonian duration is thus absent both in the outcrop and well geological record. Gap is also suspected at the Cenomanian/Turonian transition. The Kulakkalnattum Formation on similar dinoflagellates (Khowaja-Ateequzzaman and Garg 2002; Prasad and Phor 2009) is considered coeval to the subsurface Bhuvanagari Formation of mostly Coniacian age as part of a second-order TST. The Late Cenomanian–Middle Turonian later half of the first-order TST is found rich in ammonoids (*Pseudoaspidoceras footeanum, Eucalycoceras pentagonum, Romaniceras ornatissum,* etc.) in Cauvery Cretaceous that suggests maximum transgression and bathymetry. The Anaipadi Formation makes a second-order intra-Coniacian TST with *Kossmaticeras theobaldinum* bearing MFS, while the overlying Saturbhujam sands are interpreted as the following second-order RST up to or near the Santonian/Campanian boundary SB. The Silakuddi Formation disconformably overlies the second-order SB. The terminal part of the unit registered maximum bathymetry with intra-Campanian MFS with rare marker *Karapadites karapadense*. The overlying Kaller conglomerates are ranged in the second-order RST interval. The Kallankuruchhi Formation limestones make the last second-order TST with MFS above a Gryphaea abundant coquina marker. Rare *Eubaculites vagina* is present near the MFS in all the subbasins. The terminal arenaceous limestone with drastically reduced marine body fossil content in density, diversity and frequency makes the early part of the terminal Cretaceous second-order RST which is followed-up by the Otakkovil-Kallamedu units as the later part of the terminal Cretaceous RST.

4.7.2 K-G Basin

The Golapille Formation is shown pre-Albian to Early Albian in Fig. 3, Late Jurassic–Barremian in Fig. 6, and Late Jurassic–Early Cretaceous in the text (Mehrotra and Aswal 2003), and Aptian–Albian in one of the wells (Mehrotra et al. 2002); Neocomian–Aptian (Prasad et al. 1995), and Neocomian (Prasad and Pundeer 1999). The base of the disconformably overlying Raghavapuram Shale Formation in the three studied wells (Mehrotra and Aswal 2003) is shown in the terminal Jurassic to Barremian, while Early Cretaceous on the surface. The Raghavapuram Formation is assigned Aptian–Albian in text, and Late Aptian to Cenomanian in table (Prasad et al. 1995), and Barremian to Early Albian (Prasad and Pundeer 1999). Obviously, all the above highly variable ages are not correct, and speak about the limited usefulness of the palynofossil groups used. The Golapille Formation (inclusive of or together with Budavada/Nellore/Krishna/Bapatla units) is here age bracketed between the first-order late Middle Oxfordian MFS and the following first-order Barremian/Aptian boundary SB. The overlying Raghavapuram Formation is age bracketed within the first-order Barremian/Aptian boundary SB and the first-order early Middle Turonian MFS. The Tirupati Formation is ranged between the first-order early Middle Turonian MFS and the second-order intra-Maestrichtian MFS. The base of the Razole (Volcanic) Formation is placed at the Early/late Maestrichtian boundary second-order MFS, while the unit makes the second-order RST in the K-G basin. The presence of the marker ammonoid *Eubaculites vagina* in the Mahadek Formation of Assam suggests widespread transgressive event in the east sector.

4.7.3 Other Indian East Coast Mahanadi, Purnea, Bengal, Assam and Andaman Basins

In the Mahanadi and Bengal Gondwana graben basins, the first-order Late Permian–Anisian TST interval non-marine sediments are disconformably overlain by the first-order terminal Middle Oxfordian to the second-order mid-Middle Albian MFS interval non-marine sediments with a large gap of the Toarcian–late

Middle Oxfordian first-order TST interval. The sediments include the interval of the episodic Rajmahal volcanism. After another gap from the mid-Middle Albian second-order MFS to the second-order Santonian/Campanian boundary SB, follows the Campanian–Maestrictian shallow marine succession. In the Assam–Arakan region, the mega-sequence includes shallow marine first-order late Middle Oxfordian–Barremian TST interval sediments along with the disconformably overlying sediments of the Campanian–Maestrichtian–Paleocene interval. In the Purnea basin, the sediments belonging to the mega-sequence are non-marine of the Late Triassic–Pliensbachian first-order RST interval.

4.7.4 The Intra-Albian and Intra-Turonian MFSs in the Indian Basins

In Spiti, the Giumal/Chikkim lithostratigraphic boundary, between the underlying shales of the Giumal unit and the overlying limestones of the Chikkim Formation at BCP 21/02 (Bertle and Suttner 2005), is interpreted in the terminal part of the Late Albian as the second-order SB. After the initiation of spreading between India and Australo-Antarctica, the graded tilting down at the passive hind margins in the east sector basins is here suggested to have stopped and reverted to tilting up, however at a slower rate than during the earlier Late Permian–late Middle Oxfordian interval. The rate of the eustatic rise was greater than the rate of tilting up with resultant increase in bathymetry in the east sector Indian basins. However, the tilting down, combined with eustatic rise, prompted rapid increase in bathymetry in High Himalaya Spiti–Zanskar basins during the Aptian–early Middle Turonian first-order TST. This is attested by the condensation of the second-order TSTs of this first-order TST interval in Spiti–Zanskar. In Kachchh, and Jaisalmer of the west sector, the slow sedimentation is however found limited to the Aptian–mid-Albian duration. This interval is ammonoid bearing in nearly all the Indian basins in east, west and north sectors. Further up, in the remainder Cretaceous interval, ammonoids are present only in the Cauvery basin, and only in the vertical vicinity of the mega order MFS as well in the Narmada and Saurashtra basins. The intra-Turonian MFS on the basis of ammonoids is thus best determined in the Cauvery and Narmada basins (Figs. 4.6 and 4.10) in the terminal part of the Karai Formation between the *Pseudoaspidoceras footeanum* bearing beds below and *Romaniceras ornatissum* being ammonoid rich sediment interval above of shales/clays with nodules in a section east of the Odium village (Ayyasami and Bannerjee 1984), and at the Hardground II between the Chirakhan and coralline limestone units at locality L3 in the Man river section near Khor (Tripathy and Lahiri 2000). The ammonoid succession in lower Narmada suggests Late Cenomanian–Santonian interval with intra-Turonian MFS as above. The MFS is also suggested in Saurashtra basin through *Proplacenticeras stantonii* in the middle explicitly marine part of the

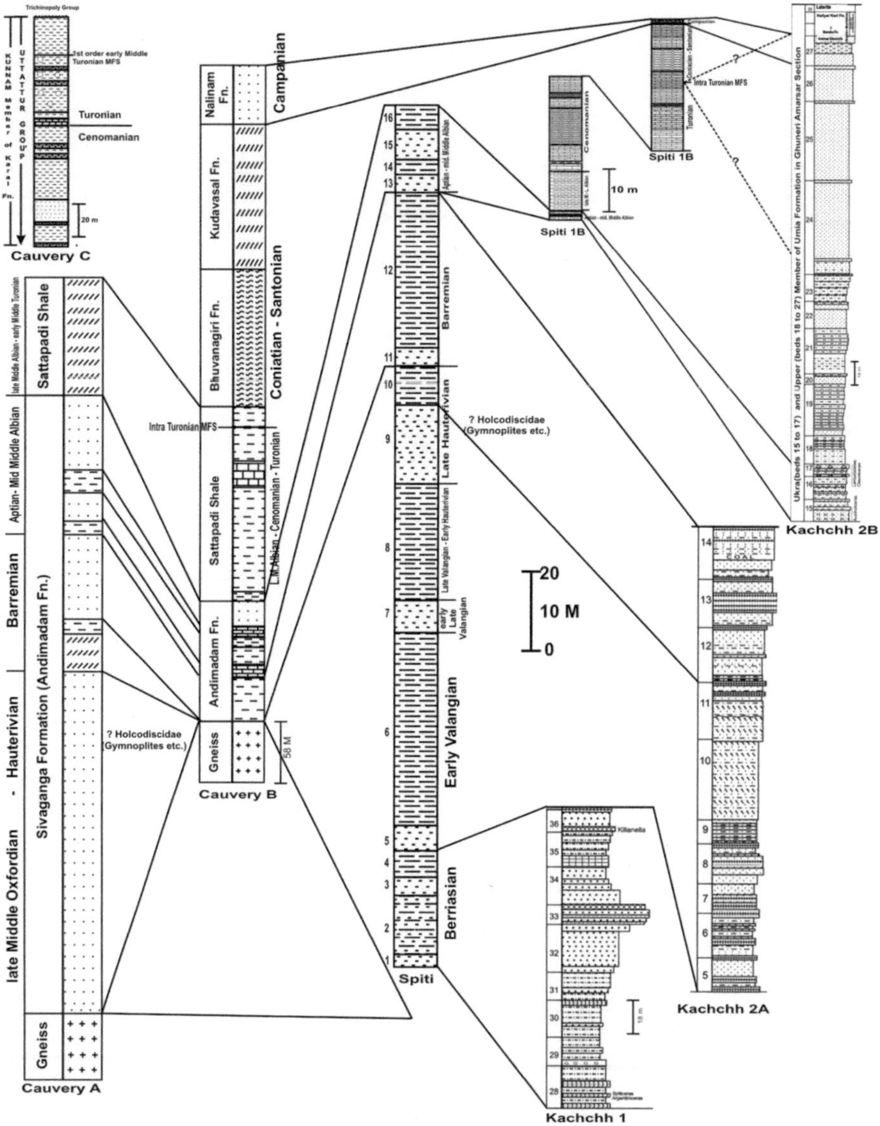

Fig. 4.10 Suggestive intra cretaceous correlations based on ammonoid/foram/palyno fossils and sequence surfaces among the Indian Cauvery, Spiti and Kachchh basins

Wadhwan Formation (Chiplonkar and Borkar 1971). The MFS is suggested at the top of the bed 23 of Kachchh.

4.7.5 Demarcation of the First-Order Intra-Cretaceous MFS

The Karai Shale in outcrops and the Sattapadi shale in wells make the middle and later part of the first-order TST with first-order MFS in the terminal part of the Karai Shale. In the K-G basin, the top of the Raghvapuram shale is interpreted as the intra-Cretaceous first-order or major MFS of early Middle Turonian ammonoid-based Turoniense Zone, planktonic foraminfer Helvetica Zone / benthonic foaminifer Turonainensis Zone nannoplankton Gartneri Zone as constrained in the Cauvery and other basins.

4.7.6 Correlations Among Cauvery, Spiti, and Kachchh Successions

The top of the ammonoid scarce bed CHG 9 of Spiti in the north sector is correlated to the *Pascoeites–Gymnoplites* bearing basal conglomeratic sands of the Golapille and coeval units of the east sector as the terminal second-order MFS of the Toarcian–Barremian first-order sequence of the Late Hauterivian Angulicosta/Ohmi zonal boundary age (Figs. 4.7 and 4.10). The event is widespread across the Neotethys, and additionally marks an anoxic event on ETM. The base of the CHG 13 is correlated to the base of the Dalmiapuram unit as the first-order intra-Cretaceous SB, and based on ammonoids in bed CHG 13 as Early Aptian. The top of bed CHG 13 in Spiti is correlated to the top of the Dalmiapuram unit in Cauvery as a second-order MFS of mid-Middle Albian age based on ammonoids in bed CHG 15. The lithostratigraphic boundary between the Giumal and Chikkim units in Spiti is integrated to the base of the Karai/Sattapadi Shale in Cauvery wells as a second-order SB of tentative Middle/Late Albian boundary or basal Late Albian age. The tentative age is supported by the suggestive start of the Cenomanian above the bed BCP 13/02 as provisionally determined in the Chichim section of Spiti within ~5 m of the lithostratigraphic boundary (Bertle and Suttner 2005).

The first-order intra-Turonian MFS is already discussed. The lithostratigraphic boundary between the normally and flaser bedded limestone of suggestive terminal Turonian age is correlated to the disconformable boundary between the Kullakalnattam Formation and the Anaipadi Formation as a second-order SB. The ~6 m thick marker flaser bedded limestone interval is interpreted as the second-order TST of mostly Coniacian age, and is correlated to the major part of the Anaipadi Formation. The lithostratigraphic boundary between the Chikkim Limestone Member below and the Chikkim Shale Member above is marked by a hard ground as a thin condensed TST of Early Santonian age. The top of the limestone unit immediately underlying the hard ground makes another second-order SB which is correlated to the disconformable boundary between the Anaipadi Formation below and Sillakudi Formation above. The top of the hard ground marks a significant Late Santonian gap. The Chikkim Shale is correlated to the Sillakudi Formation of Campanian age, which is here correlated to the gap above the Sillakudi Formation. The intra-Campanian second-order MFS is determined in the upper part of Sillakudi Formation followed-up by the Kaller Conglomerate, however, it is difficult to determine the same MFS within the Chikkim Shale unit. In the Chichim section, the terminal second-order sequence equivalent to the Kallankuruchhi Limestone–Kallamedu Sandstone T/R couplet is absent, and perhaps removed by erosion, however, the terminal Maestrichtian–early Paleogene part of the Late Permian–early Paleogene mega sequence is present in the distal Zanskar part of the Spiti-Zanskar basin.

The correlation between the Kachchh and Spiti Cretaceous successions is good in the ammonoid bearing Giumal Formation and the combine of the Basal–Ghuneri–Ukra units of Kachchh up to the mid-Middle Albian second-order MFS (Fig. 4.10). There after, the relatively slow sedimented fine textured remainder late Middle Albian to early Middle Turonian part up to the first-order MFS suggests similar development, and so the correlation is relatively easier, although only suggestive. The lithological succession farther up during the first-order RST is highly contrasting in the two basins, and the extension of the second-order surfaces from Spiti to Kachchh attempted here is at present only tentative as a working model for future precision (Figs. 4.10 and 4.11).

4.7.6.1 Suggestive to Speculative Interpretation of Significant First- and Second-Order MFSs In Kachchh

The Kachchh column includes three significant intervals within the Upper Member of the Umia Formation in between the well-dated Ukra Member below and Deccan traps above which allow an alternative somewhat speculative correlation. The oldest is the relatively fine textured shale dominated interval of the beds 18–23, the middle is the almost equally thick coarse sandy part of beds 24–27b, while the youngest—the third much thinner interval of shales inclusive of green glauconitic shales at the top with FT age of ~89/92 ma (Krishna 1983b). The FT age is considered reliable in view of the other tested FT ages which were additionally ammonoid constrained (Fig. 1.13). Since 1983 radiometric ages have undergone continuous revision. The FT age now suggests a possible range of the Turonian–Santonian interval of ~8 my which may include two significant MFSs of the early Middle Turonian first-order MFS and the second-order terminal Santonian MFS close to the Santonian/Campanian boundary. The speculative interpretations are also supported by the homotaxiality between the guide fossil-lacking post-Albian exposed Kachchh and guide fossil-rich Spiti Cretaceous. The boundary between the above mentioned shale/sand intervals as the top of the glauconitic shales are interpreted as significant MFSs. In context of the FT age range, the two surfaces, the older between the beds 23 and 24, and the younger the top of the bed 27 are here related to the early Middle Turonian first-order MFS and the terminal Santonian second-order MFS.

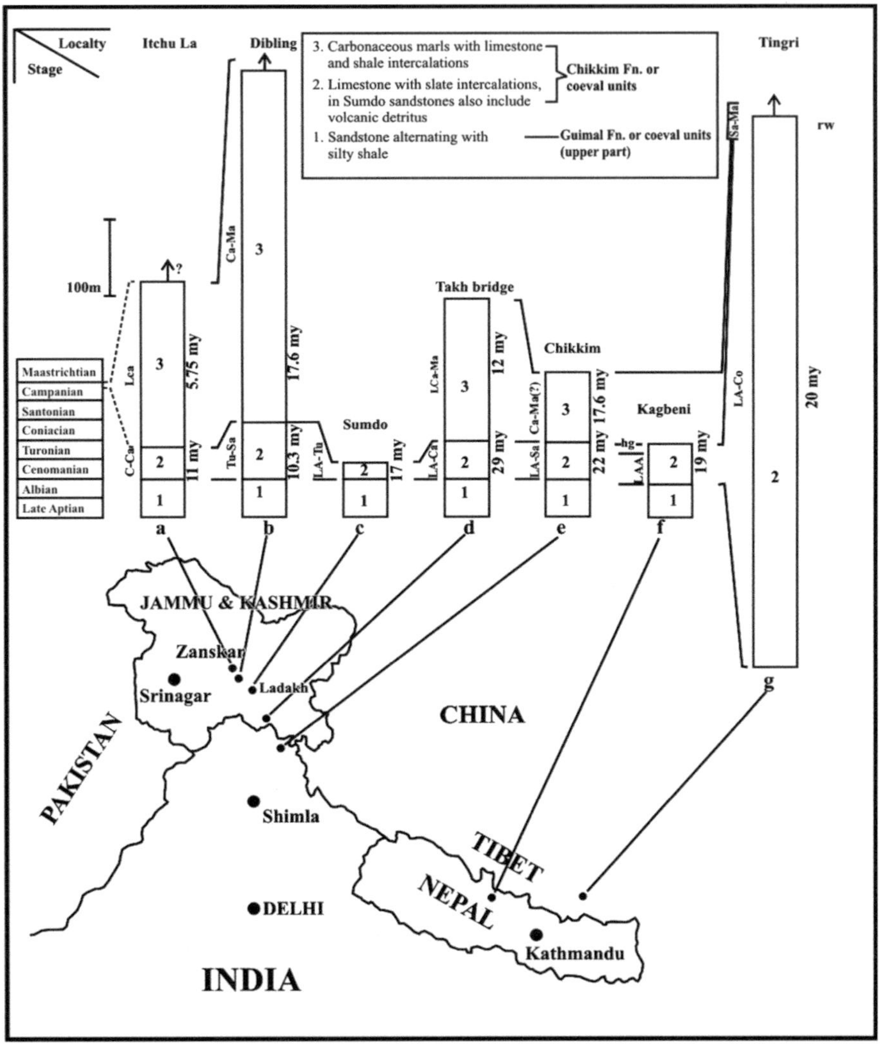

Fig. 4.11 Interpretation of sedimentary condensation in the lithostratigrapic columns of Chikkim Formation between Ladakh–Zanskar and Nepal–South Tibet during the Cenomanian–Campanian interval modified after Bertle and Suttner (2005)

4.7.7 Second-Order Surfaces in Between the Mid-Middle Albian MFS and the Above Interpreted First-Order MFS in Spiti and Kachchh

Within the Lower Member of the Chikkim Formation (Bertle and Suttner 2005) is interpreted a second-order SB in BCP 7–BCP 3 limestone interval of Reichelli Zone age near the transition of the Early/Middle Cenomanian. This is suggestively corresponded to the bed 20 of Kachchh. The one to two m thick BCP 2 flaser bedded limestone is interpreted as another second-order TST of the Middle/Late Cenomanian boundary Cushmani Zone age, which is suggestively corresponded to the bed 21 of Kachchh. The beds 30 and 29 of the Lower Member of Chikkim Limestone are interpreted as the first-order TST of the terminal Cenomanian–early Middle Turonian Archaeocretacea Zone–Helvetica Zone interval age, and are correlated to the bed 23 of Kachchh.

4.8 Salient Other Aspects of the Indian Mesozoic

4.8.1 Evaluation of the Lithobiostratigraphic Data of Jaisalmer and Barmer Basins

The subsurface Karampura Formation of the Jaisalmer basin is ranged within the first-order Late Permian–terminal Anisian TST of the GTM sequence framework. The disconformably overlying Lathi Formation (inclusive of the 'Shumerwali formation' of wells) in both the basins is age bracketed in the first-order Ladinian–Pliensbachian RST interval. The conformably overlying Jaisalmer Formation is ranged in the first-order Toarcian–late Middle Oxfordian TST interval. The disconformably overlying Bhadasar Formation and Parihar Formation in Jaisalmer and coeval Sarnu Formation in Barmer are age ranged in the first-order terminal Middle Oxfordian–Barremian RST interval. The Abur Formation in Jaisalmer is précised between the first-order Barremian/Aptian boundary SB and the second order mid-Middle Albian MFS. The Goru Formation in wells is inclusive of the Abur Formation is ranged up to the first-order early Middle Turonian MFS. The Parh Formation in Jaisalmer is ranged from Coniacian to Santonian. The Cretaceous planktonic foram succession in Jaisalmer wells is much the same as in Cauvery and Spiti. The Fatehgarh Formation disconformably lying over the Sarnu Formation in Barmer is age bracketed between the Santonian/Campanian boundarysecond-order SB to either the Early Maestrichtian second-order MFS or to intra-Paleocene and may be held homotaxial to either the Lameta Formation or the combination of Lameta and Deccan Trap of Narmada. On the lower side in absence of any guide fossils, it may even extend down to the intra-Turonian MFS or even farther below. The unit with extension in Paleocene in Barmer, lately, has assumed importance as a producing hydrocarbon reservoir. Igneous dykes and sills are found common in the unit.

In the rather marginal Barmer and Bikaner–Nagaur basins, the Late Permian–intra-Paleocene mega-sequence is either entirely absent or at least a major part of Cretaceous is found missing (Bladon et al. 2015). This in turn suggests extensional uplift related to the Barremian/Aptian boundary volcanism all round the Indian plate in east, north, west and south sectors.

The magmatism through both intrusive and extrusive expression is shown from near Barremian/Aptian boundary first-order SB to the intra-Paleocene first-order SB. The Aptian–intra-Paleogene first-order sequence began as also closed with igneous events. The basal one here is linked to the Rajmahal event. The Rajmahal event of the east sector is already evidenced well in the north, west, and even south sectors, and so its expression in the Barmer basin is well in accord with the tectono/igno-stratigraphic framework all around the Indian plate. The terminal phase with post early Middle Turonian Marion/Reunion plume expressions is the additional regional expression in Barmer. The same accords well to the India/Madagascar separation from ~92 to ~84 ma. The Ar40/Ar39 ages of the terminal event, however only range from ~70 to ~63 ma. In this context, the ONGC (1993) stratigraphic table with multiple levels of volcanic rocks within the Fatehgarh unit, may depict all the successive three volcanic events (Rajmahal, Marion, and Reunion). The evidence of the Aptian–early Middle Turonian TST interval as yet has not emerged. In tectonostratigraphic context, the absence of evidence differentiates the Barmer stratigraphic framework from the neighbouring somewhat distal Jaisalmer basin.

4.8.2 Comparative Evaluation of the Jurassic Member Units of Kachchh and Jaisalmer

The non-marine Lathi Formation is corresponded to the non-marine Nirona Formation exclusive to Banni and Nirona wells in the Ladinian–Pliensbachian age bracket. The Lathi/Jaisalmer contact is broadly equated to the Nirona/Patcham boundary in Banni well of Pliensbachian/Toarcian boundary first-order SB. The Hamira Member is correlated to the beds 1–28 of the LSM (=early and middle part of the Dingy member) of the Patcham Formation of Toarcian (Fig. 4.12). The sand dominating Joyan Member is held as the stratigraphical equivalent of the bed 29 of the LSM of the broad early Aalenian age. The early sandy part of the Fort Member is corresponded to the sandy beds 30–36 of the terminal part of the Dingy member of the LSM. The upper limestone part of the Fort Member is considered stratigraphically equivalent to the members Am to Em at Jhura of the ULM of the Patcham Formation of Bajocian age. The early sandy part of the Badabag Member is corresponded to the upper part of the LSM (beds 48–56, i.e. BCSm) at Patcham of the broad Early to mid-Middle Bathonian age. The upper limestone part of the Badabag Member is corresponded to the ULM at Goradongar (combine of GYFm + GSm + RLm) as to the ULM at Jumara of late Middle to close of Bathonian. The terminal Kuldhar Member of the Jaisalmer Member is considered stratigraphically homotaxial to the Chari Formation inclusive of the terminal Jhadasa Member at Kantkote of Early Callovian Madagascariensis Zone to the late Middle Oxfordian Orientalis Zone Schilli Subzone age bracket.

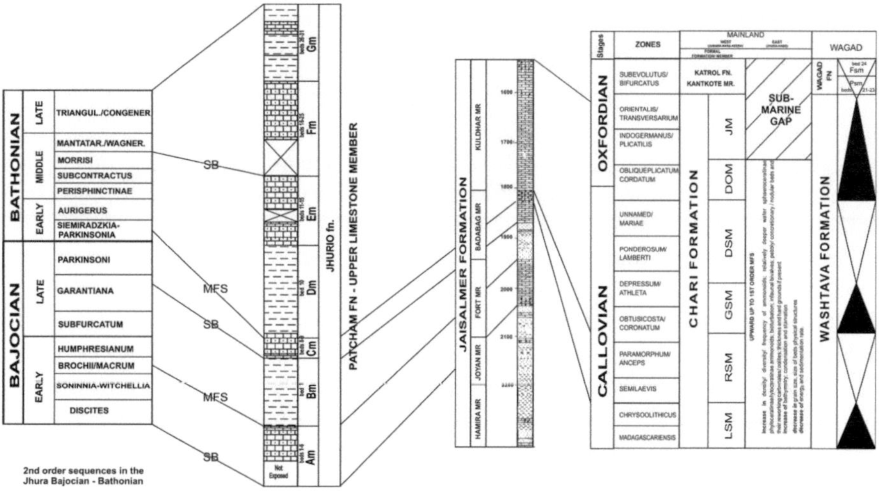

Fig. 4.12 Intra to interbasinal correlations vis-à-vis first- and third-order sequences during Callovian–Oxfordian interval between Kachchh and Jaisalmer basins

The sediment interval 1 of the Bhadasar Formation is corresponded to the terminal Middle Oxfrodian Rotoides Zone to the early part of Late Kimmeridgian Katrolensis Zone age bracket up to the second-order MFS in Late Kimmeridgian. The sandy unit 2 is corresponded to the second-order terminal Kimmeridgian upper part of the Katrolensis Zone. The units 3 and 4 are broadly corresponded to the second-order TST at the Pottingeri Zone / Virgatosphinctoides Zone boundary or within the Virgatosphinctoides Zone up to the bed 19 of Ler section. The sandy sediment interval 5 is corresponded to the sandy second-order RST interval of beds 20–26 up to the close of the Virgatoosphinctoides Zone. The sediment intervals 6–8 of the Bhadasar Formation are broadly correlated to the Late Tithonian–Early Berriasian up to the second-order MFS of the Early Berriasian sequence.

The Parihar Formation of the Jaisalmer basin is considered stratigraphically homotaxial to the Ghuneri Member of the Umia Formation of Kachchh. The relatively fine textured dominantly shaly sediment interval 1 of the Parihar Formation is correlation to the terminal part of the second-order Early Berriasian TST of bed 28 up to the top of a thin green glauconite bed. The sandy sediment interval 2 of the Parihar Formation is correlated to the dominantly second-order RST sandy beds 29–32 of Kachchh of suggestive late Early Berriasian age. The sediment interval 3 of the Parihar Formation is correlated to the second-order TST interval of beds 33–35 up to the top of the *Kilianella* bearing grits of Late Berriasian age. The sandy shale bed 4b is suggestively corresponded to the second-order TST beds 5–7 of the Early to basal Late Valanginian Verrucosum Zone age. The sediment intervals 4c–9 are only tentatively corresponded to the beds 8–14 of Kachchh.

4.8.3 Need of Formal Indian Working Groups for Each International Stage

The micropaleontological tools are serviceful, if not stretched outside their respective utility domains, and also if applied in conjunction with other tools instead of application in isolation. In this context, looking to the rapid progress in Europe in the area of stratigraphic refinement, there is dire need of formal working groups with expertise in diverse groups of guide fossils, tools and parameters for each Indian Phanerozoic stage with regular intragroup meetings and monitoring of progress and target realizations. The Indian geological record is splendidly endowed by nature, and over and above there have been early foundation studies by the Europeans, yet we have lacked in realizing appropriate degree of refinement in the Indian geological record, the basic requisite to any focused fundamental or applied geological investigation.

4.8.4 Sequence Stratigraphy-Based Revised Ages to Formations and Members of the Kachchh Jurassic

The high-resolution scale and sequence framework developed in Chaps. 2 and 3 in combination, has allowed exceptionally well-refined intrabasinal correlation, and precise age assignment to almost every cm of sediments throughout the length and breadth of the basin. In most publications, spread over the last many decades, ages have been assigned to the units along with intrabasinal correlation, without citing concrete evidence. Also, ages of the units differed in different works. Even in the same publication, ages in the text and tables are found contradictory. Cited here by way of example are three publications (Fursich et al. 1994; Alberti et al. 2012; Pandey et al. 2012a, b).

Having had a summarized treatment as above on the status of the integrated biostratigraphy in the Indian Mesozoic, the age discrepancies of the Kachchh Jurassic units are exampled in context of the several fold better ammonoid ages. Accordingly, the ages given earlier stand revised. Only gross discrepancies of academic interest are discussed in some detail for the education of students and young researchers who are carried away by renowned names and their unintended lack of objectivity.

In this context, the age data of Rajnath (1942), Agrawal (1957), Biswas (1977, 1993), Deshpande and Merh (1980), Pandey and Dave (1993), Alberti et al. (2011, 2012, 2013), Prasad (2011), Prasad et al. (1995), Mehrotra and Aswal (2003), Mehrotra et al. (2002), Pandey et al. (2012a, b), Raju (2011), Rai and Jain (2012, 2013), Prasad and Phor (2009), Raju (2008, 2011), and many other works stand revised and updated.

Mention is made in particular of Biswas (1993), Mehrotra and Aswal (2003), Mehrotra et al. (2002), Pandey and Dave (1993), Raju (2011), Prasad (2011) in view of their affiliation to ONGC, and also inclusion of their versions of interpretations, ages and geological models in the official documents/dockets circulated to hydrocarbon companies for exploration objectives.

4.8.4.1 On Ages and Correlations of Biswas (1977, 1993 and 2012) (Figs. 2.19–2.26)

Revised Ages of the Oldest Sediments as also the Oldest Jurassic Ammonoid from Kachchh

The oldest exposed sediments in the Kachchh basin are known from Patcham and Khadir of 'Island belt', and Jhura of East Mainland, and assigned by Biswas to Bathnoian. Later the earliest ammonoid-based age was extended down to at least Late Bajocian from Patcham region (Jaitly and Singh 1983). The oldest ammonoid now comes from Early Bajocian in Sadhara region of Patcham from the basal most bed there through recent record of Early Bajocian *Calliphylloceras* (Pandey et al. 2013b). The so-called *Leptosphinctes* has been revised as a Perisphinctinae and the level reassigned to early Middle Bathonian Progracilis Subzone. Additionally, nannoplankton-based Toarcian–Aalenian age has been recently suggested to the Dm of LSM at Dingy hill of Kala Doongar (Rai and Jain 2012).

'Jhurio Formation'

It is an abandoned synonym of the upper calcareous part of the Patcham Formation that has been differentiated as ULM. The unit is here understood as ULM at Jhura. It was assigned earlier by Biswas as Middle Bathonian to Callovian. The unit is here revised as Early Bajocian to Late Bathonian in sequence stratigraphic context detailed in Chap. 3. The youngest sediments of the unit on ammonoid evidence turn out to be Late Bathonian age instead of Callovian. Thus, both the base and top of ULM at Jhura are lowered down.

'Jumara Formation'

It is an abandoned synonym of the Chari Formation at Jumara, assigned earlier by Biswas as Callovian to Late Oxfordian. It is here revised as from Early Callovian to early Middle Oxfordian. Late Oxfordian is not present any where in Mainland Kachchh, and instead is included in the large submarine gap between Chari and Katrol Formation.

'Jhuran Formation'

The unit earlier assigned as Early Kimmeridgian to Early Neocomian by Biswas is précised from latest Early Kimmeridgian to early Late Valanginian, since early to mid-Early Kimmeridgian interval also all over the Mainland Kachchh is included in the submarine gap.

'Bhuj Formation'

Biswas assigned Neocomian to Santonian age to the unit in Mainland Kachchh. The unit is here included in Umia Formation, and also extended to the entire basin. It undisputedly starts as early as Late Oxfordian–Early Kimmeridgian in Wagad where 'Gamadu Sandstone member' (GSm) is held as the oldest expression of Umia Formation, rightly so in the proximal most exposed part of the basin. It may be noted that Biswas himself (1993, p. 165) suggested resemblance to near physical identity of the 'Washtawa formation' with 'Jumara formation', Kanthkot Member with 'Jhuran formation', and GSm with 'Bhuj Formation'. The unit has its earliest record in Mainland at Jhuran of Late Tithonian. Umia Formation is well known to rapidly prograde from margin to basin.

'Kaladongar Formation'

It is abandoned in favour of Patcham Formation. Biswas tentatively based on bivalves dated it as Bathonian. Here it is ranged from Early Toarcian to as young as early Late Bathonian in context of Toarcian–Aalenian coccoliths in the early part of its oldest informal member (Dm) and Late Bathonian ammonoids in the basal part of the immediately overlying ULM in different domes of the 'Island belt (Patcham in the west to Chorar in the east).

'Goradongar Formation'

It is suppressed in favour of the senior synonyms Patcham Formation (for early 3 members of 'Goradongar Formation') and Chari Formation (for the youngest member). Biswas ranged it from Early Callovian to Oxfordian. Here it is ranged from late Middle Bathonian to early Middle Callovian on the basis of late Middle Bathonian ammonoids in its basal informal member (GYFm) and early Middle Callovian ammonoids in sands in Mainland that are considered correlative to sands in the youngest informal member of the unit in discussion.

'Khadir Formation'

It is abandoned in favour of its senior synonyms Patcham Formation for the early part and Chari Formation for the later part. Biswas assigned it from Bathonian to Oxfordian, while here it is revised as from at least Early Bajocian to late Middle Oxfordian.

'Washtawa Formation'

The 'Washtawa formation' is abandoned in favour of the Chari Formation, also it does not include any Late Oxfordian as thought by Biswas. Instead, the latest Middle Oxfordian marks the base of the next younger lithostratigraphic unit in Wagad of Kantkote Member that in turn is included as the Basal Member of the Katrol Formation with exclusive presence in Wagad. 'Washtawa Formation is thus here précised in Early Callovian to late Middle Oxfordian in view of the presence of Early Callovian ammonoids in the early part and late Middle Oxfordian ammonoids in the terminal part of the 'Washtawa formation' in the type locality (Krishna unpublished).

'Wagad Formation'

It is suppressed in favour of the physically near identical undifferentiated Umia Formation in East Mainland and Wagad. Biswas assigned it to Late Oxfordian–Early Cretacous age bracket. Here, based on ammonoids, it is ranged from the latest Middle Oxfordian to Early Kimmerdgian.

Similarly, age assignments by Biswas to a few of his member units in different parts of the basin have been found grossly incorrect and imprecise.

Highly Diachronous Correlation Tried Unsuccessfully with Trigonids

T. ventricosa or for that matter any *Trigonia* can not be claimed to define the base of the Cretaceous or for that matter any chronostratigraphic boundary as wrongly claimed by Biswas (1993). Trigonids are facies fossils and have long ranges with recurrence in time and space, particularly in cyclic sediments. The base of the Cretaceous is on ammonoid evidence included in the glauconite beds of the Basal Member of Umia Formation in West Mainland. The famous trigonid bearing sediments in West Mainland are distinctly younger in the same section, although other not that famous trigonid beds have recurred a few times in the same section. It is strange that Biswas (1993) criticized the correlations based on trigonids, yet only a few pages later in the same work made the same folly when he, using trigonids, gross wrongly correlated the Wagad and Mainland sections, otherwise having a time difference of more than 25 my (Krishna et al. 1998).

Faulty time Correspondence of Mainland and Wagad Units

Broad time correlation of the Mainland succession of the abandoned 'Jumara', 'Jhuran' and 'Bhuj' units of Biswas to units in Wagad (in Biswas 1993, p. 165) has been found grossly wrong. The top of Dhosa Oolite in Mainland Kachchh is early Middle Oxfordian and not any younger while top of the 'Khadir formation' and of 'Washtawa formation' is late Middle Oxfordian, i.e. younger by several subzones. Age equivalence of 'Jhuran formation' of Mainland to Kantkote Member of Wagad emphasized upon by Biswas is grossly incorrect. Kantkote Member ranges from Latest Middle Oxfordian to Early Kimmeridgian where as 'Jhuran formation' only begins in Late Kimmeridgian and goes up to Late Valanginian. Nothing to speak of their being time correlatives, there is absence of even any small overlap between the two units of Biswas. 'Gamadu member' of Wagad similarly has been gross incorrectly suggested as time correlative of 'Lower Bhuj member'. The upper age limit of 'Gamadu member' should not be younger than Kimmeridgian. On the contrary 'Lower Bhuj member' in its stratotype in the neighbourhood of Bhuj township maynot start earlier than Late Ttithonian ranging up to Barremian. There is undisputed gross facies similarity between 'Washtawa formation' and Chari Formation, that is, why 'Washtawa Formation' is abandoned and succession included as Chari Formation in Wagad. The Kankote Member is included in the Katrol Formation and 'Gamadu formation' included in Umia Formation in the integrated unified lithostratigraphic framework.

On Intrabasinal Correlations of Biswas

In his intrabasinal correlations, Biswas (1977, 1993 and later) could use merely 5–6 correlative time lines or broad intervals. But one or two, all of them are found here diachronous, e.g. the trigonid-based correlation between Wagad and Mainland which actually deferred by ~25 my. Wagad *Trigonia* is Early Kimmeridgian while Jara *Trigonia* is Late Berriasian. So it is found distinctly diachronous physically similar Late Early to Early Middle Callovian RSM and Middle Oxfordian sands at Gangta above the marker Kakindia Limestone. Even the correlation between the Dhosa Oolite, i.e. the top of Chari Formation in Mainland of early Middle Oxfordian with Kankote ammonoid bed at the top of the Chari Formation in Wagad of late Middle Oxfordian is incorrect by 3–4 subzones. The actual bed of Wagad correlative to Mainland Dhosa Oolite is ~65 m below the top of the Chari Formation in Wagad. The correlation of the so-called 'Athleta zone' sands of Jumara as suggested to RSM in Jhuran is untenable with difference as well of a few subzones. Jumara level is decidedly post Middle Callovian and Jhuran level for sure older than late Middle Callovian.

Once the ages of the formations and members are wrong, the intrabasinal correlations based on such ages obviously involved compounding of errors, as only gilempsed above with only a few examples.

4.8.4.2 On the Ages of Fursich and Coworkers

Fursich and coworkers in most of their publications excepting those of the last 5 years, spread over last two decades or more, included ages to all the units along with intrabasinal correlation, yet hardly ever provided evidences, their own or others, either of the ages of the units or the basis of their intrabasinal correlation. Also, ages of the units differed in different works without any explanations to doing so. Even in the same publication, different ages are indicated in the text and table. Cited here by way of example are three publications (Fursich et al. 1994; Alberti et al. 2012, Pandey et al. 2012a, b) as testimony to the above comments such arbitrary age assignments (Figs. 3.21, 3.22, 3.23, 3.24 and 3.25)

Intra-LSM Ages and Correlations (Early Toarcian to Mid-Middle Bathonian)

Starting from the oldest unit, Fursich's group assigned Dm of Dingy and Kuar bet and CLm of Sadhara of LSM first to Late Bajocian in 1994 and later to basal Bajocian, however, on recent coccolith evidence (Rai and Jain 2012), it is age bracketed in the Toarcian–Aalenian span as repetitively anticipated earlier (Krishna 1994; Krishna and Pathak 1994, and onward). KSm is shown indecisively against mid-Bajocian while in the present work in sequence stratigraphic context, it is best assigned to Early Bajocian. BCSm includes LPR in its upper part instead of being followed by LPR in succession. The abandoned 'Sadhara units' are rightly shown against KSm and BCSm.

Intra-ULM Ages and Correlations (Bajocian–Bathonian)
Badi limestone', i.e. Am at Jhura is placed against Late Bajocian and correlated to basal fossiliferous lime/marl/clay of BCSm at Kaladoongar and Sadhara, while here it is dated as Early Bajocian and correlated to the basal fossiliferous limestones of KSm both at Kaladongar and Sadhara. These limestones have recently yielded the gastropod *Heterohelix* (Pandey et al. 2012a, b). GYFm of late Middle to early Late Bathonian age is wrongly placed in the table against Early Bathonian both in Mainland and Patcham, and yet in the text on page 183 is indicated to include late Middle to early Late Bathonian ammonoids *Micromphalites* and *Gracilisphinctes*. GYFm in the present work is included in ULM as extension of ULM from Mainland (Habo, Jhura, Nara and Jumara). GSm is either placed against late Middle or early Late Bathonian although it is definitively younger as mid-Late Bathonian in view of the early Late Bathonian ammonoids in the immediately underlying GYFm. GSm is here extended to Jhura as Fm (same as the abandoned 'Purple sandstone'). RLm is rightly correlated to Gm (same as the abandoned 'Sponge limestone') since the former unit is extension of the latter from Mainland.

Intra-Chari Formation Ages and Correlations up to RSM (Early to Early Middle Callovian)
The LSM (inclusive of 'Keera golden oolites') is only vaguely shown against Early Callovian interval without precise time lines while the same unit is precisely ranged here from Transitorius Subzone to Diadematus Subzone of early to mid-Early Callovian. The RSM is broadly equated to 'Anceps beds' of Spath (1927–33) by Fursich and coworkers, in turn, broadly to Middle Callovian. Based on included ammonoids, it is precisely ranged from late Early Callovian Semilaevis Subzone to early Middle Callovian Paramorphum Subzone in the entire basin. 'Modar hill sands' are held extension of the RSM at Gora doongar. In Khadir and Bela the 'Gadadha sandstone' (GSm) is considered extension of LSM and RSM combine of Mainland, and precisely ranged from Early Callovian Madagascariensis Subzone to early Middle Callovian Paramorphum Subzone, while Fursich and coworkers have indicated the top of the 'Gadadha Sandstone' roughly in the middle of GSM. The GSM, a third-order TST of a younger sequence can in no way have any age overlap with the 'Gadadha Sandstone' third-order RST of the immediately older sequence.

Post-RSM Chari Formation Ages and Correlations
The GSM is always assigned by Fursich and coworkers exclusively Late Callovian age, whereas it actually begins in Late Middle Callovian and also terminates in early Late Callovian. It is strange that the top of the 'Bambhanka/Gangta beds' have been wrongly shown to be older in age to the top of the Dhosa Oolite Member. Pandey et al. (2012a, b) also have gross wrongly correlated the top of the Dhosa Conglomerate to the top of the Khadir Formation without providing any evidence to that effect. The top of Dhosa Conglomerate is precisely of early Middle Oxfordian Indogermanus Zone, while the top of the 'Banbhanka, Gangta and Washtawa units' of Khadir, Bela and Wagad is of late Middle Oxfordian Schilli Subzone, having substantive interval in between three subzones, i.e. more than one my. On the other

hand, Pandey et al. (2012a, b) have also incorrectly placed the top of the 'Khadir formation' older to the top of the abandoned 'Washtawa Formation' while the tops of the two units are precisely correlative and of late Middle Oxfordian Schilli Subzone of Orientalis Zone (=ETM Transversarium Zone). The base of the 'Bambhanka/Gangta unit' has been wrongly shown by Alberti et al. (2012) in the middle of the GSM. For reasons detailed earlier, this base can in no way have an age overlap with the underlying third-order sequence with RSM/GSM contact as the terminal third-order SB. Alberti et al. (2012) have incorrectly shown the base of the GSM at the base of Late Callovian whereas it is of late Middle Callovian Obtusicosta Subzone. In bela and Chorar, unlike as shown by Fursich et al. 2001, The 'Bambhanka shale' is absent in Bela, and the 'Gadadha sand' in Chorar. The 'Bambhanka shales include the post RSM interval of the Chari Formation in Khadir while the 'Gadadha Sandstone member'makes the pre-GSM interval of the Chari Formation that are separated by a second-order SB as detailed in Chap. 5.

The almost ammonoid devoid DSM vis-à-vis the immediately underlying ammonoid rich GSM, also has been imprecisely ranged from the terminal Late Callovian to early part of Cordatum Zone. The DSM in fact ranges from the upper part of Horizon A-III of the Depressum Subzone of early Late Callovian Athleta Zone to at the most Early Oxfordian, i.e. Mariae Zone interval of the ETM. Even in the proximal Gangta bet, there are hardly any thick noticable sand beds in the fossiliferous 'Banbhanka shale' between 'Gadadha Sandstone' and 'Kakindia Limestone', although Biswas has mentioned about the sandy middle part of Bambhanka shale. In this context, mention here may be made of a succession of three somewhat thicker sandstone beds in the early part of the sediment interval IX which have been suggestively assigned to the Antecedens Subzone between the ammonoid rich early part of the Indogermanus Zone below and the ammonoid rich Parandieri Subzone above. These sands are included here in the basal part of the Jhadasa Member. It is extremely unfortunate that Biswas (1977, 1993) correlated these sands to the physically similar looking RSM of Mainland which are at least ~5 my older. Nevertheless, if upper limit of the DSM is extended up to the base of the DOM, its upper limit would include as well the early part of Early Oxfordian. Thus, extending the DSM up to the early half of Cordatum Subzone is imprecise and incorrect.

Alberti et al. (2012) have incorrectly assigned the top of the Dhosa Conglomerate as of Transversarium Zone age and also younger to the top of the abandoned 'Washtawa Formation' (Chari Formation in Wagad). The top is only of the Kachchh early Indogermanus Zone age (= to Vertebrale Subzone of Plicatilis Zone) which has its place in the standard zonation scheme in between Cordatum Zone and Transversarium Zone. Never ever in the past any worker has used Transversarium Zone as inclusive of the immediately underlying Plicatilis Zone. Even the basalmost part of the Transversarium Zone is not represented in the Dhosa Conglomerate. Pandey et al. (2012a, b) around the same time have shown the top of the DCM older to the top of the 'Washtawa Formation' which is correct, however, yet they have neither précised them at the subzone level nor provided any supportive ammonoid age evidence.

Katrol Formation Ages and Correlations

At the top of the table, the Katrol Formation has been again wrongly assigned in Alberti et al. (2012) to Alterneplicatus Zone, while in Mainland, the Katrol Formation begins at the base of late Early Kimmeridgian Alterneplicatus Zone and continues up to early Late Tithonian Communis Zone. Around the same time, the same combination of authors in Pandey et al. (2012a, b) have incorrectly correlated the top of the Katrol Formation to the top of the abandoned 'Gamadu Sandstone member' (early part of Umia Formation at Wagad). The top of the Katrol Formation, same as the base of the Umia Formation, is diachronous from Wagad to East Mainland to West Mainland. In the stratotype at Katrol, the top of the Katrol Formation is way younger by not just several zones but even more than a stage, i.e. difference in age of ~5 my.

The 'Upper Astarte beds' of Wagad are gross wrongly age corresponded to early part of the Katrol Fromation of Mainland by Pandey et al. (2012a, b). The early part of the Katrol Formation in Mainland ranges in age from late early Kimmeridgian Alterneplicatus Zone to mid-Late Kimmeridgian Bathyplocus Zone, while the 'Upper Astarte beds' of Wagad are of mid-Early Kimmeridgian Giganticus Zone. Nothing to speak of being age correlatives the units do not even have the slightest of overlap.

The base of the **Adhoi member** has been wrongly shown in terminal Oxfordian, while according to the revised ISJS understanding, the base of the Adhoi member marks the base of Kimmeridgian sensu lato and is precisely of basal Kimmeridgian Kachchhensis Zone based on the FAD of *Orhtosphinctes* as of the subfamily Ataxioceratinae in Kachchh. The top of the Adhoi member in Pandey et al. (2012a, b) has been wrongly correlated to the base of the Katrol Formation in Katrol. There is in fact distinct time gap between the two lithostratigraphic markers, i.e. the base of the Katrol Formation in Mainland and the top of the Adhoi member in Wagad.

The ages assigned to the succession of nine short interval lithostratigraphic units by Pandey et al. (2012a, b) are decidedly better than their ages discussed in detail in the preceding pages, however, yet imprecise and incorrect. The DOM > 20 cm below the Dhosa Conglomerate bed (DCb) is only of early part of Bukowski Zone of Cordatum Subzone. The later part of the DOM is exclusively of later part of the Cordatum Subzone, instead of including early part of the Plicatilis Zone. The DCb similarly exclusively belongs to the early part of the Indogermanus Zone (=ETM Plicatilis Zone).

Reworking in Dhosa Conglomerate bed

The reworked ammonoids in the DCb almost exclusively belong to the early Middle Oxfordian Indogermanus Zone. Their suggested inclusion even in the basal Transversarium Parandieri Zone is incorrect, and unsubstantiated on the evidence provided. Even Enay (1966) faltered with regard to the inclusion of younger species in his older zones, e.g. true *P. rotoides* Ronchadze in sediments older than Rotoides Subzone. The citation of the Madagascan zones in their support by Pandey et al. (2012a, b) fails to cut any ice, since not a single ammonoid there was collected with any precision. In fact, Collignon's soldiers use to collect bulk of ammonoids from a locality, i.e. through a thick succession of beds involving several zones at least more than one. The zonal index *Gregoryceras transversarium* is unrecorded by Pandey et al. (2012a, b) from Mainland. *P. antecedens* belongs to the Plicatilis Zone.

The Khadir ammonoids recorded in Pandey et al. (2012a, b) are also exclusively of the early Middle Oxfordian Indogermanus Zone (=Plicatilis Zone of ETM), and none of them signal the presence of even the basal Parandieri Subzone of the Transversarium Zone.

The Kantkote section from its exposed early part in the centre of the dome up to the 'Kantkote ammonoid bed' represents the later part of the Chari Formation. The older sand dominating part is not at all rich in ammonoids as the terminal part, yet the scarcely enclosed ammonoids suggest presence of the ETM Parandieri and Lucaeformis subzones. These subzones make the early half of the four-subzone Transversarium Zone. In the ammonoid zonation of the Kachchh Oxfordian, the zones are named after the characteristic perisphinctin species, and the late Middle Oxfordian zone is named as Orientalis Zone. It may be noted that the ETM index of this zone–*Gregoryceras transversarium* and other associated *Greogoryceras* are warm water low latitude ammonoids that only make the secondary/auxiliary part of the ammonoid fauna of this zone both on ETM and GTM. There are not present any exclusive elements of the Indogermanus Zone in the lower part of the Kantkote section as incorrectly suggested by Pandey et al. (2012a, b). The late Early and early Midle Oxfordian which is extremely condensed in the distal Mainland, is better exposed in the Gangta bet section in distinctly greater thickness, and accordingly has been zoned there.

The central part of the Kantkote section which Pandey et al. (2012a, b) find the richest in ammonoids is the terminal part of the Chari Formation, and has been named Jhadasa Member. It has been gross wrongly included in the early Late Oxfordian Bifurcatus Zone by Pandey et al. (2012a, b). Based on rich *Larcheria schilli* and *subschilli* together with *Gregoryceras* cf. *devauxi.* It is undoubtedly of late Middle Oxfordian Schilli Subzone. It also marks the first-order intra-Jurassic MFS not just in Kachchh but on the entire GTM. Pandey et al. (2012a, b) have again gross wrongly dated 'their terminal part' of the Kantkote ammonoid bed as of late Bifurcatus Zone, which again undoubtedly belongs to Schilli Subzone, and does not at all include even the basal Late Oxfordian.

The 'Patasar Shale member' makes the basal beds of the Kantkote Member, and represent the Rotoides Subzone of late Middle Oxfordian Transversarium Zone which is the youngest ammonoid subzone of the Transversarium Zone as the youngest subzone of the Middle Oxfordian. Even this unit on the basis of a single example of *Perisphinctes* cf. *besairiei* of unknown range has been wrongly included in the Bifurcatus Zone by Pandey et al. (2012a, b). To conclude, PSm hardly includes any part of Late Oxfordian, and the question of any part of Kimmeridgian in this unit simply does not arise.

4.8.4.3 On the Intra-Callovian Disconformity in the Jumara Section, as Succession of the Six Non-Existent Inter-Stage Unconformities of Pandey and Dave

Pandey and Dave (1993) presented a graphical model that included a large intra-Callovian disconformity of almost the entire Callovian duration in the major part of the exposed basin, as also in many wells. They, also tried to provide supportive evidence to such a non-existent unconformity through an altogether impractical graphical model (Fig. 17 of Pandey and Dave (1993), p. 111 reproduced here as Fig. 4.13) in the Jumara section to ascertain the non-existent disconformity between the Late Bathonian limestones below which they mistook as Early Callovian, and a Middle Callovian oolitic bed above, also, mistaken for the Early Oxfordian Dhosa Oolite. In their Fig. 17 on pp. 105–106, was graphically exhibited the non-existent disconformity between the Patcham on the exposed margin and distally located Sanadra and Suthri wells in Mainland almost of equal duration. In the same figure, was also indicated another similar large disconformity throughout the basin above the DOM. To begin with, there is shown deposition of beds 1–8 (bed 1 of Late Bathonian limestones while beds 2–8 of Early, Middle and Late Callovian age of shales and intervening sand). According to them, the Jumara sediments were first folded and differentially uplifted. There followed a phase of removal of several beds of the Middle and Late Callovian age (beds 3–8) selectively/locally only from the north limb of the implanted anticline at Jumara, while merely part of the bed 8 is removed on their non-existent south limb. They implied submergence again of the area for the subsequent deposition of the Early Oxfordian (beds 9 and 10) on the eroded surface. The non-existant unconformity

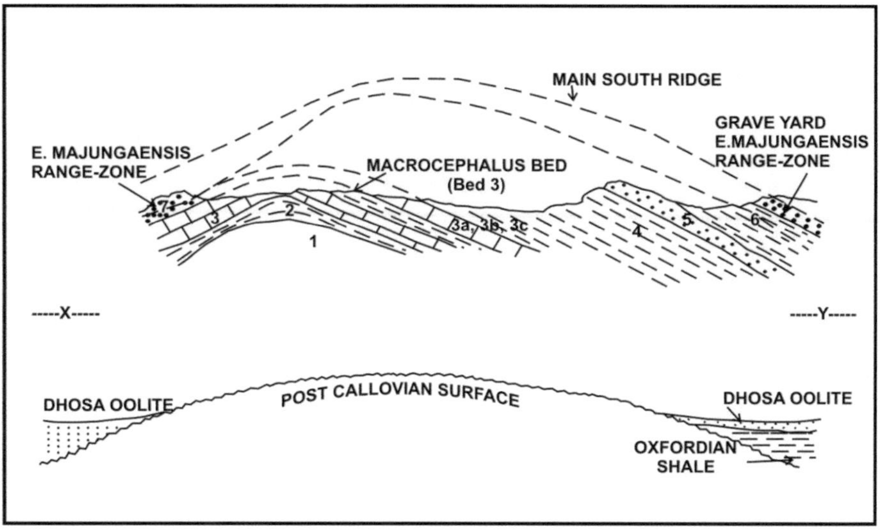

Fig. 4.13 Diagrammatic cross section of Jumara Dome, Kachchh after Pandey and Dave (1993)

surface is also shown folded in the graphic. Thus, the long phase of erosion near the Bathonian/Callovian boundary, and two phases of intra-Callovian folding were invoked by Pandey and Dave (Fig. 18).

The two large subaerial disconformities indicated by the authors below and above the Dhosa Ooilte, if at all present, must have varied in magnitude, at least conceptually, in a graded manner. The larger one should have been in proximal sites decreasing gradually towards the basin to make the succession conformable at some distal site along with transformation of the disconformity into correlative conformity and disappearance of the gap. On the ammonoid evidence, almost uninterrupted Callovian is present in the entire Mainland, Wagad and even in Khadir region of 'Island belt', not even a subzone being absent. In the sequence stratigraphic formulation, is here interpreted the tectonically influenced uplift of Patcham, Bela and Chorar at the intra-Callovian second-order SB near the close of the mid-Middle Callovian Paramorphum Subzone. It may also be emphasized here that uplift or submergence invariably occur as intrabasinal to regional events that are expressed through the sequence stratigraphic surfaces, and certainly neither locally nor at any arbitrarily chosen time as in this case close to the Callovian–Oxfordian boundary. It does neither relate to any basin wide nor regional event, nor expressed through a sequence stratigraphic surface. More over, the uplift as invoked by the authors should invariably follow shallow sands instead of the subtidal shales as in the present case. The Early Oxfordian makes part of the second-order TST during which bathymetry is increasing not merely in the basin but on the GTM at large, and any uplift/emergence that too in the distal Jumara, can not even be thought of, since that would imply disconformity of much greater duration east of Jumara increasing in magnitude up to the basin margin which is not the case.

Thus, it is concluded that the older disconformity is unintendedly implanted. During Early Callovian, the arm of the Neothys, it is expected, had submerged the entire basin to subtidal and yet greater depths. The ammonoid data indicates presence of the entire Callovian all over the Mainland as in Wagad and Khadir. Pandey and Dave, presumably, mistook sands of the DSM lying immediately below the oolitic limestones of the DOM as sands of the RSM, and also mistook silt/shale in the distal sections of the DSM as that of the GSM. Somewhat physically similar hard, resistant dark maroon rocks of Early Callovian were mistaken for the Dhosa Oolite in their photo plates. Such, not that uncommon misunderstanding of beds, probably, landed them to imagine the large disconfirmity which simply did not exist. Even the legendary Kachchh geologist Biswas similarly faulted in correlating the Oxfordian sands of Gangta bet of the JM with the distinctly older Middle Callovian sands of the RSM, without any revision all these last four decades. Jumara is a relatively basinal or distal locality of the exposed Kachchh shelf, and conceptually can not be exposed in the early part of a second-order TST above the RSM/GSM second-order sequence boundary during a second-order phase of increase in bathymetry. There can not be suspected even a remote possibility of this unconformity being present as well in the distal wells (Fig. 15 of Pandey and Dave 1993). Any subaerial unconformity if present in distal wells must increase in magnitude towards the margin, which is simply not the case here in view of the uninterrupted Callovian deposition. Apart from the disconformity being absent, Figs. 17 and 18 even conceptually stand improbable and impractical.

The younger 'subaerial' disconformity above the DOM also of large duration and present throughout the Mainland, Suthri and Sanadra wells as indicated in their Fig. 17 instead of being subareial as thought all along by one and all has been in a radical revision held as submarine (Krishna et al. 1998 and later), also discussed in detail earlier in this work, also in earlier publications (Krishna et al. 1998 onwards). If the above non-existent unconformity was not enough, Pandey and Dave additionally imagined as many as six ficticious inter-stage unconformities in the Jurassic of Kachchh inclusive of the distally located well sections. No other Kachchh Jurassic researcher has found these except the one interpreted earlier as submarine. Near continuos sedimentation is well supported by the presence of all the ammonoid zones/subzones of the Middle and Late Callovian in the Jumara section. Thus, the major and the sole unconformity present is wrongly interpreted as subaerial, while as many as five non-existent unconformities, unintendedly, are gross wrongly interpreted in the entire basin. The model diagram invoked by these authors appears to the present author as structurally impossible.

4.8.5 *On the Intra-Oxfordian Uplift and Emergence of Mainland*

Alberti et al. (2012) came out with a unique but somewhat improbable model in which the Mainland is uplifted and emerged during the large gap of several subzones between the underlying Chari Formation and the superjacent Katrol Formation from mid./late Indogermanus Zone (=early Middle Plicatilis Zone) to mid-Early Kimmeridgian Alterneplicatus Zone (=Divisum Zone of ETM). This model is violative of the basics of geology (Fig. 4.14). The sediments from the very moment after their exposure start acting as provenance to the neighbouring depo-centres, which is not the case. The model is unacceptable for the following reasons:

1. There are present reworked older Toarcian–Aalenian nannoplanktons in the terminal Callovian of Mainland and Wagad (Rai and Jain 2012) on account of the uplift/emergence of Patcham during Middle Callovian, and follow-up weathering and erosion of Patcham. Much unlike, and even to the contrary, in interpreting uplift and emergence of Mainland as envisaged during the large gap, Alberti et al. (2013) have neither provided any evidence of erosion of sediments, nor of the presence in Wagad of reworked older fossils in younger sediments. In Wagad any older Bajocian–Callovian reworked microfossils are absent in the younger Early Oxfordian sediments. Thus, the uplift and emergence of Mainland any time in late Early to Middle Oxfordian is not substantiated.

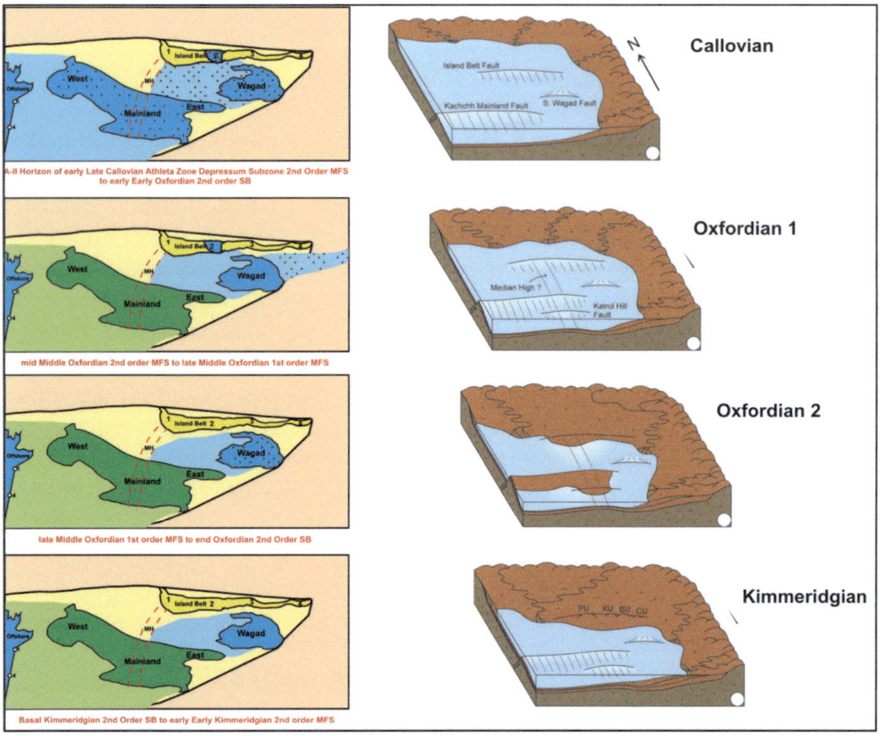

Fig. 4.14 Contrast of the present bathymetry and depositional model with that of Alberti et al. (2012) in the Oxfordian–Early Kimmeridgian of Kachchh

2. The interpreted uplift and emergence of Mainland would have resulted in the reversal of the paleoslope/paleodrainage/paleocurrent to due east instead of the earlier due west in the Wagad succession of this interval. Thus according to their model, the pre-gap succession of Wagad should have had altogether opposing paleoslope/paleodrainage/paleocurrent as different provenance which is not the case.
3. The sedimentary succession of the second-order mid-late Middle to latest Callovian is characterized by the geographically restricted GTM shallower water preferring perisphinctid ammonoid subfamily Kinkeliniceratinae. The RST sediments of this sequence are made up of the ammonoid scarce to ammonoid devoid sand/silt of the Dhosa Sandstone Member, which also suggests lesser bathymetry corresponding to the immediately underlying and overlying second-order sequences without any evidence of erosion.
4. Both, the under and overlying second-order sequences are characterized by relatively less shallow water preferring stephanoceratid subfamilies; Macrocephalitinae in the underlying sequence and Mayaitinae in the overlying sequence. It in turn suggests greater bathymetry in the basin inclusive of

Mainland in Oxfordian up to the late Middle Oxfordian Orientalis Zone Schilli Subzone first-order MFS. Thus any emergence and uplift of any part of the basin is neither supported by sedimentology nor by the enclosed ammonoids.

5. The Oxfordian interval up to the MFS of the first-order sequence includes evidences of sedimentary and ammonoid faunal condensation/starvation that are absent in the underlying and overlying sequences. These features also unambiguously support increased bathymetry during the otherwise interpreted uplift/emergence of the Mainland.
6. There is observed maximum shifting of the shoreline in the interval of uplift/emergence of Mainland, which suggests increasing transgression and increasing bathymetry up to the first-order MFS. Thus uplift and emergence is rejected by the extraordinary rapid marginward shift in the shoreline.
7. There is absence of any evidence of vertical tectonics of such magnitude in the basin. What was the timeframe and scale of the uplift? Was it sudden **or** graded during this interval? Is it present in any other Indian Jurassic basin for example in the neighbouring Jaisalmer basin of India and Indus basin of Pakistan?. Extensional tectonic events like the one in discussion conceptually can not take place in one part of the basin, and be absent elsewhere in the basin. There is hardly any similar model existing in the literature at least not in my knowledge.
8. The immediately overlying and underlying sediments are either fine textured or condensed, ammonoid rich, and suggest subtidal depths which can not be the case if Mainland was emerged. The much shallower above wave base depths should have prevailed according to the model with resultant ammonoid scarce sands which is just not the case.
9. Early part of the terminal Late Callovian Lamberti Subzone is evidenced in West Mainland at Nara (Kanjilal and Singh 2012), and Keera (Krishna and Ojha 1996) through ammonoids as part of a widespread third-order TST within DSM. The corresponding third-order RST is included in the upper/major part of DSM and thus any emergence until the close of Callovian is not supported.
10. The only possibility of emergence if so is in the early part of Early Oxfordian, and not there after that too limited to Wagad and Khadir, and at the most to East Mainland up to the MH. However, in an ongoing rift framework, uplift is hardly possible. Moreover, the deeper Mianland can not be uplifted without affecting the shallower part.

4.9 Intrabasinal Correlations with Temporal and Spatial Extent of the Kachchh Mesozoic Formations and Members

The lithostratigraphic units of the Kachchh Jurassic as originally erected by Stoliczka are described, and thereafter applied to develope the high-resolution ammonoid stratigraphy in Chap. 2. Chapter 2, in-turn, has been used to formulate the succession of the first- to second-order sequence stratigraphy in the Kachchh basin in Chap. 3. Thus could be extended the isochronous sequence surfaces in the entire basin even to the stratigraphic sections which earlier in absence of guide fossils could neither be precisely dated nor correlated to the ammonoid bearing precisely dated sections. Based on the quantum biochronologic and sequence stratigraphic progress registered in Chaps. 2 and 3, are addressed here the intra-basinal correlations, temporal and spatial extent of the Kachchh Mesozoic stratigraphic units.

4.9.1 Patcham Formation

The Patcham Formation, since its very creation by Stoliczka, has been known to include a lower, rather thick, sand dominated part and an upper, much less thick, limestone dominating part with shale interbeds. However, this unit of Stoliczka as such was never subdivided into member units by later stratigraphers. Instead, the later stratigraphers preferred to create new formations viz. Kaladongar formation, Goradongar formation, Khadir formation and other units in the 'Island belt', and Jhurio formation in Mainland, which in part or full are here suppressed as junior synonyms. The Patcham Formation is organized into two members in the Chap. 2 with formal recognition to the two lithologically contrasting lower and upper parts; as the LSM developed exclusively in the 'Island belt' and the ULM (Figs. 2.25, 2.26, 4.15 and 4.16) developed in 'Island belt' as in Mainland. The pre-Callovian pre-Chari units of Biswas (1977, 1993) and others created in the 'Island belt' and Mainland are suppressed and abandoned as junior synonyms of the Patcham Formation. The member level units in the Patcham are informally adopted, however, in view of their large number, the same are here reorganized into the above two formal Members as informal members.

The LSM is the same as Biswas' Kaladongar formation, although of member rank, with presence in Patcham, Khadir and Bela, Mouvana and Chorar in full or part. Initially, the LSM was conceived as 'Kaladongar Member', but was not proposed due to possible mix-up and confusion with the already existing Kaladongar formation and its Kaladongar Sandstone member in literature. The 'Kaladongar formation' is suppressed as junior synonym for the lower part of the Patcham Formation, while its three members are informally adopted. 'Goradongar formation' (only up to the Raimalro Limestone) is suppressed as junior synonym of the Patcham Formation, and instead included in ULM of the Patcham Formation with widespread presence in 'Island belt' and Mainland. The member units of the suppressed 'Kaladongar formation' are informally retained within the LSM as Dingy Member (Dm), Kaladongar Sandstone Member (KSm) and Babia Cliff Sandstone Member (BCSm), so the suppressed 'Goradongar formation' up to the Raimalro Limestone member within the ULM as Goradongar Yellow Flagstone member (GYFm), Gadaputa Sandstone member (GSm) and Raimalro Limestone member (RLm). The GYFm in the 'Island belt' is exclusive to Patcham. This in turn

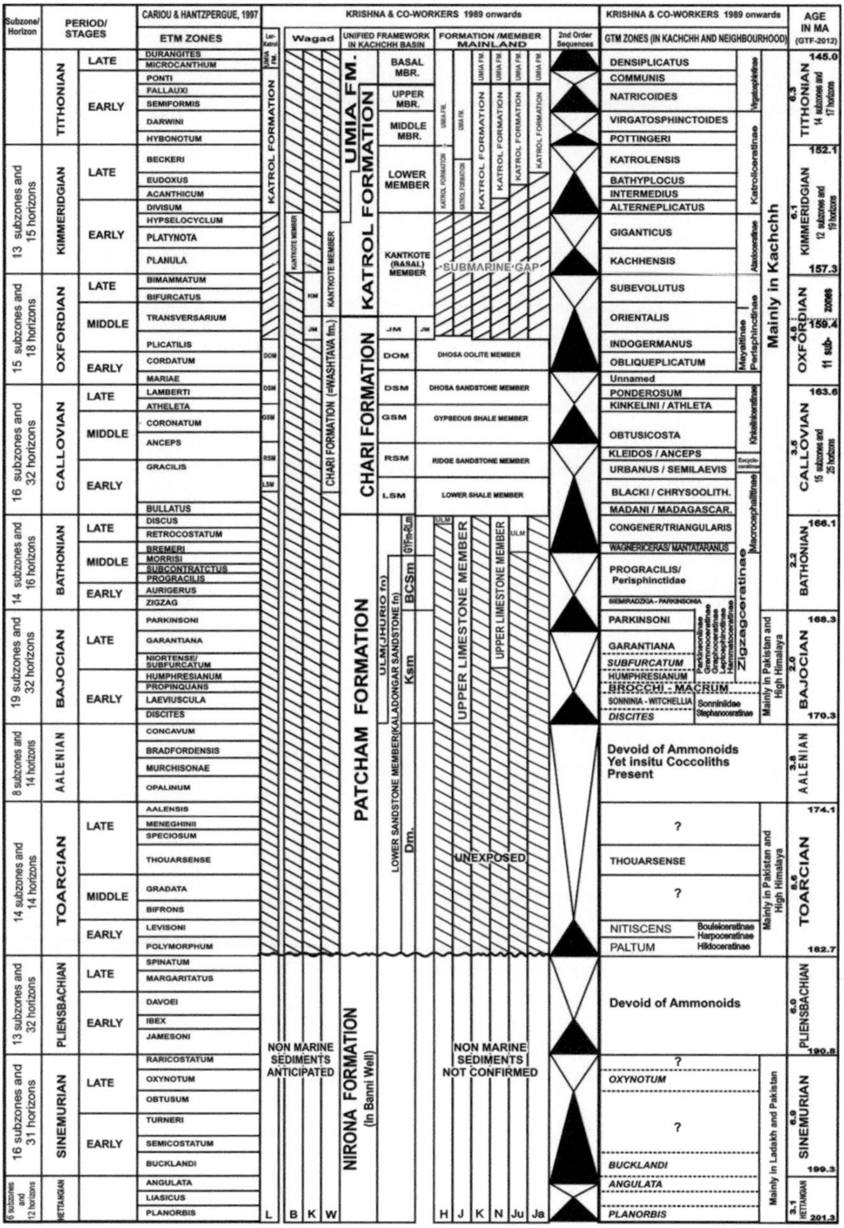

Fig. 4.15 Second-order sequence framework and Stoliczka's Jurassic scheme extended to Wagad under precise ammonoid control

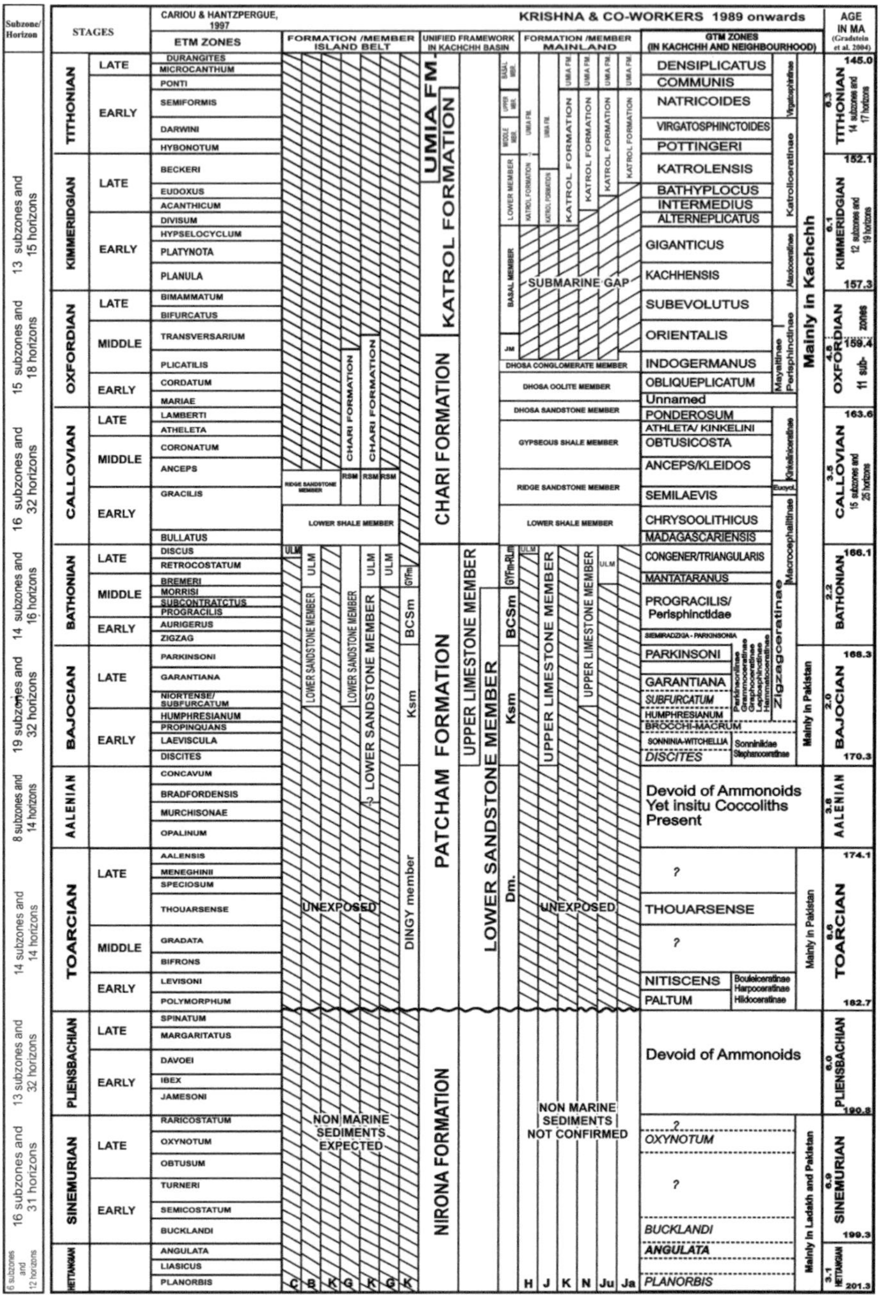

Fig. 4.16 High-resolution intrabasinal integration of Kachchh units and ammonoid zones

makes the GSm undifferentiable from the older BCSm, and in these localities, the BCSm or for that matter the LSM is understood as continuing up to the base of the Raimalro Limestone member, and is also inclusive of the GSm.

The ULM, however, has its best and full development at Jhura. The Jhura section, in view of the presence of the best and longest development of the unit, as the best lithological representative with widespread presence in Mainland at Habo and Jhura in the east and Nara and Jumara in the west in addition of 'Island belt', is here considered the informal stratotype of the ULM. The ULM in the 'Island belt' except for Patcham is limited to the RLm (=HSTm at Khadir). The 'Khadir Formation' is abandoned as junior synonym of the Patcham Formation for its lower part up to RLm, while the post RLm part is also suppressed and considered as synonym of the Chari Formation. However, the 'Hadibhadang Shale' and 'Hadibhadang Sandstone' at Khadir are used in the discussion as abandoned members, respectively as 'HSm' and 'HSTm'. The member units 'Chappar Bet member', 'Pachhampir member', 'Kharidongari member', 'Taga member', 'Kakkar member' at Goradongar (Pandey and Agrawal 1984) and corresponding member units at Sadhara ('Coral Limestone member', 'Red Sandstone member', Lower Yellow Flagstone member', 'Upper Sandstone member', 'Goradongar Yellow Flagstone member' and 'Gadaputa Sandstone member') and Jhura ('Purple Sandstone member', 'Jumara Coral Limestone member' 'Jhura Golden Oolite member', 'Badi Golden Oolite member' and 'Badi White Limestone member' within ULM of Patcham Formation (Fursich et al. 1994 and later) are also suppressed as junior synonyms of the adopted informal members of Biswas (1977, 1993) in the stratotype.

Within the complete Patcham Formation with its two formal members, the LSM is exposed only in 'island belt', possibly on account of the relatively increased depths and the associated lithological change from sands to carbonates. The ULM in Mainland includes only the later part of the unit as mixed carbonate/siliciclastics. The Jumara section includes the mid-Middle to latest Bathonian interval with the younger three informal members. Nara section perhaps includes the complete Bathonian, may also have terminal Late Bajocian. The Jhura section includes both the Bajocian and Bathonian, and the Habo section only the youngest Bathonian. The oldest informal member of the Patcham Formation is Dm, which is based on coccoliths (Rai and Jain, 2012, 2013) is here age constrained in the Toarcian–Aalenian interval. The youngest informal member is RLm, which, based on the rare presence of *M. triangularis* Spath and *M. madagascariensis* Lemoine, is well constrained in Late Bathonian (Pandey et al. 2013a). Thus temporally, the Patcham Formation is precisely ranged from the Early Toarcian to the Late Bathonian Triangularis Zone, and is of $\sim$16.6 my duration.

Member Units of the Patcham Formation

Among the four Jurassic formations, the most difficult to precisely date are the informal members of the Patcham Formation due to the near absence of guide fossils in them. The density, diversity and frequency of ammonoids in Patcham Formation are least of the four formations. Nevertheless, using the sequence stratigraphy concepts and formulations together with scarce data on ammonoids and nannoplanktons, it is now possible to invariably date the members of the Patcham Formation with increased precision as under:

Lower Sandstone Member (LSM)

The member units of Biswas are useful and here informally used; the Dingy member as Dm, the Kaladongar Sandstone member as KSm and so under the LSM. Biswas (1977) included the beds 1–38 in Dm, however, the Dm is revised here as from the beds 1–36. Ammonoids so far are not discovered. The characteristic *Corbula–Gervilla* bivalve assemblage in the early part of the Dm at Dingy/Kuar bet (beds 15–24) represents the oldest such assemblage in the Kachchh Jurassic of the Early Toarcian age. Based on recent coccolith records, the Dm is suggested as of Toarcian–Aalenian age. Thus the composite Dm exposed at Kuar bet and Dingy hill is, if not exclusively, at least largely assigned to Toarcian, while the inclusion of some or whole of Aalenian is not ruled out. According to tentative correlations of the younger member units in the 'Island belt', there are only slim chances of even the older/early part of the 'Cheria bet conglomerate' at Khadir being age equivalent of the younger/later part of the Dm, however, in absence of ammonoids and also in absence of coccolith studies in 'CBCm', the chances cannot be totally ruled out. Thus, the Dm is here held as exclusive to Patcham, and of suggestive Toarcian–Aalenian age.

Based on the homotaxiality of the ammonoid succession in Patcham and Nara sections along with the third-order sequence framework, the basal fossiliferous fine textured limestone/marl/shale beds of the BCSm are correlated to the ammonoid rich terminal Bajocian to Early Bathonian Zigzag Zone second-order TST. This in-turn suggests terminal Late Bajocian to mid-Middle Bathonian age of the BCSm. The KSm also thus indirectly gets age bracketed between the Toarcian–Aalenian Dm below and the terminal Late Bajocian to mid-Middle Bathonian BCSm above as of mostly Bajocian age but for the terminal Late Bajocian, although it may also extend down to part of the Aalenian. The 'HSm' of the Khadir section is here broadly correlated mostly to the collective interval of the LSM and GYFm, while the overlying 'HSTm' to the collective interval of the GSm and RLm of the ULM.

The GYFm is also correlated to the lithologically similar early part of the Em of Jhura section, which in-turn allows the broad correlation of the ULM at the Goradongar section to the informal members Em to Gm at Jhura (Figs. 3.18, 3.19, 3.20 and 3.21) The informal members Am to Dm of Jhura are correlated to the collective interval of the KSm and BCSm of Patcham. The Informal members Fm and Gm correlate well to GSm and RLm to the 'HSTm'. The Am + Bm combine is correlated to the KSm, and the Cm + Dm combine to the BCSm. On similar lines, the Am is correlated to the the 'Basal Coral Limestone', the Bm to the 'Red Sandstone', the Cm to the 'Lower Flagstone', the Dm to the 'Upper Sandstone', and the Em to the 'Upper Flagstone' between Jhura and Sadhara. The base or start of the ULM is highly diachronous between the Early Bajocian at Jhura to Late Bajocian at Nara, late Middle Bathonian at Jumara and Patcham and Late Bathonian at Habo, Khadir, Bela, Mouvana and Chorar. The start of the LSM is also similarly diachronous, and same as age of the oldest beds in 'Island belt' that is already discussed earlier.

Upper Limestone Member (ULM)

Precise ammonoid ages are available in the ULM, i.e. in the upper member of the Patcham Formation. The RLm of Goradongar, the Gm of Jhura, the terminal part of the 'HSm', the terminal part of the Habo, Jumara and Nara sections on the evidence of *M. triangularis* Spath in Jumara, Nara, Jhura and Goradongar are assigned to the Late Bathonian Triangularis Zone (Figs. 2.25 and 2.26). The GYFm of Goradongar, the 'Upper Flagstone' of Sadhara, the Em of Jhura, the basal part of the Jumara section are best assigned to the late Middle Bathonian GTM Mantataranus Zone (=Bremeri Zone of ETM). The early ~18 m limestone/shale strata of Nara, and the Bm + Cm +Dm combine of the Jhura section are precisely age bracketed in the Late Bajocian–mid-Middle Bathonian on the evidence of Late Bajocian to early Middle Bathonian ammonoids at Nara. The Am of Jhura, the 'BCLm' of Sadhara, the basal fossiliferous part of the KSm of Kaladongar are only indirectly assigned at least to the Early Bajocian as a second-order TST with MFS/MFI at the top of the unit. Yet lowered down, the Dm is ranged through the Toarcian–Aalenian interval on the coccolith evidence of Rai and Jain (2012).

The parallel sets of units at such close distances are considered impractical, and so avoided. The correlation is tentative and requires further strengthening. Pandey et al. (2013b) recovered the oldest ammonoid of the Kachchh basin—a single septate specimen of Early Bajocian *Calliphylloceras* from the yet older basal beds of the Sadhara section of the possible Early Bajocian age in the basal part of a second-order TST. There is a significant not so thick 7–9 m thick hard somewhat brownish pebbly fossiliferous sandstone bed in the early part of the sediment interval II at Jumara which is here corresponded to the Fm. The Fm in-turn is overlain by white to grey thick nodular limestone beds (Gm) which have yielded *Macrocephalites* spp. At Khadir, the locally erected 'HSTm' is here correlated to the combined sediments of the GSm and RLm of mid- and Late Bathonian age, the underlying 'HSm' with the combined sediments of BCSm and GYFm of the Late Bajocian to late Early Bathonian age. In this framework the 'CBCm' is suggestively placed in the Aalenian–Early Bajocian age bracket. It is some what difficult to precisely date or correlate the lithostratigraphic boundary between the 'HSm' and 'HSTm' in Khadir, Bela and Mouvana. It is also difficult to determine the correlative sediment interval in these sections as to whether of the terminal part of the 'HSm' or early part of 'HSTm'.

The entire BCSm, inclusive of the terminal thick sands above the beds 49–54, is younger than the Late Bajocian. Homotaxially, the beds equivalent to the basal beds of the BCSm in Nara section have been on ammonoid evidence firmly dated as Early Bathonian Zigzag Zone Macrescens Subzone (Fig. 1.11). On the other hand, in both the sections, the younger beds have yielded late Middle Bathonian ammonoids. The thick ammonoid devoid and even mega invertebrate scarce sands are sandwiched between the late Middle Bathonian ammonoid bearing yellow flagstones of the GYFm above and the beds 49–54 below.

The GYFm is considered a good example of the carbonate deposits in the Kachchh basin which is widely present in the 'Island belt' and also in the Mainland at Jhura, Nara and Jumara. Also is interpreted a third-order SB between the GYFm and the GSm at Kaladoongar, also picked up at Jumara either between the sediment intervals I and II, or in the early part of the sediment interval II. Also possibly, the basal beds of the sediment interval I at Jumara (beds 1–4 of the sediment interval I) may correlate with the top beds of the GYFm as the MFS/MFI of the GYFm sequence. In Sadhara, the same MFS/MFI is determined in the ammonoid bearing interval of E 2(d-h) of Fursich et al. (1994) a few meters below the GYFm/GSm boundary SB. The youngest sands plus the $\sim$5 m thick interval of the GYFm may constitute the early part of the following RST.

The GSm includes the guide fossil-devoid sands of the bed 2 of the 'Goradoongar formation', and is indirectly age restricted between the underlying GYSm and the overlying RLm in the Late Bathonian between the early to Late Bathonian around the Retrocostatum Zone/Discus Zone transition (Figs. 2.27–2.31). These rather thick significant sands in between the two relatively much thinner carbonate units are present in all the four uplifts as either GSm or 'HSm'. There are not observed any such sands in the Bathonian either at Jumara, Nara or Jhura, and the interval obviously is included in the limestone-marl succession. Thus, the GSm can not be recognized outside the Island belt.

The RLm, the youngest member of the Patcham Formation, has found majestic expression in all the four uplifts. It is very characteristic in the field with pebbly fabric in the lower part, rich in chert nodules along the bedding of the blocky/jointed limestones in the middle part, and the bedded and pelletic upper part. Bivalves, gastropods, and belemnoids are observed along with the extremely rare *Macrocephalites* aff. *triangularis* Spath and *M. madagascariensis* Lemoine, as a fresh third-order transgression which continues further up with change of lithology from the yellow brown carbonates to the Early Callovian shale/sand alternation. It is precisely placed in the latest Bathonian Triangularis Zone age (Pandey et al. 2013a). Both lithologically and paleontologically, the RLm correlates well with the greyish white nodular limestones that represent the youngest sediments of the Patcham Formation in the Jhura (Gm) and Jumara (beds 31–63 of sediment interval II) sections of the Mainland (Fig. 2.27).

The members Fm to Gm of the Patcham Formation at Jhura are correlated to the sediment interval II (bed 22 of Rajnath) while the Em to the sediment interval I (=beds 23–26 of Rajnath 1932) of the Patcham Formation at Jumara. In context of the formal six-member classification of the Patcham Formation, the sediment intervals I and II combinely are correlatable to the collective succession of the younger three members (GYSm, GSm and RLm). In view of the correlations suggested above, based mostly on sequence stratigraphic formulations developed and detailed in this work, the Am to Gm broadly correspond to the sediment succession collectively of the five younger members of the Patcham Formation in the stratotype at Patcham exclusive of the basal Dingy Member which is exclusive to the 'island belt'. Gm to RLm, Fm to GSm and the Em best correspond to the GYFm, members Cm and Dm to the BCSm, and the Bm + Am combine to the

KSm. There is large variation in facies from mostly thick sands at near the margin at Kaladoongar and Goradoongar to lime/marl and golden oolites at Jhura, Nara and Jumara. In Habo, only the RLm is present, at Jhura, the younger five members are present, and at Nara younger three members are present along with age equivalent sediments of the later part of the BCSm.

4.9.2 Chari Formation

The Chari Formation has its stratotype in Keera dome. The Jumara locality also includes another well studied section (Rajnath 1932; Spath 1927–33; Biswas 1977, 1993) of the Chari Formation which is considered here a good reference section. The Chari Formation is the only Kachchh Jurassic formation which is present in part or full in all the three principal regions, i.e. Mainland, 'Island belt' and Wagad. The large expanse of the unit is principally in view to its being the later half of the first-order TST. The unit is easily recognized all over the basin, however, its base is not exposed in Wagad. The 'Jumara formation', 'Khadir formation', and 'Washtawa formation' of Biswas with identical base, and their shale dominant over all content as of the Chari Formation, as present all over the Mainland, Island belt, and Wagad are suppressed as junior synonyms. The top of the Jhadasa Member marks as well the top of the Chari Formation. Thus, the complete Chari Formation is present only in Khadir with the exposed base and, also at Wagad but without the exposed base. Beds 14–39 of the Khadir formation above the Raimalro Limestone represent the Chari Formation in Khadir. In Bela the early part of the Chari Formation includes the youngest *Corbula–Gervilla* assemblage of the Island belt. The youngest beds in Bela are thick sands of the RSM of the Chari Formation. The LSM begins in the Madagascariensis Subzone while the top of JM is of the Schilli Subzone age. The top of the Chari Formation is diachronous between Wagad and Mainland in view of the varying submarine gap above the unit. Thus, the Chari Formation precisely ranges from the Early Callovian Madagascariensis Subzone Chrysoolithicus Zone to the late Middle Oxfordian Orientalis Zone Schilli Subzone. Time span involved in the Chari Formation is $\sim$6.4 my.

Lower Shale Member

It includes beds 01–92 in the Keera section and ranges from early to mid-Early Callovian (Madagascariensis Subzone to Diadematus Subzone). In the reference section at Jumara, the unit includes beds 1–9 of Biswas (1977, 1993). The complete Chari Formation is developed only in the Khadir uplift. Merely, the early to mid-Early Callovian dominantly fine textured shaly part represents the unit in Patcham to Bela of 'Island belt' in different names—early part of 'Modar Hill Member' (beds 4–6) in Patcham, and early part of 'Gadadha Sandstone' up to the base of the marker Pholadomya bed in Bela. The Chari Formation is almost absent in Chorar, where it has been eroded away except for some basal 'Gadadha' type sands which are here considered as part of the Lower Shale Member of the Chari Formation. Instead, the youngest sediments in Chorar are of RLm of Patcham Formation. In Khadir, the unit, conformably, overlying the calcareous cliff forming

'HSm' sands of the Patcham Formation, starts with shales of the early part of the abandoned GSm. These names whether 'MHm' or 'GSm' are informal, restricted only to the 'Island belt', and are not included in the integrated lithostratigraphic framework of the Kachchh basin. These are safely included in the collective interval of the Member I and Member II of the Chari Formation in the 'Island belt'.

Ridge Sandstone Member

The unit is dominated with ridge forming sands, particularly, east of Jhura, island belt and Wagad where it has received different informal local names. It includes beds 93–217 (sediment interval II to VI) in the Keera section, and the beds 2–5 in Jumara of late Early to early Middle Callovian age. Obviously, the RSM is diachronous from relatively less shallow Mainland to more shallow margins of either 'island belt' in the north or Wagad in the east as the second-order regressive component of the latest Late Bathonian Triangularis Zone–late Middle Callovian O-I Horizon Obtusicosta Subzone Obtusicosta Zone sequence representing the middle part of the first-order TST of the first-order Kachchh sequence. The RSM may have an older base and younger top at the margin in Island belt and Wagad, however, never temporally younger to either the late Middle Callovian O-I Horizon or older to O-II Horizon of Obtusicosta Zone. The top of the RSM in Washthava also makes the top of the 'Km' of Deshpande and Merh (1980). It may be pointed out that Biswas (1977, 1993) has correlated the Ridge Sandstone Member of this second-order sequence with the much younger similar looking Oxfordian sands of Wagad and Khadir belonging to another younger second-order RST (Figs. 2.25 and 2.26). Similar looking sediment packages, occurring cyclically at different stratigraphic levels, thus, need be cautiously studied for correlative purposes. The RSM extends in Khadir and Bela as the upper part of 'GSm'.

The Late Callovian–Early Oxfordian age to the top of the 'Modar hill Member' (Biswas 1993, p. 163) is unsubstantiated, and is being here revised to the early Middle Callovian with difference of one full stage of ~5 my duration. So is also considered the age of the RSM of Mainland. Both the 'Modar Hill Member (MHm)' and the RSM sands represent the second-order RST of the late Middle Bathonian–early Middle Callovian second-order sequence that is present through out the basin (Krishna et al. 2010, 2011), and thus any way can not be younger than O-I Horizon of Middle Callovian.

Gypseous Shale Member

It includes beds 218–372 in the Keera section from the base of the here revised Obtusicostites Zone to the late Depressum Subzone of Athleta Zone. In the reference section at Jumara, the GSM is represented by bed 6. The base of the 'BSm' of the 'Khadir formation' represents the base of the GSM at Khadir, so the base of the informal 'NSm' of Deshpande and Merh (1980) of the Washthava Formation in Wagad. This Member is absent in Patcham, Bela and Chorar.

Dhosa Sandstone Member (DSM)

It includes ~3 m thick sands and silt as its early part (beds 373 onwards up to bed 392 of Callovian, i.e. sediment interval X and further) in Keera section of the terminal part of the Depressum Subzone of the Athleta Zone to at most up to the close of the Early Oxfordian (Figs. 2.30, 2.31, 2.32, 3.25 and 3.26). It includes beds 7 and 8 in the reference section at Jumara. Broadly, the DSM extends to Khadir as middle part of the abandoned 'Bambhanka Shale' which includes significant sands in its middle part. The top of the DSM is better expressed at Gangta bet up to the sediment interval V, i.e. up to the bed 7 at Gangta bet of Biswas. The base of the DSM is here assigned to the terminal part of the Horizon A-III of the terminal part of the Depressum Subzone of the Athleta Zone of early Late Callovian age, i.e. the base of the bed 6 at Gangta bet of Biswas with approximate total thickness of 36 m.

Dhosa Oolite Member (DOM)
It includes more than one ammonoid abundant hard resistant nodular conglomeratic oolitic limestone beds, and is present all over the Mainland. Thc youngest of them is the most characteristic. The oolites are more and dense in distal sections while somewhat sandy and less oolitic, also less in ammonoids, in the proximal sections. The youngest but one of these oolitic limestone beds marks the top of the Obliqueplicatum Zone of the late Early Oxfordian age (sediment interval G-VII, beds 8–10 of Biswas). The DOM is absent in the 'Island belt' except at Khadir and Gangta bet. In Khadir (bed 36) and Gangta bet (beds 8–10), is expressed the major part of the DOM but for the youngest conglomerate bed as the 'Kakindia Limestone' with the same ammonoid assemblage and age. The youngest oolitic conglomerate includes ammonoids of early Middle Oxfordian age assigned to the Indogermanus Zone together with reworked ammonoids of late Early Oxfordian in the sediment interval G-VIII and the basal part of G-IX and beds 11–13 of Biswas.

Jhadasa Member
It includes most of the sediment interval G-IX at Gangta bet, i.e. beds 14–15 and younger of Biswas at Gangta bet (Fig. 2.33). The bed 14 is presumably ammonoid devoid but suggestively assigned to Antecedens Subzone in view of ammonoids of the early Indogermanus Zone below and ammonoids of the Parandieri Subzone above. The bed 15 is rich in ammonoids (Gangta ammonoid bed of Biswas), and encloses *Dichomosphinctes* cf *buckmani. D.* aff *buckmani* is also present in the K-B II bed 04 that marks the base of the Orientalis Zone in the Kantkote–Bharodia section. Thus, the early part of the JM is present in Gangta bet while the terminal ammonoid rich part in Desalpur and Kantkote dome (Figs. 3.27, 3.28 and 3.29). The JM starts at the base of the ammonoid devoid bed 14 of Biswas at Gangta bet, and includes the beds K-B II 01–12, and the beds K-B III 01–04 up to the top of the Chari Formation in the Kantkote section which is abundantly rich in ammonoids. The youngest ammonoid rich part is also known as Kantkote ammonoid beds. The Jhadasa Member, thus, on the basis of ammonoids is maintained to precisely range from Antecedens Subzone of the early Middle Oxfordian Indogermanus Zone to the Schilli Subzone of the late Middle Oxfordian Orientalis Zone compared to the recent broad assignment to Middle Oxfordian (Rai et al. 2015a). The unit does neither include the basal nor the terminal ammonoid subzones of the Middle Oxfordian.

4.9.3 Katrol Formation

The submarine gap between the shale dominated Chari and Katrol units interpreted earlier in the Mainland makes them incomplete in the entire Mainland. The early part of the Katrol Formation is exposed outside the Mainland only in the Kantkote–Bharodia section of Wagad conformably above the terminal part of the Chari Formation. The early part is here given the status of the Basal Member or the Kantkote Member of the Katrol Formation, which in all has four members. The oldest is the Kantkote Member while the youngest is the Upper Shale Member. The Kantkote Member is exclusive to Wagad, while other members exclusive to Mainland. The unit ranges from the late Middle Oxfordian Orientalis Zone Rotoides Subzone to the mid-Early Tithonian Natricoides Zone. The unit is absent in the 'Island belt', the sea having already withdrawn from there by the late Middle Oxfordian Orientalis Zone Schilli Subzone. In view of the margin to basin increase in the duration of the submarine gap, the base of the Katrol Formation is diachronous in Mainland from Jhura to Lakhapur. The total duration of the Katrol Formation is ~ 11.64 my.

Kantkote Member
It includes the beds 10–20 in the Bharodia section of Wagad (Fig. 3.32). It ranges from the latest Middle Oxfordian Orientalis Zone Rotoides Subzone to the mid-Early Kimmeridgian Giganticus Zone in its stratotype at Kantkote. The KM is absent in both Mainland and 'Island belt'.

KM has been organized by Deshpande and Merh (1980) in three divisions as 'Patasar Shale' (PSm), 'Fort Sandstone' (FSm) and 'Adhoi beds' (Am). The PSm based on its ammonoid content is assigned to the Rotoides Subzone (Fig. 4.1). The FSm is indirectly ranged between the ages of the PSm below and Am above to the middle and late part of the Subevolutus Zone. The Am, based on enclosed ammonoids, ranges from the Kachchhensis Zone to the Giganticus Zone of early to mid-Early Kimmeridgian.

Lower Katrol Member
It has its stratotype in Ler on the Hajapir Kukma highway, and includes beds 1–13 (Krishna et al. 1996d). The member is the best developed in the East Mainland in both north and south belts at Jawaharnagar, Ler, Walakhwas, etc. The unit is also present in the West Mainland at Keera, Jumara, sections, although not well studied. It includes the early part of the Katrol Formation at Ler of the late Early Kimmeridgian Alterneplicatus Zone. It disconformably overlies the Dhosa Oolite Member up to the major regressive thick cross-bedded sands of latest Kimmeridgian Katrolensis Zone Katrolensis Subzone (Figs. 2.36, 3.33 and 3.34).

Middle Katrol Member
It ranges from the Early Tithonian Pottingeri Zone to the mid-Early Tithonian Virgatosphinctoides Zone. It comprises marl/silts/sands of the beds 14–24 that form small to moderate monds or low ridges in between the low plains of LKM on the older side and the hard resistant cross-bedded sandy relatively more high hills of the

Umia Formation on the younger side at SW Ler (Figs. 2.39 and 3.35). The youngest Early Tithonian Natricoides Zone in India is developed and confirmed only in Jaisalmer. In IEAP, it is also present in Pakistan. However, in Ler, the Natricoides Zone is not confirmed. As such, the Virgatosphinctoides Zone may either include the Natricoides Zone in its youngest subzone, or there is a subaerial disconformable gap between the Middle and Upper members of the Katrol Formation.

Upper Shale Member
The thick shaly/silty ammonoid devoid relatively low lying expanse of the USM of the major part of the thick, yet further undifferentiated, bed 25 is not that distinctly visible from distance between the MKM and Umia Formation. The member is either ammonoid devoid or ammonoid scarce through the indeterminate incomplete whorl fragments. The USM is suggested to range from the mid-Early Tithonian Natricoides Zone (created in Jaisalmer basin and not recognized in Kachchh) to the late Early Communis Zone (Fig. 3.36).

4.9.4 Umia Formation

The Umia Formation is a sand dominating succession above the Katrol Formation, so defined in the Umia region of the West Mainland. Both the base and top of the Umia Formation are highly diachronous between the exposed margin to the exposed basin along the east–west transect. The Umia Formation is organized into five Members. It is present in Mainland and Wagad. In Wagad, it has been earlier given the name of the 'Gamadu Sandstone member' by Biswas which is here abandoned as a synonym of the Umia Formation. The base of the unit conceptually may extend down up to the intra-Oxfordian First-order MFS, however, it is here begun at least in late Early Kimmeridgian above the Early Kimmeridgian Giganticus Zone in Wagad in the proximal most exposed east, and may extend up to at least Albian or suggestively still younger up to the Santonian/Campanian boundary in the Ghuneri–Amarsar section in the distalmost exposed west (Figs. 2.41, 4.15 and 4.16). Both the base and top become younger and younger from margin to basin. The span of the Umia Formation that falls in the Jurassic is ~ 13.0 my. There is a large overlap with the Katrol Formation of $> \sim 11.0$ my. In fact both the Katrol and Umia units begin simultaneously at the start of the latest Middle Oxfordian Rotoides Subzone of the Orientalis Zone. The unit has been found differentiable only in West Mainland west of MH.

The 'Gamadu Sandstone member' ('GSm') is considered facies equivalent of the Ghuneri Member of Umia Formation, and assigned to at least as old as Early Kimmeridgian in view of the underlying ammonoid bearing mid-Early Giganticus Zone. In context of the facies, this cross-bedded dominantly sandy unit with thin interbeds of bioturbated sandstones is considered extension of the Umia Formation in Wagad of Early Kimmeridgian age. It makes the basal/oldest part of the Umia

Formation in Wagad. The base of the Umia Formation is highly diachronous in the basin from proximal to distal locations from Late Oxfordian or at least Early Kimmeridgian in Wagad to Late Cretaceous in the extreme west.

Age of the Basal Member of the Umia Formation It begins at the base of the Late Tithonian Densiplicatus Zone with termination in Early Cretaceous in Mainland. The Jurassic part includes the beds 16a–19d in the Mundhan section and beds 25–27 in the Ler section with ammonoids of the Communis Zone and Densiplicatus Zone.

4.9.5 Cretaceous

The biozonation and sequence stratigraphy of Kachchh Cretaceous is not detailed in Chaps. 2 and 3, and so the ages of the Kachchh Cretaceous units are summarized here. The ages given to the Indian Cretaceous units either on insufficient or **no** evidences are here revised and précised. The Cretaceous units in Kachchh are the abandoned 'Bhuj formation', and 'Katesar member' of Biswas (1993) besides the accepted Ghuneri, Ukra and Upper Member units of Biswas as members of the Umia Formation of Stoliczka and Amarsar Member of Krishna.

'Katesar Member'
It is the abandoned youngest Member unit of the abandoned 'Jhuran Formation' which is present only in the West Mainland. Its upper age without fail is early Late Valanginian Verrucosum Zone. Its base may vary from the Early Berriasian at Umia and Lakhapur to Late Berriasian in Mundhan and Artara section farther west in the basin based on the youngest Early Berriasian *Andesites* in Lakhapur and Late Berriasian *Kilianella* in Mundhan and Sahera.

Ghuneri Member
It has a diachronic base from the Late Tithonian in East Mailand to Late Berriasian in West Mainland, while the upper limit is Late Barremian. The subsurface Bibat Formation in Banni well may range from the Late Tithonian to Late Barremian.

Ukra Member
Biswas (1993) assigned it to Aptian. Krishna (1980, 1983a) dated it as Early Aptian to mid-Middle Albian. The Early Aptian is based on the then $\sim$112 ma fission track age of the basal green shale/sand of the UM in the Katesar–Ghuneri section. Farther up in the section are also collected *Deshayesites, Dufreyonia,* etc. of Early Aptian age. *Lemuroceras* is morphologically very close to *Deshayesites* but found to co-occur with the Early Albian index *Douvilleiceras mammilatus* in Jaisalmer and Pakistan. It may be that *Lemuroceras* is evolved from the Early Aptian *Deshayesites*, however, extends farther up independently in the region. The upper limit of the mid-Middle Albian is here given on the basis of a single *Lemuroceras* in the youngest bed of the Ukra Member in the Ghuneri section, as on the age of the second-order regional MFS in mid-Middle Albian.

Upper Member
Its age is interpreted from the late Middle Albian to the Santonian/Campanian boundary as discussed later in Chap. 7 through sequence surface-based correlation with the Spiti and Cauvery basin Cretaceous.

Amarsar Glauconite Bed
The thin green shaly unit is included in the Upper Member as its terminal marker bed which is encountered exclusively in the Ghuneri–Amarsar section. It is supposedly conformably underlain by the rest of the Upper Member, and overlain by the Deccan Traps. The unit has provided fission track age of ~89 ma (Krishna 1983b, revised as ~92 ma as per 2015 scale), and is interpreted to mark the first-order regional MFS either the early Middle Turonian first-order MFS or the second-order MFS near the Santonian/Campanian boundary.

4.9.6 Mundhra Formation

Defined in well SP 1-1, it comprises sands and subordinate shales (with occasional carbonaceous streak). It could be assigned the latest Santonian to early Campanian age as the second-order TST. It suggests shallow marine paralic environment. The unit also includes thin basalt layers. The coccolith-based Late Santonian/Early Campanian age (Raju 2011) suggests its broad equivalence to the exposed glauconitic shale below the Deccan traps. The thin basalt layers suggest Late Santonian volcanism distinctly older to the later Deccan volcanism and, possibly coeval to the volcanism at St. Mary Island of ~94–84 ma age.

4.9.7 Kori and Naliya Units

The Kori and Naliya units, both of mid-outer shelf to upper bathyal environment and of Maestrichtian age in the respective wells disconformably overly the Upper Member of the Umia Formation. These represent the youngest, the deepest and the farthest offshore sediments of the Kachchh Mesozoic. These are interpreted as transgressive, presumably corresponding to the Kallankuruchi Formation of the Cauvery basin. These units in-turn are unconformably overlain by Late Maestrichtian Deccan traps.

4.9.8 Recent Progress in the Age Assignments of the Oldest and Youngest Sediments in Different Parts of the Kachchh Basin

It has been known almost since the start of the studies of the Kachchh Mesozoic that the four uplifts in the north mostly include sediments older to those of

Mainland but for some overlaps. The density, diversity and frequency of the stratigraphically significant ammonoids compared to the Mainland are extremely low. If, at all, the marine body fossils were present, they were benthic bivalves, brachiopods, corals, gastropods, echinoids, sponges, etc. in view of the relatively much shallower marginal depositional sites. Moreover, the Island belt sections in sequence stratigraphic context included the relatively early part of the first-order transgressive interval.

Oldest Exposed Sediments in the Three Principal Sectors of the Basin
The oldest sediments were considered Bathonian in age (Rajnath 1932; Biswas 1977), however, as studies progressed, rare Bajocian ammonoids were discovered (so-called Late Bajocian *Leptosphinctes* by Jaitly and Singh (1983) which now stands revised taxonomically as with regard to its age, and more recently Early Bajocian *Calliphylloceras* by Pandey et al. 2013b). The claimed *Leptosphinctes* as per its stratigraphic position is decidedly of Bathonian age between the Early Bathonian Macrescens Subzone and the Middle Bathonian Bremeri Zone, and thus cannot even be a Leptosphinctinae. The only definite Bajocian ammonoid now left is the recently described *Calliphylloceras* from the base of the Sadhara section (Fig. 3.21). There were yet thick sediments underlying the Bajocian which unfortunately so far did not yield any ammonoids. However, Toarcian–Aalenian coccoliths have been recovered close to the basal beds in Kaladoongar at Dingy and Kuar bet, thus, extending down the age of the oldest sediments to Early Toarcian–almost to the start of the first-order marine transgression on the entire GTM (Gondwanian Tethyan Margin) from North Africa to NW Australia. This is so interpreted on the presence of the Early Toarcian ammonoid *Bouleiceras* in North Africa, Madagascar and Pakistan [Krishna (2012), Krishna et al. (2011). Krishna (in Bajpai et al. 2012)], unconformably, above the sands that are body fossil-devoid. Rai and Jain (2012) suggested the presence of marine Pliensbachian, however that is ruled out in view of the start of regional transgression only later in Early Toarcian. However, Pliensbachian, could be present in the subsurface non-marine sediments of the Banni well. These Early Jurassic sediments carry particular stratigraphic and geological significance in view of involving at least two first-order sequence surfaces, so interpreted through large nonconformities between the Precambrian basement and the non-marine Triassic in the Banni well, and the other between the palynomorph bearing non-marine sediments of the Banni well and the overlying marine Dm of the Patcham Formation. The oldest exposed beds in Goradongar at present seem extending down only up to Early Bajocian, at the most including some part of Aalenian.

The oldest exposed bed 'Cheriya Bet conglomerate' at Khadir appears younger than the oldest at Patcham. Compared to the 'Cheriya Bet conglomerate', there are two significant carbonate levels in Patcham (beds 24 and beds 32–35), both the levels being within the Dm which are interpreted to mark part of two successive third-order TSTs (Figs. 2.25 and 2.26). The younger one of Beds 32–35 is here only tentatively correlated with the 'Cheria Bet conglomerate' in view of similar *Corbula–Gervilla* biofacies in two localities only 16 km apart from each other.

Thus, the Khadir column may also include at least Aalenian and may be part of Toarcian as well. At Gangta bet the oldest beds are of the early Middle Callovian RSM, and at Kakindia bet of LSM of early and mid-Early Callovian as well with *Macrocephalites* spp. In view of the absence of Patcham Formation in Wagad, the oldest sediments belonging to the Chari Formation are of Early Callovian age. Early *Macrocephalites* has been discovered by the author several years ago (unpublished). At Bela and Mouvana, the oldest beds at least should be of the Early Bajocian age while at Chorar the oldest age is the latest Late Bathonian of RLm.

Compared to Patcham, the oldest age in the Mainland is younger by two stages Toarcian and Aalenian (i.e. ~11.4 my). These are so far not known in the exposed sediments of the Mainland, which begin in Early Bajocian in Jhura. In Jumara the oldest exposed sediments are late Middle Bathonian carbonates. Elsewhere in Mainalnd the oldest Jurassic sediments invariably are Early Callovian at Keera, Ler, Walakhwas, Jhuran, Lakhapur. West of Lakhapur, the oldest sediments are Late Kimmeridgian to Tithonian. In Habo there are present thin basal carbonates of Late Bathonian RLM of Patcham Formation. In Nara, the oldest age so far known was Middle Bathonian, however, recent studies by author's associates also suspect presence of Early Bathonian through ammonoids in a section (close to the igneous plug). Thus, the oldest sediments in Nara appear to straddle the Bajocian–Bathonian boundary with possibility of Late Bajocian as well. The igneous may be coeval to igneous event in Kachchh and Saurashtra wells of first-order Pliensbachian/Toarcian boundary SB age.

In Wagad the oldest age is Early Callovian in Washtawa dome, late Early and early Middle Oxfordian in Kantkote and Chitrod domes, Early Callovian in Dedarva dome, Late Oxfordian in Gon-Mardak and late Early Kimmeridgian in Manfara dome.

The Youngest Jurassic Sediments

The youngest Jurassic sediments in the Mainland are of the latest Late Tithonian Densiplicatus Zone. In the Island belt, the youngest sediments are from Khadir of undisputed late Middle Oxfordian Orientalis Zone Schilli Subzone (Krishna et al. 2000). The youngest sediments apart from Khadir are of late Early to early Middle Callovian age in Patcham and Bela and Mouvana of RSM and at Chorar only of LSM. Thus, the youngest Jurassic in the Island belt is older to Mainland by more than two stages (Latest Late Oxfordian, Kimmeridgian and Tithonian), i.e. older by ~12 my. In Wagad, the youngest Jurassic sediments are of the basalmost Member of Umia Formation (=the suppressed 'Gamadu Sandstone') of Early Kimmeridgian age.

The Youngest Marine Cretaceous Sediments

The Crteceous sediments are neither present in the 'Island belt' nor Wagad, nor in the East Mainland north belt. In the East Mainland south belt south of the Katrol hill range, the youngest ammonoid-based age is Early Berriasian, and the ammonoid devoid bivalve bearing part may range up to the intra-Valanginian Verrucosum Zone second-order TST. The youngest Cretaceous immediately west of the MH is ranged up to at least mid-Midde Albian coccolith record from near Jakh temple. The youngest ammonoid bearing sediments are the Late Berriasian Kilianella bearing sediments up to Mundhan and Sahera. Based on pollen/spores and forams, Mathur et al. (1970) have ranged the otherwise marine megafossil-devoid sediments all over the Mainland irrespective of the MH divide up to the Early Santonian. It implies that in the studied offshore wells, either the marine sedimentation is rather continuous in well SP-1-1 with conformable contact between the Bhuj and Mundhra units up to the close of Cretaceous, or is present only a small post Santonian duration gap. So may be the case between the Bhuj and Naliya units in the well GK-23-1. There is interpreted uplift, emergence, and withdrawl of shoreline from Banni any time after the Late Kimmeridigian.

The Youngest Non-Marine Cretaceous Sediments
The Cretaceous sediments are exclusive to Mainland. The youngest Cretaceous sediments are of the Anjar Formation of non-marine inter-trappeans of Late Maestrcihtian age. Only the first three intertrappeans are of Cretaceous, and the youngest of the three-the third has yielded dinosaurian remains of *Megalosauras, Antarctosauras, Titanosauras* and others (Ghevariya and Srikarni 1996). The dinosaurian bearing 3rd intertrappean of Anjar also marks the K/T boundary and extinction of the dinosaurs. Age constraints from the over and underlying flows have given Ar38/Ar40 age of ~65. 4–65.5 ma. The first two intertrappeans of the Anjar Formation did not yield dinosaurian remains. The intertrappean beds are also present widely in the West Mainland and dated Late Maestrichtian–Paleocene as well there, however, there is no correlation suggested of the third intertrappean at Anjar with the West Mainland intertrappeans.

References

Agrawal SK (1957) Kutch Mesozoic: a study of the Jurassic of Kutch with special reference to Jhura Dome. J Pal Soc Ind 2:119–130

Alberti M, Fursich FT, Pandey DK (2013) Deciphering condensed sequences: a case study from the Oxfordian (Upper Jurassic) Dhosa Oolite Member of the Kachchh basin, western India. Sed Geol 60:574–598

Alberti M, Fursich FT, Pandey DK, Ramkumar M (2012) Stable isotope analysis of belemnites from the Kachchh basin, Western India: paleoclimatic implications for the Middle to Late Jurassic transition. Facies 58:261–268

Alberti M, Pandey DK, Fursich FT (2011) Ammonites of the Genus Peltoceratoides Spath 1924 from the Oxfordian of Kachchh, western India. N Jb Geol Palaont Abh 262:1–18

Ayyasami K (2011) Recognition of the Cretaceous stage boundaries in Southern Indian Cretaceous based on invertebrate fauna, 78–87. In: Jaitly AK, Singh AD, Pandey B, Nath S (eds) Palaentology and stratigraphy: basics to applications. Banaras Hindu University, Varanasi

Ayyasami K, Bannerjee RK (1984) Cenomanian—Turonian transition in the cretaceous of Southern India. Bull Geol Soc Denmark 33:21–30

Bajpai S, Prasad GVR, Prasad V, Krishna J, Sarkar A (2012) Recent advances on phenorozoic biodiversity, bioevents and climates in India. PINSA 78(Spec. Issue):445–456

Bertle RJ, Suttner TJ (2005) New biostratigraphic data from the Chikkim Formation (Cretaceous Tethys Himalaya, India). Cret Res 26:882–894

Bhargava ON, Tangri SK (1996) Late Jurassic – Early Cretaceous Lingshi Formation, Bhutan Himalaya. J. Geol. Soc. Ind. 37:113–119

Bhargava ON, Krystin L, Balini H, Lein R, Nicora A (2004) Revised litho and biosequence stratigraphy of the Spiti Triassic. Albertiana Supplement 30:21–38

Bhatia SB, Jain SP (1969) Dalmiapuram formation: a new Lower Cretaceous Horizon in South India. Bull Ind Geol Ass 2:105–108

Bhatt DK, Joshi VK, Arora RK (1999) Conodont biostratigraphy of the Lower Triassic in Spiti Himalaya. J Geol Soc Ind 54(2):153–167

Biswas SK (1977) Mesozoic rock-stratigraphy of Kutch. Quart J Min Met Soc Ind 43 (3and 4): 1–52

Biswas SK (1981) Basin framework, palaeo-environment and depositional history of the Mesozoic sediments of Kutch basin, Western India. Quart J Geol Min Met Soc Ind 53:56–85

Biswas SK (1982) Rift basins in western margin of India, and their hydrocarbon prospect, with special reference to Kutch. Am Ass Petr Geol 66:1497–1513

Biswas SK (1987) Regional tectonic framework, structure and evolution of the western marginal basins of India. Tectonophysics 135:307–327

Biswas SK (1990) Correlation of nonmarine and estuarine Cretaceous of Western India, Pres. Address. In: Sem-cum-Workshop IGCP 216 and 245, Chandigarh, pp 1–13

Biswas SK (1993) Geology of Kutch. K D Malaviya Institute of Petroleum Exploration, Dehradun, vol 1, p 450.

Biswas SK (1999) A review of the evolution of rift basins of India during Gondwana with special reference to Western Indian basins and their hydrocarbon prospects. PINSA 65(3):261–283

Biswas SK (2005) Tectonic style and sediment dynamics of rifted basins, their bearing on petroleum habitat – examples from Satpura and Kutch, Basins, India. Ind J Petrol Geol 14(1):1–29

Biswas SK (2012) Stratigraphy of Kutch – An overview, Status talk (in abstract vol.). In: Thakkar MG (ed) Geology of Kachchh Basin Western India, Bhuj, Kachchh, pp 11–13

Biswas SK, Bhasin AC, Ram J (1993) Classification of Indian sedimentary basins in the framework of plate tectonics. In: Proceedings of the second seminar in petroliferous Basins of India 1:1–46. Indian Petroleum Publication

Bladon AJ, Stuart MC, Stuart DB, Hazel B, Whiteley N, Kothari V, Mohapatra P (2015) Previously unrecognised extensional tectonics exposed in the Barmer Basin, Northwest India. Search and Discovery Article #30426

Bose MN (1884) Geology of the lower Narmada valley between Nimawar and Karwant. Mem Geol Surv Ind 21(1):1–72

Chhabra NL, Kumar S (1984) Record of Carnian conodonts from the top of Kalapani Limestone, Kumaon Himalaya, India. J Pal Soc Ind 29:88–92

Chhabra NL, Sahni A (1981) Late lower triassic and early middle Triassic conodont fauna from Kashmir and Kumaon Himalaya. J Pal Soc Ind 25:135–147

Chidambaram L (2000) Middle Turonian—Santonian foraminiferal biostratigraphy of Trichinopoly group, South India. Mem Geol Soc Ind 46:173–204

Chiplonkar GW, Borkar VD (1971) Fossil algae from the Wadhwan Formation. In: Proceedings seminar Paleopalynology and Indian stratigraphy, 140–149

Deshpande SV, Merh SS (1980) Mesozoic sedimentary model of Wagad Hills, Kutch, Western India. J Geol Soc Ind 21:75–83

Diener C (1908) Ladinc, Carnic, and Noric faunae of Spiti, vol 15/5, 157pp. Springer, Germany

Diener C (1912) The triassic of the Himalayas. Mem Geol Surv Ind 36:1–176

Enay R (1966) L'Oxfordien dans le moitie sud de Jura Francais Etude stratigraphique. Nouv Arch Mus Hist Nat Lyon Stratigraphie 324 p

Fursich FT, Pandey DK, Oshman W, Callomon JH, Jaitly AK (1994) Contribution to the Jurassic of Kachchh, Western India: II Bathonian stratigraphy and depositional environments of the Sadhara dome, Pachcham Island. Beringeria 12:95–125

Garg R, Ateequzzaman K, Krishna J, Jain KP (2003) Biostratigraphic potential of dinoflagellate cysts recovered from the Late Jurassic ammonites of the Tethys Himalaya, India. J Pal Soc India 48:41–58

Gaur KN, Sisodia AK (2000) The age and biostratigraphic significance of Jurassic Foraminifera from Keera Hills, Kutch, India. Bull ONGC 37(1):1–8

Ghevaria ZG, Srikarni C (1996) Dinosaur fauna from Mesozoic rocks of Western India. Gondwana 9(2):143–164

Govindan A, Ravindran CN and Rangaraj MK (1996) Cretaceous stratigraphy and planktonic foraminiferal zonation of the Cauvery basin, South India. Memoir 37, Geol Soc Ind 155–188

Govindan A, Yadagiri K, Ravindran CN, and Kalyansundaram R (1998) The Trichy field guide. Int Sem Rec Advan Study Cret Sections 1–53

Graciansky PC, Hardenbol J, Jacquin T, Vail PR (1998) (eds) Mesozoic and Cenozoic sequence stratigraphy of European basins. Soc Sed Geol (SEPM), Tulsa (OK) 60:445–466

Gradstein FM, Gibling MR, Sarti M, Rad UV, Thruro JW, Ogg JG, Jansa LF, Kaminski MA, Westermann GEG (1991) Mesozoic Tethyan strata of Thakkhola, Nepal: evidence for the drift and break-up of Gondwana, 88:193–218

Haq BU, Hardenbol J, Vail PR (1987) Chronology of fluctuating sea-level since the Triassic. Science 235:1156–1167

Helby R, Morgan R, Partridge AD (1987) A palynological zonation for the Australian mesozoic. Memoir Assoc Australasian Palaeontol 4:1–94

Herngreen G, Randrianasold A, Verbeek JW (1982) Micropaleontology of Albian to Danian strata in Madagascar. Micropaleontology 28:97–109

Hu X, Garzanti E, An W (2015) Provenance and drainage system of the Early Cretaceous volcanic detritus in the Himalaya as constrained by detrital zircon geochronology. J Paleogeogr 4(1): 85–98

Jafar SA (1982) Nannoplankton evidence of the Turonian transgression during Narmada valley, India, and Turonian – Coniacian boundary problem. J Pal Soc Ind 27:17–30

Jafar SA (1996) The evolution of marine Cretaceous basins of India: Calibration with nannofossil zones. Memoir 37, Geol Soc Ind 1–34

Jafar SA, Rai J (1989) Discovery of Albian nannoflora from type Dalmiapuram Formation, Cauvery basin, India, and palaeooceanographic remarks. Curr Sci 58(7):358–363

Jain KP, Garg R, Kumar S, Singh IB, Singh SK (1980) Dinoflagellates and radiolarians from the Tcthyan sediments. Malla Johar area Kumaon Himalaya. A preliminary report. J Pal Soc India 21.22, 116–119

Jain S (2008) Integrated Jurassic biostratigraphy: a closer look at nannofossil and ammonite evidences from the Indian subcontinent. Curr Sci 95(2):326–331

Jaitley AK, Ajane R (2013) Comments on Placenticeras mintoi (Vredenburg 1906) from the Bagh beds (Late Cretaceous), Central India with special reference to Turonian Nodular Limestone horizon. Geo Soc Ind 81:565–574

Jaitly AK, Singh CSP (1983) Discovery of late Bajocian Leptosphinctes Buckman (Jurassic Ammonitina) from Kachchh, Western India–Neues Jahrbuch Fur Geologie und Palaontlogie, Monatshefte, 91–96

Kale AS, Phansalkar VG (1992) Calcareous nanofossils from the Uttatur Group Trichinopoly District, Tamil Nadu india. J Pal Soc Ind 37:85–102

Kale AS, Lotfalikani A, Phansalkar VG (2000) Calcareous nannofossils from the Uttatur group of Trichonopoly Cretaceous, South India. Mem Geol Soc Ind 46:213–227

Kalia P, Chowdhary S (1983) Foraminiferal stratigraphy, geography and environment of the Callovian sequence, Rajasthan, Northwestern India. Micropaleontology 29(3):223–254

Kalmalkar NR, Duraiswami RA, Jonalagadda MK, Griffin WL (2014) Mid-cretaceous lamproite from the Kutch region, Gujarat, India. Gond Res 26(3–4):942–956

Kanjilal S, Singh SR (2012) Arrival of Quenstedtoceras Hyatt: a late Callovian ammonite in Kachchh and its significance. Geology of Kachchh Basin, p 27

Khowaja-Ateequzzaman and Garg, Garg (2002) Dinoflagellate cyst evidence for the age of Kulakkalnattam Sandstone Member, Garudamangalam Formation, Cauvery basin, Southern India. Palaeobotanist 51:129–143

Kossmat F (1895–97) Untersuchungen uber die sudindische creide formation. Beitr. Pal. Geol. Ungarns des Orients, 9(11–12):1–217

Krishna J (1980) Uncoiled ammonites of Middle Albian (Lower Cretaceous) age from Habur Series, Jaisalmer, Rajasthan. J Pal Soc Ind 23 and 24, 49–54

Krishna J (1983a) Callovian - Albian ammonoid stratigraphy and paleobiogeography in the Indian subcontinent with special reference to the Tethys Himalayas. Him Geol 11:43–72

Krishna J (1983b) Reappraisal of the marine and/or "mixed" Lower Cretaceous sedimentary sequences of India; paleogeography and time boundaries. Cretaceous of India, pp 94–119

Krishna J (1994) Environmental reconstruction studies of the Mesozoic sedimentary sequences in Western Kachchh (ONGC Sponsored Project), (Technical Report), 1–143, pts, 1–10

Krishna J (2012) Litho-bio-chrono-magneto-igno-sequento-stratigraphic integration in the Kachchh Mesozoic. In: Thakkar MG (ed) Status talk – Mesozoic Biostratigraphy, In Abstract Volume, Geology of Kachchh Basin, Western India, pp 21–25

Krishna J, Ojha JR (1996) The Callovian ammonoid chronology in Kachchh (India). Geo Res Forum 1–2:151–166

Krishna J, Melendez G, Pandey B, Pathak DB (1996a) Middle Oxfordian ammonites (Perisphinctinae) from Kachchh (India): biostratigraphic and paleobiogeographic implications. Revista Espanola De Paleontologia, Numero Extraordinario, pp 140–147

Krishna J, Melendez G, Pandey B, Pathak DB (1995a) Characterisation of the ammonoid genus Larcheria (Middle Oxfordian) in Kachchh (India): paleontology, biostratigraphic and paleobiographic evaluation in the context of North-Tethyan occurences. CR Acad Sci Paris 321, lla, 1187–1193

Krishna J, Pandey B, Pathak DB (1996b) Ammonoid chronology in the Tithonian of Kachchh (India). GeoResearch Forum, Switzerland 1–2:205–214

Krishna J, Pandey B, Pathak DB (2009a) Characterization of Dichotomoceras in the Oxfordian of Kachchh (India). J Geol Soc Ind 74:469–479

Krishna J, Pandey B, Pathak DB (2010) Update on the Jurassic stratigraphy of the Indian subcontinent with focus on the tectonically controlled regional transgressive–regressive couplets. Earth Science Frontiers, China 17:138–140

Krishna J, Pandey B, Pathak DB (2011) Current status of the Jurassic ammonoid stratigraphic framework in the Indian subcontinent with focus on the tectonically controlled regional transgressive–regressive couplets. Geol Soc Ind Mem 78:140–176

Krishna J, Pandey B, Ojha JR, Pathak DB (2000) Transgressive sediment intervals in the Late Jurassic of Kachchh, India. GeoResearch Forum 6:321–332

Krishna J, Pandey B, Ojha JR, Pathak DB (2009b) Reappraisal of the age framework, correlation, environment and nomenclature of Kachchh Mesozoic lithostratigraphic units in Wagad. J Sci Res Banaras Hindu University, Varanasi 53:1–20

Krishna J, Pathak DB (1994) Stratigraphic, biogeographic and environmental signatures in the ammonite bearing Jurassic-Cretaceous in the Himalaya on the south margin of the Tethys. J Him Geol 4(2):151–167

Krishna J, Pathak DB, Pandey B (1995b) Kimmeridgian–Tithonian (Upper Jurassic) ammonoid zones in Kachchh (Western India) and their correlations. Proc Recent Res Geol Western India, Badodara, pp 323–347

Krishna J, Pathak DB, Pandey B (1996c) Quantum refinement in the Kimmeridgian ammonoid chronology in Kachchh (India). GeoResearch Forum 1–2:195–204

Krishna J, Pathak DB, Pandey B (1998) Development of Oxfordian (Early Upper Jurassic) in the most proximal exposed part of Kachchh basin at Wagad outside the Mainland Kachchh. J Geol Soc Ind 52:513–522

Krishna J, Singh IB, Howard JD, Jafar SA (1983) Implications of new data on the Mesozoic rocks of Kachchh, India. Nature 305:790–792

Krystyn L, Balini M, Nicora A (2004) Lower and Middle Triassic stage and substage boundaries in Spiti. Albertiana 30(Supplement):39–52

Mathur YK, Soodan KS, Mathur K, Bhatia ML, Juyal NP, Pant J (1970) Microfossil evidences on the presence of Upper Cretaceous and Paleocene sediments in Kutch. ONGC Bull 7(2): 109–114

Mehrotra NC, Aswal HS (2003) Late Jurassic – Cretaceous dinoflagellage cysts. Pal Ind 7, *ONGC*, Dehradoon, 1–108

Mehrotra NC, Venkatachala BS, Swami SN, Kapoor PN (2002) Palynology and hydro-carbon exploration (the Indian scenario). Mem Geol Soc Ind 48:1–161

Mehrotra NC, Shanmukhappa N, Babu R et al (2012) Development of palynology in fossil fuel exploration in India with emphasis on recent significant contributions from Western offshore. Krishna-Godavari Basin and Frontier Areas PINSA 78(3):457–473

Nagendra R, Kamalnarayanan BV, Gargi S, Gilbert H, Bakkalraj D, Nallapa-Reddy A, Jaiprakash B (2011) Sequence surfaces and paleo-bathymetric trends in Albian to Maestrichtian sediments of Ariyalur area, Cauvery basin. J Mar Petr Geol 28:895–905

Narayanan and Raju DSN (1996) Chronostratigraphic subdivisions of Uttatur of Blanford. Memoir 37. Geol Soc Ind 209–212

ONGC (1993) Lithostratigraphy of Indian petroliferous basins documents. KDMIPE, Dehradun

ONGC (2008) Pteroleum systems sequence stratigraphy. KDMIPE, Dehradun

Pal AK (1971) A note on the biozonation of the Bagh Group of Madhya Pradesh, Abstract. Proc Ind Sci Congr 333

Pandey B, Krishna J, Pathak DB, Kumar A (2013a) Ammonoid biostratigraphy of Bathonian succession at Jumara, Kachch, Western India. J Ind Geol Cong 4(2):7–18

Pandey B, Pathak DB, Krishna J (2013b) Calliphylloceras heterophylloides (Oppel, 1856) from basal most Jurassic succession of Sadhara dome, Kachchh, India. J Pal Soc Ind 58(1):61–65

Pandey B, Pathak DB, Jaitly AK, Krishna J, Venkateshwarlu M (2012a) Record of Tethyan gastropod genus Astrohelix Szabo, 1984 from Late Bajocian (Middle Jursassic) of Kachchh, western India. Ind J Geol 66(1):65–68

Pandey D K and Agrawal S K (1984) Bathonian–Callovian Molluscs of Gora Dongar, Pachchham "Island" District Kachchh, Gujarat). Q J Geol Min. Metallur Soc India, 56(4): 176–196; Calcutta

Pandey DK, Alberti M, Fursich FT (2012b) Ammonites of the genus Persphinctes Waagen 1869 from the Oxfordian of Kachchh, western India. Revue de Paleobiologie, Geneve 31(20): 139–174

Pandey J, Dave A (1993) Studies in Mesozoic foraminifera and chronostratigraphy of Western Kutch, Gujarat. Pal Ind 1:1–221

Pathak DB, Krishna J, Pandey B (2011) Differentiation of the significant Late Valanginian (Early Cretaceous) transgressive event in the Spiti Himalaya, India. J Asian Earth Sci 42:1226–1231

Prasad B (2011) Palynology: Principles, methodology and applications in biostratigraphy and oil exploration. In: Jaitly AK, Singh AD, Pandey B, Nath S (eds) Palaeontology and stratigraphy: basics to applications, pp 16–35

Prasad B, Phor (2009) Palynostratigraphy of the subsurface Gondwana and post-Gondwana Mesozoics of the Cauvey basin, India. J Pal Soc Ind 54(1):41–71

Prasad B, Pundeer BS (1999) Biostratigraphy of the exposed Gondwana and Cretaceous rocks of the Krishna–Godavari basin, India. J Pal Soc Ind 44:91–117

Prasad B, Jain AK, Mathur YK (1995) A standard palynozonation scheme for Cretaceous and pre Cretaceous subsurface sediments of Krishna-Godavar Basin, India. Geo Sci J 16(2):151–232

Prasad B, Kumar A, Raju DSN (1996) Chronostratigraphic subdivisions of pre-Aption (Pre-Uttaturaian) marine sediments in the Krishna-Godavari basin, India. Memoir 37, Geol Soc Ind 193–208

Prasad GVR (2011) Vertebrate palaeontological research in India: recent advances and future prospects, 49–58. In: Jaitly AK, Singh AD, Pandey B, Nath S (eds) Palaeontology and stratigraphy: basics to applications

Rai J (2006) Discovery of nannofossils in a plant bed of the Bhuj Member, Kutch, and its significance. Curr Sci 91–94:519–526

Rai J (2009) in response to Jain S (2008) Integrated Jurassic biostratigraphy: a closer look at nannofossil and ammonite evidences from the Indian subcontinent. Curr Sci 95(2):326–331

Rai J (2014) When did the twain meet? The epieric ethiopean gulf with Western Indian craton; coccoliths, the beacon light (abstr), 9th Internat. Congr Jur Syst Jaipur Ind 150

Rai J, Jain S (2012) Early Jurassic Gondwana break-up—a nanno-fossil story. In: Geology of Kachchh Basin, Western India (abstract), 31–32

Rai J, Jain S (2013) Pliensbachian nanno-fossils from Kachchh: implications on the earliest Jurassic transgressive event on the western Indian margin. Zitteliana, A 53:105–120

Rai J, Garg R, Kumar S (2002) Early Campanian nannofossils from Chakrud, near Zeerabad, Bagh area: stratigraphic and palaeoenvironmental implications. (Abstract) National Conference on Biodiversity—Past and Present, Lucknow, p 53

Rai J, Garg S, Gupta M, Singh A, Pandey D K, Fursich F T, Alberti M, Garg R (2015a) Integrated biostratigraphy of the Jurassic strata of the Wagad uplift, Kachchh, Western India, Vol. Jur. XIII(2):55–80

Rai J, Gupta M, Garg S et al (2013/14) Nannofossil biostratigraphy of the Chari Formation, Jumara dome, Kachchh, Western India. In: 9th International Congress Jur. Syst., Jaipur, Ind., 153–154

Rai J, Krishna J, Ojha JR (2005) Late Callovian nannofossils from Jara Dome, Kutch, Western India: biochronology and calibration by ammonites. XXth CMS, Visakhapatnam, 40

Rai J, Patel SJ, Singh A et al (2014) Middle Jurassic age calcareous nannofossils from Habo dome, Kachchh, Western India. In: 9th International Congress on *the* Jurassic System in Jaipur, 155–157

Rai J, Pathak DB, Pandey B (2015b) Early Campanian calcareous nannofossils from Chikkim Formation, Spiti Valley, Tethys Himalaya, India. In: 30th Himalaya-Karakoram-Tibet Workshop Dehradun, India, pp 32–33

Rai J, Singh A, Garg R (2013b) Calcareous nannofossils of Albian age from Tanot well 1, Jaisalmer basin, Rajasthan, and its palaeobiogeographic significance. J Pal Soc Ind 58(1): 67–73

Rajnath (1932) A contribution to the stratigraphy of Kutch. Quart J Min Met Geol Soc Ind 4(4):161–174

Rajnath (1942) The Jurassic rocks of Kutch, their bearing on some problems of Indian geology. Pres Address, 29th Ind Sci Congr 2, 93–106.

Raju DSN (2008) Stratigraphy of India; mega—charts 1–2. Bull. ONGC, Spl. Issue, 43 (1), 1-44 and mega charts

Raju DSN (2011) Cretaceous stratigraphic record on Indian plate: an overview. Mem Geol Soc Ind 78:177–228

Sahni A, Chhabra NL (1974) Middle Triassic conodonts from Kalapani Limestone, Malla Johar, Kumaon Himalaya. Him Geol 4:259–295

Sahni A, Prakash I (1973) Rhaetic condonts from Niti pass region, Kumaon Himalaya. Curr Sci 42(6):218

Sharland P, Archer R, Casey DM et al (2001) Arabian plate sequence stratigraphy. Geoarabia 9(1):199

Singh MP (1979) Foraminifera from Kuti Shale (Triassic), Malla Johar area, Kumaun Himalaya, U.P. Himalayan Geol 9(1):324–338

Singh NP (2006) Mesozoic lithostratigraphy of the Jaisalmer basin, Rajasthan. J Pal Soc Ind 51(2):1–25

Spath LF (1927–33) Revision of the Jurassic Cephalopod fauna of Kach (Kutch). Paleont. Indica, Calcutta 9(2):1–945

Srinivasan MS (1978) New chronostratigraphic subdivision of the Andaman–Nicobar Late Cenozoic rocks. Rec Res Geol. Hindustan Publishing Corporation, Delhi, 22–36

Stolickzka F (1861–63) Cretaceous fauna of South India, Cephalopoda. Pal. Ind 1:1–216

Subbotina NN, Dutta AK, Srivastava BN (1960) Foraminifera from the upper Jurassic deposits of Rajasthan (Jaisalmer) and Kutch, India. Bull Geol Min Met Soc Ind 23:148

Thierstein HR (1974) Calcareous nannoplankton, Leg 26. Deep Sea Drilling Project Intial Reports 26:619–667

Tripathi SC, Lahiri TC (2000) Marine oscillation event stratification: an example from Late Cretaceous Bagh carbonate sequence of Narmada valley India. Mem Spl Geol Soc Ind 46:15–24

Turner S, Bean LB, Dettmann M et al (2009) Australian Jurassic sedimentary and fossil succession: current work and future prospects for marine and non-marine correlation. GFF 131:49–70

Upadhyaya R, Rai J, Sinha AK (2005) New record of Bathonian-Callovian calcareous nannofossils from the eastern Karakorum block, a possible clue to understanding the dextral offset along the Karakorum fault. Terra Nova 17:149–157

Valdiya KS (2010) The making of India; geodynamic evolution, 1st edn. McMillan, India, p 816

Valdiya KS (2016) The making of India; geodynamic evolution (2nd edn). Springer, p 924

Vijaya (1997) Benchmark Permian–Triassic palynoevents in peninsular India and northwest Himalaya. J Pal Soc Ind 42:117–126

Vijaya (2009) Palynofloral changes in the Upper Paleozoic and Mesozoic of the Deocha-Pachami area, Birbhum coalfield, West Bengal. Sci China Ser D-Earth Sci 52(12):1932–1952

Waagen W (1873–75) Jurassic fauna of Cutch, Cephalopoda. Paleont, Indica, Calcutta 9(1):1–247

Yadagiri K, Govinadan A (1997) A record of algal oncolites of lower Permian age in Cauvery basin, Tamilnadu, India. J Geol Soc India 50:739–742

Chapter 5
Applicability of the Sequence Framework in IEAP and GTM, with Brief Comments on the Hydrocarbon Prospects in the Indian Basins

Abstract The validity of the Indian first- and second-order sequence stratigraphic framework is first demonstrated in India and IEAP through the isochronous character of as many sequence surfaces as possible. It is realized through investigation of the formulated sequence surfaces in the relatively better studied time slices in specific basins of the different sectors (Fig. 5.1). Subsequently, detailed comparison of the developed Indian succession of sequence surfaces is made, particularly, with the Arabian Triassic–Jurassic, and Madagascar Jurassic–Cretaceous (Figs. 5.1, 5.2, 5.3, 5.4, 5.5 and 5.6), as also mutually among the Jurassic development in India, Madagascar and Arabia (Fig. 5.4). Comparison of the Indian sequence framework is also made with regions/basins east of India at the first-order level with Dampier (Fig. 5.7), West Irian (Fig. 5.8), Timor/Sula (Fig. 5.9), and Timor Gap (Fig. 5.10), as also at second-order level with Dampier and West Irian basins. The comparisons exhibit substantive concordance of the large number of first- and second-order surfaces on the GTM both east and west of India. Comparison is also made of the Indian framework with the ETM framework, and salient differences are brought out. The comprehensive exercise reaffirms the earlier held view of the author of the need of independent development of the Indian framework. In comparison with ETM, almost all the Indian first-order surfaces (SBs/MFSs) invariably disagree in ages with the corresponding ETM surfaces by several ammonoid zones, and the obvious influence of regional tectonics in GTM is brought home well. The discordance is excellently exampled by the GTM intra-Jurassic first-order MFS in late Middle Oxfordian Orientalis Zone Schilli Subzone and the corresponding ETM first-order MFS in late Late Kimmeridgian Beckeri Zone. In contrast to the corresponding diachronous first- and second-order surfaces, the third order and finer high frequency low magnitude sequences are considered punctuated by synchronous sequence surfaces across the globe, since being governed by the earth's orbital dynamics which could not be distorted by low-frequency and high-magnitude first- and second-order extensional tectonics. The chapter is concluded with summarized sequence stratigraphy-based evaluation of the hydrocarbon prospects in the

© Springer Science+Business Media Singapore 2017
J. Krishna, *The Indian Mesozoic Chronicle*, Springer Geology,
DOI 10.1007/978-981-10-2477-1_5

mega-sequence geological record of the Indian basins that reflect an over all positive outlook. An elaborate Indian hydrocarbon prospect analysis being outside the scope of the present work is planned separately in near future.

Keywords Depositional sequences · Validation · Comparison · India · IEAP · GTM · Arabia · Madagascar · East Africa · Dampier · West Irian · Timor · Sula · Timor gap · Concordance on GTM · Discordance with ETM of first/second order · Third-order globality · Hydrocarbon prospects

5.1 Comparison with the Earlier Indian Frameworks

5.1.1 Initiation of Sequence Differentiation in India

The initial differentiation of the Indian geological record into some kind of sequences based on sharp lithological changes, stratigraphic gaps, unconformities, and other parameters in a limited context and broad manner in High Himalaya is noted in Krishnaswami and Swaminath (1965), Srikantia (1977), Bhargava (1987, 2011 also editorial preface), Bhargava and Bassi (1996), Bhargava et al. (2004), and others. Elaborate regional endeavours are made in Shanker et al. (1996), Raju (2008), ONGC (2008), and Dave (2011). The said studies, however, fall short of formal depositional sequences with appropriately defined SBs and MFSs. Sequence stratigraphic differentiation in recent years is also realized in individual basins in different sectors of India.

5.1.2 Ammonoid-Based Sequence Stratigraphic Studies in the Indian Mesozoics

The Jurassic succession in Kachchh has been in parts since the mid-1980s studied, and understood as per basic sequence stratigraphic concepts, and also organized in context of the Vailean (Haq et al. 1987 and others) succession of sequences (Krishna 1983, 1987a, b, 1994), Krishna and Cariou (1986, 1993), Krishna and Pathak (1989, 1991, 1993, 1994), Cariou and Krishna (1988), Cariou et al. (1990). The first comprehensive regional organization of the Indian Jurassic–Cretaceous succession was presented in 1994 (Krishna and Pathak 1994; Krishna unpublished project report 1994), and thereafter progressively strengthened over the last two decades through several publications (Krishna 2005 and 2012; Krishna et al. 1998, 2000, 2008, 2010, 2011; Krishna in Bajpai et al. 2012). The validity of the Kachchh Mesozoic tectono-stratigraphy, in particular of the Jurassic depositional sequences and sequence stratigraphic surfaces, in Jaisalmer and Pakistan in the west sector, and Spiti and Nepal in the north sector has been dealt with in some measure, particularly in Krishna and Pathak (1994) and Krishna et al. (2011).

5.1.3 Sequences Without Précised SBs and MFSs

The present mega-sequence broadly corresponds to the sequences VII to X of Shanker et al. (1996). The first first-order sequence spanning from the Late Permian Kazanian/Guadalupian stage or even earlier in Early Permian to mid/late Early Jurassic Pliensbachian stage broadly corresponds to the combined interval of sequences VII and VIII of Shanker et al. (1996).

Shanker et al. (1996)		Here
1. Late Permian to Carnian Sequence	VII	1. Kazanian to Pliensbachian
2. Norian to Early Jurassic Sequence	VIII	
3. Callovian to Albian	IX	2. Toarcian to Barremian
4. Cenomanian to Maestrichtian	X + XI	3. Aptian to intra-Paleogene
Without mention of MFSs or SBs		With precisely determined MFSs and SBs

Without mentioning the concept of MFS or SB, Shanker et al. (op. cit) suggested peak intra-Triassic transgression in Anisian Nodular Limestone unit. Bhargava (oral communication) favours a younger radiolaria bearing intra-Carnian level. In comparison, the intra-Triassic first-order MFS is précised in the ammonoid-rich terminal Anisian Himalayan Muschelkalk Member in the Kellnerites Zone in the present work which is 6 plus metre above the Nodular Limestone, and way older to the intra-Carnian level. The second first-order sequence of the present work spanning from the late Early Jurassic basal Toarcian to the close of late Early Cretaceous Barremian stage, broadly, corresponds to sequence IX of Shanker et al. (1996). The first-order basal SB of sequence VII below the Kuling Formation is the same as in the present work, however, the terminal SB of the sequence VIII in the present work is placed between the Kioto Limestone below and the Laptal/Tagling Formation above at the Pliensbachian/Toarcian boundary instead of the contact of the Laptal Formation and the Ferruginous Oolite Formation. The first-order MFS of this sequence is précised above the early Late Jurassic late Middle Oxfordian Orientalis Zone (=ETM Transversarium Zone) Schilli Subzone Jhadasa Member succession present exclusively in Kantkote locality of the Kachchh basin. These sequence surfaces differ distinctly from those of Shanker et al. (1996) placed below Callovian and above Albian. Krishna and coworkers in several earlier publications have repeatedly stressed on a large duration, several million year long, regional intra-Jurassic submarine gap of varied span not just in the Indian geological record, but as well all over the GTM from Arabia to NW Australia. The sequences X and

XI of Shanker et al. again correspond broadly to the third first-order sequence of the present work, however, the involved first-order sequence surfaces differ markedly. The basal first-order SB in the present work is placed near the base of the Aptian within the terminal part of the Guimal Formation much unlike that of Shanker et al. above the Guimal Formation.

5.1.4 ONGC Sequence Framework

The ONGC frameworks used here in the comparison include its formal version (ONGC 2008) as also other works by ONGC scientists (Raju 2008), Mehrotra and Aswal (2003), Dave (2011, 2012) and others.

5.1.4.1 Broad Comparison with the ONGC Framework

The ONGC outlined (without details) the status of the sequence stratigraphic framework separately in several Indian basins (ONGC 2008; Dave 2011), as also without demonstrating the interbasinal validity of the sequence surfaces of different orders, occasionally inclusive even of diachronous sequence surfaces, thus abandoning the basic premise of the regionally isochronous sequence surfaces. There also exist several papers on differentiation of sequences in the Indian Cretaceous geological record, mostly from well data, however, without ordering and precision determination of the sequence surfaces.

The ONGC framework (2008) includes four first-order SBs—MI 50 to MI 80 (respectively, of the Hauterivian/Barremian boundary, Albian/Cenomanian boundary, Santonian/Campanian boundary, and Campanian/Maestrichtian boundary age (Fig. 4.5), while the Jurassic/Cretaceous boundary is indicated as MI-40. Two second-order SBs are also included—MII 60 of the terminal Late Hautervian, and MII 70 of the terminal Late Barremian age. Additionally, two second-order MFSs are also indicated, respectively, of the intra-Berriasian and intra-Hauterivian ages. A crude comparison suggests broad correspondence of the ONGC first- and second-order sequences in duration with the second-order sequences of the present work. Irrespective of the order of the SBs, the ages of as many as four of the six ONGC first/second-order SBs do not match with those of the present work (except those of the MI 70 and MI 80), although the involved sequence surfaces may mostly be the same. So also, the ONGC first-order SBs in the Jurassic do not correspond to those of the present work except for the MI 30 of the Kimmeridgian/Tithonian boundary age. Even the determination of the Kimmeridgian/Tithonian boundary may be different in the present work and ONGC work, since details are not spelled out. Another significant aspect of the comparison is the outcrop and ammonoid-based more precise age control of the present work to not that precise microfossil control in the wells. The intra-Albian MFS of the ONGC, featured with *Planomalina buxtorfi*, *Hedbergella and Planispira*, may be the same as the second-order mid-Middle Albian Dentatus Zone MFS determined in Jaisalmer, Kachchh, Cauvery, Spiti and age precised through the included ammonoids in Jaisalmer. The other ONGC second-order MFS is placed at the Cenomanian/Turonian boundary between the planktonic foraminifer assemblages of *Rotalipora cushmani*, *R. reichelli*, and *R. appeninica* below, and *Whiteinella arcaeocretacea*, *M. Helvetica*, *and M. sigali* above. This MFS also seems close to the early Middle Turonian first-order MFS précised on ammonoid evidence in Cauvery exposures in the terminal part of the Karai Shale Formation above the concretion bearing interval ~20 m below the base of the Trichonopoly Group east of the Odhium village. The objective behind such comparison is to compliment and integrate the diverse micro/mega floral/fauna paleontological data towards objective age constraining of at least the better understood principal SBs and MFSs in the Indian Mesozoic geological record.

5.1.4.2 Comparison with Other Sequence Frameworks

The developed sequence differentiation is compared with the compilation of Raju (2008) realized in collaboration with several other Indian stratigraphers. Raju (op. cit) highlighted a large stratigraphic gap in the Jurassic without any mention of its subaerial or submarine character. This gap was determined by Mehrotra and Aswal (2003) in the K-G subsurface, and also by others (Prasad 2011) in the Indian Gondwana basins, however, without relating the gap to either regional tectonics or sequence stratigraphy. Mehrotra and Aswal (op. cit), almost in each well studied by them for palynology, have differentiated SBs and MFSs without any ordering, as also without demonstrating their intrabasinal or interbasinal applicability.

The intra-Triassic gap as per their table/chart is of the Ladinian–Carnian interval from 236 to 216 ma in the non-marine successions. The said interval falls in the early part of the first-order RST in the Spiti basin. Small duration gaps have been also suspected in the Spiti Triassic within the Late Anisian part by Bhargava et al. (2004) which are here considered submarine.

The intra-Jurassic gap in the non-marine Gondwana, and shallow marine east coast basins is made out of a large duration of ~50 my, and is ranged from almost the base of the Jurassic up to the close of the Late Jurassic Kimmeridgian stage (Prasad 2011 and many others). Large but, lesser gap is present in the marine succession of Axial belt of Pakistan, as also in the marine successions elsewhere on GTM. The said gap in the marine successions is interpreted as submarine (Krishna et al. 2011). The said gap in the non-marine Gondwana basins in sequence stratigraphic context is here corresponded to the first-order intra-Jurassic Toarcian–Middle Oxfordian TST interval. The likely explanation has been discussed in Chap. 7.

The intra-Cretaceous major drop in sea level in Raju (2008) accords well with the determination of the intra-Mesozoic bathymetry maximum above the Turoniense ammonoid zone (=Helvetica foram Zone). The bathymetric high appears more or less maintained in the Middle Albian–Middle Turonian span. The recognition of five cycles in Aptian–Albian, and 7–9 MFSs during Late Albian–Late Maestrichtian interval mentioned in Cauvery basin wells, are not detailed in the text by Raju (2008). Also any ordering of either the sequences or of MFSs is not provided. Comparison, therefore, of such differentiation with the first to third order Cretaceous sequences developed in the present work is difficult. Similarly 13 parasequences have been differentiated by Nagendra in the Cauvery basin in the Albian–Maestrichtian interval, 7 major T/R cycles by Raju and Ramesh (in Raju 2008). Four major transgressions are recognized by Govindan and Ravindran (1996) in Albian, Coniacian, Campanian and Maestrichtian in Cauvery basin. Ramkumar et al. (2005), and Nagendra et al. (2011) have made good contributions to sequence framework in the Cauvery basin.

5.1.5 On the Efforts in the Kachchh Basin

In the Kachchh basin, sequence stratigraphic interpretations have also been made by Biswas (2005, and earlier/later) by way of broad correspondence of his sequences to the framework of Haq et al. (1987) although without precise ages of SBs, and without mentioning the MFSs. Fursich et al. (2001, 2004, 2013) and Fursich and Pandey (2003) in their publications have recognized sequence surfaces in specific sections, as also widely recognizable marker beds without either any ordering or comprehensive effort spatially or temporally on the succession of sequence surfaces even within the Kachchh basin.

5.2 Validation of the Present Sequence Framework in India, IEAP and GTM

The ammonoid-based stratigraphic refinement in the Indian subcontinent except that in Kachchh, in spite of good ammonoid record, continues to be rather poor, and certainly, far from what is desired in such comparison studies. Nevertheless, here is undertaken discussion and demonstration of all the first/second-order sequence surfaces for the first time ever in these IEAP basins (Fig. 5.1). In absence of the desired degree of stratigraphic refinement, the validation here is indirectly approached. Drastic lithological and paleontological changes with gaps in ammonoid successions when found in proximal localities have been indirectly interpreted with enough caution and conjunctive evidences as subaerial, and inclusive of SB. Similar changes when noticed in distal sections are held as submarine, and inclusive of MFS. Even in a continuous section, lithological and faunal changes have been interpreted as suggesting possibility of sequence surfaces at the contact across which the changes occurred.

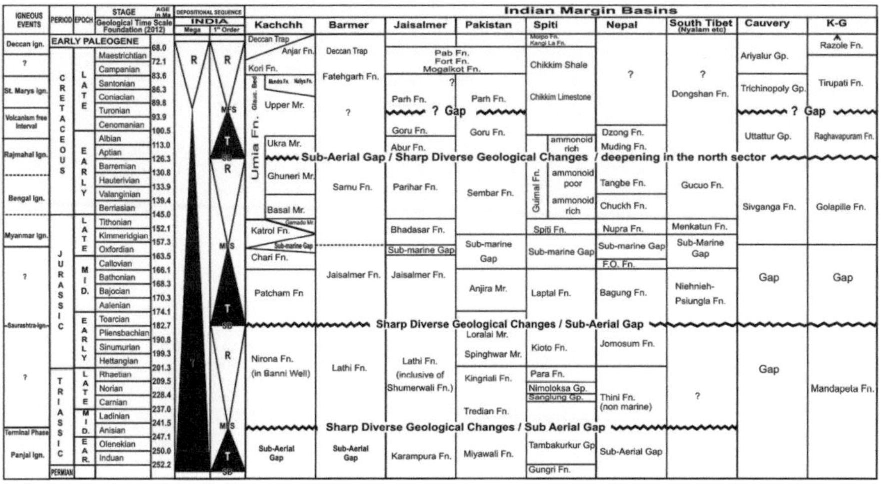

Fig. 5.1 Validation of GTM sequence framework among Mesozoic basins of the Indian subcontinent

5.2.1 *Broad Correspondence of the Indian and Arabian MFSs*

The broad correspondence of the succession of Indian MFSs with the only other comprehensive effort on GTM in Arabia (Sharland et al. 2001) is presented (Fig. 5.2).

5.2.2 *Basal Intra-Permian SB on IEAP and GTM*

Major rifting/drifting is well known to have occurred, and affected the then paleotethyan Indo-Australian margin of the erstwhile Indian subcontinent. The intra-Permian origin of the Neotethys is reflected in the geological record of the High Himalayan belt, Pakistan, Jaisalmer, Bikaner–Nagaur, Spiti, Nepal, S. Tibet, Cauvery, and K-G, in fauna, flora, facies, and stratigraphic gaps, etc. There is change of fauna and facies from Middle to Late Permian, particularly, with regard to presence of ammonoids in the latter interval. Jaisalmer, for example, is known for its disconformity and associated floral change between the Late Permian–Early Triassic Karampura Formation (∼the Late Permian part of 'Bhuvan Formation') and the underlying Proterozoic–Cambrian Birmania Formation. In Bikaner-Nagaur basin the Proterozoic–Cambrian Nagaur Formation is disconformably overlain by the Early Permian Bap and Badhaura units which are in turn disconformably overlain by the Paleocene sediments. In Kachchh, the presence of the reworked non-marine Late Permian–Early Triassic bracket sediments, is interpreted as

Periods	MFS	Arabian Plate	Number	Age	Lateral extent	Radiometric Ages	
						Sharland et al 2002	Haq et al 1987
early Paleogene		AP10	Pg20	latest Ypresian	1	49	49
		AP10	Pg10	late Palaeocene	1	58	56.5
CRETACEOUS	C12	AP9	K180	mid-Maastrichtian	1	68	69.5
	C11	AP9	K170	middle Campanian	1	78	79.5
	C10	AP9	K160	Santonian	2	85	86
	C10	AP9	K150	early Coniacian	2	88	89
	C9	AP8	K140	early Turonian	2	93	91.5
	C8	AP8	K130	middle Cenomanian	2	95	94.7
	C8	AP8	K120	early Cenomanian	1	98	95.75
	C8	AP8	K110	late Albian	1	101	97
	C7	AP8	K100	middle Albian	1	106	104
	C7	AP8	K90	early Albian	2	111	107
	C6	AP8	K80	mid-Aptian	2	116	?111
	C6	AP8	K70	early Aptian	1	120	?111
	C5	AP8	K60	late Barremian	1	123	114
	C5	AP8	K50	early Barremian	1	126	116.5
	C4	AP8	K40	late? Hauterivian	1	129	118
	C3	AP8	K30	early Valanginian	1	136	127.5
	C2	AP8	K20	late Berriasian	1	138	128.75
	C1	AP8	K10	early Berriasian	1	143	132.5
JURASSIC	J14	AP8	J110	middle Tithonian	2	147	137
	J13	AP7	J100	late Kimmeridgian	1	150.75	n/a
	J13	AP7	J90	late Kimmeridgian	1	151.25	n/a
	J12	AP7	J80	late Kimmeridgian	1	151.75	n/a
	J12	AP7	J70	late Kimmeridgian	1	152.25	138.5
	J11	AP7	J60	early Kimmeridgian	1	154	144.5
	J10	AP7	J50	middle Oxfordian	1	156	150
	J8-J9	AP7	J40	middle Callovian	1	162	153.5
	J6-J7	AP7	J30	early Bathonian	1	168	163.5
	J5	AP7	J20	early Bajocian	1	175	170
	J1-J2-J3-J4	AP6	J10	middle Toarcian	1	185	183.5
TRIASSIC	T8-T9-T10	AP6	Tr80	Norian	2	215	223
	T7	AP6	Tr70	Carnian	1	222	225.5
	T6	AP6	Tr60	Carnian	1	226	229.5
	T4-T5	AP6	Tr50	Ladinian	1	233	235
	T3	AP6	Tr40	Anisian	3	238	238
	T2	AP6	Tr30	mid-Scythian	2	245	241
		AP6	Tr20	Scythian	2	246	243
	T1	AP6	Tr10	Scythian	1	248	245.5
		AP6	P40	latest Tatarian	1	249	n/a
		AP6	P30	late Tatarian	1	250	n/a
		AP6	P20	late Kazanian	2	252.5	n/a

Fig. 5.2 Indian (first and second order) and Arabian (unordered) MFSs

disconformable to the non-marine Ladinian–Pliensbachian sediments of the Banni and Nirona wells.

In all these basins after the large break, resumption of sedimentation in either Early Permian or Late Permian is witnessed. Subsequently, the sediments are again disconformably overlain by either Late Permian–Early Triassic or Late Triassic–Early Jurassic or Middle Jurassic–Cretaceous, or Paleogene. In Pakistan, are observed intra-Permian disconformable breaks along with lithological and faunal changes. The evidences as above suggest intra-Permian basal SB of the Neotethyan mega-sequence within Early Permian or near the base of Late Permian in the entire IEAP (Fig. 2.4). The causative geological event is also evidenced with similar gap, lithological/faunal change, and igneous activity in non-marine Gondwana basins. Similar diverse geological changes are also well known west of IEAP in East Africa, Seychelles, Madagascar and Arabia, and also east of IEAP in NW Australia and neighbourhood as possible consequences of extensional tectonics (Timor, Timor Gap, New Guinea, West Iriyan, Sula, etc.). There is complete accord among stratigraphers about the basal intra-Permian sequence boundary of the mega-sequence.

5.3 Comparison of the Indian Triassic, Jurassic, and Cretaceous Sequence Surfaces

Broad comparison of the Indian Triassic first- and second-order sequence surfaces is outlined with regions east of IEAP, within IEAP, and west of IEAP in that order. The preliminary comparison of the Jurassic sequence surfaces is limited to the first order and a few selective second-order surfaces, while the comparison exercise in the Cretaceous is exclusively of the first-order surfaces.

Triassic

5.3.1 Comparison with Wombat Plateau ODP Well Sequence Framework off the NW Australian Margin Sites 759, 760, 761 and 764 East of IEAP

A succession of four SBs and four MFSs are determined of 211 and 211.5, 215 and 218, 220 and 223, and 224 and 225.5 ma in Late Triassic syn-rift succession of the Wombat Plateau (Von Rad et al. 1989, 1992). Out of the said four pairs, as many as three pairs of SBs and MFSs have been suggested without evidence as concordant to the ETM framework of Haq et al. (1987), but for the 218–220 ma pair of MFS and SB. The corresponding MFSs are indicated at 211.5, 218, 223 and 225.5 ma. Three of these four Wombat plateau MFSs appear close to Spiti MFSs (Fig. 5.3), but for one at the Carnian/Norian boundary. Thus, the limited comparison suggests

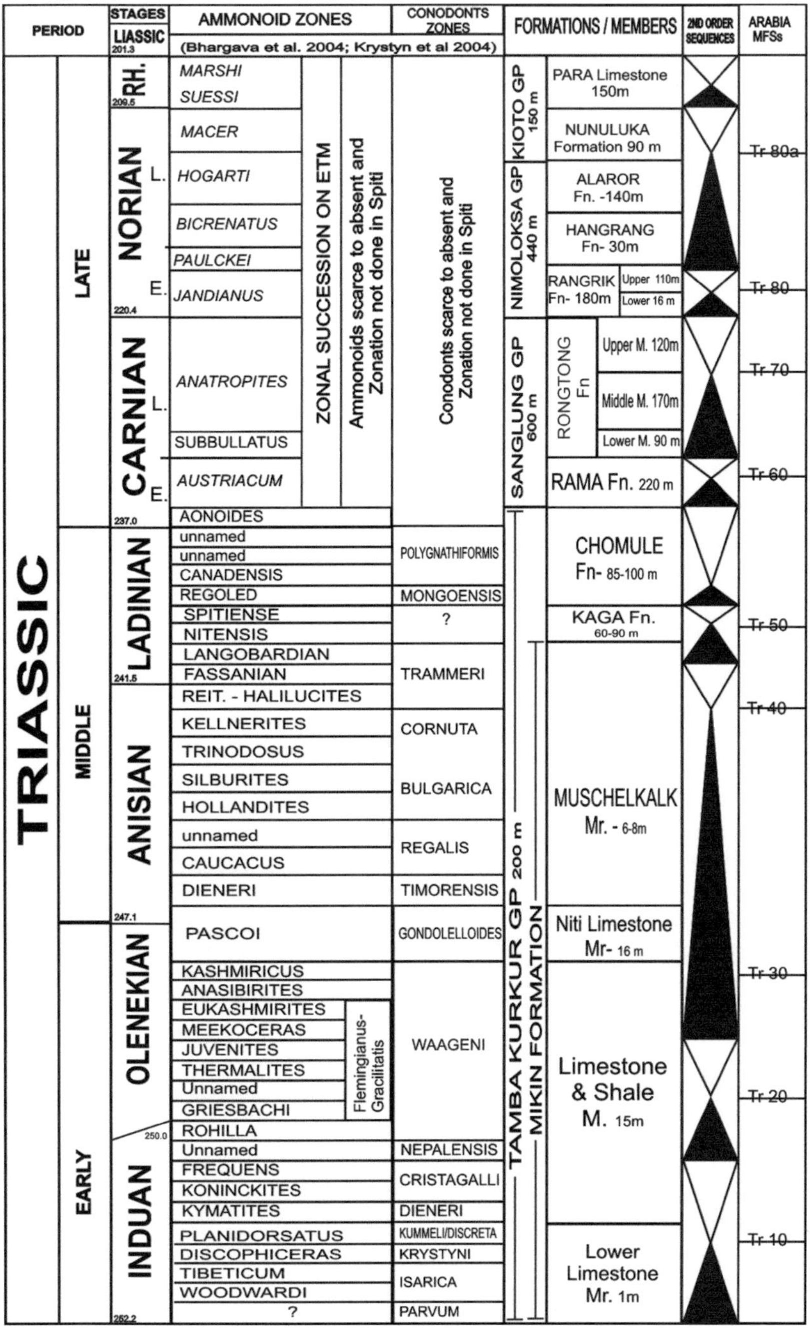

Fig. 5.3 Triassic lithostratigraphic units, ammonoid zones of Spiti and GTM second-order sequences vis-a-vis Arabian plate MFSs

substantive concordance of the Late Triassic second-order sequence stratigraphic surfaces of Spiti and Wombat Plateau on GTM.

The Late Triassic succession is followed up by the intra-Oxfordian post-breakup unconformity of the entire Late Jurassic prior to the succeeding Berriasian succession. The imprecise broad intra-Oxfordian age of the breakup unconformity is speculated here as of the Kachchh intra-Oxfordian first-order late Middle Oxfordian MFS age.

5.3.1.1 Intra-Triassic First-Order Late Anisian Hollandites Zone–Kellnerite Zone MFS on IEAP

This MFS is determined in the Spiti basin (Krishna et al. 2010, 2011), approximately in beds 34–40 of the terminal part of Muschelkalk below its contact with the Nodular limestone unit. The MFS marks drastic lithological, sedimentological and paleontological changes across the units (Fig. 2.9). In Jaisalmer, the change of microflora between the Late Permian Karampura Formation and the Late Triassic–Early Jurassic Lathi units is interpreted at the intra-Triassic MFS. In the Pakistan Indus basin, is interpreted a disconformity along with lithological and faunal changes between the ammonoid-rich Mianwali Formation below and the ammonoid- devoid Tredian Formation above. The Mid-Triassic lithological and paleontological contrast between the underlying and overlying sediments, thus, is widely documented both in marine, and non-marine successions. Even in inland non-marine Gondwana basins of Satpura, South Rewa, Damodar, Rajmahal and Pranhita–Godavari, the Middle/Triassic gap, facies and floral changes are widely known. Middle Triassic intra-Muschelkalk volcanics are recorded in Pakistan (Calkins et al. 1975; Wadia 1934). Mid-to early Late Triassic infra-rift event has since long been suggested in the West Australian Perth basin (Veevers 1984). The Arabian plate also witnessed rifting and tilting down north of the north margin along with the widest intra-Triassic flooding, however, the event is dated in Ladinian (Sharland et al. 2001).

5.3.1.2 Comparison of the Spiti Triassic and Arabian Sequence Surfaces West of IEAP

The first-order Arabian sequence that includes Triassic is AP 6. AP 6 includes a succession of 8 MFSs without any order while as many as ten age precised second-order sequences have been differentiated in the present work along with their ten MFSs and SBs (Figs. 3.13 and 5.3). The Tr 10 MFS of Arabia is assigned to the Induan playfordi palynozone (El-Bishlawy 1985 in Sharland et al. 2001). It suggests correspondence to the more precisely dated ammonoid based T-I MFS of the Induan Discophiceras Zone of Spiti (Figs. 5.2 and 5.3).

The Tr 20 MFS of mid-Schythian age of Arabia is also précised in Spiti in the Olenekian Flemingites griesbachi Zone of Early Olenekian. The Tr 30 MFS is loosely assigned to mid-Schythian in Arabia, which does not match in age to any MFS of Spiti. Tr 40 of Arabia is broadly assigned to Anisian and is of local extent. In comparison, the T-III MFS of Spiti is of wide extent and précised in the late Late Anisian Hollandites Zone–Kellnerites Zone interval. The Tr 50 of Arabia is assigned to Daonella-bearing Ladinian age, which may match the early Late Ladinian Daonella bearing MFS of Spiti that is included in early Late Meginae Zone of British Columbia. The Tr 60 MFS in Arabia is placed in Carnian which may correspond to the Early Carnian Spiti MFS in shale/siltstone alternation in the later part of Rama Formation. The Tr 70 MFS of Arabia is also assigned to Carnian which matches well with the mid-Late Carnian T-VII MFS of Spiti. It is strange, also incorrect and contradictory that the coeval and well-correlated Tr 70 MFS of Arabia of Late Carnian age has been included in the A. parvispinosus palynozone. This palynozone in the Indian Gondwana Triassic ranged exclusively within Schythian to early Anisian interval. Most likely, the range of the said species is incorrect. However, if the range in Arabia is also true, *A. parvispinosus* must not be used as a palynozonal index either in Early to basal Middle or early Late Triassic Carnian in view of its range throughout the Early, Middle and early Late Triassic through the first five stages of the Triassic. The rather long supra-stage ranges of the Indian Mesozoic palynozones have been also commented upon elsewhere. The terminal Tr 80 MFS in Arabia is placed in Norian which is here corresponded to either T-VIII MFS of Spiti of Early Norian age within the Rangrik Formation or the Late Norian T-IX MFS of Spiti in the upper part of the Alaror Formation.

The Arabian intra-Triassic MFSs, in spite of lacking precise ages as per above discussion, closely match the précised ages of the MFSs of Spiti Triassic. This in turn allows to safely infer the broad validation of the Spiti second-order framework in Arabia located at the west extreme of the GTM.

5.3.2 Comparison with ETM Second-Order Sequence Framework

Haq et al. (1987) have a total of ten MFSs in the Triassic out of which five are shown more important than the other five. The first ETM MFS is of Late Induan age while the MFS in Spiti is of Early Induan Discophiceras Zone. The second ETM MFS and second Spiti MFS are determined in Early Olenekian and may be of the same age. The third ETM MFS is placed in Late Olenekian which does not have any match in Spiti. The fourth ETM MFS is determined in Early Anisian while the third MFS in Spiti is in late Late Anisian. There is, thus, interpreted distinct discordance. The fifth ETM MFS is assigned to basal Ladinian while the Spiti one is determined in late Early Ladinian. The sixth ETM MFS is dated in Late Ladinian while the Spiti one in early Late Ladinian. The eighth ETM MFS is dated at the Carnian/Norian boundary while the Spiti one is in mid-or late Early Norian. The continuation of the discordance is more than obvious. The last ETM MFS is in Late Rhaetian which may or may not match the rather imprecise intra-Rhaetian MFS in Spiti.

Jurassic Sequence Surfaces in IEAP

5.3.3 Intra-Jurassic First-Order early Early Toarcian SB

In Spiti, this SB is readily determined between the ammonoid devoid and bivalve bearing terminal Tagling/Kioto Limestone of Pliensbachian age and the overlying Laptal marls/shales (Fig. 5.1) which are known to have yielded the solitary Early Toarcian *Alocolytoceras* (Stoliczka 1866), and Early Bajocian *Stephanoceras* in the younger beds. In Nepal also, the SB is interpreted at the contact of the Jomosum Formation and Bagung Formation, and also in other basins of High Himalaya (Gradstein et al. 1989; Fig. 48 of Gradstein et al. 1991). In Jaisalmer, this SB is interpreted between the Late Permian–Early Triassic Bhuvan Formation and the disconformably overlying Lathi Formation. Across these units, are observed the significant microfloral changes. In basins of Pakistan, the *Bouleiceras*-bearing beds either disconformably overly the unfossiliferous sands or conformably overly the ammonoid devoid sediments. The evidences as above confirm the intra-Jurassic first-order SB (Fig. 5.1).

5.3.4 Intra-Jurassic First-Order late Middle Oxfordian Schilli Subzone MFS

In Jaisalmer, Spiti, Nepal and Pakistan, the MFS is not exposed in view of a large submarine gap and associated disconformity between the Early Oxfordian belemnoid dominant shales and the ammonoid-rich nodular/concretionary shales of Spiti Formation of Late Kimmeridgian–Early Tithonian age (Krishna et al. 1982). The large submarine gap is an indirect evidence of the inclusion of the MFS in the gap. In Jaisalmer also, there is marked large gap between the Jaisalmer Formation of the Callovian–Early Oxfordian age and the Rupsi Member of the late Early Tithonian age (Pandey and Krishna 2002) as an indirect evidence of the inclusion of the said MFS in the gap which is here interpreted as submarine gap (Fig. 5.1). In Pakistan Axial belt, the submarine gap is much larger from the Early Bajocian sediments of the Windar/Anjira beds below to the Late Kimmeridgian start of the Sembar Formation above. The submarine gaps in all these basins on account of rapid increase in bathymetry provide indirect evidence of the MFS falling/included in the gap, however, the age cannot be précised.

Cretaceous First-Order Sequence Surfaces

5.3.5 First-Order Intra-Cretaceous Barremian/Aptian Boundary SB

There is observed lithological and paleontological change between the ammonoid scarce Guimal sands and the ammonoid-rich marls/limestones in the terminal part of the Guimal Formation with Early Aptian ammonoids in Spiti, similarly also in Jaisalmer, Nepal, Pakistan and many other Indian basins. The contact is here determined as the first-order intra-Cretaceous Barremian/Aptian boundary SB. Almost, identical change is interpreted between the ammonoid devoid Parihar sands of the Berriasian–Barremian age and the overlying Early Aptian (*Deshayesites*, *Dufreynia* etc.)–early Middle Albian (*Douvilleiceras*, *Lemuroceras*, *Hamites*, *Brancoceras*, *Oxytropidoceras*, *Lyelliceras* etc.) ammonoid-rich marls/cocquina of the Abur Formation with associated lthological and paleontological changes (Fig. 5.1). The rifting event is also recognized recently in the Barmer basin (Bladon et al. 2015). The multifaceted changes on account of the extensional tectonic event at the said SB are widespread in basins in India between the Sivaganga and Dalmiapuram units in Cauvery, and between the Golapille and Raghavapuram units in the K-G, between the Nimar sensu stricto and the Nodular Limestone units in Narmada, between the Dhrangadhara and Wadhwan units in Saurashtra. So is the case outside India in the east, e.g. between the Kagbeni Formation and the Muding Formation, and in the west between the Chichali Formation below and the Lumshiwal Formation above of Early Aptian–Middle Albian in Western Kohat, Kalachitta Nizampur and Hazara in Pakistan. In one section of Kohat, even Late Cretaceous ammonoids are also present in the basalmost beds of the Kawagarh Formation above the Lumshival Formation.

5.3.6 Intra-Cretaceous early Middle Turonian Turoniense Zone MFS

In Spiti, this MFS is not well determined, but included in the middle to terminal part of the Chikkim Limestone on the basis of the planktonic foraminifera of Helvetica Zone (Bertle and Suttner 2005) in beds 30 and 29. In Jaisalmer, the exposed succession, is not younger than Albian. Turonian planktic forams and pollen/spores are present in subsurface wells near the contact of the Goru and Parh units between shales below and limestones above in Pakistan and Jaisalmer wells (Singh 2006). There is obvious suggestion of bathymetric decrease at the contact. The contact has been picked up in seismic profiles as well. In the neighbouring Narmada basin, the MFS is readily determined near the top of the ammonoid-rich hard-grounded nodular limestones (Fig. 4.6). It is also indicated in the Wadhwan unit through

Proplacenticeras stantoni in the middle fossiliferous part. In the Cauvery basin of the east sector, the said MFS is interpreted in the terminal part of the Karai Formation, and also between the Raghavapuram and Tirupati units in the K-G wells (Figs. 3.37–3.40 and 4.10). The tectonic event is widely manifested in several basins of the GTM (Fig. 5.1) through a marked tectonic uplift.

Thus as above, the validity of all the first-order sequence surfaces (MFSs and SBs) of the GTM developed in Kachchh is in general validated in IEAP, and the sequence surfaces are found invariably isochronous.

Jurassic second-order sequence surface comparison within IEAP

5.3.7 *Early Bajocian Laeviuscula Zone MFS*

As per rich record of the Early Bajocian ammonoids in Pakistan (Fatmi 1986), Nepal (Gradstein et al. 1989), and S. Tibet together with the solitary *Stephanoceras* from Spiti (Stoliczka 1866), the limestone development (units M-N) of Bagung Formation over the thickly bedded sandy limestone is interpreted as a second order TST, and suggestively assigned to Early Bajocian Laeviuscula Zone (Figs. 3.22 and 3.23). It may be speculatively correlated to Am in the Jhura section, *Calliphylloceras*-bearing basal coral limestones in Sahara section, and fossiliferous beds 37–38 at the base of the Kaladongar sands (KSm) of the Lower Sandstone Member (LSM) of Patcham Formation at Kaladongar in Patcham. In Jaislamer, the marine Jurassic succession is also speculated to have begun in Early Toarcian in the basal shales of the Hamira unit disconformably over the non-marine Lathi Formation of the Late Triassic–mid-Early Jurassic Pliensbachian age. In Pakistan, the gap above the Early Bajocian to Late Kimmeridgian is large, and the same has been interpreted as submarine gap.

5.3.8 *Early Bathonian Macrescens Subzone MFS*

The ~40 m thick limestone/shale alternation after the ~80 m thick monotonous shales in Nepal (Fig. 14 of Gradstein et al. 1989) is here interpreted as a second-order TST, and speculatively assigned to the late Late Bajocian–early Early Bathonian Zigzag Zone age after the recent discovery of this MFS in Kachchh with *Parkinsonia* and *Siemiradzkia* (Pandey and Pathak 2015). It is here speculated that the thin limestone within an otherwise sand/shale alternation of Joyan unit may mark the Early Bathonian Macresecens Subzone MFS.

5.3.9 Mid/late Middle Bathonian Basal Bremeri Zone SB

In Nepal, Gradstein et al. 1989 observed change of lithology between the units O below and P above. Units P-Q include thin coquina beds which are capped by a thick marker fossil rich hard-grounded ferruginous oolite. The base of P is here interpreted as a second-order SB, and for the first time based on homotaxial stratigraphic framework between Kachchh and Nepal is assigned to Bremeri Zone. The unit P makes the early part of a second-order TST of mid/late Middle Bremeri Zone to late Late Bathonian Triangularis Zone age. Like Kachchh in Nepal too, the later part of the same TST is in iron oolitic facies of mid/late Early Callovian age. The units have yielded mainly *Macrocephalites* spp. *Clydoniceras* is also known in Jaisalmer, which attests the presence of mid/late Middle Bathonian to early Middle Callovian second-order sequence in Jaisalmer. The Bada Bag and Fort units are here speculated as representing the mid/late Middle to Late Bathonian on the basis of the homotaxial stratigraphic framework with Nepal and Kachchh. In Spiti, the Laptal sediments above the Kioto Limestone and below the Ferruginous Oolites are of Toarcian–Bathonian age. The Toarcian–Aalenian sequence is here separated from the overlying Bajocian–Bathonian part from the level which to Stoliczka yielded the Early Bajocian *Stephanoceras*.

5.3.10 Mid/late Early Callovian MFS

In Spiti, there is a thin slow-sedimented condensed ammonoid-bearing unit of the hardened Ferruginous Oolite Formation between the underlying Laptal Formation of Toarcian–Bathonian age and Early Oxfordian–Tithonian Spiti Shales. The Ferruginous Oolite Formation in Spiti is of mid/late Early Callovian age based on the presence of *Macrocephalites* and *Indosphinctes* (Krishna unpublished). In Jaisalmer, the contact between the Fort Member and Kuldhar marks the Bathonian/Callovian boundary. In the Callovian Kuldhar Member, the Chrysoolithicus Zone/Semilaevis Zone contact can be easily picked up, and suggests relative decrease of bathymetry. This contact is determined as the second-order MFS. This MFS is well expressed in the Kuldhar section between the Chrysoolithicus Zone and Semilaevis Zone.

5.3.11 On the Younger Jurassic Sequences in Jaisalmer and Spiti

The Kuldhar Member farther up includes sediments of the next second-order sequence with *Reineckeia*, *Collotia*, etc. and Middle and early Late Callovian ammonoids which prove there the presence of the late Middle Callovian to early Early Oxfordian sequence. The terminal sequence boundary of this sequence is determined in the Jajiya section underneath the hard-grounded richly fossiliferous Obliqueplicatum Zone age conglomeratic condensed slow sedimented nodular oolitic limestone of the next sequence. Subsequent to this ensued the large submarine gap. The two Kimmeridgian second-order sequences are absent and lost in Jaisalmer in the submarine gap between the Jajiya Member of Early Oxfordian age and Rupsi Member of the post-Pottingeri Zone Tithonian. These Kimmeridgian sequences are also absent in the Axial belt of Pakistan.

The Spiti Shale unit all over High Himalaya in Spiti, Nepal, Kumaon and elsewhere comprises monotonous succession of about 250 m thick grey to black shales. Shales are nodular to concretionary inclusive of either ammonoid or belemnoid or other invertebrate fossils as nucleus. Owing to the splintery and fragile nature of the shales, the detailed measurement is difficult, and ammonoid collection with stratigraphic precision is also difficult. Silt gets included in the shales in the upper part after ~25–30 m from the base. The most precise collections in Spiti have been made by Pathak and Krishna (1993), Krishna and Pathak (1994), Pandey and Pathak (2016) and in Nepal by Enay and Cariou (1999). Multiple gaps are suspected in the Late Oxfordian–Early Kimmeridgian interval. Even Late Oxfordian–Early Kimmeridgian sediments have been claimed to be present based on foraminifera (Gradstein et al. 1989) in Nepal. Nepal compared to Spiti is considered more proximal hence less deep and also with lesser/smaller gap/s. Ammonoids of the late Middle and Late Oxfordian ages are absent in Spiti that otherwise include the first-order MFS. Absence of Middle to Late Oxfordian *Gregoryceras*, *Dichotomoceras*, *Epipeltoceras*, as also of the Early Kimmeridgian *Orthosphinctes* and *Ataxioceras* accords well with a large submarine gap. Late Kimmeridgian Katroliceratinae and Aspidoceratinae are present in Spiti, and so the Bathyplocus Zone/Katrolensis Zone boundary MFS is present although the Late Kimmeridgian on expected lines is quite thin. Ammonoids of the mid-Early Tithonian Pottingeri Zone, and of Late Tithonian Microcanthum Zone are also present in good number, so also of Early and Late Berriasian second-order MFSs. The youngest ammonoids of the discussed first-order sequence are of early Late Valinginian Verrucosum Zone. Thus, guide ammonoids of the first-order RST from the latest Middle Oxfordian to early Late Valinginian in ample measure validate the Neotethyan first- to second-order sequence framework in High Himalaya and Jaisalmer. Jaisalmer being less deep or more shallow attests only the Berriasian second-order MFS. There is hardly any doubt about these neighbouring basins being controlled by the same extensional tectonic events that produced the said MFSs.

In spite of the quality of the Jurassic stratigraphic framework of Jaisalmer and High Himalaya not of desirable level, the discussion as above supports the isochronous character of the many first-and second-order sequence surfaces in the said basins to those developed in Kachchh. In case of many other sequence surfaces, the data supports the possibility and probability of their being isochronous with those of Kachchh. On the other hand, the present lithological and paleontological data does not suggest any disagreement or discordance to the validation of the Kachchh first-and second-order framework in Jaisalmer and Spiti.

5.4 Validation in Detail of the Indian Jurassic–Cretaceous Sequence Stratigraphic Framework on GTM West of India

The sequence stratigraphic framework was first developed in the Kachchh Jurassic (Krishna 2005; Krishna 1994; Krishna and Pathak 1994). Its validation on the GTM became difficult in view of the poor quality of the ammonoid stratigraphic framework in most of the basins on GTM. Moreover, the comparison of the Kachchh and ETM frameworks projected distinct diachrony of the first- and second-order sequence surfaces. Under such circumstances, it became rather desirable to verify and demonstrate the skeletal succession of sequences of the Kachchh Jurassic on GTM. Initially, the broad but preliminary comparison exercise in the foregoing pages, has been limited to few key surfaces on both sides of India on the GTM with the objective of a broad concordance. However, with the start of the twenty-first century, publications with considerably improved stratigraphic refinement emerged on the sequence stratigraphy in Arabia (Sharland et al. 2001) and Madagascar (Geiger unpublished 2004; Geiger and Schweigert 2006). These in turn allowed to test the validity in detail of the Kachchh Jurassic sequence surfaces west of India on the GTM, particularly in Mdagascar and Arabia.

The comparison studies are aimed at: (1) investigation of the concordance of the third-order sequence surfaces on a global scale, (2) validation of the regionalism of the second, first and bigger sequences on account of the overwhelming influence of regional extensional tectonics over the global eustatic influences and (3) Attestation of the discordance/diachrony of the sequence surfaces from one region to another for example between ETM and GTM. It is in this backdrop and rationality, is undertaken here the comparison of sequence surfaces, first among basins of GTM (between Kachchh and Madagascar, Kachchh and Arabia, Kachch and SW Pacific) and subsequently between GTM and ETM.

5.4.1 Comparison with Composite East Africa

The Jurassic ammonoid successions of the East African regions but for the small local/regional differences of their own are by and large identical to Kachchh, and also their sequence surfaces as summarized below.

5.4.1.1 Toarcian–Aalenian

In most East African areas, the marine Jurassic is diconformably underlain by the non-marine sands (Lohemboro Formation, Lohego Formation, Luwego Formation, Mkuju Formation and Mdaba Formation in different East African basins) of Late

Triassic to Hettangian–Pliensbachian age as in Kachchh (Fig. 3.1). In few areas, the Jurassics begin with the Early Toarcian *Bouleiceras* bearing marls/limestones (Arkell 1956) (Ngerengere beds and equivalent units in Mandawa, Jubaland, and Somaliland). The intra-Jurassic first-order SB is interpreted at the non-marine/marine contact and the same is easily dated as early Early Toarcian. The non-marine sands in turn are unconformably underlain by the Late Permian–Early Triassic sands/shales of the Songea Group (Hotambulo Formation). These are unconformably overlain on the Precambrian basement. These pre-Jurassic and pre-Triassic disconformities are interpreted as of, repectively, the Mid-Triassic and mid-/Late Permian ages, which in turn signal the intra-Triassic first-order MFS and intra-Permian SB of the Indian mega/super sequence.

5.4.1.2 Bajocian–Bathonian

The Toarcian–Aalenian sediments, in a few East African basins, particularly Mombasa and Mandava, are followed up by the Bajocian–Bathonian limestones/marls (Kambe limestones, Ngerengere beds, Matumbi beds, Mtumbei and kidugallo units in Tanzania, Hamanlei unit in Somalia and equivalents) with Early Bajocian ammonoids at the base as in the Axial belt region of Pakistan, so also in Kachchh but without guide ammonoids but for the recently discovered *Calliphylloceras* (Pandey et al. 2013). The most interesting fauna is of Mombassa of the deeper water phyllo-lytoceratids (Arkell 1956) somewhat similar to the Early Bajocian ammonoid fauna of the Axial belt of Pakistan (Fatmi 1986). It provides a good testimony to the transformation of the trans-Eritrean linear shallow marine arm into somewhat deeper water open sea that extended up to Kachchh in consequence to the spreading between Somalia and North Madagascar as part of due south transform sliding of India inclusive of Madagascar vis-à-vis East Africa along the Davies ridge. The Early Bajocian Laeviuscula Zone / Propinquans Zone second-order MFS is readily interpreted on the basis of the ammonoids of this zone. The Bathonian sediments are invariably ammonoid devoid and only bivalve rich.

5.4.1.3 Callovian

It is everywhere rich in ammonoids with Early, Middle and late Callovian ammonoids inclusive of the Indo-East African Macrocephalitinae, Eucycloceratinae and Kinkeliniceratinae. There is every possibility and reason to determine the mid/late Early Callovian second-order MFS in a detailed future study in the East African sections. At present, the development of the similar ammonoid-rich Callovian suggests similar intra-Callovian second-order sequence surfaces.

5.4.1.4 Late Early–early Middle Oxfordian

In the East African basins, the Callovian is followed up by the more rich Oblqueplicatum Zone and the basal Indogermanus Zone ammonoid fauna of late Early to basal Middle Oxfordian. The second-order SB of the Kachchh sequence framework is easily discerned at the base of the said Oxfordian fauna.

5.4.1.5 Late Middle Oxfordian–Early Kimmeridgian Gap

Invariably, the late Middle and Late Oxfordian and Early Kimmeridgian fauna is absent as in Mainland Kachchh which allows to determine the gap in the East African basins as submarine gap that is otherwise widely present almost on the entire GTM.

The Oxfordian ammonoids are followed up by the Late Kimmeridgian ammonoids as in Mainland Kachchh. This allows to infer the first-order late Middle Oxfordian MFS in the gap. The Early Kimmeridgian Ataxioceratinae ammonoids (*Orthosphinctes*, *Discosphinctes*, *Idoceras*, etc.) in Mombasa allow to interpret the Oxfordian/Kimmeridgian boundary second-order SB at the base of the Early Kimmeridgian ammonoid-bearing beds, so also the early Early Kimmeridgian Kachhensis Zone second-order MFS within the Early Kimmeridgian sediments.

The Callovian–Tithonian succession in a few East African basins is clubbed together as Pindiro shales (in Mandawa and many other basins) on the pattern of the Duvalia marls of the Morondova basin of Madagascar, representing the later part of the TST and early part of the first-order RST.

In the East African basins, the geographically restricted Late Kimmeridgian IEAP Katroliceratinae ammonoids are invariably found to disconformably overly the late Early–early Middle Oxfordian after the submarine gap. The base of the Katroliceratinae bearing beds marks the second-order SB of the late Early Late Kimmeridgian sequence with rich IEAP Katroliceratinae and ubiquitous Neotethyan *Hybonoticeras*, *Aspidoceras*, *Streblites*, *Taramelliceras*, etc. Apparently, the detailed study of this Katroliceratinae sedimentary succession should as well allow the determination of the second-order post-Bathyplocus Zone MFS.

The next and the terminal Jurassic ammonoid fauna is of Communis Zone–Micracanthum Zone Late Tithonian ammonoids which includes another second-order SB at the base of the Micracanthum Zone in Tanganayika and few other East African basins.

In many basins of East Africa and Arabia, the marine Kimmeridgian–Tithonian ammonoid-bearing sediments are unconformably overlain by the marine ammonoid-bearing Early Aptian sediments. It in turn provides strong evidence of the first-order intra-Cretaceous SB near the Barremian/Aptian boundary.

Thus as above, in spite of the poorly resolved Jurassic in East Africa, the succession of the first- and second-order MFSs and SBs is amply well demonstrated in the East African basins on GTM that has been developed in the Kachchh basin.

So is also strengthened the validity of the sequence framework developed in Kachchh widely in the East African basins on GTM.

5.4.2 Comparison with the Morondova Basin of Madagascar

The comparison and correlation of the lthostratigraphic units and ammonoid successions of Kachchh (India) and Morondova (Madagascar), unambiguously, demonstrate the validity in Morondova of the first- and second-order sequences developed recently for the Gondwanian Tethyan Margin (GTM) in Kachchh within the Indian Mesozoic mega-sequence (Fig. 5.4). It is here emphasized that even the recent detailed works of Geiger (2004), and Geiger and Schweigert (2006) could not develop there the desired level of ammonoid stratigraphic boundaries and refinement as developed in Kachchh. Ammonoids are only tentatively determined without description and illustrations, while formations and members not age precised. All the seven first-order sequence surfaces of the Indian mega-sequence are identified and found isochronous in the sedimentary succession of Morondova and/or elsewhere in Madagascar. In addition, the second-order succession of Kachchh sequences is also demonstrated in the Jurassic part of the Toarcian–Barremian first-order sequence of Madagascar.

5.4.2.1 Madagascar as Part of the Indian Plate Since Precambrian in East Gondwana

Madagascar was part of the Indian plate since the Precambrian in the south hemisphere as a constituent of the erstwhile Gondwanaland (Fig. 1.7, also discussion in Chap. 7). It was located with its east margin besides the present Indian west coast which had then not formed, while the west margin similarly paralleled the present East African coast along future north–south striking fracture zones due 40 E and 50 E. The Indian west coast formed much later only during Cretaceous with the start of the spreading between India and Madagascar (Krishna 1987a, 2005, 2012).

5.4.2.2 Basal SB of the Neotethyan Mega Sequence in Madagascar

In Madagascar, the Late Carboniferous–Early Permian Sakoa Group below and the Late Permian–Early to early Middle Triassic Sakamena Group above are separated by a supra-order disconformity. The Sakamena Group here is included in the Indian Neotethyan mega-sequence as above defined, and the disconformity separating the

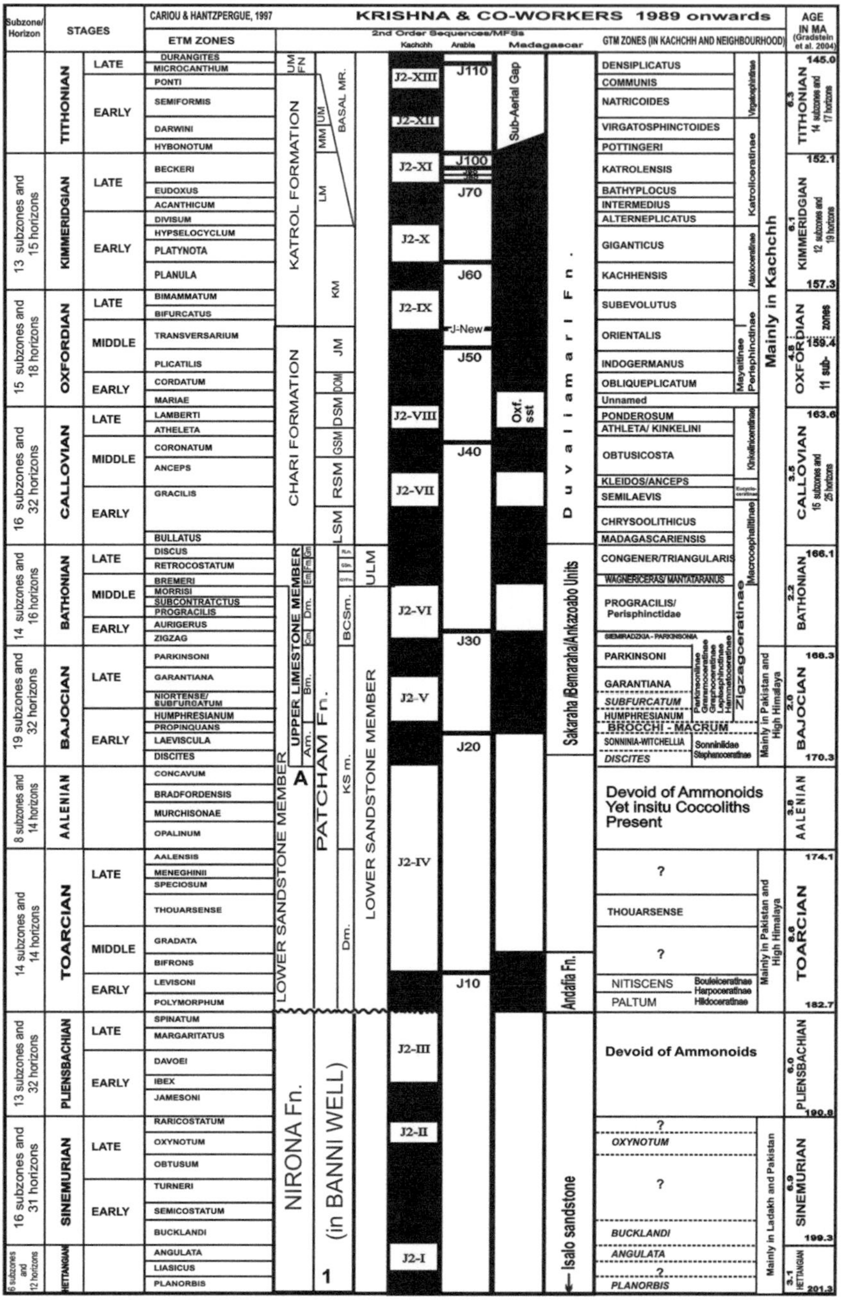

Fig. 5.4 Comparison of second-order sequences/surfaces among Kachchh, Arabia, and Madagascar on GTM

two super lithostratigraphic units in Madagascar is interpreted as the mega-order intra-Permian basal SB of the Indian Neotethyan mega-sequence.

5.4.2.3 Intra-Triassic Tectonic Event in Madagascar Near the Terminal Anisian Kellnerites Zone First-Order MFS

The Mid-Triassic lithological and associated changes between the Late Permian–Early Triassic Sakamena Group below and the Late Triassic Isalo Formation above are attributed to the widespread extensional tectonic event that occurred at the first-order intra-Traissic MFS above the late Late Anisian Kellnerites Zone as determined in the Spiti Triassic (Krishna 2012). The best observable and easily recognizable drastic change is in lithology to medium and coarse cross-bedded sands of Isalo Formation that are devoid of marine mega invertebrate guide body fossils.

5.4.2.4 Interpretation Through Tectonically Induced Uplift

The discussed change in lithology is here interpreted to have taken place on account of the tectonically induced uplift of the source area of Isalo sediments. The intra-Triassic event left widespread signatures on the GTM between Arabia and NW Australia. The said event is distinct from the later Andafia rift (Geiger 2004). It did not cause the breakup between Africa (Somalia) from Madagascar, and instead it got abandoned at the terminal Anisian Kellnerites Zone MFS. It was the Andafia rift which let the breakup between Africa and Madagascar.

5.4.2.5 Intra-Jurassic First-Order Pliensbachian/Toarcian Boundary SB

The Late Triassic unfossiliferous sands, sometimes even inclusive of the Hettangian–Pliensbachian interval, are unconformably overlain by the fine textured fossiliferous early Toarcian sediments. The Pliensbachian/Toarcian unconformable contact is easily picked up in Madagascar as in Pakistan.

5.4.2.6 Near Parallelism of the Tectono-Sequento-Stratigraphy of Kachchh and Morondova

The Toarcian–Aalenian sequence of Morondova is well age constrained through the Toarcian and even doubtful Aalenian ammonoids (De Jongh 1990 in Geiger 2004). It is found somewhat similar to the ammonoid devoid sequence in Kachchh. It is interpreted as a second-order sequence. The lithostratigraphic units of the Bajocian–Bathonian age in Morondova are age refined, revised and précised in Morondova of the like of Bemaraha, Sarkaraha, and Ankazoabo units, along with the demonstration of the Kachchh second-order sequence surfaces.

The Bathonian/Callovian sequence boundary (SB) between Ankazoabo sands and Dualia marls of Geiger and Schweigert (2006) is attributed to a latitudinal-cum-climatic change at an inferior-order sequence surface in favour of the older mid/late Middle Bathonian and younger Middle Callovian second-order SBs. Had the said lithological boundary marked the termination of a significant regression, it, as an alternative, could have been considered a second-order SB without any origin/extinction of ammonoids. However, the preceding marker carbonate unit (RLm) instead of being regressive, represents an inferior order TST in the middle of a second-order TST. Nevertheless, the said boundary signals a widespread marker event in Kachchh and Jaisalmer. The mid/late Early Callovian second-order MFS determined in Kachchh is précised also in Morondova in section XIII in place of the Middle Callovian Coronatum Zone second-order MFS of Geiger and Schweigert (op. cit) in their Fig. 3. The early Early Oxfordian second-order SB differs between Kachchh and Morondova by just one subzone, which may be due to the absence of guide ammonoids across the said SB in Morondova. The intra-Oxfordian MFS between the Schilli Subzone and Rotoides Subzone is also recognized by Geiger and Schweigert (2006). The same is strengthened in Madagascar as the most significant MFS in the Jurassic of Morondova, so of first order. The basal Kimmeridgian second-order TST is also recognized in the terminal part of the Morondova Jurassic Succession.

For the present, no comments are made on the five inferior-order (possibly third order) sequences which are not adequately described in the text, since the comparison of third-order sequences is planned in a later publication.

5.4.2.7 Basal Bajocian Second-Order SB

The pericratonic rift basin of Kachchh is one of the resultant manifestations of the pre-breakup intra Gondwana rift of the terminal Anisian age. As indicated earlier, the thermal subsidence and appreciable spreading only began later at the start of Bajocian. Kachchh interpreted a second-order sequence boundary near the said base of Bajocian which is otherwise found conformable in the 'Island belt'. Further, in Kachchh, the extent of the non-marine subsurface Nirona Formation due south and west remains yet unknown possibly on account of migration of the rift locus farther south. May be, the ULM is unconformably overlying the basement in Mainland Kachchh. Thus, there could be an earlier step of shallow-localized Patcham rift that deepened and widened into the more important Kachchh rift of basal Toarcian age, the two successive rifts belonging to two successive first-order cycles—the Spitian sequence followed up by the Kachchian sequence.

5.4.2.8 Start of Marine Sedimentation in Kachchh and Morondova in early Early Toarcian at the First-Order SB

The Kachchh basin originated in late Late Anisian Kellnerites Zone (Krishna et al. 2010), although yet earlier intra-Permian initiation is not ruled out, while the Morondova basin is of Late Paleozoic origin. In Morondova also, as in Kachchh, the marine sedimentation began in early Early Toarcian at the base of the Andafia Formation at the first-order intra-Jurassic SB as determined in Pakistan and Madagascar. The Andafia Formation in Morondova is primarily known in seismic record, while only a few metres thick presence is indicated in the Miandrivazo region. However, in the Majunga basin, the coeval Beronono Formation includes more than 100 m thick Toarcian sediments. The widespread change on GTM from non-marine to marine resulted from the renewal of rifting, e.g. between Somalia of Africa and North Madagascar of India and eustatic rise at the start of a new first-order sequence.

5.4.2.9 Similar Non-marine Sediment Fill Below the Marine Succession in Kachchh and Morondova

The marine Jurassic in Kachchh is underlain by ~600 m clastic sediments of the non-marine pollen/spore yielding Nirona Formation in Banni well subsurface of Ladinian–Liassic age (Koshal 1975) while the marine Jurassic in Morondova is immediately underlain by the non-marine sands of the Isalo Formation. In both the basins this non-marine/marine disconformity between the Islao Formation and Andafia Formation in Morondova, and between the Nirona Formation and Lower Sandstone Member (LSM) of the Patcham Formation at Patcham in the Kachchh basin is the first-order intra-Jurassic SB of early Early Toarcian age. Thus, the said SB is recognized and interpreted for the first time in Madagascar in Morondova, Majunga and other Madagascar basins as the first-order intra-Jurassic SB, so extended from Kachchh. The sand dominated succession exposed at Kuar bet and Dingy hills of Patcham region of the Kachchh basin included in the Dm (up to bed 36 of Biswas 1977) is here correlated to the mud dominated Andafia Formation inclusive of the overlying Aalenian sands. Dm up to the bed 36 is interpreted as a second-order sequence with MFS above the mid-Early Toarcian Nitiscens Zone interpreted in the beds 18–21 of Dm. The so-determined MFS is different from the one zone younger second-order MFS on ETM above the late Early Toarcian Bifrons Zone. There is also extinction of the subfamily Bouleiceratinae above the Nitiscens Zone MFS both in Pakistan and Madagascar.

5.4.2.10 Facies and Other Changes on Account of Extensional Tectonic Event at the Intra-Jurassic First-Order SB

In basins already undergoing marine sedimentation, there is change of facies from Carbonate to non-carbonate in Spiti and Nepal also in Axial belt of Pakistan, change in floral/faunal content at the start of Toarcian above the Kioto Limestone, so indicated there through Early Toarcian *Alocolytoceras*. The beds 22–36 of Dm at Patcham on margin of the Kachchh basin are sand dominant, and their upper part beds 28–36 are suggestively correlated to Aalenian sands that are included in the Andafia Formation as a distinct lithounit. The Dm second-order sequence is informally organized into five third-order sequences (details need no discussion here) corresponded to the second-order GTM sequence developed in Kachchh. *Bouleiceras* so far has not been found either in Kachchh or Morondova although having a rich presence in neighbouring Baluchistan region of Pakistan (Fatmi 1986) and also in Majunga basin of Madagascar.

5.4.2.11 Spreading Between Somalia and Madagascar and Development of Carbonate Ramp at the Start of Bajocian

The rifting gave place to spreading between Somalia and North Madagascar at the start of Bajocian on account of the earlier interpreted sliding down of Madagascar vis-à-vis East Africa along the Davies fracture (Rabinowitz et al. 1983; Krishna 1994). At the same time there developed carbonate ramp in Morondova and Kachchh as Upper Limestone Member (ULM) of Patcham Formation at Jhura, and Bemaraha and Sakaraha units in Morondova of Madagascar which also suggests that morondova and Kachchh had closely similar latitude of early 20° in south hemisphere.

Mixed Carbonate—Siliciclastics (Sarkaraha Formation) in Bajocian and Exclusive Siliciclastics (Ankazoabo Formation) in Bathonian The Sarkaraha mixed carbonate siliciclastics look quite similar to the lower (Bajocian) part of ULM at Jhura in Mainland Kachchh, while the Ankazoabo Formation similar to upper (Bathonian) part of ULM at the margin in Kaladongar and Goradongar sections of Patcham that are dominantly in siliciclastic facies. In the Sadhara section, there is present the only Bajocian ammonoid *Calliphylloceras heterophylloides* in the basal thin coral limestones, which is assigned to Early Bajocian (Pandey et al. 2013a) Propinquans Zone second-order MFS. The RST is made up of shale/sands, which is more shaly at Jhura and more sandy in Kaladongar and Goradongar sections of the basin margin.

5.4.2.12 Development of Carbonate Ramp During Bajocian–Bathonian Both in Kachchh and Morondova

Carbonate ramp developed presumably at the start of the Bajocian in Kachchh (Habo, Jhura and Jumara exposed sections in Mainland Kachchh) as in Morondova which continued and extended up to the close of Bathonian in Patcham on the north margin of the basin, but for development of thick Early Bathonian sands in Patcham. At present it is not possible to know the precise upper limit of the sands and the same are only suggestively corresponded here to Aalenian. It is not sure to date the start of the second second-order sequence on GTM, i.e. for present it is difficult to date the SB between the said two sequences in absence of guide ammonoids as to whether the sequence starts within Aalenian or at the start of Bajocian. Any way, the second second-order sequence in such uncertainty as the simpler alternative is begun at the base of Bajocian. However, the second-order MFS is somewhat better determined in Early Bajocian Laevieuscula Zone which is expressed as the basal fossiliferous oolitic or other limestones in Jhura of Kachchh Mainland, and Sadhara and Kaladongar sections of 'Island belt'. The Sadhara has recently yielded to author's team *Calliphylloceras heterophylloides* (Pandey et al. 2013) from basal coral limestones, and from basal limestones of Jhura gastropod *Heterohelix* of suggestive Early Bajocian Laeviuscula Zone. Sediments included in Am of ULM at Jhura make the second-orderTST while sediments included in Bm as the corresponding RST. However, thick cross-bedded sands are included in the same RST at Sadhara and Kaladongar. The correlation of the marginal thick sands with the mixed carbonate-siliciclastic facies Bm of less proximal Jhura by Geiger (2004) seems fine. Thus, the early part of Sakaraha Formation (oolitic limestones) makes the second-order TST of Early Bajocian while the middle Sarkaraha mudstones the corresponding RST of early late Bajocian age. The transition/contact between early and middle part of Sarkaraha unit is interpreted as the second-order MFS of Early Bajocian Laeviuscula Zone. The second second-order sequence may range up to the early Late Bajocian instead of Bajocian–Bathonian boundary.

The Bajocian–Bathonian interval in Morondova as above includes two second-order sequences plus early part of second-order TST of the third second-order sequence. The Bathonian/Callovian boundary only makes a third-order MFS much unlike an SB as interpreted by Geiger and Schweigert (2006).

5.4.2.13 Replacement of Collignon's Madagascar Ammonoid Zones by the Recently Developed Zonation Scheme of Kachchh

References are very often made to Collignon's ammonoid zones in publications on GTM Jurassic ammonoid stratigraphy. However, Collignon's collections in bags housed in Dijon (France) mention only localities without details of beds. He did record some information in his field diaries, which, however, could not be included in the collection bags. In essence, neither Collignon ever undertook proper lithostratigraphic differentiation in the sections in which he collected the ammonoids, nor ever collected ammonoids even with any useful stratigraphic details. As such the ammonoid zonal succession of Collignon is not based on field data. The author has carefully examined Collignon's unpublished diaries along with his ammonoid collections housed in Dijon (France) in order to bring out meaningful stratigraphic information. Some times in his diaries, he did make crude differentiation of his sections in more than one stratigraphic interval, however, the collections from such intervals at one locality were put together and mixed in the collection bags. The collection bags now refer only to localities. In contrast, the zonal scheme developed in Kachchh is based on high-resolution lithostratigraphic differentiation on cm scale with precise positions of all the ammonoids collected. The Kachchh zonation as above has withstood litmus test in Morondovo, after having been demonstrated earlier in Jaisalmer, Spiti, Nepal and Pakistan.

5.4.2.14 Ankazoabo Formation Sands Mostly of Middle to Late Bathonian Age

The Ankazoabo unit of South Morondova has been tentativey correlated with late Middle to Late Bathonian ammonoid-bearing sands of central Morondova which have yielded *Clydoniceras* sp., *Micromphalites hourcqi* (Collignon), *Procerites hians* (Waagen), *Delecticeras anjohense* (Collignon) and *Parachoffatia* sp. (Luger et al. 1994). However, there are thick sands in the marginal Sadhara and Kaladongar sections of Patcham which are age constrained in part of late Early and early Middle Bathonian duration. The MFS of the third second-order sequence is age constrained in the Early Bathonian Macrascens Subzone of Aurigerus Zone in Kachchh on the basis of cooccurrence of *Siemiradzkia* and *Parkinsonia* in the relatively distal Nara section (Pandey and Pathak 2015). Thus, there is good reason to suggest the correlation of Ankazoabo sands with calcareous sands or sandy limestones of Cm and Dm at Jhura section, which in turn have been correlated with the thick upper sands in the Sadhara section and thick sands of BCSm at Kaladongar. The early and middle thicker part of Ankazoabo Formation is here interpreted as the second-order RST in between the Early Bathonian Macrescens Subzone MFS below and mid-Middle Bathonian SB above of mid/late Middle Bathonian age.

5.4.2.15 Upper Body Fossil Bearing Part of Ankazoabo Formation as late Middle to Late Bathonian

The author is agreeable only in part with Besaire (1973) and Luger et al. (1994) about the correlation only of the terminal ~25–30 m body fossil bearing part (with solitary *Parchoffatia)* of Ankazoabo sands of south Morondovo with sands of central Morondovo which include extremely rare to scarce late Middle to Late Bathonian ammonoids. The older late Early to early Middle Bathonian part makes massive development of mega-marine invertebrate devoid sands as second-order RST. However, the younger late middle to Late Bathnian part includes noticeable shales as well in sands with several levels of marine mega body fossils as early part of a second-order TST that is as well deposited at distinctly slower rate in lesser energy framework and improved bathymetry. The older of the two Bathonian ages is favoured for the early and middle thicker part of Ankazoabo and Besabora sands. As in Bela section of Kachchh, the sandy facies seems to continue up to close of Bathonian in Morondova.

5.4.2.16 Second-Order late Late Bajocian–Mid-Middle Bathonian Sequence with MFS in Early Bathonian Macrescens Subzone

The second-order SB of this sequence in Kachchh is suggestively placed above the early Late Humphrsianum Zone. In Morondovo this SB can be placed at the contact of the clastics below and carbonates above in the Analamanga section. The ~15 m thick carbonates, overlying the contact are here interpreted as the slow sedimented second-order TST of the next sequence. The MFS of this sequence in Kachchh has been determined above the Early Bathonian Macrescens Subzone in the recently discovered Early Bathonian sediments in the Nara section. The determined MFS is based on the co-occurrence of *Parkinsonia* and *Siemiradzkia* (Pandey and Pathak 2015). In absence of any guide ammonoids in Morondovo, the MFS there too is suggestively dated above the Macrescens Subzone. Thus, the terminal ~15 m carbonates of Sarkaraha Formation extend into Early Bathonian up to the Macrescens Subzone. The entire Ankazoabo Formation is of post-Macrescens Subzone age, and looks similar to the clastics dominated post-Macrescens Subzone parts of Kaladongar and Goradongar sections. These thick sands of the Ankazoabo Formation are safely dated here as late Early to Late Bathonian. The intra-Bathonian SB of the second-order sequence is tentatively placed below the lone record of the *Parachoffatia* ~25–30 m below the top of the Ankazoabo Formation. Thus, quite similar lithostratigraphic development in Morondova and Kachchh together with the recent second-order sequence formulation in Kachchh has helped the assignment of safe reliable ages than any time earlier to even ammonoid devoid Sarkaraha and Ankazoabo units.

5.4.2.17 Second-Order late Middle Bathonian Mantataranus Zone–early Middle Callovian Chrysoolithicus Zone Sequence with MFS Above the Mid-Early Callovian Chrysoolithicus Zone Diadematus Subzone

The Late Bathoian *Parachoffatia* (Uhmann 1996) has been recorded within the sands of the Ankazoabo Formation about 20–25 m below the top in the Tongobory section. In several sections of Kachchh (Jhura, Nara, and Jumara of Mainland, and Goradongar of Patcham) have also been known late Middle and Late Bathonian ammonoids inclusive of the first Macrocephalitinae through *M. mantataranus* (Boehm) which have been interpreted as the early part of a fresh second-order TST of the late Middle Bathonian–mid-Early Callovian age with second-order MFS above the mid-Early Callovian Chrysoolithicus Zone Diadematus Subzone (Krishna et al. 2010, 2011). The presence of *Parachoffatia* in Tongobory section of Morondova, and of *Clydoniceras* and *Micromphalites* in the Central basin, and *Procerites hians* (Waagen) and *Delecticeras anjohense* (Collignon) south of the Onilahy river support well the late Middle to Late Bathonian age bracket for the terminal part of the Ankazoabo unit in Tongobory, so also elsewhere in Morondova and other basins of Madagascar. This ~25 m terminal part of the Ankazoabo Formation on the pattern of Kachchh is also interpreted here as the early part of the second-order TST. Thus, Ankazoabo Formation precisely ranges from Recinctus Subzone of Early Bathonian to late Late BathonianTriangularis Zone (Pandey and Pathak 2015). Unlike the interpretation by Geiger and Schweigert (2006) on the start of transgression at the base of Callovian at the contact of Ankazoabo sands below and Dualia marls above, the transgression may have started earlier ~25 m below the lithological transition. The Bathonian/Callovian boundary in Kachchh does not make a second-order SB. The lithological contrast/change at the Bathonian/Callovian boundary is on account of the drowning of the carbonate ramp presumably at a third-order SB.

5.4.2.18 Age of the Sakanavaka Formation

In spite of absence of any definite evidence, Geiger and Schweigert (op. cit) correlated the guide fossil devoid Sakanavaka Formation with the Bathonian Ankazoabo Formation. The only clue cited is the superposition of both these units by Dualia marls in different sections which unquestionably on ammonoid evidence began in basal Callovian. There perhaps are reasons to at least suggest an alternative on two counts. The first is the superposition of Sakanavaka sands by Dualia marls of late Middle Callovian age since enclosing *Obtusicostites ushas* of late Middle Callovian Obtusicosta Zone of the Indo-East African province as developed in Kachchh (Ojha 1996a, b, Krishna and Ojha 1996). The same suggests that Sakanavaka sands either range at least up to the close of early Middle Callovian or the contact is disconformable with absence of Early and early Middle Callovian age part of the Dualia marls. The second reason is the unproven stand of Geiger and Schweigert of the uninterrupted vertical spread of Dualia marls from base to the top of Callovian without any intervening thick sands of the like of Sakanavaka.

5.4.2.19 Presence of Intra-Callovian Sands in Kachchh and Morondova

So far, the Toarcian–Bathonian succession of Kachchh and Morondova have proved having similar lithostratigraphic build up, particularly the positioning and ages of sandy units like Aalenian sands, Bajocian mixed carbonates and siliciclastics, and Bathonian sands. It is quite logical to expect the said similarity in the rest of the Jurassic as well. The Callovian in Kachchh is dominantly shaly/marly but with thick marker hard resistant Ridge Sandstone Member (RSM) of late Early Semilaevis Zone to Middle Callovian O-I Horizon of Obtusicosta Zone age in between. The age framework of RSM supports well the possible minimum upper limit of the Sakanavaka sands, however, yet there is no clue about the lower limit. There are no sections in which Dualia marls are found overlain by sands. Nevertheless, the alternative suggested may be further explored. Based on homotaxial nature of the Jurassic developments in Kachchh and Morondova, the suggested alternative appears more plausible.

The ?younger Sakanavaka sands (younger than Ankazoabo sands) are overlain by mudstones with early to mid-Early Callovian ammonoids in different Morondova sections for example *Macrocephalites* cf *apertus* in the Antainakanga section and *Indosphinctes besavoesis* (Collignon) in Amparambaoto section, and *Subgrossouria wennerai* (Collignon) in Ankazemihova section. In Kachchh, the corresponding mud/shales make the later part of the second-order TST of Lower Shale Member (LSM) of Chari Formation of early and mid-Early Callovian Chrysoolithicus Zone. On the upper side, the Sakanavaka sands are found overlain by mudstones with *Obtusicostites ushas* (Spath) in Mamakiala section of late Middle Callovian age Obtusicosta Zone. The Sakanavaka sands thus on one hand appear to be of younger Bathonian age in a few sections while in Mamkialia section, the unit ranges up to early Middle Callovian, i.e. Late Bathonian to early Middle Callovian. On the other hand there are thick noticeable late Early Callovian sands in section Via, and thick bioclastic oolitic limestones in VIb of around the same age which total ~ 15 m in between Dualia marls on both sides. Thus, like the Ridge Sandstone Member (RSM) of Chari Formation of Kachchh, the Sakanavaka Formation or similar looking sands are age constrained between late Early Semilaevis Zone and early Middle Callovian Anceps Zone. Sakanavaka or similar sands could be somewhat broadly age equivalent of RSM of Kachchh. The fourth second-order sequence in Morondova includes Ankazoabo Formation of late Middle Bathonian to Late Bathonian as early part of the second-order TST, while later part of the TST is made up of the early part of Dualia marls.

5.4.2.20 Age of the Intra-Callovian MFS, Above Mid-Early Chrysoolithicus Zone or Above the late Middle Coronotum Zone

Geiger and Schweigert (op. cit) in their Fig. 3 without adequate evidence and discussion indicated the intra-Callovian MFS in the later part of late Middle Callovian Coronatum Zone in Rota Subzone, which may have been influenced by the widespread Coronatum Zone MFS on ETM. In the Kachchh sequence framework, the second-order intra-Callovian MFS is determined above the mid-Early Callovian Chrysoolithicus Zone Diadematus Subzone. Apart from the sedimentological evidences in favour of the maximum intra-Callovian bathymetry at this point of time, there is also maximum frequency, diversity and density of ammonoids at the determined surface, also extinction of the relatively deeper water preferring inflated Chrysoolithicus lineage of Macrocephalitinae. In Morondova and other basins of Madagascar, there are as well widely noticed similar Macrocephalitinae ammonoids. The contact of the basal ammonoid-bearing Dualia marls with the overlying sands can be reasonably well assigned to Chrysoolithicus Zone, and further interpreted as the second-order MFS. On the upper side, the contact between Sakanavaka or similar looking sands with the overlying *Obtusicostites*-bearing marls in the Mamakiala section can be dated as early/late Middle Callovian O-I/O-II horizontal boundary within the Obtusicosta Zone above the Paramorphum Zone / Obtusicostites Zone boundary, and further interpreted as the second-order SB.

5.4.2.21 Ammonoid Constraints on Bathymetry

The marls that include the inflated Macrocephalitinae suggest grater depth than the superjacent Kinkeliniceratinae-bearing marls. The stephanoceratid Macrocephalitinae in context of functional morphology are considered relatively less shallow than the perisphinctid Kinkeliceratinae. Moreover, the frequency, diversity and density of ammonoids are greater in Macrocephalitinae-bearing marls below the thick sands than the Kinkeliniceratinae-bearing marls above the sands. The argument as above allows here to interpret much stronger propriety of the Early Callovian second-order MFS, than the Middle Callovian as suggested earlier by Geiger and Schweigert (2006).

5.4.2.22 Start of Ammonoid Faunal Differentiation as Indo-East African Faunal Province on the GTM

The late Middle Bathonian–early Middle Callovian sequence could also be termed as Macrocephalitinae sequence, so named after the dominant indigenous ammonoids in the sequence. It is not mere coincidence, that the sequence framework so far has been nearly concordant on ETM and GTM with the same succession of either precisely or nearly isochronous sequence surfaces (SBs and MFSs); intra-Toarcian Early Toarcian Nitiscens Zone/Bifrons Zone second-order MFS, intra-Bajocian Early Bajocian Laeviuscula Zone second-order MFS, intra-Bathonian Macrascens Subzone second-order MFS, Morrisi Zone second-order SB. The first unambiguous discordance is noticed in the intra-Callovian MFS, in Early Callovian in the Kachchh basin of GTM and Middle Callovian in ETM. It is here interpreted that the start of the explicit discordance is possibly related to regional tectonic influence exclusive to GTM at the second-order SB which continued for the rest of the Jurassic, i.e. for a large duration. In this interval, the first- and second-order sequence surfaces exhibit distinct discordance with those of ETM. This is in contrast to the global concordance of the third- and finer-order sequences of duration less than ~ 2 my. In general the mega and, first-order sequences are controlled by regional tectonics at regional level of the like of GTM or ETM, whereas provincial tactonics like within the Indo-East African province governs the second-order sequences.

5.4.2.23 Late Middle Callovian Obtusicosta Zone–early Early Oxfordian Second-Order Sequence with MFS Above the Early Late Callovian Athleta Zone Depressum Subzone Sequence

This sequence starts at the base of the Gypseous Shale Member (GSM) above the RSM sands in Kachchh with its dominant ammonoid content of geographically restricted Kinkeliniceratinae in the Indo-East African faunal province. Kinkeliniceratinae ammonoids also dominate in this interval in Morondovo and other basins of Madagascar. The younger mudstones of Morondova with base in the late Middle Callovian Obtusicosta Zone, are precisely age equivalent of the Gypseous Shale Member (GSM) of the Chari Formation. These mudstones constitute another second-order TST. The MFS in Kachchh is at the top of the GSM above the early Late Callovian Depressum Subzone.

Section XIII suggested the start of this sequence above the Sakanavaka sands (coeval to RSM of Kachchh) as the early part of the second-order TST, while the RST is interpreted in the so-called Oxfordian sands. In Kachchh, the RST comprises sands of the Dhosa Sandstone Member (DSM). The MFS is determined at the GSM/DSM contact which also marks the extinction of Kinkeliniceratinae in the IEAP. In Kachchh, DSM ranges from the early Late Callovian Athleta Zone to either close of early Early Oxfordian or restricted within the Late Callovian itself in absence of the early Early Oxfordian Mariae Zone ammonoids in the sands. The same age duration is here very logically extended to the Oxfordian sands of Moronodovo and elsewhere in Madagascar which can be interpreted there as well as the second-order RST. This second-order sequence could also be designated as the Kinkeliniceratinae-bearing sequence with terminal SB either near the Callovian/Oxfordian boundary or at the most extending up to early/late Early Oxfordian Mariae Zone. In global context, similar sands are restricted to late Late Callovian with second-order sequence boundary near the Callovian/Oxfordian boundary on the Callovian side.

These late Late Callovian–?early Early Oxfordian sands include substantive silt in the distally located sections of the like of Keera, Lakhapur etc. In Keera, the RST is divisible in three parts, among which the middle part includes ammonoids. The other two parts so far have not yielded any ammonoids. Based on the ammonoids, the middle part of the sand dominant RST may range up to the late Late Callovian Ponderosum Zone. The details as above allow us to interpret the approximately dated 'Oxfordian sands' of Morondova as a second-order RST of late Athleta Zone–?early Early Oxfordian Mariae Zone.

5.4.2.24 Late Early Oxfordian Obliueplicatum Zone–late Late Oxfordian Bifurcatus Zone Second-Order Sequence with MFS Above the late Middle Oxfordian Orientalis Zone Schilli Subzone

This sequence starts at the base of the Dhosa Oolite Member (DOM) above the sands of DSM. The TST comprises succession of hard-grounded ammonoid-rich slow sedimented condensed nodular conglomeratic limestones. These make the most characteristic marker unit of the late Early Oxfordian Obliqueplicatum Zone to basal Middle Oxfordian Indogermanus Zone age (DOM) in Mainland. Near the east margin in Wagad, the marker lithology is of Middle Oxfordian Antecedens Subzone of Indogermanus Zone (equivalent of the ETM Plicatilis Zone) to schilli Subzone of the late Middle Oxfordian Orientalis Zone (equivalent of the ETM Transversarium Zone). In Mainland Krishna et al. (2000 and later) interpreted a large submarine gap from early Middle Oxfordian Indogermanus Zone to mid-Early Kimmeridgian Giganticus Zone of several my. In Morondova, the ~8 m succession of the yellow bioclastic iron oolitic limestones is equivalent of the DOM and JM as the second-order TST.

The said ammonoid-rich marker Oxfordian mudstones in the Dangovato section include multiple marine mega-invertebrate-rich hardgrounds as in the Dhosa Oolite Member (DOM) in Mainland and Jhadasa Member (JM) at Kantkote in Kachchh. The maximum frequency of the hard grounds and associated thinning of the intervening mudstones/shales in the Dangovato section is in the youngest part of the hard-grounded succession. There are two such ~4 m thick intervals of hard ground succession there which are separated by less than 2 m thick mudstone bed. These have been dated as latest Early Oxfordian and Middle Oxfordian, respectively. Immediately above of the marker unit, follows the ~10 m thick mudstone interval with three thin hard fossiliferous beds which have been assigned to Late Oxfordian and basal early Kimmeridgian. Thus, the hard-grounded ammonoid-rich lower part includes near the top *Enayites* of Lucaeformis Subzone, while the oldest date from the overlying shales is Bifurcatus Zone. It is quite possible that the contact between the two contrasting lithologies in Madagascar represents the Schilli Subzone/ Rotoides Subzone boundary first-order MFS. In context and comparison with Kachchh sections, the Morondova appears to be proximal to margin farther South. The presence of late Middle Transversarium Zone and Late Oxfordian ammonoid elements suggests greater similarity to the continuos succession at Kantkote of Jhadasa Member of Chari Formation and Basal Kantkote Member of the overlying Katrol Formation. The said succession of Kantkote is absent in Mainland Kachchh and is instead included in the large submarine gap of the early Middle Oxfordian Indogermanus Zone to mid-Early Kimmerdgian Giganticus Zone.

5.4.2.25 Intra-Oxfordian First-Order late Middle Oxfordian Schilli Subzone MFS

The most visible and evident feature of the Oxfordian succession of Morondovo is the lithological, sedimentological and paleontological contrast between the older part of the section with two thick succession of hard grounds with thin mudstones in between below and the younger succession with absence of hard grounds within the thick beds of mudstone. The absence of hard grounds in the younger succession and decrease in frequency of body fossils are interpreted here as improvement in sediment accumulation rate on account of decrease in bathymetry. Thus, the said contact is interpreted here as the first- or second-order intra-Oxfordian MFS, which is tentatively assigned to Middle/Late Oxfordian in Morondova. With large scale similarity of sedimentary succession, ammonoid succession and sequence stratigraphic framework, it is logical to extend the date of the Kachchh intra-Oxfordian first-order MFS to Madagascar, i.e. the late Middle Oxfordian Schilli Subzone/Rotoides Subzone boundary.

5.4.2.26 Second-Order Dominantly Shaly/Muddy Latest Middle to late Late Oxfordian Subevolutum Zone RST in Morondovo

The RST of this sequence in Kachchh is made by major part of the Kantkote Member of the latest Middle Oxfordian Rotoides Subzone of Orientalis Zone to latest Late Oxfordian Subevolutus Zone which is exposed only in Kantkote at Wagad at the proximal-most exposed east margin of the basin. The RST in Kachchh except for its early shaly part is dominated by sands. In Morondovo too, the preserved RST of this sequence is made up of ~35 m mudstone with drastically reduced frequency, diversity, density of the ammonoids as also of other body fossils across the MFS. It is here assigned to Late Oxfordian.

Geiger and Schweigert (2006) in the text do not provide explicit evidence of either continued increase in bathymetry during Late Oxfordian and Early Kimmeridgian or decrease in bathymetry at the Middle/Late Oxfordian transition, while in Fig. 3 they have decisively indicated the MFS at the above discussed contact. Instead, there are good evidences of decrease in bathymetry for placement of the MFS at or near the Middle/Late Oxfordian transition. The above discussed second-order sequence could be termed Mayaitinae sequence or Mayaitinae–Perisphinctinae sequence in view of the two dominant ammonoid lineages in the sequence of the geographically restricted IEAP Mayaitinae and the cosmopolitan Perisphinctinae.

5.4.2.27 Basal Kimmeridgian Kachchhensis Zone Second-Order TST in Kachchh and Morondova

Capping the Jurassic succession in Morondova is the ~10 m thick shale with thin intercalations of conglomeratic sandstone. These shales have yielded *Subdiscosphinctes* of basal Kimmeridgian age. As in Kachchh, the shaly unit is here interpreted to signal the start of a fresh second-order sequence, however, only the early part of the TST is preserved in Morondova in contrast to the presence of the entire sequence at Kantkote in Kachchh. The sequence in both Kachchh and Morondova marks the origin of Ataxioceratinae on GTM.

5.4.2.28 Early Aptian Transgression in Morondova and Kachchh

In the relatively distal West Mainland, there is continuity of sedimentation from Jurassic into Cretaceous. As indicated in Chap. 3, the terminal first-order sequence begins with a wide spread transgression not simply in Kachchh but in several basins of India and GTM. This Early Aptian transgression is already known in basins of Madagascar. The same is interpreted here as the start of the first-order terminal Cretaceous sequence in Madagascar.

5.4.2.29 First-Order Intra-Cretaceous MFS/MFI in the Early and early Middle Turonian in India and Madagascar

The Early and early Middle intra-Turonian first-order MFS/MFI as discussed elsewhere in this work has been recognized widely in several Indian basins, so also in basins of Madagascar.

The detailed discussion as above has convincingly demonstrated the presence of all the first-order Indian intra-Mesozoic SBs and MFSs of the Neothyan mega-sequence in Morondova and/or elsewhere in Madagascar. The basal intra-Permian SB at the base of Kazanian is determined between Sakoa Group and Sakamena Group, first-order intra-Triassic MFS above the late Late Anisian Kellnerites Zone between Sakamena Group and Isalo Formation, the first-order intra-Jurassic SB. In basal Early Toarcian Andafia Formation, the first-order intra-Jurassic MFS above the late Middle Oxfordian Orientalis Zone Schilli Subzone within the Dualia marls between the Middle Oxfordian hard-grounded richly ammonoid-bearing slow-sedimented condensed oolitic bioclastic limestones, and not so slow sedimented hard ground scarce, not so ammonoid-rich Late Oxfordian mudstones in Dangovato section of Morondova (Fig. 12 of Geiger and Schweigert 2006). The first-order intra-Cretaceous SB is determined again in Dangovato section of Morondova and in other basins of Madagascar of basal Aptian age between the terminal part of Basal Kimmeridgian Dualia marls below and the disconformably overlying Early Aptian bioclastic conglomerate.

5.4.3 Comparison of the Kachchh and Arabian Plate Sequence Stratigraphic Frameworks

Prior to the present effort of sequence stratigraphic formulations in Kachchh on GTM, there already existed on GTM the comprehensive study realized in Arabia (Sharland et al. 2001) by a competent group of petroleum geology specialists of the Arabian region. It is in fitness of things that a comparison is undertaken to attest the regional applicability of the two frameworks of GTM, with focus on their similarities in order to compliment each other towards minimizing the differences (Fig. 5.5). The exercise also aims at providing explanations of the differences as also relooking at the respective frameworks towards further strengthening and improvement in our quest for finding more and more hydrocarbons.

5.4.3.1 Hydrocarbon Context

The Arabian plate includes proven high-quality hydrocarbon source and reservoir sediments to the extent of 50 % of total global reserves, and at present shares about 40 % of total global production. Further, ~65 % of the hydrocarbon source sediments are skewed in the Mesozoics. Among the Arabian plate sequences, AP7 sequence has been by far the most productive. The reserves and the production are heavily skewed in favour of AP7. Most of the Arabian giant oil fields belong to Mesozoic sequences. Among the 12 giant hydrocarbon reservoirs in global context, six are included from the Mesozoic of Arabia. Among these 12, the Ghawar field of Saudia is undisputedly considered the giantest of all (Lindsay et al. 2006). Again within AP7 there is heavy skewing towards J50 (having maximum proven source sediments) and J60 (good source sediments but less than J50) as source sediments in intra-shelf basins (Sharland et al. 2001, Table 3.18, p. 103). Even the biggest reservoir of the Ghawar field—the Arab–D—is of Kimmeridgian age with source sediments of Oxfordian age in close vertical proximity of the first-order MFS of late Middle Oxfordian. The Kachchh first-order sequence corresponding to AP7 has also proven source sediments in Middle and Late Jurassic, although as per published data (Biswas 2012a and others), the reservoirs are in Early and Late Cretaceous. Basins in India (K-G and Cauvery in the east and Kachchh and Jaisalmer in the west), and as well in neighbouring Pakistan, are either producing hydrocarbons or shall start producing soon in the late Middle Jurassic to Early Cretaceous, and Early–Late Cretaceous petroleum systems. Moreover, India has large thicknesses of Jurassic and Cretaceous sediments underneath the latest Cretaceous Deccan traps in Kachchh, Cambay, Narmada, Saurashtra, and even Deccan basins. Thus, in hydrocarbon context as well, it is desirable to compare the two sequence frameworks, for future complimentary collaborations in the hydrocarbon exploratory geoscientific researches.

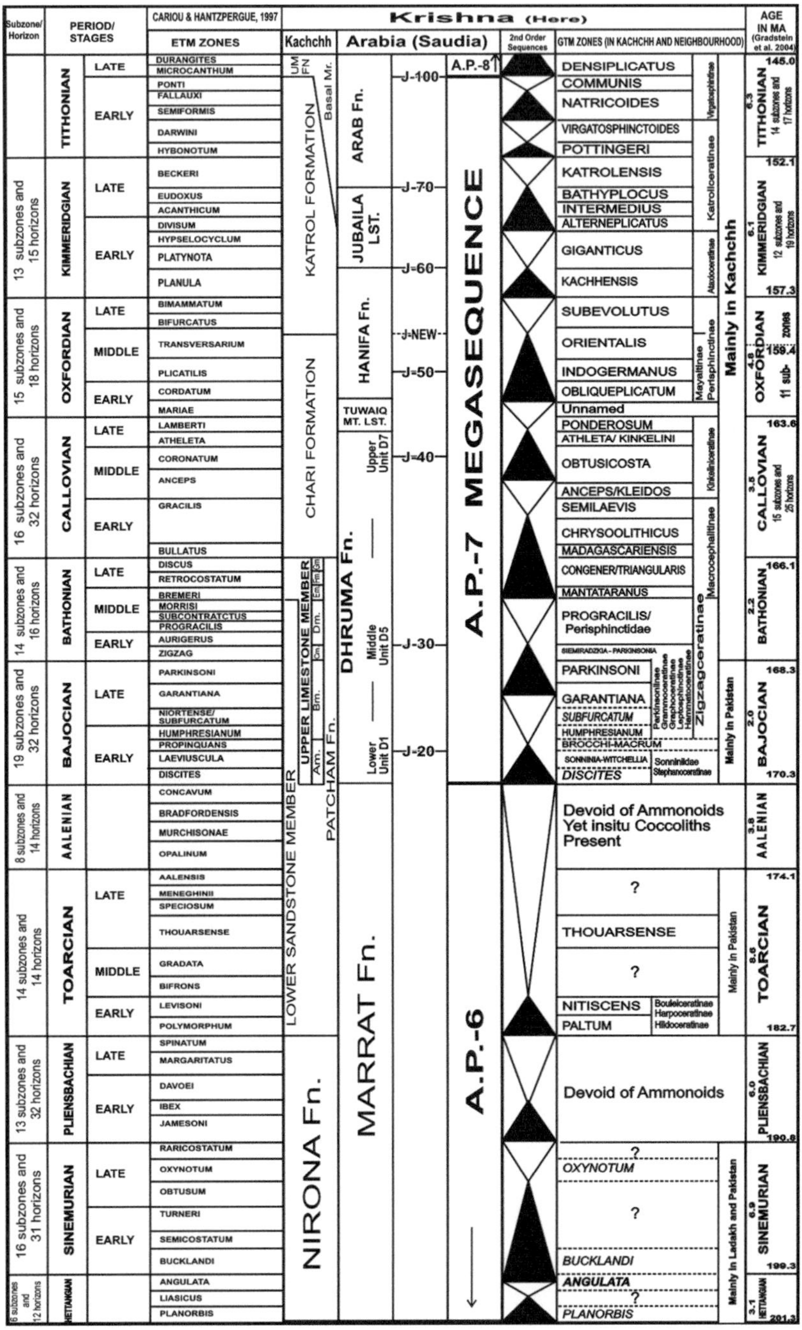

Fig. 5.5 Comparison of second-order sequence surfaces (SBs/MFSs) between Kachchh and Arabia on GTM

5.4.3.2 Intra-Jurassic Rifting History of the Two Regions

In Sharland et al. (2001), the origin of the Minjur clastics has been interpreted as sedimentary response to rifting in north (Khan 1989). Further, it is commented upon that the said rifting may have been a relatively localized attempted breakup of the extreme plate margin, which did not extend far inside the plate. There is also drastic change in lithology and environment between the Minjur and the overlying Marrat units in Arabia, and also in Kachchh at the same time from non-marine to marine. Further according to them, the Arabian intra-shelf basins developed on account of the renewed rifting mainly between J20 and J50 that prompted the submergence of the said basins there after at a fast rate. Kachchh has broadly similar intra-Jurassic rifting framework. It also witnessed renewed rifting and fast submergence since the start of Bajocian up to late Middle Oxfordian, i.e. between ages of J20 and J50. The initiation of the major regional GTM rifting event is here interpreted in early Early Toarcian which affected the divergent fracture zones in the India/Somalia sector in the west as also India/Australia sector in the east, and even in the north sector on GTM. It is equally easy to pick up in Arabia the older rifting related Minjur/Marrat unconformity even in well logs in gamma ray, porosity and density curves (Fig. 5.6). May be, Sharland et al. (2001) were overwhelmingly obsessed with ETM framework of Haq et al. (1987), in which the corresponding sequence has begun in Late Toarcian above the Marrat Formation instead of the opinion here of the start in early Early Toarcian below the Marrat Formation. Unlike Arabia, no younger intra-Toarcian unconformity is observed in Kachchh. The pre-Marrat unconformity is more widespread and as well involves larger gap. The available Jurassic rift history of Arabia justifies preference for the pre-Marrat SB as in Kachchh instead of the post-Marrat SB as the basal SB of the first-order sequence of Sharland et al. (2001). This has also been proved on a regional scale on the GTM in subsequent studies. The Nitiscens Zone Early/Middle Toarcian boundary ammonoid fauna in view of being relatively richer and more diverse fauna than the Middle Toarcian Bifrons Zone fauna, is considered better for the first second-order MFS of the so emended basal SB of AP7.

5.4.3.3 Sediment Package of J50–J60 Interval

The sediments corresponding to J50 and J60 are in part present in Mainland Kachchh (late Middle Callovian to early Middle Oxfordian), however, the complete and continuous succession is only at Khadir, Gangta bet and Wagad (late Middle Callovian to mid-Early Kimmeridgian). These belong to the younger 4 Members of the Chari Formation and the early part of the Kantkote Member of the Katrol Formation. The mid-Middle Oxfordian to mid-Early Kimmeridgian sediments are absent in Mainland due to a submarine gap (Krishna et al. 2000 onwards). The sediments corresponding to the gap are exclusively present in Khadir, Gangta bet and Wagad. However, the present author is clueless, if this mid-Middle Oxfordian to mid-Early Kimmeridgian succession, absent in the rest of the relatively distal

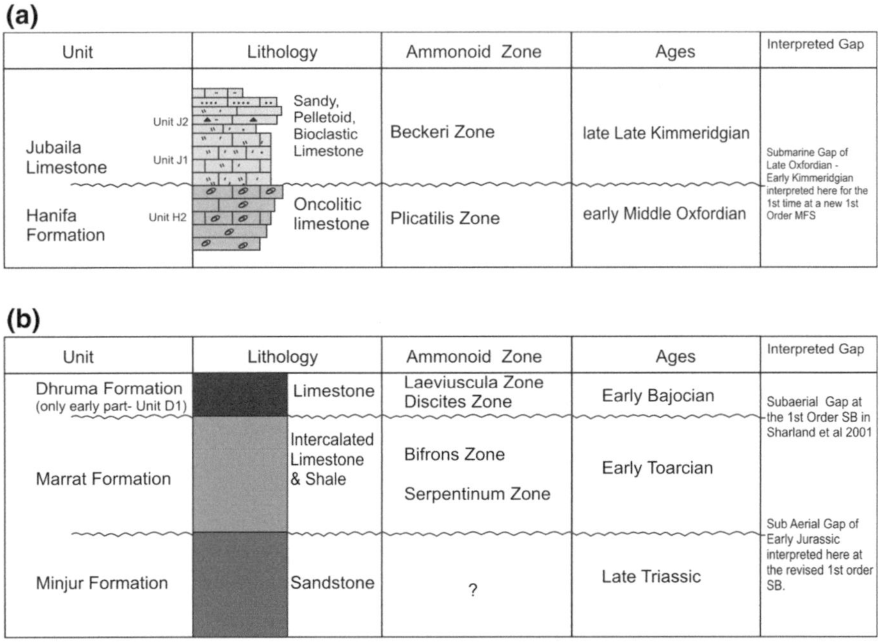

Fig. 5.6 **a** Submarine Gap of at least Late Oxfordian to Early Kimmeridgian interpreted between Unit H2 of Hanifa formation below and Unit J1 of Jubaila formation above at the new first-order Intra-Jurassic MFS in Central Saudi Arabia between J5 and J6 and else where on GTM; and **b** subaerial Gap of at least part of Early Jurassic Interpreted between Minjur Formation below and Marrat Formation above at the first-order Intra-Jurassic SB in Central Saudi Arabia and else where on GTM

shelf in Mainland Kachchh, could at all be compared with the Arabian intra-shelf successions. This proximal part of the shelf has an inner subsurface part (between east margin of Wagad and Radhanpur) of ~50 km stretch and an outer equally long exposed part up to the west margin of Wagad (Fig. 1.5). The inner stretch between Radhanpur and the exposed east margin of Wagad, covered underneath the recent sands, is interpreted to include sediments only of mid-Middle to latest Oxfordian. The exposed mid-Middle Oxfordian–mid-early Kimmeridgian sediments exclusive to Wagad, Khadir and Gangta bet have approximate thickness of ~400 m. Surface geochemical studies in the exposed part have suspected possible presence of hydrocarbons. In context of the first-order or principal intra-Jurassic MFS of late Middle Oxfordian, the said exclusive sediments in part belong to the terminal part of the first-order TST (two third-order sequences from terminal part of the first-order TST from early Early Oxfordian onwards) and early part of the first-order RST (one and half third-order sequences of the early first-order RST up to the close of Oxfordian). In context of the hydrocarbon prospects, this interval in relatively deeper offshore having enough overburden of the thick pile of younger sediments may be considered promiseful.

5.4.3.4 Location of Both Arabia and Kachchh on GTM

Both Arabia and Kachchh in the Jurassic were located on the same margin of the Neotethys—the GTM; Arabia almost on the west extreme of GTM and Kachchh located somewhat in the middle of GTM. These are separated from each other by a few thousand km. Arabia was located in low near equatorial latitudes and the resultant deposits are mostly carbonates. Kachchh on the contrary occupied sub-temperate high latitudes for the late Middle and Late Jurassic with dominance of clastic facies. In view of the different climatic framework, the geographically relatively cooler water restricted Indo-East African province Jurassic ammonoid subfamilies (early Middle Callovian Eucycloceratinae, late Middle and Late Callovian Kinkeliniceratinae, Early and Middle Oxfordian Mayaitinae, Late Kimmeridgian Katroliceratinae and Virgatosphinctinae) failed to spread to the warm and arid Arabian region. In this context, it may be noted that Fig. 3.26, p. 98 of Sharland et al. (2001) has incorrectly depicted Kachchh around 13°–14° in subequatorial tropical latitudes.

5.4.3.5 Pre-Jurassic History of the Two Regions

Kachchh is a pericratonic east–west striking rift basin which originated in the Middle Triassic. Until its origin, this Precambrian terrain had been undergoing constant weathering and erosin, and had reached near peneplation. It was located in proximity of India (inclusive of Madagascar)/Africa divergent fracture zone. The basin had a resultant southwesterly slope. The infant non-marine basin, subsequently, was inundated by an arm of the Neotethys near basal Toarcian. The exposed Jurassic shelf from Ghuneri in the west to Wagad in the east measures about 300 km, however, there are unexposed parts of the shelf on either side of the exposed shelf as discussed above. There is, obviously, large variation in facies observed between the distal sections in the west to proximal sections in the east. The initial gradient of the shelf would have been presumably small, say less than a degree of the rather extra-wide shelf. Unlike Kachchh, the Arabian basins have a long geological record since the Precambrian.

5.4.3.6 Data Base, Outcrops/Sections and Sequence Framework

The exercise in Arabia is based on voluminous data of exposed sections, well logs, and seismic profiles of a wide region. However, the effort in Kachchh here is heavily based on the exposed Jurassic ammonoid stratigraphic data, that too, generated over the last three decades by the author himself together with associates.

Kachchh is endowed with splendid outcrops and stratigraphic sections with a continuous succession of ammonoids in the ~24 my long Bathonian–Tithonian interval, which facilitated the development of highly refined ammonoid zonation (Krishna et al. 1983, 2010, 2011). The Kachchh GTM framework includes the ordering of sequences along with their ammonoid inclusive precisely dated MFSs and SBs. In comparison, the Arabian exercise does not provide any such ordering, yet it has taken note of the significance and future desirability of such ordering. The Kachchh framework includes a succession of plus 13 second order longer than ~1.5 my sequences in the ~37. 3 my long Toarcian–Tithonian intra-Jurassic interval, and about 30 less than ~1.5 my long third order, along with examples of fourth-order sequences as well. Thus, Kachchh includes ~38 MFSs inclusive of ten second order and one first order. Among the second-order Kachchh MFSs, five are included in the first-order TST and four in the Jurassic part of the RST.

5.4.3.7 Subaerial Gap in Early and Submarine Gap in Late Jurassic

The first-order TST in Kachchh is age constrained by either the underlying non-marine sediments or big subaerial early Jurassic gap around the basal SB. It is also featured with a substantive and distinct early Late Jurassic submarine gap close to the MFS above of several subzones (Figs. 2.20 and 2.21). Similar large ~20 my plus early Jurassic subaerial gap is also known in Arabia below the Marrat Formation and coeval units (Krishna 2005 onwards, also in Sharland et al. p. 202). On the younger side, Sharland et al. (2001, Table 4.35) have indicated a disconformity between the Hanifa and Jubaila units (Fig. 5.6). The ammonoid evidence suggests definite early Plicatilis Zone J50 age below and Late Kimmeridgian J70 age above the disconformity, which Krishna since 2005 (Krishna 2005 and later) have all along interpreted as mid-Middle Oxfordian to Early Kimmeridgian gap. With fine textured organic-rich slow sedimented low energy decidedly subtidal units below and above, there can be little doubt about this substantive gap being submarine. Further, Krishna et al. (2011) interpreted the intervening succession between the said two major gaps as the first-order Jurassic TST almost exactly as in Kachchh and in several other basins of GTM.

In comparison, the Arabian AP7 sequence provides a succession of 9 MFSs in AP7 roughly in the same or lesser interval (Figs. 5.4 and 5.5). Among the nine MFSs, five are included in Kimmeridgian (~5.2 my), and one each in Oxfordian (~6.2 my), Callovian (~2.6 my), Bathonian (~2.2 my) and Bajocian (~2.0 my), i.e. highly skewed in favour of Kimmeridgian. It is clear, that compared to the succession of the 30 MFSs in Kachchh (4 in Bajocian, 6 in Bathonian, 6 in Callovian, 10 in Oxfordian, 6 in Kimmeridgian and 6 in Tithonian), several MFSs seem to have been missed in Arabia, which obviously require to be determined in near future, particularly in the hydrocarbon source-rich Oxfordian. They also did not determine any principal or first-order MFS among the nine MFSs but for passing reference to Vail who interpreted J20 and J50 as second-order MFSs along with subaerial exposure in between J40 and J50. Krishna (2012) and Krishna et al. (2011)

organized the mega-sequence in India into three first-order sequences with their SBs and MFSs. The first-order intra-Jurassic MFS was determined above the late Middle Oxfordian Orientalis Zone Schilli Subzone near the close of the Chari Formation in the Kantkote section of Wagad. The intra-Triassic first order MFS above the early Middle Triassic late Anisian Kellnerites Zone, and the Cretaceous first order MFS in the early Late Cretaceous late Middle Turonian Turoniense Zone (Krishna et al. 2011). The mega sequence began with the origin of the Neotethys in Permian in Kachchh as well as Arabia.

5.4.3.8 Sequence Stratigraphic Frameworks in the Jurassic

In tune with the universally understood definition of tectono-stratigraphic depositional mega-sequences, Sharland et al. (2001, p. 13) differentiated 11 TMSs in Arabia that are separated by major unconformities, i.e. subaerial stratigraphic gaps. The author in Kachchh also followed the same basis for differentiation of the sequences. The first-order sequence in Kachchh which includes the Jurassic, according to the above criterian, has its basal SB above a major regional subaerial unconformity in early Early Toarcian between the late Triassic–early Jurassic marine fossil devoid cross-bedded thick sands and the fine textured higly fossiliferous clays/marls/limestones with the early Early Toarcian index ammonoid *Bouleiceras*. This unconformity or its correlative conformity is regionally present, particularly, exhibited in Arabia, East Africa, Madagascar and Pakistan. In case the unconformity also includes Toarcian, the first-order TST begins with Bajocian ammonoid-bearing sediments. This large unconformity is between the Minjur and Marrat units in Qatar, Saudia, Kuwait, etc. (Figs. 4.35–4.38, Sharland et al. 2001). In spite of this large regional unconformity, Sharland et al. (2001) presumably under the influence of the ETM sequence framework (Haq et al. 1987) based the AP7 sequence on a younger unconformity in Arabia above the Marrat Formation.

5.4.3.9 Closure of AP7

Sharland et al. closed their AP7 sequence prior to the early Late Tithonian Micracanthoceras micracanthum Zone MFS of AP8, i.e. prior to the end of the Jurassic along with their claim of the origin of the proto Indian ocean at the AP7/AP8 SB event. Instead, a much later origin of the Indian ocean is interpreted in India, and also continued quasi-inclusion (without appreciable spreading) of Arabia on the GTM until the close of Barremian the $\sim$20 my later, i.e. the terminal SB of the first-order sequence.

5.4.3.10 Comparison of the Arabian and Kachchh MFSs

The oldest AP7 MFS in the basal Bajocian Laevieuscula Zone matches well with the second order early Early Bajocian MFS interpreted at Patcham in the Sadhara and Kaladongar sections. However, the interpreted Early Bajocian in Kachchh has so far yielded only a single fragmentary *Calliphylloceras* near the base of the Sadhara section. Early Bajocian ammonoids are otherwise well known in Pakistan and Madagascar. In Kachchh, the first ammonoid-based second-order MFS is determined in basal Early Bathonian Zigzag Zone Macrescens Subzone. There are present guide ammonoids from ?terminal Late Bajocian to basal Early Bathonian in Kachchh at Nara (Pandey and Pathak 2015). The MFS is age constrained in Nara above the bed 3 of ~1.5 m thickness (Fig. 1.11) is in Zigzag Zone in complete accord with J30 early Early Bathonian Zigzag Zone MFS of Arabia as also of ETM. The carbonate ramp development is significant in the entire Kachchh basin in Bajocian–Bathonian interval, whereas such ramp is present in Arabia for most of the Jurassic with exceptionally rich hydrocarbon reservoirs. In Kachchh, it is also interpreted in Patcham and Jhura. At Patcham, this MFS is determined above bed 55, while at Sadhara, it is determined at the top of 'Lower Flagstone' (LFm), i.e. in early part of BCSm of ULM. In Jhura, it is determined at the base of Cm.

J40 intra-Coronatum Zone Arabian MFS is a third-order MFS in vertical/temporal proximity of the second-order intra-Callovian SB above the early Middle Callovian Anceps Zone. On the other hand, the intra-Callovian second-order MFS determined in Kachchh is above the mid-Early Callovian Chrysoolithicus Zone Diadematus Subzone.

J50 early Middle Oxfordian Plicatilis Zone MFS is a third-order MFS in Kachchh also of early Middle Oxfordian Indogermanus Zone which is widely recognized across the Kachchh basin. It is noteworthy, that this third-order MFS is followed up in the succession by a widely present large submarine gap (Krishna et al. 2000). It is two MFSs older to the principal or first-order Jurassic MFS determined in the proximal Kankote section of Wagad. A somewhat similar gap is interpreted here between the Hanifa Formation or equivalent units and the Jubaila unit or equivalents (Arkell 1956; Sharland et al. 2001, Fig. 4.35, p. 200). Almost, eight to nine ammonoid subzones appear to be absent. Further support is available from absence of the guide ammonoids (*Gregoryceras*, *Dichotomoceras*, *Larcheria*, *Epipeltoceras*, etc.) that are diagnostic in the concerned interval in Arabian basins (Arkell 1956). It is rather strange that there is just a single MFS in almost equally long Oxfordian as compared to five in Kimmeridgian from J.60 to J100. Based on above revelations, the late Middle Oxfordian first-order MFS of Kachchh is strongly suggested in Arabia as well in the proximal exposed sections between J50 and J60. The well sections are invariably distally located, where the said principal Jurassic MFS is lost in the submarine gap, hence cannot be recognized in well logs and seismic profiles. The terminal part of the Jubaila Formation and the superposed Arab D Member of the Arab Formation compositely have provided the giantest hydrocarbon reservoir of the world of Kimmeridgian age, while the hydrocarbon source sediments are included in the underlying Hanifa and early part of the Jubaila units of mostly Oxfordian age and Early Kimmeridgian age. The Anhydrite seal sediments of Arab C and D Members are here assigned to latest part of late Late Kimmeridgian Beckeri Zone. More so, it is interpreted here that the hydrocarbon source sediments include lower part below the first-order MFS determined in

Kachchh, i.e. the terminal part of the first-order TST, and an upper part above the first-order MFS, i.e. the early part of the first-order RST. J 60 MFS of Early Kimmeridgian Baylei Zone age of Arabia corresponds well with the Early Kimmeridgian *Orthosphinctes* age second-order MFS in Kachchh. The discussion on the age of J70 is based on the mixed ammonoid evidence of Early Kimmeridgian (*Discosphinctes gredgingensis* in Toland et al. 1995) and Late Kimmeridgian Beckeri Zone ammonoids (Howarth and Morris 1998). In comparison, in Kachchh in addition of the *Orthosphinctes*-based second-order MFS, there are two younger third-order MFSs of mid-Early Kimmeridgian Giganticus Zone (=ETM Platynota Zone) age MFS recognized in the proximal Bharodia section of Wagad, followed up by the late Early Kimmeridgian Alterneplicatus Zone MFS in the relatively distal Ler–Katrol section. Further, in Late Kimmeridgian, there are determined three MFSs of third-order early Late Kimmeridgian Acanthicum Zone, second-order Linguiferous Subzone of Bathyplocus Zone MFS (=Eudoxus Zone/Beckeri Zone boundary), and third-order Pressulum Subzone of Beckeri Zone MFS. The youngest Late Kimmeridgian MFS of Kachchh may or may not be isochronous with Arabia. However, in absence of the precise ammonoid-based ages there is no confirmation of the ages of the older two Late Kimmeridgian Arabian MFSs.

5.4.3.11 First-Order Sequence Surfaces

In Arabia are differentiated two successive MFSs K70 and K80, respectively, of Late Barremian and Early Aptian. The SB between them must be close to or at the Barremian/Aptian boundary with good match to the first-order Barremian/Aptian boundary SB in Kachchh. This SB is otherwise excellently expressed widely in several Indian basins. Further up, in Arabia is determined K140 of Early Turonian Archaeocretacea Zone which matches closely to the first-order intra-Cretaceous MFS of early Middle Turonian Helvetica Zone.

5.4.3.12 Second-Order Sequence Surfaces

The Late Berriasian K20 MFS of Arabia matches well the Late Berriasian Kilianella Zone of Kachchh, however, absence of common guide ammonoids in Arabian basins and Kachchh prevents precise correlation. *Fauriella boissirie* of Berriasella paramimouna Subzone in Iraq at this MFS suggests broad correlation of Kilianella Zone with the Late Berrisain B. boissirie Zone. The second-order Verrucosum Zone MFS determined in Spiti (Pathak et al. 2011) of early Late Valanginian matches closely the Arabian MFS K30, and signals a major global transgressive event. The Arabian K70, K80 and K90 also seem to match well with Jaisalmer MFSs within the transgressive condensed nodular Abur Formation. The slow sedimented often condensed fine textured ammonoid-rich Karai Formation of Cauvery basin includes three MFSs which broadly correspond well with the Arabian MFSs between K100 and K140. These MFSs in Arabia are bereft of guide ammonoids. Comparison of sequence surfaces between Arabia and India is not extended farther up in view of different tectono-stratigraphic history after complete oceanic isolation of India from rest of the Gondwana constituents.

5.4.3.13 Salient Remarks

In spite of the different latitudinal setting, exclusive carbonate lithology, warm climate, different pre-Jurassic geological history in Arabia, different geographically restricted ammonoids, the comparison as above decidedly brings out that all the nine Arabian AP7 MFSs are isochronous with Kachchh. Three of these, in comparison, are isochronous with second-order MFSs of Kachchh. These are J30, J60 and J100 which conceptually are understood as second-order MFSs. One second-order MFS between J50 and J60 is interpreted as lost without physical record in the submarine gap between J50, and J60. J20, J70, J80 and J90 are here considered third-order MFSs. The above interpretations call for fresh search for second-order MFSs in Arabian basins above J20, between J30 and J40, and between J40 and J50. The differentiation between second- and third-order MFSs as above should pave way towards some sort of preliminary ordering of the sequences in Arabia.

There is similarity between the rifting history and rift-related submergence of Arabia and Kachchh. It is apparent, particularly, in intra-shelf basins of Arabia from near Callovian/Oxfordian boundary second-order MFS within the J40–J60 interval and also in Kachchh along with rift climaxing and aborting at the first-order late Middle Oxfordian MFS. There is evidence in both these regions of an earlier phase of rifting-related submergence at the base of Bajocian related to J20, besides initiation of rifting near basal Toarcian.

The 20 my plus regional subaerial gap above the Minjur Formation in early Jurassic along with drastic change of lithology and environment is already explicitly mentioned in Sharland et al. (2001). Most of the Middle and Late Oxfordian are absent in Arabia and also in Kachchh Mainland. The suggested widespread Late Middle Oxfordian disconformity in Arabia as a regional submarine gap requires to be critically examined by the Arabian region specialists.

The characterization in Kachchh of each second-order sequence by ammonoid lineage, and origin/extinction of the successive ammonoid lineage at first- or second-order sequence surfaces (MFSs and SBs) may serve as good clues to near future determination of second-order MFSs in Arabia as well.

The revision of basal and terminal SBs of AP7 may be given a fresh look through reinterpretation of data including new information.

Multifaceted similarity, demonstrated above between Arabia and Kachchh, attests excellently, the regionality of the high-magnitude and low-frequency longer than 2 my sequences on GTM, along with their net governance by regional extensional tectonics.

The Jurassic sequence stratigraphic framework of Kachchh (India) on being compared with Arabia (located on the same Gondwanian Tethyan Margin—GTM), is found to exhibit substantive similarity in (1) MFSs, (2) similar rift induced subsidence history, and (3) similar large pre-Toarcian subaerial gap near the base, and (4) similar substantive submarine gap above the early Middle Oxfordian MFS of several ammonoid subzones, (5) drastic change of lithology and environment across the Minjur/Marrat units in Arabia, and also in Kachchh from non-marine to marine at the same time, (6) absence of guide ammonoids of mid. Middle to Late Oxfordian in Arabia as in the distally located Kachchh Mainland. Thus is demonstrated regionality of large duration sequences, along with their control by high-magnitude low-frequency extensional tectonics. Low magnitude smaller sequences are interpreted as controlled by globally isochronous sea level changes.

5.4.3.14 Dire Need of Collaborative Studies

Dire need of collaborative studies between the geoscientists of these two far apart regions is here stressed upon, which in the Jurassic were neighbours and tectono-stratigraphically belonged to the GTM. Such studies will be of mutual benefit and facilitate further refinement of the geology and sequence stratigraphic formulations on GTM, that are so crucial for petroleum exploratory geological studies.

5.5 Validation of the Kachchh First- and Second-Order Sequence Framework on GTM East of India

5.5.1 Comparison with Wombat–Exmouth Plateaus and Dampier Basin Well Sections

The Mid-Triassic change of facies, environment, and micro/mega flora and fauna with or without gap is widely known in all the above-mentioned areas of GTM east of India. Similar evidences are present even up to the Wombat and Exmouth plateaus east of the NW Australian coast (Hocking 1990; Hearty et al. 2002; Von Rad et al. 1989; Jablonski 1997; Jablonski and Saitta 2004; Turner et al. 2009; Geoscience Australia 2014), so also in Dampier basin (Westphal and Aignar 1993) up to the east margin of GTM inclusive of Iriyan Jaya, Sula, Timor, and Timor Gap (Figs. 5.7, 5.8, 5.9, 5.10, 5.11 and 5.12). The Mid-Triassic age is not precise, but interpreted here as the intra-Triassic first-order MFS of late Late Anisian Hollandites Zone–Kellnerites Zone interval. In Dampier basin, this boundary is claimed as SB1 (Westphal and Aignar 1993). This is the time some Greater India microblocks started spreading from NW Australia. Permian–Early Triassic sediments are found truncated below SB1. The structural style that developed is of half grabens bounded by listric faults. The Dampier basin sequence stratigraphic framework is based on huge amount of geophysical data of over 600 km of seismic lines and huge number of wells. In total over a dozen second-order sequences have been differentiated among which seven sequences belong to late Paleozoic, as many as seven sequences in Mesozoic, two in Triassic and five in the remainder Mesozoic. The SB1, between the truncated Late Permian–Early Triassic below and the Ladinian–Norian Locker Formation above, is of Mid-Triassic late Late Anisian first-order intra-Triassic MFS. It is précised in Spiti basin for the entire GTM. The onset of rifting is placed at SB4 as in several other basins on GTM near Pliensbachian/Toarcian boundary which otherwise is the basal SB of the second first-order sequence on GTM. This SB marks the onset of a major transgression on GTM in several basins of Arabia, East Africa, Madagascar, Kachchh, etc. The said major transgression is so dated by Early Toarcian *Bouleiceras* bearing sediments either over non-marine sands with a large subaerial gap in between or over ammonoid devoid marine sediments. The pre-rift stage continues up to the breakup angular unconformity at SB5 which is claimed as intra-Callovian by Westphal and Aigner (1993), and breakup unconformity / intra-Oxfordian unconformity / main intra-Jurassic unconformity (Geoscience Australia 2014 and others earlier) major flooding and many others although without good reliable guide fossil evidence. Instead, the age of the said breakup unconformity could also be intra-Oxfordian late Middle Oxfordian MFS of ~159 ma age (Fig. 5.7) as determined for the entire GTM in the Kachchh basin of India. At this point of time began the spreading away of the Argo abyssal plain (Krishna 2016, unpublised/communicated). It is associated with ceasure of rifting in the Barrow–Dampier basin as in Kachchh. The SB5 is between the Learmouth Formation and Biggada Formation. The oldest oceanic crust at DSDP site 765 of ~155 ma age was first assigned to Callovian by Australian geologists. However, later interpretations based on palaeontological data ranged the same from Callovian to Valanginian (Exxon and Coldwell 1994). Gascoyne and Cuvier abyssal plains, based on magnetic anomaly data (Gibbons et al. 2013), are interpreted to have been created between ~130 and ~120 ma which is placed here between ~136 and ~126 ma in the Verrucosum Zone second-order intra-Valanginian MFS to the first-order Barremian–Aptian boundary SB. Gibbons et al. (2013) based on new ocean-floor and magnetic anomaly data have age constrained the initial breakup event in mid-Late Jurassic around ~155 ma.

The post-rifting continued well beyond the Mesozoic. The fifth second-order sequence of Late Tithonian–Early Berriasian duration, is understood here as a part of the first-order RST, and unconformably overlies the fourth second-order sequence with Oxfordian–Kimmeridgian gap in Dampier basin in relatively distal wells much in the same manner as in Kachchh, Spiti, and other basins of GTM. Recently, Stilwell et al. (2011) have recorded a Late Tithonian *Blanfordiceras* in a wild cat well of Browse basin. The Callovian–Ttihonian gap is interpreted as a submarine gap in the vertical vicinity of the first-order intra-Jurassic MFS. The submarine gap in part is below and in part above the MFS as in Kachchh. The Sixth second-order sequence includes Valanginian sediments. The sixth sequence between SB 6 and SB 7 is here interpreted as middle part of the first-order RST with submarine gap above the first-order MFS and subaerial gap below the first-order

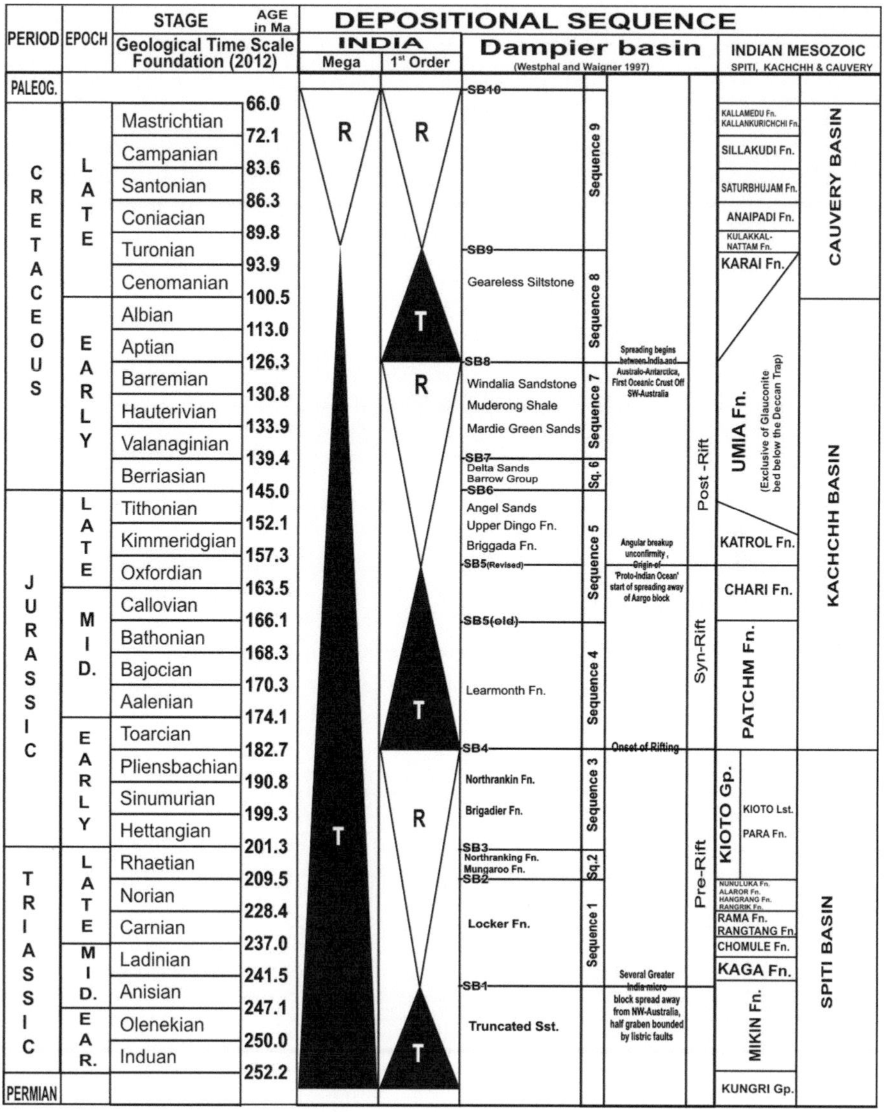

Fig. 5.7 Comparison of first-order sequences of India and Dampier basin on GTM

SB. The Hauterivian–Barremian gap in the Dampier basin is here interpreted as a subaerial gap in the terminal part of the first-order RST. The first-order intra-Cretaceous SB of Barremian/Aptian boundary is placed at SB 8 between Windalia sands and Geareless silts. At this point of time began the spreading between India and Australo-Antarctica with first oceanic crust off SW Australia. Another major early Turonian marker surface is present in Northwest Australia that corresponds well to the major intra-Cretaceous MFS in IEAP, particularly in the Indian basins. All the nine seismic marker surfaces (Geoscience Australia 2014) accord well with the first- and second-order surfaces of the Indian mega-sequence.

The discussion as above unmistakably marks accord of four SBs (SB1, SB4, SB5 and SB8) out of a total of eight between India and Dampier basin off NW Australia in Mesozoic up to the close of Barremian. These are here considered first-order SBs as elsewhere on GTM, while the remaining four are found to accord with second-order SBs of GTM as determined in India. The SB2 is intra-Triassic near Norian/Rhaetian boundary between the Locker Formation/Mungaroo Formation below and Brigadier Formation above in the Dampier basin which corresponds to the second-order SB between the Alaror Formation and the Para Formation in Spiti basin of India (Bhargava et al. 2004). The next Dampier basin second-order SB3 is at Rhaetian/Hettangian boundary between the Brigadier Formation below and North Rankin Formation above which accords well with the second-order SB of Spiti basin between Para Formation and Liassic part of the Kioto Formation. The next younger second-order SBs of the Dampier basin, respectively, are between the Angel sands and the Barrow Group (Figs. 5.7 and 5.11), and between the Barrow Group and the overlying condensed green sands accord well with the Kachchh second-order SBs of Early/Late Tithonian, respectively, between the Communis Zone and the Micracanthum Zone, and between the Basal Member and the Ghuneri Member of the Umia Formation.

As in Dampier basin, the Wombat and Exmouth plateaus also should include the breakup unconformity at the intra-Oxfordian first-order GTM MFS of late Middle Oxfordian age. The Berriasian sediments in these plateaus have a Callovian–Tithonian gap below, and Valanginian–Barremian above. The Berriasian thus is disconformably followed up by the Aptian–Albian transgressive succession. The said gaps are even of greater duration than in the High Himalaya. The gaps above or below the breakup unconformity, i.e. at the MFS, are interpreted as submarine as earlier interpreted by Sarti (in Von Rad et al. 1992). The relatively shallow Berriasian early Late Valanginian sediments are considered as part of the first-order RST. The Late Valanginian–Barremian gap is considered subaerial prior to the first-order Barremian/Aptian SB. The deep water post-Aptian sediments constitute part of the next first-order TST after the start of spreading between India and Australo-Antarctica.

The above comparison permits differentiation of the Dampier basin tectono-stratigraphic framework in reference of the Kachchh first-order sequence framework. The succession below SB1 of Late Permian–Anisian interval of the Truncated Sandstone unit represents the pre-rift first-order TST. The sediments included between SB1 and SB4 constitute the infra-rift part of the Locker Formation to North Rankin Formation. Thus, the succession from the Truncated Sandstone Formation to North Rankin Formation corresponds to the first-order sequence of Kachchh. The contact of the Truncated Sandstone Formation with Locker Formation at SB1 makes the first-order MFS, while the base of the Truncated Sandstone Formation and the top of the North Rankin Formation represent the basal and terminal SBs. The succession between SB4 and SB5 included in the Learmouth Formation constitutes the first-order TST with first-order intra-Jurassic MFS at SB5 at the contact of Learmouth Formation below and Briggada Formation above with revised age of late Middle Oxfordian Orientalis Zone Schilli Subzone as determined in Kachchh for the entire GTM. The first-order RST comprises sediments from the base of the Briggada Formation to the top of the Windalia Formation which, respectively, are interpreted as the basal and terminal SBs of the first-order sequence. The Gearele Formation makes the TST of the last first-order sequence of the mega-sequence with its intra-Turonian top as the MFS. Farther up the sequence eight between SB8 and SB9 is interpreted as the terminal first-order RST of the mega-sequence. As a consequence of the above validation of the developed Indian sequence framework, it is possible to correlate lithostratigraphic units of the Dampier basin on the east extreme of GTM with the Arabian lithostratigraphic units on the west extreme of the GTM. For example the succession between the base of the Locker Formation to the top of the Northrankin Formation, i.e. between SB1 and SB4 of the Dampier basin is correlated with the Minjur Formation of Saudi Arabia, to Isalo Sandstone of Madagascar, and to the Adigrat Sandstone of NE Africa in the Ladinian–Pliensbachian first-order RST age bracket. The said lithostratigraphic units in absence of guide fossils, nothing to speak of mutual correlation, have remained mostly only vaguely dated.

5.5.2 *Iriyan Jaya*

The ammonoid-based stratigraphic framework of S. Tibet, Iriyan Jaya, Sula, and Papua New Guinea hardly has any basic difference from that of the High Himalaya. There is basal Toarcian transgression and change with or without gap, evidenced and dated by Early Toarcian *Hammatoceras* in the basal part of the Innawatan unit invariably unconformably over the Late Triassic–early to mid-Liassic, even Late Permian in wells. In Iriyan Jaya (Figs. 5.8, 5.9 and 5.10), the Early Permian sediments are unconformably overlain by the Late Permian–Early Triassic sediments of Aifam Group. The unconformable contact at the base of the Aifam Group evidences the first-order basal SB of the mega sequence. In Timor Gap, the Early Permian acritarch bearing marine shales are unconformably overlain by the Late Triassic–Liassic non-marine sands.

The changes across the Pliensbachian/Toarcian transition (with or without gap) are here interpreted to mark the first-order intra-Jurassic SB. In Exmouth plateau, the Rhaetian carbonates are followed up by early Liassic volcanics that are in turn unconformably overlain by the Toarcian–Callovian coal bearing succession. The volcanics may be coeval to the initial phase of Karoo igneous activity. The age framework is based on palynomorphs. Rich Early Bajocian ammonoid record having an ETM imprint is evidenced in the exposed sections of S. Tibet, Sula, Iryan Jaya and Papua–New Guinea as in Pakistan and the Nepal High Himalaya. However, in wells, the Bajocian–Bathonian interval is indicated as part of a large gap that is here interpreted as submarine gap. In Kachchh, the newly discovered

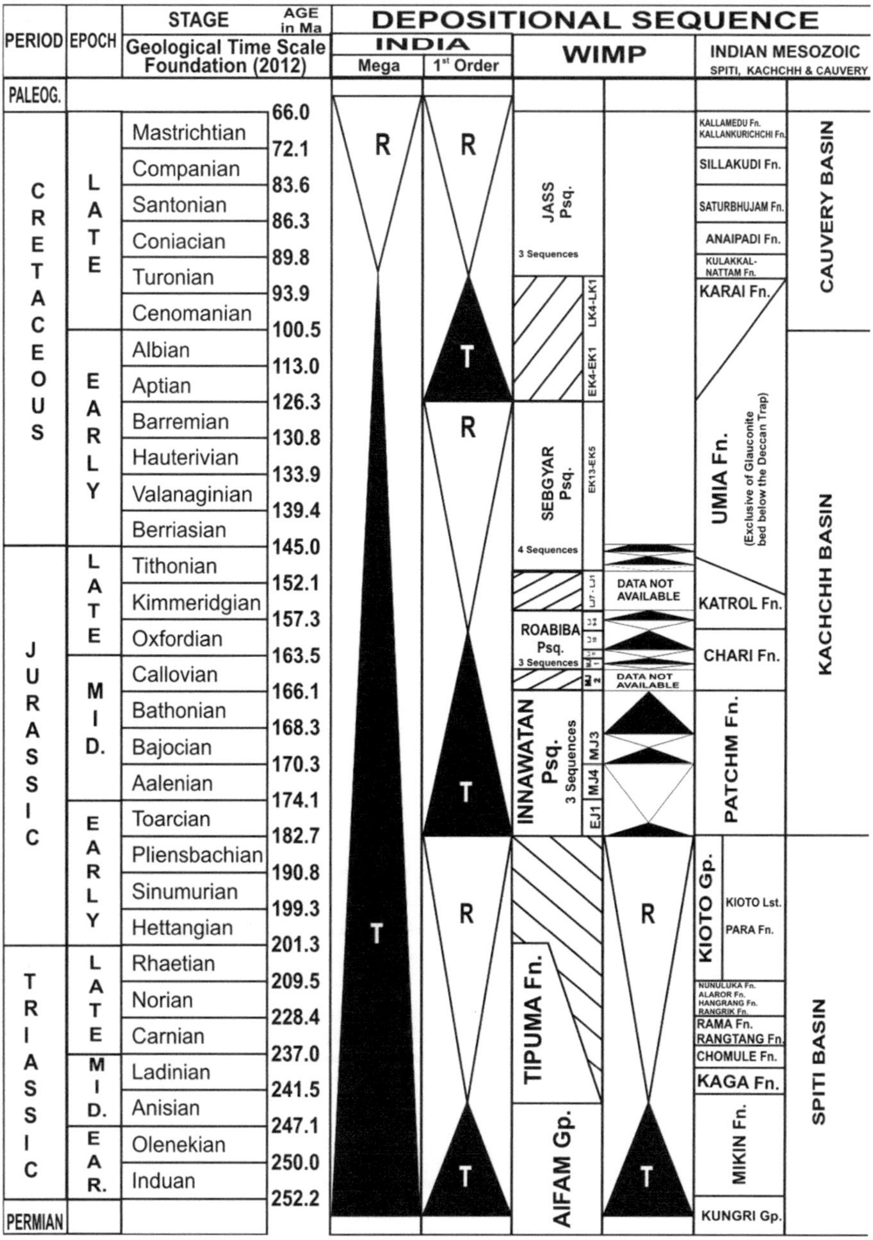

Fig. 5.8 Comparison of first-order sequences of India and West Iriyan Jaya Microplate (WIMP) on GTM

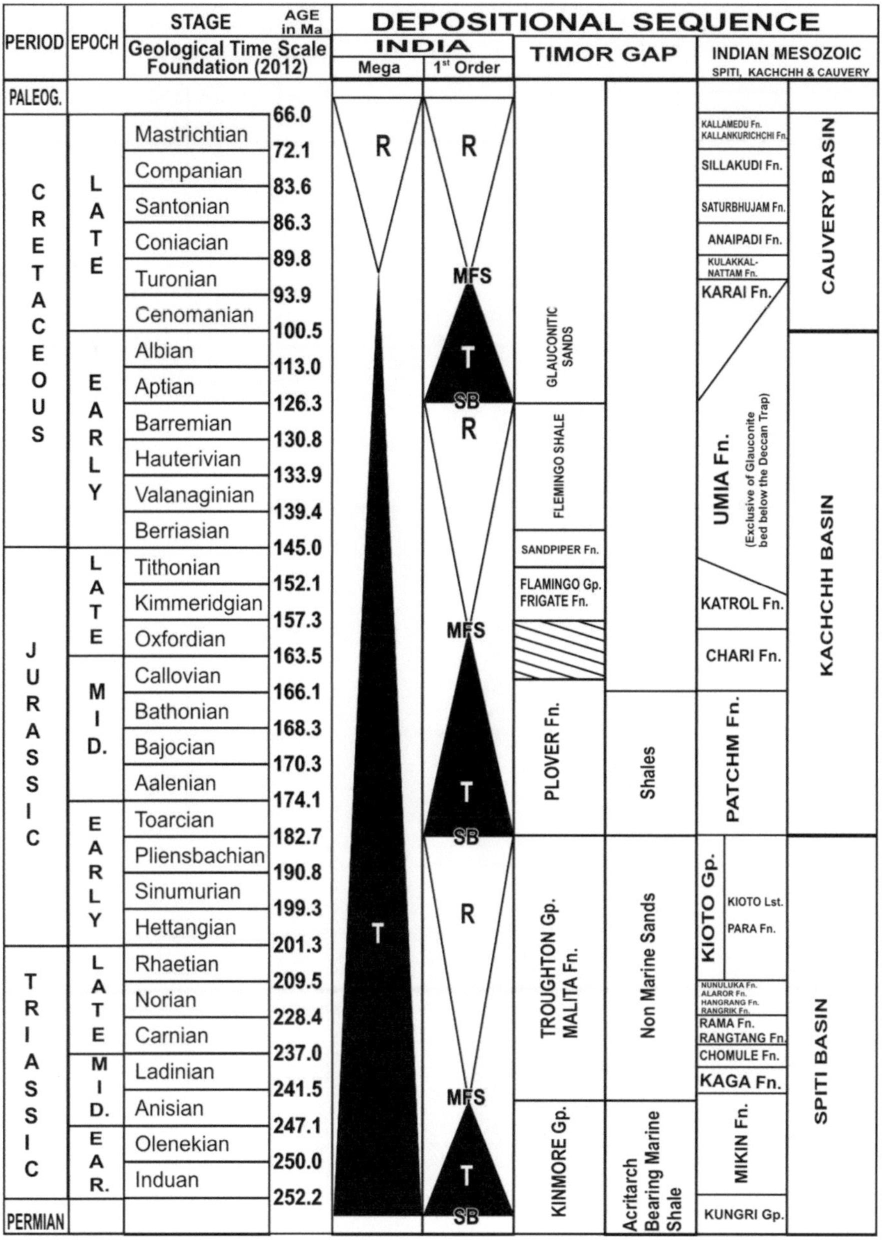

Fig. 5.9 Comparison of first-order sequences of India and Timor Gap on GTM

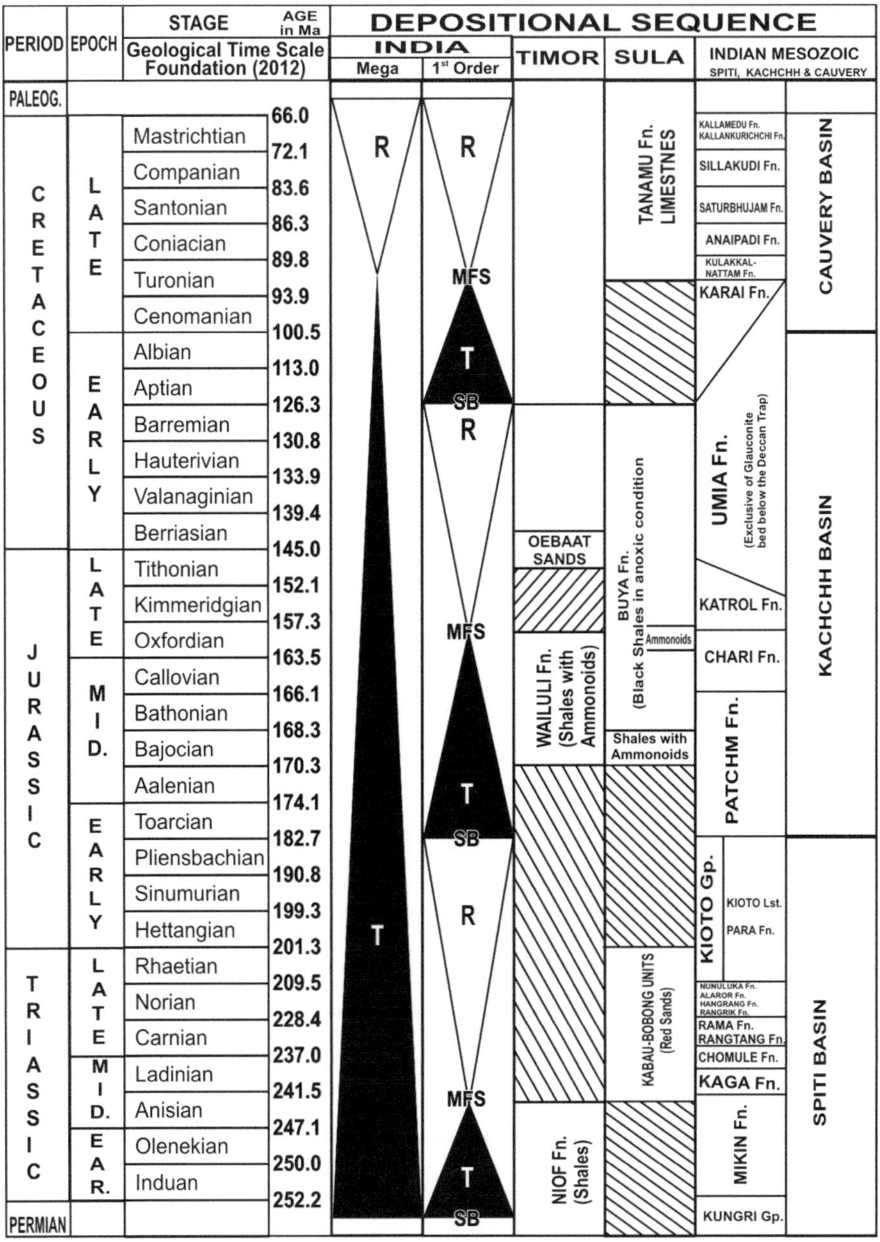

Fig. 5.10 Comparison of first-order sequences of India with Timor and Sula on GTM

Early Bathonian ammonoids also have an ETM impress. Late Middle to Late Bathonian and Early Callovian Macrocephalitinae are quite similar in Kachchh, Sula and Iriyan Jaya. Early and Middle Oxfordian ammonoids are particularly similar in context of their domination by Mayaitinae. Thus, a characteristic GTM ammonoid fauna begins in Middle Bathonian. Late Kimmeridgian is known through *Paraboliceras* followed up by/Early Tithonian *Kossmatia*. The classical Kimmeridgian–Early Tithonian ammonoids (*Orthosphinctes*, *Ataxioceras*, *Idoceras*, *Aspidoceras*, *Streblites* and others) are absent which may imply a large submarine gap. The succession terminates with Late Tithonian ammonoid elements that are well known from both ETM and GTM. The substantive similarity of the ammonoid succession demonstrated as above suggests likely presence of similar subaerial and submarine gaps as also similar first- and second-order sequence framework in the said exposed areas east of High Himalaya.

There are four polysequences and several second/third-order sequences organized in Iriyan Jaya, and all claimed as Gallovian sequences by Fraser et al. (1993); Innawatan (Toarcian to Aalenian from zonal interval EJ1 to MJ 4 further divided into three cycles, one as EJ1 and the following two as MJ4 and MJ3), Roabiba (Middle Callovian to early Early Kimmeridgian from MJ1 to LJ8 further splitted into three cycles, one in MJ1, one in LJ 11-10, and one in LJ 9 to LJ 8)) and Sebgyar (Late Tithonian to early Early Valanginian LJ 2 to EK10 further organized in four cycles, two in LJ1 to EK13, two in EK 12-10.) and Jass (Coniacian to Maestrichtian). Gallovian sequences range from MFS to MFS, however, polysequences that are separated by gaps, conceptually can neither begin at MFS nor end at MFS, and need be better considered neither Vailean nor Gallovian, instead just sedimentary packages with intervening gaps, in which MFSs/SBs need be determined as per usual criteria. There is clear-cut discrepancy between the ages/range of Innawatan parasequence in the stratigraphic table and text, while in this work, the Innawatan sequence is ranged from EJ1 to MJ3, i.e. from early Early Toarcian to early Early Bathonian.

5.5.2.1 Comparison of the First-Order Sequences

The framework east of India up to NW Australia appears largely of deeper water sites and corresponding facies from Middle Tithonian onwards. The first large gap is observed between Roabiba and Sebyar polysequences. In comparison with Kachchh and High Himalaya, this large gap is in the vertical vicinity of the first-order intra-Jurassic MFS of late Middle Oxfordian age of at least Late Kimmeridgian to Early Tithonian. Callovian is widespread compared to Oxfordian and Kimmeridgian. Early Kimmeridgian is present only in a maximum of 4 wells out of a total of 13 wells. The pattern suggests submarine gap, followed up by development of relatively deeper facies up to Berriasian. This is quite similar to the framework in Pakistan. The first-order basal SB in WIMC is here interpreted in early Early Toarcian and the first-order MFS in late Middle Oxfordian. It is also suspected that what is palynologically considered Late Oxfordian to Kimmeridgian might be Middle Oxfordian. In this context it may be noted that even any non-in situ/float Late Oxfordian–Early Tithonian ammonoids could never be found in WICM and Papua New Guinea (Sukmato and Westermann 1992). It is here suggested that the top of LJ 11 is of late Middle Oxfordian of first-order MFS age. MJ1 makes the second-order sequence from late Early Callovian Semilaevis Zone to Depressum Subzone of Athleta Zone of early Late Callovian. The LJ 11–LJ 10 Gallovian sequence ranges from post-Depressum Subzone Athleta Zone to late Middle Oxfordian Schilli Subzone, and the third Gallovian sequence of LJ 9–LJ 8 interval from latest Middle Oxfordian Rotoides Subzone early Early Kimmeridgian Kachchhensis Zone. These match excellently with Kachchh. The author has made appropriate shifting/revision/correction of palynological boundaries by one subzone less or more. It may be clarified that the chronologic resolution of the palynological assemblages/zones is not very refined or is not that refined to even precisely mark stage or substage boundaries in the Jurassic. The first-order intra-Cretaceous SB is more explicit near Barremian/Aptian boundary at the base of Aptian limestones.

There is just one sequence of Toarcian–Aalenian age recognized both in WIMC and Kachchh. There is also facies change from coarse to fine clastics at the Pliensbachian/Toarcian sequence boundary. This gap is widely presented on GTM, and is here considered the first-order SB between the Tipuma and Innawatan polysequences.

5.5.2.2 Comparison of Second-Order Sequences

The Toarcian–Aalenian early Innawatan sequence succession matches well with Kachchh as also with numerous basins of GTM. It is unconformably overlain by the Early Bajocian to early Early Bathonian sediments of late Innawatan polysequence which correspond well to the Early Bajocian to early Early Bathonian second-order sequence of Kachchh. Thus, the Innawatan polysequence is interpreted to include only two second-order Gallovian sequences instead of three as proposed by Fraser et al. (1993). One sequence of late Early Toarcian to Early Bajocian Propinquans Zone and the other from late Early Bajocian Humphresianum Zone to early Early Bathonian Zigzag Zone. May be one of the three Gallovian sequences is a third-order sequence. The Roabiba PSq ranges from MJ1 to LJ 8 interval from late Early Callovian Semilaevis Zone to early Early Kimmeridgian Kachchhensis Zone which in turn is unconformably overlain by Middle/Late Tithonian–Early Berriasian sediments of Jass PSq. The Late Oxfordian to Early Tithonian is interpreted as a large submarine gap. Similar gaps are particularly present in High Himalaya, so also in many other areas of GTM. In Agung-1 well, the Late Tithonian–Barremian sediments are overlain by Aptian limestones as in Kachchh and numerous other basins of GTM. It attests Early Aptian transgression of the terminal first-order Mesozoic sequence over the earlier first-order sequence.

Clues to smaller order sequences are found within the above major sequences. There is noted a regional gap between Late Bajocian and Middle Callovian in well Roabiba IX. This gap is here interpreted as a submarine gap around the two successive second-order MFSs of Early Bajocian Propinquans Zone and Early Callovian Chrysoolithicus Zone Diadematus Subzone. Such gaps have been also interpreted on similar arguments in the Jurassic succession in Pakistan. The late Early Callovian–Early Kimmeridgian interval of Roabiba polysequence corresponds well to the Kachchh late Early Callovian Semilaevis Zone–early Early Kimmeridgian Kachchhensis Zone second-order sequence. The small diachrony of sequence surfaces is most likely on account of much less precise boundaries of palynological zones. There are three second-order Gallovian sequences in Kachchh as in WIMC in the said Roabiba interval.

Again the Jurassic part (LJ4 to LJ1) of the Sebgyar polysequence in WIMC corresponds well with the late Early Tithonian Virgatosphinctoides Zone to Densiplicatus Zone interval of Kachchh. Almost two second-order Gallovian sequences are included in WIMC, and the same number in Kachchh which match well. The intervenig Natricoides Zone is best correlated with Semiforme Zone of Europe. Fraser et al. (1993) have merely given the number of second/third-order sequences and mentioned the Gallovian nature of their sequences without precise boundaries of SBs and MFSs. Gallovian nature of their sequence in turn implies that the bounding surfaces of the sequences are MFSs. However, in this work, the polysequence bounding surfaces have been found as per above discussion closely matching to first- and second-order MFSs.

The large gaps collectively are of much large duration than the collective duration of sedimentation in WIMC which clearly suggests dominant influence of extentional tectonics over eustatics.

5.5.2.3 Bathymetric Framework

There is in general graded increase in bathymetry in WIMC wells from Innawatan to Raoabiba to Shebyar polysequence. The maximum bathymetry is interpreted in Shebyar sequence and minimum near shore conditions during the interval of Innawatan sequence. There is a mention of the first significant tectonically induced subsidence at the start of Bajocian in Kachchh. In the normal circumstances, maximum bathymetry should have been in the terminal part of Roabiba polysequence which in turn suggests extensional tectonics and rapid down faulting during the Shebyar sequence. The same might have outweighed and negated the eustatic fall. Alternatively, the bathymetry at MFS was quite deep to allow deeper conditions even during the early part of RST.

Thus, it is quite satisfactorily demonstrated that the WIMC sequence framework at first-order level accords well with the Kachchh framework. It is, however, not possible to confirm the accord at the second-order level in absence of precision dating of the WIMC palynostratigraphic zonation of Davey 1987 and Helby et al. (1987). The number of second-order cycles in WIMC is eight which refers only to the duration of their two plus polysequences while the succession of second-order Kachchh sequences in the Toarcian–Tithonian duration is ten. At least a few of the eight WIMC sequences must be third order. The Roabiba polysequence of Middle Callovian to Early Oxfordian of three cycles, appear to match the three second-order sequences in the same interval in Kachchh. However, it is not possible to compare the sequence surfaces in absence of precise dates in WIMC. It is opined that the small age discrepancies if any are on account of much less precise palynological zonation in WIMC in comparison to high-resolution ammonoid zonation developed for the entire GTM in Kachchh.

Comparison with Timor Gap, Timor, and Sula

5.5.3 Timor Gap

In Timor Gap (Fig. 5.9), the Jurassic succession is part of the Triassic–Cretaceous sedimentation. The Mesozoics, unconformably, overlie the Early Permian acritarch bearing marine shales of Kinmore Group. These are unconformably followed up by the Late Triassic–Liassic non-marine sand dominating sediments of the Malita Formation of the Troughton Group. These in turn are overlain by the Toarcian–Early Callovian near shore Plover Formation shale dominating sediments. There is suggested general increase in bathymetry. In comparison, in the age duration of the Plover Formation, there are three to four second-order sequences in Kachchh. The contact of the Malita Formation with the overlying Plover Formation is interpreted as the first-order intra-Jurassic SB. The upper surface of the Plover Formation is found close to the second-order mid-Early Callovian MFS and mid-Middle Callovian second-order SB. These are again unconformably followed up by black shales of the Frigate Formation of the Flamingo Group of Kimmeridgian–Tithonian age. These shales suggest maximum bathymetry of the exposed succession above the large Callovian–Oxfordian gap which is interpreted as a submarine gap inclusive of the level of the first-order MFS. The shales are superposed by the Tithonian–Early Berriasian marine sands of the Sandpiper Formation of the Flamingo Group. Farther up, are the Berriasian–Barremian shales of the Flamingo Formation. The Flamingo shales are uncoformably overlain by Early Aptian glauconitic sediments.

The first-order intra-Triassic MFS is interpreted between the Kinmore Group and the Malita Formation of the Troughton Group. The acritach bearing Early Triassic Kinmore Formation shales make the first-order TST, and the Late Triassic–Early

Jurassic Malita Formation non-marine sands represent the RST. The first-order intra-Jurassic SB is interpreted between the Malita Formation and the Plover Formation near the Pliensbachian/Toarcian boundary. The first-order intra-Jurassic MFS is interpreted to fall in the large gap between the Plover Formation of Toarcian–Callovian age below and the Frigate Formation above of the Kimmeridgian–Tithonian age. The first-order intra-Cretaceous SB is placed between Berriasian and Barremian shales of Flaminngo Formation below and the slowly deposited, condensed glauconitic shales/marls of Aptian age near the Barremian/Aptian boundary. At this sequence boundary initiated the spreading between India and Australo-Antarctica, which now delimits the GTM in the east only up to Nepal and S. Tibet.

5.5.4 *Timor*

In Timor (Fig. 5.10), there are Early Triassic shales of Niof Formation (with lateral equivalents of the Aitutu Formation pelagic limestones and the Babulu Formation turbidites and sands of the same age). These are unconformably followed up by the Wailuli Formation ammonoid-rich Middle Jurassic shales which include Callovian sands at the top. As in Timor Gap, the Callovian sands are unconformably superposed by the Oebaat Formation Tithonian–Berriasian sands that are similar to the Sandpiper sands of Timor Gap. The succession is similar to that of Timor Gap although in deeper water Middle Jurassic facies, and thus accords well with the first-order sequence surfaces of Kachchh.

5.5.5 *Sula*

In Sula, the Late Triassic non-marine red beds (Kabaau–Bobong units) are unconformably superposed by marine guide fossil-bearing Early Toarcian shales. These are conformably followed up by the Buya Formation black shales formed under restricted circulation of Toarcian–Barremian age as in High Himalaya. These black shales in turn with a large Aptian–Coniacian gap are uncoformably overlain by carbonates of the Tanamu Formation. Three ammonoid-rich intervals are recorded of the Late Bajocian–Early Callovian, Early and Middle Oxfordian, and farther up of Late Tithonian. Absence of Aalenian, Middle and Late Callovian, and Late Oxfordian to Middle Tithonian ammonoids suggests inclusion of multiple submarine gaps or ammonoid devoid yet deeper facies. The ammonoid succession and facies are similar to High Himalaya, and thus there is good accord with the GTM sequence framework developed in India, in Kachchh with supplements from Spiti to Nepal High Himalaya, and Pakistan within the Indian subcontinent. The first-order intra-Jurassic SB is included in the subaerial gap between the red sands (Kabaau–Bobong units) and overlying black shales of the early part of Buya Group. The first-order intra-Jurassic MFS is interpreted between the ammonoid-rich Early Oxfordian interval and ammonoid-rich Late Tithonian interval.

Summary of Comparison of Sequence Framework on GTM
The summarized comparison of the first-order sequence framework of GTM areas east of High Himalaya as above in unmistakable manner testifies well the applicability in the said areas of the GTM sequence stratigraphic framework developed in Kachchh. The only short coming is the lack of precise ages of the first-order sequence surfaces east of High Himalaya because of invariable absence of ammonoids, and instead the same being mostly palynomorph based. However, the available data does not negate the precise age data from Kachchh. Obviously, there should not be any objections to the extension of precise ages of the Kachchh sequence surfaces, respectively, to areas on GTM both in the east and west up to NW Australia and Arabia. The identified Arabian sequences (Sharland et al. 2001) neither have any defined SBs nor any ordering of MFSs; Madagascar sequences (Geiger and Schweigert 2006) neither have any ordering nor precise ages of sequence surfaces; Dampier basin (Westphal and Aignar 1993), Exmouth–Wombat plateaus (Von Rad et al. 1989, 1992), Iriyan Jaya (Fraser et al. 1993), Sula, Timor Gap, Timor regions (Brown 1992) neither provide precise ages of their SBs beyond stage level surfaces, nor define their MFSs. However, the comparison as above with the Indian basins allows extension of precise ages to at least ammonoid zonal resolution of the first-and second-order SBs and MFSs to entire GTM, and also ordering of sequences and sequence surfaces from mega to third order. It is a breakthrough progress in the validation of the outcrop-based sequence framework to seismic profile-based framework in wells. Further studies are in progress on correlation of not only first to third order but even fourth-order sequences identified in outcrops to well logs along with demonstration of ~400 ky and finer, ~200, ~100, ~40 and ~20 ky sequence cyclicities also in well logs wherever possible (Figs. 5.11, 5.12 and 5.13).

5.6 Sequence Stratigraphic Comparison of the Gondwanian Tethyan (GTM) and Eurasian Tethyan (ETM) Margins

All the 30 plus Kachchh third-order MFSs, so also as many as nine AP7 Arabian Jurassic MFSs are isochronous/concordant with ETM, which in turn suggests their control by global sea level changes. However, nine out of the twelve Kachchh

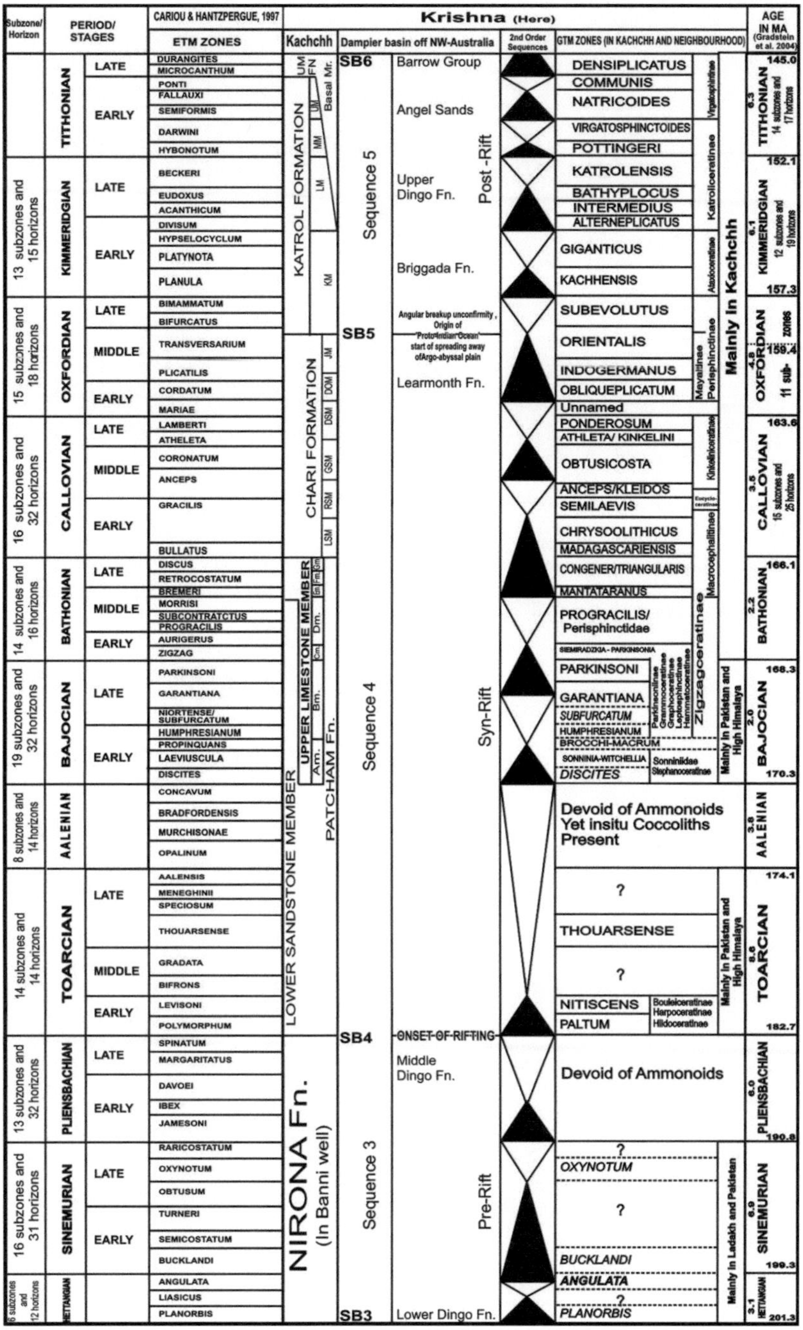

Fig. 5.11 Comparison of second-order sequences between Kachchh and Dampier basin off NW Australia on GTM

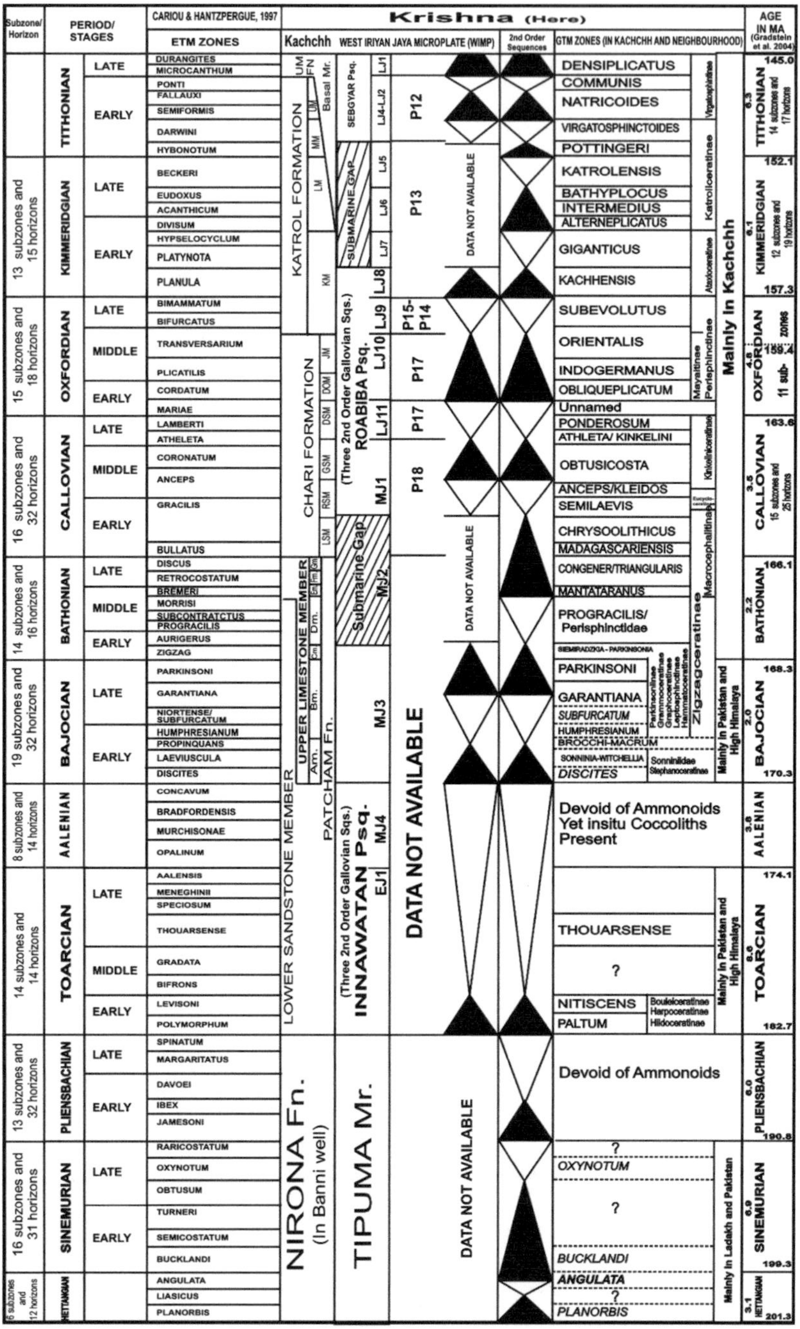

Fig. 5.12 Comparison of second-order sequences between Kachchh and WIMP on GTM

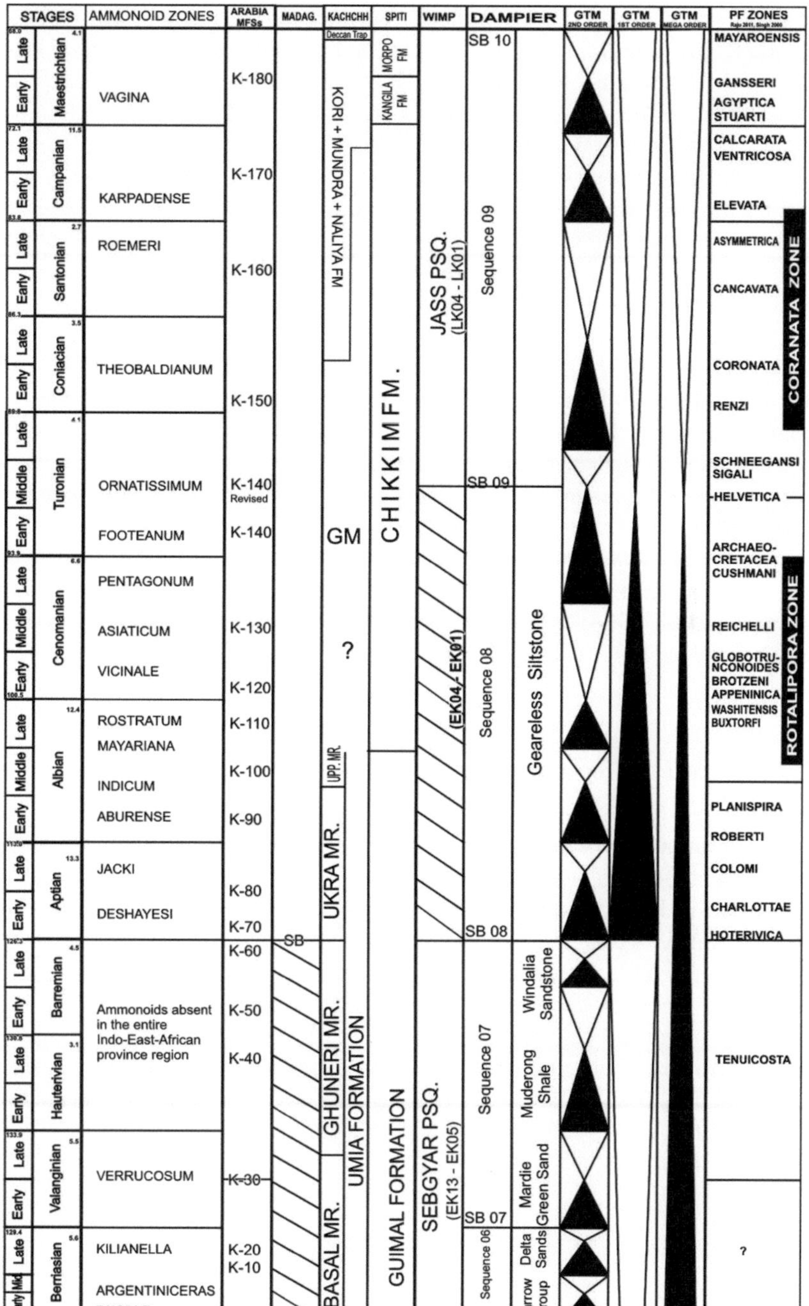

Fig. 5.13 Comparison of second-order sequences on GTM determined in India up to Arabia in the west and up to NW Australia in the east

second-order SBs and MFSs indicate pronounced diachrony with ETM. The first-order intra-Jurassic MFS in Kachchh is determined in late Middle Oxfordian Schilli Subzone while on the ETM, it is several zones higher up near late Late Kimmeridgian Eudoxus Zone / Beckeri Zone boundary. The basal and terminal SBs also are diachronous. Drastic changes of lithology and environment near the Pliensbachian/Toarcian boundary along with presence of a large ~20 my plus gap do not match on the ETM. Also the large submarine gap around late Middle Jurassic first-order MFS does not accord with ETM. The rifting and subsidence history is also chronologically different. There is a marked diachrony/discordance in the origin and extinction of the corresponding, some what parallel, ammonoid lineages, e.g. the GTM Mayaitinae and ETM Cardioceratinae at the respective first/second-order SBs/MFs. The above-mentioned age differences, markedly separate the second-and bigger order Jurassic sequences of longer than ~2 my duration between GTM and ETM.

Over the last few decades, there has been an increasing realization that objective, and precision comprehension of geology and geological events are possible only through sequence stratigraphy. In spite of the conceptualization of the subject hardly half a century ago, it has evolved rapidly across the globe. Here is undertaken comparison of the mega/super order sequences of ETM (Haq et al. 1987; Graciansky et al. 1998; Jacquin et al. 1998) and of GTM (Krishna 2005; Krishna et al. 2010, 2011).

5.6.1 *Basal SB of the Mega-Sequence*

Both ETM and GTM have their basal SB within Permian. On ETM, the basal SB is determined at the Kazanian/Tatarian boundary (Fig. 5.1), while on GTM, in Arabia, the basal SB is determined a little older in early Early Kazanian, in the Himalaya the same SB is placed at a still older level; certainly not above the base of Kazanian. Even yet older Sakmarian/Artinskian boundary geological and radiometric evidences have emerged (Sharland et al. 2001; Shellnut et al. 2009), which seem to suggest the protracted character of the origin of the Neotethys along its long stretch from west to east. This SB marks the origin of Neotethys, as also the start of the mega-sequence. The small differences in age are only academic and may still imply the same event, unless the event is of protracted nature.

5.6.2 *Terminal SB of the Mega-Sequence*

The terminal SB on the GTM marks the closure of the Neotethys which is also subjective, however, since long mostly accepted in Paleogene. Nevertheless, multidisciplinary recent opinions favour the protracted closure of the Neotethys in multiple steps between Paleocene to Oligocene/Miocene boundary between ~51 to ~34 ma.

In Haq et al. (op. cit), it is assigned ~58.5 my for the maximum compared to ~60-59 ma in India. Thus, again there seem close proximity of the ages.

5.6.3 MFS of the Mega Sequence

MFS is précised in Haq et al. (op. cit) at or very close to the base of Turonian in the basal Archeocretacea planktonic foraminiferal Zone. In the Indian Cretaceous geological record, the author has précised this MFS in the planktic foraminiferal Helvetica Zone or upper part of the nannoplanktic Gartneri Zone or Turoniense ammonoid Zone of early Middle Turonian, i.e. one zone younger than on ETM. The well-accepted eustatic curves, the recent time scales, and other data depict a protracted eustatic high almost throughout Turonian. The mid-Middle Albian eustatic high is only slightly behind the Turonian highs by only a few metres only with nearly entire Cenomanian in between of distinctly lower sea levels. If sedimentation rates are any guide, the Early Aptian–mid-Middle Albian TST suggests slower sedimentation rates compared to any time during Turonian. The basal or intra-Turonian MFS is linked to Marion plume volcanism in the west sector while the intra-Albian MFS is related to Kerguelen plume volcanism in the east sector. The author personally considers the east sector event more important and of wider influence than the west sector event. Yet more significant is the intra-Oxfordian volcanism.

5.6.4 First-Order Intra-Triassic MFS

Before taking up the detailed comparison of the Jurassic sequence stratigraphic framework of ETM and GTM, is here scruitnized the first-order intra-Triassic MFS. In Haq et al. (op. cit), it is determined in Late Carnian Welleri ammonoid Zone or Polygnathiformis conodont Zone. In Arabia (Sharland et al. 2001), there are two MFSs within the Carnian (Tr. 60 and Tr. 70), one in Ladinian of plate-wide extent while the exclusive intra-Anisian MFS is of local extent (Sharland et al. op. cit) (Fig. 5.3). It is also commented that the local extent is possibly due to nondeposition in other areas. It may be noted that the principal MFS among a few close ones invariably has lesser extent, since, in relatively distal sites it gets included only in submarine break, e.g. the first-order intra-Jurassic MFS in Kachchh is preserved only in a proximal site (Krishna et al. 2000 and later). In absence of any ordering, the intra-Anisian Arabian MFS is here interpreted as the first-order intra-Triassic MFS for reasons given above. Krishna et al. (op. cit) have in Spiti Himalaya this MFS precised above the intra-Anisian Kellnerites Zone above the beds 34–40 in Spiti of Krystyn et al. (2004). Krystyn et al. (op. cit) have interpreted the same surface as a significant SB.

5.6.5 First-Order Intra-Jurassic MFS

On ETM, it is précised at the Eudoxus Zone/Beckeri Zone boundary in Late Kimmeridgian. In Kachchh, it is determined within the late Middle Oxfordian Orientalis Zone (=ETM Transversarium Zone) between Schilli Subzone and Rotoides Subzone which is older by some four zones. In ETM, the GTM first-order intra-Jurassic MFS makes a second-order MFS while on GTM, the ETM first-order MFS is interpreted as a second-order MFS. Krishna (2005), attributed the said discordance of the first-order intra-Jurassic MFS on ETM and GTM to the mutually differing regional tectonics, which is also reflected in the evolution of their respective ammonoid subfamilies/lineages. The ETM subfamily Cardioceratinae became extinct at the ETM first-order MFS in Late Kimmeridgian, while the morphologically corresponding GTM subfamily Mayaitinae vanished at the GTM first-order intra-Jurassic MFS in late Middle Oxfordian. This in turn demonstrates that origin/evolution/radiation of ammonoid subfamilies and lineages on ETM and GTM are controlled by their differing/discordant regional extensional tectonism.

Much unlike the ETM first-order MFS in the early part of the Late Kimmeridgian Beckeri Zone near the Eudoxus Zone/Beckeri Zone Boundary, the Kachchh first-order MFS is about eight ammonoid subzones earlier above the Schilli Subzone Orientalis Zone (Fig. 5.4). The ETM second-order MFS preceding the first-order MFS is three zones earlier.

5.6.6 First-Order Intra-Jurassic SB

The intra-Jurassic SB on ETM is difficult to determine in the chart of Haq et al. (op. cit), however, the best suggestion is in the early Early Aalenian Opalinum Zone. The SB older to this is indicated in the late Late Pliensbachian. Graciansky et al. (1993) determine it in the basal Toarcian Paltum Zone. In Arabia, it is indicated in the Late Toarcian. The present author's interpretation of this SB widely on GTM inclusive of Arabia is in the early Early Toarcian near the Pliensbachian/Toarcian boundary at the base of the Paltum Zone. It is different with that of Haq et al. while in agreement with Graciansky et al. Review of the Arabian data as discussed elsewhere calls for a revision of the said SB in favour of the early Early Toarcian between the Minjur and the Marrat units in place of the SB above the Marrat.

5.6.7 First-Order Intra-Cretaceous SB

The SB on ETM is in the Early Valanginian near the top of the Otopeta Zone in Haq et al. Widely in India, this SB is précised in the early Early Aptian (Fig. 5.5).

In Arabia, this SB instead of being intra-Cretaceous is intra-Tithonian. Perhaps, the broad homogeneous character of the GTM is lost due to renewed Late Jurassic rifting in Arabia which made it different from the Indian plate region.

To summarize, the basal and terminal SBs of the mega-sequence on ETM and GTM are for sure not isochronous and instead differ markedly. The mega intra-Cretaceous MFS on ETM and GTM differ by one or more ammonoid or planktic foraminifer zone. The first-order intra-Triassic and intra-Jurassic MFSs on ETM and GTM differ in age by one or more stages. The first-order SBs also are highly diachronous on ETM and GTM. The over all evidence thus suggests lack of agreement at mega to first-order sequences.

5.6.8 *Broad Comparison of the Sequence Fameworks of Kachchh on GTM with ETM Frameworks of Haq et al. (1987), Graciansky et al. (1993, 1998)*

The exposed Toarcian–Tithonian part of the Kachchh Jurassic first-order sequence is organized into ten plus second-order transgressive-regressive cycles. These range in age from the Early Toarcian to Late Tithonian. On ETM on one hand there is the classic detailed framework of Haq et al. (1987) and on the other hand, the relatively more recent works of Jacquin et al. (1998) and Graciansky et al. (1993). The latter are more detailed at the third order while the former is more detailed at the second order level. These factors do bring in some subjectivity in comparison at second order with the GTM framework developed in Kachchh.

5.6.9 *Long-Term TST on ETM and GTM*

The base of the long intra-Jurassic TST of ETM and GTM are of different ages. In Haq et al. it is the Late Bathonian. In Jacquin et al., it is at the base of the Bremeri ZoneMiddle/Late Bathonian boundary. In Kachchh, it marks at least the base of the Bajocian, could even be the base of Toarcian (Fig. 5.6). Haq et al. have two first-order sequences in the Mesozoic mega-sequence viz., (1) Absaroka of Late Permian to early Early Aalenian up to close of Opalinum Zone of $\sim$79 my duration, and (2) Zuni from start of Early Aalenian Murchisonae Zone to early Late Maestrichtian of $\sim$109 my duration. These in turn include eight second order A 1 to A eight, respectively, and are followed up by eight second order Z 1 to Z 8 sequences. The comparison is possible only in the ammonoid bearing precisely dated Early Bathonian–Tithonian interval. There are two second-order sequences (LZA 3 and LZA 4 plus part of LZA 2 on the lower side and part of LZB 1 on the upper side) on the ETM in the interval under comparison. However, these are of relatively much longer duration compared to the second-order sequences developed in Kachchh. Therefore, even the third order ones longer than 2 my, are considered together with second order which now total nine. Jacquin et al. have just two second-order sequences 6 and 7, and there is no way to increase this number as done in case of sequences of Haq et al. Graciansky et al. (1993) have its second-order MFSs common with Jacquin et al. (1998), however, the SBs differ by a few zones in the Sinemurian–Aalenian interval. In comparison, Kachchh has 7 plus second-order sequences (Fig. 5.7).

5.6.10 Second-Order MFSs of Haq et al., Jacquin et al., and Granciansky et al.

The MFSs of the ETM sequences of Haq et al. and Jacquin et al. are invariably of different ages. Haq et al. have successive MFSs at the base of Early Toarcian Bifrons Zone, near the close of the Laeviuscula Zone, the basal Early Bathonian Macrascens Subzone, top of the Early Callovian Macrocephalus Zone, late Middle Callovian Coronatum Zone, the basal Kimmeridgian Baylei Zone, and mid-Tithonian, while Jacuin et al. (op. cit) have the corresponding MFSs during the compared interval of basal Early Bathonian, in basal Oxfordian Mariae Zone Scarburgense Subzone, and above the mid-Late Kimmeridgian Eudoxus Zone.

5.6.11 Diachrony and Differences in Second-Order MFSs Between ETM and GTM

The basal Early Bathonian second-order MFS is found common among Haq et al. (1987), Jacquin et al. frameworks of ETM and that of Kachchh (Figs. 5.5, 6.3 and 6.5). Two other second-order ETM MFSs of Haq et al. and Jacquin et al. of basal Kimmeridgian and the mid-Late Kimmeridgian Eudoxus Zone / Beckeri Zone match with GTM second-order MFSs determined in Kachchh. The other second-order MFSs do not match between ETM and GTM, however, a few differ in age by only one ammonoid zone or so. The Early Toarcian Nitiscens Zone, the Early Bajocian Propinquans Zone, the mid-Early Callovian Chrysoolithicus Zone Diadematus Subzone, the early Late Callovian Athleta Zone Depressum Subzone, the mid-Early Tithonian Pottingeri Zone, and the mid-Early Tithonian close of Natricoides Zone are the GTM MFSs which differ with corresponding ETM MFSs by about a zone or so. Thus only three second-order MFSs are common while six are diachronous although only by a zone or so. The intra-Tithonian diachrony of the second-order MFSs could also be due to extreme provincialism and absence of good correlations among the provinces.

5.6.12 Diachrony and Differences in Second-Order SBs of ETM and GTM

Comparing the second-order SBs, Haq et al. has placed the second-order SBs at the base of intra-Bathonian Discus Zone, intra-Cordatum Zone, the Early Kimmeridgian, the late Early Tithonian, while Jacquin et al. have their second-order SBs at the base of intra-Bathonian Bremeri Zone, the intra-Oxfordian Berrense Subzone. The Kachchh second-order SBs in the compared interval have been précised at the base of the Bremeri Zone, the intra-Anceps Zone, early/late Early Oxfordian, at the revised Oxfordian/Kimmeridgian boundary in the Tethys, the late Early Kimmeridgian Alterneplicatus Subzone, the Kimmeridgain/Tithonian boundary, at the base of the late Early Tithonian Natricoides Zone, and at the base of the early Late Tithonian Densiplicatus Communis Zone. Out of eight only two SBs précised at the base of intra-Bathonian Bremeri and early/late Early Tithonian Natricoides Zone are common between GTM and ETM framework of Jacquin et al. (1998) (Figs. 6.3 and 6.5). Two other second-order SBs intra-Cordatum Zone, and Berrense Zone also are close to GTM second-order SBs. Thus four second-order SBs are common and other four are diachronous between ETM and GTM.

5.6.13 Intra-Bathonian Second-Order MFS

Haq et al. has one second-order MFS in Early Bathonian Zigzag Zone. Graciansky et al. has also one second-order MFS in the Early Bathonian Macrescens Subzone almost at the same time as in Haq et al. Jacquin et al. also have one second-order MFS in the Early Bathonian Macrescens Subzone. In Kachchh too, the lone intra-Bathonian second-order MFS is determined in Early Bathonian Macrescens Subzone. Thus, there is complete identity among the known ETM and GTM sequence frameworks with regard to intra-Bathonian second-order MFS.

5.6.14 Intra-Bathonian Second-Order SB

Haq et al. (1987) have one intra-Bathonian SB at the base of the Late Bathonian Discus Zone. Gracianski et al. place the same above the Morrisi Zone. Jaquin et al. have the same at the base of Middle Bathonian Bremeri Zone. Kachchh too has the same SB at the base of Bremeri Zone. The difference in the age of the SB is thus of one to two zones with Haq et al. while close concordance with Graciansky et al. (1993) and Jacquin et al. (1998).

5.6.15 Intra-Callovian MFS

Haq et al. have two intra-Callovian MFSs, one above the Early Callovian Macrocephalus Zone and the other in the late Callovian Athleta Zone, above the Coronatum Zone. Graciansky et al. (1993) have one in the Late Callovian Lamberti Zone, while Jacquin et al. do not include any second-order MFS in Callovian. Krishna et al. in Kachchh envisage two MFSs above the mid/late Early Callovian Chrysoolithicus Zone Diadematus Subzone and above the early Late Callovian Depressum Subzone. The first of Haq et al. may be isochronous or close to the first of the Kachchh framework. While the second Kachchh Callovian MFS is one subzone younger too that of Haq et al. and about one zone older to that of Graciansky.

5.6.16 Intra-Callovian SB

Haq et al. have one in late Early Callovian Calloviense Zone in comparison to early Middle Callovian Paramorphum Zone. These have difference of age of about one zone.

5.6.17 Intra-Oxfordian MFS

Haq et al. (1987) do not have any second-order MFS in Oxfordian. Jacquin et al. have interpreted it in Early Oxfordian Scarburgense Subzone. In Kachchh it is determined in late Middle Oxfordian Orientalis Zone between Schiili Subzone and Rotoides Subzone. Thus, there appear substantive differences with regard to this MFS.

5.6.18 Intra-Oxfordian SB

Haq et al. (1987) include it in Early Oxfordian. Jacquin et al. determine in Berrense Subzone. Krishna et al. have two. The earlier one in between early/late Early Oxfordian is older by one Subzone to that of Haq et al., while the younger one near the Oxfordian/Kimmeridgian boundary seems very close to that of Jacquin. The ETM SBs appear very close to those of Kachchh.

5.6.19 Intra-Kimmeridgian MFS

Haq et al. placed it in basal Kimmeridgian Baylei Zone. Jaquin et al. determine it in Late Kimmeridgian between Eudoxus Zone and Beckeri Zone. Krishna et al. in Kachchh have two MFSs in basal Kimmeridgian as also another in Late Kimmeridgian between Bathyplocus Zone and Katrolensis Zone. Thus the two ETM MFSs have the same ages as in Kachchh.

5.6.20 Intra-Kimmeridgian SB

Haq et al. (1987) include just one SB only above Baylei Zone in Early Kimmeridgian. Jacquin et al. do not include any second-order SB in Kimmeridgian. Krishna et al. in Kachchh place it near the Oxfordian/Kimmeridian and another second-order SB in Early Kimmeridgian between Giganticus Zone and Alterneplicatus Zone. Thus, there is evident distinct diachrony of second-order SBs between ETM and GTM.

5.6.21 Intra-Tithonian MFS

Haq et al. (1987) determined one in Early and two in Late Tithonian. Jacquin et al. do not include any second-order MFS in Tithonian. Krishna et al. interpret the MFS in Early Tithonian above the Pottingeri Zone and also above the Natricoides Zone. There could be closeness of ages, however, in view of provincialism and absence of correlation, fingers need only to be kept crossed.

5.6.22 Intra-Tithonian SB

Haq et al. have two SBs, one in Early Tithonian and the other in latest Late Tithonian. Jacquin et al. interpret the said SB also at the close of Tithonian while Krishna et al. determine one above the Virgatosphinctoides Zone, and the other above the Communis Zone. There is good agreement between GTM and ETM in late Early Tithonian SB, while no agreement with regard to the Late Tithonian although the difference may be of about one zone.

5.6.23 Comparison of Principal Ammonoid Lineages

The comparative study also reveals near parallel or corresponding existence of ammonoid stocks/subfamilies/lineages on ETM and GTM. For example the Macrocephalitinae lineage terminates in extremely depressed whorl section on both GTM and ETM. On GTM, the morphologically similar Macrocephalitinae lineage terminates at the second-order MFS above the mid-Early Callovian Chrysoolithicus Zone Diadematus Subzone, while on ETM the corresponding Macrocephalitinae lineage terminates in Anceps Zone immediately below the coronatum Subzone second-order MFS. Another good example is the parallel existence on GTM of Mayaitinae and on ETM of Cardioceratinae. Mayaitinae terminates at the first-order MFS above the late Middle Oxfordian Schilli Subzone on GTM, while Cardioceratinae terminates on ETM at the first-order MFS in Late Kimmeridgian. Such higher order sequence framework-related origin and extinction of parallel lineages on ETM and GTM allows the interpretation that origin, extinction, and evolution of principal ammonoid lineages on GTM and ETM are genetically linked to sequence stratigraphic surfaces SBs or MFSs. This in turn is the consequence of respective yet distinctly diachronous regional extensional tectonic events of ETM and GTM.

5.6.24 Summary of Comparison of Sequence Frameworks of GTM and ETM

The detailed comparison as above brings out, (1) pronounced discordance of the first-order framewok, and moderate but distinct diachrony in second-order framework and sequence surfaces (SBs and MFSs) between GTM and ETM. Diachrony is somewhat greater in SBs than MFSs. The pronounced or moderate discordance has been attributed to regionality of the sequences, in turn on account of their net governance by high-magnitude low-frequency regional extensional tectonics, (2) near complete accord of the third-order sequence framework having third-order SBs/MFSs isochronous between GTM and ETM, which in turn has been interpreted on account of their globality and net control by earth's orbital dynamics inclusive of globally low-magnitude and high-frequency synchronous sea level fluctuations, (3) the high-magnitude and low-frequency regional extensional tectonics does not interfere with low-magnitude and high-frequency globally synchronous sea level fluctuations, and (4) the origin, extinction and evolution of somewhat or near-parallel ammonoid lineages are genetically linked to differing/diachronous/discordant regional extensional tectonics, so also the resultant expressions at the first- and second order sequence surfaces, respectively, on GTM and ETM.

5.6.25 Comparison of Third-Order Sequences

There has been constant improvement and refinement on sequence differentiation on ETM since Haq et al. (1987) to Graciansky et al. (1993) to Jacquin et al. (1998). The number of third-order sequences has increased in the compared interval of the Jurassic from 13 to 28. The number of third-order MFSs determined in Kachchh and neigbourhood on GTM is about 38, and these are found invariably isochronous to those of ETM.

5.7 Hydrocarbon Prospects in the Mega-Sequence Geological Record in India in Sequence Stratigraphic Context with Focus on the Kachchh Basin

The scope of evaluating the hydrocarbon prospects in the mega-sequence is immense and beyond the scope and focus of the present study, and so planned separately for a later occasion. However, in context of the importance of applied economic aspects involved, a crude evaluation is outlined as an edifice to a future focused effort.

The Indian Mesozoic basins number around 25, and are known to include huge onshore and offshore areal expanse. Time to time, almost since the establishment of the ONGC, hydrocarbon prospects have been evaluated in the Indian sedimentary successions. The first comprehensive evaluation was done in the 1983 publication on the petroliferous basins of India, and has continued on an annual or biennial basis mainly in east and west sector basins (Biswas 2012b).

5.7.1 Conceptual Remarks in Sequence Stratigraphic Context

Krishna (2005) in context of his sequence, stratigraphic studies demonstrated strongly similar broad geology and stratigraphy on GTM constituent basins during the said interval. In sequence stratigraphic context, hydrocarbon source sediments are located in the vertical proximity of MFSs of different orders at a particular site on the shelf on account of maximum bathymetry at that site, in turn allowing maximum/densest presence of nanno/microorganisms in its spatial proximity. The densely packed organisms in specially favoured conditions are transformed into hydrocarbons. There is off course a minimum bathymetry for the growth of such organisms, burial depth and appropriate heat for transformation. Restricted depositional conditions are most favoured for source rocks. Reservoirs and seal rocks are the other two important elements of the petroleum systems. Also, certain minimum overburden and depth (from the surface) are required to provide the necessary pressure and temperature for the transformation of the included microorganisms into hydrocarbons.

There is a minimum bathymetry for the growth of such organisms. The better quality Arabian hydrocarbon source sediments in the said interval are further restricted mostly in the Oxfordian–Kimmeridgian interval that is also the closest on either to the intra-Oxfordian first-order MFS. The said first-order MFS in Arabia on GTM and also in Kachchh is recently determined in Kachchh above the late Middle Oxfordian Orientalis Zone Schilli Subzone bed I–O8 in the Kantkote section of Wagad outside the Mainland Kachchh (Krishna et al. 2000 and later) (Fig. 5.14).

Conceptually, the best source sediments need be concentrated in the younger second-order TSTs of the first-order TST being relatively time wise closer to the first-order MFS, e.g. the Bajocian–Middle Oxfordian interval of mid-shelf and deeper environments. These second-order TSTs are Early Bajocian, mid-Middle to Late Bathonian to mid-Early Callovian, late Middle Callovian to early Late Callovian, and late Early to late Middle Oxfordian. The relatively early second-order TSTs of the first-order RST also make good source sediments, which include most of Late Oxfordian, Kimmeridgian, and Late Tithonian to Valanginian interval second-order TSTs. Even in the above second-order TSTs, the TSTs of the third-order sequences make better source sediments than the corresponding third-order RSTs (Fig. 5.15).

Even the organically rich fine textured sediments of the older second-order TSTs of the first order TST in farther offshore make good source sediments. Example may be cited of the organically rich shaly facies that is time equivalent of the early part of the Dingy Member of Toarcian age as good prospective source sediments.

Similarly, the following first-order RST up to nearly the close of Barremian should make the major reservoir. As in the case of source sediments, the second- and first-order RSTs of the first-order RST make better reservoirs compared to the second and first-order RSTs of the first-order TST. The later parts of the first-, second- or third-order RSTs make better reservoirs than the earlier parts of the same RSTs.

Third-order TSTs of the second-order TSTs make better hydrocarbon source sediments compared to the corresponding third-order TSTs of the corresponding second-order RSTs. Thus for example, the Middle Oxfordian shales of a third-order TST of the terminal second-order TST of the first-order TST should make better source sediments than early Late Oxfordian shales of the basal third-order TST of the first-order RST.

5.7.2 Tectono-Stratigraphic Framework

India in essence is made up of four prominent cratons from south to north: the Dharwar, Bastar, Singbhum, and Bundelkhand, along with intervening Narmada–Son–Damodar, Pranhita–Godavari, and Son–Mahanadi linear weak zones. Apart from thick non-marine Permian–Pliensbachian, terminal Middle Oxfordian–

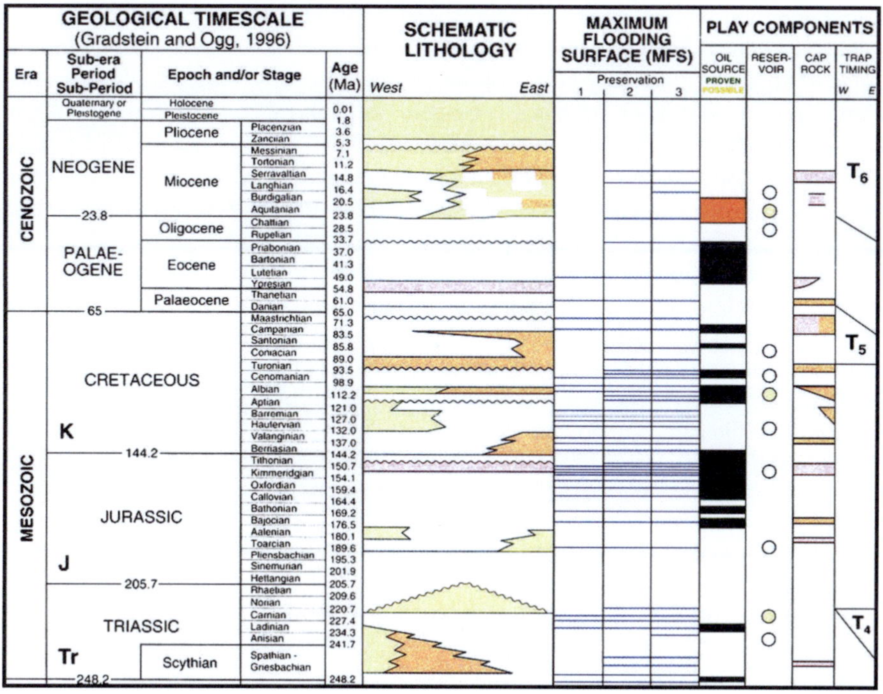

Fig. 5.14 Major TST intervals as prospective source sediments in the Kachchh Mesozoic vis-à-vis Arabia Source rock distribution through the Phanerozoic modified after Sharland et al. (2001)

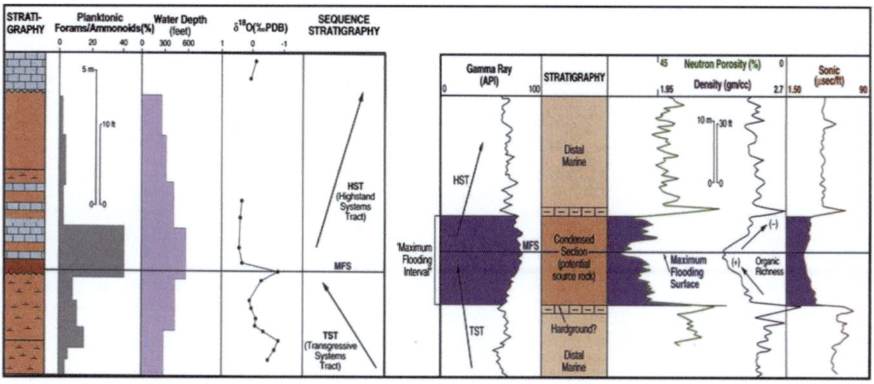

Fig. 5.15 Delineation of source rocks in wells as also exposed sections based on stratigraphic correspondence of maximum density/diversity/frequency of ammonoids, planktic forams, condensed intervals, oxygen isotope, marked shifts in the logs during maximum bathymetry, maximum transgression/flooding after Sharland et al. (2001)

Barremian, and Aptian–Maestrchtian sediments in the said weak zones, the Indian shield is rimmed by substantively thick marine and non-marine successions in umpteen Indian sedimentary basins on the external margins of the said four cratons. These are Cauvery, K-G, Mahanadi, Bengal, Meghalaya, Assam, Purnea, Sikkim, Kumaon–Garhwal, Spiti, Kashmir–Ladakh, Jaisalmer, Barmer, Kachchh, Cambay, Kachchh, Saurashtra, Narmada, Deccan, and Konkan–Kerala basins. A few of these basins in the west and east sector are already either proven petroliferous basins or producing hydrocarbons. Temporally, all the six systems involved in the mega-sequence from Permian to Paleocene include, both hydrocarbon source and reservoir rocks, in some or the other Indian basins, and a few have been producing hydrocarbons. The sediments of the mega-sequence in the north sector in High Himalaya are not being actively explored on account of preserving the ecology and fragile environment there. The sedimentary succession both in the west and east are governed by the same tectno-sedimento-stratigraphic framework and is punctuated by the same succession of first- and second-order sequence stratigraphic surfaces that are resultants of the GTM tectonic and igneous events in divergent passive margin setting, but for transient convergent interludes. Only a few later studies have their evaluations in broad sequence stratigraphic backdrop.

5.7.3 Paleogeographic Perspective

In the paleogeographic context, somewhat as a 'Central' unit, it was bordered on both sides by divergent regions in the west by Africa of West Gondwana, and in the east by the East Gondwanian Australo-Antarctica, which presumably also influenced the source/reservoir sediment framework.

5.7.4 Differentiation of Marine and Non-marine Regimes Within the Mega-Sequence in India

The Late Permian–Pliensbachian first-order sequence interval in both east and west sectors is mostly non-marine but for transient marine interludes during the Late Permian–Anisian first-order TST. The Toarcian–late Middle Oxfordian sediments are fully marine in the west sector, however, the said interval makes important and long stratigraphic gap in the east sector. The terminal Middle Oxfordian–Barremian sediments are essentially marine in both the sectors. The Aptian–early Middle Turonian sediments are also largely marine in both the sectors. The Late Turonian–Paleocene sediments again in both the sectors is largely marine.

5.7.5 Temporal Organization into Syn-Rift, and Rift/Drift Intervals

The rift-drift stage in a conceptual context was realized first in the west sector progressively in the Toarcian Barremian interval, in contrast to distinctly later rift/drift transition in the east sector at the Barremian/Aptian boundary first-order SB. The drifting picked up appreciably only since the mid-Middle Albian MFS in both the sectors. The Late Permian–Pliensbachian interval in the rift basins of both the sectors exclusively include pre-rift, infra-rift and syn-rift sediments. The Kachchh basin includes rift/drift sediments for the rest of the mega-sequence. The Saurashtra and Kerala–Konkan basins should include syn-rift sediments even up to mid-Middle Albian. The east sector basins make syn-rift sediments progressively up to late Middle Oxfordian/Early Valanginian/Late Barremian, and post-rift sediments from Aptian, particularly from mid-Middle Albian onwards.

5.7.6 Regional Igneous Events Among the Required Heat Sources in the Hydrocarbon Source Sediment Kitchen

The present study has time ranged the near continuous igneous activity in the east sector from terminal Oxfordian to mid-Middle Albian, particularly in the Aptian–Albian span in the east sector. The available heat is expected to have had enough time to adequately cook the hydrocarbon sources during the terminal Late Oxfordian to mid-Middle Albian interval, to in turn let the resultant spelled hydrocarbons start moving for storage in the overlying younger reservoirs during the Late Cretaceous to Neogene. The heat source for the post-Albian hydrocarbon source sediments is interpreted as non-igneous residual source underneath. The terminal Cretaceous Deccan event or its coeval igneous manifestations are known even in the east sector.

In the west sector, the suspected ~ 183 ma Saurashtra and Kachchh volcanism may have primarily served as good heat source to the ~ 272 to ~ 183 ma ranged Late Permian–Pliensbachian hydrocarbon source sediments in the west sector. The younger ~ 183 to ~ 92 ma spanned Toarcian–early Middle Turonian hydrocarbon source sediments may have had similar heat source in the widespread ~ 92 to ~ 84 ma Turonian–Early Campanian volcanism in the India–Madagascar region. The Campanian–Early Maestrchtian sediments e hydrocarbon source sediments had the Deccan volcanics as their prime cooking source.

India in context of its interpreted travel during the span of the mega-sequence yoyaged due south and north along transforms involving altogether different latitudinal framework between 35° south to 35° north across the equator from subtemperate to tropical to subtropical zones. In this respect, it differed much with Arabia which during the span of the said mega-sequence maintained subtropical to tropical framework. The Arabian source and reservoir rocks are mostly carbonates while India is beset with carbonates only in the early Middle Jurassic, locally in the Middle Cretaceous, and then again in Maestrichtian to early Eocene.

Indian west sector basins were neighbored to Afro-Arabia, yet in distinctly higher subtemperate latitudes during the mega-sequence span, but for Bajocian–Bathonian, and Campanian–early Early Eocene. The east sector basins almost throughout maintained subtemperate to temperate latitudes except for Maestrichtian–early Early Paleocene.

5.7.7 Availablility of Restricted Depositional Environments

Conducive environment for organic matter as source sediments is available in both the sectors from Late Permian to mid-Middle Albian in context of only limited spreading in both the sectors on account of acute space constraints.

5.7.8 Context of Oceanic Anoxism Through the Mega-Sequence

On the GTM in general, anoxic events suggestively occurred in some or other basin throughout the mega-sequence right since the start at the second-order Permian/Triassic boundary SB, Middle Triassic MFS, Early Toarcian MFS, late Middle Oxfordian MFS, intra-Valanginian second-order MFS, terminal Hauterivian second-order MFS, intra-Barremian second-order MFS, first-order Barremian/Aptian Boundary SB, mid-Aptian second-order MFS, mid-Middle Albian MFS, terminal Albian MFS, early Middle Turonian MFS, and intra-Campanian second-order MFS. The ones in the east sector are genetically linked to Late Oxfordian–mid-Middle Albian volcanism, while those of the west sector to Karoo, Marion, and Deccan volcanism. Thus a genetic link is suggested among hydrocarbon source sediments, major volcanism, ocean anoxicism, and sequence surfaces. Supposedly, the OAEs followed the volcanism with some time lag, and invariably lasted for the duration of an ammonoid zone or that of a small third-order sequence. The OAEs in the geological record are signaled through short duration ~one million year and less long cyclically alternating thinly bedded condensed organically rich dark to black shales particularly close to first- and second-order MFSs near thermal maxima. Such sediments are quite common offshore Australo-Antarctica. The phytoplankton-rich oceanic waters of juvenile nascent oceans like the Indian ocean in the east sector during Early Aptian–mid-Middle Albian TST interval spread the OAEs far and wide at finely tuned eustatic highs by way of upwelling, and even reach into the shallow marine depositional frameworks as splendidly exampled in the exposed Grey Shale unit of the Karai Formation (Jafar and Rai 1989; Kale et al. 2000). The Grey Shale unit suggestively formed between mid- to outer shelf where the clastics failed to reach, possibly because of the distance from the provenance. Such units, formed under anoxic framework, are known to make good source rocks. Non-condensed moderate to slow sedimented mega body fossil devoid glauconitic and other green shaly/silty coeval intervals are known in Spiti, Pakistan, Jaisalmer, Kachchh, and Narmada basins in the intra-Cretaceous TSTs particularly in Aptian–Albian, Late Cenomanian–Early Turonian intervals that are considered coeval to OAEs in marginal setting. Condensed to starved sedimentation (Thierstein 1974) is also interpreted in the east, south, and west sectors of the Indian ocean at sites 241, 249, 258, 259, 260, 261, 268 of Legs 26 and 27. In Madagascar as well, the condensed marls, glauconites, and pelitic clays of Aptian–early Middle Turonian first-order TST interval are followed up by rapidly deposited shallow water cross-bedded sands In Madagascar, Pakistan, Jaisalmer, and Kachchh. Dark shales and/or glauconitic marl/silt/sand with or without body fossils are also evidenced in the Berriasian–Early Valanginian interval, followed up by lagoonal coaly carbonaceous shale locally in Kachchh of Late Barremian age.

5.7.8.1 Example of the Widely Known Mid-Middle Albian OAE

In the Aptian part of the major Aptian–mid-Middle Albian TST, the juvenile Indian ocean could not immediately develop deeper bathyal and abyssal depths possibly because of acute space constraints all round the Indian plate. In the east sector of the Indian ocean, this second-order TST interval ($\sim$CC 7–CC 8) after the initiation of oceanic spreading and associated volcanism at the first-order SB is evidenced to have developed stagnant anoxic black shale facies at several sites and even sub-basins with episodic availability of the nutrient-rich waters leading to cyclic carbonaceous and calcareous laminations as in the Cauvery basin as an strong evidence of OAE. It is not just the slow expansion of the Indian ocean, the South Atlantic also followed a similar time frame of origin, and similar early slow growth under similar space constraints in association of episodic volcanism. The ammonoid fauna is South Pacific, and the juvenile ocean maintained intermittent communication with South Pacific which facilitated the intermittent mixing of the cooler South Pacific waters with the relatively warmer juvenile Indian ocean. In fact, the Dravidian shallow marine corridor from the Cauvery basin to the South Pacific margin of the Gondwana acted first as an immediate praecursor arm of the South Pacific, and subsequently advanced northward to unite with the Neotethyan arm of the sea that progressed from north to south in the east sector divergent zone as the Dravidian corridor. During Aptian, the corridor had extended up to K-G basin as evidenced by common ammonoid, and in a graded manner, yet later, transformed itself into the new Indian ocean during the course of the second-order TST interval through punctuation marks of the third-order intra-Aptian MFSs.

5.7.9 Global and East Arabian / Northwest Australian Context

According to current estimates, about 60 % of hydrocarbon source sediments in global context are restricted in Late Jurassic Early Cretaceous interval of earth's history. During this interval India was a constituent of the Gondwana in Southern hemisphere in rather subtemperate latitudinal framework. Its north and northwest margins then were under the influence of south waters of the Neotethys as part of the Gondwanian Tethyan margin (GTM). The hydrocarbon-rich Arabia was also a part of GTM in extreme west. Arabia includes about 60 % of the total hydrocarbon source and reservoir rocks of the world in Late Jurassic–Early Cretaceous interval. Also, 60–70 % of the Arabian hydrocarbon source sediments are in Late Jurassic–Early Cretaceous interval. Similarly, the NW and West Australia also within the mega-sequence in the east sector include substantive hydrocarbon source sediments (Geoscience Australia 2014, and others), giant gas reservoirs in the Permian–Pliensbachian units, and oil in the Toarcian–Barremian sequence. The Ladinian–Pliensbachian units hold huge gas reservoirs in addition of oil source sediments (Geoscience Australia 2014). The NW Australian discoveries are made only recently in the last decade since 2008–09 which makes a positive hope in similar Indian tectono-sequento-stratigraphic setup. Thus, prior to the oceanic separation of India from Australo-Antarctica that began at the first-order intra-Cretaceous SB, both the Spitiian (Anisian–Pliensbachian) and Kachchhian (Toarcian–Barremian) first-order sequences have substantive to near identical tectono/sediment/sequentostratigraphic framework with NW and West Australia, and suggests positive hydrocarbon outlook in the Indian basins (Fig. 5.16).

A cursory scan of the Phanerozoic succession of the unordered Arabian MFSs reveals, that the hydrocarbon source sediments are temporally age constrained in the GTM first-order TSTs as interpreted in the present work namely Early to Middle Cambrian, Middle Ordovician–Middle Silurian, early Late Permian, Early Bajocian Late Kimmeridgian, Aptian–Middle Cenomanian, Maestrichtian–early Early Paleocene. Among these, the most important intervals are Early Bajocian–Kimmeridgian, Aptian–Cenomanian, and Early Paleocene in the mega-sequence under consideration, which, obviously, deserve extensive and intensive studies.

5.8 Summarized Evaluation of the Indian Hydrocarbon Prospects in the Mega-Sequence

In an over all global scenario, among the temporally organized five Phanerozoic mega-sequences, the intra-Permian–intra-Paleogene mega-sequence has yielded the major chunk of the hydrocarbons. The remark holds good for proven and prospective source rocks to produce in near future until exhaustion.

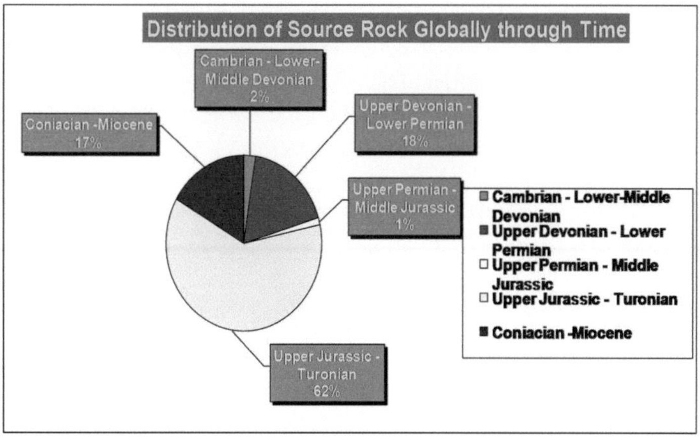

Fig. 5.16 Skewing of hydrocarbon source rocks in the Late Jurassic–Turonian interval

Among the three first-order sequences of the mega-sequence, the rich source sediments are higly skewed in the ~57 my long Toarcian–Barremian span. It includes many giant oil and gas fields in Arabia inclusive of the giantest Ghawar oil field of Saudi Arabia. East of India, the major reservoirs are in the ~60 my long Ladinian–Pliensbachian span in NW Australia.

5.8.1 *On the Successive Second-Order Intervals of the Mega-Sequence*

The 37–40 odd second-order TSTs within the Indian basins in the said mega-sequence in terms of biostratigraphic zones are in the Raniganjensis, Woodwardi–Discophiceras, unnamed-Griesbachi, Meekoceras–Kellnerites, Langobardian–Nitensis, Regoled, early Austriacum, Subbulatus–early Anatropites, early Jandianus, Paulcki–Hogarti, Suessi zonal/subzonal intervals in the Triassic; Planorbis–Liasicus, Bucklandi–Obtusum, Margarata–Nitiscens, Discites–Propinquans, Garantiana–Macrascens, Mantataranus–Diadematus, Obtusicosta–Depressum, unnamed–Schilli, Kachchhensis, Alterneplicatus–Bathyplocus, Pottingeri–Rajnathi, Natricoides, Densiplicatus–Jacobi, early Boisseri, Pertransensis–Verrucosum, Radiatus–Balearis, Vandeheckii–Sartousiana, Oglanlensis–Nolani, Tardefurcata–Dentattus, Lautus–Dixoni, Geslinianum–Turoniense, Deverianum–Serratomarginatus, Biodorsatum–Delawarensis, Tridens, P0–P5 zonal intervals. Among these more important ones of the first-order TSTs number fourteen (Fig. 7.2).

The following first-order RSTs (Ladinian–Pliensbachian, terminal Middle Oxfordian–Barremian, and late Middle Albian–early Early Eocene) should make the major reservoirs. As in the case of source sediements, the second- and first order RSTs of the first-order RST make better reservoirs compared to the second- and first-order RSTs of the first-order TST. The later parts of the first-, second- or third-order RSTs make better reservoirs than the earlier parts of the same RSTs.

The assessment of hydrocarbon prospects in the west sector up to Madagascar–India separation during ~98 to ~84 ma is discussed in the Arabian context, while that in the east sector in the NW and West Australian context up to India–Australo-Antarctica separation at ~107 ma in spite of the broad tectono-stratigraphic similarity all over the GTM.

5.8.2 *Late Permian–Anisian First-Order TST Interval*

Arabia

Right since the start of the mega-sequence, the stratigraphic units Khuff D and C mostly as second-order TST of Guadalupian–Lopingian represented the early part of a major widespread transgression consequented by the post-rift thermal subsidence. India and Arabia continued their intra-plate relationship during this interval. The cyclic carbonate sediments of the said units D and C of Saudi Arabia, Iran, Oman, Abudhabi, etc. bounded on both sides by unconformities have been considered as good prospective source rocks (Sharland et al. 2001). The proven Arabian Induan and Ladinian source sediments are in the close temporal vicinity, respectively, of the second-order intra-Induan, and first-order terminal Anisian MFSs.

India

The present study under similar tectono-stratigraphic framework in the east and west sectors features thick Late Permian–Anisian non-marine sediments in Gondwana basins, particularly K-G in the east, and Jaisalmer, Kachchh, Saurashtra, possibly also, Deccan, and Kerala–Konkan basins in the west. The K-G basin is already a proven source rock and producing basin in this first-order TST interval from Early Triassic reservoirs. Even the coeval inland Gondwana sediments, particularly, the Panchet and coeval units require exploration in this context.

Indian West Sector

The Guadalupian–Anisian first-order TST interval in the west sector is suggested in Jaisamer, Saurashtra, and Kerala–Konkan in moderate non-marine sediment thickness at the base of a thick Mesozoic–Cenozoic column in subsurface wells. Like the presence of proven/interpreted hydrocarbon source sediments in Arabia, East Africa–Madagascar, Seychelles, K-G, and NW/W Australia, the Guadalupian–Anisian interval in the Indian west sector as well is anticipated to include moderate grade source sediments under similar tectono-sedimento-sequento-stratigraphic framework.

Indian East Sector
In the K-G basin, the first-order Late Guadalupian/Kungurian–Anisian Kommugudem unit makes the main source sediments. In the Triassic, the early part of the Mandapeta Formation of Anisian–Ladinian age in the temporal vicinity of the first-order MFS includes good source sediments.

5.8.3 Ladinian–Pliensbachian First-Order RST Interval

Arabia
Only the early part of the said interval up to Early Norian includes proven/prospective source rocks, however, even this interval is not considered significant in the over all Mesozoic hydrocarbon scenario in Arabia.

India
The sediments in this interval are again quite thick in all the east and west sector basins of largely non-marine framework. These have already been considered quite promising in Seychelles, Madagascar, Somalia, Tanzania, and Mozambique basins in the west sector, and also on the NW and West Australian margins, and should make interesting near future targets particularly along the vast Jaisalmer, Kachchh, Saurashtra, Deccan, and Kerala–Konkan tract in the west, and even in the Gondwana inland basins. These are considered highly promising but have not been adequately explored on account of their basaltic covers.

Indian East Sector
The upper part of the Mandapeta Formation in the first-order RST interval is interpreted to include moderate source sediments, and hydrocarbons are being produced that were generated in the said first- and second-order TST source intervals. However, in the Northwest Australian context, similar non-marine Ladinian–Pliensbachian with huge expanse and thickness units in both the Indian sectors are interpreted as highly positive.

5.8.4 Toarcian–late Middle Oxfordian TST Interval

Arabia
This first-order TST interval is the most productive of the earth's history in Arabia and elsewhere on the globe, particularly, its late Middle Callovian–late Middle Oxfordian part (Sharland et al. 2001). The source sediments of the world's biggest

hydrocarbon field—the Ghawar field of Saudi Arabia—are in this interval. Even the second- and third-order RSTs of this interval include high-quality source sediments. However, the source sediments mostly are housed in a specialized setting of intra-shelf basins. The source sediments are in the Early Bajocian, late Middle Callovian–early Late Callovian, and late Early to late Middle Oxfordian successive second-order TSTs in the later part of the first-order TST. The best quality source sediments with maximum spatial expanse are age constrained in Middle Oxfordian inclusive of the world's giantest Ghawar field.

Source Sediment Intervals
Mostly, fine textured siliciclastics mixed carbonates of ULM of Patcham Formation and exclusive fine siliciclastics of Chari, Katrol and Umia units of second-order TSTs of the first-order TST, i.e. Early Toarcian (early part of Dm of LSM of Patcham Formation), Early Bajocian (at Patcham and Jhura of ULM of Patcham Formation), Bathonian to mid-Early Callovian (ULM of Patcham Formation and LSM of Chari Formation at Patcham, Jhura and Jumara), late Middle Callovian early Late Callovian (GSM of Chari Formation (all over Mainland, Wagad and Khadir), early Early Oxfordian to late Middle Oxfordian (JM of Chari Formation) compared to second-order TSTs of the first-order RST, i.e. most of the Kimmeridgian, and Late Tithonian to Valanginian (early part of Katrol Formation and early part of Umia Formation) should make good source rocks. Among the above, the Oxfordian–Kimmeridgian shales/clays should make the best source rocks, followed by Callovian and Late Tithonian to Valanginian shale/silt as the next best, and equally good are Bajocian–Bathonian carbonates.

West Sector
The said TST interval is present in Jaisalmer, and Kachchh, and also suggested in Saurashtra, and Kerala–Konkan. In the relatively proximal part of the unusually wide Kachchh shelf, the early Toarcian–Bathonian part is much thicker under relatively restricted depositional environment than the younger Callovian–Middle Oxfordian part. The Bajocian–Bathonian carbonates as in Arabia include good hydrocarbon source rocks. The 'Dingy member' of the Lower Sandstone Member (LSM) of presumable Toarcian second-order TST should also include good source intervals in context of the widespread basal Toarcian OAE. Similarly the shaly second-order TST interval of GSM (Gypseous Shale Member of Chari Formation) of late Middle–early Late Callovian, and other younger shaly intervals of Oxfordian age also need exploration down the dip even in the exposed basin. The sediments of this interval in Kachchh have found the widest expanse up to the extreme margin of the basin near the Radhanpur arch. Restrictive depositional conditions are interpreted north of the Kachchh Mainland fault up to Khadir and east of Mainland. Biswas (2012a, b) interpreted partially restricted environment and large thickness in the half graben of the Gulf of Kachchh south of the East Mainland. The sediment thickness in the interval is known to increase in the offshore wells.

Similarly, the interval in Jaisalmer and Saurashtra is considered highly promising in the Arabian context, and pursuance of diverse exploratory studies is highly desirable on high priority basis. The sediments are in marine framework at the north extreme of the west sector, but possibly non-marine to marginal marine in Saurashtra, Deccan, and Kerala–Konkan. The marine framework in Kachchh and Jaisalmer is considered more interesting in Arabian context than the non-marine in Saurashtra, Deccan, and Kerala–Konkan. This interval is unfortunately absent in the east sector coastal and inland basins. The better quality Arabian hydrocarbon source sediments in the said interval are further restricted mostly in the Oxfordian–Kimmeridgian interval that is also the closest to the intra-Oxfordian first-order MFS. In Arabia, the intra-shelf basins in Iraq, Kuwait, Central Saudi Arabia, and rift basins in Yemen are known to have the best quality source rocks in Early Bajocian, Middle Callovian, and Middle Oxfordian (Sharland et al. 2001) which correspond to the second-order TSTs of the first-order Toarcian–late Middle Oxfordian TST, and thus primarily provide strong support to the above outlined conceptual source intervals in the second-order TSTs of the first-order TST. The East Mainland, and Wagad during the second-order TSTs of the first-order TST suggest restricted circulation framework somewhat similar to that of the rather interiorly located intra-shelf source rock rich basins of the Arabian plate. The said first-order MFS in Arabia on GTM is recently determined in Kachchh above the late Middle Oxfordian Orientalis Zone Schilli Subzone bed I–O8 in the Kantkote section of Wagad outside the Mainland Kachchh (Krishna et al. 1998). The anticipated Toarcian–Kimmeridgian sediment in the vast proximal tracts in Jaisalmer, Kachchh, Saurashtra, Kerala–Konkan basins appear interesting in this context.

The best source sediments, thus, should be concentrated in the younger second-order TSTs of the first-order TST being relatively time wise closer to the first-order MFS, i.e. in the Bajocian–Middle Oxfordian interval of mid-shelf and deeper environments. These second-order TSTs are Early Bajocian, early Early Bathonian, mid-Middle to mid-Early Callovian, late Middle to early Late Callovian, late Early to late Middle Oxfordian. Second-order TSTs of the first-order RST also make good source sediments, which include early to mid-Early Kimmeridgian, late Early to mid-Late Kimmeridgian, early Early Tithonian, and late Late Tithonian to early Late Valanginian interval second-order TSTs. Even in the above second-order TSTs, the TSTs of the third-order sequences make better source sediments than the corresponding third-order RSTs. Even the organically rich fine textured sediments of the older second-order TST of Early Toarcian age of the first order TST in farther offshore should make good source sediments.

Indian East Sector

The said extremely important interval in hydrocarbon exploration context of proven high-quality source rocks in Arabia is unfortunately absent in the east sector.

5.8.5 Terminal Middle Oxfordian–Barremian First-Order RST Interval

Arabia
The early part of this first-order RST interval up to close of Berriasian includes good source sediments particularly during the second-order TSTs, while the later Late Valanginian–Barremian part is almost bereft of any source sediments. The NW Australian basins also include substantive oil and gas reservoirs.

Indian East Sector
The interval is almost fully or partly present in the Cauvery, K-G, Bengal, Upper Assam, Jharkhand. The terminal Middle Oxfordian–Early Valanginian interval of the Golapille and coeval units in the K-G basin also includes good source rocks. In the Cauvery basin, the Late Oxfordain–Early Berriasian early part of Andimadam and coeval units include good source sediments. In the inland east sector basins, studies in the concerned interval require more exploratory initiatives. The sediments in this interval formed in a new paleogeographic framework in the east sector that is quite different from that of the west sector. A new ocean—the proto-Indian ocean—has come into being particularly in the Indo-Australian margin east sector with slow due south growth in the north–south-directed rift zone. The Indian East coast basins have originated, and offer better and more conducing passive margin tectono-stratigraphic framework in Bengal, Mahanadi, K-G, Cauvery, and Mannar basins all along the long east coast.

Indian West Sector
The interval is widely present in the west sector in Jaisalmer, Kachchh, Saurashtra, Narmada, Deccan and Konkan–Kerala basins. In Jaisalmer, Kachchh and Saurashtra basins, the related stratigraphic units in their early Oxfordian–Early Valanginian part in successive second-order organically rich shaly intervals include good source rocks, and the interval in these are considered highly promising targets for intensive exploration. The said interval in the west sector offshore is equally promising. Both oil and gas are already being produced in the Indus basin of Pakistan. The Katrol Formation–Ghuneri Member succession of Kachchh, and Parihar Formation in Jaisalmer basins also are known for hydrocarbon generation. Lately, even commercially viable hydrocarbon pools are confirmed in the Kachchh Late Jurassic–Cretaceous. Dhrangadhara Formation in Saurashtra, and unnamed coeval units in Kerala–Konkan. The setting in the west sector below the traps defies detailed subsurface studies.

5.8.6 Early Aptian–Mid-Middle Albian TST Interval

Arabia
The interval in Arabia includes good quality source sediments.

Indian East Coast
The interval is included in the upper part of the Andimadam unit in the Cauvery basin, and early part of the Raghavapuram unit in the K-G basin. These are known to include moderate quality source rocks.

Indian West Coast
The interval is included in the Abur unit of Jaisalmer, Ukra unit of Kachchh. The interval is also suspected in Cambay, Saurahtra, Narmada, Deccan, and Konkan–Kerala, and over all the TST interval is considered promising.

5.8.7 Cenomanian

Indian East Coast
The interval is included in the Dalmiapuram Limestone, and its facies equivalent Karai Formation in the Cauvery basin, middle part of Raghavapuram Formation in the K-G basin, and encloses moderate quality source rocks.

Indian West Coast
The interval hardly includes any source sediments in outcrops or wells in Kachchh. The interval includes the upper part of the Goru Formation in wells of the Jaisalmer basin and deserves attention.

5.8.8 Late Cenomanian–early-Middle Turonian TST Interval

The sediments both in east and west sectors include interesting source intervals particularly in the Karai Formation in Cauvery, Raghavapuram Formation in K-G, Abur/Lower Goru units in Jaisalmer, Ukra and coeval offshore sediments in Kachchh, Wadhwan Formation in Saurashtra, and coeval sediments in Narmada, Kerala–Konkan appear extremely promising targets.

5.8.9 Late Turonian–Intra-Paleocene First-Order RST Interval

It includes source sediments of the already producing reservoirs in Bombay High, and Gujarat, as also in east coast basins.

Indian East Coast
The interval is included in the upper part of sand dominating Tirupati Formation and early part of Razole Formation. Good quality source sediments are known in the Razole Formation.

Indian West Coast
The interval in Kachchh for sure is included in wells in the Mundra/Naliya/Kori units. The Kori unit carbonates are anticipated to include hydrocarbon source sediments. The other west sector Indian basins of Saurashtra, Narmada, Deccan, and Konkan–Kerala include the said interval in facies and environment that should also be conducive for hydrocarbon source sediments.

5.8.9.1 Late Santonian–Coniacian

Indian East Coast
The interval as per ammonoid evidence includes the Trichinopoly Group (Kulakkalnuttam and Anaipadi units) in the Cauvery exposures. Good source rocks are expected as in Pakistan.

Indian West Coast
The interval for sure includes the Parh Formation in the Jaisalmer basin. The interval is also included in Saurashtra, Narmada, Deccan and Konkan–Kerala basins. Source sediments are known in the Parh unit. In Pakistan, the unit is known to have generated hydrocarbons.

5.9 Concluding Remarks

The scenario within the mega-sequence in sequence stratigraphic context in this preliminary exercise is over all quite positive.

The best source sediments are expected in the Bajocian–Bathonian carbonates of the Upper Limestone Member of the Patcham Formation, and Callovian–Oxfordian–Kimmeridgian shales of second-, third- or fourth-order TSTs of Chari Formation (LSM, GSM, and JM) to early part of Katrol Formation of second, third and fourth-order TSTs, and even in the Berriasian–Valanginian shales of the basal Member of Umia Formation again of second-, third and fourth-order TSTs, and also shaly facies of the Toarcian–Bathonian of LSM down the basin in offshore of KSm and BCSm. The source rock prospects in the vast expanse of the west sector are considered equally positive as in Kachchh. The terminal Oxfordian–Barremian sediments in east sector are considered equally promising.

The Aptian–Early Maestrichtian sediments should also provide hydrocarbon source sediments but of inferior order than those of the preceding first-order sequence. Best intervals are Aptian to early Middle Albian (Ukra Member), Turonian–Santonian, if present in subsurface wells, and late Late Turonian to Coniacian (Mundra Formation in wells), Early Campanian and Early Maestrichtian (Naliya/Kori Formation also in wells). Cenomanian and Campanian–Maestrichtian sands should make good reservoirs. The coeval sediments in the vast west and east sectors on similar considerations deserve extensive exploratory studies.

References

Arkell WJ (1956) Jurassic geology of the world. Oliver and Boyd, London, pp 1–806

Bajpai S, Prasad GVR, Prasad V, Krishna J, Sarkar A (2012) Recent advances on Phanerozoic biodiversity, bioevents and climates in India. PINSA 78(Spec. Issue):445–456

Bertle RJ, Suttner TJ (2005) New biostratigraphic data from the Chikkim formation (Cretaceous Tethys Himalaya, India). Cret Res 26:882–894

Besaire H (1973) Precis de geologies malgache. Annales Geologiques de Madagascar, Tananarive, XXXVI, pp 1–141

Bhargava ON (1987) Stratigraphy, microfacies, and paleoenvironment of the Lilang Group (Schythian-Dogger), Spiti valley, Himachal Himalaya, India. J Pal Soc Ind 32:92–107

Bhargava ON (2011) Early Paleogene paleogeography, basin configuration, paleoclimate, and paleotectonics on the Indian plate. Mem Geol Soc Ind 78:69–99

Bhargava ON, Bassi UK (1996) The Spiti Kinnaur Tethyan basin, Himachal Himalaya, and regional sedimentological pauses, Proceedings of Symposium. Himalaya and foredeep. Sp. Publ. Geol Surv Ind 21(1):19–27

Bhargava ON, Krystin L, Balini H, Lein R, Nicora A (2004) Revised litho and biosequence stratigraphy of the Spiti Triassic. Albertiana Suppl 30:21–38

Biswas SK (1977) Mesozoic rock-stratigraphy of Kutch. Quart J Min Met Soc Ind 43(3and 4):1–52

Biswas SK (2005) Tectonic style and sediment dynamics of rifted basins, their bearing on petroleum habitat—examples from Satpura and Kutch, Basins, India. Ind J Petrol Geol 14(1):1–29

Biswas SK (2012a) Stratigraphy of Kutch–an overview, abstract, 11–13. Geology on Kachchh Basin Western India, Bhuj, Kachchh

Biswas SK (2012b) Status of petroleum exploration in India. PINSA 78(3):475–494

Bladon A, Stuart J, Clarke M, Stuart DB, Beaumont H, Nicholas W, Kothari V, Mohapatra P (2015) Previously unrecognized extensional tectonics exposed in the Barmer Basin, Northwest India. Search and Discovery Article #30426

Brown S (1992) The Mesozoic stratigraphy of the Timor Gap and its bearing on the hydrocarbon potential of Eastern Indonesia. Proc Indones Petrol Ass 553–561(92–11):21

Calkins JA, Offield TW, Abdullah KM, Ali ST (1975) Geology of the Southern Himalaya in Hazara Pakistan and adjacent area. USGS Professional Paper, 716 C, p 29

Cariou E, Krishna J (1988) The Tethyan Reineckeiinae of Kachchh and Jaisalmer (West India): systematic, biostratigraphic and biogeographic implications. Paleontographica A 203(4–6):149–170

Cariou E, Krishna J, Enay R (1990) Dynamique biogeographique des ammonites tethysienne au Callovien inferiour et correlations stratigraphique entra marges indiennese et europeennese. C R Acad Sci Paris 311(2):373–379

Dave A (2011) Recent developments in sequence stratigraphy and depositional sequences of Indian petroliferous basins. In: Jaitly AK, Singh AD, Pandey B, Nath S (eds) Palaentology and stratigrahy: basics to applications. Banaras Hindu University, Varanasi, pp 88–112

Dave A (2012) Depositional sequence stratigraphy and petroleum systems of Kachchh basin, Status Talk–Hydrocarbon Potential and offshore geology. In: Thakkar MG (ed) Abstract volume, Geology of Kachchh basin, Western India, 94–97

Davey RJ (1987) Palynological biozonation of the Lower Cretaceous and Upper and uppermost Middle Jurassic in the northwest Papuan Basin of Papua New Guinea. Geological Survey PNG Memoir #13

El-Bishlawy SH (1985) Geology and hydrocarbon occurrence of Khuff Formation in Abu Dhabi UAE. In: Proceedings of the 4th society of petroleum engineers middle east oil technical conference, Paper SPE 13678, pp 601–606

Enay R, Cariou E (1999) Jurassic ammonite faunas from Nepal and their bearing on the palaeobiogeography of the Himalayan belt. J As E Sci 17:829–848

Exxon NF, Coldwell JB (1994) Geological history of Northwest Australia—a synthesis. J Austr Geol Geoph 6:1–29

Fatmi AN (1986) Lower and Middle Jurassic ammonites from windar group of Phusi Jhal (Kanrach) and Sand (Windar Nai) sections of Lasbella, Baluchistan, Pakistan. Mem Geol Surv Pak 16:1–17

Fraser TH, Bon J, Samuel L (1993) A new dynamic Mesozoic stratigraphy for the West Iriyan micro-continent Indonesia, and its implications. In: Proceedings of Indonesia petrol association, 2006 22nd annual meeting, IPA, 93-1.1-024, pp 707–761

Fursich FT, Pandey DK (2003) sequence stratigraphic significance of sedimentary cycles and shell concentrations in Upper Jurassic–Lower Cretaceous of Kachchh, western India. Paleogeogr Paleoclimat Paleoec 193:285–309

Fursich FT, Alberti M, Pandey DK (2013) Stratigraphy and paleoenvironments of the Jurassic rocks of Kachchh (field guide). Berringeria 7(Spl. Issue):1–174

Fursich FT, Oshmann W, Pandey DK, Jaitley AK, Singh IB, Liu C (2004) Paleoecology of Middle to Late Jurassic macrofaunas of the Kachchh basin, western India: an overview. J Pal Soc Ind 49:1–26

Fursich FT, Pandey DK, Callomon JH, Jaitly AK, Singh IB (2001) Marker beds in the Jurassic of the Kachchh basin, western India: their depositional environment and sequence stratigraphic significance. J Pal Soc Ind 46:176–198

Geiger M (2004) Sedimentary and stratal patterns in Jurassic successions of Western Madagascar —facies, stratigraphy, and architecture of Gondwana break up and drill sequences, Ph.D thesis, Uni. Bremen

Geiger M, Schweigert S (2006) Toarcian–Kimmeridgian depositional cycles of the southwestern Morondova basin along with rifted continental margin of Madagascar. Facies 52:85–112

Geoscience Australia (2014) Regional geology of the Northern Carnavon basin, Austraian Government Department of Industry, 33p

Gibbons AD, Whittaker JM, Muller RD (2013) The break-up of East Gondwana: assimilating constraints from Cretaceous ocean basins around India into a best fit tectonic model. J Geophy Res, Solid Earth, pp 808–822

Govindan A, Ravindran CN (1996) Cretaceous biostratigraphy and sedimentation history of Cauvery basin, India. In: Pandey et al (eds) Contribution to XV Indian Colloquium on micropaleontology and stratigraphy, 19–32

Graciansky PC, Dardeau G, Dumont T, Jacquin T, Marchand D, Mouterde R, Vail PR (1993) Depositional sequence cycles, transgressive-regressive facies cycles and extensional tectonics: example from the southern subalpine Jurassic basin, France. Bull Soc Geol France 164(5): 709–718

Graciansky PC, Hardenbol J, Jacquin T, Vail PR (1998) (eds) Mesozoic and Cenozoic sequence stratigraphy of European basins. Soc Sed Geol (SEPM), Tulsa (OK) 60:445–466

Gradstein FM, Gibling MR, Jansa LF, Kaminski MA, Ogg J, Sarti M, Thurow JW, Von Rad U, Westermann GEG (1989) Investigation of Mesozoic sediments in Thakkhola, Central Nepal, vol 1, 60, 48 figs. and plates. Special Report, Centre for Marine Geology, Dalhousie University

Gradstein FM, Gibling MR, Sarti M, Rad UV, Thruro JW, Ogg JG, Jansa LF, Kaminski MA, Westermann GEG (1991) Mesozoic Tethyan strata of Thakkhola, Nepal: evidence for the drift and break-up of Gondwana. Palaeogeogr Palaeoclimatol Palaeoecol 88:193–218

Haq BU, Hardenbol J, Vail PR (1987) Chronology of fluctuating sea-level since the Triassic. Science 235:1156–1167

Hearty DJ, Ellis GK, Webste KA(2002) Geological history of the Barrow subbasin; implications for hydrocarbon entrapment at Woollybutt and surrounding oil and gas fields. In: Proceedings of symposium on sedimentary basins of Western Australia, Perth, pp 577–598

Helby R, Morgan R, Partridge AD (1987) A palynoogical zonation for the Australian Mesozoic. Memoir Assoc Australas Palaeontologists 4:1–94

Hocking RM (1990) Carnarvon basin, geology and mineral resources of Western Australia. Mem West Austral Geol Surv 3:457–495

Howarth MK, Morris NJ (1998) The jurassic and lower cretaceous of Wadi Hajar, Southern Yemen. Bull Nat Hist Mus Geol Ser 54(1):1–32

Jablonski D (1997) Recent advances in the sequence stratigraphy of the Triassic to Lower Cretaceous succession in the northern Carnarvon basin. APPEA J 37(1):429–454

Jablonski D, Saitta AJ (2004) Permian to Lower Cretaceous plate tectonics and its impact on the tectonostratigraphic development in the Western Australia margin. The APPEA J 44(1): 287–327

Jacquin T, Dardeau G, Durlet C, Graciansky PC, Hantzpergue P (1998) The North Sea cycle; an overview of 2nd-order transgressive/regressive facies cycles in Western Europe. In: de Graciansky PC, Hardenbol J, Jacquin T, Vail PR (eds) Mesozoic and Cenozoic sequence stratigraphy of European basins. Soc. Sed. Geol. (SEPM), Tulsa (OK), 60:445–466

Jafar SA, Rai J (1989) Discovery of Albian nannoflora from type Dalmiapuram Formation, Cauvery basin, India, and palaeooceanographic remarks. Curr Sci 58(7):358–363

Kale AS, Lotfalikani A, Phansalkar VG (2000) Calcareous nannofossils from the Uttatur group of Trichonopoly Cretaceous, South India. Mem Geol Soc Ind 46:213–227

Khan AA (1989) Stratigraphy and hydrocarbon potential of Peromo-Triassic sequence of rocks in the state of Kuwait. In: Proceedongs of deep formation, organization of Arab patroleam exporting countries, Kuwait, pp E3-E29

Koshal VN (1975) Palynozonation of Mesozoic subsurface sediments of Banni Kutch Gujarat. Quart J Min Met Soc Ind 47:79–81

Krishna J (1983) Callovian–Albian ammonoid stratigraphy and paleobiogeography in the Indian subcontinent with special reference to the Tethys Himalayas. Him Geol 11:43–72

Krishna J (1987a) Biological evidence for better appreciation of Indian Gondwana. Paleobotanist 36:268–84

Krishna J (1987b) The Jurassic–Cretaceous ammonoid geography in the Indian subcontinent vis-à-vis marine seaways and plate-tectonics in the Indian ocean region, Sp. Publ. Geol Surv Ind 11(1):453–482

Krishna J (1994) Environmental reconstruction studies of the Mesozoic sedimentary sequences in Western Kachchh (ONGC Sponsored Project), (Technical Report), 1–143, pts, 1-10

Krishna J (2005) Update on kachchh Jurassic ammonoid zonation: summarised Intra-basinal to Global Perspectives. Spl. Publ.-1. Assoc Pet Geol, pp 63–68

Krishna J (2012) Litho-bio-chrono-magneto-igno-sequento-stratigraphic integration in the Kachchh Mesozoic, Status Talk–Mesozoic Biostratigraphy. In: Thakkar MG (ed) Abstract volume, Geology of Kachchh basin, Western India, pp 21–25

Krishna J (2016) Initiation of Indian ocean at $\sim$159 ma late middle oxfordian orientalis zone schilli subzone MFS event (abstr). In: 35th International Geological Congress, Capetown

Krishna J, Cariou E (1986) The Callovian of Western India: new data on stratigraphy, biogeography of the ammonoids and correlation with Western Tethys. Newsl Stratigr 17(1): 1–8

Krishna J, Cariou E (1993) The Tethyan Macrocephalitinae: evolutionary, environmental and dispersal strategies. Geobios 15:227–236

Krishna J, Ojha JR (1996) The Callovian ammonoid chronology in Kachchh (India). GeoResearch Forum 1–2:151–166

Krishna J, Pathak DB (1989) Kimmeridgian in Ler–Katrol area of Kachchh, Western India: ammonoid systematics and biochronology. In: Abstracts 28th International Geological Congress, Washington, 228–229

Krishna J, Pathak DB (1991) Ammonoid biochronology of the Kimmeridgian stage in Kachchh, India. J Pal Soc India 36:1–13

Krishna J, Pathak DB (1993) Late Lower Kimmeridgian – Lower Tithonian virgatosphinctins of India: evolutionary succession and biogeographic implications. Geobios 15:217–226

Krishna J, Pathak DB (1994) Stratigraphic, biogeographic and environmental signatures in the ammonoid bearing Jurassic-Cretaceous in the Himalaya on the south margin of the Tethys. J Him Geol 4(2):151–167

Krishna J, Kumar S, Singh IB (1982) Ammonite stratigraphy of the Spiti Shale (Upper Jurassic), Tethys Himalaya, India. Neues Jahrbuch für Geologie Paleontologie. Abhandlungen 10:580–594

Krishna J, Pandey B, Ojha JR (2008) Gregoryceras in the Oxfordian of Kachchh (India): diverse eventful implications. Geobios 42:197–208

Krishna J, Pandey B, Pathak DB (2010) Update on the Jurassic stratigraphy of the Indian Subcontinent with focus on the tectonically control regional transgressiv–regressive couples. Earth Sci Front 17(Special Issue):138–140

Krishna J, Pandey B, Pathak DB (2011) Current status of the Jurassic ammonoid stratigraphy framework in the Indian Subcontinent with focus on the tectonically control regional transgressive—regressive couplets. Memoir Geol Soc Ind 78:140–176

Krishna J, Pathak DB, Pandey B (1998) Development of Oxfordian (Early Upper Jurassic) in the most proximal exposed part of Kachchh basin at Wagad outside the Mainland Kachchh. J Geol Soc Ind 52:513–522

Krishna J, Pathak DB, Pandey B, Ojha JR (2000) Transgressive sediment intervals in the Late Jurassic of Kachchh, India. GeoResearch Forum 6:321–332

Krishna J, Singh IB, Howard JD, Jafar SA (1983) Implications of new data on the Mesozoic rocks of Kachchh, India. Nature 305:790–792

Krishnaswami VS, Swaminath J (1965) Himalayan and Alpine geology: a review. Wadia Mem Vol, Min Mettalurg Inst Ind 172–192

Krystyn L, Balini M, Nicora A (2004) Lower and middle Triassic stage and substage boundaries in Spiti. Albertiana 30(supplement):39–52

Lindsay RF, Cantrell DL, Hughes GW, Keith TH, Mueller III HW, Russell SD (2006) Ghawar Arab-D reservoir: Widespread porosity in shoaling-upward carbonate cycles, Saudi Arabia. In: Harris PM, Weber LJ (eds) Giant hydrocarbon reservoirs of the world: From rocks to reservoir characterization and modeling: AAPG Memoir 88/SEPM Special Publication, pp 97–137

Luger P, Groschke M, Bussmann M, Dina A, Mette W, Ulhman A, Kallenchbach H (1994) Comparison of the Jurassic—Cretaceous sedimentary cycles of Somalia and Madagascar: implications for the Gondwana break-up. Geol Rundschau 83:711–727

Mehrotra NC, Aswal HS (2003) Late Jurassic–Cretaceous Dinoflagellate Cysts. Pal Ind 7, ONGC, Dehradoon, pp 1–108

Nagendra R, Kamalnarayanan BV, Gargi S, Gilbert H, Bakkalraj D, Nallapa-Reddy A, Jaiprakash B (2011) Sequence surfaces and paleo-bathymetric trends in Albian to Maestrichtian sediments of Ariyalur area, Cauvery basin. J Mar Petr Geol 28:895–905

Ojha JR (1996a) The Callovian at Keera, Kachchh, India; ammonoid systematic, chronology and geography. Unpublished Ph.D thesis, Banaras Hindu University, Varanasi

Ojha JR (1996b) The Callovian at Keera, Kachchh, India: ammonoid systematics, chronology and geography. J Sci Res BHU 46:233–244

ONGC (2008) Petroleum systems sequence stratigraphy. KDMIPE, Dehradun

Pandey B, Krishna J (2002) Ammonoid biostratigraphy in the Tithonian (Late Jurassic) of Jaisalmer, Western India. Geophytology 30(1 and 2):17–25

Pandey B, Pathak DB (2015) Record of Early Bathonian ammonoids from Kachchh, India: biostratigraphic and paleobiogeographic implications. J Pal Soc Ind 60(1):33–44

Pandey B, Pathak DB (2016) The possibility of the oceanic anoxic events (OAEs) study in the Indian marine Jurassic–Jurassic outcrops. J Geol Soc Ind 87:261–267

Pandey B, Pathak DB, Krishna J (2013) Calliphylloceras heterophylloides (Oppel, 1856) from basal most Jurassic succession of Sadhara dome, Kachchh, India. J Pal Soc Ind 58(1):61–65

Pathak DB, Krishna J (1993) Preliminary remarks on the biostratigraphic relevance of the new ammonite collection from Spiti Shale Formation, Tethys Himalaya, India. J Him Geol, India 4(2):169–183

Pathak DB, Krishna J, Pandey B (2011) Differentiation of the significant Late Valanginian (Early Cretaceous) transgressive event in the Spiti Himalaya, India. J Asian Earth Sci 42:1226–1231

Prasad B (2011) Palynology: principles, methodology and applications in biostratigraphy and oil exploration. In: Jaitly AK, Singh AD (eds) Palaeontology and stratigraphy: basics to applications. Pandey B, Nath S, 16–35

Rabinowitz PD, Coffin MF, Falvey D (1983) The separation of Madagascar and Africa. Science 220:67

Raju DSN (2008) Stratigraphy of India; mega—charts 1–2. Bull ONGC, Spl. Issue 43(1):1–44 and mega charts

Ramkumar M, Subramaniam V, Steuben D (2005) Deltaic sedimentation during Cretaceous period in the northern Cauvery basin, South India: facies architecture, depositional history and sequence stratigraphy. J Geol Soc Ind 66:81–94

Shanker R, Singh G, Kumar G (1996) Sequence stratigraphy in Himalaya and its reflections in Gondwana sedimentation. In: Proceedings of 9th Gondwana symposium, Spl Publ Geol Surv Ind, Hyderabad, Indi, pp 1159–1173

Sharland P, Archer R, Casey D M et al (2001) Arabian plate sequence stratigraphy. GeoArabia, Sp. Pub. 2:371

Shellnut JG, Bonit GM, Brookfield ME, Jaba BM (2009) No link between the Panjal traps (Kashmir) and Permian mass extinction. Geophys Res Lett 38:219308. doi:10.1829/2011GL059052

Singh NP (2006) Mesozoic lithostratigraphy of the Jaisalmer basin, Rajasthan. J Pal Soc Ind 51(2):1–25

Srikantia SV (1977) Sedimentary cycles in the Himalaya and their significance on the orogenic evolution of the mountain belt. Inte Colloq Centre Nat Res Sci (CNRS) 268:395–407

Stilwell JD, Benedikt MD, Gamarra LS (2011) J/C boundary ammonite Blanfordiceras (Mollusca Cephalopoda) from Fortissimo–1 wildcat well, Brown basin, NW Shelf, Australia. J Palaeontol 85(3):549–552

Stoliczka F (1866) Geological section across the Himalayan Mountains in Spiti. Mem Geol Sur Ind 5(1)

Sukmato R, Westermann GEG (1992) In: Westermann GEG (ed) Jurassic of Circum-Pacific. Cambridge University Press, New York, pp 181–193

Thierstein HR (1974) Calcareous nannplankton, Leg 26, Deep Sea Drilling Project, Intial reports, 26, 619–667

Toland C, Simmons MD, Walkden GM (1995) A new sequence stratigraphic reference section for the Upper Jurassic of Southern Yemen. In: Al-Hussaini (ed) Middle East Petroleum Geosciences, Geo 94 Gulf Petro Link, Bahrain 2, pp 891–899

Turner S, Bean LB, Dettmann M et al (2009) Australian Jurassic sedimentary and fossil succession: current work and future prospects for marine and non-marine correlation. GFF 131:49–70

Uhmann A (1996) Sedimentologische und fazielle entwicklung im Jura und Kreide im Morondova becken (Sudwest Madagascar)—untersuchengen zur mikrofazies tongeologie und schwermineralogie. Verlag fur Wissenschaft und bildung, Berlin 129 p

Veevers JJ (1984) (ed) Phanerozoic Earth history of Australia, Clarendon press, Oxford, 418p

Von Rad U, Exon NF, Boy R, Haq BU (1992) Mesozoic paleoenvironments of the rifted margin off NW Australia (ODP legs 122–123 American Geophysical Union). Geophysical Monograph 70:157–184

Von Rad U, Thurow J, Haq B, Gradstein F, Ludden JN (1989) Triassic to Cenozoic evolution of the NW Australian continental margin and the birth of the Indian Ocean (preliminary results of ODP Legs 122 and 123). Geologische Rundschau 78(3):1189–1210. doi:10.1007/BF01829340

Wadia DN (1934) The Cambrian- Triassic sequences of N W Kashmir. Rec Geol Surv Ind 68:121–167

Westphal H, Aignar T (1993) Seismic stratigraphy and subsidence analysis in the Barrow–Dampier basin, Northwest Australia. Am Assoc Petrol Geol B V 81:1721–1749

Chapter 6
Crucial Links Among Evolution, Extensional Tectonics, Ammonoid Provincialism and Sequence Surfaces

Abstract The first- and second-order sequence surfaces are considered to mark the origin, extinction, drastic reduction, and sudden expansion of ammonoid subfamilies on GTM in general, and in IEAP in particular, even farther to other regions, as held consistently over the last three decades in context of ammonoid faunal cuts and expansion events in the Jurassic (Figs. 6.3, 6.4, 6.5 and 6.6). During the second-order TSTs (particularly near the MFSs) of the first-order TSTs, the faunal similarity within IEAP, as also with ETM, SE Pacific, and SW Pacific is found distinctly greater than in the second-order RSTs and TSTs of the first-order RST. The spatial distribution within the basin, in IEAP, and even outside is wider during the second-order TSTs and MFSs than during any other interval. The temporal punctuation of the infraspecifc, intra-generic, intra-lineage evolution (Figs. 6.7, 6.8, 6.9 and 6.10) occurred at second-order sequence surfaces, and is well attested by the heterochronic ammonoid evolutionary processes of unidirectional paedomorphism in the Macrocephalitinae chrysoolithicus lineage (Fig. 6.8), and unidirectional peramorphism in Katroliceratinae lineage (Figs. 6.8, 6.8, 6.9 and 6.10). The temporal replacement of one lineage by another, also, occurred near exclusively at the sequence surfaces. The explicit control above the subfamilial ammonoid evolution in IEAP by extensional tectonics encouraged further investigation at higher (superfamily/suborder/superorder/subclass/class) level, and the same provided strong links of the entire evolution of the subclass Ammonoidea, and also of the class Cephalopoda, seemingly controlled by extensional tectonics (Fig. 6.2). The origin of Cephalopoda appears linked to the first-order intra-Cambrian MFS, and also of Ammonoidea at the first-order intra-Devonian sequence surface. The origin of Ceratitina, and extinction of Goniantitina may have occurred at the Carboniferous/Permian second-order SB. The origin of Ammonitina is held at the first-order intra-Triassic MFS, and extinction of Stephanoceratina at the first-order intra-Jurassic MFS. The graded decline of the subclass Ammonoidea began at the major intra-Turonian MFS, with the follow up final extinction at the K/Tboundary

© Springer Science+Business Media Singapore 2017

J. Krishna, *The Indian Mesozoic Chronicle*, Springer Geology,
DOI 10.1007/978-981-10-2477-1_6

second-order MFS. The interpretations require confirmation through further investigation. On similar lines are also speculatively related the major evolutionary landmarks, and transformations of marine and non-marine animals and plants to first-order sequence surfaces (Fig. 6.1). The fossil fuel/energy resources of the GTM in general and India in particular are also age-constrained to sequence framework.

Keywords Origin · Extinction · Drastic reduction · Extensional tectonics · Major evolutionary landmarks · Animals · Plants · Marine · Non-marine · Micro · Mega · Mega/lineage/micro evolution · Seedless · Seed bearing · Flowering · Dinosaurs · Birds · Mammals · Planktic forams · Coccoliths · Ammonoids · First/Second-order surfaces · Heterochrony · Paedomorphism · Macrocephalitinae · Peramorphism · Katroliceratinae · Mass extinctions · Phyletic extinctions · First/Second-order SBs/MFSs · Biotic density · Frequency · Diversity

6.1 Fossil Fuel/Energy Resources Vis-à-Vis Sequence Stratigraphic Surfaces

Somewhat out of place, yet, the exploration of the possible links of the prime Indian and GTM energy resources to sequence stratigraphic surfaces, in economic context is considered significant. The origin and evolution of the discipline of sequence stratigraphy has kept pace primarily due to the application of the discipline in hydrocarbon exploratory geological studies. The start and closure of the mega-sequence incidentally is in the midst of Permian and Paleogene coal, lignite, and associated coal bed methane resources that have continuously sustained the human civilization through ages immemorial. The quantum change in the pace of the societal material progress over the last century and half has been catalyzed through the discovery and use of fossil hydrocarbon resources that are age-constrained mostly in the middle part of the mega-sequence during the Late Jurassic–Early Cretaceous time span.

6.1.1 Permian and Paleogene Coal

During the Permian and Paleogene, there formed rich coal deposits in India and elsewhere within, prior to, as also immediately after the duration of the mega-sequence in question. The said energy resources at first sight appear to be

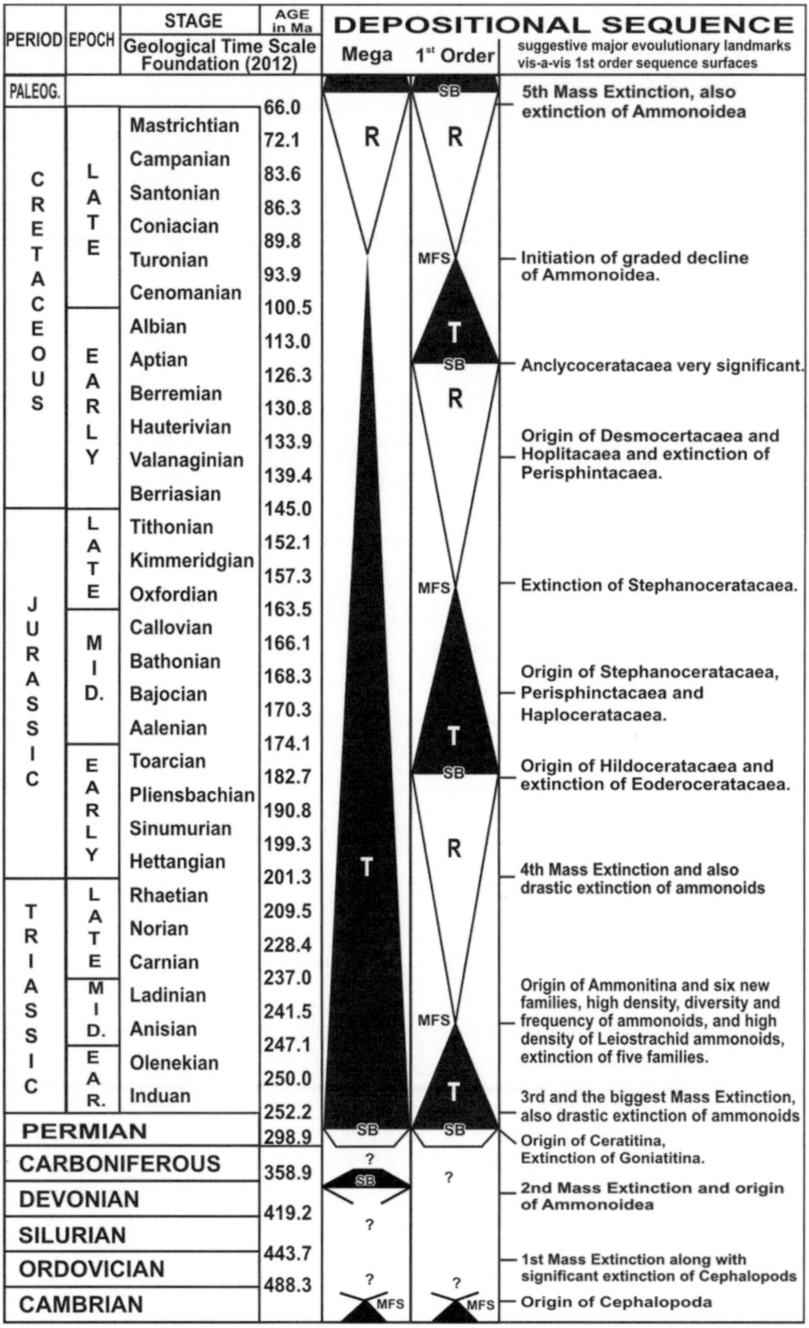

Fig. 6.1 First-order sequence surfaces on GTM, mass extinctions, and ammonoid evolution

related to the geological events which facilitated the availability of favoured environments. The coal formation in India here interpreted to have begun at the second-order Carboniferous/Permian boundary SB. The P/T boundary second-order SB marked the closure of the coal-rich regime. There is also marked mid-Permian break in coal formation, supposedly, at the basal SB of the mega-sequence. The climate changed from wet subtemperate to dry, arid and subtropical. The luxuriant plant life that followed the basal Permian glaciation transformed into rich coal deposits in India under rapid amelioration of the climate. The Early Permian Panjal and coeval volcanism may also have indirectly facilitated the rapid favourable climatic change in the inland coal-rich basins. Similarly, the Paleogene coal is also age-constrained soon after the closure of Deccan outpouring between the first- and second-order surfaces mostly in Paleogene soon after the closure of the mega-sequence, additionally facilitated by early Paleogene thermal maxima.

6.1.2 Hydrocarbon Source/Reservoir Sediments

All the five systems involved in the mega-sequence from Late Permian to Early Paleogene include hydrocarbon source/reservoir sediments in some or other Indian basin. The mega-sequence includes about 60 % of the global reserves. Further, the GTM includes about 50–60 % of the global reserves and production. The best source sediments are stratigraphically located close to the first- and second-order MFSs. The major share of the above already skewed in favour of the GTM, particularly, its Arabian plate, is in the Late Jurassic–Cretaceous interval of the mega-sequence. Thus, the link of the first- and second-order surfaces of the mega-sequence to hydrocarbons is more than obvious.

6.2 Cephalopod and Ammonoid Evolutionary Events

In course of the study, the ammonoids have been primarily applied for development of high-resolution scales of the succession of zones/subzones/horizons in the three Mesozoic systems. The developed high-resolution ammonoid scales were in turn used for the formulation of the succession of first- to third-order sequence stratigraphic surfaces. In this process, information emerged on genetic linkages of the ammonoid origin, evolution, expansion, sudden drastic reduction and extinction on one hand to supra and interregional extensional tectonic events, and on the other hand to mega, first- and second-order sequence stratigraphic surfaces (SBs and MFSs) (Krishna 1983a, b, 1987, 2005, 2012; Krishna and Cariou 1986, 1990, 1993; Krishna and Ojha 1996; Krishna and Pathak 1989, 1991, 1993, 1994; Krishna and Thierry 1987; Krishna et al. 1983, 1994, 1995a, b, 1996a, b, c, d, 1997, 1998; Pandey et al. 2010 and others) that were the resultants of the tectonic events themselves. Salient highlights of said relationships are discussed.

6.2.1 *Global Distribution of Higher Grade Cephalopod/Ammonoid Taxa Vis-a-Vis Geographically Restricted Regional Spread of Ammonoids*

The ammonoid taxa largely have had Pan-Tethyan distribution until the start of the second first-order sequence of the mega-cycle in Early Toarcian. The geographical provincialism in ammonoids suggestively showed up from Toarcian onwards invariably at the family/subfamily level. The ammonoid record in the Neotethys, apart from Pan-Tethyan and globally spread ammonoids, also, exhibits somewhat parallel development and evolution of subfamilies/lineages independently of each other on GTM and ETM of geographically differentiated ammonoids.

6.2.2 *Simultaneous Mass Extinctions Across the World and Diachronous Regional Extinctions*

It is noticed that the higher grade taxa (class to superfamily) exhibited nearly isochronous extinctions simultaneously across the globe, without any provincialism, e.g. the well-known five mass extinctions. In contrast, there are also records of diachronous regional inferior order extinctions on GTM and ETM of lower grade taxa (super to subfamilies). The extinction record of the ammonoids in the studied Late Permian–Early Paleogene mega-sequence vis-à-vis sequence surfaces is reviewed here.

6.2.3 *Five Major/Mass Extinctions*

Among the five major/massive global extinctions, the younger three occurred during the said mega-cycle. The first extinction of the mega-cycle at the Permian/Triassic boundary, has been the most catastrophic in which only 5 % of the existing marine species could survive. In terms of number of ammonoid genera, merely the solitary genus *Sageceras* out of a total of 50 plus ammonoid genera in the Permian could get through into Triassic. In comparison, during the other four global extinctions that took place during the Early Paleozoic and Mesozoic, the survival percentage was about 25–35 %. Apart from the most catastrophic Permian/Triassic boundary extinction event, two additional extinction catastrophies happened during the mega-cycle in question, respectively, near the Triassic/Jurassic boundary of ~200 ma age, and the Cretaceous/Paleogene boundary of ~66 to 65 ma age.

6.2.4 *Mass Extinctions and Evolution of Cephalopods, Particularly, Ammonoids*

In context of the cephalopods/ammonoids, not only the three younger of the five globally spread mass extinctions affected their evolution, but even the first two extinctions—the terminal Ordovician and intra-Devonian) (Fig. 6.1) also explicitly governed them. The first major extinction of cephalopods occurred near the terminal Ordovician, while another similar extinction characterized the ammonoids within the Devonian. According to the treatise on Ammonoidea (Moore 1957), during the first two mass extinctions of the mega-cycle, merely a solitary ammonoid subfamily could get through from near complete extinction. The ammonoids on both the occasions had to make herculean efforts to again adapt, radiate and diversify. The single subfamily Medlicottinae of the suborder Prolecanitina through merely the solitary genus *Sageceras* survived the P/T boundary extinction event. The Jurassic ammonoid world did not include a single known survival from the Triassic, although supposedly having evolved from the single three genera family Discophyllitidae (Moore 1957). From Discophyllitidae evolved, radiated and diversified all the Jurassic and Cretaceous ammonoids. However, at the K/T boundary mass extinctions, the subclass Ammonoidea as a whole could not carry on any further and eventually perished lock, stock and barrel.

6.2.5 *Origin of the Nectic Cephalopods Near the First-Order Intra-Cambrian Sequence Surface*

The younger four of the five mass extinctions, thus, punctuated the origin, evolution, radiation and extinction of ammonoids. Even the first one played a big role in the evolution of the class Cephalopoda. The origin of the nectic Cephalopoda near the Middle/Late Cambrian boundary from benthic mollusks marked an extremely innovative solution in the fierce struggle for nutrients and space. On one hand, the sudden skeltalisation at the start of the Cambrian prompted longer ontogenic span, while on the other hand the acquiring of the protective skeleton resulted in sudden several fold multiplication of the individual skeletal organisms on the sea floor. During the first-order eustatic rise, until near the end of the Middle Cambrian, the struggle for space on the sea floor may have been slow and not that severe due to the increased availability of the additionally encroached land surface as part of the shelfal sea floor. However, during the following first-order RST, the struggle for space as also for nutrition on the seafloor is here interpreted to have become yet fiercer. To get over the fast-increasing space and nutrition constraints on the shallow sea floor came the innovative solution of the nectic way of life through the origin of the nectic cephalopods who could live above the sea floor between the water surface and the land/water interface. The origin and/or sudden radiation of the cephalopods thus presumably took place near the Middle/Late Cambrian boundary first-order MFS.

6.2.6 *Cephalopod Evolution and Extinctions at Sequence Surfaces*

The first major extinction of cephalopods coincided with the first globally spread terminal Ordovician Hirantian stage mass extinction near a first-order SB. The second significant cephalopod extinction is also interpreted to have occurred at the mega-order intra-Devonian SB. The third mass extinction at the P/T boundary occurred at a second-order SB which again caused near complete extinction of ammonoids but for the solitary genus *Sageceras*. The fourth mass extinction took place at the Triassic/Jurassic boundary second-order SB. Finally, the fifth mass extinction near the K/T boundary second-order SB completely wiped out the ammonoids from the world oceans.

All the mass extinctions on the planet earth, thus, explicitly affected the nectic nautiloids and ammonoids. Based on the precise ages provided by the ammonoid taxa, all the five major mass extinctions occurred at the mega, first- or second-order SBs. To put it more objectively, the major SBs have been determined at such exemplary catastrophic geological instants of the Earth history.

6.2.7 *Pre-Jurassic Ammonoid Evolution and Sequence Surfaces*

There are only two pre-Jurassic first-order sequence stratigraphic surfaces in the Late Permian–intra Paleogene mega-sequence. Major superfamily level extinction occurred at the older of the said two sequence surfaces at the intra-Permian mega-order SB. Only one suborder, i.e. Goniatitina is present in Permian. Out of the nine superfamilies of the suborder Goniatitina, as many as six became extinct by the close of Middle Permian (Moore 1957), i.e. at the close of the preceding or prior to the start of the Late Permian–Early Paleogene mega-sequence (Fig. 6.1). Without going into details, it may be mentioned that the biggest intra-Triassic decline took place near the younger of the two pre-Jurassic sequence surfaces of the Anisian/Ladinian boundary age, i.e. at the first-order intra-Triassic MFS. Not just ammonoids, the mid-Triassic extinctions in between the end-Permian and end-Triassic mass extinctions affected the overall diversified life. Five ammonoid families became extinct near the first-order intra-Triassic MFS, while six other ammonoid families registered their origin near the same MFS (Fig. 6.1).

6.3 Origin/Extinction of Subfamilies at Second-Order Sequence Surfaces

Examples of familial origin/extinction linked to sequence surfaces from the start of the Jurassic interval on GTM:

6.3.1 *Early Jurassic Psiloceratidae and Others*

Psiloceratidae began at the second-order SB at the base of the Hettangian and possibly terminated at the second-order intra-Sinemurian MFS. Ectocentritidae and Schlotheimiidae also had similar origin and extinction or drastic reduction (Fig. 6.2). **Arietitinae** started at the second-order intra-Hettangian MFS and terminated at the end-Sinemurian second-order SB. **Oxynoticeratinae** had its origin at the second-order intra-Sinemurian MFS while it became extinct at the end-Sinemurian second-order SB. A few other familial/subfamilal origin and/or extinctions at second-order surfaces are depicted. The representation of the above families/subfamilies in India as also IEAP is extremely scanty in density, diversity

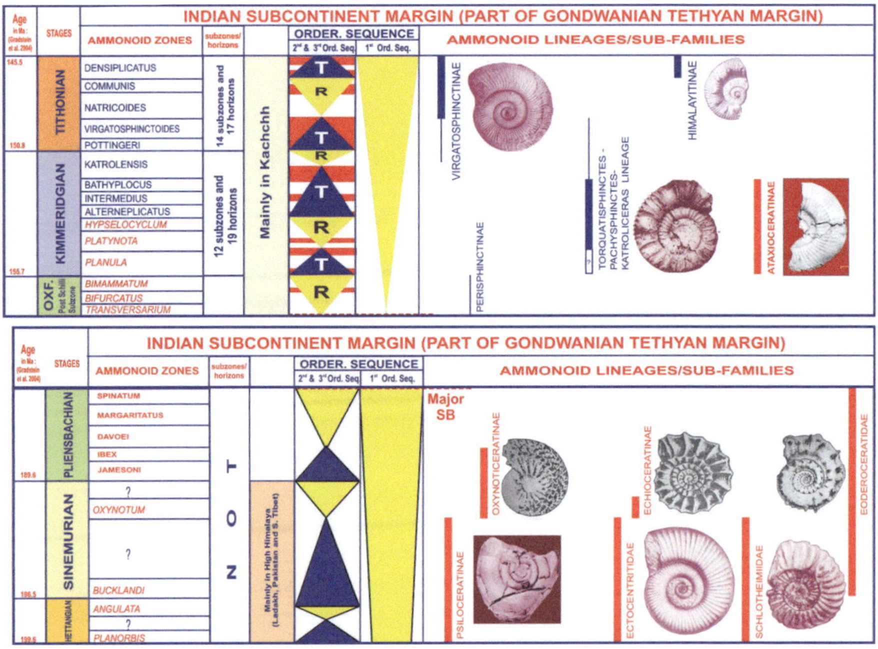

Fig. 6.2 Origin/extinction drastic reduction of ammonoid families/subfamilies vis-a-vis first- and second-order sequence surfaces

and frequency, and data is also poor, and so the data used is mostly from the treatise (Moore 1957). Further detailed treatment is neither desirable nor possible.

Biogeographic/climatic differentiation began near the first-order intra-Jurassic SB with **Bouleiceratinae** at the first-order basal Toarcian SB (Figs. 6.2 and 6.4). Bouliceratins became extinct during the second-order intra-Toarcian MFS in Pakistan, Madagascar and Arabia (Blaison 1967; Fatmi 1986; Fatmi et al. 1986). It is a short ranged (early to mid-Early Toarcian), and presumably latitudinally/climatically restricted subfamily with low latitude warm water distribution in Arabia, Spain, Portugal, East Africa, Madagascar and Pakistan during the early to mid-Early Toarcian. Its rich presence in the relatively deeper Axial belt and Indus basin localities of Pakistan and the absence in India is attributed to both relatively shallower bathymetry and not that low latitudes.

6.3.2 Middle Jurassic Macrocephalitinae and Others

The advent of the **Stephanoceratinae** on GTM took place at the second-order basal Bajocian SB, and had its extinction in the next second-order SB. Thus it ranged through a single second-order sequence. **The Stephanoceratidae** of the Early Bajocian–mid-Late Bathonian duration exhibits near pandemic distribution. Besides the ETM and both the East and West Pacific margins, the family is found to be widely present in Arabia, East Africa, Madagascar, Pakistan and NW Australia on the GTM. There is recently discovered even in Kachchh at Nara doubtful/ suggestive fragment of *Telermoceras* (Krishna unpublished report 2014; Pandey and Pathak 2015). The rich Early Bajocian ammonoid presence has been long known in Pakistan and Madagascar. The second-order TST on GTM is ranged from the base of Bajocian to the Early Bajocian Propinquans Zone.

6.3.2.1 Zigzagiceratinae

It started in basal Bathonian at a second-order MFS and extincted at the second-order intra-Bathonian SB at the close of the Bullatimorphus Subzone (Cariou and Hantzpergue 1997)

6.3.2.2 Macrocephalitinae Chrysoolithicus Lineage (late Middle Bathonian Mantataranus Zone–mid-Early Callovian Chrysoolithicus Zone)

The subfamily began in the late Middle Bathonian Mantataranus Zone in SW Pacific and later expanded to IEAP region as across the Neotethys to ETM. The Chrysoolithicus lineage is however restricted to IEAP region of GTM with best development in Kachchh. The Macrocephalitinae of the Bathonian age are found in Jhura, Nara and Jumara in Mainland and 'Island belt'. As expected, the maximum frequency, diversity and density is observed in the relatively distal and deeper Jumara. The oldest Macrocephalitinae so far known presumably are from Jumara and Nara. The origin of the GTM Macrocephalitinae chrysoolithicus lineage is suggested at the second-order SB between the GYFm above and the BCSm below, which can be easily traced throughout the 'island belt' and also in Jhura and Nara at or closely beneath the first Macrcocephalitinae record there. The lineage is found to terminate at the close of the Chrysoolithicus Zone Diadematus Subzone (Krishna and Ojha 1996) (Figs. 6.3, 6.4 and 6.10), between the sediment interval I and the sediment interval II, i.e. between the LSM and RSM.

The Macrocephalitinae Semilaevis lineage began at the second-order MFS in mid-Early Callovian (Krishna and Ojha 1996).

6.3.2.3 Eucycloceratinae

This started at a third-order SB above the Semilaevis Subzone in the terminal part of the Early Callovian and became extinct at the third-order MFS at the close of the Eucyclum Subzone in early Middle Callovian (Krishna and Ojha 1996).

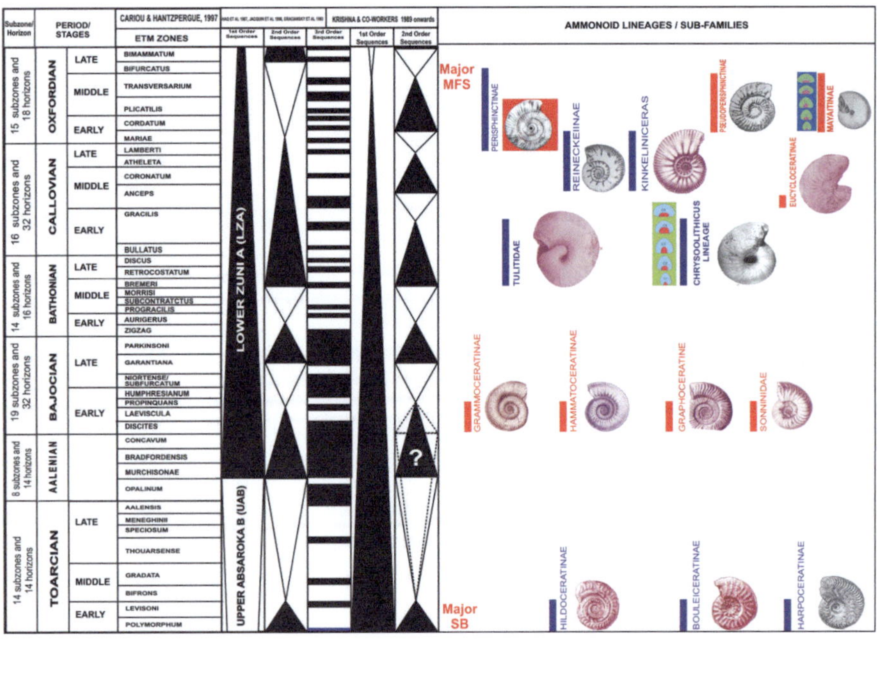

Fig. 6.3 Origin/extinction drastic reduction of ammonoid subfamilies vis-a-vis first- and second-order sequence surfaces

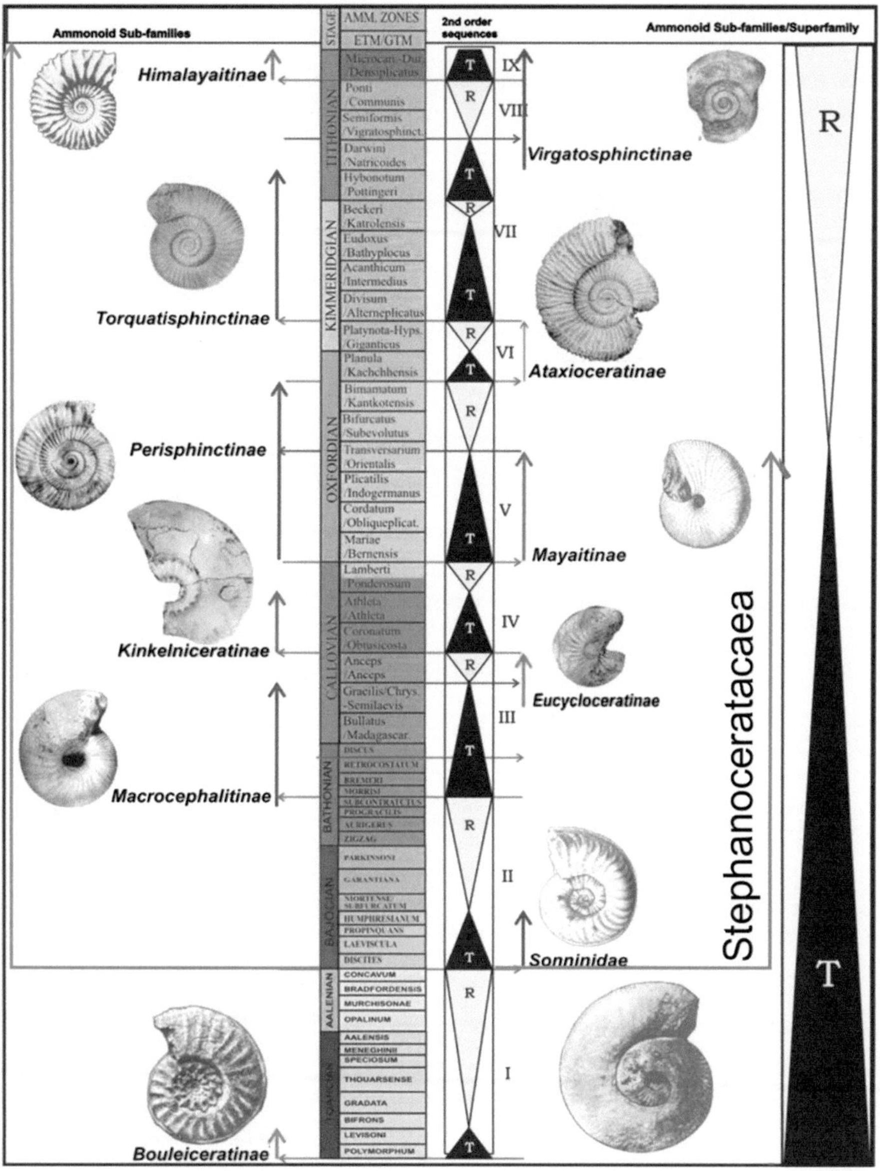

Fig. 6.4 Subfamilial origin/extinction in successive second-order sequences on GTM (ammonoid illustrations not to size)

6.3.2.4 Kinkeliniceratinae

This had its origin at the second-order SB in the Middle Callovian and near extinction at the second-order MFS in Late Callovian (Krishna and Ojha 1996).

6.3.2.5 Late Jurassic Mayaitinae and Others

Mayaitinae lineage began at a second-order SB above the Early Oxfordian unnamed zone and became extinct at the close of the first-order Schilli Subzone MFS in Middle Oxfordian (Krishna et al. 2000b, 2008).

6.3.2.6 The Ataxioceratinae

Originated at the second-order SB (Fig. 6.3) at the base of the Kachhensis Zone in Early Kimmeridgian and extincted at the second-order SB in Early Kimmeridgian itself.

6.3.3 Late Jurassic Katroliceratinae and Others

6.3.3.1 The Katroliceratinae

It appeared at the second-order SB at the base of the Alterneplicatus Zone (Krishna and Pathak 1991). The principal **Katroliceratinae** lineage registered drastic reduction at the close of the Linguiferus Subzone second-order MFS in Late Kimmeridgian

6.3.3.2 The Virgatosphinctinae

It started near the close of the second-order SB above the Virgatosphinctoides Zone, and extincted near the close of the late Late Ttithonian Opelli Subzone second-order MFS (Krishna et al. 1996b; Pandey and Krishna 2002) (Figs. 6.3 and 6.4)

6.3.4 Appearance/Disappearance of Superfamilies Linked to Bathymetry

In the successive second-order sequences, the change in bathymetry is found linked to the corresponding change in the dominant ammonoid lineage from one superfamily to another. For example, the Macrocephalitinae of the superfamily Stephanoceratacaea flourished during relatively deeper water duration of the late Middle Bathonian to early Middle Callovian. It was replaced by the relatively shallower water Kinkeliniceratinae of the superfamily Perisphinctaceae in the next younger sequence of decreased bathymetry (Figs. 6.1 and 6.3). Thus is unambiguously demonstrated that the origin and extinction of ammonoid lineages is genetically linked to the changing bathymetry in the region which in turn is governed by regional vertical tectonics. Moreover, the interpretation is further strengthened through appropriate change as well in the proportion of the associated secondary elements that expanded from other regions/subregions/provinces. From deeper to shallower bathymetry, there is decrease or absence of ammonoid elements that are common (pan-tethyan in context of the two tethyan margins—ETM and GTM) with other broad tectono-stratigraphic homogeneous regions like ETM and vice versa.

6.3.5 Globally Concordant Third-Order Sequences

The third-order TSTs are found age-constrained even at zonal to horizonal level (Fig. 6.3), thus unambiguously demonstrating either concordant or at the most slightly overlapping ages of the TST's in Kachchh on GTM with the Graciansky et al. (1993) and Haq et al. (1987) ETM frameworks. The concordance in turn strongly affirms the intercontinental/interrealmal to global applicability of the third-order framework as also their being explicitly driven by earth's orbital dynamics such as sea-level cyclicity. The demonstration and affirmation is particularly significant, since it is for the first time exampled from outside Euro-America. Several ammonoid devoid yet bivalve inclusive third-order older than late Middle Bathnoan TSTs have been differentiated and broadly included in the Toarcian–Bajocian interval with their suggestive ages. The suggestive ages of these third-order TSTs are based on the global to supra-regional to subrealmal age-concordance of the third-order MFSs.

6.3.6 Replacement of Lineages at the First/Second-Order Sequence Surfaces

The Kachchh Jurassic ammonoid succession has been nearly all through dominated and distinguished by the subfamilies/lineages geographically restricted to the entire GTM or substantive part of it along with the secondary/subordinate presence of the ETM elements (Fig. 6.5). The succession of the subfamilies/lineages is invariably without any gaps. Once a lineage becomes extinct, it is without any time lag replaced by a better adapted one. However, as an exception, there are small time

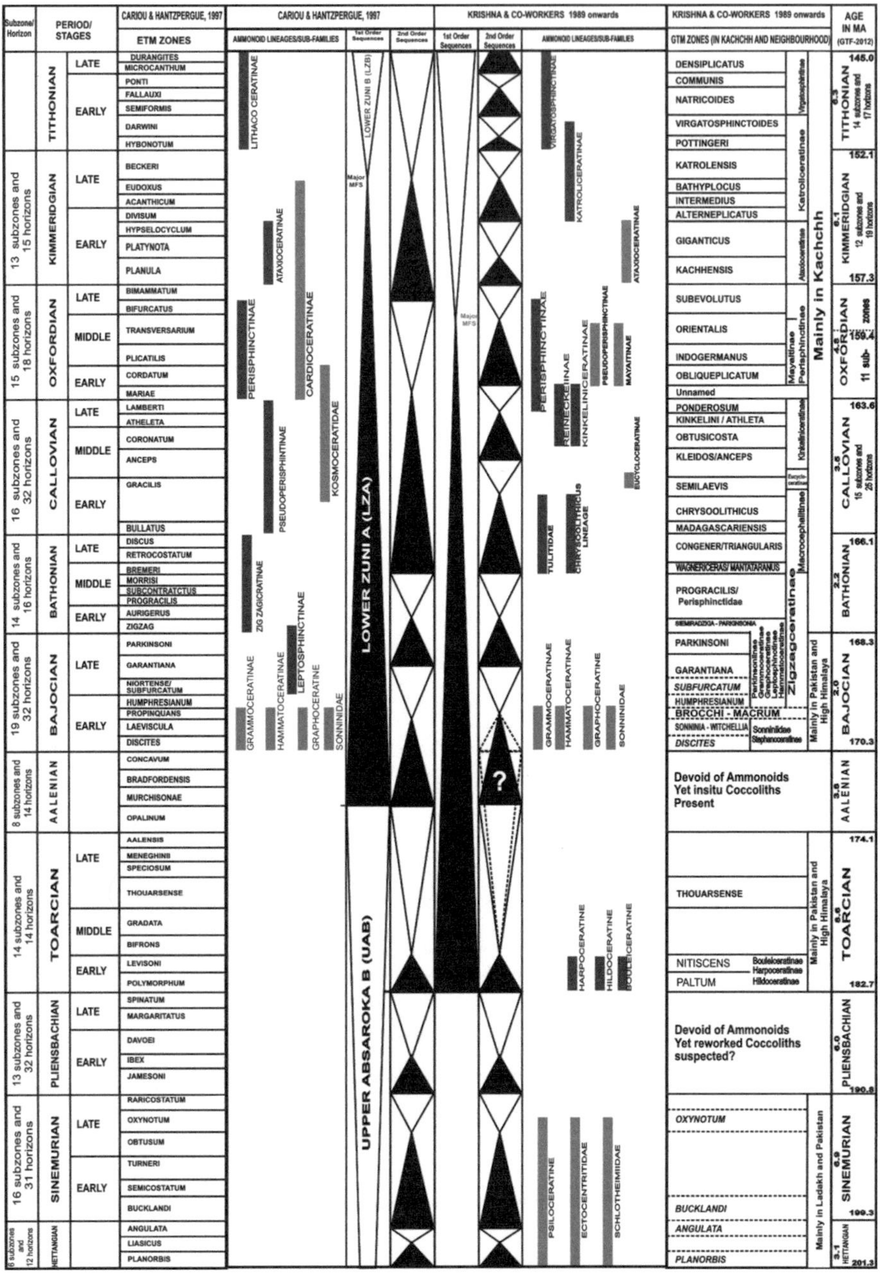

Fig. 6.5 Origin/extinction drastic reduction of ammonoid subfamilies vis-a-vis first- and second-order sequence surfaces on GTM and ETM (modified after Krishna 2005)

lags of one or two zones, e.g. there is replacement of the Kinkeliniceratinae by the Mayaitinae with some time lag near the Middle/Late Jurassic widespread climatic and other perturbations. The origin/extinction or drastic reduction of the lineages takes place at or close to the first- or second-order sequence surfaces.

6.3.7 Subfamilial Origin/Extinctions in the Subboreal Region

In the context of the above revelations, it is noteworthy that the first- and second-order sequence surfaces are manifestations of the tectonically influenced minimum or maximum bathymetry at any depositional site on the basin shelf. In consequence of the revelation of such important relationship between the ammonoid evolution and basin bathymetry in Kachchh, it became all the more desirable to scan the similarly high latitude constrained NW European record. Drastic depletion to complete termination of the **Leptosphinctinae** is confirmed at the basal Bathonian second-order MFS of Zigzag Zone, of the **Kosmoceratinae** at the close of late Late Callovian Lamberti Subzone immediately below the second-order MFS above the basal Oxfordian Scarburgense Subzone and of the **Cardioceratinae** at or near the close of the Late Kimmeridgian Eudoxus Zone in the Subboreal Britain at either first- or second-order MFSs or SBs (Fig. 6.5). Both the NW Europe Subboreal and Kachchh are examples of subtemperate to subtropical latitudinal setting for the greater part of the Jurassic with domination of the relatively low diversity ammonoid assemblages in comparison to high diversity picture in the Submediterranean European region of the ETM. As a consequence, the successive ammonoid faunas are found moderately to strongly skewed in favour of the geographically restricted families/subfamilies.

6.4 Genetic Link Between the Origin/Extinction of Ammonoid Taxa and Extensional Tectonics

The extensional tectonics in view of the resultant sudden changes in bathymetry, shelfal accommodation, nutrients, etc. that it induces, conceptually should also drastically affect the origin, evolution and extinction of bathymetry sensitive nektonic biotic lineages as ammonoids happen to be. In this context is examined the conduct of the ammonoid suborders/superfamilies/families/subfamilies vis-à-vis the first- and second-order sequence stratigraphic surfaces in the Neotethys, particularly on GTM.

6.4.1 Origin/Extinction of Multiple Subfamilies and Superfamilies Triggered by Rift-Tectonism and Volcanism at First/Second-Order Sequence Surfaces

In addition to the control exercised by extensional tectonics on origin, evolution, drastic reduction to near extinction of subfamilies, there is evidenced even the origin and extinction of the superfamilies as well linked to such tectonics.

Out of the six, as many as four superfamilies of the suborder **Goniatitina** became extinct near the start of the mega-sequence. Five families of the superfamily **Otoceratacea**e extincted near the terminal Anisian first-order intra-Triassic MFS, while six ammonoid families originated at the same MFS, two of the **Otoceratacea**e and four of the **Noritaceratacea**e. The ammonoid stratigraphic range chart of Krystyn et al. (2004) of the Muschelkalk Member of the Mikin Formation includes a total of 37 ammonoid genera, among which 25 ammonoid genera are present in the first-order intra-Triassic MFI between the beds 34–40 (Fig. 2.9).

The extinction of the superfamily **Eoderocerataceae** in Late Pliensbachian prior to the start of the first-order intra-Jurassic SB is quite noteworthy (Fig. 6.2). The radiation is particularly noted at specially eventful moments like the basal Toarcian of the origin of the **Hildocerataceae** (Fig. 6.3) and at the basal Bajocian of the origin of the **Stephanocerataceae**, **Perisphinctaceae** and **Haplocerataceae** (Fig. 6.2) so also of the extinction of the **Stephanocertaceae** above the late Middle Oxfordian on the GTM and at the late Late Eudoxus Zone / Beckeri Zone boundary on the ETM at their respective first-order MFSs.

The above exampled origin and extinction of ammonoid families and superfamilies unambiguously demonstrate the influence of the extensional tectonics on the bigger as also smaller order taxa from superfamily to genus (Figs. 6.1, 6.2, 6.3, 6.4, 6.5 and 6.6).

Majority of the **Stephanocerataceae** are interpreted to have dominated and maintained themselves around mid-shelf depths during the middle and terminal part of the first-order TST, while most **Perisphinctaceae** dominated during the early part of the first-order RST that fell in the later part of the Jurassic in lesser depths. The high energy above normal wave base depths did neither favour any ammonoids to habit nor get fossilized.

Even the superfamilies with secondary much less reduced presence in relative proportion compared to the dominant superfamily at any point of time, as also in density, diversity and frequency, too, followed similarly age-constrained pattern of origin, extinction and drastic reduction. The **Perisphinctaceae** which dominated the successive ammonoid faunas during the late Middle Oxfordian–Barremian first-order RST interval, became drastically reduced at the second-order intra-Valanginian MFS with complete extinction later at the Barremian/Aptian boundary first-order SB. The same happened to the **Haploceratacacea** as the

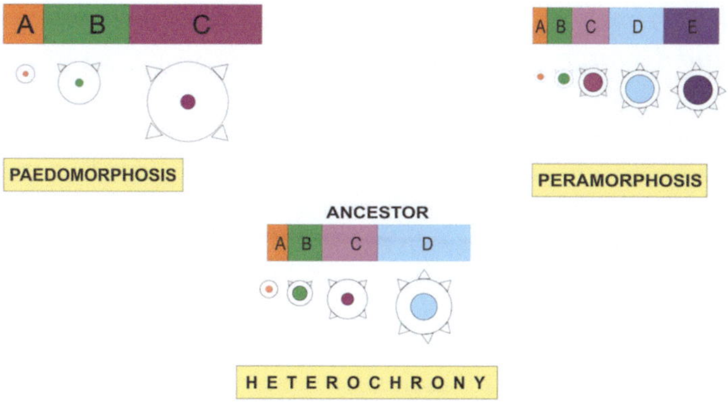

Fig. 6.6 Types of heterochronic processes (modified after McNamara 1986)

copatriat of Perisphinctaeae in the said interval. At the same intra-Valanginian second-order MFS, the **Desmoceratacea** filled in the niches occupied by the **Perisphinctacea**. The rapid diversification of the **Desmoceratacea** picked up at the start of the last Neothyen mega-sequence at the first order Barremian/Aptian SB and continued up to the extinction of ammonoids at the K/T boundary. The **Hoplitaceae** shared part of the shallow marine niches with the big brother **Desmoceratacea** and had its origin near the second-order intra-Valanginian MFS. It underwent similar rapid radiation near the first-order Barremian/Aptian boundary SB. The distinct but graded decline in density, diversity and frequency of the said three ammonoid families began at the mega-order early Middle Turonian MFS. The specialized **Ancycloceratacea** on GTM are largely restricted in the second-order Early Aptian–mid-Middle Albian TST with drastic decline there after.

6.4.2 Tilt of the Median High at the Second-Order Intra-Callovian MFS

The regional vertical tectonic event is interpreted to have tilted up the median high northward. The tilting up resulted in the uplift of Patcham and the emergence with final withdrawal of the sea by the RSM/GSM second-order SB, so also from the entire 'Island belt' except Khadir. Khadir presumably subsided as compliment to the uplift of Patcham being located next to it. Once emerged, Patcham supplied sediments during the next younger second-order sequence. This is attested by the reworked Toarcian–Aalenian nannoplanktons recovered from the Late Callovian sediments of the Lakhapur section by Rai (2006). Similar uplift of Jhura resulted in the distinct facies differentiation even between the neighbouring sites of Jhura, one site lying on the median high and the other site otherwise in proximity yet not on the median high.

6.4.3 *Rift Abortion Related Inversion of Bathymetry, and Change of Paleoslope*

The initiation of oceanisation on the Indo-Australian margin and associated reorganization of plates triggered rifting between India and Australo-Antarctica with tilting due east and southeast of the Indian Plate. This in turn caused tilting up in the west and northwest. The ongoing rifting aborted in Kachchh along with shallowing in Mainland and emergence of Khadir and Wagad. The resultant decrease in bathymetry caused the extinction of the relatively less shallow water Mayaitinae

6.5 Crude Suggestion of Relative Depths Based upon Spatio-Temporal Distribution of Ammonoid Sub/Superfamilies in the Basin

The development of a high-resolution scale based on evolutionary succession of ammonoids coupled with bathymetric distribution of ammonoid lineages/subfamilies/superfamilies allows estimation of relative bathymetry. Among the orders/suborders present are relatively deeper water Lytoceratina and Phylloceratina and relatively shallower water Ammonitina. There are two basic premises in this respect; (1) at the same site, the bathymetry increases from SB to MFS in the TST, and decreases between MFS and SB in the RST, and (2) in the same geological time, relative proportion of relatively deeper water preferring ammonoid taxa increases from margin to basin.

6.5.1 *Maximum Bathymetry Based on the Suborder Level Distribution*

It has been interpreted at the first-order MFS above the late Middle Oxfordian Schilli Subzone. The Schilli Subzone sediments exposed in Wagad at Kantkote, Khadir, Gangta bet, Desalpur and neighbourhood in ammonoid rich hard grounded sediments include noticeable proportion of the Lytoceratina–Phylloceratina. The maximum bathymetry of $\sim$50 to 60 m in these sediments has been estimated earlier (Krishna et al. 2000a, b). Somewhat lesser depths in the same stratigraphic sections are thought of in the immediately underlying late Early to early Middle Oxfordian up to the Parandieri Subzone and Lucaeformis Subzone in which the Lytoceratina–Phylloceratina presence is much less. The post-Schilli Subzone Jurassic ammonoid fauna in Wagad does not at all include even a single Lytoceratina–Phylloceratina and instead is exclusively made of the Ammonitina, that too with drastically reduced density, diversity and frequency. Thus is proven that the drastic reduction/extinction/change in the temporal distribution of the crucial suborders has taken place across the MFS as also from MFS to SB in the TST, so also is proven in turn the control of the same by extensional tectonics.

6.5.2 Depth Estimates Based on the Distribution of Superfamilies

Another parameter is the relative presence/dominance of the relatively less shallow water Stephanocerataceae to the shallow water preferring Perisphinctaceae within the suborder Ammonitina. There is not only relatively increased dominance of the Stephanocerataceae up to the first-order MFS but the superfamily also became extinct at the first-order MFS. The early part of the first-order RST up to the close of the Jurassic exclusively includes the Perisphinctaceae in Wagad. This is in tune with the interpretations of bathymetry made otherwise in the respective pre- and post-MFS intervals on the abiotic sedimentological, lithological and non-ammonoid biotic parameters. The Stephanocerataceae is present and dominant for over ~10 my upper part of the first-order TST except for a small interval of the ~1.5 my of the dominance of shallower water preferring Kinkeliniceratnae of Perisphinctaceae and complete absence of any Stephanoceratacea. Lesser depths in the said ~1.5 my interval than the intervals below and above are ascertained through the distinctly faster rate of sediment accumulation in the said interval compared to that in the underlying and overlying intervals. In the post-Schilli Subzone Jurassic interval of well over ~12.0 my, the complete absence of the Stephanoceratacea gets explained by the absence of enough depths for the superfamily. Thus again the extinction of the Stephanocerataceae and the associated increase in the relative proportion of the Perisphinctceae has occurred across the MFS, so also is ascertained the control of the superfamily level distribution of ammonoids by extensional tectonics. The given examples also apply amply well in other transects of the exposed Kachchh Jurassic.

6.5.3 Estimates Based Even at the Subclass Ammonoidea Level

The sheer absence or presence, and varying spatio-temporal distribution of the subclass Ammonoidea provide good clues to the relative depth estimates. At any site in the basin, be that most proximal exposed or most distal exposed or in between, the maximum density, diversity and frequency is found at the first-order MFS, and then at the second- and third-order MFSs. Another premise is the minimum depth of ~20 to 25 m, i.e. around the storm wave depths necessary for the presence of ammonoids.

6.5.3.1 'Island Belt'

The sedimentary columns in the exposed north margin of the basin in 'Island belt' are sand dominated, by and large unfossiliferous, but for thin fine textured invariably benthic fossil inclusive lime/marl/shale interludes with a total thickness of ~600 to 800 m spanning through ~22.0 my long Toarcian–late Middle Oxfordian interval. The first-order MFS sediments are present only in Desalpur (here included in the 'Island belt because of closeness with Khadir), and ammonoids are present in good density, diversity and frequency. Ammonoids are also present in the Kakindia limestone marker bed in Khadir and Gangta bet at or vertically close to a third-order MFS obviously in much less density, diversity and frequency. Yet lowered down, occasional ammonoids have also been collected at or in vertical proximity of the second-order MFS of early Late Callovian age in Gangta bet. In Patcham, ammonoids are best present in GYFm at the third-order MFS. Rare examples in RLm, and near second-order MFS in Early Callovian, yet rarer and solitary in the terminal part of BCSm at the third-order MFS of early Middle Bathonian, and recent solitary discovery of Early Bajocian *Calliphylloceras.* The cited examples of ammonoid distribution are all in thin intervals, and maximum bathymetry suggested is ~25–30 m except for the first-order MFS occurrence of somewhat greater depth of ~40 m. Elsewhere in 'Island belt', the depths range from the terrestrial coastal influence to ~15 m near shore environments.

6.5.3.2 Wagad

The sedimentary columns in the exposed east margin of the basin in Wagad are also quite thick of the order of ~600 m spanning through ~9.0 my long Callovian–Early Kimmeridgian, which are by and large ammonoid devoid but for thin intervals as in 'Island belt' at or vertically near the first- to third-order MFSs. Maximum density, diversity and frequency is present at the first-order MFS in Kantkote in the stream section discovered by the author and coworkers in 1994. Immediately below and above there is observed drastic depletion of density, diversity and frequency. The next best density, diversity and frequency of ammonoids are at the early Early Kimmeridgian second-order MFS above the first-order MFS in the Bharodia section. There are at least two levels with rare ammonoids in Washtawa near the second-order MFSs respectively in early Late Callovian and Early Callovian. The depth suggested at the first-order MFS in Kantkote is c 50–60 m in late Middle Oxfordian, at Bharodia ~30 to 35 m at the second-order MFS in Early Kimmeridgian, ~20 to 25 m in Early and early Late Callovian second-order MFSs in the Washtawa dome. However, by and large, the entire column of sediments was deposited between zero and normal wave base ~15 m depth.

6.5.3.3 East Mainland

In East Mainland localities proximal to KMF are Habo, Jhura and Jhuran, the sedimentary succession includes ULM, i.e. upper part of the Patcham Formation, the Chari Formaton, the Katrol Formation and the Umia Formation with a total thickness of ~1000 m. It includes ~25.3 my long interval from Early Bajocian to end Tithonian. Ammonoids are present in Em to Gm of ULM of late Middle to Late Bathonian second-order TST. In Callovian, ammonoids are present almost throughout, yet the significant presence is only in the second-order TSTs at Jhura and Habo. The late Early to early Middle Oxfordian interval of the Dhosa Oolite also includes appreciable presence of ammonoids. The Katrol Formation includes good presence of ammonoids in Late Kimmeridgian TST as the youngest ammonoid bearing interval in the East Mainland proximal to KMF. In East Mainland immediately south of the Katrol fault, ammonoids are additionally present in Tithonian as well in the second-order TSTs. Members Am to Dm without ammonoids suggest 10–20 m depth, Em to Gm indicate increase in bathymetry of 15–25 m. 15–35 m depth in early and mid-Early Callovian later part of second-order TST, mostly 5–15 m in late Early to early Middle Callovian, 15–30 m in late Middle to early Late Callovian, 0–20 m late Late Callovian to early Early Oxfordian, 55–65 m in late Early to basal Middle Oxfordian, 10–35 m in Kimmeridgian.

6.5.3.4 West Mainland

Significant presence of ammonoids is observed in the terminal Late Bajocian–basal Early Bathonian interval at Nara, and nearly continuous from the late Middle Bathonian to early Middle Oxfordian at Jara and Jumara. In Jara, ammonoids are also present in the late Early to Late Tithonian. In Nara there is scarce to moderate yet stratigraphically extreme significant presence of ammonoids in the late Late Bajocian–mid-Middle Bathonian which is perhaps the only ammonoid bearing interval of the said age in the entire IEAP between East Africa to NW Australia. The richest ammonoid presence is in DOM. There is graded increase in the ammonoid presence from the terminal Late Bajocian onwards up to DOM in each successive second-order TST up to Late Tithonian. Ammonoids are also present in almost all the RSTs. Obviously, the maximum depths realized in DOM may be 80–100 m while depths in the submarine gap may have reached even a maximum of ~120 m, i.e. deeper than mid-shelf in late Middle Oxfordian which is almost the double or more to that in Kantkote between the most proximal Kantkot and Jara during the submarine gap.

6.6 Heterochronic Evolutionary Framework of the Dominant IEAP Ammonoid Subfamiles

The broad ammonoid evolutionary heterochronic framework along with the ammonoid stratigraphic data is analyzed in the context of regional sequence stratigraphy. Salient heterochronic evolutionary features of the Macrocephalitinae (late Middle Bathonian Manatataranus Zone to mid-Early Chrysoolithicus Zone), Eucycloceratinae (late Early Callovian Semilaevis Zone to early Middle Ramosa Subzone), Kinkeliniceratinae (early Middle Paramorphum Zone to early Late to mid-Late Callovian Depressum Subzone to late Late Callovian), Mayaitinae (late Early Obliqueplicatum Zone to late Middle Oxfordian Schilli Subzone), Katroliceratinae (late Early Kimmeridgian Alterneplicatum Zone to mid-Early Tithonian Virgatosphinctoides Zone) and Virgatosphinctinae (Early Tithonian Natricoides Zone to late Late Tithonian Densiplicatus Zone) are considered.

6.6.1 Approach/Methodology to the Heterochronic Evolutionary Framework

According to Gould (1977) and McNamara (1986), the heterochrony refers to changes in relative time of appearance and rate of development of characters already present in ancestors in either direction forward or backward. It includes two main processes Peramorphosis and Paedomorphosis (Fig. 6.7). Peramorphosis is characterized by the introduction of new characters at the terminus in the descendent taxon along with centripetal extension of existing features. It is in turn divided into three types: (a) Acceleration (reduction in size), (b) Predisplacement (without change in size) and (c) Hypermorphosis (increase in size). Paedomorphosis is featured by the removal of the existing characters from the terminal most part along with centrifugal extension of the remaining ontogenic stages from the ancestral to the descendent taxon. It again includes three types; (a) Neoteny (increase in size), (b) Post-displacement (without change in size) and (c) Progenesis (reduction in size).

The individual ammonoid spiral shells are uniquely favoured for heterochronic evolutionary studies since the spirals include all the successive ontogenic stages. The significance and application of heterochrony to evolution, particularly in ammonoids, has been dealt recently in several works (Alberch et al. 1979; House 1985; Dommergues et al. 1986; Dommergues 1987; Krishna and Cariou 1990, 1993; Krishna and Pathak 1991, 1993; Krishna et al. 1996a, b; Pandey unpublished 1993; Pandey et al. 2010).

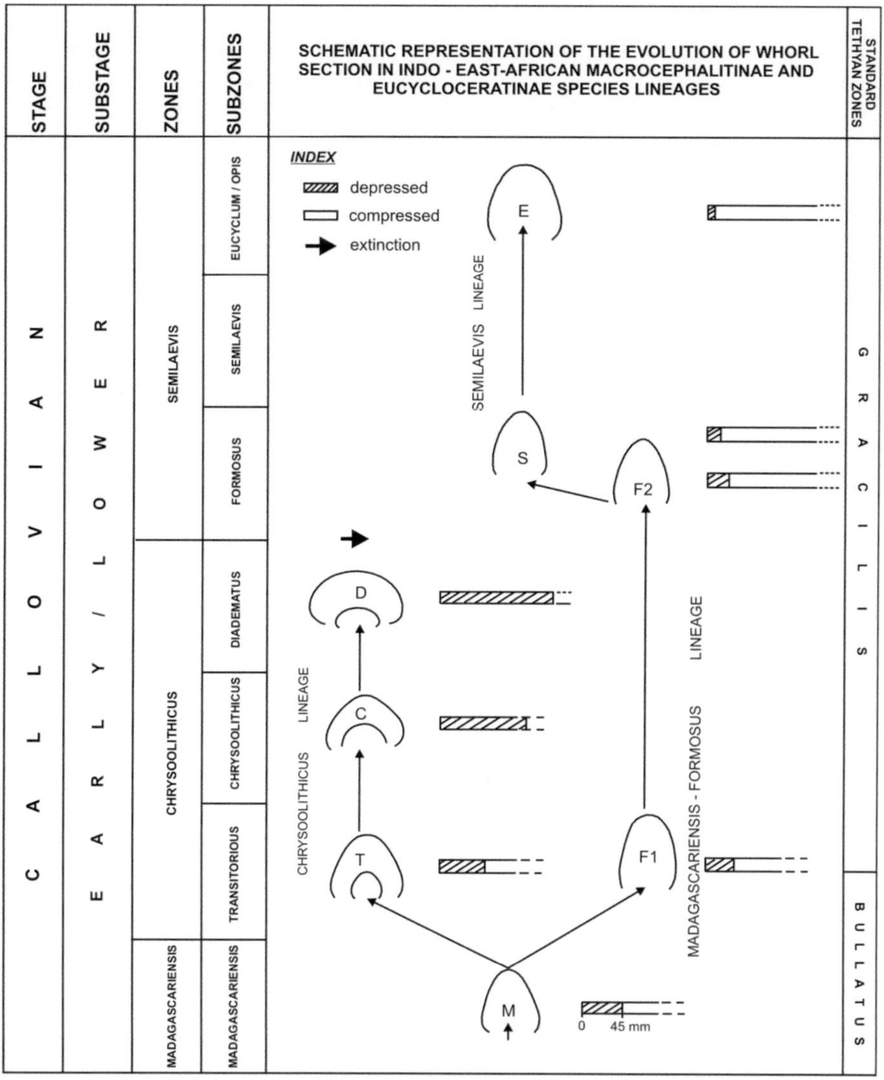

Fig. 6.7 Schematic representation of the evolution of whorl section in Indo-East African Macrocephalitinae and Eucycloceratinae species lineages

6.6.2 *Macrocephalitinae and Its Chrysoolithicus Lineage*

Among the GTM areas, the record of Macrocephalitinae is the best in Kachchh. The successive assemblages are relatively homogeneous, whether monospecific compressed or inflated. The best understood lineage consists of a succession of six species. The lineage is known as Chrysoolithicus lineage after the important species —*M. chrysoolithicus* (Waagen). The lineage ranges from the start of early Late Bathonian Mantataranus Zone to the close of mid-Early Callovian chrysoolithicus Zone (Fig. 6.7). The lineage up to Madagascariensis Zone dominantly comprises fine-ribbed compressed species *M. mantataranus, M. triangularis* and *M. madagascariensis*. During the remainder of the lineage span, it ranged up to Diadematus Subzone of the Chrysoolithicus Zone through inflated species. The heterochronic evolution of the lineage is understood through increase in size, prolongation of the otherwise short depressed stage by neoteny. The evolution is considered unidirectional as a paedomorphocline (Krishna and Caroiu 1993; Ojha 1996, and others).

6.6.2.1 Morphological Characteristics of the Macrocephalitinae Paedomorphocline

There is centrifugal unidirectional progressive evolution of the morphological features in the successive species of the lineage at the same growth stage (Fig. 6.7). The older species of the lineage have either compressed (higher than wide) or less depressed whorl section. The whorl section in successively younger species at the same size is found either (wider than high) depressed or less compressed than the immediately older species of the lineage. In the species evolution, the progressive change is such that the rate of increase of whorl width during ontogenic growth is more compared to that of whorl height, resulting into progressively inflated shell.

6.6.3 Evolutionary Inversion of Macrocephalitinae at the 2nd Order MFS

At the MFS when the progressive increase in bathymetry stopped and reversed to unidirectional decrease, the youngest species of the lineage which had become extra-specially adapted to unidirectional graded increase in bathymetry failed to adapt to sudden reversal and became extinct. Thus origin, evolution and extinction of the lineage is found directly and genetically linked to bathymetric changes and to first- and second-order sequence stratigraphic surfaces that in turn are governed by intrabasinal to interregional tensional tectonics. The heterochronic morphological evolution of the ammonoid lineages unambiguously demonstrates its genetic link to regional tectonics. As a corollary, the sudden changes in morphological features of ammonoids provide excellent clues to precision dating of sequence surfaces to resolution of c 200–400 ky, and in turn to such precise dating of tectonic events that could not be even thought of using any other tool.

6.6.3.1 Morphofunctional Features Linked to Bathymetry

In the lineage range, the earlier three compressed species in context of morphofunctional studies are best adapted to shallow shelf while the inflated species in the remainder of the lineage range are found more at home with relatively less shallow shelf. Thus, there has been a gradual unidirectional increase of bathymetry from the start to close of the lineage range. Interestingly, in context of the developed second-order sequence framework, the lineage range has been interpreted as a phase of unidirectional steady increase in bathymetry during a second-order TST. The lithological framework during the said interval also lends support to such an interpretation. It is also noticed that the compressed species in the different realms are relatively widely distributed in context of the episodal faunal expansion events. The widespread distribution of the compressed species in turn suggests their relative greater ability for dispersal by way of enlargement of shallow platforms during the phases of eustatic rise.

It is interpreted that the lineage from its origin at the basal SB to MFS of the second-order sequence became progressively inflated by environmental response and morphological adaptation to progressive increase in basin bathymetry. There is as well increase in size of the successive species in the lineage.

6.6.4 *Eucycloceratinae*

The ammonoid subfamily Eucycloceratinae is known to range from the late Early Callovian Semilaevis Zone to the early Middle Callovian Ramosa Subzone of Anceps Zone. The earliest species of the subfamily (*Ammonites opis* Sowerby) was described from Kachchh in 1840. Eucycloceratins differ from their ancestors (macrocephalitins) in their protracted simplified septal sutures and the particularly coarse ribbed regressing body chamber (Ojha 1996). Eucycloceratinae derived from Macrocephalitinae by considering *Macrocephalites semilaevis* directly connected to *M. formossus* (Sowerby) on the lower side and to *E. eucyclum* on the upper side. Ojha (unpublished thesis 1996) proposed the eucycloceratin macroconch lineage M. formosus–E. eucyclum–S. ramosa. There is unidirectional increase in size of the umbilicus, spacing and coarseness of ribs in the evolution of the lineage. The heterochronic evolutionary process is interpreted as peramorphosis in a peramorphocline.

6.6.5 *Kinkeliniceratinae*

It was created by Krishna and Ojha (1996) and Ojha (unpublished 1996) for the IEAP 'proplanulitins' through differentiation and separation from the ETM Proplanulitinae for reasons of geographical segregation, mutual stratigraphic exclusion and evolutionary differentiation. The apparent morphological similarities were considered on account of evolution under similar environments. The phylogenetic derivation of Kinkeliniceratinae is yet not clear, although the same is imperative from some perisphinctid ancestor during the Late Bathonian–Early Callovian, e.g. *Indosphinctes*. The principal Kinkeliniceratinae lineage of *K. kleidos–K. paramorphum–K. kinkelini–K. catillus* exhibits unidirectional decrease in umbilicus, increase in whorl section, increase in ovateness and decrease in size. The lineage ranges from mid-Middle Callovian Paramorphum Subzone to mid-Late Callovian Depressum Subzone.

6.6.6 Mayaitinae

The late Early to late Middle Oxfordian subfamily Mayaitinae is considered homeomorphic to late Middle Bathonian–Early Callovian Macrocephalitinae in view of the stratigraphic exclusion of early Middle Callovian Kleidos Subzone to early Early Oxfordian unnamed zone between the LAD of Macrocephalitinae and FAD of Mayaitinae. The sudden rapid rise in IEAP bathymetry is interpreted on account of cumulative effect of the extensional downfaulting along with eustatic rise. The bathymetry in IEAP is even much deeper for Mayaitinae in Oxfordian than for Macrocephalitinae in Bathonian–Callovian. The speculation is prompted by the revised age assignment of the Middle to Late Bathonian Macrocephalitinae in SW Pacific earlier considered as Early Callovian on account of the absence of any stratigraphically precise SW Pacific material coupled with long held opinion on ETM of the origin of Macrocephalitinae in basal Callovian. It may be that like Macrocephalitinae, Mayaitinae too may have longer total range to make up for the above-mentioned stratigraphic exclusion. The problem can be approached to directly obtain guide/marker dinocysts or coccoliths from the complete morphological range of mayaitins from Iriyan Jaya and Papua–New Guinea.

The heterochronic evolutionary framework of monogeneric subfamily Mayaitinae has not been worked out in the absence of the establishment of the succession of species in the Mayatinae lineage. In general, in the view of the subfamily's range restricted to a second-order TST as that of morphologically nearly identical Macrocephalitinae, it is anticipated that the heterochronic evolution of the Mayaitinae species lineage would have been a paedomorphocline. In this paedomorphocline is anticipated increase of size, increase of inflation of whorl section from inner to outer whorls in the macroconch succession of species. The start of post-Schilli Zone RST, marked the replacement of Mayaitinae by Grayiceratinae on the similar pattern of replacement of Macrocephalitinae by Eucycloceratinae.

6.6.7 *Katroliceratinae*

Katroliceratinae is known to range from the late Early Kimmeridgian to late Late Tithonian in IEAP. Currently, this subfamily is restricted to include the genera *Torquatisphinctes, Pachysphinctes, Katroliceras, Indodichotomoceras and Aulacosphinctoides* (Krishna and Pathak 1991, 1993; Pandey unpublished 1993; Pandey et al. 2010). The comprehension of the phyletic transformation is also strengthened under heterochronic framework. The morphologically close genera *Torquatisphinctes, Pachysphinctes and Katroliceras* have been precisely differentiated from one another in terms of heterochronic framework as observed in their respective genotypes (Fig. 6.8). The generic differentiation is mostly based on the ontogenic positioning of their respective specialized rib sculpture from later part of phragmocone to the body chamber (Krishna and Pathak 1993).

6.6.8 *Virgatosphinctinae*

The subfamily Virgatosphinctinae is here exclusively retained for the genus *Virgatosphinctes*. The entire morphological range of *Virgatosphinctes* whorl

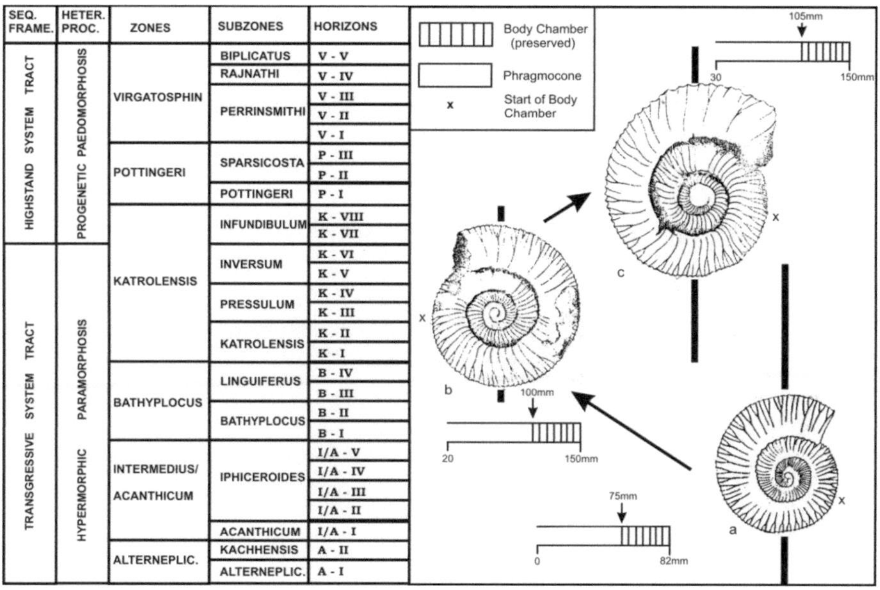

Fig. 6.8 Schematic illustration of the evolution of Kachchh Katroliceratinae microconch lineage Torquatisphinctes–Pachysphinctes–Katroliceras; *a* to *c* respectively indicate the successive species Torquatisphinctes intermedius Spath m, Pachysphinctes linguiferus Spath m and Katroliceras katrolensis (Waagen) m after Pandey et al. (2010)

section is retained instead of creating a separate genus *Malagasites* for ones having compressed whorl section as done by Enay (2009). Similar examples can be cited of Macrocephalitinae, Mayaitnae where the entire range of whorl section is included under the same genus. The author here firmly considers the genus and subfamily as exclusively GTM, primarily even of IEAP, which could only occasionally spread to east and west only on GTM during a phase of eustatic rise in late Late Tithonian. The occasional mention of microconch *V. transitorius* Oppel on ETM is attributed to mere homeomorphic resemblance of microconchs. Such resemblance of microconchs in altogether phylogenetically unconnected lineages that undergo parallel evolution on two margins of the Neotethys is ascribed to the availability of extremely limited range of whorl section and rib sculpture.

6.6.9 Examples of Paedomorphic and Peramorphic Heterochronic Evolution

The most observed heterochronic evolutionary process is peramorphosis in both microconch and macroconch lineages. It is characterized by the addition of new characters at the terminus in the descendent taxon along with centripetal extension of existing features and distinct increase in size. This trend of evolution is found to range for a substantive interval of the total range from A-I Horizon of Alterneplicatus Zone to K-VI Horizon of Katrolensis Zone (~3.4 my), and later again from V-V Horizon of Virgatosphinctoides Zone to D-III Horizon of Densiplicatus Zone (~3.0 my) while paedomorphosis is observed relatively for much less distinctly shorter interval from K-VII of Katrolensis Zone to V-IV Virgatosphinctoides Zone (~2.5 my) (Fig. 6.8).

6.6.9.1 Example of the Intra-generic Heterochrony

Torquqtisphinctes intermedius–T. torquatus–T. bathyplocus microconch lineage (Fig. 6.9): *T. intermedius* Spath m appear in A-I Horizon of Alterneplicatus Zone and is found to range until K-I Horizon of Katrolensis Zone. *T. intermedius* Spath m is known through incomplete specimens only in the collections, however, the holotype of Spath (Spath 1927–33) includes body chamber. Next in the succession appears *T. torquatus* (Sowerby) m in I/a-IV Horizon of Intermedius/Acanthicum Zone and ranges up to K-III Horizon of Katrolensis Zone. The evolution of *T. Torquatus* appears through the extension of the existing rib sculpture by way of acceleration (peramorphosis). *T. bathyplocus* evolved next through distinct increase in size and centripetal extension of the existing rib sculpture by way of hypermorphic peramorphosis (Krishna and Pathak 1993; Pandey et al. 2010).

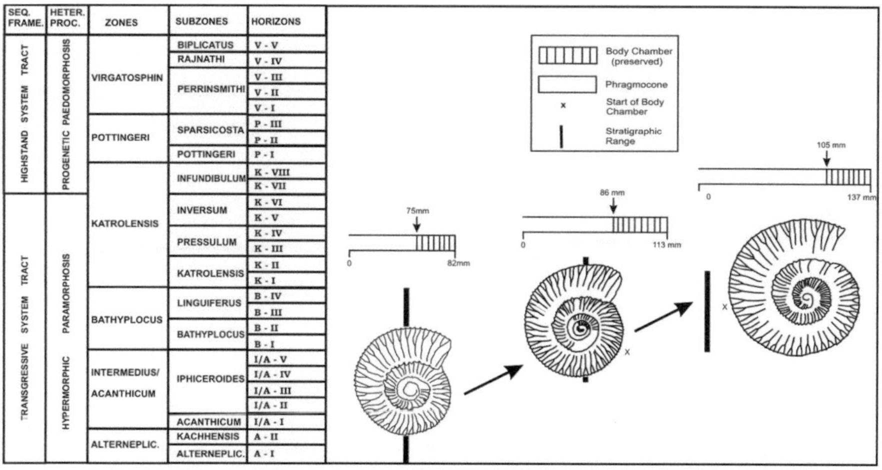

Fig. 6.9 Infraspecific heterochrony in Torquatisphinctes bathyplocus (Waagen) after Pandey et al. (2010)

6.6.9.2 Example of the Microevolutionary Intra-species Heterochrony

In specially favoured situations (Fig. 6.10), there are even unidirectional infraspecific microevolutionary heterochronic modifications within the species range. It is the first such effort of this kind on the GTM Jurassic ammonoid succession. This allows the resolution of the Bathyplocus Zone into four (B-I to B-IV) and Katrolensis Zone into eight (K-I to K-VIII) horizons (the older four horizons K-I to K-IV based on successive transient morphs within the range of *K. katrolensis* while the younger four horizons K-V to K-VIII based on the first appearances of successive younger species *I. inversum, A.infundibulum, Hybonoticeras ornatum* and last appearance of *K. katrolensis* (Waagen) m (Krishna and Pathak 1993; Pandey et al. 2010). By way of examples, the case history of infraspecific heterochrony in *Torquatisphinctes bathyplocus* (Waagen) m is discussed (Fig. 6.9). The evolution is featured through the gradual centripetal extension of the first triplicate rib from near the close of the body chamber to near the start of the body chamber through intermediate positions in the successively morphs becoming young through the species range. This in turn allows unambiguous yet easy differentiation of the Bathyplocus Zone interval into four smaller intervals (horizons) characterized by the successive evolutionary transient morphs of the species as morph I to morph IV in ascending order. These morphs (morph I to morph IV) of *T. bathyplocus* (Waagen) m are found to appear respectively at the base of B-I Horizon with first triplicate rib in the last quarter, at the base of B-II Horizon with first triplicate rib in the end of the second quarter, at the base of B-III Horizon with first triplicate rib in the beginning of the second quarter and at the base of B-IV Horizon of Bathyplocus

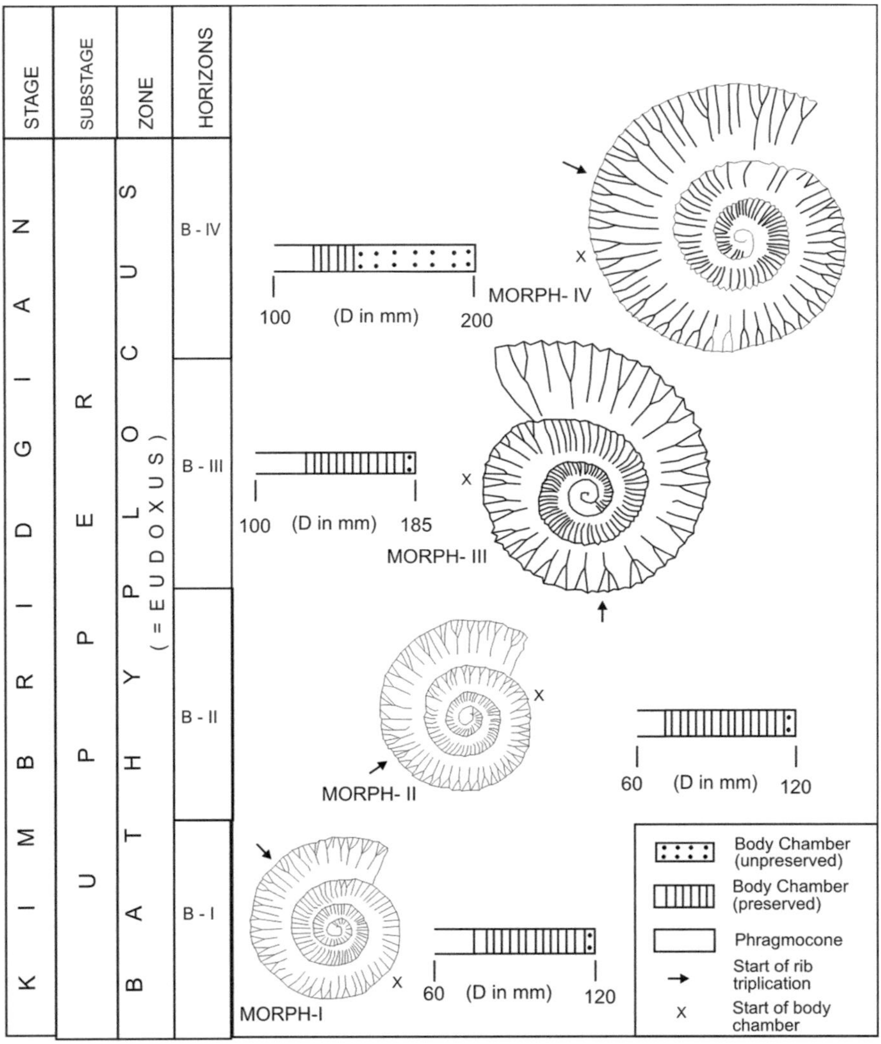

Fig. 6.10 Infrageneric heterochrony in Torquatisphinctes (Spath) after Pandey et al. (2010)

Zone with first triplicate rib in the first quarter of the body chamber. The differentiation of Bathyplocus Zone into four and Katrolensis Zone into eight smaller intervals results in a chronological resolution, respectively of the order of ~200,000 years and ~100,000 years per horizon. It compares well with the highest pre-quaternary resolutions ever achieved in any part of the world.

6.6.10 Suggestive Genetic Link Between Heterochronic Evolutionary Changes and Sequence Stratigraphic Framework (Fig. 6.5)

The studied ~56.4 my long first-order sequence in this work includes major part of the Jurassic from near basal Toarcian of ~37.7 my and early to middle part of the Early Cretaceous up to close of Barremian of ~18.7 my. The heterochronic evolutionary framework has been studied in bits only during the ~20.3 my long late Middle Bathonian–end-Tithonian interval. Within this heterochronically studied part, the ~7 my late Middle Bathonian–late Middle Oxfordian duration represents the terminal one-third part of the first-order TST, while the superjacent ~14 my long terminal Middle Oxfordian–end-Tithonian interval constitutes less than half of the first-order RST as its early part. It is found that the paedomorphic evolutionary framework has prevailed from the late Middle Bathonian to the late Middle Oxfordian Schilli Subzone, i.e. during the ~7 my long later half of the first-order TST up to MFS but for short interlude of peramorphism. In terms of subzones, the two large phases of paedomorphism total to 19 subzones while that of the short peramorphic interlude is only 6 subzone long. Immediately after the MFS up to the late Early Kimmeridgian, heterochronic framework could not be studied in the absence of any exclusive IEAP subfamily/ies. In the following middle part of the first-order RST from late Early Kimmeridgian to late Late Tithonian, the dominant and distinctly longer ranging evolutionary framework is peramorphism with intervention in between of a relatively short phase of paedomorphism. In terms of chrostratigraphic horizons, the duration of two large peramorphic intervals totals 27 while the short paedomorphic phase is only 10 horizon long. As such, only two and that too tentative yet cardinal interpretations are made here. The first is possible suggestive relationship, respectively, of the first/second-order TSTs with paedomorphism and of RSTs to peramorphism as observed in Kachchh. The other one is the heterochronic evolutionary inversion and/or cut at significant MFSs. Further the short interludes within the principal heterochronic patterns are here suggested to be related to second-order RSTs within the first-order TST or second-order TSTs within the first-order RSTs.

The heterochronic evolutionary change from the peramorphocline takes place at the start of the MFI (Maximum Flooding Interval) in B-III Horizon. It is interpreted that the MFI maintains stable maximum bathymetry from B-III Horizon to K-VI Horizon, while the precise MFS (Maximum Flooding Surface) is near the base of the K-II Horizon. The part of the MFI below the MFS registers extremely slow increase in bathymetry, so also extremely slow decrease of bathymetry in the part of MFI above the MFS.

The second-order sequence stratigraphic framework suggests possible genetic links between the two phenomena of ammonoid heterochronic evolutionary and sequence stratigraphic surfaces (Krishna et al. 2000a, b; Krishna 2005). In view of the possible governance of the second-order sequence stratigraphic framework by regional extensional tectonics, the major heterochronic evolutionary changes appear to be heavily influenced by regional extensional tectonics. Within the paedomorphic interval (corresponding to RST), there is also observed drastic reduction of ammonoid density, diversity and frequency; coarsening of texture, increase in sediment accumulation rate, decrease of bathymetry and increase of energy framework. However, similar genetic links require to be investigated in other intervals of the Jurassic on GTM (Gondwanian Tethyan Margin) and other parts of the world. Heterochronic evolutionary studies not only facilitate high resolution refinement of time, but also provide interesting leads to governance of sedimentation pattern, sequence surfaces and biological evolution.

6.7 Ammonoid Distribution, Provincialism and Migration to and Fro IEAP

The qualitative biogeographic exercise of comparison of the ammonoid faunas of IEAP with neigbouring provinces/regions is summarized during the successive second-order sequence intervals or their multiples from the early Early Toarcian to the latest Tithonian ~22.2 my long duration. The views held earlier are reinforced in a second-order sequence stratigraphic context without elaborate discussion that is already detailed (Krishna 1983a, 1987; Krishna and Pathak 1994). In addition of the GTM and ETM as super-biogeographic units, five paleogeographic units are used in the exercise; Submediterranean and Mediterranean Europe as (A) Arabia inclusive of North Africa as (B), East Africa, Madagascar, Pakistan, India and the High Himalaya as (C), SW Pacific as (D), SE Pacific as (E). These include regions in the neighbourhood of GTM west and East of IEAP as also regions of ETM.

The earlier ammonoid geographic studies (Krishna 1983a, 1987; Krishna and Pathak 1994) included High Himalaya, West India, West Pakistan, Madagascar and East Africa into Indo-East African ammonoid faunal province (IEAP) region in the Middle and Late Jurassic. It was characterized and dominated by a low diversity geographically restricted indigenous ammonoid rich fauna during the successive substage intervals. For short intervals, its characteristic ammonoid elements also expanded to high Himalaya, Sula and New Guinea. At other short intervals, extraneous ammonoid elements also spread to the Indo-East-African faunal province. The ETM elements were always present as minor/secondary constituents, which facilitated precise correlation of most of the Kachchh zones assemblages with the ETM scale. At the time of proposing the IEAP, neither did exist there any semblance of ammonoid zonal scale nor even any thought of developing a comprehensive first- to third-order sequence stratigraphic framework in Kachchh on GTM. In the present biogeographic review the comprehensive ammonoid zonal scale and sequence stratigraphic formulations with focus on the Kachchh basin have been made full use of. The IEAP during successive 2nd order sequence intervals is found dominated by geographically restricted ammonoid subfamilies as revealed in Kachchh. It is observed that the similarity of the IEAP faunas with others in comparison is maximum at and near the MFSs and minimum at the SBs since interregional communication is maximum and minimum respectively at MFSs and SBs. However, at the moment, there is no way to undertake such comparison at the MFSs or SBs, since such data does not exist in areas other than IEAP.

6.7.1 Salient Ammonoid Distribution Features During the Toarcian–Barremian 1st Order Sequence Interval with Focus on 2nd Order Sequences

Bouleiceratinae is considered of IEAP origin although without any definitive evidence. During the Toarcian–Aalenian 2nd order sequence interval at the very start, the geographically restricted Bouleicertinae fauna is found to have spread from IEAP to Arabia and Mediterranean. The said spread suggests some kind of low latitude warm climatic control on the distribution of *Bouleiceras*. At the next second-order MFS in late Early Bajocian Propinquans Zone Macrum Subzone, the ammonoid faunas on GTM and ETM have complete identity, which suggests improved communication under increased bathymetry. There is however, slight diachrony of MFSs, above the Laeviuscula Zone on ETM and Propinquans Zone Macrum Subzone on GTM. At the next younger second-order MFS in Early Bathonian Zigzag Zone Macrescens Subzone, there is again complete generic similarity on both ETM and ETM which suggests graded increase of bathymetry in Kachchh, Pakistan and High Himalaya through each successive MFS. In the Mediterranean and Submediterranean provinces of ETM, the bathymetric scenario is not much different. GTM extensional tectonics and related submergence seem to have contributed to this framework, however, only through enhancing the rate of increase of bathymetry compared to ETM. Except for the Toarcian–Aalenian interval, the IEAP ammonoid faunas are expansions from the ETM Mediterranean and Submediterranean faunal provinces until late Middle Bathonian. The ammonoid provincialism remained very explicit, yet the GTM/ETM zone by zone correlation did not suffer. It instead even became stronger until the intra-Jurassic first-order MFS in the late Middle Oxfordian. In this duration, bathymetry registered rapid increase due to the combined effect of eustatic rise and rift related basinal subsidence on GTM. At the late Middle Oxfordian MFS, in spite of ammonoid faunal provincialism, the over all similarity of the ammonoid faunas between GTM and ETM was maximum of the Jurassic.

Immediately after the first-order MFS, correlation between GTM and ETM became increasingly difficult possibly due to sudden drastic decrease of ammonoid density, diversity and frequency, particularly during the successive second-order

RSTs. In the two successive second-order Kimmeridgian sequences, the *Paraboliceras–Kossmatia* fauna from SW Pacific spread only to deeper High Himalaya of IEAP and not any farther due west to Kachchh. This was most possibly on account of difference of bathymetry, perhaps not sufficient in the shallower Kachchh for the said relatively deeper water fauna. The TSTs also continued to decline in ammonoid diversity, density and frequency, particularly, in the proximally exposed parts of the IEAP basins. The GTM/ETM correlations soon reached their rock bottom in the early Early Tithonian second-order sequence when indigenous IEAP acquired 100 % provincialism with nothing in common between GTM and ETM. The next two sequences again seem to have developed moderately good ammonoid faunas, only, in the relatively distal deeper sites in the Berriasian–Valinginian duration. The said deeper sites until then were lying deep under starved conditions, and only resumed sedimentation in late Early Tithonian Natricoides Zone after having remained starved of sediments for a long interval since Late Bajocian, in any case since mid-Middle Oxfordian. This is the reason that the GTM and ETM faunas hardly exhibit any provincialism in the late Early Tithonian–Valinginian duration. The GTM faunas are best expressed in this interval in Pakistan and High Himalaya which constituted the relatively distal deeper parts of the composite Indus shelf and the neighbouring High Himalaya shelf. The said locations of the Indian margin had remained starved during the long submarine gap without any geological record until mid-Early Tithonian. The Kachchh, being relatively proximal compared to the Axial belt of Pakistan and High Himalaya, had already started receiving sediments in East Mainland (Ler, Jhuran, Walakhwas localities) on account of the rapid decrease of bathymetry and the associated fast receding shore line. The ammonoid faunas, however, until the early Early Tithonian, in view of shallower depths in East Mainland, remained indigenous and geographically restricted within the IEAP.

6.7.1.1 Relatively Deeper Water Macrocephalitinae and Mayaitinae

The subfamily Macrocephalitinae did expand to both the Tethyan margins but its Kachchh Chrysoolithicus lineage could not expand out of IEAP. Macrocephalitins at the time of entrance in IAEP were relatively smaller in size with highly compressed to triangulate whorl section. These early forms of Macrocephalitinae for expansion to ETM around Mid/Late Bathonian transition, needed only marginal improvement in bathymetry. The said Kachchh lineage grew unidirectionally with increase in size and inflation with graded increase in bathymetry. The ancestral small compressed Macrocephalitinae species of the like of *Macrocephalites mantataranus* managed to rapidly spread widely throughout the Neotethys, even to Boreal England. Later in deeper Early Callovian waters, there developed independent lineages in different regions, e.g. GTM and ETM. The descendant Early Callovian species adapted to deeper waters could not intermigrate across the equator because of the possible intervening shallow water bottlenecks along the coastal regions between GTM and ETM. The question arises as to why there is provincialism within the duration of the first-order TST of the relatively deeper water preferring Macrocephalitinae, and Mayaitinae even close to the first- and second-order MFSs between the Early Callovian and the Middle Oxfordian, so also even in the early part of the RST of the relatively shallower water preferring perisphinctid Katroliceratinae. The probable answer is that the provincialism in the relatively deeper water TST may have been possibly on account of the intervening shallow water depth barriers during the relatively shallower water RST intervals due to temperature-cum-depth barrier. Around the first-order late Middle Oxfordian MFS in the terminal second-order TST, the much deeper water Early and Middle Oxfordian subfamily Mayaitinae, unlike its possible ancestors (Late Bathonian to Early Callovian Macrocephalitinae of moderate depths) remained exclusive to IEAP during the terminal part of the TST on account of surrounding/intervening shallower water bottlenecks. During the RST, the Arabo-African low latitude warm climate prevented the spread of relatively high latitude much less warm water ammonoids. Expansion around the East and South African coast was not possible since the Gondic corridor had not come into existence at that time. Eastward expansion of the relatively shallower water ammonoids into SW Pacific may have been prevented by the deeper water open sea conditions.

In spite of the origin in Early Toarcian at the start of the first-order sequence, the IEAP could receive identity of its own only from late Middle Bathonian onwards with the arrival of Macrocephalitinae from SW Pacific. Possibly, for the first time sufficient bathymetry developed in the basins of this province for development of its indigenous lineages. However, the other elements in the fauna are all from ETM, which arrived here via Middle East.

6.7.1.2 Significant Inferences

It emerges that starved to condensed conditions prevailed in the relatively distal and deeper parts of IEAP from as early as Late Bajocian in Pakistan, from mid. Middle Oxfordian in Kachchh, from Late Callovian in Jaisalmer, from Middle Callovian in Spiti, similarly so also in South Tibet. The starved conditions continued even during the first-order RST from late Middle Oxfordian onwards up to the intra-Valanginian second-order MFS. The resultant rather slow sedimented post MFS succession mostly, included low diversity, low density and low frequency IEAP ammonoid faunas.

The ammonoid faunal provincialism since the late Middle Bathonian is because of the deeper depth and subtropical less warm temperature requirements of Macrocephalitinae and Mayaitinae whose expansion was prevented by the intervening shallower warm water barriers.

The late Middle to Late Callovian IEAP Kinkeliniceratinae is interpreted due to the uplift of IEAP part of GTM at mid/late Early Callovian MFS during a long phase of increase of bathymetry on GTM.

6.7.2 Toarcian–Mid-Middle Bathonian

The presence of several Toarcian–mid-Middle Bathonian subfamilies along with lyto-phylloceratids in Madagascar and Pakistan, more than one Bajocian–Bathonian genus in Nepal, S. Tibet, and solitary *Calliphylloceras* in Kachchh demonstrate rapid improvement in bathymetry in IEAP. The faunas suggest second-order MFSs in the Early Toarcian Nitiscens Zone and around the mid-Early Bajocian late Laeviuscula Zone to early Propinquans Zone. The suggested MFS in GTM is only slightly younger to that of Laeviuscula Zone of ETM. The faunas are also quite similar without any marked provincialism but for the Early Toarcian *Bouleiceras*.

6.7.3 Late Middle Bathonian–Early Middle Callovian

The subfamily Macrocephalitinae has its origin in SW Pacific province with Pan-Tethyan expansion, also even outside to Boreal realm. However, the lineages are independently developed on GTM and ETM.

6.7.4 Late Middle Callovian–Late Early Oxfordian

The second-order sequence surfaces—the MFS, and both the SBs—indicate distinct discordance between GTM and ETM. The subfamily Kinkeliniceratinae is exclusive to IEAP and attests strong provincialism for the first time at subfamily level. It is here interpreted as influenced and controlled by regional extensional tectonism in at least IEAP part of GTM.

6.7.5 Late Early to Latest Late Oxfordian

The second-order sequence surfaces of GTM continue to exhibit distinct discordance to corresponding second-order ETM sequence surfaces. The subfamily Mayaitinae is again exclusive to IEPA which suggests continuation of the provincialism under the control of regional extensional tectonics.

6.7.6 *Basal to Mid-Early Kimmeridgian*

The provincialism has somewhat decreased with continuation only at the species level. The appropriate development of the lineages of the Pan-Tethyan subfamily Ataxioceratinae could not be realised within IEAP. It is almost a highly scarce single subfamily fauna of low diversity, low density and low frequency. It is surmised that in the adjoining deeper starving sites, there would be present a richer Ataxioceratinae fauna, which unfortunately because of the starved conditions could not leave any record.

6.7.7 *Late Early to Latest Kimmeridgian*

The provincialism is strong with each further step of decrease in bathymetry and associated retreat of the shore line during the 1st order RST. The subfamily Katroliceratinae is dominant and exclusive to IEAP. The fauna in spite of being dominated by Katroliceratinae includes ETM elements in good numbers because of better depths at Ler–Katrol sites than Bharodia of Wagad, where was exclusively preserved the record of the immediately preceding sequence, in spite of distinct decrease of bathymetry in the basin. The decrease in bathymetry also caused the emergence of Wagad, the site of the previous second-order sequence. The margin during the time of the present sequence is formed by East Mainland at Jhuran, Jhura, Habo, etc. while Ler–Katrol down the basin was located somewhat deeper.

6.7.8 *Early Early to Mid-Early Tithonian*

The provincialism extends with continued exclusiveness of Katroliceratinae in IEAP. There is further decrease of bathymetry in the basin, even at Ler–Katrol, with resultant low diversity, low density IEAP elements. The ETM elements disappeared presumably by moving basinward to relatively deeper sites that were still starving of sediments.

6.7.9 *Late Early to late Late Tithonian*

The provincialism disappears during late Early Tithonian. The cosmopolitan subfamily Virgatosphinctinae dominates during Late Tithonian with presence both on GTM and ETM, however, lineages are independent. In spite of the reduced provincialism, the fauna suggests low density, low diversity and low frequency. The subfamily Virgatosphinctinae continues to dominate with IEAP exclusive

V. densiplicatus. Richer fauna is present in Pakistan, High Himalaya and Jaisalmer, being deeper and more distal, and yet having somewhat starved to condensed conditions.

6.8 Speculative Linkage of Prime Evolutionary Landmarks to Major Regional/Global Geological Events and Their Precision Dating Through Sequence Surfaces

Having exampled and demonstrated genetic links among ammonoid evolution at both mega and micro-level, regional extensional tectonics, and the resultant sequence surfaces in a general manner in the Phanerozoic, and somewhat focused manner in the Jurassic, the author inquisitively, scanned through the timing of the major landmarks of the organic evolution. In spite of neither being a biologist nor an evolutionary connoisseur, the author became awe-struck with the broad correspondence of the major organic evolutionary landmarks in the global geological record to tectonic events, and to first-order sequence surfaces. This correspondence by any means cannot be mere coincidence, and deserves serious investigation. The said tectonic events in turn have been dated through the mega, first-, and second-order sequence surfaces in this work. Unable to fully understand the causal relationship, the said evolutionary leaps, nevertheless, are here enumerated for readers to ponder over for developing understandable process/response links (Fig. 6.11).

6.8.1 Intra-Permian Basal SB of the Mega-Sequence

6.8.1.1 Evolutionary Leaps by Land Plants

The seed-devoid vegetation on land during the terminal Early Permian phase of the previous mega-cycle was by and large replaced by the seed-bearing plant life in the intra-Permian–intra-Paleogene mega-sequence. This significant evolutionary landmark is here interpreted as consequence of the quantum change from cold and wet to warm and dry climatic conditions along with widely known significant drop in delta C13 (Sarkar et al. 2003) in the early Late Permian–Early Triassic first-order TST interval near the P/T boundary second-order SB.

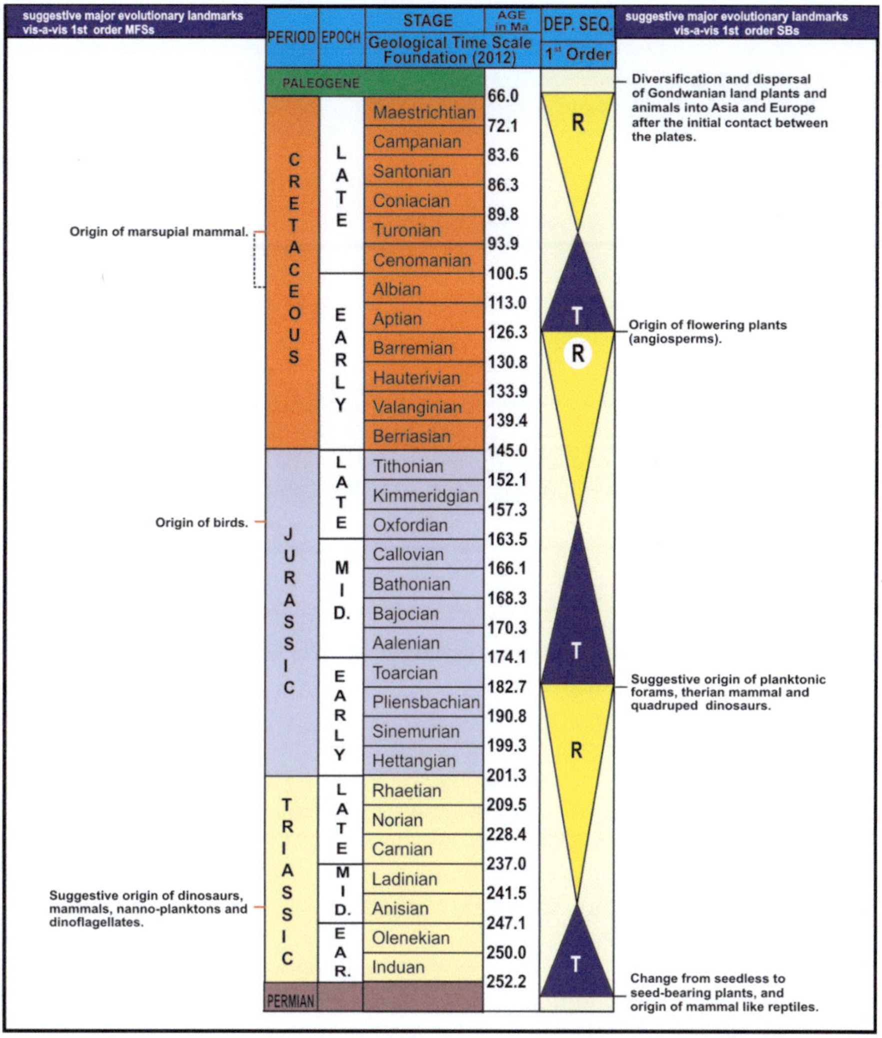

Fig. 6.11 Clade diagram for precise chronicle of separation of Successive micro/mega plates/blocks during the dismemberment of Gondwana

6.8.1.2 Possible Origin of the Mammal like Reptiles at the Basal SB of the Mega-Sequence

Among the animals, there proliferated the early Amphibians in the Late Permian–Early Triassic interval during the first-order TST interval. The reptiles of mammalian affinity, also, may have advented at the basal mega-sequence surface. The origin of the mammal like reptiles from the amniote egged amphibians was an extremely significant evolutionary step in the organic evolution. These flourished and diversified particularly during the first-order TST interval up to the terminal Anisian in the Gondwana inclusive of India as evidenced through the recovered remains in the Indian Gondwana from the Kamthi, Maleri and Panchet units of the Damodar, Narmada and Godavari basins.

6.8.2 *Intra-Triassic First-Order MFS*

6.8.2.1 Dawn of Dinosaurs at or Near the Intra-Triassic First-Order MFS

One of the most important evolutionary landmarks on the planet—the origin of the dinosaurs—is here interpreted near the intra-Triassic first-order MFS. Dinosaurs were not big sized in their early evolutionary history, however, these organisms in the later phases of the mega-cycle acquired giant size—the largest of the organic world even of several metres, a few sized up to 40 m × 30 m or even more, in their evolutionary acme prior to the sudden extinction at the K/T boundary second-order SB.

6.8.2.2 Evolution of the Warm Blooded Mammals at or Near the Intra-Triassic MFS

Subsequent to the end-Permian environmental degradation prior to the mass extinctions, the origin of the mammals is interpreted at the mid-Triassic first-order MFS. The diversification presumably picked up with the improvement in climate and environment during the first-order RST up to the close of Pliensbachian into egg laying triconodonts and reproducing ones as symmetrodonts. The latter ones abound in the Kota and Tiki units. The early mammals were also of small size, like that of mices.

6.8.2.3 Mega-Floral Evolutionary Step at the Mid-Triassic First-Order MFS

The South Rewa mega-floral assemblages are clearly differentiated into pre-Ladinian, and post-Anisian divisions. The pre-Ladinian flora has a distinct Lower Gondwana impress up to the Panchet unit of Late Permian–Anisian age bracket of the first-order TST interval vis-à-vis the floral range of the remainder Gondwana units.

6.8.2.4 Possible Origin of the Nannoplanktons and Dinoflagellates Near the First-Order MFS

The oldest, rather rare, nannoplanktons are known from the Late Triassic sediments. In India, the only Late Triassic nannoplanktons have been recovered from the

Ladakh region of High Himalaya (Rai et al. 2004). The oldest Ladinian dinocysts are recorded from Australia. The origin of the nannoplanktons and dinocysts is here linked to the mid-Triassic first-order MFS. It has been since long hypothesized that organic innovation has been recurrently catalyzed in the open marine deep extremely stressed environments. After the end-Permian all time massive volcanism, regressions, drastic deterioration of the climate and the resultant extinctions, it took time for the amelioration of the climate. The extreme stressed environments, presumably, catalyzed the origin of the said planktonic micro-organisms in the oceans from marine benthic ancestors, also of coral reefs, and of warm blooded reptiles and mammals on the coastal tracts at the mid-Middle Triassic MFS.

6.8.2.5 Distinct Palyno-Change in the K-G Subsurface

The non-marine Late Permian–Triassic sediments are known in K-G subsurface wells. Prasad and Phor (2009) have given a succession of six palynozones during the Triassic in which the marked intra-Anisian change is noticed between the Fossulatus and Isolates palynozones within the Mandapeta Formation which is here linked to the terminal Anisian first-order MFS event.

6.8.3 Suggestive Origin of the Planktonic Forams at the First-Order Intra-Jurassic SB

As a major organic evolutionary step in the oceanic cradle of planktonic forams, it is suggestively linked from their benthonic ancestors near the Pliensbachian/Toarcian first-order SB. Their fossil record is rather scarce in the Toarcian–Barremian first-order sequence, however, in the succeeding Aptian–intra-Paleogene first-order sequence, the planktonic forams acquired immense significance as an unparalleled biostratigraphic tool, particularly, in the hydrocarbon exploratory studies. Also is suggested the origin of irregular echinoids at the first-order boundary SB.

6.8.4 The Origin of Birds—The Flying Dinosaurs Near the First-Order Intra-Jurassic MFS

The first significant graded decline of dinosaurs occurred during the later phase of the first-order TST through the graded reduction of the low lying coastal tracts that were inundated by the graded sea-level rise and the resultant marine encroachments. Near the first-order intra-Jurassic intra-Oxfordian MFS and associated volcanism, although geographically restricted in spite of evidence in all the sectors on the Indian margin, both the front and hind ones, the survivors faced strong evolutionary stress and resultant distinct decline. Significant organic extinctions and origins are here linked at the said major intra-Oxfordian first-order event. As one of the all time most spectacular feats of the organic evolution, occurred the origin of the flying organisms—in the evolutionary line of the fast moving therapods, as we all know today as the beautiful world of birds. The unbearable limiting environmental stresses prompted one line of dinosaurs to seek escape and better sustenance in the atmosphere.

6.8.5 Barremian/Aptian Boundary First-Order SB Evolutionary Marks

The origin of the colourful world of the flowering plants may have occurred at the said SB or even earlier at the intra-Jurassic first-order MFS.

The more obvious examples, as above summarized, deserve further examination of possible linkages among the regional to global geological events and evolutionary landmarks in a process response framework.

Only a few examples are outlined above to invite the geoscientists across the world for further inquiry to precise dates of similar innumerable quantum evolutionary landmarks. Major extinctions and drastic reduction of organisms on a mass scale are causally better linked to igneous events and associated major sequence surfaces, however, origin and/or sudden radiation requires to be inquired into as to whether also linked to tectonics and igneous events and associated major sequence surfaces as catalysts or triggers.

References

Alberch P, Gould SJ, Oster GF, Wake DB (1979) Size and shape in ontogeny and phylogeny. Paleobiology 5:296–317

Blaison J (1967) Etude biometrique et revision de l'espece Protogrammoceras (Thevnin) 1908. Ann Sci Univ Besancon, Geologie 3:59–115

Cariou E, Hantzpergue P (1997) Biostratigraphie du Jurassique ouest-europeen et mediterraneen. Bulletin des Centres de Recherches. Elf Explor Prod Mem 17:1–440

Dommergues JL (1987) L'evolution chez les Ammonitina du lias moyen (Carixien, Domerien basal), University Claude Bernard, pp 1–297

Dommergues JL, David B, Marchand D (1986) Les relations ontogenese—phylogenese: applications paleontologiques. Geobios 19(3):335–356

Enay R (2009) Les faunes d'ammonites de l'Oxfordien au Tithonien et la biostratigraphie des Spiti-Shales (Callovien Superieur – Tithonien) de Thakkhola, Nepal Centrale, Docum. Lab Geol Lyon 351

Fatmi AN (1986) Lower and Middle Jurassic ammonites from Windar group of Phusi Jhal (Kanrach) and Sand (Windar Nai) sections of Lasbella, Baluchistan, Pakistan. Mem Geol Surv Pak 16:1–17

Fatmi AN, Hyderi IH, Anwar M, Mengal JM (1986) Stratigraphy of Zidi Formation (Ferozabad Group) and Parh Group, Khuzdar district, Baluchistan, Pakistan. Rec Geol Surv Pak 1:1–32

Gould SJ (1977) Ontogeny and Phylogeny. Harvard University Press, Cambridge, Massachusetts, 1–501

Graciansky PC, Dardeau G, Dumont T, Jacquin T, Marchand D, Mouterde R, Vail PR (1993) Depositional sequence cycles, transgressive-regressive facies cycles and extensional tectonics: example from the southern subalpine Jurassic basin, France. Bull Soc Geol France 164(5): 709–718

Haq BU, Hardenbol J, Vail PR (1987) Chronology of fluctuating sea-level since the Triassic. Science 235:1156–1167

House M (1985) Major factors in the evolution of Cephalopoda, (abstr.). In: 2nd International cephalopod symposium Tubingen

Krishna J (1983a) Callovian–Albian ammonoid stratigraphy and palaeobigeography in the Indian subcontinent with special reference to the Tethys Himalayas. Himalayan Geol, India 11:43–72

Krishna J (1983b) Reappraisal of the marine and/or "Mixed" Lower Cretaceous sedimentary sequences of India; Paleogeography and time boundaries, Cretaceous of India, 94–119

Krishna J (1987) The Jurassic-Cretaceous ammonoid geography in the Indian subcontinent vis-à-vis marine seaways and plate tectonics in the Indian ocean region, In 3 decades of development in Paleontology and Stratigraphy in India. Spl Publ Geol Surv Ind 11(1):453–482

Krishna J (2005) Update on Kachchh Jurassic ammonoid zonation: summarised intra-basinal to global perspectives. In: Raju DSN et al (eds) Spl. Publ.-1, Ass Petr Geol, 63–68

Krishna J (2012) Litho-bio-chrono-magneto-igno-sequento-stratigraphic integration in the Kachchh Mesozoic, Status Talk—Mesozoic Biostratigraphy. In: Thakkar MG (ed) Abstract volume, Geology of Kachchh basin, Western India, 21–25

Krishna J (2014) Ammonoid—sequence-magneto stratigraphic integration in the Kachchh Jurassic. Unpublished DST project report, 60 p

Krishna J, Cariou E (1986) The Callovian of Western India: new data on stratigraphy, biogeography of the ammonoids and correlation with Western Tethys. Newsl Stratigr 17(1): 1–8

Krishna J, Cariou E (1990) ammonoid faunal exchanges during Lower Callovian between the Indo-East African and submediterranean provinces: implications for the long distance east-west correlations. Newsl Stratigr 23(2):109–122

Krishna J, Cariou E (1993) The Tethyan Macrocephalitinae: evolutionary, environmental and dispersal strategies. Geobios, MS 15:227–236

Krishna J, Ojha JR (1996) The Callovian ammonoid chronology in Kachchh (India). GeoResearch Forum 1–2:151–166

Krishna J, Pathak DB (1989) Kimmeridgian in Ler–Katrol area of Kachchh, Western India: ammonoid systematics and biochronology, (abstr.). In: 28th international geological congress, Washington, pp 228–229

Krishna J, Pathak DB (1991) Ammonoid biochronology of the Kimmeridgian stage in Kachchh, India. J Pal Soc Ind 36:1–13

Krishna J, Pathak DB (1993) Late Lower Kimmeridgian–Lower Tithonian virgatosphinctins of India: evolutionary succession and biogeographic implications. Geobios 15:217–226

Krishna J, Pathak DB (1994) Stratigraphic, biogeographic and environmental signatures in the ammonoid bearing Jurassic- Cretaceous in the Himalaya on the south margin of the Tethys. J Him Geol 4(2):151–167

Krishna J, Thierry J (1987) Discovery of the Middle Callovian ammonoid genus Erymnoceras from Kachchh. Western India. Newsl Stratigr 17(2):71–78

Krishna J, Singh IB, Howard JD, Jafar SA (1983) Implications of new data on Mesozoic rocks of Kachchh, Western India. Nature 305:790–792

Krishna J, Pathak DB, Pandey B (1994) New ammonoid evidence on the Jurassic/Cretaceous boundary in Kachchh (India). Geobios, MS 17:327–335

Krishna J, Melendez G, Pandey B, Pathak DB (1995a) Characterization of the ammonoid genus Larcheria (Middle Oxfordian) in Kachchh (India): paleontology, biostratigraphic and paleobiographic evaluation in the context of North-Tethyan occurences, CR Acad Sci Paris, 321 lla:1187–1193

Krishna J, Pathak DB, Pandey B (1995b) Kimmeridgian–Tithonian (Upper Jurassic) ammonoid zones in Kachchh (Western India) and their correlations. Proc Rec Res Geol Western India, Badodara, pp 323–347

Krishna J, Melendez G, Pandey B, Pathak DB (1996a) Middle Oxfordian ammonites (Perisphinctinae) from Kachchh (India): biostratigraphic and paleobiogeographic implications. Revista Espanola De Paleontologia, Numero Extraordinario, pp 140–147

Krishna J, Pandey B, Pathak DB (1996b) Ammonoid chronology developed in the Tithonian of Kachchh (India). GeoResearch Forum 1–2:205–214

Krishna J, Pathak DB, Pandey B (1996c) Ammonoid age control in the Mesozoic succession of Wagad (Kachchh Basin) outside the Mainland Kachchh. Geophytology 26(1):63–68

Krishna J, Pathak DB, Pandey B (1996d) Quantum refinement in the Kimmeridgian ammonoid chronology in Kachchh (India). GeoResearch Forum 1–2:195–204

Krishna J, Sinha AK, Upadhyay R (1997) The first south Tethyan Hettangian Psiloceras from the Indus Suture, Ladakh Himalaya: diverse implications. J Him Geol 18:45–51

Krishna J, Pathak DB, Pandey B (1998) Development of Oxfordian (Early Upper Jurassic) in the most proximal exposed part of Kachchh basin at Wagad outside the Mainland Kachchh. J. Geol. Soc. Ind. 52:513–522

Krishna J, Pandey B, Ojha JR, Pathak DB (2000a) Transgressive sediment intervals in the Late Jurassic of Kachchh, India. GeoResearch Forum 6:321–332

Krishna J, Pandey B, Ojha JR, Pathak DB (2000b) Ammonoid–Foraminifer Zonal integration in the Jurassic of Kachchh, India. ONGC Bulletin 37(1):9–21

Krishna J, Pandey B, Ojha JR (2008) Gregoryceras in the Oxfordian of Kachchh (India): diverse eventful implications. Geobios 42:197–208

Krystyn L, Balini M, Nicora A (2004) Lower and Middle Triassic stage and substage boundaries in Spiti. Albertiana 30(Supplement):39–52

Mcnamara KJ (1986) A guide to the nomenclature of Heterochrony. Paleobiology 60(1):4–13

Moore RC (ed) (1957) Treatise on invertebrate palaeontology, Part L, Ammonoidea, Geol. Soc. Amer. and Univ. Kansas Press, LI-490

Ojha JR (1996) The Callovian at Keera, Kachchh, India; Ammonoid systematic, Chronology and Geography. Unpublished Ph.D thesis, Banaras Hindu University, Varanasi

Pandey B (1993) The Kachchh Jurassic Virgatosphinctinae, high resolution evolutionary succession and chronology. Unpublished Ph.D. thesis, Banaras Hindu University, Varanasi, pp 1–473

Pandey B, Krishna J (2002) Ammonoid biostratigraphy in the Tithonian (Late Jurassic) of Jaisalmer, Western India, Geophytology 30(1 and 2):17–25

Pandey B, Pathak DB (2015) Record of Early Bathonian ammonoids from Kachch, India: biostratigraphic and paleobiogeographic implications. J Pal Soc Ind 60(1):33–44

Pandey B, Krishna J, Pathak DB (2010) Review of the subfamily Virgatosphinctinae with special reference to its evolutionary succession in the Indian subcontinent on the Gondwanian margin, J Sci Res BHU 54(1–2):1–20

Prasad B, Phor (2009) Palynostratigraphy of the subsurface Gondwana and post-gondwana Mesozoics of the Cauvery basin, India. J Pal Soc Ind 54(1):41–71

Rai J (2006) Reworked Pliensbachian – Aalenian nannofossils from Jara dome Kachchh; early Jurassic palaeobiogeography of western India. In: XIth International Nannplankton Association, Nebraska, pp 84–85

Rai J, Upadhyay R, Sinha AK (2004) First Late Triassic nannofossil record from the Neo-Tethyan sediments of the Indus- Tsangpo Suture Zone, Ladakh Himalaya, India. Curr Sci 86(6): 774–777

Sarkar A, Trivedi JR, Yoshioka H, Ebihara M, Naraoka H (2003) Geochemical and organic Carbon isotope studies across the continental Permo-Triassic boundary of Raniganj basin, eastern India. Palaeogeogr Palaeoclimat Palaeoecol 191:1–14

Spath LF (1927–33) Revision of the Jurassic cephalopod fauna of Kach (Kutch). Paleont Indica, Calcutta 9(2):1–945

Chapter 7
High-Resolution Intrabasinal to Inter-regional Geodynamic Chronicle During the Span of the Intra-Permian–Intra-Paleogene Mega-Sequence in and Around India on the GTM

Abstract This chapter chronicles the geodynamic evolution of India and associated dismemberment history of the Gondwanaland (Figs. 7.43 and 7.44). The narration is through the succession of geological events and their interdisciplinary manifestations that are uniquely punctuated with the first- and second-order physically feelable and visually observable sequence surfaces. These surfaces are considered products of the causative tectonic/igneous events. In the process, the entire ~540 my long Phanerozoic history in general and the intra-Permian–intra-Paleogene Indian/GTM mega-sequence geological record in particular inclusive of the surface and offshore wells is, intrabasinally to interregionally, correlated, integrated and interwoven at a distinctly improved average resolution of ~3.5 my involving a succession of ~40 odd second-order sequence surfaces. Within the Kachchh basin, the precision tell-tale has been realized at yet finer resolution of ~400 ky or multiples with the help of a series of graphics. The mega-sequence began with Panjal and coeval igneous events that are particularly manifested in the north sector. Its differentiation into three first-order sequences (Spitiian, Kachchhian and Dravidian) is punctuated by major igneous events in the west near the ~183 ma Pliensbachian/Toarcian boundary in the Kachchh and Saurashtra wells (coeval to Karoo volcanism), and later in the east near ~the 126 ma Barremian/Aptian boundary first-order SB Rajmahal and coeval igneous events. Major volcanic events are evidenced in the east near the ~243 ma first-order Middle Triassic MFS in Papua New Guinea, and again at the first-order intra-Jurassic ~159 ma MFS with spreading away of the Argo block from Greater India at the Indo-Australian divergent margin. Madagascar separated from India during the ~94 to ~84 ma span. Both the start and end of the span mark significant sequence surfaces. The initial impingement of the drifting India with the Kohistan–Ladakh oceanic island arc is interpreted around the ~51–50 ma late Early Eocene Late Cuisian planktonic foram P8 Zone sequence surface, well after the intra-Paleocene basal/intra P3 Zone SB ~60–59 ma closure of the mega-sequence. The stratigraphic gaps in the Indian record inclusive of the break-up unconformities are better resolved and differentiated as subaerial and submarine, and also are precisely ranged through the succession of sequence surfaces. The succession of the ~40 highly time-precised

© Springer Science+Business Media Singapore 2017
J. Krishna, *The Indian Mesozoic Chronicle*, Springer Geology,
DOI 10.1007/978-981-10-2477-1_7

sequence surfaces, in a process-response relationship, has, precisely, dated as many rift, igneous, eustatic, anoxic, T/R events and the multifaceted collage of other geological phenomena and expressions. The Mesozoic litho- and biostratigraphic units in outcrops and wells, marine and non-marine, guide fossil bearing or devoid, carbonates or clastics, shallow or deep of the Indian basins across the length and breadth of India have been precisely age-constrained and mutually correlated through the succession of sequence surfaces. Major evolutionary landmarks of organic evolution are also speculatively age-constrained, since conceptually influenced by tectonic and magmatic events with or without igneous expressions. The principal energy resources of India, hydrocarbon, coal, and atomic minerals, are temporally and spatially better constrained, and understood through the succession of the second-order sequence surfaces.

Keywords Neotethys · Indian plate · GTM · ETM · IEAP · Arabian · SW Pacific and SE Pacific ammonoid provinces · Event integration · Marine/non-marine correlation · Sequence surfaces · Sequence-ordering · Rift · Drift · Volcanic island arcs · Convergence · Subduction · Hotspots · Karoo · Ferrar · Rajmahal · Deccan · Volcanism · Transform sliding · Marine corridors · Major evolutionary landmarks · Ammonoid ages · Magnetic anomalies · Radiometry · Origin · Extinction · Stratigraphic gaps · Condensation/starvation · Emergence/submergence · See-saw tilt-tectonics · Slope reversal · Spreading ridges · Gondwana dismemberment · Cyclicity · Cycle-periodicity · ~400 ky resolution

7.1 Backdrop to the Geodynamic Chronicle

7.1.1 Transformation from the Origin to the Closure of the Late Permian–Early Paleogene Mega-Sequence

India, at the intra-Permian start of the mega-sequence (Fig. 7.1), was a constituent of the Gondwanaland with high altitude areas in the east and south of the Narmada–Son–Damodar fracture in the south hemisphere. In the east and southeast, it was in union with Australo-Antarctica, and on the west contiguous with West Gondwana (Fig. 1.1). Glacial ice sheets were seated on the then Indian highlands in the south and east. The coastal low lands faced intermittent onslaught of the just created Neotethys in the west, northwest, north, and northeast. The mega-sequence started with the origin of the Neotethys and associated graded shrinking of the Paleotethys (Fig. 2.2). It was followed up later by the initiation of the Indian and Atlantic Oceans in the middle of the mega-sequence. In the west and east divergent sectors on either side of India, the origin of the Indian Ocean was precursored, respectively, by the transient Gondic and Dravidian shallow marine corridor seaways, which in

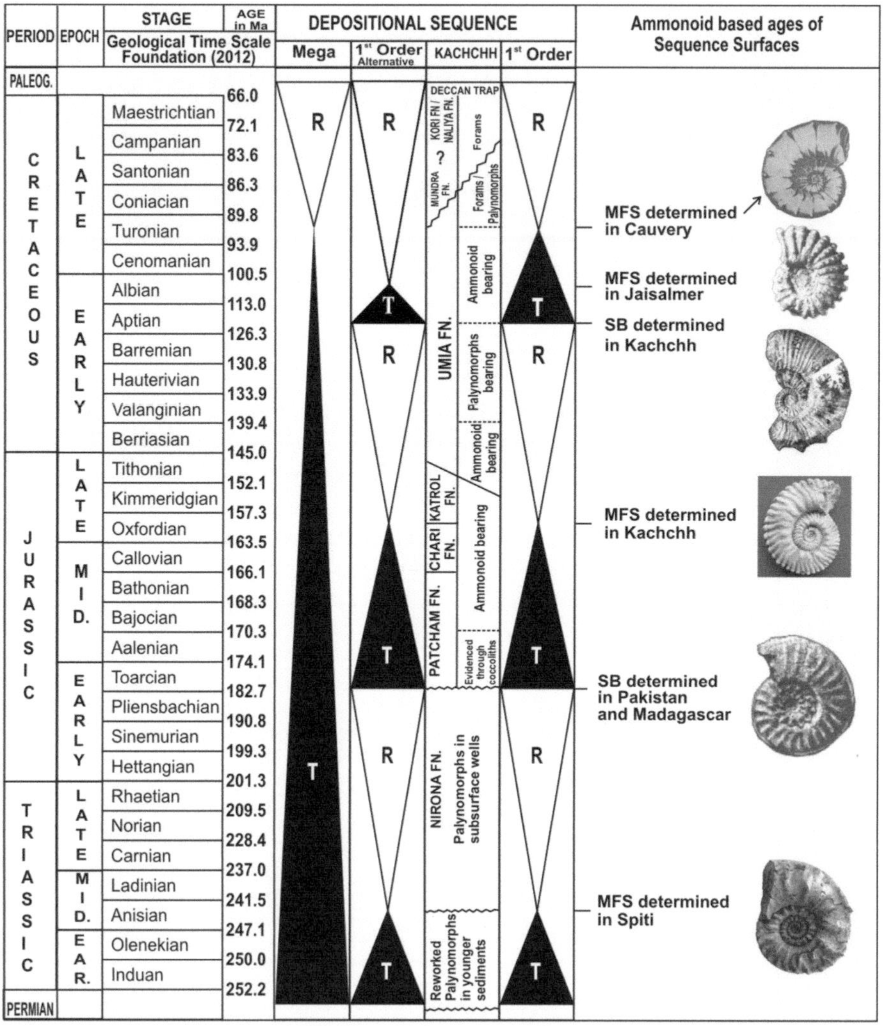

Fig. 7.1 First-order sequences on GTM determined in Indian subcontinent with focus on the Kachchh succession

turn were preceded by transform sliding in both sectors in mutually opposite directions (Fig. 1.3). The said oceans later grew substantively to their near present sizes by the close of the mega-sequence. The Indian Ocean at the close of the mega-sequence, much unlike the position of the Neotethys, was placed in the west, east and south of the Indian plate. The Neotethys shrank in stages during the span of the mega-sequence, and eventually disappeared. High lands and glacial sheets now are seated and form the north margin of the Indian plate in High Himalaya. The slope/drainage/current is towards east and south. Near the terminal and post

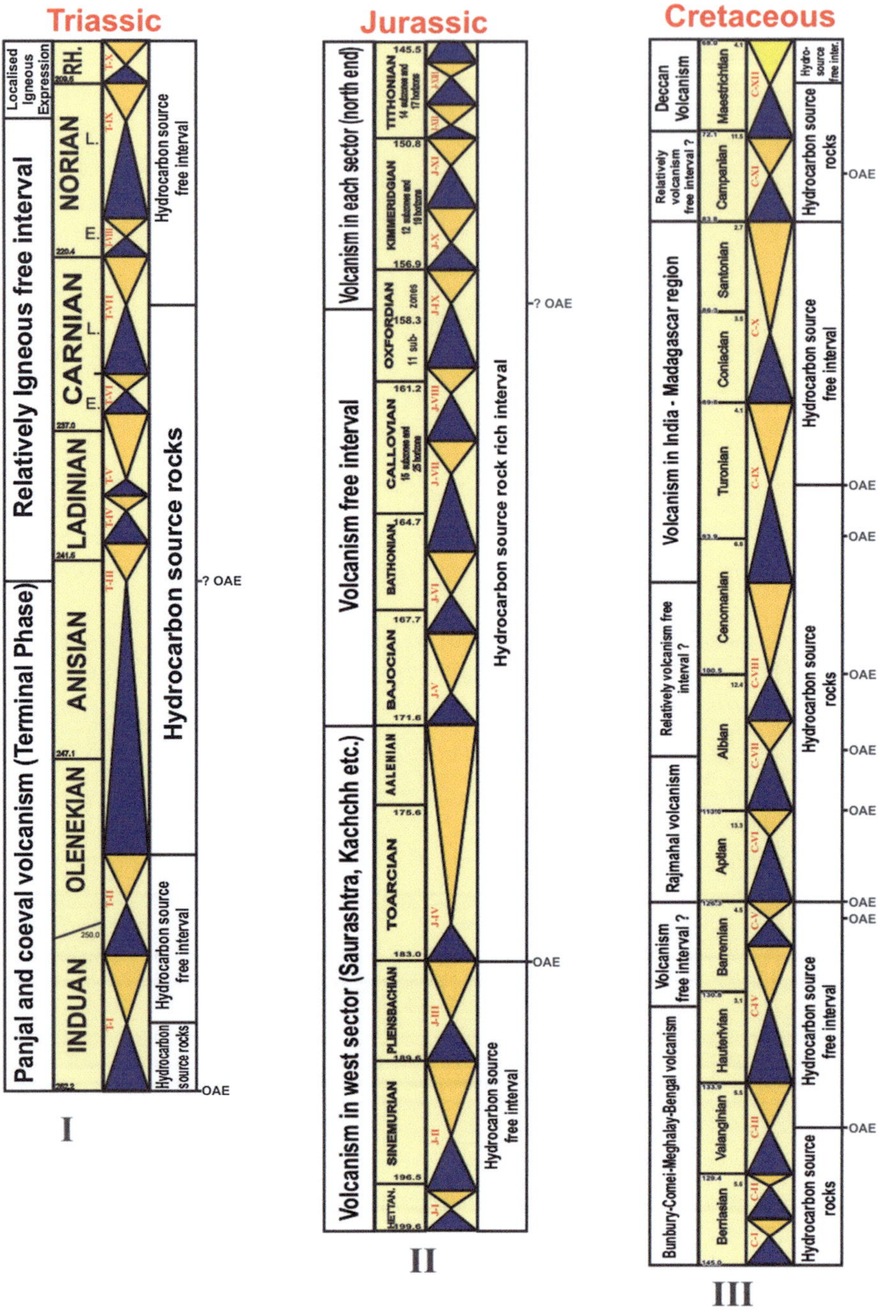

Fig. 7.2 Succession of second-order sequences in the Indian Mesozoic vis-a-vis igneous events, OAEs, and economic mineral resources

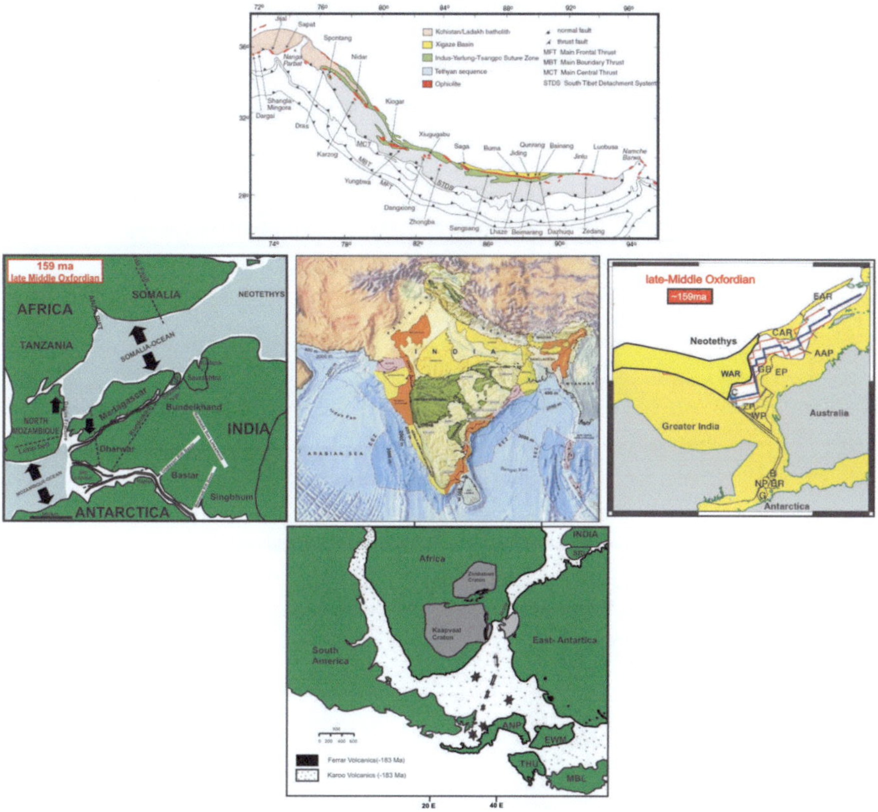

Fig. 7.3 Indian sedimentary basins along with neighbouring regions in the north, east, south and west (Modified after Reeves 2014; Gibbons et al. 2013; Zhu et al. 2009)

mega-sequence time, India is brought in the north hemisphere as a constituent of Asia. India, thus, is interpreted to have travelled a distance of several thousand kilometres (Fig. 1.1). Its neighbours prior and post mega-sequence are altogether different. Important regional to global tectonic and igneous events manifested themselves both near the start, and terminus, and as well during the span of the mega-sequence in and around the Indian plate (Figs. 7.2, 7.3, 7.4 and 7.5). Major hydrocarbon, coal, and other energy resources, sea level changes and anoxic events are found temporally related to sequence surfaces (Figs. 7.6, 7.7 and 7.8). Important evolutionary landmarks are also speculatively linked to major sequence surfaces of the mega-sequence (Figs. 7.9 and 7.10). Indian coastlines, both in the east and west, were sculpted in steps at sequence surfaces. To summarize, the scenario during the span of the mega-sequence underwent drastic changes that altogether not only changed the face of India, but also of the globe at large. It brought us closer to the broad geographical framework, it exists today, particularly contextual to the position of India vis-à-vis rest of Asia (Fig. 1.2).

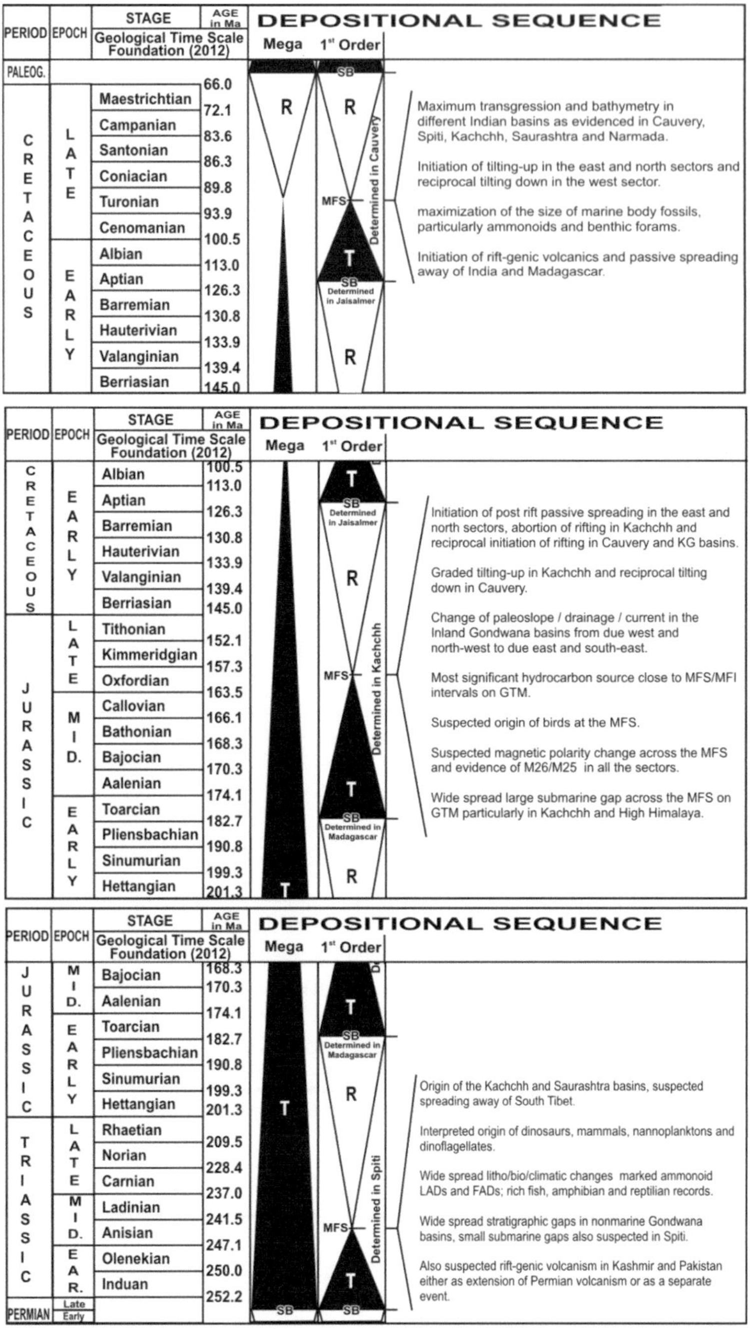

Fig. 7.4 Salient features/events at the first-order sequence surfaces of the mega-sequence

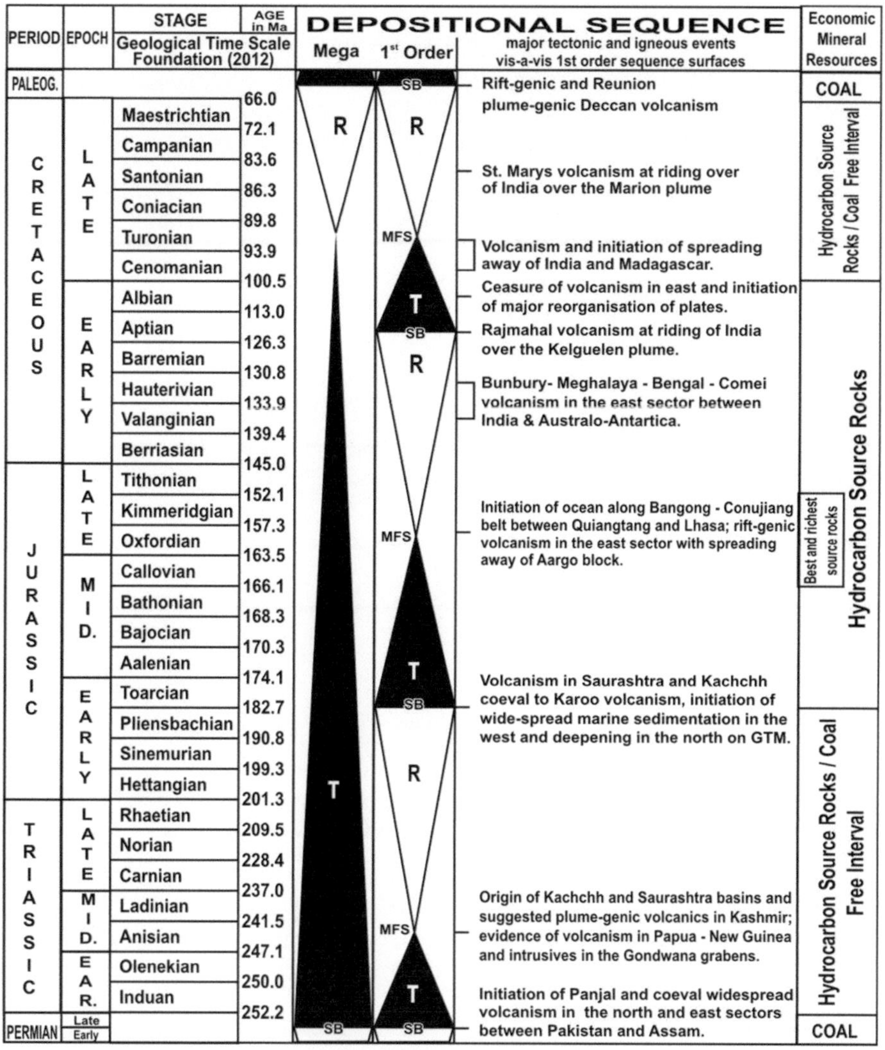

Fig. 7.5 First-order sequences surfaces vis-a-vis salient tectonic/igneous events, and economic deposits of hydrocarbon source sediments and coals

7.1.2 Sequence Surfaces as Regional Eventful Additional Timelines to Date Events of One Sector in Some Other Distant Sector

The era, period, stage and zones as classical chrono-units, since their establishment in Europe in the mid-ninteenth century, served well the cause of chronologic refinement and correlations. Yet, with ever-increasing refinement, it became

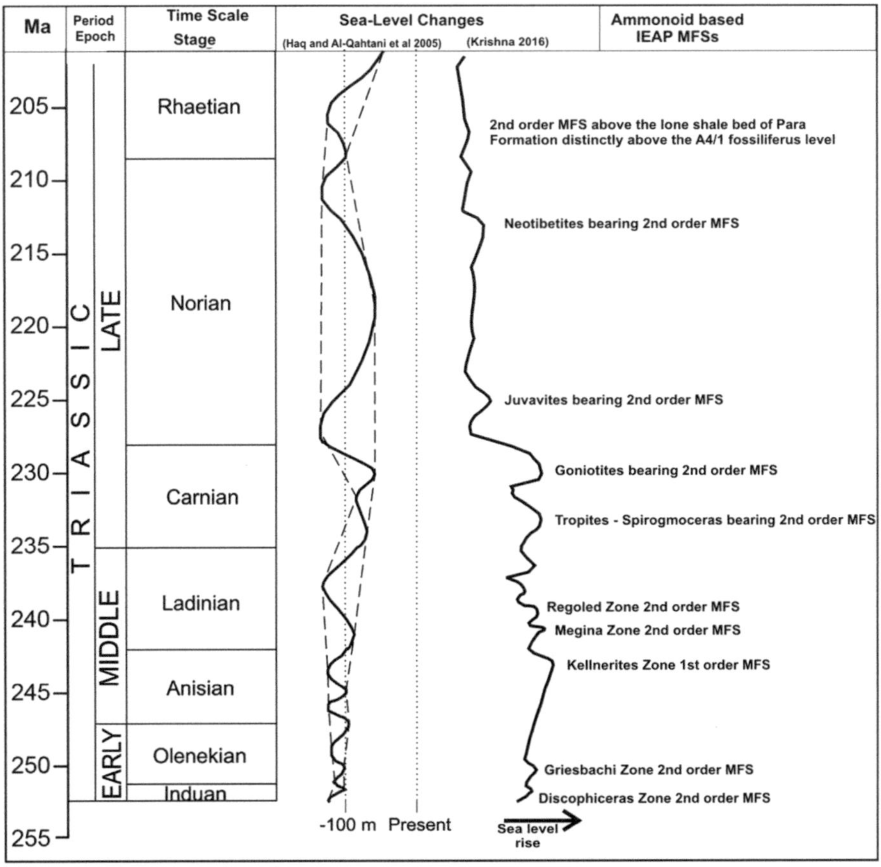

Fig. 7.6 Comparison of sea level curves during Triassic (modified after Ruban 2014)

increasingly clear that the important geological events and changes invariably occurred in between the boundaries of the said chrono-units, rather than at them. Also, in spite of the explicit definition of these standard chrono-units in localities and regions of their original establishment in Europe, their boundaries have been difficult to recognize and correlate across distant lands, e.g. India, China or America. The geological successions and events in regions away from Europe, thus, could only be tentatively correlated using the standard time scale based on the evolutionary succession of guide fossils. The accuracy of such correlations has been mostly supra-zonal to supra-stage, depending upon the evolutionary rates of the guide fossils, if so available, and studied in the correlated lithostratigraphic column. It is in the above backdrop that the integrated multidisciplinary high-resolution comprehension of the holistic geological dynamics is approached through the additional application of the succession of regionally to globally isochronous

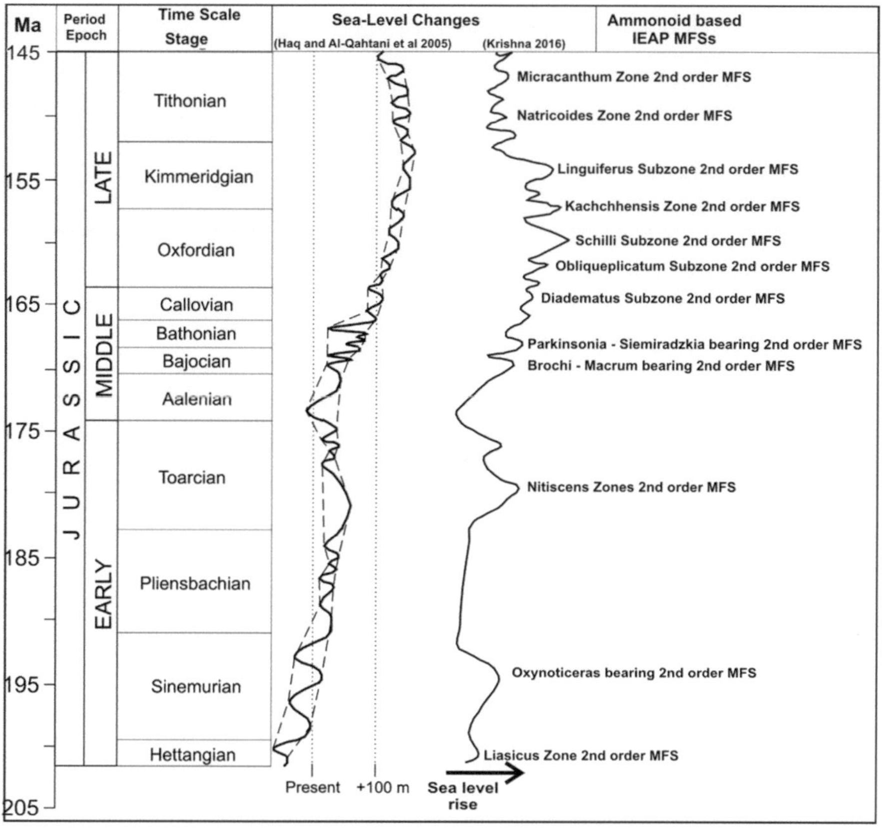

Fig. 7.7 Comparison of sea level curves during Jurassic (modified after Ruban 2014)

depositional sequences, and the visually observable sequence stratigraphic surfaces that characterize them, also the discipline of sequence stratigraphy encompassing them.

7.1.2.1 Sequence Surfaces as the Common Binding Tools

The sequence surfaces, in spite of varied manifestations in individual basins or regions, unite and bind together the diverse geological expressions to a common geological/tectonic/igneous event that is best expressed in a single basin or region. Also, thus, the resultant expression of any tectonic event in one basin or region can be precisely dated in some other distant basin of the region with the help of better guide fossil record in that distant basin compared to the basin having the best and near exclusive resultant expression of the regional tectonic event, e.g. the dating of the spreading away of the Argo block from the Indo-Australian margin in the

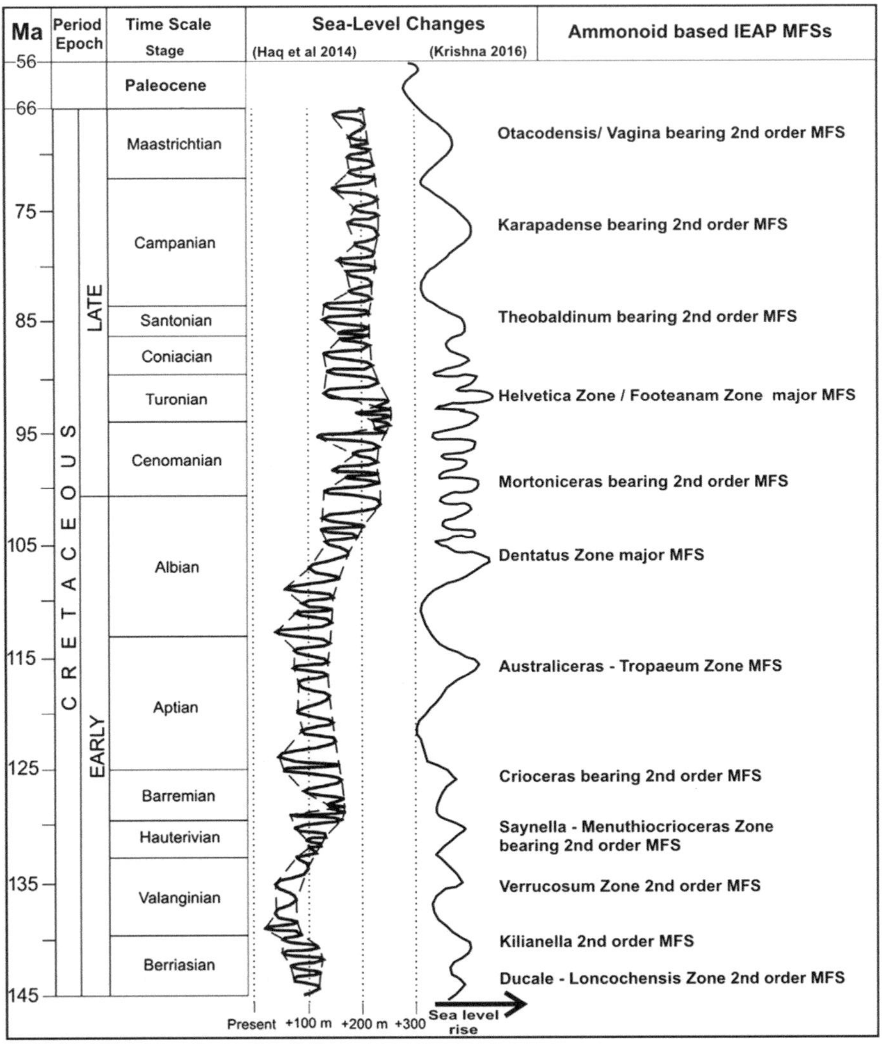

Fig. 7.8 Comparison of sea level curves during Cretaceous (modified after Ruban 2014)

~3000 km or more away Kachchh basin of India through the ~159 ma late Middle Oxfordian Pan-Neotethyan Schilli Subzone, discussed later in detail.

As a fundamental shift, particularly so in the Indian context, the first and second-order sequence stratigraphic surfaces, genetically linked to extensional tectonic events in the earlier chapters, are used for the intrabasinal to inter-regional geodynamic chronicle. The said extensional tectonic events are spread in a broad tectono-stratigraphically homogeneous region—the GTM. As a corollary, the extensional tectonic events of substance are postulated to have occurred only at

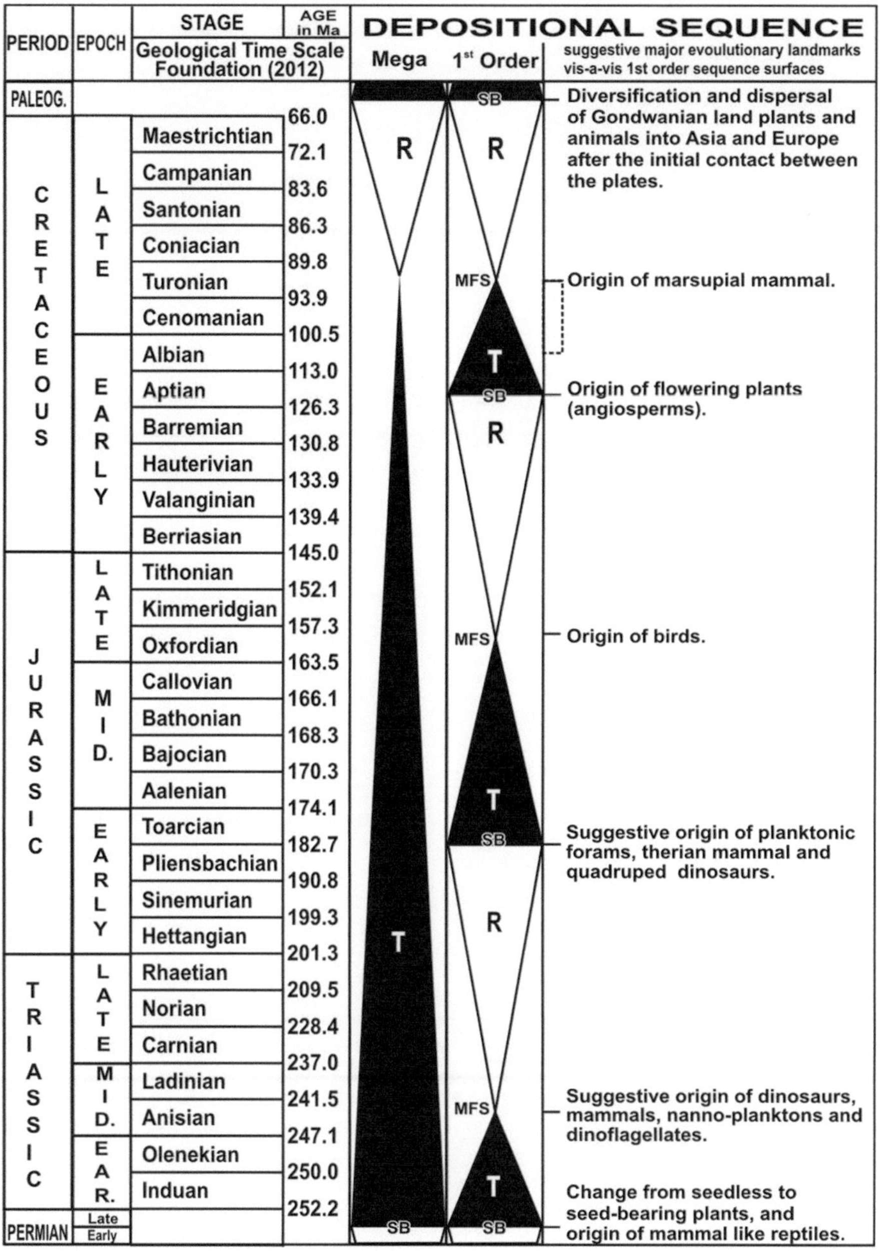

Fig. 7.9 First-order Sequences and major organic evolutionary landmarks

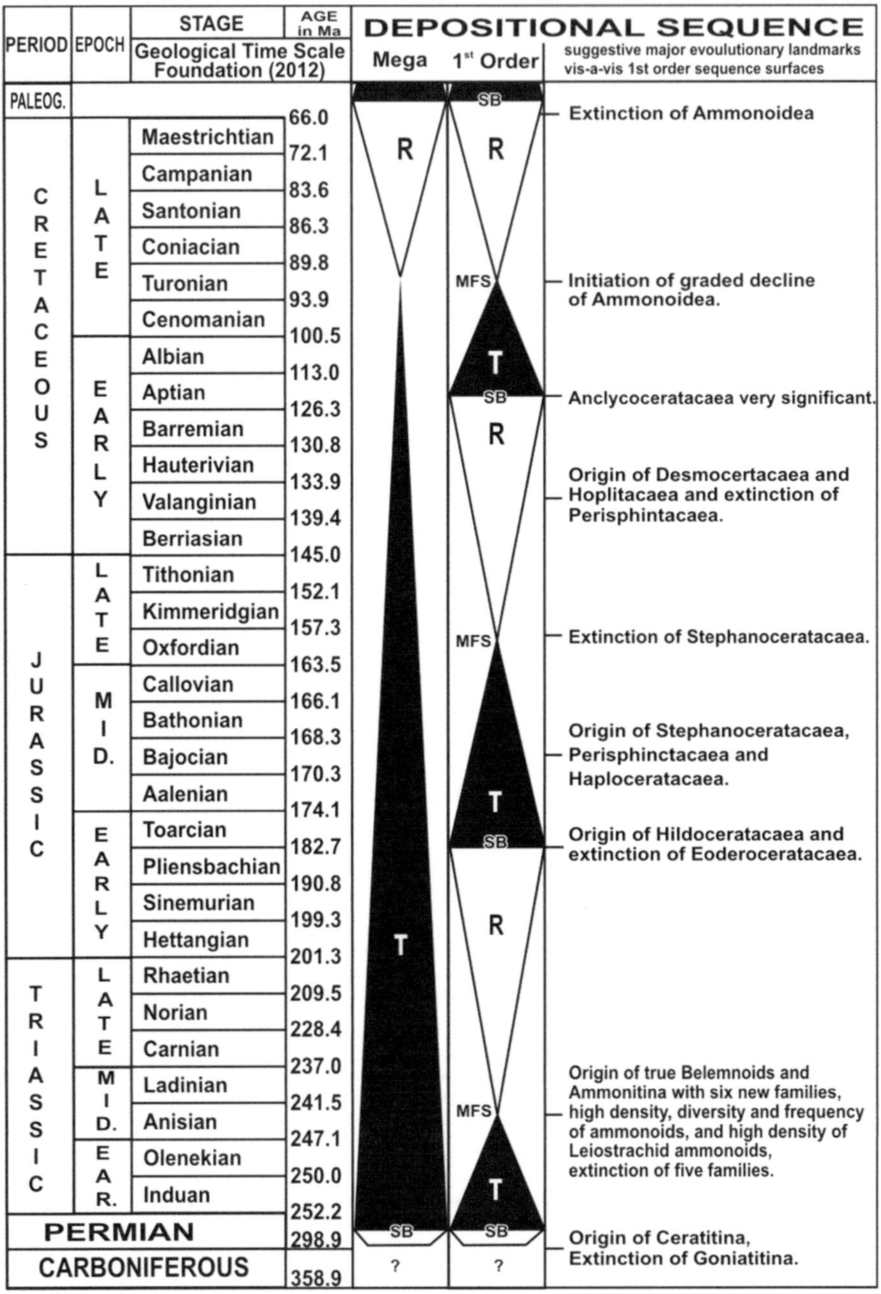

Fig. 7.10 First-order Sequences and ammonoid evolution on GTM

specific geological times along with the genetically linked resultant first/second-order sequence surfaces. There is origin, peaking, vaning and closure of relatively protracted events at the time the said sequence surfaces were produced, while their influence is invariably not considered as appreciable in the geological record in between the sequence surfaces. The sequence surfaces manifest the episodically punctuated framework of the extensional tectonics. The sequence surfaces have also allowed correlation of the lithostratigraphic units of the Indian basins, marine or non-marine, carbonate or clastic, shallow or deep water, high or low latitude, into a unified scheme as a new, and not earlier tried working model.

The sequence surfaces (SBs and MFSs) carry extra-meaningful correlative significance, being practically isochronous within the basins, across the basins, across big regions and even across the globe, irrespective of lithology, structure, texture, fauna, flora, environment, bathymetry, climate, latitude and many such parameters. Only the mega to second-order sequences have intrabasinal to regional extent in view of their net governance by the regional tectonics. In contrast, the third-order and yet finer and smaller sequences are isochronous across the globe in view of their net control by Earth's orbital dynamics and associated global manifestations, e.g. the eustatic sea level changes. Eustatically guided third-order and yet finer/smaller sequences have high frequency and low magnitudes, and escape any interference or distortion by low-frequency high magnitude second to mega/super-order sequences, and hence their globality. The mega to second-order sequence stratigraphic surfaces of the Gondwanian Tethyan margin (GTM—extending from Arabia to NW Australia on the Neotethyan margin of Gondwana) need neither be necessarily nor be ever isochronous to those of the Eurasian Tethyan margin (ETM—extending from Spain to Tibet). Obviously, the dire need in such individual regions and large countries such as India to develop and establish its succession of DSs inclusive of SBs and MFSs, cannot be over-emphasized for more objective comprehension of the high-resolution holistic geological dynamics and application of the same to more efficient and cost-effective hydrocarbon and other economic exploration ventures in the Indian Phanerozoic geological record.

7.1.3 *Predictable Spatio-Temporal Distribution of Biotic and Abiotic Parameters Across the Sequence Surfaces as also from Proximal Margin to Distal Basin on the Individual Surfaces*

The low-frequency high magnitude regional extensional tectonics causes sudden marked changes in environmental parameters, particularly bathymetry, shelfal accommodation and nutrients in shallow marine systems that lack globality. Such extensional tectonics in many cases overrides the eustatic effect. However, both eustatics and tectonics combinely influence the spatio-temporal distribution and migration of both biotic and abiotic elements in the depositional system, particularly so in the shallow marine geological record, and also get imprinted in sequence stratigraphic context through punctuation marks of sequence surfaces of different orders, frequency, magnitudes and extent. In this work, apart from the lithofacies variation on predictable lines, the biotic changes are also found specially revealing in the bathymetry sensitive guide fossil and other groups of ammonoids, other molluscs, corals, forams, etc. Not simply the origin, evolution, and extinction, but even their frequency, density and diversity and spatio-temporal distribution in the basin seem to suggest predictable patterns as exampled later in the Kachchh Jurassic.

Chapters 3, 5 and 6 and this chapter, respectively, deal with the same succession of the first- to second-order surfaces of the mega Indian Neotethyan sequence in respective focused contexts of the (1) formulation of the succession of well-defined and well-structured depositional sequences in Chap. 3, (2) demonstration of the applicability of the said sequence surfaces in the Indian, IEAP and finally GTM basins from Arabia to NW Australia in Chap. 5, (3) demonstration of the genetic links of the origin/radiation/extinction/depletion of ammonoid subfamilies of the Indo-East African ammonoid faunal province on GTM with extensional tectonic events in Chap. 6 and finally (4) culminating in the intrabasinal to inter-regional to global geodynamic multidisciplinary interrelationships of the geological record, events and sequence surfaces in this chapter. As a consequence, some overlapping, even occasional repetition, has crept in, inspite of all possible efforts towards minimization. Alternatively, the said four chapters could have been synthesized in just a single chapter at the cost of the independent development of each, yet mutually interrelated aspects separately.

7.1.4 Chronicle of the Events at the Subzonal Resolution of ~400 ky or its Multiples

The major geological events have been age-precised here, invariably, to subzonal/horizonal precision of ~400/200 ky or multiples, or of even yet finer precession/obliquity/eccentricity cycles, hardly ever exceeding 1 my. The premise is, that the first and second-order sequence surfaces are products of the regional tectonic events, and so the ages of the successive sequence surfaces provide the most precise chronicle of the tectonic events that produced them (Krishna 1987a). It is observed that the tectonic events taking place at the first-order surfaces are invariably more widespread compared to those at the second-order sequence surfaces. The basinal/regional tectonic events responsible for the first and second-order Kachchh Mesozoic sequence architecture are the same that structured and punctuated the successive phases of the drifting history of the Gondwana dismemberment, origin and growth of the Indian Ocean and rapid due north drift of India for graded inching, initial impingement, collision in multiple steps and final assimilation into Asia.

7.1.4.1 Hierarchical Order, Cyclicity and Duration Framework

According to spatio-temporal magnitude, like any other type of chronostratigraphic units, the DSs have hierarchical ordering and spatial extent; mega-order of ~60 my to a few hundred my, first-order of a few ten my, second-order of ~1.5–2 to 5–15 my, third-order of multiples of ~ 400 ky, and fourth-, fifth- and sixth-order sequences of yet smaller durations. However, these durations are subjective, and occasional exceptions are not excluded, and are already discussed in Chap. 3. The Mesozoic span of ~185 my is organized into ~35 second-order sequences. The ~56 my long Jurassic includes about 45 third-order sequences with an average of ~1.2 my duration as multiples of ~400 ky cycles. The durations of the different orders of sequences in this work differ substantively from the standard ordering, particularly, with that practiced in India by the ONGC. May be, the outcrops provide consideration of parameters at smaller thickness, time, and space. The durations of the sequences of different orders are distinctly different, i.e. the present outcrop-based sequences are more refined and better resolved. For example, the ONGC has differentiated the Indian Cretaceous sequences through a maximum of four first-order SBs—termed as MI 50 to MI 80 or a total of eight sequences in the Indian Mesozoics differentiated through MI 10–MI 80. Thus, the first-order sequences of the present work are of bigger durations compared to the first orders of the ONGC (2008), while the second-order sequences of the present work have durations on an average at least twice more refined in duration than those of the ONGC. This has been possible because of their being outcrop based, and several fold better resolving power of ammonoids—the fastest evolving marine organism of the planet's organic history.

7.1.4.2 Relative Resolution of the Principal Tools of Age Determination of Geological Events

The data on initiation of new ocean floor is obtained through deep sea ocean floor studies under programs like DSDP, ODP, etc. The principal tools are magnetic anomalies, radiometric ages of the oldest ocean floor basalts and planktonic microfossils that form the oldest sediments on the new ocean floor close to the mid-oceanic ridge. The actual absolute geological moment of the initiation of oceaniszation is conceptually distinctly older to the ages realized in the said three directions. The ages of the magnetic anomalies as also of planktonic microfossils for convenience are linked and quantified to the constrained radiometric ages. A good number of radiometric ages are obtained, followed by averaging of the closest/densest clusters. Improvement in radiometric techniques and sea-floor data results in regular revision and refinement of the geological time scale with ever-increasing precision. An extreme example of significant changes in the revision is the base of Aptian, which used to be placed in the 1980s at ~112 ma (Fig. 1.13), for example, in the highly used Haq et al. (1987) Science paper. It was revised to ~125 ma in the 1990s. In the 2010s, it is at ~126.3 ma (2015 GTS). So, also, is necessitated near continuous revision in the correspondence of the radiometric ages of the standard stages, and magnetic anomalies. On the other hand, the ages obtained through the above three directions are related to the stages/substages of the Mesozoic scale based primarily on the evolutionary succession of ammonoids—the fastest evolving marine organisms of not merely Mesozoic, but of the entire Phanerozoic. In comparison to the significant change/revision/refinement in absolute ages, there is far less change in the age definition of the base of Aptian or for that matter the base of any of the 30 Mesozoic stages and ~75 substages based on the succession of ammonoid zones/subzones/horizons, so also their resolution in the Mesozoic. Ages of the biostratigraphic zones of parallel scales based mostly on microplanktons are integrated with the primary Mesozoic scale based on ammonoids. The ammonoid-based scale in turn has been perfected in the shallow marine sedimentary successions. The resolution of correlation of the principal regional ammonoid scales is invariably of zone level, which is of the order of ~800 ka. The resolution in the best resolved individual regions/sections is of horizonal grade, i.e. ~200 ka. The subzonal sediment interval, although further indivisible in context of the included ammonoid taxa, invariably includes physically/visually countable number of cycles/sequences known to us as milankovitch cycles mostly based on the earth's orbital dynamics. This in turn allows further refinement in the age of a bed or surface in context of the precession/obliquity/eccentricity cycles.

7.1.4.3 Precise Dating of Sequence Surfaces Vis-à-Vis not so Precise Radiometric Dating

The punctuated individual stages/steps of the geodynamic evolution are invariably dated through radiometric ages of the rift and plume-genic igneous rocks, and ophiolites, respectively, of divergent and convergent regions, magnetic anomalies and polarity reversals, and ocean floor volcanism, which in spite of the technological progress over the decades still include large plus-minuses and margin of errors of a few to several million years. For example, the Rajmahal traps in the literature are ranged from the basal Berriasian to close of Albian, i.e. between ~145 and ~100 ma, while the individual dates have a margin of ~1–2 my between the minimum and maximum. The dismemberment history also includes movement along the transforms, and across the spreading axis, as also shifts/jumps of the spreading axes, which again are interpreted through magnetic anomalies and radiometry. Also, based on the included different groups of mega- and micro-guide fossils, an event marker bed or sediment interval of a lithostratigraphic unit provides highly varying ages. Such age data in turn does not allow workers to discriminate, filter, or converge to an objective precise age to the event. Vast and significant new geophysical information from deep in the Indian Ocean inclusive of sequences of magnetic anomalies has in recent years poured in, however, with near exclusive radiometric age constraints. Even the magnetic anomalies in these recent studies are exclusively assigned radiometric age references, and hardly referred even to the involved standard international stages. It is perhaps side-tracked, that the radiometric ages have changed with refinement over the years, while the stages have had more or less unchanged definitions at least over the last few decades. Even the correlations of strata and included geological events are beset with similar divergence of opinions and ages. In such a fluid scenario, the rift and plume-genic igneous expressions are here genetically tied to the extreme precisely age-constrained sequence surfaces. The most refined U-Pb shrimp and Ar39/Ar40 ages in absence of conceptual linkage to some geological event hardly fulfil the desired objective. In this backdrop, single highly precise sequence surface based ages are offered as an alternative plausible innovative approach. The precisely dated regionally isochronous sequence surfaces, also, facilitate much better age constraints and correlations in both marine and non-marine successions.

7.1.4.4 Fuller Record in Structurally Undisturbed Basins

In the divergent and convergent zones, most of the information is, invariably, lost for posterity, while the little that is left is extremely difficult to retrieve, and when retrieved is extremely difficult to date with precision. Even in subsurface wells, in spite of the constantly and progressively improved multidisciplinary geophysical tools, the resolving power hardly has been beyond 5–10 m. Paradoxically, on the other hand, major geological events often occur in the extremely slow sedimented condensed to starved sediment intervals that require centimetre level precision, e.g. the first-order intra-Oxfordian oceanization event or the K/T boundary event. When a new ocean floor basalt sample, or a volcanic flow sample, or a magnetic anomaly recognized in some region/basin/section is radiometrically dated, one is hardly ever sure of the same being the oldest or the youngest of a particular event. Added to that is the resolving limitation of the radiometric system used. On the other hand, the first- and second-order major/regional, and the third and, yet refined sequence stratigraphic surfaces produced by such geological events are precisely dated in the ammonoid rich structurally undisturbed exposed successions at extremely high resolution. The physical determination of the sequence stratigraphic surface of no thickness or interval of a small thickness once determined is highly reliable and involves much less subjectivity of interpretation. The involved diverse readily observable drastic changes across such surfaces provide excellent evidence of the tectonic event. The same surfaces and intervals are also readily open to later checks and verifications any time by any individual or groups of specialists. Ages of such ammonoid dated sequence stratigraphic surfaces are then conceptually extended to the diverse resultant facets of the said event in the eventful convergent/divergent theatres as, apparently, unrelated produce of a common geological event. It is this innovative approach that is applied in this work with several fold better resolved ages of the geological events for the objective consideration of the multidisciplinary geoscience community across the world. The combination of ammonoid-based and sequence surface-based timelines, thus, seemingly hold the key to precise absolute chronicling of regional to global tectonic and igneous events.

7.1.4.5 Relative Scales of Operation of Magmatic, Sedimentation and Ammonoid Evolutionary Processes

The magmatic events are few and geologically protracted. The entire Indian Phanerozoic span of ~540 my is organized into a succession of 11 periods, however, the number of magmatic events is any way distinctly less than that. Even the individual flows of a single magmatic event are found alternating with sediment intervals that invariably include multiple fine order milankovitch eccentricity, precession and obliquity sedimentary cycles/sequences. The ammonoid heterochronic infraspecific evolutionary succession allows to precisely date the ~200 ky Jurassic subzones/horizons, and within the individual subzone/horizon approximates ages of the visible individual cycles of fifth and yet finer order. Example is cited of the ~400 bed 15 zone 25 horizon ~3.4 my Callovian in the Keera section of the Kachchh Kachchh, so also are the other highly resolved Jurassic stages in Kachchh for the entire IEAP and GTM, that in turn allow the dating of the GTM first and second-order sequence surfaces at subzonal/horizonal and Mialncowich cycle precision.

Mere two wide spread regional magmatic events are included in the Jurassic in the Gondwanaland that so far remain only broadly dated, respectively, in the Early and Late Jurassic vis-à-vis over ~300 ammonoid-based precisely resolved horizons. Thus, the regional tectonic and associated magmatic events can be several fold better age précised in the structurally undisturbed copy book type ammonoid bearing successions located away from the extension/collision theatres.

7.1.4.6 Diverse Geological Aspects Already Covered in the Earlier Chapters

In the foregoing chapters, the diverse resultant facets of the geological changes and events involving lithology, facies, texture, structure, mega- and micro-flora/fauna, stratigraphic gaps, tectonic, igneous, glacial, radiometric, isotopic, geomagnetic, climatic, anoxic, latitudinal and many others have been detailed, synthesized, and used in the formulation of the GTM mega, first-, second-, third-order and yet finer sequences. In this geodynamic tell-tale, the plethora of the names of the lithostratigraphic units of the individual basins are largely done away with that the students and even other professional scholars inclusive of the hydrocarbon industry, invariably, find extremely strenuous to remember and cope up with. These lithostratigraphic units, otherwise, are bounded together by the succession of geological events common to the entire GTM that occurred at sequence stratigraphic surfaces. The succession of the geodynamic events of the mega-sequence is developed first in Kachchh, and subsequently followed in India, IEAP and GTM, across the Neotethys and finally across the Gondwana as the biggest tectonostratigraphically somewhat homogeneous unit of the globe. As an example the near Pan-Gondwanian manifestation of the first-order intra-Oxfordian ~159 ma MFS volcanism is cited on the one hand on the GTM Indo-Australian margin (Krishna 2016 communicated) while on the other hand in the Patagonia–Africa–Antarctica region on the Southeast Pacific margin of the Gondwana (Pankhrust et al. 1999) located several thousand km away from each other. Most surprising is the precise determination of the date neither by zircon nor Ar40/Ar39 nor U/Pb systems of the volcanics at either place of eruption but in distant remotely located Kachchh in an ammonoid abundant sedimentary column bereft of any volcanism to a precision of ~200 ky, even yet finer.

7.1.5 *Unified Integrative Geodynamic History of the Indian Plate, and the Ammonoid Provinces, Corridor Seaways and the Related Oceans Included in the Tell-Tale*

The Indian plate margin at the start of the mega-sequence compositely included the west margin between India (inclusive of Madagascar) and Africa, north margin along the High Himalaya and east margin between India and Australo-Antarctica. These margins in turn were parts of the divergent fracture zones in the north, south, east and west sectors. The parts of the GTM between the two divergent sectors constitute the IEAP. The east and west margins on either side of India, eventually, were united in the south in the vicinity of South Africa, Antarctica, and South America. In addition to the basins on the GTM, there are significant inland non-marine basins, e.g. the Gondwana basins. The succession of the first and second-order sequence surfaces is applied to develop an integrated evolution of the Indian plate during the progress of the mega-sequence through demonstration of the near isochronous geological happenings in all the sectors of the Indian plate margin as also on the entire GTM. The geodynamic evolution as well dwells on the ammonoid geographic units of India and neighbourhood. The shallow shelfal margin between East Africa and NW Australia makes the IEAP. The Arabian and SW Pacific provinces border the IEAP, respectively, in the west and east, while due south of IEAP is SE Pacific. The Gondic and Dravidian corridors (Fig. 1.3) run due south from the Neotethys to Pacific and unite in the SE Pacific. The chronicle of the events as well bears on the five oceans in the vicinity of the Indian plate. Paleotethys vanished at the start of the mega-sequence. Neotethys originated at the origin of the mega-sequence. Pacific existed before the start of the mega-sequence and continues to be extant, while the Indian and Atlantic Oceans originated in the later part of the sequence on either side of the Axial Gondwana and are also extant. The disappearance of the Neotethys marked the closure of the mega-sequence.

7.1.5.1 Suggestive Integration and Dating of the Volcanism-Devoid Rift, Rift-Genic and Plume-Genic Igneous and Associated Sedimentary Rocks Through Sequence Surfaces

The crustal thinning, necking, stretching, active and passive, narrow and wide rifting, preceding or following the volcanism, transform sliding, drifting and riding over plumes make the most important stages of the geodynamic evolution and plate-tectonic history of a region of the like of GTM. During the course of the Gondwana dismemberment and drift, the plate-tectonic history of the Indian plate is considered frequently punctuated invariably with a succession of as many as six rift and plume-genic igneous events almost in each successive period, which are intervened by relatively volcanism-free intervals in the Late Permian to Early Paleogene span of the mega-sequence.

It is conceived that the individual igneous events, for that matter all kinds of geological events, invariably include three differentiable stages. There is the start and the initial phases that are invariably mild in context of the defining parameters of the igneous event. The start of the event after a distinct lag of time peaks or climaxes, and finally the volcanism wanes to termination. The peaking includes bulk volcanism invariably within one to three million years, and is considered to include the deepest melts. Also, the spread of the volcanics at the climaxing is the farthest. The time lag from initiation to peaking, or from peaking to closure may differ from event to event, may be large, or the start in no time may reach to climax, and so also from climax to closure. If, influenced, or catalyzed, or triggered by a hot spot/plume, the start may synchronize with the arrival/positioning of the plate within the circle of influence of the plume.

The Panjal and Deccan igneous events occurred near the start and closure of the mega-sequence with their initiation, peaking and closure dated suggestively through sequence surfaces. The Rajmahal, St Marys, Mannar volcanism occurred during the span of the mega-sequence. Although, the principal igneous events through the mega-sequence have their main focus in a single sector, coeval expressions are invariably evidenced as well from other sectors. The Panjal event is focused in northwest, the Saurashtra event in southwest, Myanmar–Meghalaya–Bengal event in the northeast, Rajmahal in the east, St. Marys–Mannar event in west to south and Deccan event in northwest. The sudden proliferation of the radiometric ages of igneous events is noted in recent decades, so triggered by refinement in techniques, particularly the more reliable zircon ages through the U-Pb SHRIMP, and Ar40/Ar39 systems. Even this spurt and refinement, in spite of substantively enhanced precision coupled with minimization of the margin of error, has hardly succeeded in preventing the relative large range of ages from a single unit. The said spurt and refinement has, nevertheless, facilitated relative statistical clustering of ages of supposedly single events, seemingly around the ages of the first-order sequence surfaces with regional to global isochronism of at least some major geological events. However, integration of the multidimensional new data in recent decades has substantively constrained the paleogeographic reconstructions, as also the order of the events, but for their precise timings.

7.1.5.2 Varied Rift Processes and Styles

The rifting, if not always, quite often, is evidenced to occur in episodes with intervals of relative quiescence in between. Also in the progress of rifting, shifts/jumps of the rift/spreading axes take place. The two episodic phenomena correlate to result in multiple rift phases/episodes. The continental break-up occurs when the rate of extension or velocity exceeds a critical point, while stretching below the critical extension velocity does not lead to break-up and instead undergoes shift/migration of the rift axis. This causes abortion of the rifting at the former locus, and in turn follows uplift of the rifted region. Small extension rates produce multiple rift phases as perhaps in the Indian west sector in Kachchh, Jaisalmer, Cambay and Narmada, while large extension rates lead to break-up, e.g. in the west divergent sector between the East and West Gondwana as exampled at the south extreme of the west sector in Saurashtra, Mozambique, South Africa and Antarctica. The multiphase rifting, under low extension velocity, results in small horizontal extension of several tens of kilometre to couple of hundred kilometre, while the break-up rifting under high extension velocity involves horizontal extension even of a few thousand km. The Kachchh–Narmada rift system is thus included under the low extension velocity with horizontal separation of less than 200 km each. There could also be involved change in extension velocity from low to high during the rift progression. The narrow rifting/necking or wide extension is also dependent on the lithospheric temperature and crustal thickness, with former related to normal crustal thickness and the latter to thickened crust. The rift-genic magmatic activity depends upon the thermal anomaly in the mantle beneath the crust. The volcanic rifted margins are invariably bound to plumes, although many do not accord with this idea. However, the volcanic rifted margins are often associated with LIPs of the like of Karoo, Rajmahal, or Deccan.

7.1.5.3 Geodynamic Chronicle Punctuated Through the Mega, First- and Second-Order Sequence Surfaces

Apart from the igneous manifestation of the geolocial events black boxed through the mega-sequence/mega-surfaces, the said surfaces also evidence and date the multifaceted geological facets. The high-resolution chronicle uses seven first-order sequence surfaces of the mega-sequence. The multifaceted geological events at each first-order surface are equally important, and have hardly been unified in the manner done here. The chronicle additionally includes about 30 somewhat less eventful second-order surfaces (Tables 7.1 and 7.2).

The successive first-order and a few unordered major sequence surfaces are the Late Permian ~272 ma first-order intra-Permian first-order SB, terminal Anisian intra-Triassic first-order ~243 ma MFS, Pliensbachian/Toarcian boundary first-order intra-Jurassic ~183 ma SB, late Middle Oxfordian ~159 ma intra-Jurassic first-order MFS (Table 7.3).

Table 7.1 Summary of Triassic second-order sequences

No. of second-order sequence	Age range of the sequence	MFS and comments
T-X	Rhaetian Para Formation	MFS in the lone shale bed of the unit
T-IX	Middle to Late Norian Entire 28 m thick Hangrang Formation plus entire 130 m Alaror Formation and entire Nunuluka Fn	MFS in the *Neotibetites* bearing shale dominated upper part of Alaror Fn of Late Norian age
T-VIII	Early Norian Entire Rangrik Formation	MFS in 8 m thick *Juvavites* bearing shales of late Early Norian age in Upper Member. The Early/Norian boundary is also précised above the *Juvavites* beds between thick sandstone/siltsone below and thick limestones above, both lying above the *Juvavites* bearing thick shale bed
T-VII	Latest Early to Late Carnian entire Rongtong Formation	MFS in *Gonionotites* bearing calcareous siltstone in Middle Rongtong Member of mid-Late Carnian age
T-VI	Mid to late Early Carnian 220 m thick Rama Formation	MFI in the siltstone/shale cyclic interval of Rama Formation enclosing *Tropites* and *Spirogmoceras* in late Early Carnian. This differs with the Early/Late Carnian boundary of Bhargava et al. (2004) within Rama Formation. The said boundary is suggested between Rama and Rongtong units
T-V	late Early Ladinian Sutherlandi Zone–Early Carnian Aon Zone. Entire Chomule Formation	MFS in or above the hard ground level BL 12 at Muth 3 at the boundary of Regoled Zone and Canadensis Zone in the terminal part of Ladinian
T-IV	Intra-Ladinian Langobardian Gredlerensis Zone to late Early Ladinian Maclerni Zone). Terminal part of Niti Limestone and entire Kaga Formation	MFS in or above bed VT 24 of early Late Ladinian Meginae Zone age inclusive of *Daonella* bearing interval
T-III	Late Olenekian Meekoceras Zone–Early Ladinian Fassinian Curionii Zone, Terminal part of Limestone + Shale Formation, entire Niti Limestone to bed 47 of Himalayan Muschelkalk	MFS/MFI in Hollandites Zone to Kellnerites Zone
T-II	Basal Olenekian Rohilla Zone–Intra-Olenekian Flemingites-Euflemingites Zone Beds 12–39 of the Muth section of Limestone + Shale Formation	MFS within the bed 15 (Intra-Olenekian Flemingites griesbachi Zone

(continued)

Table 7.1 (continued)

No. of second-order sequence	Age range of the sequence	MFS and comments
T-I	Late Permian–intra-Induan Frequens Zone (Limestone + Shale Member) Beds 1–11 of the Muth section	MFS either within or at the top of Discophiceras Zone (below bed GU 4/4)

Table 7.2 Summary of Jurassic second-order sequences

No. of second-order sequence	Age range of the sequence	MFS and comments
J-XIII	Natricoides Zone–Densiplicatus Zone Sequence of ~3.6 my. **Bed 25 at Ler to bed 18 at Mundhan**	MFS above the Natricoides Zone
J-XII	early Early Tithonian Pottingeri Zone–Virgatosphinctoides Zone Sequence of ~2.8 my. **Beds 14–24 at Ler**	MFS above bed 19b in Rajnathi Subzone of Virgatosphinctoides Zone
J-XI	late Early Kimmeridgian Alterneplicatus Zone–late Late Kimmeridgian Katrolensis Zone of ~4.0 my. **Beds 1–13 at Ler**	MFS above bed 3 in Ler of the Bathyplocus Zone
J-X	early Early Kimmeridgian Kachchhensis Zone–mid. Early Kimmeridgian Giganticus Subzone of ~2.0 my. **Sediment intervals X–XX at Bharodia**	MFS above sediment interval XV of the Kachchhensis Zone
J-IX	Obliqueplicatum Zone–Subevolutus Zone Sequence of ~4.8 my. **Beds 25–105 at Gangtabet**	MFS above bed I-08 of the late Middle Oxfordian Schilli Subzone of Orientalis Zone
J-VIII	Late Middle Callovian O-II Horizon Obtusicosta Subzone Obtusicosta Zone–early Early Oxfordian of ~2.1 my. **Em to Fm of ULS in Jhura Section**	MFS above bed 392 in Keera section in A-II Horizon Depressum Subzone Athleta Zone of early Late Callovian age
J-VII	late Middle Bathonian Mantataranus Zone/Bremeri Zone–O-I Horizon of late Middle Callovian Obtusicosta Zone Obtusicosta Subzone sequence of ~2.40 my **Sediment intervals I to II at Jumara to bed 217 at Keera**	MFS above bed 92 in Keera section of the Diadematus Subzone of Chrysoolithicus Zone
J-VI	Late Bajocian Garantiana Zone–mid. Middle Bathonian Morrisi Zone of ~1.5 my **Beds 49–58 of LSM at Kaladongar**	MFS above bed 9 in Nara section of the Early Bathonian Macrescens Subzone
J-V	Intra-Aalenian / Basal Bajocian–mid-Late Bajocian of ~3.6 to ~1.7 my **Beds 37–48 of LSM at Kaladongar**	MFS above bed 47 at Kaladongar in the Early Bajocian Propinquans Zone
J-IV	Early Toarcian–? Intra or end Aalenian second-order sequence of < ~12.4 my **Beds 1–36 at Kaladongar**	MFS above bed 24 in Kaladongar section in Nitiscens Zone

(continued)

Table 7.2 (continued)

No. of second-order sequence	Age range of the sequence	MFS and comments
J-III	Latest Sinemurian to end Pliensbachian of ~8 my **Beds 6–9 in Spiti**	MFS above bed 8 in Spiti of Ibex Zone
J-II	Hettangian–Sinemurian boundary to Rauricostatum Zone of ~8.5 my **Beds 4–5 in Spiti**	MFS above bed 6 in Spiti of Oxynotum Zone
J-I	Hettangian of ~2.0 my **Beds 1–3 in Spiti**	MFS at the top of bed 2 in Spiti of the Middle Hettangian Liasicus Zone

Table 7.3 Summary of Cretaceous second-order sequence

No. of second-order sequence	Age range of the Sequence	MFS and comments
C-XII	Maestrichtian	MFS above the Early Maestrichtian Tridens Zone. The sequence in the Cauvery basin includes the Kallankuruchhi Formation up to the top of the rich *Gryphaea* bearing sediment interval as TST and the Ottakovil Formation/Kallamedu Formation as the corresponding terminal part of the RST
C-XI	Late Campanian	MFS above the? Early Campanian Delawarensis Zone. The sequence in the Cauvery basin includes the Sillakudi Formation
C-X	Coniacian–Santonian	intra-Coniacian MFS. The sequence includes the Kullakalnattum Formation/Anaipadi calcarenite as the TST at the early/late Late Coniacian boundary while the sandy Saturbhugam Formation makes the following RST
C-IX	Late Cenomanian–Late Turonian	MFS above the early Middle Turonian Turoniense Zone/Helvetica Zone/Gartneri Zone. The sequence includes the Karai Formation with MFS in the terminal part of the Karai Formation
C-VIII	Late Middle Albian–Middle Cenomanian	MFS near the Albian/Cenomanian boundary
C-VII	Early to mid-Middle Albian **Beds 17 and 18 of the Ukra Member**	MFS above the early Middle Albian Dentatus Zone at the top of the bed 17
C-VI	Early Aptian Deshayesites Zone–late Aptian Furcata Zone **Early part of the Ukra Member (beds 15 and 16)**	MFS within the bed 16 at the top of the green glaucontic shale/sands within the Ukra Member of Weissi Zone age
C-V	Late Barremian **Beds 13–14 of the Ghuneri Member) in the Ghuneri–Amarsar section**	MFS in the terminal Berremian at the top of bed 13

(continued)

Table 7.3 (continued)

No. of second-order sequence	Age range of the Sequence	MFS and comments
C-IV	Hauterivian–Early Barremian **Beds 10–12 of the Ghuneri Member) in the Ghuneri–Amarsar section**	MFS in the terminal Hauterivian at the top of bed 11
C-III	Valanginian **Beds 5–9 of the Ghuneri Member) in the Ghuneri–Amarsar section**	MFS in the early Late Valanginian Verrucosum Zone with *Olcostephanus* in the Spiti basin tentatively correlated to the top of the bed 7
C-II	Late Berriasian **Beds 33–36 at Lakhaper**	MFS above the *Kilianella* bearing bed 35 in the Ghuneri section
C-I	Middle Berriasian **Beds 28–32 at Lakhaper**	MFS at the transition of Jacobi Zone/Occitanica Zone with *Argentiniceras* and *Spiticeras* above the basal Green Glauconites of the bed 28 of the Basal Member of the Umia Formation in the Lakhapur section

MESOZOIC Marker Ammonoids of First/Second-Order TSTs/MFSs

The 12 plates include important ammonoid genera that have facilitated delineation of the second-order MFSs as also their precise dating. The ammonoid figures included in a single plate are neither to size nor maintain uniform reduction.

Plates I–XII

TRIASSIC (Plate-I–III)	*PLATE-I*
	Figure 1: *Otoceras woodwardi*, T-I second-order sequence TST, basal Induan Woodwardi Zone
	Figure 2: *Discophiceras* sp, T-I second-order sequence MFS, Early Induan Discophiceras Zone
	Figure 3: *Flemingites griesbachi*, T-II second-order sequence MFS Late Induan Rohilla/Flemingianus Zonone
	Figure 4: *Kellnerites trinodosus* T-IV second-order sequence MFS, terminal Anisian Trinodosus Zone; and
	Figure 5: *Hollandites* sp T-IV MFS
	All the five figures ammonoid specimens from Spiti and are sourced from Krystyn et al. (2007)
	PLATE-II
	Figure 6: *Maginoceras* sp T-V second-order sequence MFS—Late Ladinian Maginae Zone
	Figure 7: *Frankites* sp T-VI second-order sequence MFS—terminal Ladinian Sutherlandi Zone

(continued)

FP. 7.1 (continued)

	Figure 3a, b: Lateral and ventral view of *Tropites subbullatus* T-VII second-order sequence MFS—early Late Carnian
	PLATE-III
	Figure 1a, b: Lateral and apertural views *Goniatites* sp—T-VIII second-order sequence MFS, Late Carnian *Goniatites* bearing bed, Middle Member, Rongtong Formation
	Figure 2a, b Lateral and apertural views, *Juvavites* sp—T-IX second-order sequence MFS, Late Norian *Juvavites* beds, Upper Member, Rongtong Formation
	Figure 3a, b Lateral and apertural views of *Tibetites* sp—T-X second-order sequence MFS, Rhaetian *Neotibetites* beds, Upper Member, Alaror Formation
JURASSIC (Plate-IV–IX)	*PLATE-V*
	Figure 1a, b: *Psiloceras planorbis* J-I second-order TST sourced from the Treatise on Ammonoidea
	Figure 2a, b: *Oxynoticeras* J-II second-order TST Late Sinemurian sourced from the Treatise on Ammonoidea
	Figure 3a, b: *Bouleiceras nitescens* J-IV second-order TST from Madagascar, sourced from the Treatise on Ammonoidea; and
	PLATE-V
	Figure 1: *Calliphylloceras heterophylloides* from Kachchh, J-V second-order sequence MFS, Early Bajocian Laeviuscula Zone
	Figures 2–4: Procerozigzag spp, 2. Septate, 3. body chamber and 4. Only early party septate all from Nara sourced from Pandey and Pathak (2015a)
	Figures 5–6: *Parkinsonia* spp body chamber fragments, both from Nara
	Figures 7–8: *Zigzagiceras* body chamber fragments
	Figure 9: *Parkinsonia* sp body chamber fragment
	Specimens 2–8 come from the second-order sequence J-VI TST/MFS interval at Nara sourced from Pandey and Pathak (2015a) plate 1
	PLATE-VI
	Figure 1a, b: Lateral and apertural views, *Macrocephalites diadematus* from Kachchh J-VII MFS, mid/late Early Callovian Didematus Subzone; and
	Figure 2a, b: Lateral and ventral views, *Kinkeliniceras depressum* from Kachchh J-VIII second-order sequence TST, early Late Callovian Depressum Subzone; and
	Figure 3: *Alligaticeras obliqueplicatum* from Kachchh J-IX second-order sequence TST, late Early Oxfordian Obliqueplicatum Zone
	PLATE-VII
	Figure 1: *Perisphinctes orientalis* from Kachchh J-IX second-order sequence TST, Middle Oxfordian Orientalis Zone
	Figure 2: *Gregoryceras gr. devauxi* from Kachchh J-IX second-order sequence MFS, late Middle Oxfordian Schilli Subzone; and
	Figure 3: Larcheria schilli from Kachchh J-IX second-order sequence MFS, late Middle Oxfordian Schilli Subzone

(continued)

FP. 7.1 (continued)

	PLATE-VIII
	Figures 1–3: *Orthosphinctes kachchhensis* from Kachchh, J-X second-order TST, early Early Kimmeridgian kachchhensis Zone
	Figure 4: *Ataxioceras giganticus* from Kachchh, J-X second-order TST, mid Early Kimmeridgian Giganticus Zone; and
	Figure 5: *Torquatisphinctes bathyplocus*, J-XI second-order sequenceTST, early Late Kimmeridgian Bathyplocus Zone
	PLATE-IX
	Figure 1: *Katroliceras rajnathi*, J-XII second-order sequence TST, Early Tithonian Rajnathi Subzone; and
	Figure 2: *Virgatosphinctes densiplicatus*, J-XIV second-order sequence TST, terminal Late Jurassic Densiplicatus Zone
CRETACEOUS (Plate-X–XII)	*PLATE-X*
	Figure 1a, b: Lateral and ventral views, *Argentiniceras locochensis* from Kachchh, C-I MFS, Early Berriasian Argentiniceras Zone
	Figure 2a, b: Lateral and ventral views, *Kilianella* sourced from the Treatise on Ammonoidea, C-II second-order sequence MFS, Late Berriasian *Kilianella* Zone; and
	Figure 3: *Olcostephanus astierianus* sourced from the Treatise on Ammonoidea, C-III second-order sequence TST, early Late Valanginian Verrucosum Zone
	PLATE-XI
	Figure 1: *Pascoeites* sp from Cauvery, C-V second-order TST,? terminal Late Hauterivian
	Figure 2: *Australiceras jacki* sourced from the Treatise on Ammonoidea, C-VI second-order sequenceTST, Late Aptian
	Figure 3: *Douvilleiceras mammillatus* from Jaisalmer, C-VII second-order sequence TST, early Middle Albian
	Figure 4: *Mortoniceras inflatum* sourced from the Treatise on Ammonoidea, C-VIII second-order sequence TST, Late Albian; and
	Figure 5: *Pseudoaspidoceras footeanum* sourced from the Treatise on Ammonoidea, C-IX second-order sequenceTST, early Middle Turonian
	PLATE-XII
	Figure 1: *Kossmaticeras theobaldinum* from Cauvery sourced from the Treatise on Ammonoidea, C-X second-order TST, Early Coniacian
	Figure 2: *Karapadites karapadense* from Cauvery sourced from the Treatise on Ammonoidea, C-XI second-order sequence TST, Late Campanian
	Figure 3: *Eubaculites otacodensis* from Cauvery sourced from the Treatise on Ammonoidea, C-XII second-order TST, Early Maestrichtian
	Figure 4: *Gunnarites* sourced from the Treatise on Ammonoidea, C-XI second-order TST, Late Campanian; and

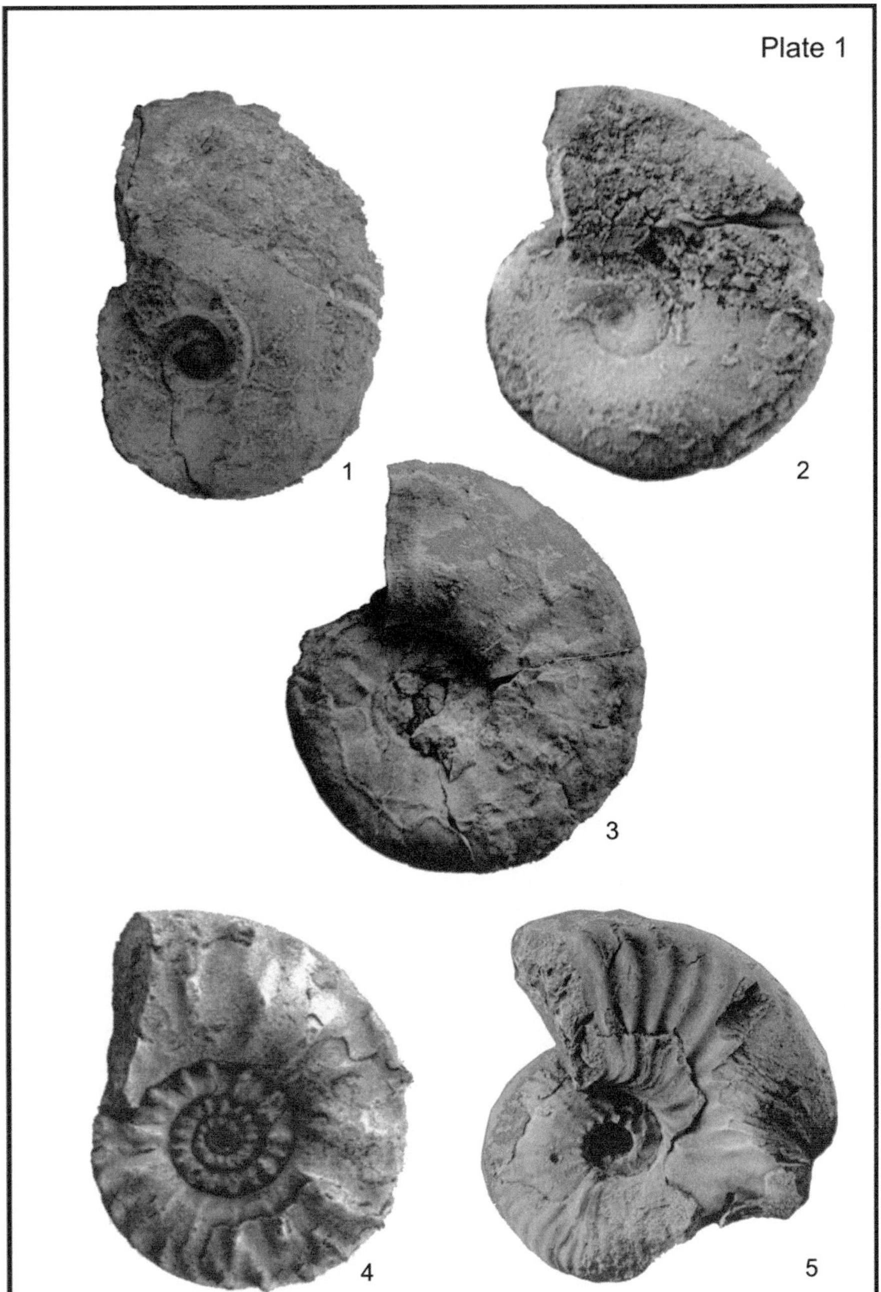
Plate 1
1
2
3
4
5

Plate 2

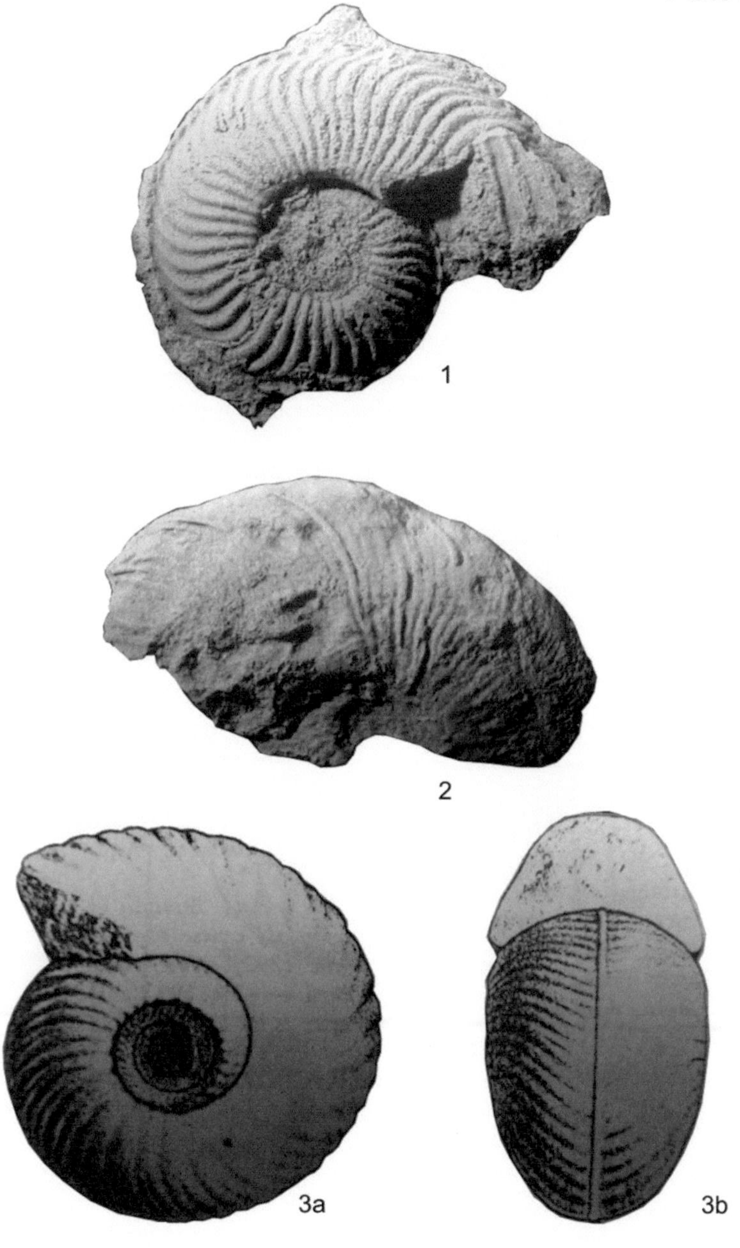

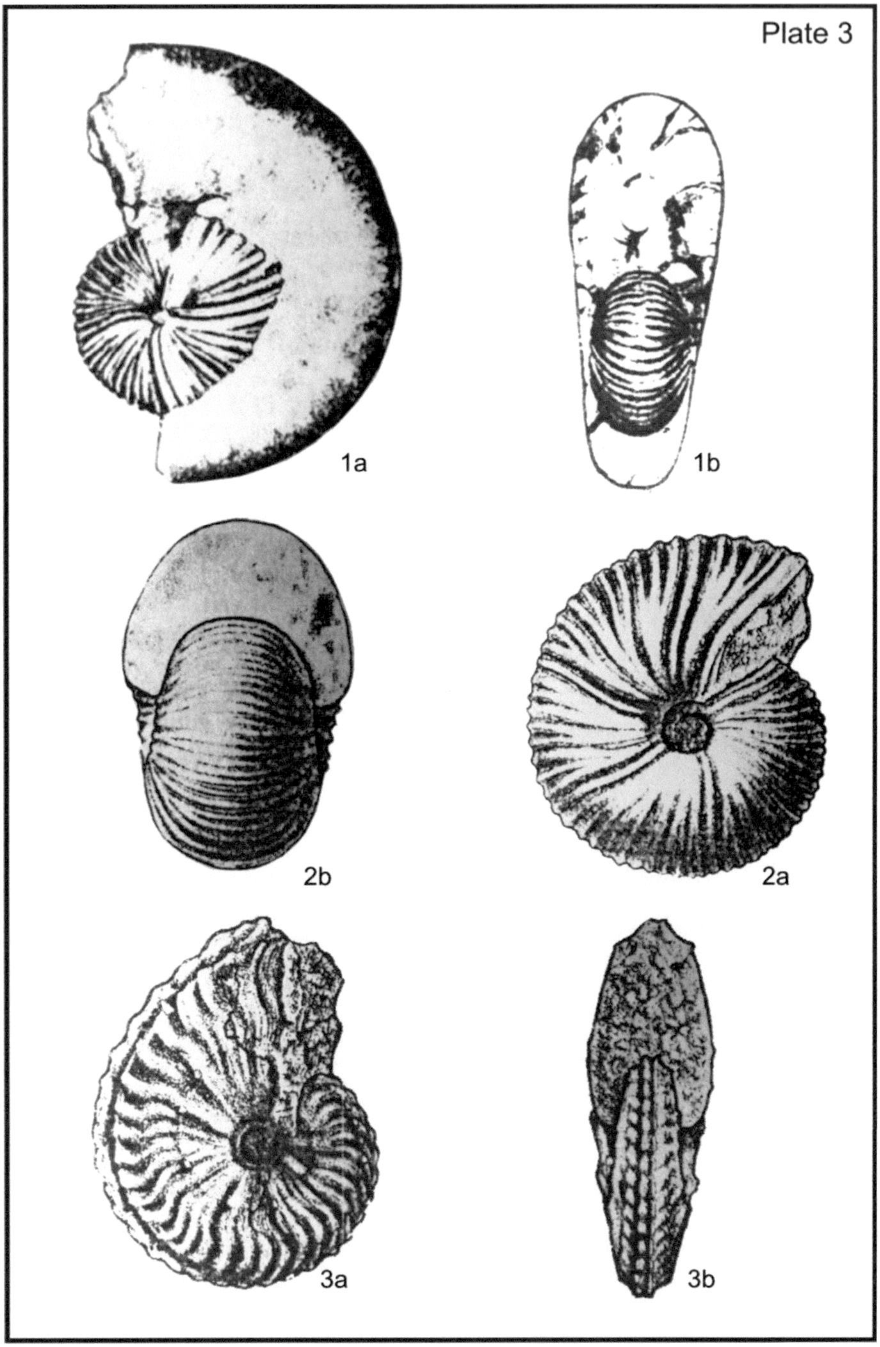
Plate 3
1a
1b
2b
2a
3a
3b

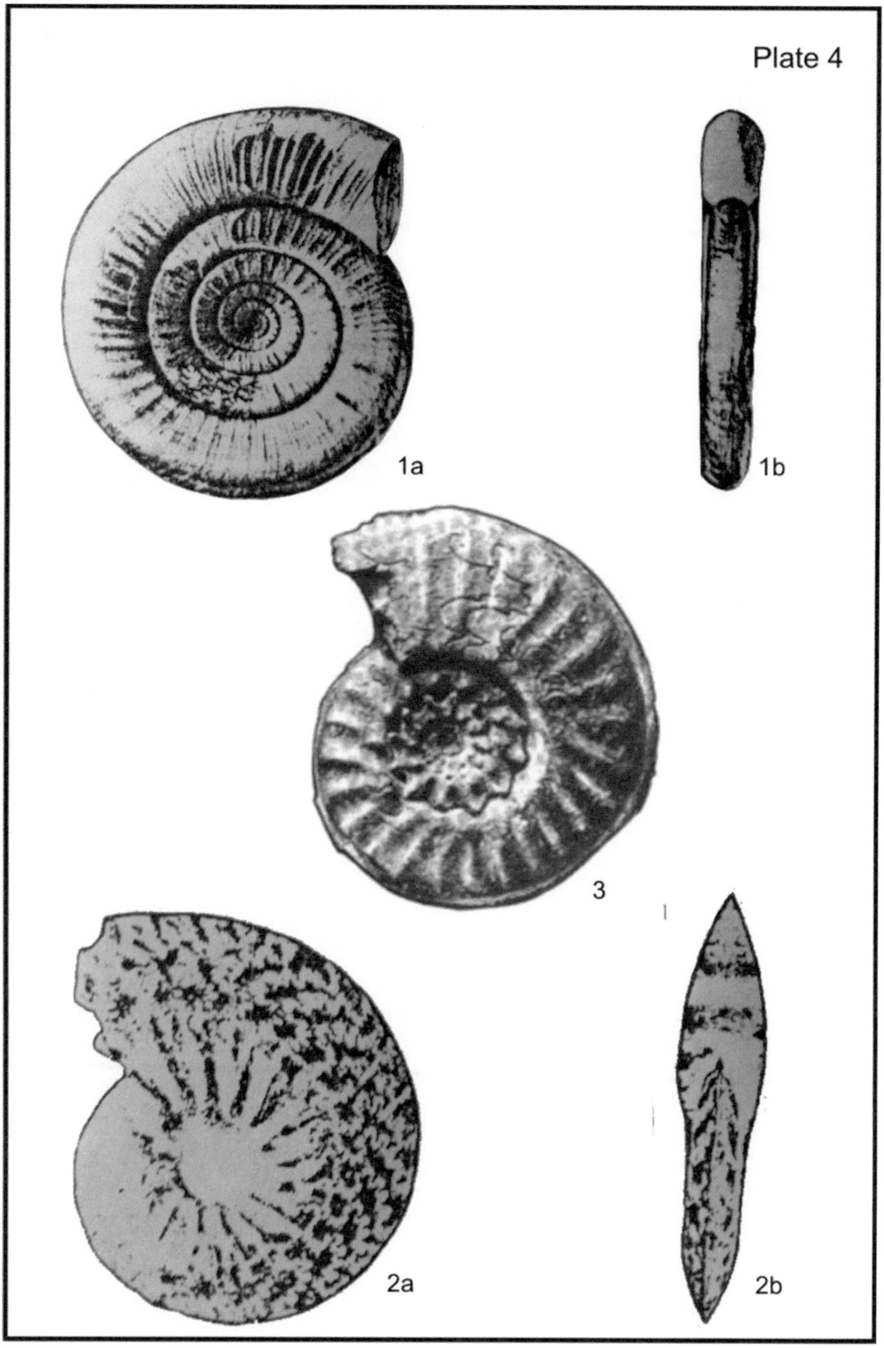
Plate 4
1a
1b
3
2a
2b

Plate 5
7
8
9
5
1
6
2
3
4

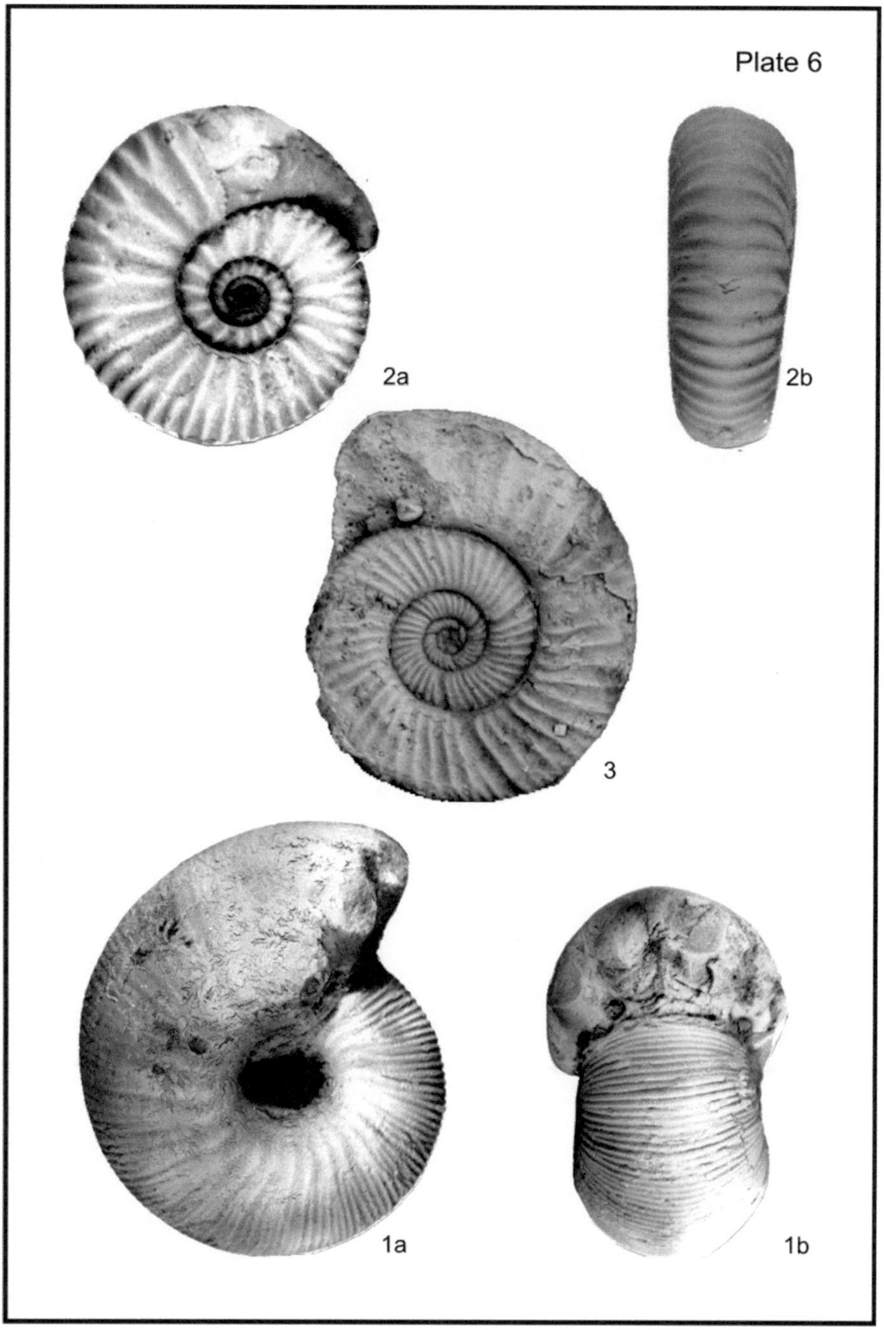

Plate 6

Plate 7

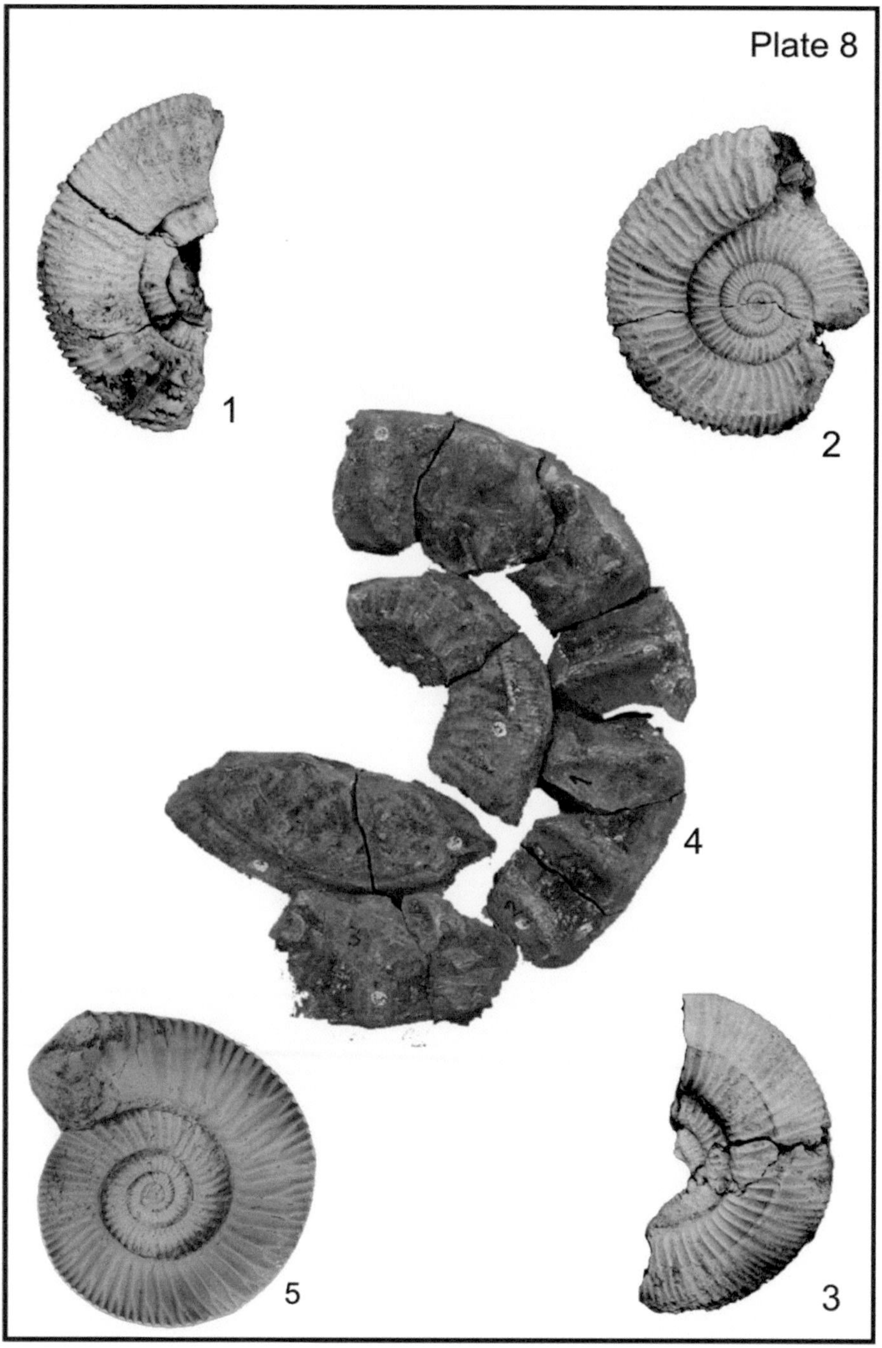
Plate 8
1
2
4
5
3

Plate 9
2b
2a
1

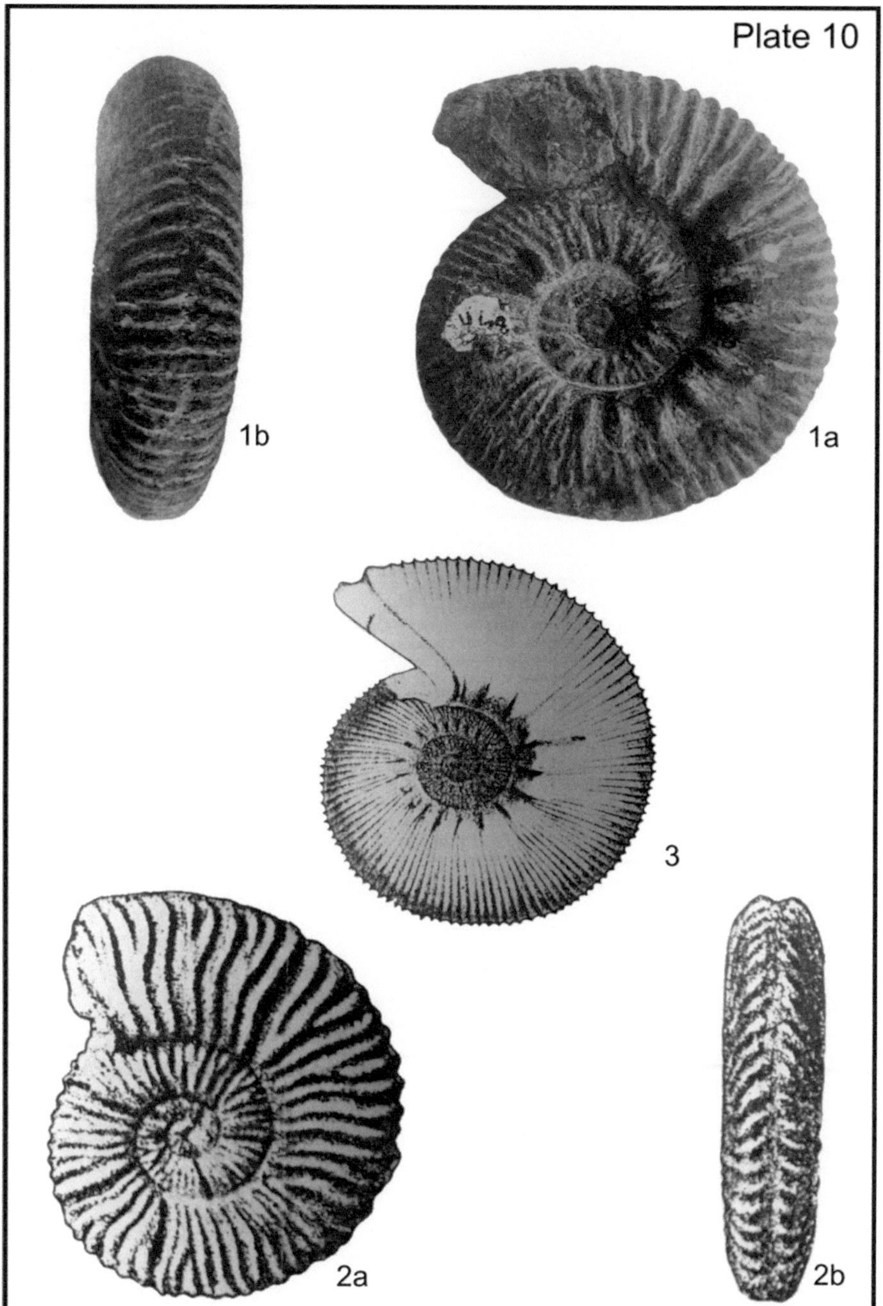
Plate 10
1b
1a
3
2a
2b

Plate 11
1
2
5
4
3

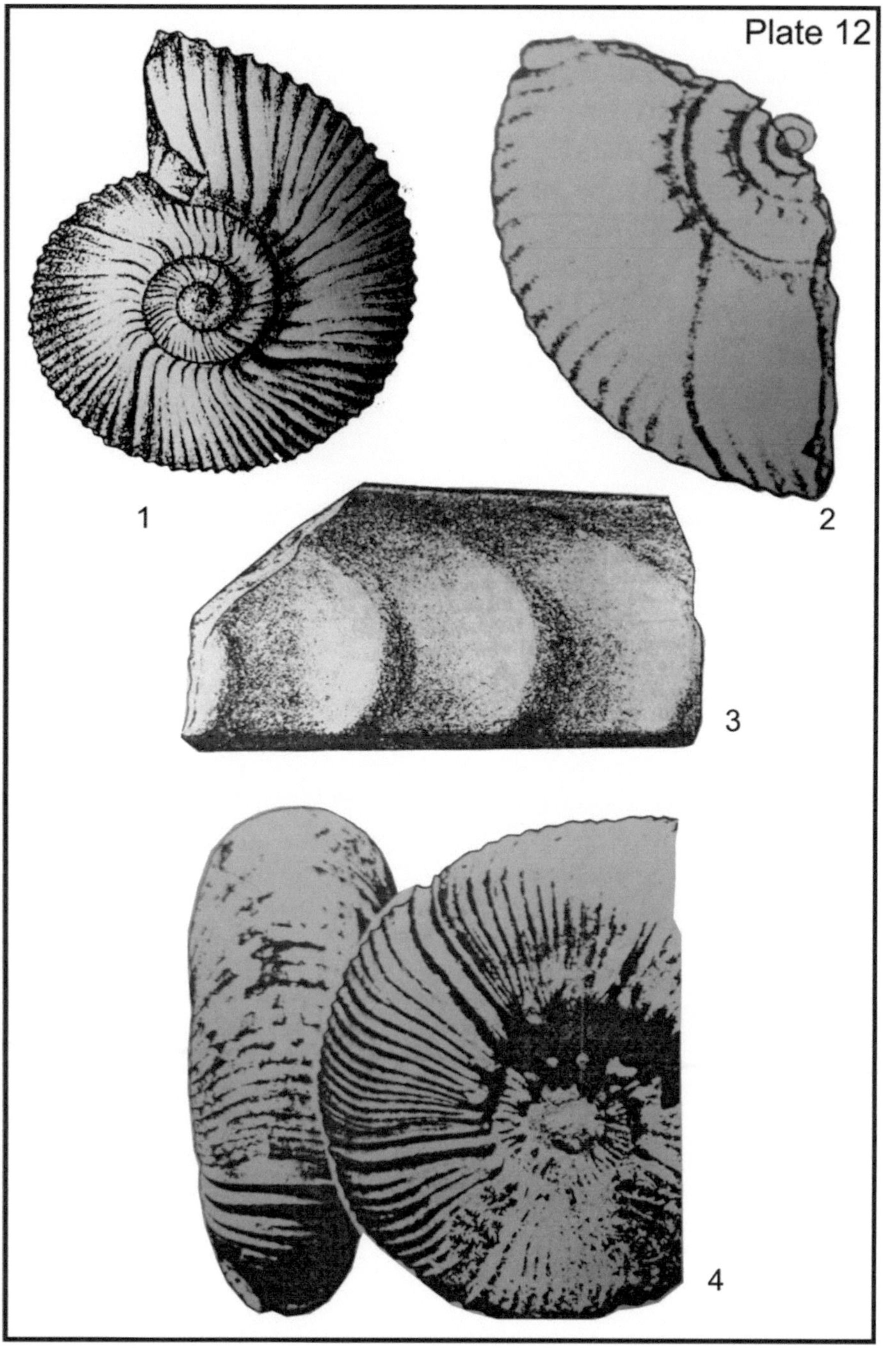
Plate 12
1
2
3
4

Intra-Valanginian ~136 ma MFS, Barremian/Aptian boundary first-order intra-Cretaceous ~126 ma SB, mid-Middle Albian ~107 ma MFS, early Middle Turonian ~91 ma first-order MFS, and intra-Paleocene first-order ~60–59 ma mega SB. The individual events are discussed in integrated geodynamic context with other manifestations in chronological order.

7.1.6 Constraints and Influence of the Inherent Precambrian Structural Trends in and Around India, and Progressive Refinement of the Size of Greater India

The relative positioning of the principal components in the east and west sectors is constrained by the inherent composite structural grains on either side of the conjugates. The Davies ridge–Gunnerus ridge runs in the NNW–SSE direction along the east face of Mozambique (Rabinowitz et al. 1983). In the east sector, another fracture runs along the east coast of Sri Lanka to east face of Cauvery between the Gunnerus ridge and Napier. In between these two is included the Mascarens ridge/axis between India and Madagascar. These three rift margins first acted as rift axes, then as a transform faults, and finally as spreading axes prior to their jumps to other locations.

7.1.6.1 West Sector

Another feature somewhat parallel to the Davies ridge is the Chennai–Mumbai Kurduvadi fracture from the Indian southeast margin to northwest margin, then farther along northeast margin of Madagascar on the west to east margin of Seychelles as also of east face of Saurashtra on the east (Rajaram et al. 2006; Reeves 2014) (Fig. 1.7). Farther along northwest, it seems to extend along the Malda fault, with alignment of NSF to MF along the south margin of Saurashtra. Mumbai High is located south of Saurashtra parallel and next to the Indian west coast (Bastia et al. 2010). The Moyar–Cauvery shear zone into Madagascar (Reeves 2014), as also the conjugation of the terrace like cut on the SW India to Madagascar (Yateesh et al. 2006) place further constraints to the India–Madagascar fit.

7.1.6.2 East Sector

The outward projecting Napier margin of Antarctica fits against the Cuddapah basin of Andhra. The Pranhita–Godavari shear zone is interpreted to extend due north into Saurashtra, due south into East Antarctica and also Mahanadi as Lambert rift. The Narmada–Son–Damodar belt is constrained through extension into the Albany and Fraser shear zones between Australia and East Antarctica. The Indian East coast up to Mahanadi is fitted against Antarctica from Gunnerus ridge to Prydz bay. The Prydz bay also appears aligned to the Son–Mahanadi zone.

Such an assembly in the neighbourhood of India inclusive of Madagascar in the India–Africa sector in the west, India–Australo-Antarctica sector in the east, India–Madagascar block prior to the start of the Late Permian–intra-Paleocene mega-sequence also included small terraines like Kachchh, Seychelles, Laxmi ridge, Bombay High and Saurashtra north of the NSL in close proximity of Northwest India and Northeast Madagascar until their progressive separation that began in early Late Cretaceous in Cenomanian–early Middle Turonian duration ($\sim$98/$\sim$91 ma MFSs).

7.1.6.3 Significant Differences on the Positioning of Madagascar Vis-à-Vis India

In the Africa–India–Antarctica region, the reconstructions of Gibbons et al. (2013) and Reeves (2014) differ quite a bit with regard to positioning of Madagascar vis-à-vis India and Sri Lanka. Reeves has constrained the same through alignment of the Moyar–Cauvery shear zone with Ranatsora zone of Madagascar, and the same is followed in the present work. The two detailed studies also differ with regard to the orientation of the Davies ridge and location of the Marion plume vis-à-vis India on the same count.

7.1.6.4 Progressive Refinement of the Size of Greater India

The size of Greater India has all along the decades since its conception way back in the 1940s varied by a few thousand kilometre, so also the magnitude and the timing of the crustal shortening involved. The recent discovery of a sliver of Late Jurassic crust amidst the Late Cretaceous ocean floor has further constrained the size of Greater India (Gibbons et al. 2012), with most of it located south of the large offset WZFZ.

7.2 Sequence Stratigraphy Based Geodynamic Chronicle

The entire Phanerozoic geodynamic chronicle is punctuated through more than 20 major events inclusive of 14 during the span of the intra-Permian–intra-Paleogene mega-sequence.

7.2.1 Brief Outline

The GTM at the start of the mega-sequence represented the Neotethyan margin of the Gondwana, which constituted more than half of the entire global land. As such, the significance of the geological events on the GTM in particular related to Gondwana in general cannot be over-emphasized. The first major reorganization of plates in the south hemisphere, initiation of the collision of East and West Gondwana and the associated origin of the Gondwanaland, also the resultant change of paleoslope from due west and northwest to due east and northeast occurred near the terminal Middle Cambrian first-order MFS. The said MFS event may have also resulted in the origin of the Molluscan cephalopods as the first necto-benthonic organisms. The Gondwana assembly was finally realized near the Cambrian/Ordovician boundary Pan-African event first-order SB. Most of India and neighbourhood except its Paleotethyan margin up to Ganga basin remained stable as positive/emerged areas up to the close of Carboniferous with the resultant major intra-Paleozoic stratigraphic gap/s of the Indian Phanerozoic geological record. The said $\sim$250 my long duration stability was shaken up by the initiation of widespread rifting in early Early Permian in Gondwana (the Karoo rifting in the India–Southeast Africa region in the vicinity of the south extreme of the India–Africa divergent sector), along with the origin of the Gondwana rift basins, and graded spreading away of Cimmeria with origin of Neotethys on India's Tethyan margin. In these just opened inland rifts of the Gondwana constituents are preserved the non-marine Early Permian–Early Jurassic sediments inclusive of the rich hydrocarbon and coal resources. The tectonic evolution during the course of the mega-sequence started with significant rifting in the east sector with origin of the Gondwana rift basins. Multiple jumps of the rift and spreading axes are included either way along east–west or north–south strikes with consequential separation of blocks in succession from different sizes of India in the east and west sectors. The appreciable spreading across the spreading axes is invariably preceded by transform sliding and rifting along the spreading axes. The first major event was the intra-Devonian origin of Paleotethys with spreading away of South China, Indo-China, etc. However, the mega-sequence in question started with spreading away of Cimmeria near the first-order basal SB event during $\sim$289 to $\sim$272 ma along with Panjal and coeval igneous event ($\sim$289 to $\sim$243 ma—Shellnut et al. 2009; Sakmarian volcanic intercalations in Chumic unit of Zanskar—Garzanti 1999; also volcanics intercalated within the Gondwana Talchir unit in M.P. Hoshangabad borehole—by Karve in Shanker et al. 1996; and in the lower part of Barakar at Ramkola, trachytic and phonolitic intrusives as also flows of Late Permian–Early Triassic age bracket in wells of the south part of the Carnarvon basin—Harris 1996). Marine signatures have also been interpreted in Barakar, Barren Measures, and Raniganj units of Permian Gondwana as evidence of more than one marine incursion in the otherwise dominantly inland non-marine Permian. The Early and early Middle Triassic marine signatures are present in North Madagascar and Pakistan; however, the same are absent within the non-marine Gondwana because of widespread uplift near the

Permian/Triassic boundary second-order SB. It is followed up by the spreading away of Lhasa, etc., at ~243 ma first-order MFS along with igneous event at the GTM in Papua New Guinea (Hill and Hall 2003), south part of the Carnarvon basin, and elsewhere in neighbourhood. Next in succession occurred, the separation of Argo block at the first-order ~159 ma MFS, origin of GAP, etc., at the second-order ~136 ma MFS, separation of CAP, PAP and EAP at ~126 ma first-order SB, and Elan bank, and South Kerguelen blocks at ~107 ma MFS in the east and northeast sectors. Madagascar separation peaked at the ~91 ma MFS and was fully realized by ~84 ma second-order SB, and finally separation of Seychelles at the second-order end Maastrichtian SB in the west sector. The significant steps of the India–Asia collision process are interpreted at the ~51–50 ma MFS in late Cuisian, and ~34 ma terminal Eocene SB events.

7.2.2 *Basal to Intra-Permian Infra-Rift Sagging to Syn-rift Transition Event in the East*

The inland Gondwana rift basins (in India, and Northwest and West Australia) originated as the outcome of the infra-rift sagging along the Precambrian intra-cratonic / intra-shield shear zones as interior sag basins of the master basin (Mitra and Rao 1987; Biswas 2005) near the second-order SB during the Sakmarian interval (~298 to ~289 ma). The Gondwana congregation in the east sector (Reeves 2014; Bastia et al. 2010) but for modifications on its Tethyan margin so continued from Middle Cambrian to close of Pliensbachian (Fig. 7.11). The transition from the infra-rift stage master sag basin to individual Gondwana rift basins is linked to the geographically restricted initiation of Neotethys as early as at ~289 ma Sakmarian/Artinskian boundary. Non-marine sedimentation in the individual Gondwana rift basins began near the Karharbari/Barakar or Barren Measures/ Raniganj lithostratigraphic transition. The next significant subsidence events in the Gondwana rift basins occurred near the Sakmarian/Artinskian, Early/Late Permian, Permian/Triassic boundary and intra-Triassic sequence surfaces with resultant stepped decrease in the quality and expanse of coal, D13Corg content, stoppage of coal formation, differential subsidence in the central and outer parts of the master basin and others. Monosaccate pollens/spores changed to disaccate at the start of Late Permian as part of the palynoevent 4 (Vijaya 1997). A minor transient transgression is also suggested in Kulti Formation. The palynoevent 4 is précised at the intra-Permian mega-sequence SB. The said progressive changes are found reflected in Spiti and other High Himalaya basins in prolific ammonoid/conodont rich slow sedimented fine textured basal member of the Mikin Formation.

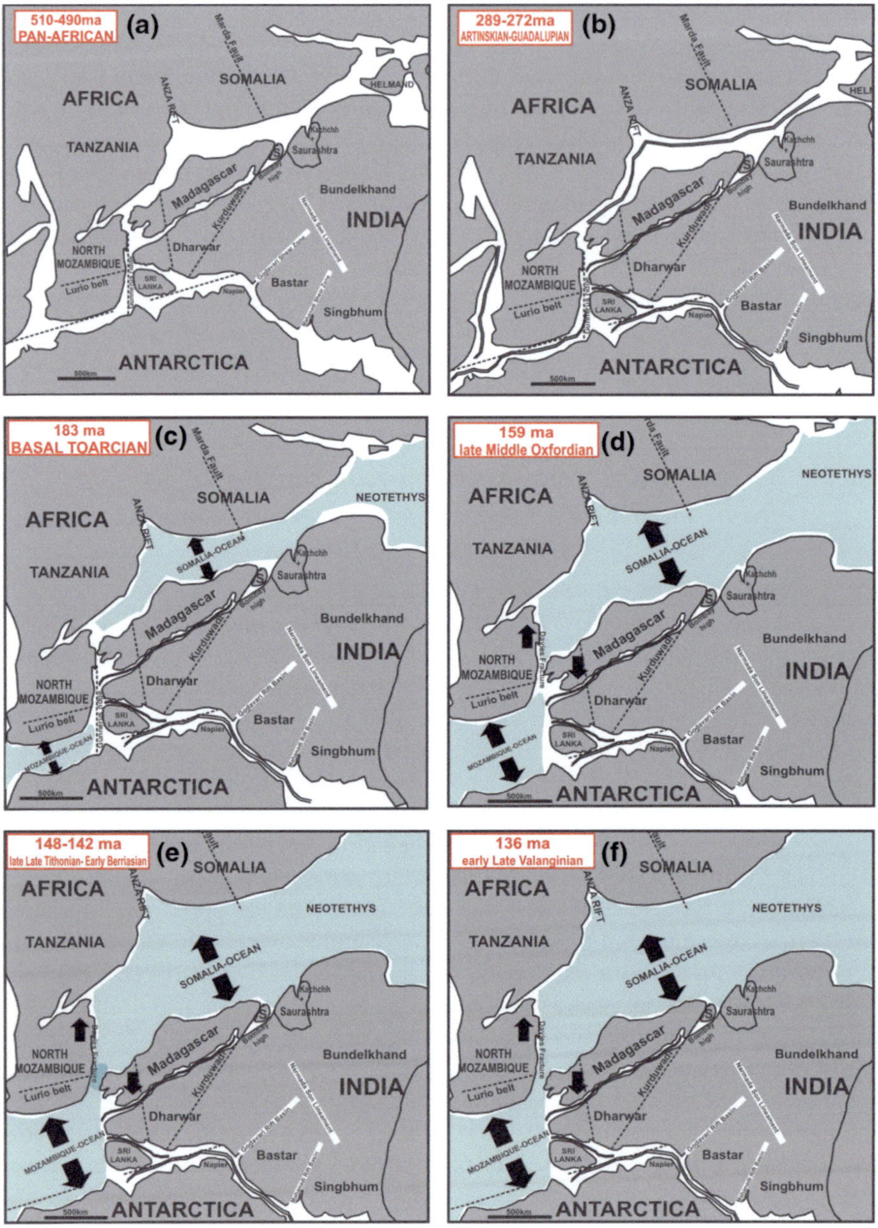

Fig. 7.11 Paleogeographic reconstructions at important first or second-order sequence surfaces near the **a** Cambro/Ordovician boundary, **b** Artinskian-Guadalupian, **c** Basal Toarcian, **d** late Middle Oxfordian, **e** late Late Tithonian–Early Berriasian and **f** early Late Valanginian with focus on the India–Africa sector (reconstructions modified after Bastia et al. 2010)

The surface, and subsurface well, geological and geophysical studies in the west sector basins of Madagascar, Indus, Jaisalmer, Kachchh, Saurashtra, Deccan and Konkan–Kerala, also, reveal similar largely non-marine Permian–Early Jurassic successions unconformably overlying the Precambrian (Chandrasekhar et al. 2011; Nathaniel 2013; Singh et al. 2000; Singh 2006). These suggest similar tectono-stratigraphic framework as that of the inland Gondwana basins of east and central India. Similar unconformities and gaps are presented prior to Permian and after Early Jurassic, as also within, of intra-Permian, and intra-Triassic ages. The diverse geological changes are also present in High Himalaya. The first true rifting was accomplished during Artinskian to Guadalupian by the Mid/Late Permian mega-order basal SB.

7.2.3 *Early Permian–Middle Triassic Igneous Event (~289 to ~243 ma) with Climax at the Second-Order Sakmarian/Artinskian MFS of the Intra-Devonian–Early Permian First-Order Sequence*

The event supposedly began in Early Permian near Sakmarian/Artinskian boundary as attested by ground geology (Bhargava 2008, 2015), further supported through the recent ~289 ma date (Shellnut et al. 2009) in the north sector of Panjal and coeval/related volcanics in Kashmir, Zanskar, Ladakh, Nepal, Assam, NW Australia, Oman, Madagascar, Pakistan, as widespread evidence of the rifting/spreading and origin of the Neotethys. So also is known the widespread Early/Late Permian Zagros rifting, and the spreading away of the Sanandraj–Sirjan terrains in the Arabian sector (Sharland et al. 2001). The related second-order event is also occasionally manifested in the Gondwana graben basins (Sarkar et al. 1996). The igneous event on the GTM in the north sector may be or not, in some manner related to the end-Permian peaking of volcanism in South China and Siberia. In the Chelyabinsk region of Siberia, the terminal phase of volcanics has been précised at ~243 ma (Reichow et al. 2009, and others) through a U/Pb date near the terminal Anisian that is somewhat similar to the intra-Muschelkalk age constraints of the termination of the volcanism in Pakistan (Calkins et al. 1975) and neighbourhood. Thus, the volcanism in the north sector, having begun in Early Permian in the west part of High Himalaya, with possible dual peaks at Mid/Late Permian, and end-Permian, may have culminated in the Middle Triassic, although more radiometric age data is desirable, apart from the known ground geological evidences. It is interpreted that the unzipping between the Cimmeria and the Neotethyan Gondwana initiated with widespread extensional tectonics on the margin as also in the interior in a restricted manner near the start of the Permian through transform sliding; however, the volcanism mainly occurred in the neighbourhood of Kashmir and Pakistan in northwest and with only localized expression elsewhere in the region, possibly in Assam and neighbourhood. The volcanism first peaked near the start of the Late Permian, and triggered the graded unzipping of the Cimmerian microcontinents from west to east. If the South China volcanism is also considered part of the same regional event, it once again peaked in end-Permian. Alternatively, the end-Permian extinctions were exclusively caused by the Siberian and the South China events. The terminal phase of the igneous event may have lasted until near the close of the first-order mid-Triassic MFS, when the Lhasa block finally broke away from the Indo-Australian Paleotethyan margin (Metcalfe 2011 and others). The history of spreading in this interval is better documented in the northeast (Metcalfe 2011) than in the northwest (Gaina et al. 2015). The spreading away of Cimmeria did not make a single event, and instead possibly ranged through a rather long intra-Permian–late Middle Oxfordian duration up to the final separation of the Argo block and the resultant origin of the Argo Abyssal Plain.

7.2.4 Possible Earlier Origin of the Neotethys

In recent years, the revised basal Permian start of the Panjal and associated volcanism (Shellnut et al. 2009; Bhargava 2008, 2015; Gibbons et al. 2012, 2015) has been interpreted as the initiation of the spreading away of the Cimmerian continent from the Gondwana along with origin of the Neotethys in Early Permian near the Sakmarian/Artinskian boundary. However, the greater number of the Cimmerian continents, supposedly, may have separated at the start of Late Permian near the Kungurian/Ufumian/Guadalupian boundary than either in Early Permian or Middle Triassic. The Early Permian volcanism and initiation of rifting in the east sector is considered as important as at the start of Late Permian. There is additional context of the reworked non-marine Late Permian pollen/spore elements in the west sector Kachchh (Venkatachala 1970) and Kerala–Konkan basins (Potonie and Sah 1960), about the mid-Permian break and pre-rift sagging event as well in the west sector. May be, both the basal and intra-Permian rifting events have found wide expression in the Gondwana. In India, the basal Permian event is better expressed in the east sector, while the intra-Permian one in the west sector. The origin and closure of the Neotethys are in reality graded events that happened at the start between the Early Permian and Middle Triassic and K/T boundary to intra-Paleogene. It is neither prudent nor practical to conceptualize either an instant start or sudden closure of such events that by their nature range through a protracted interval.

The chronicle is realized through a series of geographic reconstructions (Figs. 7.11, 7.12 and 7.13) focusing on the principal events:

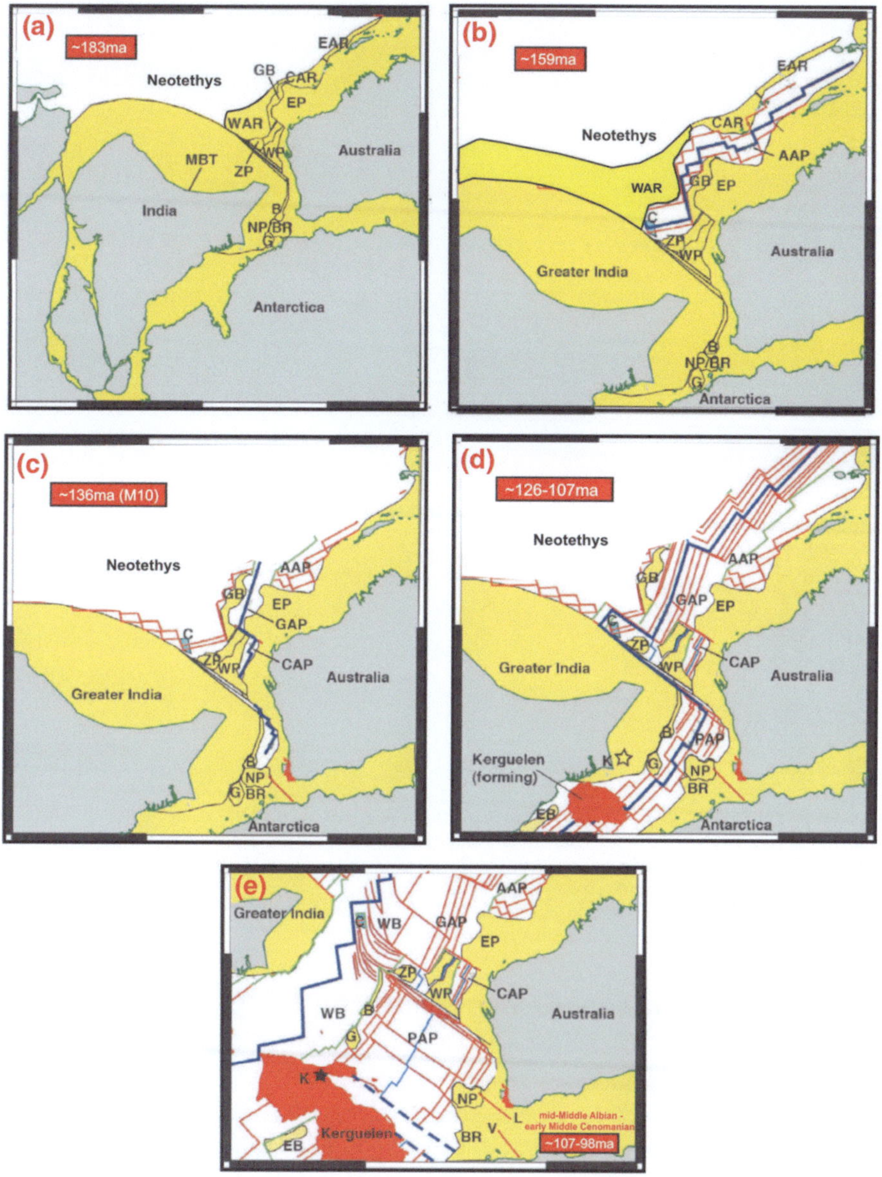

Fig. 7.12 Paleogeographic reconstructions at important first or second-order sequence surfaces near the **a** Pliensbachian—toarcian boundary **b** late Middle Oxfordian **c** early Late Valanginian **d** early Aptian–mid-Middle Albian and **e** mid-Middle Albian–early Middle Cenomanian with focus on India–Australo-Antarctica sector (modified after Gibbons et al. 2015)

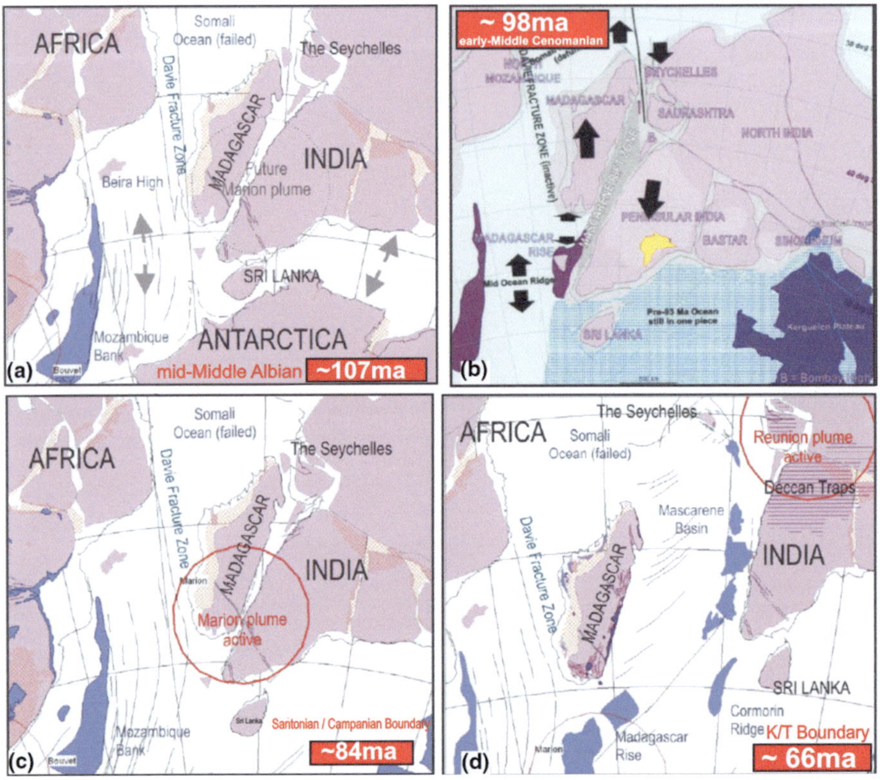

Fig. 7.13 Paleogeographic reconstructions at important first or second-order sequence surfaces near the **a** mid-Middle Albian, **b** early Middle Cenomanian, **c** Santonian/Campanian boundary, **d** Cretaceous/Cenozoic boundary; with focus on India–Madagascar separation (modified after Bastia et al. 2010)

7.3 The Spitiian First-Order Sequence

(Late Permian–Pliensbachian)

The first-order sequence began with igneous activity (Panjal and coeval volcanics) mainly in the north sector. The sequence also witnessed igneous activity in Papua New Guinea in northeast at the mid-Triassic MFS, and closed with igneous activity in the west sector (Karoo and coeval volcanics) at the terminal SB. The sequence is underlain by a large pre-Permian or pre-Late Permian stratigraphic gap in most basins of India, so also followed up by a large Toarcian–Middle Oxfordian stratigraphic gap in inland non-marine Gondwana basins. The geological record of the first-order sequence span, except on the Neotethyan margin, is largely non-marine in the linear inland Indian Gondwana rift basins.

7.3.1 *Early/Late Permian (~272 ma) Basal Mega SB Geological Event*

The start of the mega-sequence besides its regional spread on the GTM inclusive of spreading away of Cimmeria (Sanandraj-Sirjan, Afghanistan, Central Iran, NW Iran, Simasu) from the Indo-Arabo-Africano-Australian Paleotethyan margin also evidenced widespread diverse geological changes, events, and gaps at the start of the mega-sequence, both in the marine and non-marine, exposed and subsurface geological record in the basins of the Indian subcontinent, in particular, in the west, north and east sectors, as also on GTM, that are already discussed in the earlier chapters. The intra-Permian mega SB and associated subaerial unconformity or its correlative conformity and associated geological contrast are interpreted in Konkan–Kerala, Deccan, Narmada, Saurashtra, Kachchh, Jaisalmer, Barmer, Indus basin, Spiti, Kashmir, Kumaon, Lesser Himalaya, Assam, Mahanadi, K-G, Cauvery and Palar; also in non-marine inland Gonwana graben basins, and so detailed in Chaps. 2–5 (Fig. 7.14). There is recorded a mid-Permian mafic peridotite K/Ar age of 268/275 ma from Richuguta along a distinct Precambrian shear zone which has been time and again reactivated (Ghose et al. 1973). Even parallel igneous radiometric evidence has over the years grown from high and lesser Himalaya localities inclusive of South Tibet. The explicit contrast between Early Permian (similar

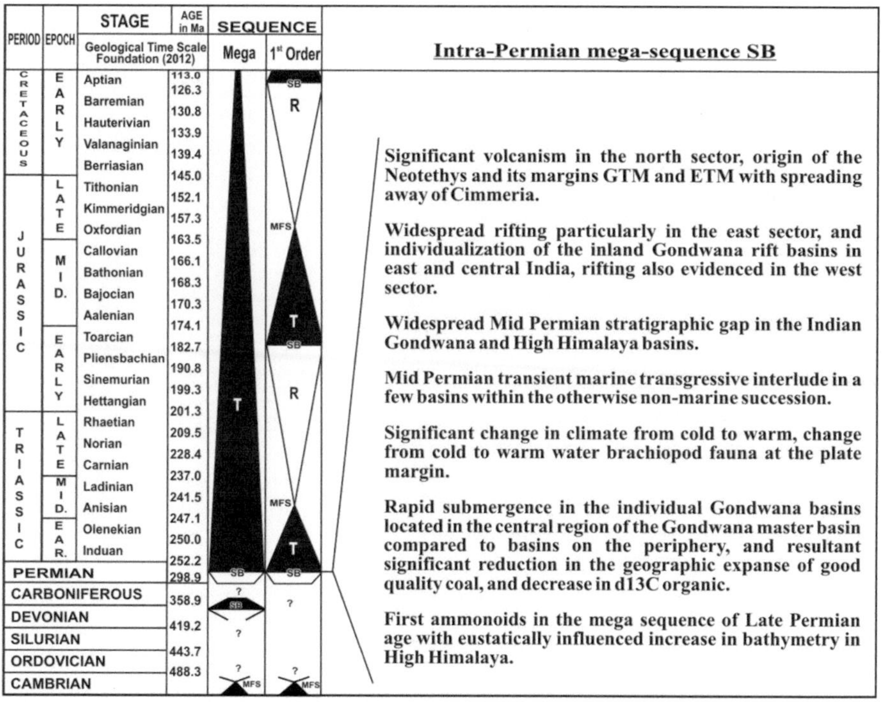

Fig. 7.14 Salient diverse geological facets at the basal SB of the mega-sequence

lithology of Talchir–Barakar units) and the post-rift Late Permian (variable lithostratigraphic development in the different individual grabens of Raniganj and coeval units) is particularly evident in the inland Gondwana basins. There is thus well demonstrated the mega-order basal SB geological event throughout the GTM. The event in the east and west sectors largely is expressed through crustal thinning, extension and rifting without any marked igneous expression. Similarly, the said event is expressed on the GTM east of IEAP in the east sector Dampier basin, Iryan Jaya, West Iriyan, Timor and Timor gap, and west of IEAP in Somalia, Kenya, Seychelles, Mozambique, and Arabia (6.3–11).

7.3.2 *Anoxic Events Vis-à-Vis Sequence Stratigraphic Surfaces*

The oceanic anoxic events (OAEs) are characterized by widely present inter-regional marked positive shift in the stable carbon (C) isotope in marine systems across the globe. The OAEs are said to be triggered by volcanic activity that in turn lead to quick succession of somewhat coeval events of increase in atmospheric CO_2, temperature, nutrient availability, productivity, rate of preservation of organic matter, etc. (Jenkyns 1980, 1985). OAEs are of short duration of an ammonoid zone/subzone of less than one my. Presence of black shales in the marine sediments facilitates the identification of the OAEs. In India, the widespread Early Permian volcanism prior to the start of the mega-sequence is marked with abrupt shift of C12/C13 ratio in Late Permian marine sediments (Shankaran 1997; Valdiya 2010). Also are noticed deficiency of O_2 and simultaneous increase of CO_2, along with the foresaid parameters in the oceans that resulted in anoxic conditions. The overall scenario was detrimental to life, and in turn prompted the end-Permian mass extinctions at the second-order SB. In this context, the geochemical studies of the Late Permian sediments of the High Himalaya linear belt may reveal evidences of anoxicity. Several other intra-Mesozoic OAEs are also found linked to sequence surfaces (Fig. 7.2).

7.3.3 *Mega-Sequence Surfaces and the Indian Energy Resources (Fig. 7.5)*

7.3.3.1 The Theatre Geotectonically Set for the Generation/Accumulation of Black Gold Particularly in Arabia in the Ensuing Neotethyan Mega-Sequence

Although, the hydrocarbon source/reservoir rocks are known throughout the Phanerozoic, even earlier in the Precambrian, the origin of the Neotethys heralded the geological stage for the generation and accumulation of hydrocarbons in huge quantity. In context of the hydrocarbon source and reservoir rocks, the duration of the mega-sequence is the most important of all the Phanerozoic mega-sequences across the globe. Most of the source and reservoir sediments were formed in the Neotethyan mega-cycle during a span of ~100 my of Late Jurassic–Early Cretaceous (Figs. 7.2 and 7.5). Approximately, 60 % of the global hydrocarbon reserves are located in Arabia on the GTM in the said time span. It is this black gold that transformed the bullock cart and horse driven world to the present rail, road, air and interplanetary space transport. Hydrocarbon source and reservoir sediments are present in all the five systems included in the mega-sequence, and also, age-constrained through sequence surfaces as elaborated upon in Chap. 5.

7.3.3.2 Start and Cessation of Coal Formation in India at Significant Second-Order C/P and P/T Boundary SBs for the Permian and K/T Boundary SB and Intra-Paleogene Terminal Mega SB for the Paleogene Coal

During the Permian and Paleogene, there formed rich coal deposits in India near the start and close of the mega-sequence. The coal formation in India is interpreted to have begun at the second-order Carboniferous/Permian boundary SB. The Early Permian coals are, however, different from the Late Permian coals, and are separated from each other by a coal-devoid interval. The P/T boundary second-order SB marked the closure of the coaliferous regime. The climate also changed as already stated from wet subtemperate to dry, arid and subtropical. Similarly, the Paleogene coal is also age-constrained between the second and first-order surfaces immediately after the terminal phase of the mega-sequence.

7.3.4 The End-Permian Mass Extinction at the Second-Order SB, and Its Varied Causes

Called the mother of all extinctions, the P/T boundary extinction event does not need any special mention. Besides the widely accepted model of volcanism and associated warming as the prime cause of the extinctions that supposedly climaxed at the P/T boundary second-order SB, a few rather unusual explanations have also been made for the mass extinctions that occurred over a short duration of a few million years. These causes include low oxygen atmospheric levels (Retallack et al. 2003; Berner 2002; Berner et al. 2007 and others), depletion of the ozone layer (Beerling et al. 2007), increase of CO_2, H_2S, methane, etc., (Meyer and Kump 2008), which provided the lethal combination for the mass extinctions. Even a combination of cooling followed by warming event has also been invoked (Korte and Kozur 2011). Short episodes of cooling have also been interpreted prior to, as also after, the principal event during the course of the steady graded warming. Brookfield et al. (2003, 2013), and Baud et al. (2014) have interpreted seismites and psunamites as part of a strong shake-up event in Kashmir. The Recent interdisciplinary investigations in Spiti (Ghosh et al. 2014; Williams et al. 2013) have found ample explicit geochemical signatures of the P/T boundary extinctions, and the associated catastrophic changes at or immediately prior to the extinctions, which are much on the same lines as elsewhere on the GTM. Bond and Wignall (2014) have lucidly discussed the relevance of diverse possible sudden causes, in particular volcanism, of the mass extinctions. Haggerty (1996) in a detailed review correlated the mass extinctions with episodes of flood basalts.

7.3.5 Major Organic Evolution Events Suggestively Linked to the First-Order Sequence Surfaces

The organic evolution through the geological time, apart from inherent genetic hereditary factors, occurred in response to the principal environmental constraints of space and nutrients. Thus, the evolutionary radiation and diversification in terrestrial life is interpreted to have been environmentally linked to marked increase in the availability of land, and nutrients, during regression of the shore line, i.e. during the first-order RST intervals. The organic evolution exhibits gradualistic as also quantum frameworks. Here is dwelt upon mainly the quantum mode involved. The drastic decline and extinction during phases of marked encroachment of land and associated decrease in the availability of land, maximized the space and food constraints (Fig. 7.9). The opposite is true for the marine life with maximum stress during the terminal parts of the RST intervals. It is emphasized that the MFSs and SBs present the most stressed environments in sedimentary columns to in turn lead to innovative evolutionary solutions.

7.3.6 Example of Quantum Evolutionary Steps in Cephalopods/Ammonoids

In case of the rapidly evolving marine organisms, e.g. ammonoids, the said genetic relationship has been demonstrated earlier in Chap. 6, and also again outlined here. The advent of skeletal organisms resulted in longer ontogenic span, as also multiplication of the individual organisms. The resultant fierce struggle for shelter, food and security prompted the intra-Cambrian origin of the extremely innovative nectic buoyant mode of marine life as the class Cephalopoda out of the ancestral shallow marine benthic Mollusca at the first-order MFS. The cephalopods occupied the shallow shelf above the bottom in the marine water columns. Significant eustatic rise during the Ordovician–Silurian interval necessitated more efficient depth negotiators than the early nautiloids, and thenceforth originated at the intra-Devonian mega/first-order surface the better swimming subclass Ammonoidea with simplified goniantitic sutures. At the intra-Permian mega SB evolved the ammonoids with somewhat more frilled ceratitic sutures. Farther up in the mega-sequence evolved at the first-order mid-Triassic MFS ammonoids with intricately frilled ammonitc sutures as more efficient depth adaptors.

7.3.7 Major Evolutionary Leaps of Higher Grade Taxa in the Animals and Plants

Similar links among the evolution and extensional tectonic events are difficult to ascertain in case of rather slow evolving land animal and plant groups. The links, however, seem conceptually possible, and preliminary enquiries into the relevant data are found encouraging. The published origin and extinction ages, are found invariably close to the ages of the first and second-order sequence surfaces. The same cannot be mere coincidence, and deserves further detailed investigation (Fig. 7.9).

7.3.7.1 Examples of Major Evolutionary Landmarks Linked to Sequence Surfaces

The more obvious examples of such linkage, already exampled in Chap. 6, include the distinct replacement of the seed-devoid vegetation on land during the terminal Early Permian phase of the previous mega-cycle by the seed-bearing plant life in the Late Permian–intra-Paleogene mega-sequence (Fig. 7.9). On the pollen–spore front, there is as well a drastic change from the monosaccate to striate and non-striate disaccate groups. The possible origin of the mammal like reptiles, from the amniote egged amphibians as an extremely significant evolutionary step in the organic evolution, is also speculated at the basal SB of the mega-sequence. On similar lines are suggested the dawn of dinosaurs on the land, of the marine nannoplanktons, dinoflagellates, reef building corals and true belemnoids near the intra-Triassic first-order MFS, of planktic forams near the first-order intra-Jurassic SB, of birds at the intra-Jurassic MFS, of flowering plants linked to the intra-Cretaceous first-order SB, and of marsupial mammals during the rather long intra-Cretaceous eustatic high in between the mid-Middle Albian and Early Turonian MFSs.

7.4 First-Order Intra-Triassic Late Anisian Hollandite Zone–Kellnerites Zone MFS Rifting Event

It is considered one of the most important events of not merely the intra-Permian–intra-Paleocene mega-sequence but of the entire Phanerozoic (Figs. 7.4, 7.5, 7.6, 7.7, 7.8, 7.9, 7.10, 7.11, 7.12, 7.15 and 7.16), and précised here at zonal precision. The significance, largely, has emerged through the correlation and integration of the otherwise separately known developments. The biggest global event suggestively linked to this surface has been the differentiation/break-up of Pangea into Gondwana and Laurasia. The classic tripartite definition of Triassic included a distinct eventful middle marine Muschelkalk division, altogether different from the underlying Bunter and the overlying Keuper non-marine units. The same exemplifies well the diverse mid-Triassic changes in the classical Germanic Triassic. The Middle Triassic igneous expressions rather appear localized as the vaning phase of the widespread Permian volcanism in the north. In India, the same is suggested through a few intrusives along the inland linear Gondwana. There is also suggested prolongation of the Permian volcanism in Pakistan up to the mid-Triassic. Mid-Triassic volcanism of 30 m thick rhyolitic lavas is evidenced in Pakistan prior to the Late Triassic limestones. In addition, the Chelyabinsk region in Southeast Siberia marks the precisely dated termination of the Siberian volcanism (Reichow et al. 2009), which is physically located not very far from the Indian Neotethyan margin. Metcalfe (2011) and many others have even preferred the spreading away of the Lhasa block from the Indo-Australian margin in Middle Triassic rather than in Permian. Mid-Triassic igneous event is also recognized in Papua New Guinea (Hill and Hall 2003). A few other recent studies separate Lhasa as late as late Early Jurassic. The start of the non-marine sedimentation in the Kachchh, and Saurashtra, even Konkan–Kerala basins is among the diverse consequences of a regionally

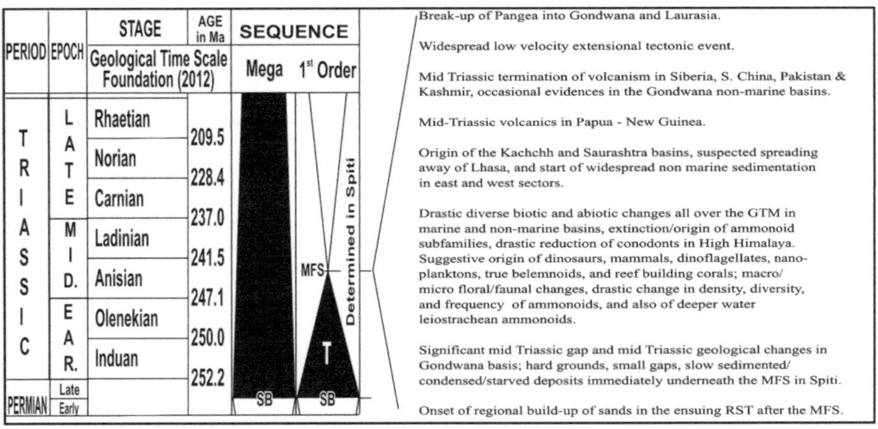

Fig. 7.15 Salient diverse geological facets at the first-order intra-Triassic MFS

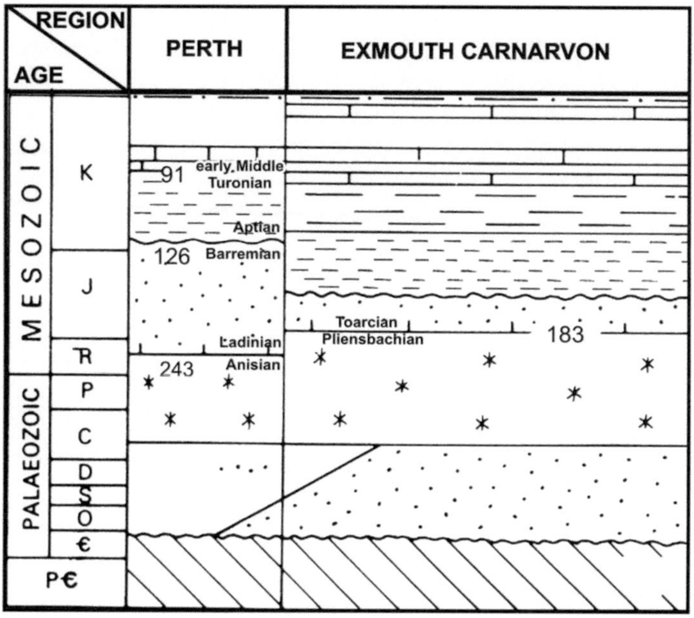

Fig. 7.16 Significant events/changes on the northwest and West Australian margin during the span of late Permian–Paleocene near the first-order sequence surfaces mega-sequence modified after Veevers 1984

spread extensional tectonic event on GTM of ~243 ma intra-Triassic MFS age. A significant rifting age in the Gondwana basins, also, occurred at the mid-Triassic first-order MFS, along with stratigraphic gap of varying durations, and associated palynological changes at the palynoevent 7 (Vijaya 1997). Major widespread mid-Triassic mega-floral changes (Awatar 2011) may also have occurred at the intra-Triassic MFS event. In this context is exampled the ~231 ma igneous date from the Lower Tianba Formation of South Tibetan region (Hu et al. 2015). The subtrappean sequence in Saurashtra is already interpreted to extend down at least into the Ladinian–Pliensbachian first-order RST age bracket, and if stratigraphic homotaxiality among Seychelles, Madagascar, Mozambique, Tanzania, Kenya, and Saurashtra is given credence, the subtrappean succession in the Indian west and east sectors may even extend down to the intra-Permian base of the mega-sequence as defined in this work. There are already known reworked Late Permian sediments in Kachchh, Kerala and Cauvery (Potonie and Sah 1960; Venkatachala 1970; Prasad and Phor 2009). The Late Permian–Pliensbachian first-order sequence has assumed particular significance in context of its highly prospective hydrocarbon source/reservoir estimates. Rawal (1995) has suggested continuation of the Panjal volcanism up to the Middle Triassic in some Kashmir localities, which has been otherwise on ground geological considerations and recent reliable radiometric age restricted to Early Permian (Bhargava 2008; Shellnut et al. 2009). Thus, the mid-Triassic first-order MFS initiation of rifting and localized igneous activity in the west sector at least in Pakistan, and Saurashtra is firmly placed. Relatively localized volcanism is also recorded of ~213–192 ma age bracket in West Australia in the vicinity of the Dampier basin in the Ladinian–Pliensbachian first-order RST. The ~215 ma intra-Norian localized rift event is also recorded in Iraq near the Arabian plate margin (Khan 1989). Similarly, a ~190 ma localized extension event is recorded at Lebombo locality of South Africa (Reeves and De Vit 2000). In East Antarctica, initiation of silicic volcanism is palynologically age-constrained in Late Sinemurian to the Sinemurian/Pliensbachian boundary followed up by mafic volcanism from Early Toarcian onward (Bomfleur et al. 2014) in tune with the Karoo–Ferrar–Kirpatrick LIP age framework. In Seychelles is recorded ~190 ma event (Zahirovic et al. 2014). There is also recorded intra-Triassic igneous Rhyodacite in Bihar of ~217 to ~214 ma (Sarkar et al. 1996) during the Ladinian–Pliensbachian RST. These may be, respectively, linked to the intra-Norian and intra-Sinemurian second-order MFSs. The ~217 to ~213 ma igneous event is suggestively correlated to the rift-related lithological change from the limestone–shale mixed Alaror unit to the Nunuluka sands in High Himalaya at the mid/late Late Norian Hogarti/Macer zonal boundary, in turn correlated to the Arabian Tr 80 MFS. The ~193 to ~190 ma intra-Sinemurian MFS event is suggested in the Kioto shale–carbonate Kioto succession.

The intra-Triassic first-order ~243 ma MFS extensional tectonic event is held rather bereft of any significant widespread igneous activity on IEAP in particular and GTM in general. The Late Permian–Early Triassic interval in the Gondwana basins is marked by pre-rift crustal thinning/attenuation/stretching/sagging along the ancient Precambrian weak linear fracture zones, otherwise mainly restricted to infra and early rift stages. The widespread diverse mid-Triassic geological changes, gaps, and events on GTM are as well known west of India in basins of Pakistan, Madagascar, East Africa, and Arabia, also east of India in East Myanmar (Bronnimann et al. 1975), on Wombat and Exmouth plateaus west off the NW Australian coast, in Dampier, Carnarvon, and Perth basins up to the east margin of GTM (Fig. 7.16). These are detailed in earlier chapters as related to the first-order intra-Triassic MFS as determined in the exposed Triassic of the Spiti basin. The mid-Triassic changes all over the Himalaya are obvious between the relatively distal deep highly condensed, hardgrounded, and even occasionally starved TST of mid to outershelf environment up to the MFS, so also in the relatively much shallower, moderately to rapidly deposited post-MFS RST. Widespread but only low extension velocity-related extension activity is interpreted in all the sectors while the rift-genic volcanism is rather localized. The diverse changes in the inland non-marine basins, in proximal Kashmir, north and south parts of Salt Range in Pakistan are somewhat different but well marked, than in the distal High Himalaya from Pakistan to South Tibet.

FP. 7.1 First-order Intra-Triassic MFS in the terminal part of the Muschelkalk member in Spiti in the field photo of Krystyn et al. 2007

7.4.1 Early Triassic Bathymetric Increase in Spiti Vis-à-Vis Eustatic Changes During the Intra-Olenekian–Intra-Carnian Interval

The Early and Middle Triassic on ETM evidence is mostly considered an interval of graded eustatic rise with the late Early Carnian maximum (Haq et al. 1987; Hornung et al. 2007), and has been linked with a regional to global crisis in carbonate productivity (Reingaben effect/event). The said event is also interpreted in Spiti near the Chomule/Rama boundary lithostratigraphic change (Figs. 2.11 and 2.12). Garzanti (1999) interpreted bathyal depths in Olenekian in Nepal, while Bhargava et al. (2004) suggested maximum of storm wave base bathymetry in the Spiti Olenekian. The carbonate productivity crisis is considered graded and protracted since the close of the Mikin Formation during successive second-order sequences with the carbonate deficient Kaga Formation in between the Muchelkalk Member and Chomule Formation. Moreover, the sudden change from carbonates to clastics appears also influenced by global climatic changes, and necessarily may not be linked to the intra-Triassic MFS/MFI as determined by multiple features of drastic biotic and abiotic changes across the determined MFS/MFI of density, diversity and frequency of ammonoids, presence and frequency of hard grounds, sedimentation rates and others. The bathymetry change during the Late Olenekian–Early Carnian span, i.e. during the formation of the Nodular Limestone–Muschelkalk–Kaga–Chomule succession may be relatively distinctly less in a phase of graded slow increase compared to the relatively rapid increase during the Induan–Olenekian interval. The ammonoid and other biotic extinctions also are linked to sharp bathymetric changes at first and second-order sequence surfaces. The Late Anisian in Spiti suggests stronger decrease and associated ammonoid extinctions than any time in Carnian. The carbonate/clastic change also appears more marked in terminal Anisian between Muschelkalk and Kaga Member than in Carnian above the Chomule Formulation. The terminal Anisian MFS is thus favoured over the Carnian MFS, and the marked intra-Carnian change considered at a second-order surface.

7.4.2 Application of Sequence Stratigraphic Concepts and Sequence Surfaces in Correlation of Marine and Non-marine Units

In the uncertain and long ranging age scenario of the palyno-assemblages, recourse is taken to the use of the first to second-order regionally isochronous sequence stratigraphic surfaces determined in the marine Triassic of Spiti basin, for suggestive correlation as a futuristic working model between the Indian marine and non-marine units as realized in the earlier chapters (Fig. 3.13). The premise has been the common control of regional tectonics on both marine and non-marine successions.

7.4.3 Regional Build-up of Sands in the Ensuing First-Order RST

The said MFS also dates the change of bio/lithofacies and fossil assemblages with or without stratigraphic gap both in marine and non-marine successions in the exposed High Himalaya and the non-marine Gondwana linear graben basins, in subsurface K-G wells, also elsewhere on the GTM (Fig. 3.13). Uniquely associated thick cross-bedded coarse textured unfossiliferous non-marine sediments are known widely on GTM, particularly in proximal sites as Minjor Sandstone in Arabia, Adigrat Sandstone in East Africa, Isalo Sandstone in Madagascar, Datta Sandstone in Pakistan, and similarly elsewhere on GTM. These sands are here related to rift-related rejuvenation of the provenance areas at the successive first-order intra-Triassic MFS and intra-Jurassic SB. Sands of presumably the same age bracket are also recently discovered in the Dhandhuka and Lodhika wells of the Saurashtra basin (Singh et al. 2000). The Nunuluka and coeval sands in High

Himalaya sandwiched between the finer textured limestone/shale units below and above suggest the climaxing of sand development in the terminal Norian as a distinct uplift event on the GTM.

7.5 The Kachchhian First-Order Sequence

(Toarcian–Barremian)

7.5.1 The Multifaceted Widespread First-Order Intra-Jurassic SB Geological Event

The multifaceted changes across the first-order SB (Fig. 7.17) have been highlighted in the earlier chapters. Also are known widespread Early Toarcian anoxic, and igneous manifestations, and even Early Toarcian hydrocarbon source rocks on GTM, particularly in the west sector in the Arabian and East African basins.

7.5.2 Suggestive Origin of the Planktic Forams at the First-Order Intra-Jurassic SB

As a major organic evolutionary step in the oceanic cradle, the origin of the planktonic forams is suggestively placed near the Pliensbachian/Toarcian first-order

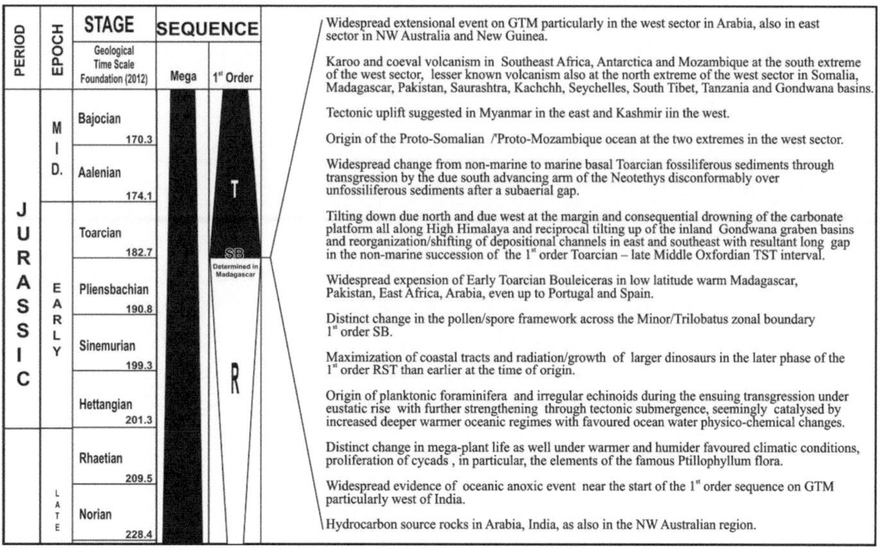

Fig. 7.17 Salient diverse geological facets at the first-order intra Jurassic SB

SB (Chap. 6). The planktonic forams subsequently acquired immense significance as an unparallel biostratigraphic tool, particularly, in the hydrocarbon exploratory studies. Also is suggested the origin of irregular echinoids, and a few other marker invertebrate evolutionary events at this first-order boundary SB.

7.5.3 Igneous Activity at the Intra-Juraasic First-Order SB Near the Start of Toarcian

Okimura and Fatmi (1988) noted intercalation of volcanic rocks in the Early Jurassic carbonate Loralai unit in Baluchistan. Ahmad (1991) observed intrusive kimberlitic lamprophyres in Early Jurassic Shirinab unit. Both the Pakistan occurrences are close to the ~183 ma SB. Extensive Karoo and associated volcanism is known in Mozambique, Tanzania, Morondova, Seychelles and South Africa of ~183 ma (Konig and Jokat 2006, and others). Shellnut et al. (2015) reported ~190 ma devitrified ash from Seychelles and related the same to Karoo volcanism. The Saurashtra Lodhika and Dhandhuka subsurface volcanics (Singh et al. 2000) may also be linked to the first-order SB. In a few Kachchh offshore wells, thick dolerite intervals have been included within or at the base of Middle Jurassic sediments (DGH report on Kachchh), which as well may be related to the first-order SB event. These may also be cogenetically linked to the Karoo event in the west sector as manifestations of the large Karoo–Ferrar–Gujarat LIP. In this context, the K-Ar ~181/185 ma dolerite dyke age from an inland Gondwana rift basin locality of Latehar (Ghose et al. 1973) assumes significance, and deserves fresh dating through improved Ar40/Ar39 system. Acidic metavolcanics in Garhwal Himalaya a few decades ago were dated by the then prevailing K/Ar system ~197 and ~186 ma (Sinha and Bagdasarian 1976), which require fresh dating by the improved new systems. In context of the regional rift-genic igneous event in Pakistan, Kachchh, Saurashtra, the igneous plug at the base of the newly discovered ammonoid bearing terminal Bajocian/early Bathonian succession at Nara is considered pre-Bathonian, and linked to the igneous event at the north end in the west sector of supposedly Karoo age. A ~177 ma U-Pb age of plagionite intrusive is mentioned in Hu et al. (2015) in the Spontang region of the north sector of the Indian Neothyan margin. Gibbons et al. (2015) have mentioned late Early Jurassic intrusives/volcanics, in Qingtang, ISZ, YTSZ, and Lhasa. Zhu et al. (2008) have determined SHRIMP zircon ages of Yeba volcanics from Gangdese (South Tibet) of ~174 to ~190 ma range, and interpreted their creation by northward subduction of Neotethyan oceanic crust in Early Jurassic. Evidence of guide fossil bearing (Megalodonts, Lithiotes, etc.) Late Triassic–Early Jurassic is mentioned through country rock xenoliths in Ladakh. In Shigatse island complex also has emerged evidence of Early Jurassic ~190 ma magmatic event. All the above apparently isolated/sporadic ages, also many uncited magmatism from west, north and east sectors, and other evidences are included in the ~190, ~170 ma age bracket. The rifting and localized origin of the intra-oceanic Shyok-Dras volcanic arc during ~180, ~170 ma (Jain 2014; Jain et al. 2012), and the above-mentioned somewhat protracted igneous activity may be suggestively linked to its climax at the first-order SB near the Pliensbachian/Toarcian boundary with initiation and closure near the ~190 and ~170 ma second-order surfaces. The latest in the chain of extremely robust evidences to mention is the Karoo–Ferrar and coeval volcanism of the Patagonia–Africa–Antarctica region which is palynologically and radiometrically age-constrained in Early Toarcian (Bomfleur et al. 2014). Rifting event of this age is also interpreted in NE Arabia, near the Mediterranean region (Sharland et al. 2001). The occasional localized oceanization in the west sector is interpreted near ~183 (Stampfli and Borel 2002). The Dongkeng basin also includes ~170 ma granite (Zhu et al. 2009). There also occurred sudden tectonically induced uplift and halt to sedimentation near the intra-Jurassic first-order SB in East Myanmar (Valdiya 2010, 2016) in the east and Kashmir in the north (Bhargava 2015). The ceasure of sedimentation in the Gondwana rift basins in India is for the first time précised through sequence stratigraphy at the Pliensbachian/Toarcian boundary first-order SB, besides localized ceasure in a few basins earlier in Middle to Late Triassic. The palynoevent 9 (Vijaya 1997) is considered linked to the intra-Jurassic first-order SB. The closure of rifting in most of the inland Gondwana basins appears linked to the onset of transform sliding of India inclusive of Madagascar vis-a-vis Africa and related Karoo igneous event.

7.5.4 *Widespread and Significant Rifting on the NW Australian Margin at the Intra-Jurassic Pliensbachian/Toarcian Boundary First-Order SB Geological Event*

The onset of rifting off the NW Australian Plate is placed at SB4 (Westphal and Aignar 1993) near the Pliensbachian/Toarcian boundary first-order SB. Zahirovic et al. (2014) support the initiation of rifting in New Guinea near the first-order SB based on the oldest age of ~187 ma. The pre-spreading stage continues up to the intra-Oxfordian break-up unconformity. In the Timor Gap, the Late Triassic–Early Liassic non-marine sand dominating Malita Formation is unconformably overlain by shallow marine shales of the Plover Formation of the Toarcian–Early Callovian age. The widespread rifting event and the resultant disconformity, thus, is interpreted at the Pliensbachian/Toarcian first-order SB.

Audley-Charles et al. (1988) interpreted crustal uplift in the Middle Jurassic and subsequent submergence near the J/C boundary in context of his studies on the rift-drift history of Asian terranes that fragmented from the Gondwana. The two poorly dated events appear linked, respectively, to the first-order intra-Jurassic SB extensional uplift and first-order intra-Jurassic MFS submergence.

7.5.5 *Suspected Coeval Volcanism in Gujarat*

The record of the thick Lodhika traps / Formation and coeval Dhanduka traps / Formation in the Saurashtra basin wells (Lodhika and Dhanduka wells) is quite significant (Singh et al. 2000). The Dhanduka igneous unit is underlain by 100 m thick sands and overlain by $\sim$150 m thick sands of Viramgam Formation in the Dhanduka well. In the Lodhika well, the igneous Lodhika Formation is overlain by the $\sim$1500 m thick Dhrangadhra Formation, which comprises conglomerates and coarse sands with thin clay/silt intercalations. Even a few wells (Suthri, and GK 22-C-1) in Kachchh (DGH website report 2003), have, also, indicated intrusives in the basal part of the Middle Jurassic sedimentary succession. The Saurashtra basin assumes specific significance in view of the suspected presence there of the late Early Jurassic, Late Jurassic, as also pre and post Lameta Late Cretaceous volcanic events, respectively, related to the Karoo igneous event at the south extreme of the west divergent sector, spreading away of Argo block at the north extreme of the east divergent sector and Deccan volcanism at the north extreme of the west sector and associated LIPs.

7.5.6 *Age of the Sediments of Pre and Post Lodhika Volcanism Interval in the Lodhika and Dhandhuka Wells of Saurashtra Basin*

The Lower Member of the post volcanism Dhrangadhra Formation is entirely non-marine without any micro/mega flora/fauna, but on stratigraphic considerations is assigned Late Jurassic age (Singh et al. 2000). The Viramgam Formation in the Dhanduka well is suggestively correlated with the Upper Member of the Dhrangadhra Formation in the said work, there by implying the absence of the Toarcian–late Middle Oxfordian Lower Member in the Dhanduka well. However, the Upper Member in a marine $\sim$10 m clay interval in the middle part has yielded foraminifera. Rich palynoflora is also recorded from the Upper Member (Shanmukhappa et al. 1995 in Singh et al. 2000). Based on the recorded microfossils, the Upper Member has been suggested Late Jurassic–Early Cretaceous age (Singh et al. 2000). The Dhrangadhra Formation inclusive of both the members is ranged in the Toarcian–Barremian first-order sequence age bracket. The pre-volcanism sands in the Dhanduka well are here assigned to the Early Jurassic in the Ladinian–Pliensbachian first-order RST age bracket. Much like the interpreted Saurashtra volcanism, a few Kachchh basin wells $\sim$200 km southeast of the said Saurashtra wells include thick dolerites at the base of the Middle Jurassic sediments (DGH website report), and so also as in Saurashtra suggestively assigned $\sim$183 ma age. However, the so suggestive ages of Kachchh and Saurashtra wells require confirmation through radiometry. The said widespread igneous activity is not evidenced farther north in Jaisalmer, and Spiti High Himalaya.

7.5.7 Stratigraphic Homotaxiality in Kachchh and Saurashtra Successions

The ~40 m thick tuff below the 2300 m depth in Saurashtra is considered interesting which Singh et al. (2000) have related to the early drift phase with marked shift in the maturation pattern at the boundary between the Lower Member and Upper Member of the Dhrangadhra Formation. At the first thought, the said tuff in a regional frame may be linked to the first-order intra-Jurassic MFS volcanic event. This makes the broad stratigraphic history of Saurashtra basin somewhat similar to that of Kachchh with isochronous origin at the intra-Triassic MFS, oldest phase of the Ladinian–Pliensbachian RST interval as coeval of the Nirona Formation, second phase of the Lower Member of the Dhrangadhra Formation of the Toarcian–late Middle Oxfordian first-order TST interval disconformably over the oldest phase, third phase of the first-order RST sediments of the Upper Member of the Dhrangadhra Formation of the terminal Middle Oxfordian to Barremian interval. The third phase includes a short transient marine incursion as evidenced by foraminifera from the thin ~10 m clay in the early to middle part of the Upper Member of the Dhrangadhra Formation, which is suggestively assigned intra-Valanginian age. The fourth phase of the first-order TST interval of the Aptian–early Middle Turonian of the Wadhwan Formation is again disconformably overlain by the Deccan volcanics. However, in spite of being much less explored compared to Kachchh, the Saurashtra basin is academically considered more important and revealing in context of its unique feature of multiple volcanic events of suggestive Early Jurassic, Late Jurassic, and terminal Cretaceous that are neither known in Kachchh, nor in any other Indian or IEAP basin of any sector, west, north or east. These three igneous events of Saurashtra basin in the regional context suggestively occurred at the Pliensbachian/Toarcian boundary first-order intra-Jurassic SB related to Karoo volcanics, first-order intra-Oxfordian MFS related to volcanics at the Indo-Australian margin, and intra-Turonian first-order MFS St Marys igneous event or the intra-Maestrichtian second-order MFS Deccan volcanism. May be, both the Cretaceous events are present as in Kachchh and Narmada. Kachchh in addition has recently evidenced the ~124 ma igneous event at Jhura (Kalmalkar et al. 2014).

7.5.8 Similarly Positioned and Age-Constrained Volcanic Units in the West Sector

The stratigraphically, similarly positioned volcanic units above the underlying sediments of the suspected Ladinian–Pliensbachian age are known in South Africa, Mozambique and Seychelles in the west sector (Singh et al. 2000; Shellnut et al. 2015, and others). Dolerites of this interval either in wells or outcrops are recognized in the Maud Island of Antarctica, in the Falkland region of S. America, and

Kachchh in India. The said suspected igneous activity in Saurashtra and Kachchh is considered part of a regional rift-genic event in the west divergent sector with similar stratigraphic frameworks to that of Mozambique below and above the igneous event. The event is coeval and similar to the not that good manifestation in the Somalia–North Madagascar region at the north extreme. The radiometric ages of over 25 samples of the Mozambique, Karoo and Ferrar igneous events by different techniques are ranged in ~184 to ~172 ma age bracket (Manninen et al. 2008; Jourdan et al. 2007; Duncan et al. 1997 and others). There is considerable difference of opinion on the rift or plume-genic character of the Karoo and related volcanics (Moore et al. 2008, and others).

7.5.9 Pre-Toarcian Similarity of the Mesozoic Successions at Mozambique, Seychelles, Madagascar, Pakistan, and Saurashtra

The Ladinian–Pliensbachian succession of the first-order RST sediments at Mozambique (Konig and Jokat 2006), Seychelles (Plummer and Belle 1995), Pakistan Indus basin (Fatmi 1986), and Madagascar (Geiger 2004) are in non-marine clastic facies. The Kachchh, Dhanduka and Seychelles wells (Fig. 7.18) also evidence pre-Toarcian clastics. In this context, the ~100 m thick sands in the Dhanduka well below the Lodhika Volcanic Formation are here interpreted to belong to the said first-order RST Ladinian–Pliensbachian age bracket.

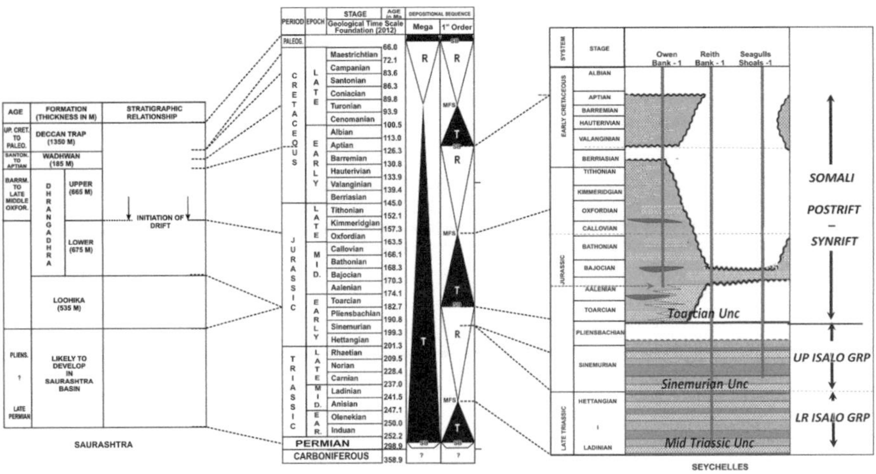

Fig. 7.18 First-order Indian Mesozoic sequence surfaces interpreted in Saurashtra (modified after Singh et al. 2000) and Seychelles (modified after Robinson et al. 2012)

7.5.10 Probable Age of the Lodhika and Dhanduka Trap Formation

Singh et al. (2000) suggested the presence of the Late Triassic to Late Jurassic sediments below the Lodhika Formation in the text. In their Fig. 2, the Lodhika formation is assigned to Jurassic with a question mark. In Fig. 4, the Lodhika formation is assigned Early to Middle Jurassic age. In Fig. 5, the Lodhika formation is suggested during Middle Jurassic age. The given highly variable ages are somewhat contradictory, and point out the absence of any firm views on the age of the Lodhika Formation. Nevertheless, a few significant age constraints have emerged. One is of the volcanism being distinctly older than the Late Jurassic Upper Member of the Dhranagadhara Formation. The second is of considerable parallelism of Jurassic succession between Saurashtra and Southeast African basins like Mozambique. The third is the ~ 184 to ~ 172 ma radiometric ages (Toarcian) of the widespread volcanism in the west divergent sector. In context of the genetic link of sequence surfaces to extensional tectonics, the Lodhika igneous event is linked to the intra-Jurassic first-order SB near the Pliensbachian/Toarcian boundary, which perhaps lasted up to the succeeding second-order MFS of ~ 170 ma early Early Bajocian. Moreover, the Lodhika is considered rift-genic rather than plume-genic near the north extreme of the said divergent sector at the first-order Pliensbachian–Toarcian boundary SB. Only more wells in Saurashtra, and radiometric dating of said volcanics may establish the physical expanse. The igneous event seems widespread all along the west divergent sector from Karoo–Ferrar regions to Pakistan, and occasional radiometric evidence is also recorded from South Tibet.

7.5.11 So Called Origin of the Proto-Indian Ocean in the East and West Sectors

Many works interpreted the said spreading between Somalia and India inclusive of Madagascar as the origin of the Proto-Indian Ocean in the west sector in Early Jurassic (Sharland et al. 2001 and others). The said interpretation is difficult to accept in broad context of the origin of the Proto-Indian Ocean and Indian Ocean distinctly later in the east sector. At the most, these may be termed as the localized Somalia/Mozambique Proto-oceans (Fig. 7.11) as the resultants of transform sliding along the Davies ridge between the East and West Gondwana at the two extremes of the west divergent margin. These are also considered, respectively, as extended Neothethyan ocean offshore Somalia, and extended South Pacific Ocean offshore Mozambique. There is evidenced mere progression of rift/drift and associated arm of the Tethys due south. The same is localized between South Somalia and North Madagascar without any appreciable spreading of India inclusive of Madagascar vis-à-vis Africa until at least the Late Jurassic. Sharland et al. (2001) also have interpreted the origin of the Proto-Indian Ocean between Arabia and India in the Owen fracture region within or above Early Tithonian. The initiation of the Proto-Indian Ocean is held having occurred in the east sector as the praecursor to the actual origin of the Indian Ocean later near the first-order Barremian/Aptian boundary SB up to the mid-Middle Albian MFS, at the final separation between India and Australo-Antarctica in the Enderby–Elan Bank sector, i.e. along the entire stretch of the east divergent sector. The early localized spreading in the east or west sectors need be differentiated from the actual distinctly later origin of the Indian Ocean.

7.5.12 Stratigraphically Similarly Positioned Marine Incursions in Mozambique and Saurashtra

The first marine incursion at the northeast margin of the Mozambique basin at Nacale based on bivalves has been assigned Late Kimmeridgian–Tithonian age (Prinz et al. 1993). This could be coeval to the first marine incursion in the Lodhika well in the middle part of the Upper Member of the Dhrangadhra Formation. This incursion is interpreted in either second-order intra-Kimmeridgian or intra-Tithoian TSTs. If that be so, it lends additional support to the earlier suggested Toarcian–late Middle Oxfordian first-order TST interval age bracket to the Lower Member of the Dhrangadhra Formation, which is followed up by the early part of Upper Member of Dhrangadhra Formation of early part of the first-order RST of the Late Oxfordian–Early Kimmeridgian age. The Saurashtra Cretaceous is considered near continuous and complete. Moreover, pre-Trappean Mesozoics are also surmised in the Cambay basin. Thus, there is a long belt of Mesozoic sediments from the Axial belt of Pakistan to Jaisalmer, Kachchh, Saurashtra, Narmada, Deccan, Konkan, Kerala in the on/offshore west sector of India for hydrocarbon exploration.

7.5.13 Origin of the Saurashtra Basin

The Lodhika well succession, on account of the pre-Toarcian non-marine clastic sedimento-stratigraphic similarity with Kachchh in spite of only suggestive presence of the overlying volcanic rocks, is interpreted to have originated as well at the intra-Triassic MFS. Similar pre-Toarcian sediments on either side of the divergent sector may extend down to Early Permian as in the East and South African basins, Indian inland Gondwana basins, and NW/W Australian basins in the classical non-marine Gondwana tectono-stratigraphic framework. In some basins like Kachchh and Saurashtra, the successions range only though the Ladinian–Pliensbachian duration due to the intra-Triassic origin of these basins, although reworked non-marine Late Permian–Early Triassic sediments have been recorded in Kachchh and Konkan–Kerala. In Kachchh, Rajnath (1934) even suspected the Early Permian transgression based on a reworked Early Permian brachiopod *Streptorhyncus* in the terminal Late Bathonian bed 22 of the Jumara section.

7.5.14 Onset of the Transgression in Kachchh and Elsewhere on GTM

The revival of rifting in the Somalia–North Madagascar divergent region coupled with eustatic rise resulted in the relative deepening of the newly opened Kachchh rift. It in turn let the southward advancing arm of the Neotethys invade the young basin at the intra-Jurassic first-order SB in early Early Toarcian (Figs. 3.5, 3.6, 3.7 and 3.8). The start of the eustatic rise at the said SB in steps continued up to the late Middle Oxfordian. The first-order TST is punctuated by several second-order transgressive/regressive couplets (Krishna et al. 2010, 2011), and associated intrabasinal extensional tectonics at the second-order sequence surfaces as attested by the strand shoreline marine sedimentation of the Toarcian–Aalenian age spread to the entire exposed basin. The origin of Bouleiceratinae, (Fig. 6.4), along with origin/extinction of a few other ammonoid subfamilies, is also linked to the first-order intra-Jurassic SB event.

The first-order intra-Jurassic SB rift event in the west sector may also be genetically linked to the ceasure of sedimentation in the Indian Gondwana basins. The ceasure may have resulted due to uplift, rejuvenation of the provenance, reorganization/lateral shifting of the river channels, and distinctly changed depositional framework. The Paleoslope during the Early Permian was presumably due northeast in context of the basal Permian marine transgression in the Gondwana basins of the east sector with present day remnants in Arunachal, Mizoram, Meghalaya, Assam, Bengal, Purnea, Daltonganj, Manendragarh, and Umaria. The possible connection of the said inroads of the Neotethys with the marine Early Permian in Pakistan and Jaisalmer via the Barmer–Sanchore–Cambay and Narmada grabens has been speculated (Krishna 1986, 1987a, 1987b), although without any conclusive evidence of marine Early Permian in the subsurface Cambay and west margin of the Narmada graben (Fig. 2.3).

7.5.15 Tectono-Stratigraphic Evolution in the Kachchh Basin in the Second-Order Sequence Stratigraphic Context

The Mesozoic geological dynamics in the Kachchh basin is discussed at length at second-order sequence surfaces through series of 2-D graphics and tabulated figures (Figs. 7.19–7.40) as the preamble to possible near-future similar detailed treatment in other Indian basins. The second-order geodynamics in Kachchh is outlined.

7.5.16 Lithological Evidence of Rift-Genic Early Bajocian, and Early Middle Bathonian Submergences at Successive Second-Order MFSs

After the initiation of the rift-genic transgression near the basal Toarcian first-order SB, the next significant submergence in Kachchh and the GTM occurred at the

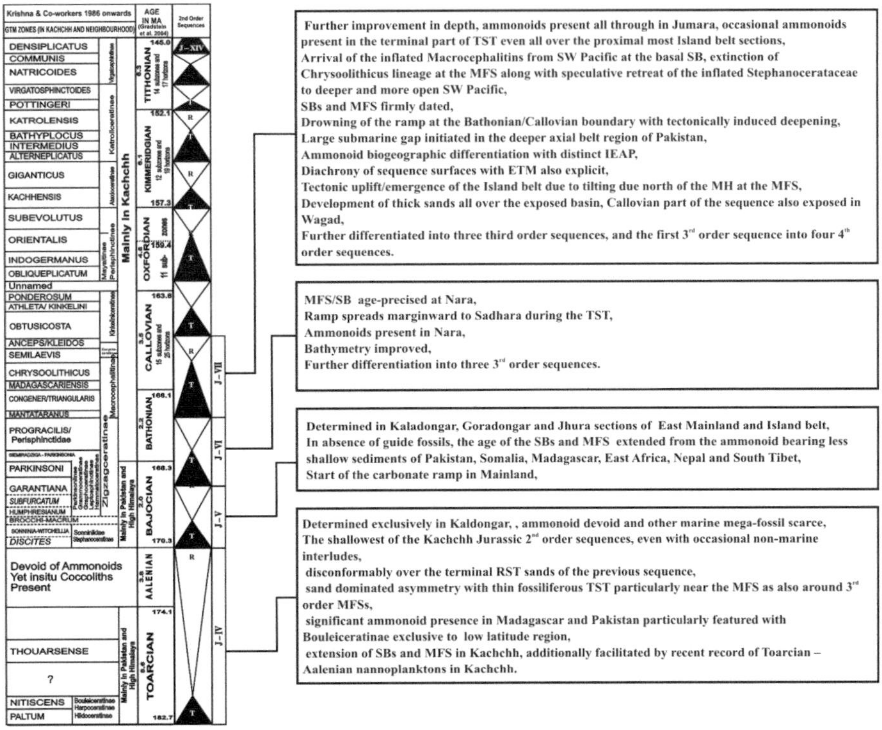

Fig. 7.19 Salient geological facets at the second-order sequence surfaces in Kachchh (sequences J-IV to J-VII)

second-order SB at the base of Bajocian (Figs. 7.19 and 7.22). It was the outcome of cumulative effect of sea level rise and another rift-related tectonic submergence. Yet, the Bajocian to early/mid-Middle Bathonian interval, particularly, in 'Island belt' remained extremely shallow. The Bajocian, as interpreted in subsurface well and seismic logs of West Mainland, is also in subtidal carbonate facies. It is significant to note that the Early Bajocian ammonoid record is in carbonate facies from Goradoongar, thereby suggestive of subtidal depths in the Early Bajocian even in South Patcham. Nothing as yet is known of the Early Bajocian record in Wagad subsurface although surmised here in view of the increased depths in the Early Bajocian as compared to the Toarcian–Aalenian. Thus, the Early Bajocian sea in Kachchh could have spread at least to the exposed limits of the basin, possibly farther in north and east. On the organic evolutionary scenario, near the Aalenian/Bajocian boundary also originated the relatively deeper water Stephanoceratacaea (Fig. 7.10). The increase in bathymetry is inferred from the Early Bajocian to the early/mid-Middle Bathonian towards the margin, i.e. from Goradoongar to ~25–30 km farther away in Kaladoongar as a progressive relatively unidirectional transgression in the Kachchh basin. The Early Bajocian submergence is expressed in the early part of KSm (beds 37–47 of Biswas), while the

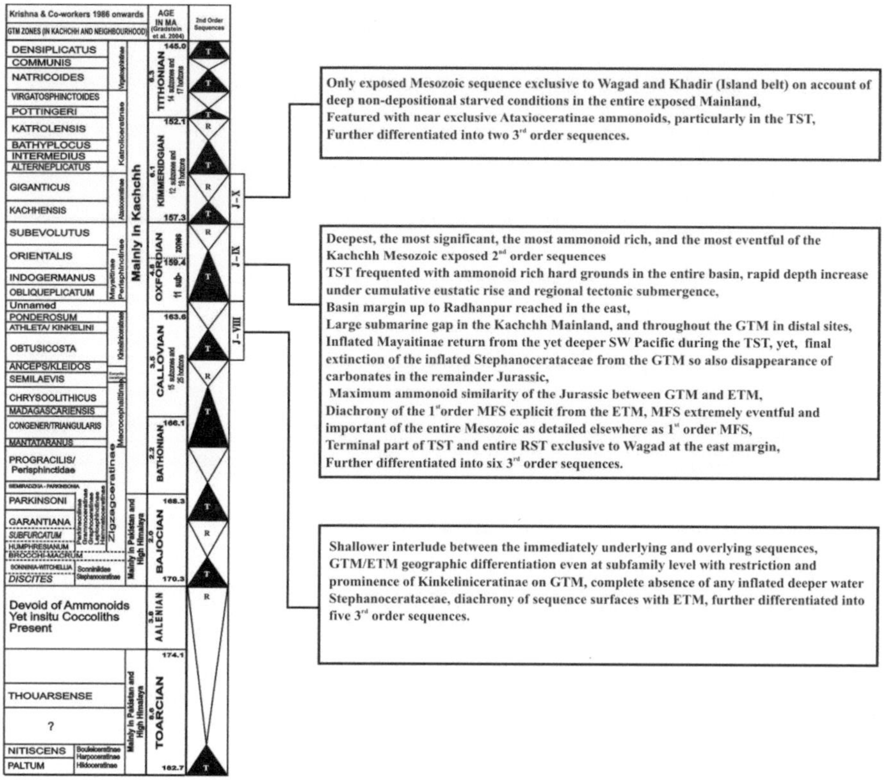

Fig. 7.20 Salient geological facets at the second-order sequence surfaces in Kachchh (sequences J-VIII to J-X)

Early Bathonian one in the basal part of BCSm (beds 49–55). The lithological succession of the two sections corroborates the same. There is additional evidence to support the increase in bathymetry in the basin in the form of the recently discovered distinctly richer ammonoid bearing sediment interval in the distally located Nara section of West Mainland of Early to mid-Middle Bathonian (Krishna unpublished report 2014; Pandey and Pathak 2015a), in particular the common *Micromphalites* and coeval ammonoids not just in Kachchh (Sadhara, Jhura, Nara, Jumara) but every where from Arabia to India in the west sector. The entire Bajocian along with basal Early Bathonian in the ETM sequence framework makes two successive second-order TSTs along with the single intervening RST. In Jhura succession, there is as well suggestion of step wise tectonically facilitated increase in bathymetry in successive second/third TSTs (Am, and Cm), respectively, of the Early Bajocian, and the basal Early Bathonian. Around the same also supposedly ceased the widespread Karoo and coeval igneous event in the west sector.

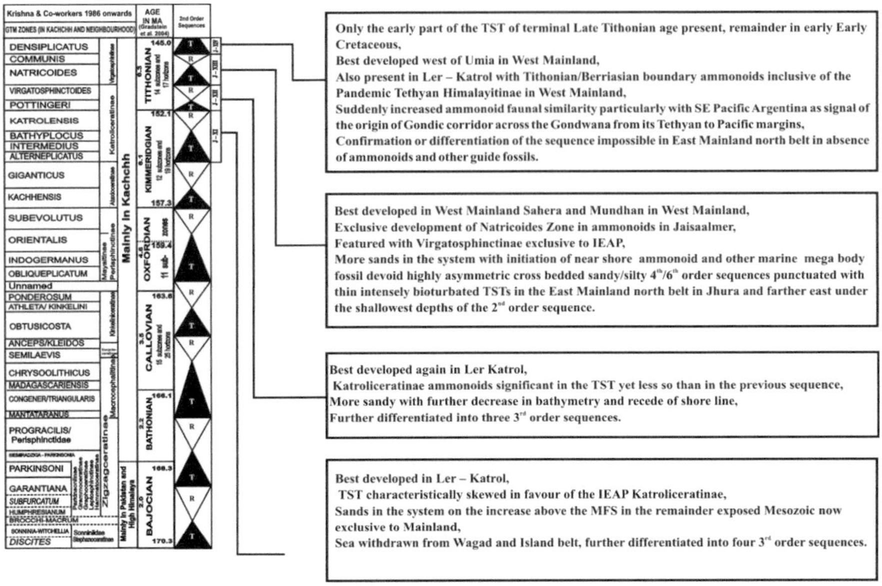

Fig. 7.21 Salient geological facets at the second-order sequence surfaces in Kachchh (sequences J-XI to J-XIV)

7.5.17 *Transform Sliding of India Vis-à-Vis Africa Along the North–South-Directed Davies Fracture Zone During the First-Order TST*

The oceanic spreading and deepening between Somalia and India (inclusive of Madagascar) of the Axial Gondwana has been interpreted through southward transform sliding (Krishna and Pathak 1994; Rabinowitz et al. 1983) in multiple steps at the second-order sequence surfaces from Early Toarcian to late Middle Oxfordian during the first-order TST (Figs. 7.19–7.20, 7.22–7.25). The next significant step as above occurred at the start of the Aalenian/Bajocian boundary (Figs. 7.19 and 7.24) of transform sliding. The said deepening/spreading/oceanization at the north extreme of the west divergent sector accords well with the interpreted oceanization at the south end between Southeast Africa and East Antarctica based on igneous event in Mozambique of ~174–172 ma age (Jourdan et al. 2007; Gaina et al. 2015). This is supported by deeper water Early Bajocian ammonoids in the Somalian basin (Arkell 1956) and the Axial belt of Pakistan (Fatmi 1986). Thus, during the Toarcian–late Middle Oxfordian first-order TST, the spreading was mostly due to transform sliding locally at the extremes of the transform boundary along the boundary instead of across the boundary. Significant spreading across the Davies ridge between India inclusive of Madagascar on one side and East Africa on the other side in essence did not begin at least until the intra-Oxfordian first-order MFS.

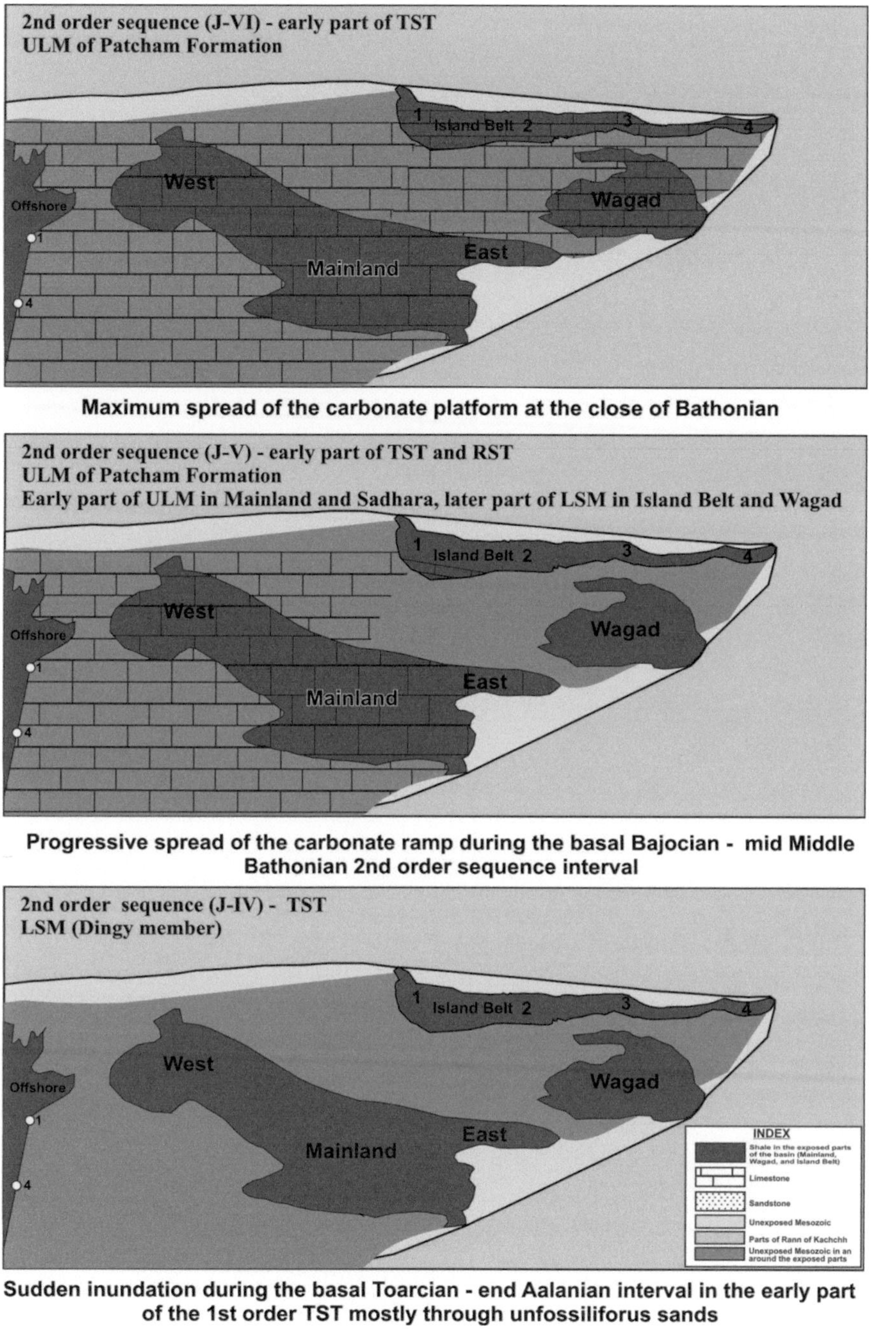

Fig. 7.22 Graphics from bottom to top respectively refer to Toarcian-Aalenian, Bajocian–mid-Middle Bathonian and top of Bathonian intervals

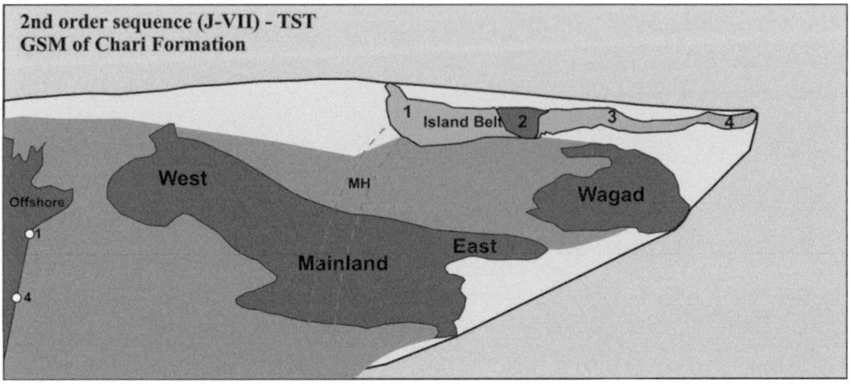

Late Middle Callovian O-II Horizon of to early Late Callovian A-II Horizon shale rich interval of Athleta Zone Depressum Subzone

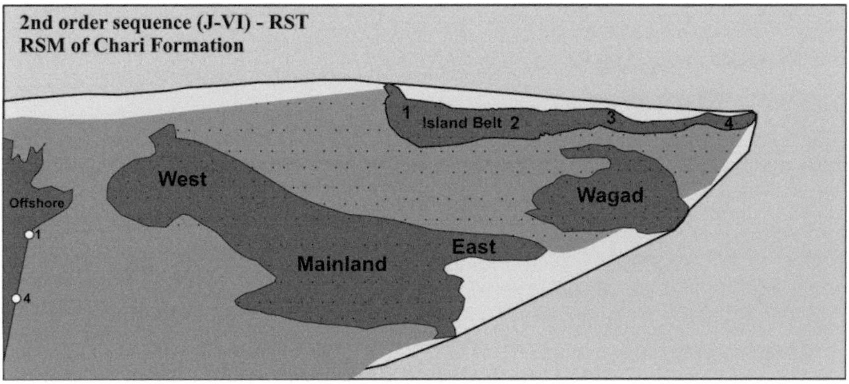

Late Early Callovian Base of Semilaevis Zone - mid Middle Callovian O- I Horizon of Obtusicosta subzone

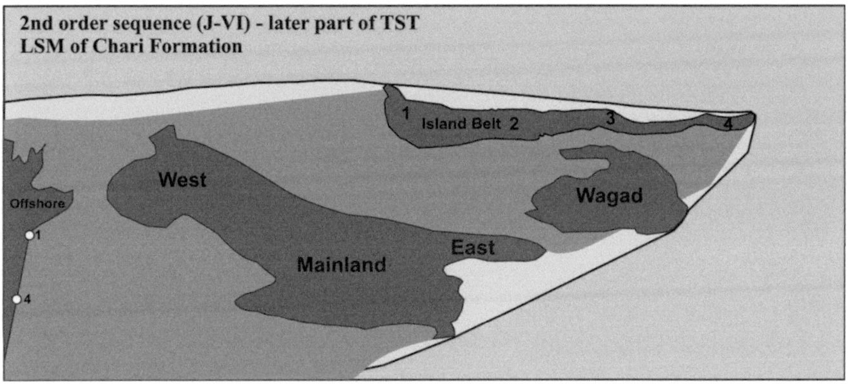

Basal to mid Early Callovian shale rich sediment interval after the drowining of the carbonate plateform

Fig. 7.23 Graphics from bottom to top respectively refer to basal to mid-Early Callovian, late Early to mid-Middle Collovian, O-1 Horizon, and O-2 Horizon to early Late Callovian A-2 Horizon interval

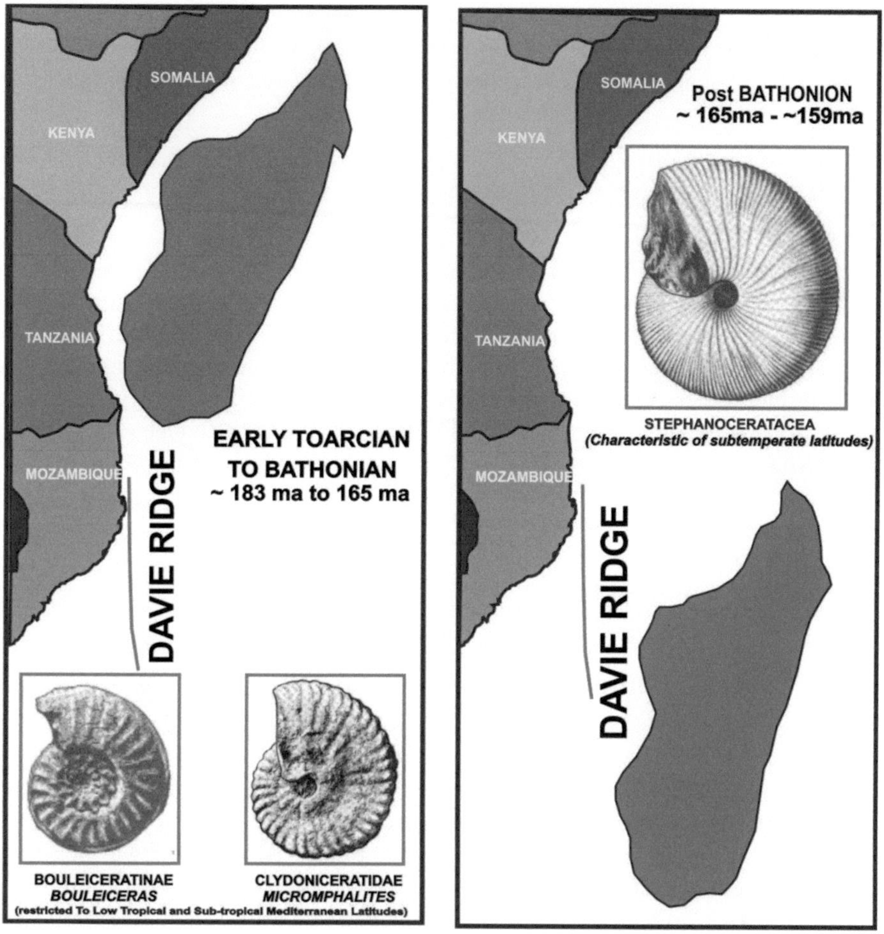

Fig. 7.24 Transform sliding due south of India (inclusive of Madagascar) vis-a-vis Africa along the Davies fracture reflected in the progressive yet distinct change from low to high latitude ammonoids during the Toarcian–late Middle Oxfordian first-order TST

7.5.18 Drowning of the Carbonate Platform in High Himalaya

The drastic lithological change from the significant pre-Toarcian carbonates to the absence of carbonates, and the associated drowning of the carbonate platform in High Himalaya is ascribed to the transform sliding of a few degrees along the Davies fracture transform boundary from subtropical to subtemperate (Figs. 7.19,. 7.22, 7.24) (Rabinowitz et al. 1983; Krishna and Pathak 1994). There is also regional change in climate from warm arid to somewhat wet humid. The over all multifaceted change in High Himalaya is suggestively linked to the widespread tectonically influenced submergence event at the first-order intra-Jurassic SB.

7.5.19 Mid/Late Middle Bathonian Second-order SB Event

The carbonate ramp began in High Himalaya at the T/J second-order SB. In Kachchh, the ramp conditions started developing since the start of Bajocian in Mainland, and extended proximally to the 'Island belt' in late Middle Bathonian (Figs. 7.19, 7.22). The ramp also got drowned in Kachchh later through the tectonically induced additional subsidence at the Bathonian/Callovian boundary. The intra-Bathonian second-order SB marked further increase in bathymetry in Kachchh along with marked lithological and faunal changes which also found explicit expression in Madagascar, Pakistan, and Jaisalmer in the west divergent sector.

7.5.20 Origin of the Geographically Restricted IEAP Ammonoid Subfamilies/Lineages at the Mid/Late Middle Bathonian Second-Order SB

The ammonoid faunal changes across the early/mid-Middle Bathonian second-order SB, from their earlier ETM imprint to later geographically restricted IEAP character, also support such interpretation (Krishna and Pathak 1994). The bathymetry continued to improve in the successive second-order TSTs, and for the first time allowed the development of the indigenous moderately rich and diverse ammonoid fauna in the relatively distal Jumara section of somewhat deeper water Macrocephalitinae, also with scarce ammonoids in Nara and Jhura above the second-order SB.

7.5.21 Arrival of Macrocephalitinae Under Increased Bathymetry in Kachchh from SW Pacifc

The Macrocephalitinae arrived in Kachchh from SW Pacific after realization of sufficient depth in Kachchh (Krishna and Cariou 1993). The near absence of ammonoids in the earlier Toarcian–Bajocian interval implied less than ~20–25 m depths in the 'Island belt' localities in the considered interval except Goradongar. Even Goradongar lacked optimum depth for the additional presence of relatively less shallow water preferring Macrocephalitinae. The late Middle to early Late Bathonian ammonoid record of Goradongar does not include any Macrocephalitinae, and instead includes mostly shallower water preferring Perisphinctinae. The said Macrocephalitinae mark the start of ammonoid provincialism and differentiation of IEAP through another step of rift-genic submergence in the region.

7.5.22 First Marked Prsesence of Ammonoids in Goradongar

The late Middle Bathonian provides the first ammonoid level of Goradongar with more than one ammonoid genera recorded ~500 m above the base. The LSM of Patcham Formation in the entire 'Island belt' so far yielded merely two ammonoid specimens (single example of Early Bajocian *Calliphylloceras* in the basal limestones at Goradongar and single early Middle Bathonian Perisphinctinae in the Kuran section of Kala Dongar). The scenario testifies well the prevalence of extremely shallow environments for ~20 my in the 'Island belt' with only rare occasional supratidal interludes.

7.5.23 Mid/Late Early Callovian Chrysoolithicus Zone Diadematus Zone Second-Order MFS Event

Almost north–south-directed MH (median high, Fig. 1.7), is interpreted to have undergone reactivation with slight tilting up due north and associated rejuvenation of the provenance at this second-order MFS. The tilt resulted in graded/stepped shallowing during the ensuing second-order RST (Fig. 7.25a–c). The shallowing in turn prompted rapid deposition of thick regressive sands as RSM of the Semilaevis Zone to O-I Horizon of Obtusicostites Subzone duration. The event also initiated differentiation of the sedimentation regimes, relatively and distinctly deeper west of MH and relatively shallower east of MH. The increase in bathymetry during the terminal part of the second-order TST of early to mid Early Callovian duration allowed for the first time presence of ammonoids inclusive of Macrocephalitinae in the depositional system even in the most proximal exposed sections of the basin at Khadir, Kakindia bet, Bela, Mouwana and Washtawa. The Macrocephalitinae required depths of at least 25–30 m which is also the maximum depth realized in Bela and Mouwana. If such depths were acquired at the second-order MFS in proximal-most locations, the depths attained at the same point of time at distally located sites in West Mainland of Keera, Nara, Jumara and Jara ~150–200 km down the basin in either of the northeast–southwest or east–west transect would have been much more. There is estimation of ~40–50 m with an assumed shelf gradient of half degree which is about double the depths estimated at the margin. In

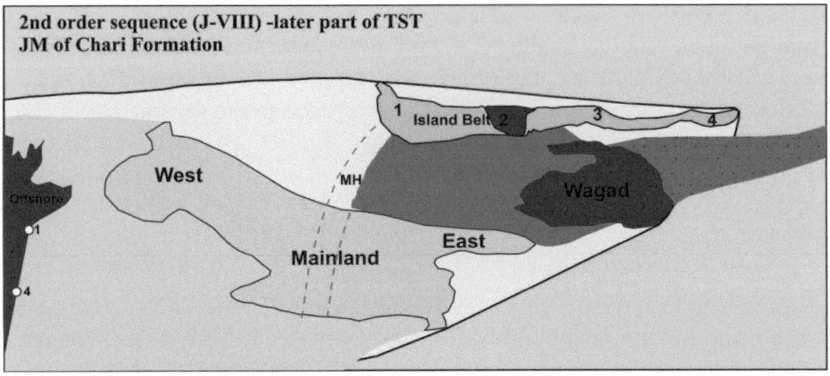

Starvation in mainland, slow sedimented ammonoid rich sediment from Wagad to Khadir during mid Middle Oxfordian 2nd order MFS to late Middle Oxfordian

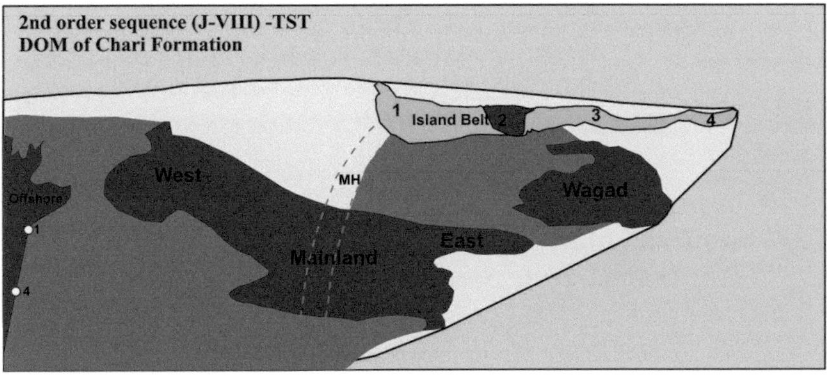

Early highly condensed slow sedimented (Mainland) TST to moderately sedimented (Wagad and Island Belt), later part in submarine gap in Mainland during early Early Oxfordian 2nd order SB to early Middle Oxfordian 2nd order MFS

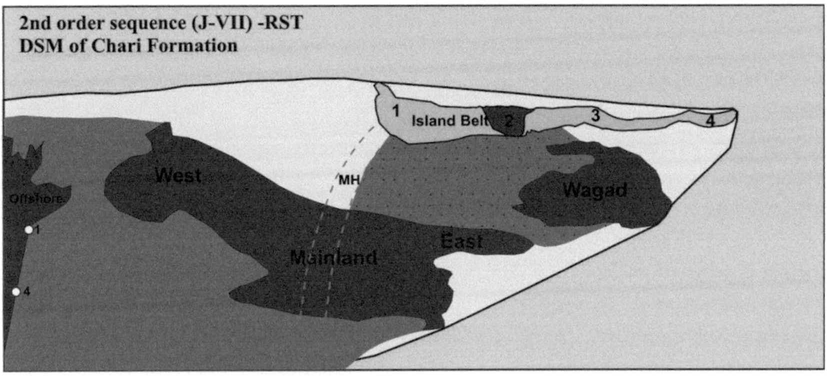

Early Late Callovian of Athleta Zone Depressum Subzone A-III Horizon 2nd Order MFS to early Early Oxfordian 2nd order SB

Fig. 7.25 Graphics from bottom to top respectively refer to early Late Callovian A-3 Horizon to early Early Oxfordian, mid Early Oxfordian to early Middle Oxfordian and mid Middle to late Middle Oxfordian intervals

the total sediment thickness of over 600 m at Khadir section, the first ammonoids occur around the second-order mid/late Early Callovian MFS ~300 m or more above the base, so also in Mouwana and Bela. In context of time, the said first ammonoid in Khadir is ~19 my above the base, and over 10 my above the base in Bela and Mouwana. It may be noted that this also is the second highest bathymetry scored in the entire exposed Mesozoic succession of the Mainland of over 1000 m thickness and ~80 my duration, and practically the highest bathymetry registered (Fig. 7.7) except for the ~four my interval of the exposed Early and Middle Oxfordian in the entire Kachchh Mesozoic with a total time span of over 110 my.

Somewhat similar ammonoid succession and graded increase in bathymetry have been interpreted as well in Arabia, East Africa, Madagascar, Pakistan, India, Nepal, South Tibet, Indonesia and New Guinea up to the second-order intra-Callovian MFS. It suggests distinct regional control over a wide area on GTM, at least in IEAP. In Paksitan, the slow sedimentation to sediment condensation, and yet further complete starvation evidences rapid deepening. The NW Australian ammonoid succession is disjunct and extremely meagre, nevertheless, Sula and New Guinea provide good testimony of somewhat similar ammonoid stratigraphic record as in Kachchh. The second-order MFS also marked the termination of the Macrocephalitinae Chrysoolithicus lineage.

7.5.24 *Middle Callovian O-I Horizon Obtusicosta Subzone Second-Order SB Event*

The deposition of RSM as the second-order RST was followed further up by uplift/emergence of the Patcham region that was located on the median high at the second-order SB. Thus, the activation of MH resulted in the cessation of marine sedimentation in the Patcham, Bela, and Mouwana localities for the rest of the Mesozoic. Sedimentation, however, continued in the Khadir region which may have undergone complimentary deepening due to the uplift of neighbourly Patcham. Khadir, now, formed relatively less shallow part of the north margin of the basin than in the earlier second-order sequence (Figs. 7.19–7.20, 7.23). The associated decrease in bathymetry prompted the Macrocephalitinae to either retreat to SW Pacific or become even extinct, already discussed at some length in an earlier chapter.

7.5.25 *Early Late Callovian Athleta Zone Depressum Subzone Second-Order MFS Event*

The tectonically produced shallowing of the previous second-order sequence in general occurred all over the Mainland, as also in Wagad. The next younger yet shallower sequence is characterized by the Indo-East African perisphinctid Kinkeliniceratinae. Obviously, the basin underwent yet another phase of tectonically produced shallowing at the intra-Late-Callovian second-order MFS (Fig. 7.25) that resulted in the deposition of, yet another, sand dominated systems tract as DSM after RSM of the previous second-order sequence. The RSM-GSM-DSM succession is interpreted either as the resultant of significant global eustatic fall or tectonically induced sand dominated interlude in the middle of the first-order TST. In this interlude, the relatively thin and fine textured ammonoid inclusive GSM as a second-order TST is found sandwiched between two thick sand units. DSM in absence of guide fossils could be restricted in age in the terminal Late Callovian more so in the proximal Khadir, Wagad, and East Mainland east of MH as a short duration distinct shallowing/cooling phase near the Callovian/Oxfordian or Middle/Late Jurassic boundary as suggested in Europe (Dromart et al. 2003) and many other regions of the world already discussed in Chap. 4. Possibility of a glacial interlude at present is discounted, however, the interval deserves deeper investigation.

7.5.26 On the Middle/Late Jurassic Boundary Cooling Event

Based on estimation of sea surface temperatures through isotopic thermometry, and other multidisciplinary evidences, Dromart et al. (2003) interpreted severe widespread cooling event at the Middle/Late Jurassic transition (MLJT) in the Jurassic. The Jurassic period otherwise has been in the past consistently considered having had all through a warm green house type of climate. The cooling event was widely interpreted in the north hemisphere inclusive of mid latitude Arabia. The event postdated widespread occurrence of organic rich sediments. The cooling climaxed in terminal Late Callovian through abrupt fall of temperature for ~0.8 my. One of the evidences cited was expansion of the Boreal ammonoids that climaxed in the Lamberti Subzone. Drop in sea surface temperature was of ~5 °C from ~25–20 °C to ~19–17 °C, i.e. a drop comparable to one between interglacial and glacial phases.

In Kachchh, the terminal Callovian is included in the middle of a second-order RST of a first-order TST. It includes sand/silt of DSM without guide fossils, and is indirectly age-constrained on the strength of underlying and overlying ammonoid bearing units. It is considered a strong regressive phase, and may include even emergence at the proximal sites of the basin. But for this ammonoid devoid interval, the entire Callovian is ammonoid rich so also the entire Oxfordian. The Oxfordian is ammonoid rich even in proximal part of the basin in context of first-order intra-Oxfordian MFS. DSM invariably understood as of late Late Callovian to early Early Oxfordian could also be considered exclusively late Late Callovian with early

Early Oxfordian as subaerial gap. In context of ammonoids, a single distorted fragment of *Quenstedtoceras* (Singh 1992, Kanjilal and Singh 2012) has been reported in one of the distal sections which is underlain a few metres below by suspected rare kosmoceratid examples. Two other ammonoid examples collected in the same bed were *Pachyerymnoceras* and *Kinkeliniceras*, and the level is assigned to Lamberti Subzone. The level is followed up by a ~3 m ammonoid devoid terminal part of the RST prior to the ammonoid rich second-order TST with the Pan-tethyan late Early Oxfordian *Properisphinctes gr. bernensis*. The almost ammonoid devoid interval is ~5 m thick that includes a single ammonoid level with mere three ammonoid specimens. The crude estimate of the second-order RST is ~800 ky to 1.0 my, while the marker level/interval under consideration may be considered ~400 ky or less. It is preceded and proceeded by ammonoid devoid intervals. As many as five third-order sequences are differentiated in the Keera section, and the Lamberti Subzone there makes the third-order TST of the fourth third-order sequence of one subzone duration, i.e. ~400 ky. Thus a transient cooling event cannot be totally ruled out and the subject requires further investigation from different angles. The eustatic fall, on the ETM prior to the onset of third-order Lamberti Subzone TST, is estimated as of at least a few tens of metres. Any tectonic event is not interpreted in Kachchh, and so similar decrease in bathymetry could be envisaged in the Kachchh basin, particularly in Mainland. Almost since the start of the Callovian, Kachchh was located in subtemperate latitudes. Similar sand rich second-order RSTs have, also, been differentiated particularly in Aalenian and late Early to early Middle Bathonian which temporally correspond well with distinct cooling phases on ETM (Dera et al. 2011; Korte et al. 2015). Otherwise, the first-order significant eustatic rise is considered between the basal Toarcian SB to late Middle Oxfordian MFS.

Alberti et al. (2012a, b) have in the above context based on stable isotope records from belemnoid, brachiopod and oyster shells interpreted decrease of sea water temperature from 18 C to 14 C between Late Callovian to Late Kimmeridgian inclusive of two phases of rise and minimum of 14 C in early Late Oxfordian. They discounted any distinct drop at the MLJT, while the minimum was linked to their suspicion of the origin of the Gondic corridor seaway (Krishna 1987a, b, c, 1996; Krishna and coworkers—several papers since 1991) by early Late Oxfordian. In biogeographic and sequence stratigraphic context, cooling phases are possible in Aalenian, late Early to early Middle Bathonian, and terminal late Late Callovian/early Early Oxfordian second-order RSTs during a long phase of graded rise in the Toarcian late Middle Oxfordian first-order TST. The decline in sea surface temperature is invariably observed during first-order RSTs, i.e. terminal Middle Oxfordian Rotoides Zone onward in the Kachchh Jurassic and post early Middle Turonian first-order intra-Cretaceous. The slight rise in parts of Early and Late Kimmeridgian is in the corresponding second-order TSTs.

The origin of the Gondic corridor (Fig. 1.3) on biogeographic evidence of ammonoids, bivalves and other numerous studies over the last few decades is not substantiated at least until close of Kimmeridgian. The definite ammonoid evidence is available in late Late Tithonian Densiplicatus Zone (Krishna 1987a, b, c onward several papers) because of the persisting land barrier of Mozambique ridge between the Somalian and Mozambique shallow seas. The ammonoid faunal similarity is low until at least Middle Tithonian. The IEAP ammonoid subfamily Katroliceratinae is not evidenced to get through the Mozambique ridge until mid-Late Tithonian, while the ammonoid faunal similarity between the two regions is found suddenly increased during the late Late Tithonian Densiplicatus Zone with sudden arrival of *Corongoceras, Durangites, Argentiniceras, Groebericeras* and others from Argentina to India through the Gondic corridor (Krishna 1996 and later). Thus, the terminal Middle Oxfordian–early Late Oxfordian drop in sea surface temperature is not linked to the origin of the Gondic corridor seaway. Arrival of the cool South Pacific or the nascent juvenile South Atlantic Ocean waters could not get through the Mozambique ridge until at least the close of Oxfordian. The origin of such corridors additionally requires tectonically caused submergence of the intervening land barrier. Much better alternative explanation can be sought in the large scale plate reorganization all round the Indian plate. The reorganization is prompted by acute space constraints on the NE and SW Gondwana margins on account of localized transform sliding and associated spreading on these two margins initiated at the ~159 my late Middle Oxfordian Schilli Subzone first-order MFS event.

7.5.27 Climatic and Bathymetric Barrier Between IEAP and Arabia at the Second-Order Intra-Callovian SB

Bathymetric framework and ammonoid succession similar to that of Kachchh are known in IEAP, e.g. Madagascar. These attest the regional character of the tectonic events, so also of the resultant first and second-order sequence surfaces on GTM. However, unlike the relatively deeper water stephanoceratid Macrocephalitinae, the relatively shallow water perisphinctid Kinkeliniceratinae could spread up to East Africa, and not farther west to Arabia because of the distinctly lower latitudes and warmer climate in Arabia. The latitudinal differences seem accentuated by the progression of transform sliding of India due south along the Davies ridge (Fig. 7.24).

7.5.28 Land Barrier Between IEAP and South Pacific Mozambique

The IEAP ammonoid subfamilies are found geographically extended due south in the west divergent sector only up to the south margin of the Morondova basin.

However, the said IEAP ammonoid subfamilies are absent in the SE Pacific basins and its extended arm up to the Mozambique sea owing to the land barrier of the possible combination of Mozambique ridge–Astra ridge–Erving bank between the two shallow marine Somalia and Mozambique regions. The Neotethyan arm through further progression due south may have eventually reached the SE Pacific basins on account of possible submergence of the Mozambique ridge during the maximized bathymetry at the first-order intra-Oxfordian MFS. The eustatic maximum at the first-order late Middle Oxfordian MFS may have combined with rift-related submergence and associated transform sliding along the Davies ridge. The relative deepening may have been facilitated by the initiation of volcanism and associated spreading between Africa and Antarctica between ~159 and ~154 of Late Oxfordian–Early Kimmeridgian (Konig and Jokat 2006). This is further supported by the recognition of M26/25 anomaly in the region (Gaina et al. 2015). The Late Kimmeridgian–Early Tithonian bivalve evidence supports the above interpretation. However, biogeographic ammonoid evidence of the origin of the Gondic corridor is absent up to the early/mid-Late Tithonian Thus, the shallow marine ammonoid Gondic corridor seaway between IEAP from the Neotethyan margin of the Gondwana to the SE Pacific margin across the Gondwana through the East/West Gondwana divergent zone became operative only in late Late Tithonian Micracanthum Zone/Densiplicatum Zone as is evidenced by the common *Corongoceras, Argentiniceras, Durangites,* etc during the terminal Tithonian second-order TST between the two shallow marine provinces (Krishna 1987a, b, c, 1996). The initiation of the Weddell Sea at ~147 ma around the terminal Tithonian (Konig and Jokat 2006) also accords well with the origin of the Gondic corridor.

7.5.29 Early/Late Early Oxfordian Second-Order SB

At this SB began the most important interval in Kachchh of relatively ~six my long phase of the tectonically induced rapid/sudden increase in the bathymetry (Fig. 7.25). The associated deepening resulted in the hard grounded, nodular, conglomeratic, extremely slow sedimented, ammonoid abundant, oolitic limestones, particularly, in the distally located West Mainland. The oolitic limestones evidence repeated reworking of sediments and ammonoids of more than one depositional episode. Slow sedimentation strengthened with time, and prompted the sedimentary-cum-biological condensation. It was followed by complete starvation on further deepening. The duration of the complete starvation is expressed as ~plus six my several subzone long submarine gap (Krishna et al. 1998, 2000) (Figs. 2.19, 2.20, 2.21, 2.22 and 2.23). The sedimentation was the slowest of the entire Kachchh Phanerozoic, may be the slowest ever of the exposed Indian geological record, as also one of the slowest ever of the world, of less than a metre in more than two my, i.e. 1 mm per 2 ky. If the starvation interval is also included in the estimation, it becomes yet slower of the-order of 1 mm per 10 ky. This was an exceptional transient intra-Oxfordian interlude on the GTM that culminated later with even more spectacular geological event at the first-order late Middle Oxfordian MFS. The rapid deepening began at the second-order early/late Early Oxfordian SB immediately after a shallow regional sandy regime which may have included even a cooling event although discounted by estimates of sea surface paleo-temperatures (Alberti et al. 2012a, b). The relatively deeper water Mayaitinae either originated or returned/re-migrated from the deeper water SW Pacific at this second-order SB from where had earlier arrived in Kachchh the similarly deeper water Macrocephalitinae near the mid/late Middle Bathonian second-order SB. As earlier discussed in Chaps. 3 and 4, emergence of any sort as interpreted by Alberti et al. (2012a, b, 2013) in late Early Oxfordian (Kachchh Obliqueplicatum Zone = ETM Cordatum Zone) or in early Late Oxfordian (Subevolutus Zone–Bifurcatus Zone) is not favoured (Figs. 3.26a, 4.14, 7.25 and 7.30), but for extremely short duration transient one less than even an ammonite horizon. Instead, emergence, if any at all, could be only thought of the proximal depositional sites exclusively near the Callovian/Oxfordian boundary during or immediately after the deposition of DSM, and not any later, as already remarked earlier. The sudden deepening is rift related around the start of a phase of regional reorganization of plates and associated major shake up involving transform sliding, rotational movements under space constraints, followed later by volcanism in the northeast and southwest margins of Gondwana at the late Middle Oxfordian first order MFS. This was perhaps the biggest, the most widespread, and the most violent reorganization shake up that the Gondwana ever underwent. It is interpreted to have been rather protracted with early/late Early Oxfordian start of shake up that later climaxed at late Middle Oxfordian first-order MFS. Frequent occurrence of earthquakes and associated seismites and tsunamites as observed at the Permo/Triassic boundary in Kashmir (Brookfield et al. 2013) are suggested in Kachchh throughout the long tectonically unstable interval from the early Middle Oxfordian Indogermanus Zone to the late Middle Oxfordian Orientalis Zone Schilli Subzone at least. The already lithified/hardened late Early / early Middle Oxfordian chunks/slabs of the oolitic limestones fragmented and rotated, even topsy-turvied upside down in the repetitive violent/explosive shake ups. Also, topsy-turvied/mixed/reworked more than one subzonal ammonoid and other mega fossil assemblages as part of the explosive/volatile event. However, neither in Kachchh nor even in any other GTM basin any emergence is interpreted at the Early/Middle Oxfordian boundary as made out by Alberti et al. (op cit). The unusual short duration highly volatile instability as precursor to the pre-oceanization tectonics may be related to the initiation of oceanisation and associated plate reorganization under acute space constraints far northeast on the Indo-Australian margin and far southwest in the West Antarctica–Africa–Patagonia margin. Detailed comments have been separately made in Chap. 4 on the said two intra-Oxfordian emergences in late Early Oxfordian Cordatum Zone and the early Late Oxfordian Bifurcatus Zone exclusively in Mainland of Alberti et al. (2013).

7.5.30 *Late Middle Oxfordian Orientalis Zone Schilli Subzone first-Order MFS Event*

The rifting climaxed in Kachchh at the late Middle Oxfordian Orientalis Zone Schilli Subzone in or above the bed I–08 of the youngest Member (Jhadasa Member) of the Chari Formation (Figs. 7.26, 7.27 and 7.28), and then aborted as evidenced by uplift related shallowing in the basin immediately after the MFS. This subzonal intra-Oxfordian precision on the abortion of rifting is much contrary to the broad and incorrect age of rift abortion in early Cretaceous by Biswas since 1977.

The abandonment of the rifting in Kachchh may have been forced by the initiation of volcanism at the north end of the east sector, as also associated revival of rifting and tilting down due east of the Indian plate all along the India–Australo-Antarctica sector. The so interpreted reorganization and vigorous shake up in the west and south parts of the Indian plate is linked to the igneous activity and onset of the spreading away of the Argo block between Greater India and NW Australia. The rift abortion in Kachchh was followed by the suspected migration of the rift axis to possible revival of rifting in the Narmada basin as evidenced there by the Nimar Sandstone of the terminal Middle Oxfordian–Barremian first-order RST age bracket. The onset of the decrease in bathymetry in Kachchh in particular and in the IEAP in general marked the extinction of the relatively deeper water Mayaitinae, so also the extinction of many other relatively deeper water ammonoid subfamilies from the IEAP and the GTM (Figs. 7.10 and 7.26). The ammonoid

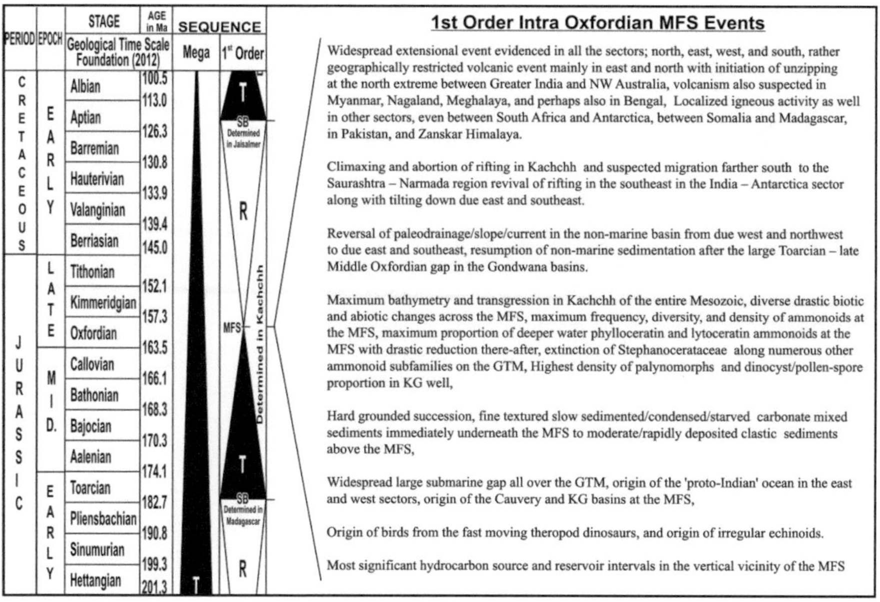

Fig. 7.26 First-order intra-Oxfordian MFS events

extinction event at the first-order intra-Oxfordian MFS may be considered the biggest ammonoid extinction event of the Indian, IEAP or even the GTM Mesozoic. In context of ammonoids, the maximum diversity is recorded at the first-order MFS of ~10 subfamilies which is drastically reduced to mere one subfamily across the MFS (Krishna and coworkers—several papers since 1998). May be the ammonoid familial extinctions at the intra-Triassic and intra-Cretaceous MFSs are of comparable magnitude but are not that well spelled out in literature. Quantitatively, the intra-Oxfordian first-order MFS extinction at the ammonoid subfamily level is ~90 %, while in context of ammonoid density the drastic reduction may be yet more severe. At the intra-Triassic first-order MFS, the extinction event is neither sudden nor well marked in the range chart of Krystyn et al. (2007 Fig. 2.9). The reduction in the number of ammonoid genera is from six to three, No such range chart or data exists in the Indian or IEAP or GTM Cretaceous.

7.5.31 Other Facets of the Intra-Oxfordian First-Order MFS Event

The intra-Oxfordian MFS provides an excellent example of precise dates to the major regional and global multifaceted geological events. The event marked drastic change in lithology, texture, structure, fauna, flora across the MFS. Mixed carbonates and siliciclastic sediments changed to exclusively clastics. Slow/condensed/starved relatively fine textured rather structure deficient hard grounded terminal part of the TST changed to rapid coarse textured cross-bedded sands across the MFS. Singh et al. (2000) locally in Saurashtra basin even interpreted distinct change in maturation pattern between the Lower and Upper units of the Dhrangadhra Formation. This aspect, in context of similar tectono-stratigraphic framework in the basins of the west sector as also similar influence of the later volcanism, requires further examination.

Also are observed Middle Oxfordian changes on ETM in the marine carbonate sedimentation pattern (Dromart et al. 2003), positive anomalies in both marine and terrestrial carbonate isotope records (Pearce et al. 2005), climatic change from cold to warm and arid (Dromart et al. 2003 and many others), reorganization of plates, reorganization of ocean current system (Rais et al. 2007), lowest values of sea water Sr87/Sr86 of the entire Phanerozoic, exceptionally high ocean floor spreading rates (Cogne and Humler 2004), more efficient circum global seaways, deepening of the east–west low latitude Hispanic seaway and associated differentiation of Pangea into Laurasia and Gondwana, mega-change and turning point in ocean chemistry, intense mid-ocean ridge hydrothermal activity, partial breakdown of the mega-monsoonal climatic system (Rais et al. 2007), and others. These changes seem to be triggered by the intra-Oxfordian silicic volcanism of the Chile–Patagonia–Africa–Antarctica LIP in the S and SE Pacific margin of the Gondwana

(Pankhrust et al. 1999) on one extreme, and the Argo block volcanism on the Neotethyan Indo-Australian margin of the Gondwana on the other extreme. Igneous events of the said age are also evidenced elsewhere in the divergent region round the Indian plate. The ocean floor volcanism along with origin of the Proto-Indian Ocean on the Indo-Australian margin is indirectly determined at the ~159 ma late Middle Oxfordian Orientalis Zone (=Transversarium Zone) Schilli Subzone first-order intra-Jurassic MFS. The argument for genetically linking the first-order MFS with the tectonic/igneous event is their being products of the same geological event. On one hand the initiation of a new ocean is the biggest expression of tectonics in the region, so is the first-order MFS the biggest manifestation of the same regional to global geological event in the shallow marine sedimentary record.

7.5.31.1 Diverse Sudden Evolutionary Changes in the Ammonoid Scenario Across the MFS

The MFS exhibited large number of large sized lyto-phylloceratid and perisphinctid ammonoids inclusive of the largest Jurassic ammonoid of the world—a lytoceratid. However, the ammonoid diversity across the MFS is suddenly reduced from ten subfamilies to just one (Figs. 3.31 and 3.32). Density is reduced several tens of fold. Similar drastic reduction is in the frequency of ammonoid levels across the MFS. The relatively deeper water superfamily Stephanoceratacaea became extinct on GTM at this MFS.

7.5.32 Maximum Phanerozoic Bathymetry Ever in the Kachchh Basin at the Intra-Oxfordian First-Order MFS

Never again in the Kachchh Phanerozoic were realized depths greater than at the intra-Jurassic first-order MFS in spite of the graded rise of sea level until the early Middle Turonian intra-Cretaceous first-order MFS. Also never again is interpreted the eastward reach of the Neotethys up to or close to the basin margin near the Radhanpur arch (Fig. 1.7). The cephalopod subfamilial extinctions at the intra-Oxfordian MFS are more marked than any other time in the Kachchh Phanerozoic. The principal factors responsible were perhaps the graded uplift over and above the thickness of sediments deposited during the ~67 my long interval between the first-order intra-Jurassic and intra-Cretaceous MFSs. The rather shallow Mainland part of the basin deepened substantively from few tens to well over one hundred metres mid to outer shelf depths, and may be even more at the MFS. On the other hand, the supply of sediments reduced, particularly, to the distant and deepened Mainland.

7.5.33 Strong Condensation and Starvation

The said deepening continued with reduction in sediment supply during the Oxfordian. This in turn resulted at first in extreme condensation during the late Early to early Middle Oxfordian, and there after complete starvation at least up to the late Early Kimmeridgian in East Mainland or even up to the mid Early Tithonian in West Mainland farther down the dip. The condensation is evidenced by ammonoid and other invertebrate rich succession of hard grounds (Fig. 3.29). The hard grounds, with time, progressively shifted towards the margin from West Mainland to East Mainland to Wagad. During the phase of the lateral marginward shifting of the hard grounds, the Mainland underwent complete starvation as evidenced by the plus six my submarine gap. The said gap for over a century had been considered subaerial. The submarine gap gradually decreased towards the margin and vice versa increased towards the basin. In the interval of condensation and starvation of mid-Middle Oxfordian to Early Kimmerdgian of maximum bathymetry in the west and north sectors, the Neotethys is interpreted to have also reached Saurashtra and Cambay as a transient marine interlude in the middle of the Upper Member of Dhrangadhra Formation.

7.5.34 Sandy Regimes During the First-Order TST

There are sandy intervals at the north and east margins, respectively, in the 'Island belt', Wagad, and in the East Mainland that were formed mostly during the successive second-order RSTs (Figs. 2.25 and 2.26). Precision dating and intrabasinal spatial differentiation of such guide fossil-devoid sands is significant in context of such sands being prospective hydrocarbon reservoir rocks. The oldest of these sandy intervals is in the Dingy Member with definite presence at Dingy and Kuar bet that is exclusive to Kaladongar locality of Patcham region. The relatively younger two such sandy intervals are spread widely in the Island belt. These sands in the Island belt are included in LSM of Patcham Formation as Dm, KSm and BCSm at Patcham, CBCm, HSm and HSTm at Khadir, Bela, and Mouwana. These range in age from the Early Toarcian to early Late Bathonian having a total duration of ~ 18 my and maximum thickness of over 600 m. Farther up, there developed RSM in East Mainland, at Wagad as Km, and at 'Island belt' (as MHm in Goradongar, GSm in Khadir and other domes). The duration of the RSM or its equivalents was greater near the margin of over four zones of entire Early and major part of the Middle Callovian up to O-I Horizon of Obtusicosta Subzone in comparison to over two zones from the late Early Callovian to O-I Horizon of Middle Callovian all over Mainland (Figs. 2.19–2.29). There is another sandy unit (DSM) either across the Callovian/Oxfordian boundary or restricted to the terminal Late Callovian of single subzone to nearly one and half zone duration in Mainland, Wagad and Khadir. The youngest sandy interval of the first-order TST is evidenced in the early part of JM which is exclusive to Washtawa, Khadir, Gangta bet and Desalpur. It is only of one subzone duration. It is observed that upward in the TST, the time duration of the successive sandy second-order RSTs has decreased as also the thickness, along with graded decrease in rate of sedimentation. The thickness and sand content, so also duration of the successive sandy RSTs, as anticipated also decrease from margin to the basin.

7.5.35 Terminal Middle Oxfordian–Mid-Middle Kimmeridgian Second-Order Sequence

During the interval of the submarine gap in Mainland, the entire Wagad region witnessed continuous sedimentation. The uninterrupted sedimentation, thus, allowed the determination of the principal intra-Jurassic Kachchh MFS exclusively in late Middle Oxfordian in Kantkote section (Krishna et al. 1998, 2000 and later). Wagad continued with the next phase of sedimentation up to the close of the revised Early Kimmeridgian (Fig. 7.30) (Kantkote Member of Katrol Formation and Gamadu sands of Umia Formation). The shore line and depositional site shifted rapidly towards the east margin of the basin at the late Middle Oxfordian first-order MFS. The Kantkote Member and the Gamadu sands are exclusive to Wagad.

7.5.36 Unique Scenario of Starvation/Non-deposition in the Distal Mainland and Shallow Marine Thick Sand Deposition in the Proximal Wagad

The scenario of sedimentation exclusively in the proximal Wagad at/near the margin appeared unique while the nearly eightfold bigger distal deeper Mainland part of the basin inclusive of the two Ranns and Banni depression underwent complete starvation (Figs. 4.14, 7.25 and 7.30). Completely emerged 'Island belt' at this time was part of this dynamic landscape to witness passively the unique happenings.

7.5.37 Origin of the 'Proto-Indian Ocean' at the North Extreme of the East Sector, with Spreading Away of Argo Block, also Similar Scenario in the Africa–Patagonia Region

The pre-drift stage on the NW Australian Neotethyan margin continued up to the break-up angular unconformity at SB5 (Figs. 5.7–5.13). The event in a regional context in Chap. 5 is reinterpreted at the late Middle Oxfordian MFS determined for entire GTM in the Kachchh basin (Krishna 1994, 2005, 2009; Krishna et al. 1998, 2000, 2010, 2011). In this context is also mentioned the origin of 'the Proto-Indian Ocean' in the west sector yet earlier at the first-order intra-Jurassic SB. The spreading in the Patagonia–Africa–Antarctica is also evidenced through volcanism of ~156 ma (Konig and Jokat 2006; Pankhrust et al. 1999; Bomfleur et al. 2014). Thus, distanting-cum-spreading at the south extreme of the west divergent sector, and at the north of the east divergent sector, presumably, started almost simultaneously at the first-order intra-Oxfordian MFS on GTM. In fact, the rifting/distanting/volcanism/spreading/reorganization in all the four sectors are interpreted as near isochronous manifestations of a single intra to Pan-Gondwanian event of different scales at the first-order intra-Oxfordian MFS. Recent evidences through refined radiometry (Aitchison et al. 2007; Badsah et al. 2000; Gaina et al. 2015; Gibbons et al. 2012; Guo et al. 2007; Jain 2014; Qi et al. 2013; Malkowski et al. 2016; Poblete et al. 2016; Polvé et al. 1997; Radhakrishna et al. 1999; Sarkar et al. 1996; Sharland et al. 2001; Shellnut et al. 2015; Shi et al. 2008, 2014; Yang et al. 2015; Yui et al. 2013 and others) support the said interpretation of a widespread intra-Oxfordian rifting tectonic-cum-magmatic event in all the intra-Gondwana divergent zones (Krishna 2016 communicated). The multiple geological expressions, specially the extensional tectonics in different sectors at the said MFS may also have been driven by the northward slab pull from subduction beneath Asia as interpreted by Zahirovic et al. (2014).

The geographically restricted 'Proto-Indian Ocean' possibly happened through rifting, transform sliding, associated volcanism and spreading between the Argo block of Greater India and Northwest Australia during the Late Jurassic at the first-order MFS of ~159 ma. Yateesh et al. (2013) interpreted it earlier than ~152 ma. Gibbons et al. (2012, 2015) provided significant new evidence of the extended Late Jurassic oceanic crust located immediately west of the Zenith Plateau. Thus, they extended the Argo block due south up to the ZP (Fig. 6a Gibbons et al. 2015). The two Ar40/Ar39 ages obtained from the Argo block of ~155 and ~153 ma (Gibbons et al. 2012) suggest Late Oxfordian–Late Kimmeridgian age bracket which are considered to post-date the initiation of the oceanization at the late Middle Oxfordian Orientalis Zone Schilli Subzone Subschilli Horizon above the bed I–08 first-order MFS as precisely determined in Kachchh. Voluminous magmatism of Rowley Terrace, part of Browse, and Scott Plateau (Symonds et al. 1998) may also be related to this first/initial stage separation event.

7.5.38 *Involvement of the Kerguelen Plume not Supported in the Initial Late Jurassic Volcanism*

The event perhaps did not involve the Kerguelen hot spot, in view of its location away from the location of AAP. The said early spreading is recognized exclusively in AAP (Hein 2002) off the NW Australian margin until the early Late Valanginian ~136 ma Verrucosum second-order MFS, when the ridge abandoned to allow distinct shift/jump into a new active spreading ridge. In a recent interdisciplinary study, the volcanic tuffs constrained in intra-Valanginian on nannoplankton evidence from the immediately underlying and overlying sediments, have provided U/Pb age of ~136 ma as the first such radiometric age of the intra-Valanginian volcanics of South Tibet (Yin et al. 2011). Gibbons et al. (2012) restrict the said initial volcanism to only AAP region. The intra-Oxfordian maximum bathymetry and transgression through the combined effect of eustatic rise and accelerated subsidence is also evidenced in the stratigraphic record of the Canning basin in the vicinity of Broom and Derby. Also, is interpreted maximum bathymetry, in the dinoflagellate rich Oxfordian–Kimmeridgian well data in Iriyan Jaya, Sula Wasi, Papua and the neighbourhood (Symonds et al. 1998). In Timor Gap, the first-order intra-Jurassic MFS is interpreted between the Plower Formation below and the Frigate Formation above with submarine gap in between of perhaps Middle and Late Callovian, entire Oxfordian and early Early Kimmeridgian of ~ten my. Geochemical signatures of back arc framework are also recorded in subduction zone of Papua New Guinea ophiolites. During the Oxfordian–Valinginian interval the Northwest Australian margin has been conceived as a combination of promontories and embayments as either extended continental crust or small compartments with oceanic lithosphere.

7.5.39 Significant Radiometric Ages of the Igneous Rocks on the Said Spreading in the East and North Sectors

The last few decades have seen a spurt in radiometric ages from the erstwhile Neotethys on its two margins (Figs. 7.27 and 7.28), e.g. ~148 ma active continental margin magmatism, and ~153 ma in Southeast Tibet (Guo et al. 2007), Late Oxfordian–Early Valanginian Dras Volcanic Formation in the Ladakh ITSZ, ~ 177 ma zircon U-Pb age of plagionites in Spontang ophiolite (Pedersen et al. 2001), shoshonite and associated rocks in a submarine oceanic island arc setting at Zedang of ~161 to ~152 ma, and associated intrusive of ~163 to ~165 ma interval (Aitchison et al. 2007) on the said oceanization events in the Neotethys. They have also interpreted Late Jurassic intrusives in KLA and Lhasa, and terminal Jurassic volcanism in YTSZ, Qingtang and associated oceanization events. Gibbons et al. have interpreted intra-oceanic KIA, TIA, and continental rifting. On the other hand ~149, and ~158 ma ages are determined in Meghalaya–Nagaland–Myanmar region (Sarkar et al. 1996), and Late Jurassic/basal Cretaceous volcanism in Bangladesh (Johnson and Alam 1991), U–Pb-based zircon ages of ~163 and ~146.5 ma (Shi et al. 2008), glaucophane age of ~158 ma Myanmar jadeitite (Qiu et al. 2009, 2010, 2013). Yui et al. (2013) interpreted subduction in the Jadeitite uplift in Myanmar during Oxfordian–Kimmeridgian at ~152.4 ma, and considered the

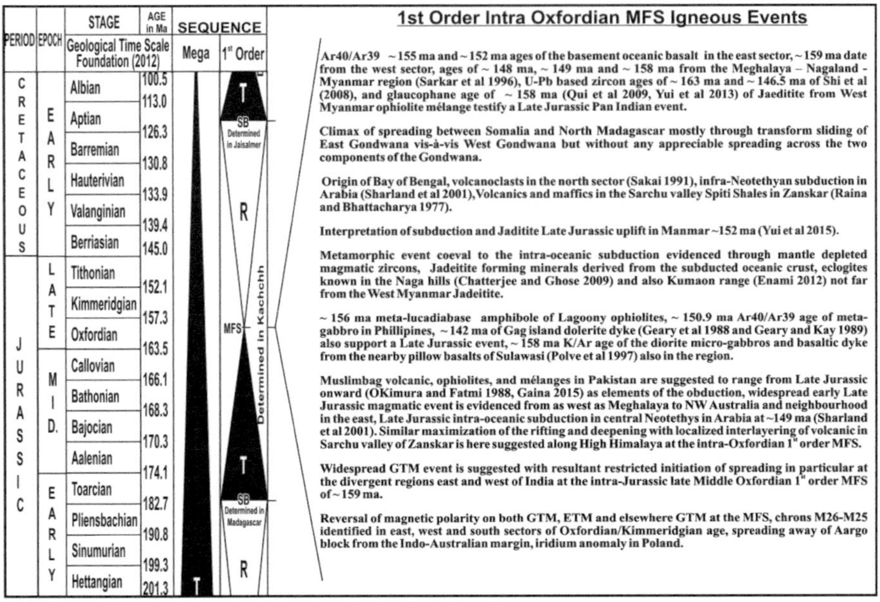

Fig. 7.27 First-order intra-Oxfordian MFS igneous events

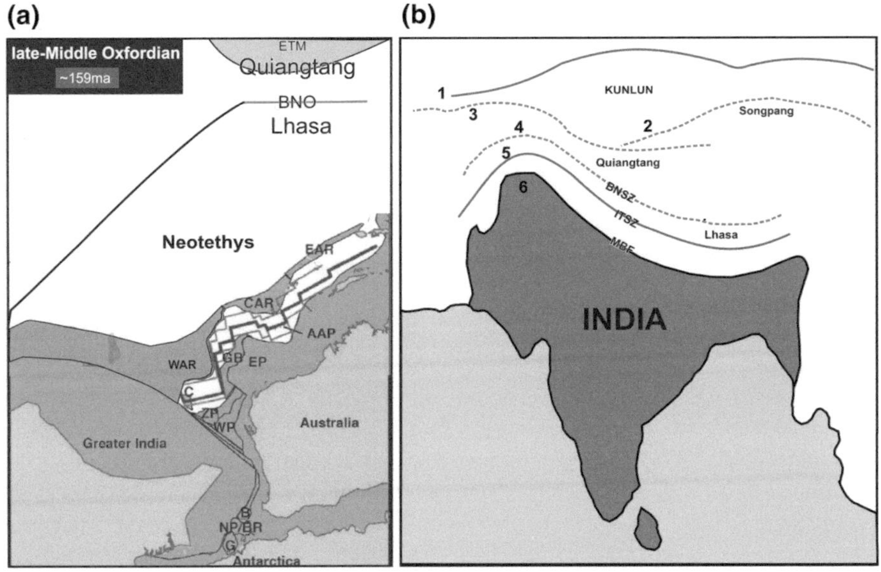

Fig. 7.28 **a** Schematic depiction of the origin of Bangong–Co-Nujiang Ocean at the same time of spreading away of Argo block (modified after Gibbons et al. 2015). **b** Closure of Bangong–Co-Nujiang Ocean along BNSZ (after Krishna 1987a, b, c): 1–6 respectively stand for the successive suture zones of Tienshan belt, Nanshan belt, Kunlun belt, Bangong–Nujiang belt (Early Aptian to mid-Middle Albian), Indus-Psangpo belt (Intra-Paleogene) and Main Boundary Fault

Jadeitite forming minerals as derived from the subducted oceanic crust. The ~156 ma age of meta-lucadiabase amphibole of Lagoony ophiolites, and ~150.9 ma, Ar40/Ar39 age of meta-gabbro in Phillipines (Geary et al. 1988; Geary and Kay 1989) are equally significant. The ~158 ma K/Ar age of diorite micro-gabbros and basaltic dyke from nearby pillow basalts of Sulawasi (Polvé et al. 1997) also are recorded in the region. Samples of basement oceanic basalts from two different locations of AAP yielded Ar40/Ar39 ages of ~155, and ~153 ma (Gibbons et al. 2012, 2015), and spreading away of AAP, etc., are here linked to the first-order MFS event. In recent literature, have emerged numerous radiometric ages in the said range, all of which are difficult to mention, and need not be cited here. Thus, the radiometric ages as above clinch the Late Jurassic oceanisation events along the Bangong–Co-Nujiang belt between the Quaingtang and Lhasa, and between the Argo block and Greater India on the GTM. The initiation of the said ocean is within the Neotethys, and is précised here at ~159 ma. These radiometric ages are here suggestively linked to the ~159 ma late Middle Oxfordian first-order intra-Jurassic MFS.

7.5.40 *Volcacnoclasts in the Tethyan Late Jurassic Sediments Related to Possible Penecontemporaneous Volcanism*

There are reports of volcanic clasts in the Jurassic sediments on the deeper shelves at several localities of GTM between South Tibet and Pakistan (Fig. 7.34). Durr and Gibling (1994) based on whole rock analysis of such volcanoclasts interpreted their provenance from the intra-plate geotectonic setting. The suspected inland volcanism in part might have been pene-contemporaneously related at least in a few places within the Neotethys itself (Valdiya 2010). Presence of such volcanoclasts in Late Jurassic turbidites that are interbedded in radiolarian chert in Ladakh and South Tibet allowed the interpretation of volcanic activity even in the Late Jurassic (Xiachi and Grant-Mackie 1993). Widespread Late Jurassic volcanism of ~161 to ~140 ma is also interpreted through volcanoclasts in Southeast Tibet (Stumpata, Tianba, Sangxiu, Jiduala, Tingri), Zanskar, Ladakh and Nepal (Garzanti 1999; Hu et al. 2015; Sakai 1984; Robertson and Sharp 1998) which is linked to the intra-Oxfordian MFS. In Zanskar, the shales in the Spiti Formation are known to include interbedded volcanics, and intrusive mafic rocks as evidence of rift-genic deepening (Raina and Bhattacharya 1977). These are here interpreted in the terminal Oxfordian–Early Valanginian age bracket, most likely intra-Valanginian. The Zanskar volcanism is significant as the farthest northwest of High Himalaya.

A Late Jurassic intra-oceanic subduction in the middle of the Neotethys is interpreted in Arabia at ~149 ma (Sharland et al. 2001). Radiolarites on the Altimur–Kabul ophiolites of Afganistan have provided Late Oxfordian age (Badsah et al. 2000; Gnos et al. 1997; Gaina et al. 2015). The Muslimbag and Bela volcanics, ophiolites, and mélanges in Pakistan are suggested to range from Late Jurassic to terminal Cretaceous (Okimura and Fatmi 1988) as elements of the obduction. The ~ 159.7 ma gabbro from Kiogar, ~ 140 ma from Nidar–Kazog (Ahmad et al. 2008), Shyok volcanics of Ladakh (Jain 2014; Jain et al. 2012) also are extended down in Late Jurassic. ~ 163 ma trachyte from Seychelles (Shellnut et al. 2015), ~ 149 ma basalt in Madagascar (Yang et al. 2015), ~ 145 ma intrusive at the south terminus of Kerala (Radhakrishna et al. 1999), undated Late Jurassic volcanic tuffs in a Saurashtra well (Singh et al. 2000), and above-mentioned ages hardly lead to convergence to either a single short range age or convergence to a single event. Instead, the said radiometric ages range between the Middle Bathonian ~ 167 ma to J/C boundary ~ 145 ma interval. The drastic geological changes in the Jurassic are observed only at the first-order sequence surfaces, i.e. ~ 183 ma SB and ~ 159 ma MFS. The initiation of oceanization in the north part of the west sector is interpreted at ~ 183 ma SB through localized spreading between Somalia and North Madagascar, while that in east sector at ~ 159 ma MFS. The above ages are here linked to the ~ 159 ma intra-Oxfordian MFS, with much less chance of occasional younger volcanism in steps during the early part of the terminal Middle Oxfordian–early Late Valanginian interval at any of the second-order sequence surfaces of the interval.

7.5.41 *Magnetic Anomaly Evidence*

The anomalies M26–M24A are present exclusively immediate east and southeast of the COB that is delineated by the Exmouth–Scott plateaus and Rowley terrace margin in the east sector. The oldest oceanic crust between Madagascar and East Africa is assigned to M26/M25 of ~ 156 to ~ 154 ma age (Coffin and Rabinowitz 1992; Segoufin and Patriat 1980). Gaina et al. (2015) suggested initiation of spreading in the North Somali and Owen basins between ~ 160 and ~ 155 ma on the basis of recognition of M26/25 which is considered younger to the ~ 159 ma first-order MFS. The recent averaged ~ 167 ma trachyte date from Seychelles (Shellnut et al. 2015) is also linked to the ~ 159 ma oceanisation event in context of its actual much longer range between ~ 167 and ~ 145 ma.

7.5.42 *Interpretation of a Single Interactive Spreading Framework on the GTM*

The near correspondence to precise matching and almost echoeing of the multiple lines of evidences of the Neotethyan early rift-drift history of the Indian plate in the east, north and west sectors is not mere coincidence on the GTM inclusive of the East Gondwana tethyan margin. The multiple evidences suggest a common unified integrated spreading system and geodynamic evolution in the region in the early part of the Late Permian–Early Paleogene Neotethyan mega-sequence up to the spreading away of Australo-Antarctica from the Gondwana. The strongest testimony to this unifying tectonic history is best provided by the succession of the precisely dated first and second-order physically visible sequence surfaces which are the direct manifestations of the tectonic events themselves.

Gaina et al. (2015) and others have also suggested a single contiguous volcanic arc framework to multiple differentiated arc/s from Iran to Gangdese arc of Tibet via Kohistan and Ladakh which is here assigned Aptian–mid-Middle Albian age. The validation of the isochronous character of the first-order intra-Oxfordiam MFS between Arabia and NWAustralia, and similar radiometric ages of the volcanics and the associated igneous expressions as above provide excellent testimony to the interactive single composite integrated spreading system on the GTM from east to west through north. The recent paleomagnetic evidence (Yang et al. 2015) from Lakang Formation volcanics also suggests the single Indian plate extending up to South Tibet until ~135 ma, i.e. intra-Valanginian second-order MFS. The relatively wider spread of volcanics in the east and north is interpreted to start at the said second-order MFS.

7.5.43 *Sequence Surface Based Precise Ages in the Backdrop of Tentativeness of Radiometric Ages*

Over the last many decades has emerged overwhelming convergence on initiation of oceanization, development of volcanic arc/s, and subsequent subduction events on the GTM and even ETM (between Quiangtong and Lhasa along the Bangong–Co-Nujiang zone) from Arabia to High Himalaya to NW Australia, in particular in the north sector in the terminal Middle Oxfordian–Barremian age bracket. Scores of excellently constrained igneous radiometric ages have been obtained, yet the tentativeness in their convergence to precise dates of the events has continued till this date. In this backdrop, the said events, using an earlier unthought of innovative alternative, are linked to relatively much less simpler expressions in some distant undisturbed geological setting, and dated there with several fold better precision.

The said tentative Late Jurassic initiation of oceanisation and volcanic island arc formation is thus précised at the ~159 ma late Middle Oxfordian Orientalis Zone Schilli Subzone bed I–08 first-order MFS age précised in Kachchh, and the subsequent subduction event at or the latest by the ~126 ma Baaremian/Aptian boundary first-order SB age précised in Jaisalmer. There seem possible oceanization and/or subduction also at the intra-Valanginian MFS.

7.5.44 *Disagreement with Late Jurassic Separation and Spreading Away of the Entire Tethys Himalaya from Greater India*

The strong similarity of the Tethys Himalaya Late Jurassic ammonoids of Spiti, Nepal, etc., with those of Kachchh does not accord with the Late Jurassic separation and spreading away of Tethys Himalaya as one of the constituents of the Argo block as advocated by Gibbons et al. (2012). Instead, the major chunk of Tethys/High Himalaya (Spiti, Zanskar, Uttarakhand, Nepal, Bhutan, Jolmo Lungma region of South Tibet) remained a constituent of India until the closure of the Neotethys in early Paleogene. There is separation and spreading away of only the distal most part of the Tethys Himalaya which by some also has been addressed as North Tethys Himalaya. The oceanization/spreading occurred within the Tethys Himalaya between its submerged middle and yet deeper distal part (Tethys Himalaya), instead of the entire block between Greater India and Tethys Himalaya as made out by Gibbons et al. (2012).

7.5.45 *On the Possibility of yet Older Localized Spreading in the West Sector*

Müller et al. (1997) suggested yet older initiation of oceanization between ~154 and ~180 ma. The oldest radiometric ages on the East African coast between Somalia and Mozambique as also on the west coast of Madagascar are ranged between ~183 and ~155 ma (Geiger 2004), i.e. between the intra-Jurassic first-order SB and first-order MFS. The progression of the oceanization is interpreted to initiate locally at the north and south extremes in the west sector near the first-order Pliensbachian/Toarcian boundary SB of ~183 ma, which spread up to the Indo-Australian margin by the first-order intra-Oxfordian MFS of ~159 ma. In the east sector, the start of the oceanization at the two extremes is at the intra-Oxfordian MFS. Based on ammonoid evidence from the region, sudden marked increases in bathymetry are observed during the early Bajocian, Bathonian/Callovian boundary, and Early to Middle Oxfordian which could be related to possible oceanisation in the region.

7.5.46 *Wide Spread Ammonoid and Other Evidences of Large Gap in India*

The maximum gap on ammonoid evidence is present in West Mainland Kachchh of mid-Middle Oxfordian to mid Early Kimmeridgian or even more as detailed in

Chaps. 2–4. In Jaisalmer, and High Himalaya inclusive of the Axial region of Pakistan, Nepal, and South Tibet, the gap is explicit, even larger with older start and younger closure than Kachchh.

7.5.47 Widespread Evidences of Large Submarine Gap from Arabia to India on GTM

As earlier dealt in Chaps. 3 and 5, a large widespread submarine gap of varied duration has been inferred on evidence of lithology, texture, structure, rate of sedimentation, ammonoids, micro-flora/fauna and others on GTM (Figs. 3.5, 3.7, and 3.9) in relatively deeper distal sites in Madagascar, East Africa and Arabia west of India on the GTM. In absence of ammonoids, the gap although present also in NW Australian basins as discussed in Chap. 5 is not that obvious as in the Indian basins. Scanty ammonoid data exists in Sula, Timor, etc., however, in absence of any recent modern revision, similar gap is surmised but without conclusive evidences. The revised submarine interpretation is highly radical revision from its long understanding for over a century as a subaerial gap.

7.5.48 GTM First-Order Intra-Oxfordian MFS Event Considered More Important in Global Context Compared to the Corresponding ETM First-Order Intra-Kimmeridgian MFS Event

The multifaceted first-order intra-Oxfordian MFS event is considered a global intra-Jurassic event, particularly so on the GTM, more significant in comparison to the corresponding Late Kimmeridgian first-order MFS event on the ETM (Haq et al. 1987 and most others). It is held one of the most important events of the Gondwana geological history since the intra-Cambiran origin of the Gondwana, and equally important as the separation of the East and West Gondwana. In fact, the effective dismemberment of Gondwana with greater rate of spreading picked up in the east sector than in the west. On the ETM, the first-order intra-Oxfordian GTM event has been interpreted as a second-order MFS, which preceded the ETM first-order terminal Kimmeridgian MFS by $\sim$5 my, i.e. little over a stage. The importance of the multifaceted event at the intra-Oxfordian first-order MFS can not be over-emphasized in view of its fundamental as also economic consequences not only all over India, but all over the GTM. Even on the ETM, the intra-Oxfordian changes as a second-order event are considered highly significant. Thus as a whole, the intra-Oxfordian event assumes global dimensions, and is held equally significant as the first-order intra-Jurassic SB event in the Jurassic. These two intra-Jurassic events are distinctly more important not only to the intra-Kimmeridgian first-order MFS event but also to other intra-Jurassic events in the over all Pan-Tethyan and even global context. The intra-Oxfordian event has been age précised, and lime lighted only recently in 1994 merely two decades ago in a new, hitherto unknown, stratigraphic section and outcrop (Krishna and coworkers–several papers since 1994), where no geologist earlier had even put feet upon over nearly last two centuries of history of geological studies in Kachchh.

7.5.49 Suspected Magnetic Polarity Reversal at the Intra-Oxfordian MFS Event

Ogg and Gutowski (1996) marked polarity reversal above the late Middle Oxfordian MFS between the pre-MFS Schilli Subzone and the post-MFS Rotoides Subzone. Preliminary studies also suspect a similar reversal in Kachchh in the Kantkote section between the said subzones (Krishna et al. 2011, Krishna unpublished report 2014 Fig. 1.12). The major Jurassic gap in the non-marine inland Gondwana basins and in east coast basins highlighted in palynological studies is for the first time interpreted as a consequence of the same tectonic event.

7.5.50 Rich Hydrocarbon Source Sediments in the Vertical Proximity of the First-Order MFS

The giant oil fields of Arabia have their best hydrocarbon source sediments in the vertical proximity of this intra-Oxfordian MFS (Fig. 7.5). The recently discovered hydrocarbon source sediments in the Indian east coast (K-G and Cauvery) and in the west sector (Kachchh and Jaisalmer) are also in vertical proximity of the said first-order MFS, so also are the hydrocarbon source sediments of the Indus basin in Pakistan. The hydrocarbon prospects are already discussed in an earlier chapter. The Nupra shales in Nepal, coeval and lithologically similar to Spiti shales of the Indian High Himalaya, have evidenced high over mature TOC of 1.7–1.8 % values of dominantly terrestrial origin along with dysaerobic episodicity immediately above the intra-Oxfordian MFS (Gradstein et al. 1991). The original pre-maturity TOC values are interpreted as double of the measured values, and gas generation stage of maturation based on organic petrology and geochemistry. This is also supported by seeps of natural gas in the form of a holy flame at Muktinath temple.

7.5.51 See-Saw Tectonics of Alternate up and Down Tilting (Down at the Active Margin with Complimentary Cogenetic Tiliting up at the Passive/Hind Margin)

The principal tectonic events in the simplest possible manner are of alternate stepped tilting down of the Indian plate margin in consequence of the renewed phase of rifting with cogentic complimentary tilting up in the opposite direction at the respective hind margin of the plate. During the Late Permian–Pliensbachian first-order sequence interval, as discussed elsewhere, the paleoslope is due east and southeast in the east sector. In the northwest, in Kachchh is due southwest and west. In Jaisalmer and Salt range, the slope is due northwest. At the Pliensbachian/Toarcian boundary first-order SB, there is rift-influenced tilting down in the west sector with simultaneous cogenetic uplift of the inland Gondwana basins. The general resultant slope became due west, and northwest, and southwest. Just the reverse/opposite happened at the late Middle Oxfordian first-order MFS to at least end Barremian first-order SB interval of the first-order RST just like a see-saw. There occurred active tilting down in the east and southeast margin of the Indian plate, while the complimentary cogenetic tilting up at the hind plate margin in the west, northwest and north. During the first of the two intervals, there is developed consequential stratigraphic gap equal to the interval of at least first-order TST in the Gondwana linear graben basins and east coast basins, while in the latter interval, shallowing is observed in the west sector, e.g. in Kachchh. Tilting down is consequented by initiation or revival of rifting at one margin and possible large gaps or shallowing at the other margin.

7.5.52 Absence of Substantial Part of Jurassic in Non-marine Basins

The basal Permian is marine and wide spread in Cauvery, K-G, Palar and Tanjore. The Late Permian to Early Jurassic is non-marine and localized in K-G basin, while the Toarcian–late Middle Oxfordian interval is absent. The paleobotanists and palynostratigraphers have repetitively emphasized about the near complete absence of the Jurassic (late Hettangian to Early Tithonian—Fig. 4.5) between the true non-marine Gondwana units of Permo-Triassic age and the disconformably overlying Late Tithonian to Late Maastrichtian marine successions in both Cauvery and K-G basins (e.g. in Prasad et al. 1995; Prasad and Phor 2009). The tectonic event which resulted in the initiation of marine transgression in the Kachchh basin, presumably, also uplifted the Gondwana graben basins in the east to put complete stoppage even to any non-marine Jurassic sedimentation in those east sector non-marine basins. The gap of the Ladinian–late Middle Oxfordian first-order Gallovian sequence duration from intra-Triassic to intra-Jurassic is also interpreted in the Bengal basin boreholes between the Barakar Formation below and Dubrajpur

Formation above. The conceived uplift as above, perhaps caused lateral shifting of the non-marine depositional systems which consequented the gap. The gap is claimed to be even greater with time extension both at the lower and upper ends of the minimum gap. The progressive/stepped tilting down of the east margin of the Indian plate later from late Middle Oxfordian onward caused increasing submergence and deepening.

7.5.53 Sequence Stratigraphic Context of the Gap in the Non-marine Gondwana Basins

The said much talked about major Jurassic gap in the east sector basins gets explained and precised in sequence stratigraphic context. The gap spans from the basal Toarcian first-order SB to the late Middle Oxfordian first-order MFS. The Kota Formation extends up to Pliensbachian in Early Jurassic prior to the gap while the Late Oxfordian is suggested in the Cauvery and K-G basins. Similarly, in the Satpura belt, Casshyap and Khan (2000) have suggested similar gap between the Denwa and Bagra units. The termination of the non-marine Gondwana succession in some basins at the close of the Early Jurassic Pliensbachian stage first-order SB could also be due to uplift, and reorganization/shifting of the fluvial/lacustrine depositional systems at the end of Pliensbachian. The associated initiation of the gap on account of the uplift of the non-marine basins at the basal Toarcian SB is in turn genetically linked to the tilting down of the Indian plate in the west sector. The initiation of the marine transgression occurred in the Cauvery and K-G basins at the intra-Jurassic first-order MFS. Thus, the resumption of the non-marine sedimentation in the east sector is also linked to the geological event at the intra-Oxfordian first-order MFS. In the past, the above-mentioned multifaceted expressions were neither precisely dated nor genetically linked to the said single geological event. The culmination of the gap at the intra-Oxfordian MFS is on account of tilting down in the east sector and associated consequential tilting up in the west sector. All the 13 first-order MFSs of the Indian Phanerozoic carry similar multidimensional significance.

7.5.54 Abortion of Rifting, and Reversal of Paleoslope/Drainage/Current at the First-Order MFS

The tilting of the Indian plate in opposite directions in turn also caused the reversal of the paleoslope/drainage/current from due west and northwest to due east and southeast near the east sector (Fig. 7.29). Rao et al. (1996) interpreted the reversal of the paleoslope in Early to Middle Cretaceous. Tewari (1998), Tewari and Casshyap

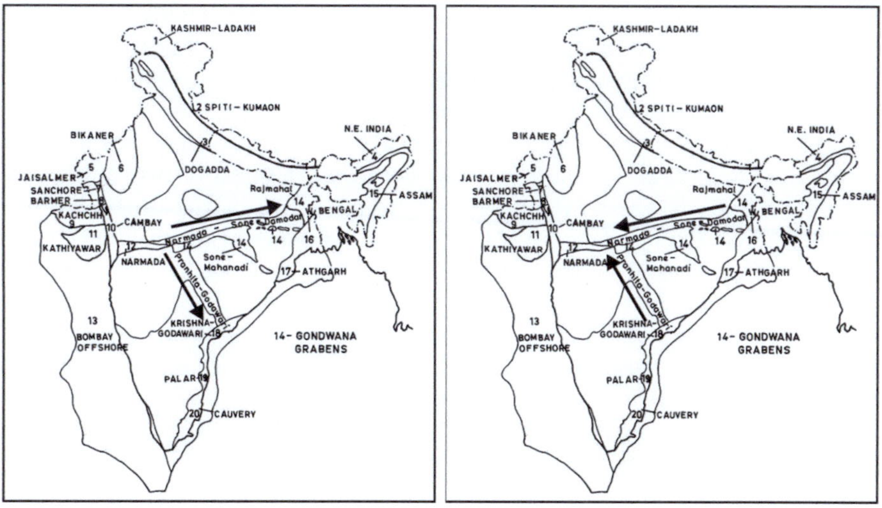

Fig. 7.29 Reversal of Paleoslope from due west and northwest to due east and southeast at the first intra-Oxfordian MFS

FP. 7.2 The 1st order ~159 ma late Middle Oxfordian Orientalis Zone Schilli Subzone MFS in the Kantkote section of Wagad in Kachchh

(1996) interpreted the change in Middle Jurassic. Veevers et al. (1996) interpret two tectonic events in the earliest Late Jurassic, and Earliest Early Cretaceous. Lakshminarayana (2002) interpreted the change in paleodrainage in Middle Jurassic. Prasad (2011) interpreted the directional change in paleodrainage near the base of the terminal Jurassic Tithonian stage. Veevers and Tiwari (1995) interpreted a unified master Gondwana drainage basin towards northwest. Valdiya (2010) sourced the non-marine Permo-Triassic Gondwana sediments of Godavari, Mahanadi and Damodar basins from Enderby–Wilkies lands of Antarctica of the then united East Gondwana in a general northwesterly direction. The rivers in the Early Permian presumably drained into the Paleotethys both in the northeast and northwest. The origin of the Neotethys thus did not bring any significant change. However, the origin of the Gondwana rift basins in the east, evidences paleodrainage directed to the rift basins, and thus due east and southeast during the Late Permian–Pliensbachian first-order sequence. This view is further supported by the Neotethyan Early Permian marine incursion in the said rift basins from the northeast. Probably, the Precambrian highlands between the east and west sectors allowed differently or somewhat oppositely directed paleodrainage during the east and west sectors at least during the Early Permian. The change in paleodrainage is only vaguely dated as above by various workers from near the start of Jurassic to Middle Cretaceous during a a span of $\sim$100 my. Conceptually, the change was caused by a single tectonic event. It required precision dating that alluded the Gondwana geoscience community in spite of innumerable studies spread over a century and half. The said change in paleodrainage is genetically linked to the widely manifested extensional tectonic-cum-magmatic event on GTM at the intra-Jurassic late Middle Oxfordian Orientalis Zone Schilli Subzone $\sim$159 ma first-order MFS. The single visually observable surface (FP 7.2) is age précised to the horizonal resolution of $\sim$200 ky in Kachchh with the time of abortion of rifting in the basin along with renewal of rifting in the India–Australo-Antarctica divergent sector inclusive of the geographically restricted initiation of oceanization and spreading away of the Argo land within East Gondwana at the $\sim$159 ma first-order intra-Oxfordian MFS.

The relative high lands in the east and south east could not any more remain high lands due to revival of rifting and associated tilting down of the Indian plate in that direction. The tilting down and tilting up is on the simple model of the children see—saw game. The game has hinge or fulcrum about which the two extremes of the plank alternatively tilt down and up. The paleoslope reversal is also linked to the above-mentioned oceanisation event.

7.5.55 *Shift of Action from West and Northwest to East and Southeast*

The theatre of action, after the rift abortion in Kachchh, shifted to the east sector. The further tell-tale of the tectonic events is focused in the east with integration of signatures in the west. These include origin of the Indian east coast basins (K-G, Cauvery and others) through revival of rifting in this sector that initiated with abortion of rifting in Kachchh on the other extreme of the Indian plate.

The abortion of rifting in the Kachchh basin in the west coincided with revival of rifting in the east, along with reversal of paleoslope/drainage/current framework. The initiation of the stepped/graded rifting and associated tilting down due east between India and Australo-Antarctica resulted either in the origin of the Indian east coast rift basins or resumption of sedimentation, e.g. K-G basin above the late Middle Oxfordian first-order MFS. It is essential to keep in mind that the K-G basin wells are located far west and south of the zone of rifting between India and Australo-Antarctica. Even transform sliding and associated localized spreading in the region may have started with the origin of the 'Proto-Indian Ocean' as also 'Proto-bay of Bengal' at one extreme of the divergent sector several hundred km or more away from the K-G and Cauvery basins. Sedimentation in the east sector suggestively began in the late Middle Oxfordian in the newly formed basins along the east coast strike, which could have been locally even non-marine in the early phase of rifting as in Sivaganga, and only later would have turned marine. The oldest marine sediments, i.e. of Late Oxfordian and Early Kimmeridgian should be present at locations far away due east and southeast of the K-G basin wells, closer to COB. There is dinoflagellate based suggestion of the Late Oxfordian–Kimmeridgian in Bengal basin (Vijaya 1997) to support the origin of 'Proto-Indian Ocean' and/or 'Proto-Bay of Bengal' at the said MFS. The oceanization–subduction cycle in the late Middle Oxfordian–mid-Middle Albian age bracket has evidenced at number of localities in South Tibet from west to east inclusive of Myanmar not far from Bengal and Jharkhand. In such a broad context, the record of more than one thin beds of oolite, occasional dinoflagellates, and multiple thin beds of grey/dark/organically rich shales, as also the clay mineral Illite of invariable marine origin in the early part of the Rajmahal Formation intertrappeans both in Bengal basin boreholes and exposed Jharkhand sections (Vijaya 1997, 2009; Ghose et al. 1996) deserves further investigation for microfossils and specific depositional environment. There could be possible extremely short duration, few hundred/thousand year reaches of the nascent juvenile ocean during the late Middle Oxfordian–mid-Middle Albian bracket volcanism in the northeast India–northwest Australia sector in the otherwise fluviolacustrine depositional systems of the Bengal–Jharkhand intertrappeans particularly during a tectonically fast subsiding Bengal–K-G–Cauvery neighbourhood. Such a remote possibility may correspond to any of the early rift deepening steps at the second-order MFSs. Relatively frequent anoxic events are recognized in the region in the Hauterivian–Albian interval, inclusive of the one in mid-Middle Albian through the organically rich shales in the Cauvery basin (Jafar and Rai 1989; Kale et al. 2000). The dark grey shales presumably may not be coeval and may represent a few successive (more than one) milankovitch type short successive eustatic maxima at cumulative effect of the eustatic rise and tectonic submergence in an extremely shallow marine or lagoonal/braksih framework not far from true marine framework with occasional/intermittent influence of marine waters. In context of time, greater possibilities are thought of in the Early Aptian–mid-Middle Albian second-order TST owing to the greater sea levels as also greater rift-related submergence than during the successive Late Jurassic second-order TSTs of the preceding first-order sequence. Radiometrically, the exposed Rajmahal stratigraphic sections constrain the said intertrappean sediment interval as Late Aptian or older.

The interpreted spreading in the east sector is understood as on account of transform sliding along the transform from Sri Lanka's east margin to Cauvery's east margin on the pattern of Davies ridge in the west sector. No appreciable spreading is conceived here across India–Antarctica but for India's transform sliding due north–northwest vis-à-vis Australo-Antarctica, and geographically restricted spreading away of the Argo block. The transform sliding for at least some interval east and west of India is in mutually opposite directions along the parallel to subparallel transforms of Davies ridge in the west and Sri Lanka–East Cauvery transform in the east with resultant counterclockwise rotation. The oldest marine sediments, i.e. in the Late Oxfordian–Early Kimmeridgian span are only anticipated at locations far away due east and southeast offshore of the K-G basin onshore wells. Even the presence of Oxfordian–Kimmeridgian sediments has been reported by Mehrotra and Aswal (2003) in one well, though not supported by others (Prasad and coworkers). This situation is not different from the somewhat similar case of the oldest sediments in the Kachchh basin of the Early Toarcian age, as repetitively suggested by Krishna (1994 and later), which could be proved only recently (Rai and Jain 2012, 2013). The oldest age interpreted in Kachchh is as many as 3 stages older to the initially known oldest age of the Bathonian, i.e. difference of ~14.4 my. In case of the K-G basin, the oldest age predicted is Late Oxfordian (Mehrotra and Aswal 2003), while the oldest age based on palynomorphs inclusive of dinoflagellates is Late Tithonian (Prasad et al. 1995; Prasad and Phor 2009). The interpreted age is older by ~12 my to the already confirmed Late Tithonian. In fact Mehrotra and Aswal (2003) in the K-G basin wells RCPM-A and MVD-A interpreted the presence of the Oxfordian and the Kimmeridgian sediments in neighbouring wells. All these wells are more or less at the same distance along dip direction from the outcrops.

7.5.56 Myanmar Shoshonite as Evidence of Tectonic-Cum-Igneous Event at ~159 ma MFS

In the last decades, considerable radiometric and other evidences (included in 7.6.8 and 7.6.10 in particular) have emerged in favour of an early Late Jurassic Late Oxfordian shoshonite event and jaeditisation in Myanmar in the neighbourhood of Nagaland, Meghalaya of Northeast India. The Shoshonite and associated jaeditisation event may be linked to the Late Oxfordian ~159 ma first-order MFS oceanization event (Krishna 2016 communicated) on the Indo-Australian margin. Also, terminal Jurassic volcanism is suspected in Bengal wells on the basis of palynological evidence (Vijaya 2009). These are geographically not as widespread in comparison to the well-known widespread Early and Mid Creteceous volcanism of the east sector. Amidst all such evidence in favour of the event, Yui et al. (2013) based on a detailed reassessment of zircon types and their ages interpreted that the earlier claimed Late Jurassic jaeditisation in Myanmar may also have alternately occurred during the Late Cretaceous subduction. According to them, the earlier Late Jurassic ages were based on incompletely recrystalized zircons. Nevertheless, the Late Jurassic zircon ages unambiguously provide the age of the igneous protolith of the Myanmar jaeditites.

7.5.57 An Arm of the Neotethys in the East Divergent Sector

Much like the arm of the Neotethys making inroads due south in the India (inclusive of Madagascar)–Africa divergent sector, there could as well be interpreted somewhat similar inroads by the Neotethys due south in the India–Australo-Antarctica sector. The advances in the west sector began at the first-order SB at the start of the sequence both in the east and west. The localized spreading, presumably, started earlier in the west. The oceanization in the northeast started at the first-order MFS, and the arm like advance of the sea resulted from the north-south transform sliding of India vis-a-vis Australo-Antarctica at the late Middle Oxfordian first-order MFS. The only significant difference west and east of India was that the Neotethys was narrow to closed in the west but wide and open in the east and possibly in communication with the SW Pacific. The initiation of the 'Proto-Bay of Bengal' may also be conceived in the terminal Middle Oxfordian–Barremian age bracket. However, the shallow east coast basins could effectively start growing and deepening only subsequent to the start of appreciable spreading between India and Australo-Antarctica during Late Aptian–early Middle Albian. It occured only after the final separation between the Inner (Africa–India) and outer Gondwana (Australo-Antarctica and South America) subsequent to spreading in the Enderby–Elan bank sector as the final leg of separation between India and Australo-Antarctica.

7.5.58 Possible Arms of the South Pacific in the South Africa–Antarctica–South America Region

Similar to the said arms of the Neotethys at the north extreme of the east and west divergent sectors, shallow arms of the South Pacific are also conceived in the South Africa–Argentina–Antarctica region. These arms were related to the origin and growth of the South Atlantic and the Indian Ocean in the south sector. The said arms, presumably, travelled and grew due north with the growth of the South Atlantic, and progression of the Pacific waters between India and Australo-Antarctica. The said arms of the Pacific may have had their origin either at the first-order intra-Jurassic SB as that of the arm of the Neotethys in the west sector or at the first-order intra-Jurassic MFS as that of the arm in the east sector. However, the appreciable growth of the arms of the Pacific is interpreted only from the first-order intra-Jurassic MFS onward.

7.5.59 Desirability of Investigating Possible Anoxic Signatures at the Intra-oxfordian first-Order MFS

The multifaceted geological event at the intra-Oxfordian MFS is considered one of the most important events of the intra-Permian–intra-Paleocene mega-sequence on the GTM, also more important than the corresponding ETM intra-Kimmerdgian first-order MFS event. It is manifested through widespread rifting/spreading and volcanism, maximum bathymetry and transgression on GTM, widespread submarine gap, lithological/sedimentological/paleontological changes, magnetic polarity reversal, distinct evolutionary origin/extinction event, rich hydrocarbon source rocks, frequent hard grounds in the terminal TST immediately underlying the MFS, fall in temperature, climatic change, seismitic, and psunamitic expressions, suspected europium anamoly event on ETM, strong condensation and starvation, black shale lithology along the entire High Himalaya. The manifestations are much the same as at other significant anoxic events, which make the MFS event suspectful for anoxic manifestation in the east and northeast.

Events During the Late Middle Oxfordian–End Barremian First-Order RST

7.5.60 Graded Withdrawal of the Neotethys from Kachchh and the Associated Recede of the Shoreline

After the abortion of rifting and associated tilting up of the Indian plate in the west, northwest and north, the sea withdrew first from Wagad by the close of the second second-order sequence above the Early Kimmeridgian Giganticus Zone ($\sim$ETM Hypselocyclum Zone), next from the KMU belt east of the MH by the close of the Early Tithonian second-order SB, from the Ler-Walakhwas belt by early Early Cretaceous and from the entire shelf up to Ghuneri–Amarsar between Cenomanian to Early Campanian. The step-by-step withdrawal is induced by all round eustatic fall with or without tectonic element through punctuation of the second-order sequences included in the first-order RST. This stepwise withdrawal, (Figs. 7.21, 7.31 and 7.32), may have been affected either by uplift of the basin or by uplift of the source area resulting in the rejuvenation of the hinterland geomorphology in turn causing increased sediment supply. The geological history in this interval in Madagascar basins is nearly identical to that of Kachchh as detailed in Chap. 5.

7.5.61 Mid/Late Early Kimmeridgian Second-Order SB

The Kachchh basin tilted up at its west extreme presumably complimentary and cogenetic to the tilting down of the Indian plate due east and southeast in the India–Australo-Antarctica sector. The so interpreted tilting of the Indian plate let the shore line recede farther towards the basin along with complete emergence of Wagad (Figs. 7.21, 7.30). Patcham, Bela and Chorar had emerged above the early Middle Callovian second-order SB. Khadir emerged subsequently above the late Middle Oxfordian first-order MFS. As part of the stepped recede of the shore line, the entire Wagad emerged above the mid/late Kimmeridgian second-order SB. Marine conditions, thereafter, remained restricted exclusively to Mainland Kachchh. The resultant sediments are designated as the Lower Member of the Katrol Formation. The event also marked the near extinction of the ammonoid subfamily Ataxioceratinae from the IEAP.

7.5.62 Relative Faster Bathymetric Increase During Oxfordian up to First-Order MFS Than Decrease up to Mid-Late Kimmeridgian Second-Order MFS

The bathymetric increase since the late Early Oxfordian up to Schilli Subzone was much faster having taken place only during less than three zones in comparison to the later much slower bathymetric decrease spanning through at least succeeding eight zones. The possible explanation could be the eustatic rise that contributed to the rift-related submergence up to Schilli Subzone. While in the post Schilli Subzone duration, there, possibly, continued the phase of eustatic rise up to the Late Kimmeridgian Eudoxus Zone, and the same had to be negated and overridden tectonically either by uplift of the basin or increased supply.

7.5.63 Unique ~Four to Six to Eight My Long Duration of Submergence in Mainland, yet Deprived of Sediments

In course of the ~four my long Late Oxfordian to mid Early Kimmeridgian, the basin landscape underwent sea-saw change. But for Patcham, and may be also Chorar, Bela, etc., the entire basin, roughly to the extent of 75–85 % of its extent, remained submerged immediately prior to the start of this duration. However, as a quantum change, the Island belt and Wagad emerged by the close of the said ~four my interval, never to be submerged again during the rest of the Mesozoic (Figs. 7.20–7.21, 7.25, 7.30). Also, the Mainland, in spite of complete submergence, remained starved of sediments during this interval. Sediments could not be fed to the Mainland on account of drastic reduction in the supply of sediments coupled with sudden distanting and recede of the source of sediments. Under such conditions of acute shortage caused by the sudden increase of bathymetry to maximum of the Mesozoic, the supply of sediments could only be maintained in Wagad, while the entire Mainland was kept hungry. The supply could only be resumed after the close of the interval, that too in graded manner, with resumption first in relatively shallow north sector of the East Mainland east of MH, then in south sector of East Mainland, and finally in West Mainland. The exposed West Mainland, thus, remained without food for the longest duration of ~6–8 my, even up to early Early Tithonian, i.e. an interval of two successive second-order sequences. The said rapid rift-genic submergence and resultant starvation in the exposed Indian plate margin areas, e.g. the axial belt region of Pakistan, Zanskar has been determined as of much longer duration of more than ~10 my duration. In the High Himalayan convergence zone, the said duration also included intra-Neotethyan oceanisation and ocean floor volcanism.

7.5.64 Mid / late Late Kimmeridgian Post Bathyplocus Zone Linguiferus Subzone Second-Order MFS

The Kachchh basin at this second-order MFS witnessed either another step of further upward tilting of the Mainland region east of the median high or of rejuvenation of the provenance or both. This was complimentary and cogenetic to downward tilting in India–Australo-Antarctica divergent sector. It caused gradual emergence of the Mainland east of the Median High and south of the KMF by the mid/late Early Tithonian second-order SB. The resultant sediments are the late Late Kimmeridgian dominantly fine and medium laminated sands of the middle part of the Katrol Formation at Ler–Katrol (Figs. 7.21, 7.30). The Mainland became the arena of dynamic depositional geological processes through resumption of sedimentation with sediment supply from 'Island belt' in the north as also Wagad in the east, the shoreline having also receded from Wagad. During this event also initiated the oldest Jurassic marine incursion in Mozambique.

7.5.65 Mid Early Tithonian Post Rajnathi Subzone Second-Order MFS

There is yet another step of tilting up interpreted in Kachchh that is cogenetic with tilting down in the east sector. It happened on account of further intensification of

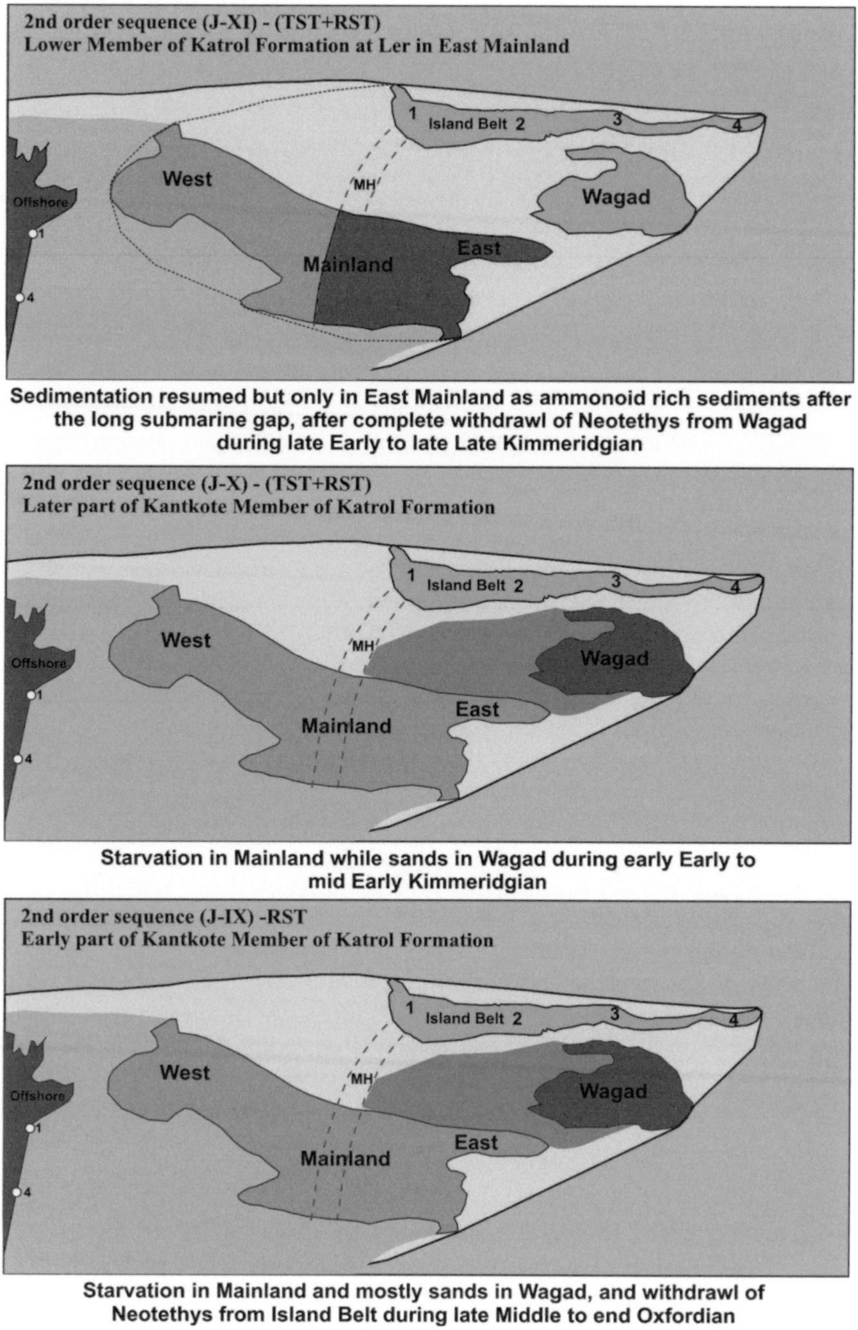

Fig. 7.30 Graphics from bottom to top respectively refer to late Middle to end Oxfordian, early-Early to mid Early Kimmeridgian and late Early to late Late Kimmeridgan intervals

the rifting and associated deepening in the Cauvery, K-G and other East Indian coast basins. The final withdrawal of the sea was thus realized from the entire north belt of the East Mainland, i.e. from Jhura, Habo, Jhuran, Jawahar nagar and associated localities east of the MH (Figs. 7.30–7.31). The second-order sequence also marked the extinction of the ammonoid subfamily Katroliceratinae.

7.5.66 Origin of the Intra-Gondwana Gondic Shallow Marine Corridor Across the Gondwana at the Intra-Tithonian Second-Order MFS

Krishna (1983, 1994, 1996 and later) interpreted the late Late Tithonian origin of the Gondic corridor across Gondwana from its Neotethyan margin areas of Kachchh, Pakistan, High Himalaya to its SE Pacific margin of Argentina round the South and East African coast between the East and West Gondwana (Fig. 1.3). The said corridor, which allowed intermigration of the shallow marine ammonoids between its two ends, opened during the terminal Late Tithonian second-order TST. Such corridor types of linear shallow marine seaways are invariably short duration transient precursor geological features to the origin of the new oceans as long duration persistent features. However, there could be involved variable time lags between the origin of the shallow corridor and the origin of the new ocean. The new ocean to have originated is the Indian Ocean in the west sector which could acquire oceanic depth and size only from the intra-Turonian mega-order MFS onward after the spreading away of India from Madagascar. The origin of the Gondic corridor could also be related to possible terminal Late Tithonian volcanic episode in the east sector. The due south progression of the arm of the Neotethys began at the intra-Jurassic first-order SB in early Early Toarcian, and was realized in steps at the second-order sequence surfaces. The corridor for sure did not operate until the end of Virgatosphinctoides Zone as attested by the absence of Katroliceratinae ammonoids in the Argentinian SE Pacific (Fig. 1.3). The oldest Jurassic marine sediments in Mozambique are suggested in the Late Kimmeridgian–Tithonian sediments (Prinz et al. 1993) although without ammonoids which supports the yet later late Late Tithonian origin of the corridor with common bivalves (Riccardi 1991), ammonoids (Krishna 1987a, b, c, 1996) and belemnoids (Mutterlose 1986).

7.5.67 Possible Volcanic Episode in Bengal Basin at the Late Tithonian / Early Valanginian MFS in the East Sector

The initial phase of the geographically restricted volcanism with spreading away of the Argoland suggestively at the intra-Oxfordian MFS apparently did not reach the

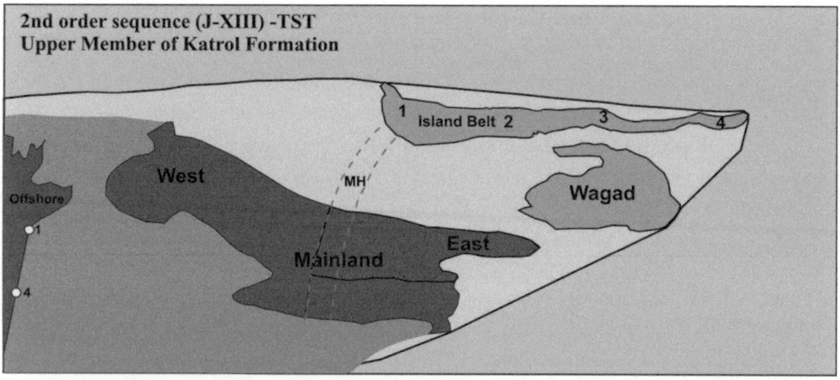

Ammonoid rich slow/condensed to moderate sedimentation in West Mainland, moderate sedimentation in South Belt of East Mainland, and marine body fossil devoid sediments in North Belt of East Mainland during post Virgatosphinctoides Zone

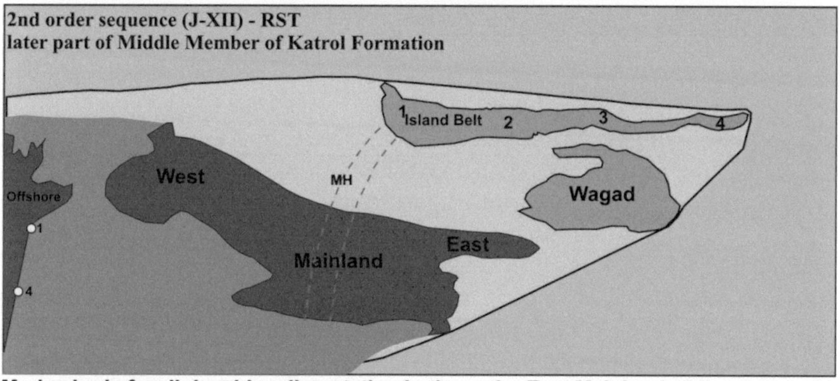

Marine body fossil devoid sedimentation in the entire East Mainland while ammonoid bearing sediments in West Mainland upto close of Virgatosphinctoides Zone

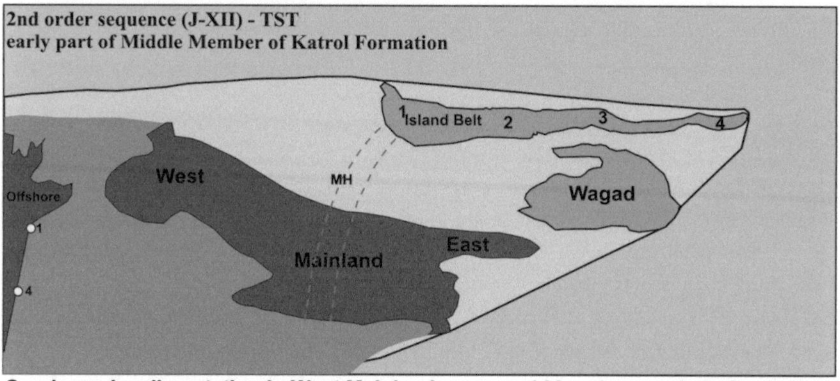

Condensed sedimentation in West Mainland, ammonoid bearing sands in South Belt of East Mainland, and ammonoid devoid sands in North Belt of East Mainland during Basal to early Early Tithonian (Pottingeri Zone to early part of Rajnathi)

Fig. 7.31 Graphics from bottom to top respectively refer to the basal to early Early Tithonian, mid Early Tithonian to mid Middle Tithonian, and late Middle to mid Late Tithonian intervals

Indian east sector of Meghalaya–Bengal in the form of trap flows (Figs. 7.27–7.28, 7.33–7.36). However, the oldest flow in the Bengal basin is suggested at the Tithonian/Berriasian transition either at the intra–Late Tithonian or Early Berriasian MFSs or even still earlier at either the intra-Oxfordian or intra-Kimmeridgian MFSs. This suggestion stems from the underlying and overlying intertrappean palynological record in the Bengal basin boreholes at Birbhum (Vijaya 2009). Vijaya (1997) in an earlier work in another Bengal basin coal field bore hole assigned the younger three of a succession of seven volcanic flows to the Valanginian–Hauterivian interval, which indirectly implied early Early Cretaceous or even older Late Jurassic ages to the older four flows. The palynotaxa and assemblages are long ranging, and thus the assigned ages to the intertrappeans lack the desired precision. The lateral spread of the flows depends upon the intensity as also the composition of the magmatic melt. This episode is not evidenced in Jharkhand. The oldest flow in the well DPD 6 has its base at 226 m depth while the base of the fourth volcanic interval has its base at 180 m. The younger 180 m thickness includes four flows, the third and fourth flow within a thickness of ~40 m between depths of 140–180 m. Thus, the thickness up to 226 m is expected to include the oldest volcanic interval above the base in the well DPD 6, which unconformably overlies the Cooksoniae Zone. The fourth volcanic interval is dated below the intra-Berriasian Australensis Zone and the oldest above the Toarcian–late Middle Oxfordian Cooksoniae Zone. The oldest and the fourth volcanic intervals are, thus, age-constrained in the Late Oxfordian–Early Berriasian interval, while the third volcanic interval is considered terminal Hauterivian. If the palynological age data is relied upon, all the three phases of volcanism differentiated earlier are present in the Bengal basin. The pre-Aptian ones are interpreted to have reached up to Bengal basin but not farther west to Jharkhand. It seems logical to interpret the progressive younging of the volcanic phases from the spreading axis to periphery, so appears at least the case between Mynmar and Jharkhand. May be, the direct or indirect evidence available at present of the oldest flows in the Myanmar–Nagaland-Meghalaya-Bengal region also include the Late Oxfordian–Kimmeridgian initiation as well. The samples of the older flows in the Bengal basin borehole stored in GSI require immediate Ar40/Ar39 dating on a priority basis if not already obtained to confirm the time of initiation of the volcanism in the east divergent sector. The younger two flows in the said well are here considered of the later post-Barremian–Aptian–Albian final phase.

May be, the volcanism in NE India also began in Late Oxfordian itself as in the AAP region offshore NW Australian margin or is restricted to the terminal Tithonian / early Early Berriasian–Early Valinginian interval as part of the said early phase of volcanism in the east. The latter interval is preferred although the Late Oxfordian is not totally ruled out in context of the extent of the interpreted oceanic crust in the AAP region up to the Zenith Plateau (Gibbons et al. 2012).

7.5.68 *Duration of the Multiple Stage Mesozoic Volcanism in the East Divergent Sector*

The volcanism in the east divergent sector is interpreted to have begun at the first-order intra-Oxfordian MFS and closed at the mid-Middle Albian second-order MFS. Evidences from radiometric, magnetic anomaly, palynology, rate of magma output, spatial spread, chemical composition of the basalts, age duration data from either side of India–Australia divergent region suggest more than one episode of pre-Rajmahal volcanism that began at the first-order intra-Oxfordian MFS. The volcanism in the east sector can be organized into three principal phases, which were supposedly punctuated by jumps of the spreading axes. The first phase began with spreading away of the Argo block but had a geographically restricted expanse. It may or may not have spread eastward to Nagaland, Myanmar and Meghalaya, and certainly not any farther. The middle phase spread to Bengal, Meghalaya and Nagaland in India, Comei in South Tibet and Bunbary in West Australia, yet did not reach Jharkhand. Possibly, the advent of Rajmahal/Kerguelen volcanism is linked to the ceasure of spreading in the west sector with another jump of the spreading axis. The Rajmahal event marked the final volcanic phase and as per radiometric dates had the widest reach in Southeast Tibet and Jharkhand. All the episodes may not have reached the Indian east and northeast sectors, so also all the phases may not be present in all the principal regions (Figs. 7.27–7.28, 7.33–7.36). It is rather difficult to assert or discount the continuity of the volcanism in the rather long Oxfordian–Albian interval. It is also not possible to link the Late Oxfordian–Early Valinginian volcanism (initial/early phase) to Kerguelen plume, which perhaps at this time was located at substantive distance from NW Australia. In context of the sequence framework, the first two volcanic phases are here age bracketed in the Late Oxfordian–Late Barremian first-order RST interval from the MFS to SB. Also, the first phase may be ranged up to the early Late Valanginian second-order MFS. The first significant shift/jump of the spreading axis at M14/M 13 (Hein 2002; Gibbons et al. 2012, 2013) is interpreted at the said intra-Valanginian MFS ($\sim$ 136 ma), which otherwise marked significant lithological/faunal change in several GTM basins. The second volcanic phase is ranged in the Late Valanginian–Late Barremian bracket from one second-order MFS to another second-order SB. The third and the final phase spans from the Barremian/Aptian boundary first-order SB to mid-Middle Albian second-order MFS after the second significant jump of the spreading axis. The third and final shift of the spreading axis occurred at the mid-Middle Albian second-order MFS. Thus, the three volcanic/spreading phases are punctuated through significant jumps in the spreading axis, while their ages are précised through the ages of the sequence surfaces determined, respectively, in Spiti and Jaisalmer. The third phase suggests genetic linkage to the Kerguelen plume. The Shyok/Dras volcanics of Ladakh and Mikir volcanics in Assam, and evidences of the igneous activity elsewhere on the GTM in this context appear to make the

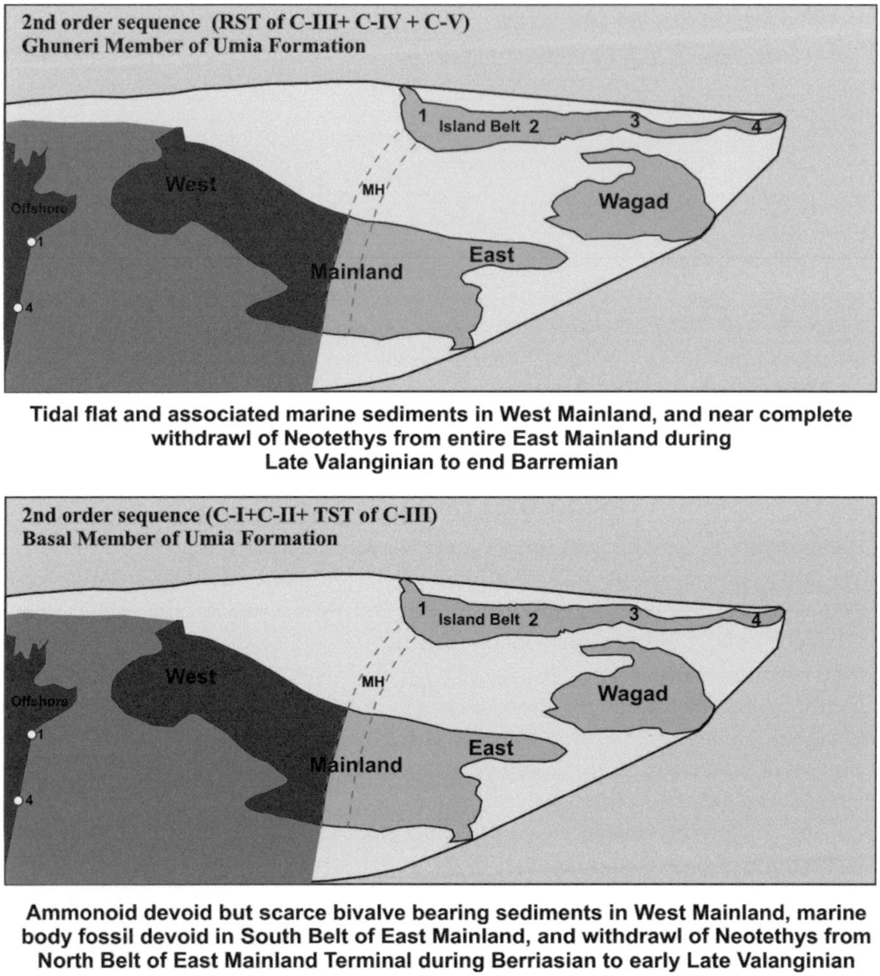

Fig. 7.32 Graphics from bottom to top respectively refer to Berriasian to Late Valanginian, and mid Late Valanginian to end-Berramian intervals

volcanic activity present not merely in High Himalaya but on the entire GTM as testimony to the integrated composite spreading system on the entire GTM.

7.5.69 Early Berriasian Second-Order MFS Event

It marks another step of tilting up due west and northwest of the Indian plate margin. There is further significant increase of cross-bedded and ripple marked sands with interbeds of shale and silt or even fine sands in the ensuing second-order RST as the Basal Member of the Umia Formation (Fig. 7.32). The thin interludes of astartid and trigonid bearing levels are understood as transgressive lag reworked benthic fauna in the West Mainland and the Ler–Katrol region of the Basal Member of the Umia Formation. By the close of the second-order RST, there occurred the final withdrawal of the sea from the north belt of East Mainland. There is further graded depletion of the ammonoid density, diversity, frequency, and spread of ammonoids as a consequence of the stepped shallowing caused by the cumulative effect of the sea level fall and tectonically induced tilting up due west and north-west. Perhaps, this is the last second-order MFS which is expressed in ammonoids both in east (at Ler–Katrol) and west Mainland (at Lakhapur), and precisely dated as Early Berriasian.

7.5.70 Late Berriasian Second-Order MFS Event

It is expressed in ammonoids only at Mundhan in west Mainland in a sand dominated succession (Fig. 7.32). The ammonoid *Kilianella* is found in coarse grits. There is further tilting up at this MFS followed up by coarsening of sands. These sands, but for Mundhan, are ammonoid devoid, though invariably continue with astartids and trigonids as part of the transgressive lag deposits. By the end of the RST is realized complete withdrawal of the sea from East Mainland east of MH.

7.5.71 Early Late Valanginian Second-Order Verrucosum Zone MFS

It is identified in the exposed succession in extremely high energy shallow tidal flat depositional system which hardly let any kind of biota inhabit the Kachchh basin (Fig. 7.32). Sequence stratigraphic concepts have facilitated dating in Kachchh of the second-order Verrucosum Zone MFS that was earlier determined in Spiti within the Guimal Sandstone. The event is significant and of global dimension in context of the widespread expansion of *Olcostephanus* (Pathak et al. 2011). The MFS also allowed correlation of sediments below and above it between basins that are located a few thousand kilometres away at the opposite margins of the Indian plate. In fact, the interpreted tilting up, that was age précised in Kachchh through ammonoid-based sequence surfaces, could also be picked up in the geological record of east coast and north sector basins. The MFS is considered additionally significant in context of its possible linkage to the first significant shift in the oceanic spreading axis in the east divergent sector at the magnetic anomaly M14/M13. The thin volcanic tuffs in the Dhrangadhra Member of Saurashtra wells may be linked to the intra-Valanginian Verrucosum Zone MFS event.

7.5.72 *Wider Expanse of Volcanism After the Shift of the Spreading Ridge*

There is interpreted shift/jump of the spreading ridge/axis at M14/M13 with slight anticlockwise rotation of the spreading ridge after the initial late Middle Oxfordian to Early Valanginian spreading (Hein 2002). At the same time also ceased spreading in the Dampier and Barrow basins. The sequence includes Valanginian sediments in Wombat and Exmouth plateaus as also in the Dampier basin followed by the Hauterivian–Barremian gap. In the Dampier basin, the SB 7 (Figs. 6.7 and 6.13) differentiates sequence 6 below of the Barrow Group Delta sands from green sands of the Mardie unit of the basal part of sequence 8. The green sands of the Mardie unit are here considered of early Late Valiniginian near the time of the shift/jump of the spreading axis. In WIMP (Figs. 5.8 and 5.12), the Sebgyar Formation in its early part includes the intra-Valanginian change of lithology and facies. In Timor Gap (Figs. 6.7 and 6.13), the change is interpreted between the Sandpiper sands and the overlying shales of Flamingo Formation. A significant post-rift subsidence with maximum bathymetry and transgression is noted all along the West Australian margin from north around Timor to Perth in south during the early Late Valanginian in several basins. M14 is the oldest magnetic anomaly in the GAP and CAP, where the spreading is attested by continued volcanism. The frequency of magnetic polarity reversals are high after the jump of the spreading ridge during the interval of M14 to M10 than during M22 to M14, which also supports the jump of the spreading ridge at the intra-Valanginian M14 second-order MFS. The peaking of the volcanism at the start of the second/middle phase at M14 early Late Valinginian ($\sim$136 ma) and associated changes east of India on the GTM are here interpreted as in the west and north sectors of India at the second-order MFS. In this context, the record of shoshonite from the oceanic crust located south of the Indus-Psangpo suture on the High Himalaya margin along with the mid-Valanginian assemblage in the overlying radiolarian chert is found interesting. The relatively farther spread of the intra-Valanginian volcanism is also supported by volcanic intercalations in Spiti Shale Formation of High Himalaya in Zanskar (Raina and Bhattacharya 1977), and shales of somewhat coeval stratigraphic units at Zedang in South Tibet (Aitchison et al. 2007). The said volcanic tuffs at Zedang have yielded U/Pb shrimp age of $\sim$137–136 ma. The said age is further strengthened by nannoplankton records in the intervening sediments. The Zanskar volcanism is particularly significant as the farthest west in High Himalaya. However, these could also be separate individual volcanic expressions, nevertheless coeval to the intra-Valanginian volcanism in the east sector. Both these records are here linked to the Pan-Neotethyan early Late Valanginian Verrucosum Zone second-order MFS so determined in Spiti (Pathak et al. 2011). Yateesh et al. (2013) placed the event at $\sim$132 ma. The volcanics suggest widespread Late Valanginian–Late Barremian second/middle phase magmatism much in contrast to the relatively restricted initiation at late Middle Oxfordian first-order MFS. The Comei volcanism of South Tibet of $\sim$132 ma age also appears linked to the said Middle phase in the early Late Valanginian–Barremian age bracket (Zhu et al. 2009). It also suggests submarine ocean island arc setting close to the suture. The said volcanic phase continued until the next jump of the spreading axis at ~126 ma. During this middle volcanic phase were created GAP, CAP, and PAP and associated transfer of the Wallby and Zenith plateaus to Australia (Gibbons et al. 2013). Gibbons et al. (2013) also suggested the unzipping in the east sector up to the south tip of India by ~126 ma, while the same is considered fully realized during the Early Aptian–mid-Middle Albian major TST (~126 to ~107 ma), so also ceasure of spreading and transfer of Elan Bank and South Kerguelen to Antarctica at the ~107 ma mid-Middle Albian MFS instead of ~115 ma (Gibbons et al. 2013; Yateesh et al. 2013).

7.5.73 *Wide Spread Terminal Dominantly Regressive Sandy Regime with Initiation at the Intra-Valanginian Second-Order MFS*

Wide spread sands with subordinate shales are evidenced all around the Indian margin in the north in Pakistan, Spiti, Nepal; in the west in Kachchh (Fig. 3.2) and Jaisalmer and Saurashtra; in the east and southeast in Mahanadi, Bengal, Pranhita, K-G, Palar and Cauvery; as also in inland Gondwana basins in Jabalpur and elsewhere. The interval is shale/silt dominated in the east sector. The regional control is more than obvious. All over the Arabian plate is present a significant Late Valanginian–Early Hauterivian subaerial gap in Bahrain, Saudi Arabia, Yemen, Oman, and Iran. Levant, Syria, and Iraq were probably also emergent. East of IEAP, in the Dampier and Barrow basins, the said interval suggests graded shallowing from Mardie green sands to Windalia sands. In a few wells in the Dampier basin, the interval marks stratigraphic gap which is interpreted as subaerial gap. Thus, major all round shallowing is evident above the second-order intra-Valanginian MFS on the entire GTM. This shallowing is tectonically driven, since Haq et al. (1987) and others have suggested sea level rise of ~80–85 m during the Hauterivian.

7.5.74 *Speculative Differentiation of A ~2.8 my Long Second-Order TST of Nearly Full Hauterivian Stage Widely on the GTM Between Arabia and India (Spiti, Kachchh and K-G)*

In Kachchh, the ~plus 13 my long mid-late Late Valanginian–Barremian dominantly sandy interval of the Ghuneri Member, includes three second-order sequences (Figs. 2.41 and 3.39), which are devoid of any guide or other marine fossils for any kind of age approximation. In the Chichhim section of the Spiti basin are also interpreted three second-order sequences with the upper two MFSs in the middle of the Giumal Formation column, respectively, at the top of the beds CHG9 and CHG 11 (Pandey and Pathak 2015b). The bed CHG 9 has for the first time yielded ammonoids whose taxonomic determination is in progress. Using the isochronous character of the sequence surfaces, the ammonoid bearing bed CHG 9 is suggested as of the late Late Hauterivian age near the Angulicosta/mortilleti zonal boundary. It may indicate the basal/early part of the terminal Hauterivian Catulloi Zone. The level besides marking a second/third-order MFS on both ETM and GTM, also marks a short-lived Faraoni anoxic event on the ETM (Baudin et al. 1999; Bodin et al. 2006), and is speculatively assigned to the terminal Hauterivian age. The event occurred one ~400 ky long ammonoid zone/subzone below the Hauterivian/Barremian boundary. Using sequence stratigraphy, the correlative of the said event in the Ghuneri column is placed between the beds 11 and 12. In Haq et al. 1987, the said level is suggested to indicate the LZB 3.2 MFS. The beds 10 and 11 of the Ghuneri Member and the beds CHG9–CHG11 of the Giumal Formation in the Chichim section are assigned to the Hauterivian stage from the base of the stage up to the base of the terminal Mortilleti/Catulloi/Ohmi Zone as a second-order TST. Thus for the first time is differentiated the significant ~2.8 my long second-order TST Hauterivian interval in India within the major ~plus 13 my long Late Valaniginian–Barremian over all regressive interval. In Madagascar, even scarce ammonoid presence (*Crioceratites, Menuthioceras* in Besaire 1973) is mentioned in a few sections within the Dualia marls. In the east sector also, the event is suggestively determined in the rather endemic *Pascoeites and Gymnoplites* (ammonoids) bearing basal conglomeratic sands of the Sivaganga Formation at the Golapille locality of the K-G basin, so also extended to other Indian basins in the east sector. The suggested MFS event has widespread manifestation all around the Indian plate.

These second-order TST sand dominated sediment intervals in Spiti and Kachchh, respectively, include glauconite grains at intermittent levels. In Kachchh are present bioturbated sands with thin extremely characteristic densely bioturbated caps and occasional oolitic limestone levels as successive inferior-order MFSs.

Guide ammonoids are absent in the numerous other Indian basins mentioned above. In Arabia two MFSs K40 of KN48/49 Zone and and K50 of KN46 Zone ages are suggested, respectively of Late Hauterivian and Early Barremian. As in Spiti, the ammonoids are absent in this interval west of India up to Arabia and in east up to NW Australia, i.e. an ammonoid devoid interval on the entire GTM. The ages of sequence surfaces in Arabia as also east of India are palynology based, however, the Arabian ages are based on coccoliths, while those of NW Australia are dinocyst based. In Haq et al. 1987, two MFSs are indicated in Early Hauterivian and near Hauterivian/Barremian boundary. The limited data as above is inconclusive, however, ages of the Arabian MFSs are here tentatively extended to Kachchh. Thus is correlated the Arabian Late Hauterivian K40 MFS of KN48/49 zonal boundary (Sharland et al. 2001) with the ammonoid-based Angulicosta/Ohmi zonal boundary in India.

7.5.75 Younging of the Successive Sandy RSTs in the First-Order Latest Middle Oxfordian to Late Barremian RST from Margin to Basin

In the Gon Mardak and Desalpur sections, the Chari Formation is conformably overlain by sands of the GSm of the abandoned 'Wagad Formation' that are here included as the Gamadu Member of the Umia Formation mostly of Late Oxfordian age, while in many other sections, the Chari Formation is conformably overlain by B (K)M of the Katrol Formation with a relatively thin capping of GSm of approximately the same time range. Thus, GSm is ranged from the latest Middle Oxfordian to mid-Early Kimmeridgian. The undifferentiated Umia Formation overlies the Katrol Formation in the north belt of East Mainland and is mostly of Tithonian age. West of Jhura, i.e. west of MH, the Umia Formation is differentiated into four Member units as the Basal, Ghuneri, Ukra and Upper Members in ascending order. In Wagad and East Mainland, the undifferentiated Umia unit may be referred to as the Umia Formation. At Wagad, the Umia Formation is expressed as the Gamadu Sandstone member (GSm) that overlies the Basal (Kantkote) Member of the Katrol Formation. In East Mainland Umia Formation is domianantly made up of sands above the Katrol Formation. In different parts of the basin, the base of the Umia Formation is highly diachronous between the Late Oxfordian and the late Late Tithonian. Its total duration is also highly variable (Figs. 2.19–2.20). It is difficult to date the top of Umia Formation in different sections in the basin. Conceptually, as per observed range of single sandy development, it is invariably only one second-order sequence above the precisely dated ammonoid-based date of the top of Katrol Formation. This is also highly diachronous. However, no guide fossils are available. The maximum thickness in Wagad is $\sim$100 m, in East Mainland of $\sim$400 m and in West Mainland of a maximum of $\sim$800 m. These sand interval are interpreted very significant as prospective hydrocarbon reservoirs, particularly in offshore.

7.5.76 Pattern of Migration of Sandy Facies in the Basin During the Toarcian–Barremian First-Order Sequence

From early Toarcian to Late Bathonian, there is spatial migration of sandy facies from basin to margin. Sandy limestones in the Jhura and Nara localities of Mainland have changed to thick sands in 'Island belt' at the margin. Late Early to mid-Middle Callovian RSM is widespread in the basin, which suggests decrease in bathymetry in the basin. There after the migration pattern of sands again becomes in the basin to margin direction. To summarize, any definite unidirectional pattern of migration of sands in the basin is not evident from base to the top of TST in early Early Toarcian–late Middle Oxfordian duration. Yet except for the duration of RSM, there is evident explicit migration of sands from basin to margin under over all increase of bathymetry in the basin. In contrast in the following first-order RST from Late Oxfordian to Barremian, sands have unidirectionally migrated from magin to basin without any change in the pattern in between under rapid decrease in basin bathymetry and recede of the shoreline.

7.5.77 Age Range of the Syn-rift Succession in the K-G Basin Subsurface Wells Through Sequence Surfaces

Prasad et al. (2000) differentiated five units; Nellore Claystone, Baptala Sandstone, Pennar Shale, Krishna Sandstone and Golapille Formation, and ranged the same from Tithonian to Barremian with unconformable contacts on both sides. They have also named the succession as Nizamapattanam Group. The Nizamapatnam Group is conceptually age bracketed within the post-MFS first-order RST interval from the latest Middle Oxfordian to Late Barremian (Figs. 7.31 and 7.32), since the K-G and other east coast basins either originated or revived sedimentation at the late Middle Oxfordian MFS. The gap between the Mandapeta Formation and the Nizamapattanam Group is also revised in the sequence stratigraphic context and précised as at least of the early Early Toarcian–late Middle Oxfordian duration which is also the duration of the first-order TST. The individual units of the Nizamapatnam Group of K-G basin have been age ranged using sequence stratigraphic surfaces in the earlier chapters (Fig. 7.33).

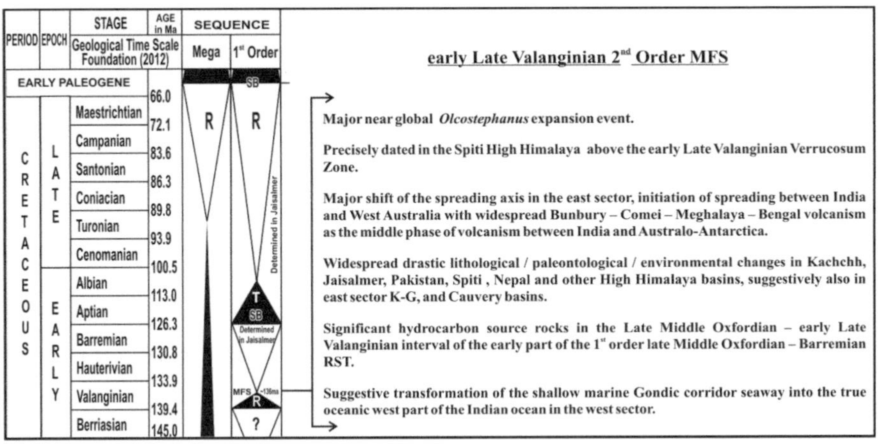

Fig. 7.33 Early Late Valanginian second-order MFS

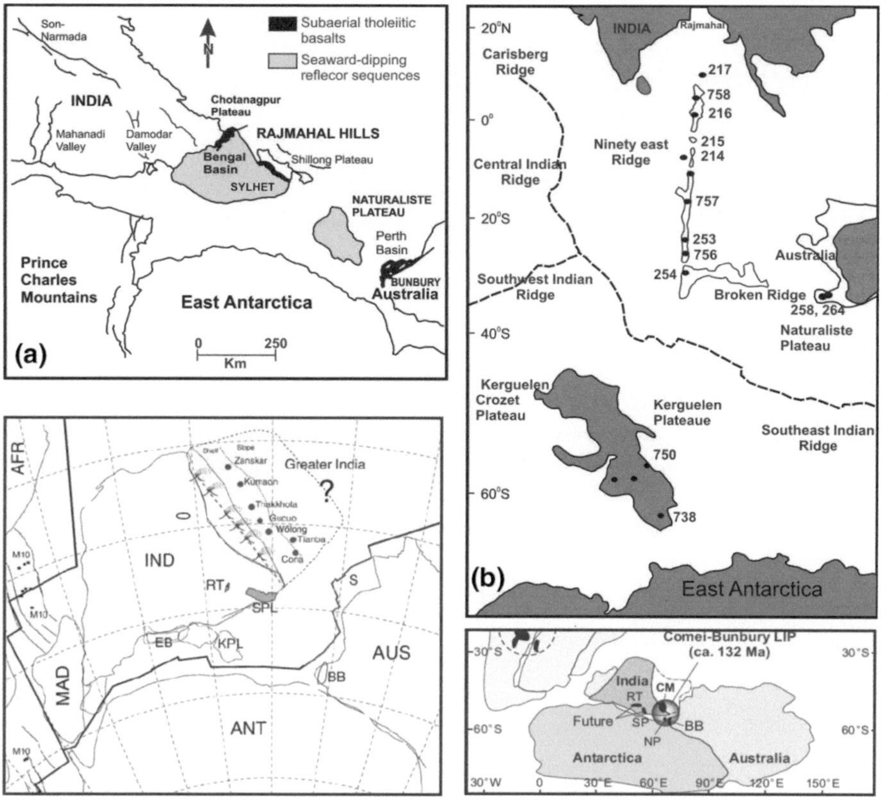

Fig. 7.34 **a** Early Cretaceous volcanic rocks in East India and southwest Australiain the east divergent sector. **b** Oceanic island arc framework between Zanskar to Tianba (modified after Zhu et al. 2009). **c** DSDP and ODP sides on Ninety east ridge, Kerguelen Plateaue and East Indian Ocean (Modified after Kent et al. 2002; Storey et al. 1992). **d** Comei–Bunbury LIP (modified after Zhu et al. 2009)

7.5.78 On the Short Convergence Interpreted Near the Hauterivian/Barremian Boundary

Gaina et al. (2015) interpreted a short duration shortening/convergence near the Hauterivian/Barremian boundary, i.e. in the terminal phase of the Toarcian–Barremian first-order sequence near its terminal Hauterivian second-order MFS, which, however, cannot be ascertained but for the indirectly interpreted anoxic event widely on both the margins of the Tethys. The convergence may have been forced under space constraints by the early phase of anticlockwise rotation of India, in turn so caused by mutually opposite pushes that India underwent during the

~ 136 to ~ 126 ma due south and east in the west sector and due north and west in in the east sector prior to the abandonment of spreading in the west sector near the ~ 126 ma first-order SB. This may have let India rotate, squeeze and eventually escape out due north. The spreading rate would have also got a substantive boost after emancipation of India from the clutches of its Gondwana captors by the mid-Middle Albian MFS event.

7.5.79 End Barremian First-Order SB Event

The marine sedimentation in Kachchh continued at the most up to the Barremian/Aptian boundary first-order SB in Ler–Katrol region as the marine mega body fossil-devoid sand dominated undifferentiated Umia Formation down the basin slope. Alternatively, the sea withdrew gradually in steps from east to west from the late Middle Oxfordian MFS to the Barremian/Aptian first-order SB prior to the start of the next first-order sequence. The TST of the next first-order sequence witnessed transgression exclusively in Mainland Kachchh west of the median high, and there by produced a distinct subaerial gap east of the median high. The subaerial gap increased in duration from basin to margin from west to east in the Kachchh basin (Fig. 7.32).

7.6 The Dravidian First-Order Sequence: Aptian–Intra-Paleogene

7.6.1 Graded Reduction in the Extent of the GTM After the Initiation of Spreading Away of Antarcto-Australia in the East and Spreading Away of the South America in the West with Respective Origin of the Indian and Atlantic Oceans

The GTM is now reduced to the loosely held together components of the Axial Gondwana. The East Gondwana is reduced just to India inclusive of Madagascar (Fig. 7.38), and West Gondwana reduced to mere Africa with India–Africa divergent zone in between. After the initiation of the spreading between India and Madagascar, the GTM reduced further in extent, and the solitary Indian plate exclusive of Madagascar represented the GTM in the terminal mid-Middle Turonian–Early Paleogene RST.

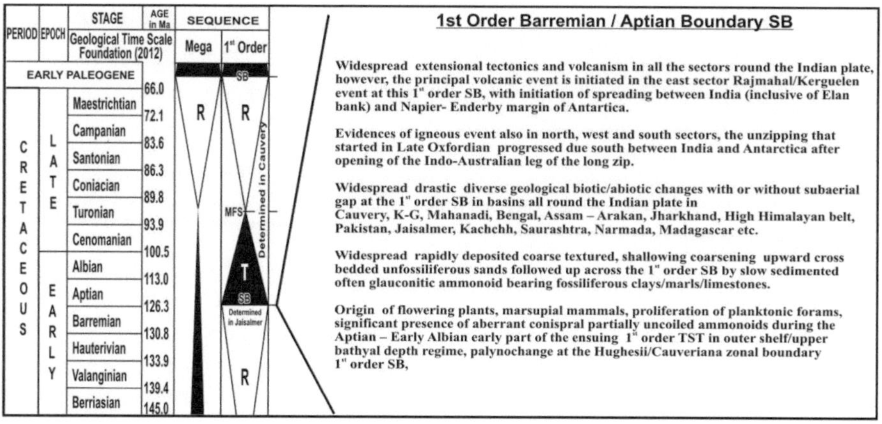

Fig. 7.35 First-order Barremian/Aptian Boundary SB

7.6.2 *Basal Early Aptian First-Order SB Event*

Distinct lithological/paleontological/physical signatures are observed in most of the Indian basins across the subaerial disconformity or its correlative conformity near the Barremian/Aptian boundary (Fig. 7.35). The said diverse drastic changes are observed between the Ghuneri Member and the Ukra Member in Kachchh, between the Parihar Formation and the Abur Formation in Jaisalmer, between the unfossiliferous Nimar Formation and the overlying fossiliferous calcareous sands in the Narmada basin, within the Giumal Formation between the ammonoid devoid sands below and the marls with Aptian–Albian ammonoids above, between the Sivaganga Formation and the Dalmiapuram Formation / Uttatur Group, between the Golapille and Raghavapuram units in the K-G basin, between the Jabalpur/Gangpur Formation and the Lameta Formation in Satpura, South Rewa, and Pranhita–Godavari basins with or without disconformity (Figs. 4.3 and 4.10). Even in the subsurface Cauvery and K-G wells, there occurred a palynoevent with change of palynoflora near the Barremian/Aptian transition between the Hughesii Zone below and the Cauveriana Zone above as the palynoevent 10 (Vijaya 1997). Several palynotaxa became extinct near the said zonal boundary. There is apparently no disconformity at this level in the relatively deeper well locations compared to the outcrops at the margin, however, distinct break and gap have been suggested in offshore Kachchh and other basins in the India–Africa sector. The physical, lithological, paleontological, bathymetric and textural changes as enumerated above in the Indian basins are more than apparent. Also, at the said first-order SB originated the shallow marine Dravidian corridor (Fig. 1.3). In context of the major evolutionary event, the origin or sudden radiation of the colourful world of flowering plants is suggested at the first-order SB.

7.6.3 Final/Late Phase (~126 to ~107 ma) of the Volcanism East of India Between the Basal Aptian–mid-Middle Albian Interval with Peak in Late Aptian (~118 to ~114 ma)

The intra-Cretaceous first-order SB suggests a fresh step of progressive due south advancement of the tectonic unzipping in the divergent sector that had begun earlier at the intra-Oxfordian MFS. The basal Aptian–mid-Middle Albian volcanism of the Rajmahal, Kerguelen, Enderby, etc., (Krishna 1987a) makes the final phase during the early part of the first-order TST of the Aptian–intra-Paleogene sequence. The initiation of the final phase is considered of relatively low key. In West Bengal and Bangladesh, the shelf started to subside near ~127 ma, i.e. near the Barremian/Aptian boundary first-order SB (Johnson and Alam 1991). The Sung valley of Meghalaya also recorded post Sylhet carbonatite complex of ~107 ma age with chemical characteristics similar to those of the Kerguelen basalts (Srivastava et al. 2005). The true initiation of the South Atlantic Ocean is also dated at ~126 ma (Nurnburg and Muller 1991). In Bengal basin and Bunbury, both pre-Aptian and post-Barremian volcanics seem to be evidenced (Vijaya 2009). In Bengal boreholes, the two distinct events seem to be separated by stratigraphic gap, whereas in Bunbury, there is compositional differentiation between the early Casuarina type basaltic flows and later Gosselin type of flows. Again in both Bengal and Bunbury, the number of flows is much smaller compared to Rajmahal. The low key Bunbury event marks slow rate of magma output (Coffin et al. 2002). May be there is a gap between the eruption of Casuarina basalts in the early phase and of Gosselin basalts in the later phase at Bunbury, so also in Bengal boreholes. The magmatic event is expressed in large areas of Jharkhand, Bengal, Bangladesh, Meghalaya, Nagaland and Myanmar, and recently in several localities of High and even Lesser Himalaya from northwest to southeast through direct igneous expressions as also indirectly through volcanoclasts, paleontologic and other evidences (Hu et al. 2015). Igneous rocks of this final phase in the east sector are also manifested in New Guinea and farther east in Australia, and Antarctica through ~115 ma lamprophyres from East Antarctica (Coffin et al. 2002), in distant Mainland Kachchh in the west sector (Kalmalkar et al. 2014), even in the west sector of the Indian Ocean (Vaillier 1974).

7.6.4 Origin of Bay of Bengal Based on Magnetic Anomalies from Indian Ocean in the East

The framework of magnetic anomalies on the floor of the Bay of Bengal, at Enderby (Antarctica) and in neighbourhood of Kerguelen hot spot in the India–Australo-Antarctica divergent region suggests the start of spreading at ~133 ma

(Ramana et al. 1994, 1997, 2001) with M 11 as the oldest anomaly. The said anomaly falls between the major jumps of spreading axis of M14 (136 ma) and MO (~126 ma). Banerjee et al. (1995) interpreted oceanic crust no older than Barremian. Based on the jump of the sprading axis in this sector in many recent studies, the causative volcanism in Bengal is favoured during the ~136 to ~107 ma interval. As conjugate of Enderby, is favoured a younger origin of Bay of Bengal during the ~126 to ~107 ma span of the final phase of volcanism in the east sector, however, the older ~136 ma origin is not ruled out. The later phase of Bunbury event is interpreted at ~126 ma near the Barremian/Aptian boundary around the start of the Deshayesites Zone. The spreading of Ceylon due south also should have become appreciable from the mid-Middle Albian second-order MFS onward.

7.6.5 Subduction of the Volcanic Island Arc in the North Linked to Final Phase of Volcanism Between Antarctica and India

Gibbons et al. (2015) have summarized evidences of the subduction of volcanic island arc/s in the Bangong–Co-Nujiang fracture region (Figs. 7.28 and 7.34), and neighbourhood in the North Tethys Himalaya, many of which have been also mentioned earlier (Krishna 1987b, and others). Thanh et al. (2012) based on data from the Shyok ophiolites have interpreted the initiation of subduction in Ladakh Himalaya on the south margin of the Karakorum block prior to ~104 ma. Gibbons et al. (2012) have suggested the slowest spreading rate in the Late Valanginian–terminal Hauterivian interval, and thus explained the absence of M14 to M11 anomalies in GAP and CAP. The oldest anomaly west of Perth is M9 (Williams et al. 2013). Somewhat improved spreading rate, yet slow, is interpreted in the M10—MO ~130 to ~120 ma interval near the Barremian/Aptian boundary MFS. The volcanic arc subduction in High Himalaya and initiation of spreading between Enderby region of Antarctica and India (inclusive of Elan bank) seem to be linked to the Barremian/Aptian boundary first-order SB event (Fig. 7.12). However, the spreading rate again changed/dropped in the late Middle Albian–early Middle Turonian interval (~107 to ~92 ma) after the oceanic separation of Elan Bank, South Kerguelen block, etc., due to space constraints and the anticlockwise rotation of the drifting India at another jump of the spreading axis from east to west of Elan Bank, etc. at ~107 ma. In the India–Australo-Antarctica sector, the spreading rate supposedly changed at ~107 ma after the realization of oceanic depths and complete/final separation of India from Australo-Antarctica at the mid-Middle

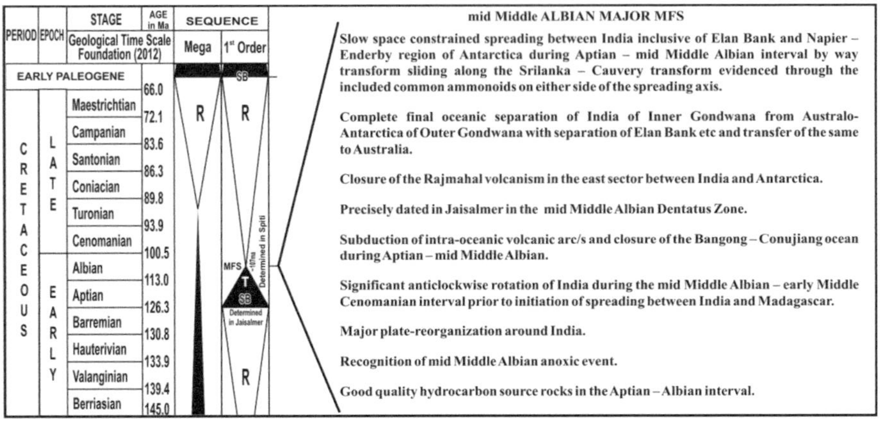

Fig. 7.36 mid-Middle Albian major MFS

Albian MFS. The variation in the spreading rates based on ammonoid distribution and space constraints is also dealt elsewhere in this chapter (Figs. 7.35 and 7.36).

7.6.6 Culmination of the Tectonic Unzipping Between the Axial and Outer Gondwana

The final unzipping in the Elan bank–Enderby sector of the India–Australo-Antarctica divergent region resulted in the complete oceanic separation of Inner/Axial Gondwana (India, Madagascar and Africa) from Australo-Antarctica and South America, which had started near the Barremian/Aptian first-order SB (Figs. 7.12–7.13 and 7.37–7.38). Until the so interpreted separation of the Inner/Axial and Outer Gondwana, the Axial Gondwana constituents (India, Madagascar, and Africa) were somewhat held together except for some spreading across India (inclusive of Madagascar) and Africa during the ~159 ma intra-Oxfordian MFS –~136 ma intra-Valanginian MFS span. The slow down to ceasure of spreading across India–Africa occurred during the ~136 to ~126 ma interval may be linked to the ridge jump in the east sector at the start of the middle phase of volcanism. The spreading between India and Africa stopped at ~126 ma. The oceanic break-up of Elan Bank out of the drifting Indian plate, was caused by shift/jump of the spreading axis from east to west of Elan Bank at the intra-Cretaceous mid-Middle Albian ~107 ma MFS. It is maintained that the major spreading within the Inner/Axial Gondwana occurred only after the complete oceanic separation between the Inner and Outer Gondwanas after the mid-Middle Albian MFS event (Krishna 2005, 2009, 2012).

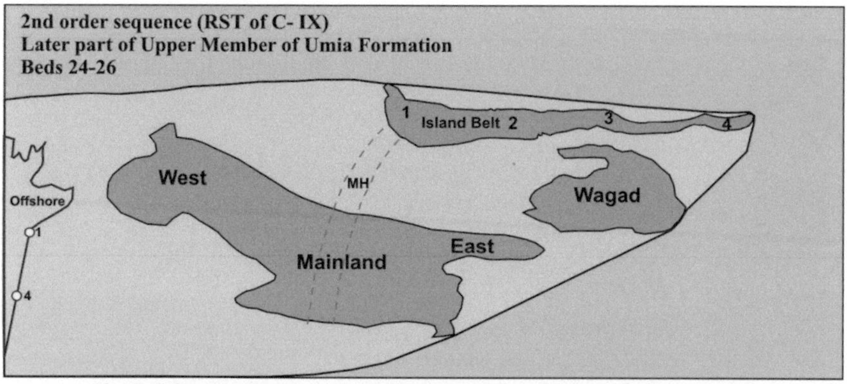

Kachch basin almost completely emerged inclusive of offshore part of the shelf during Late Turonian to Late Santonian

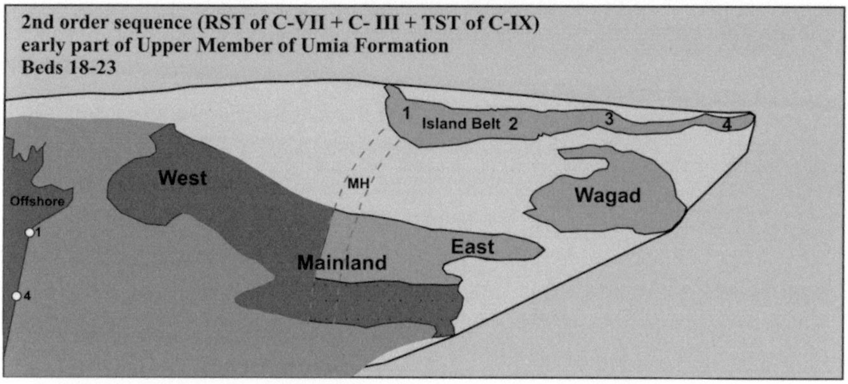

Marine mega body fossil devoid densly bio-turbeted sediments in West Mainland, presence also suspected in South Belt of East Mailand during Cenomanian to early Middle Turonian

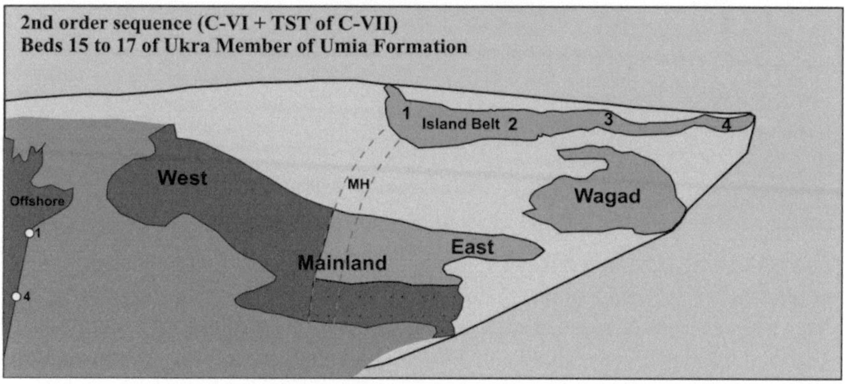

Ammonoid rich subtidal Glauconite bearing lime/marl/silt in West Mainland west of the Median High during Early Aptian to mid Middle Albian

Fig. 7.37 Graphics from bottom to top respectively refer to the basal Aptian to mid Middle Albian, Cenomanian to early Middle Turonian, and Late Turonian to late Santonian intervals

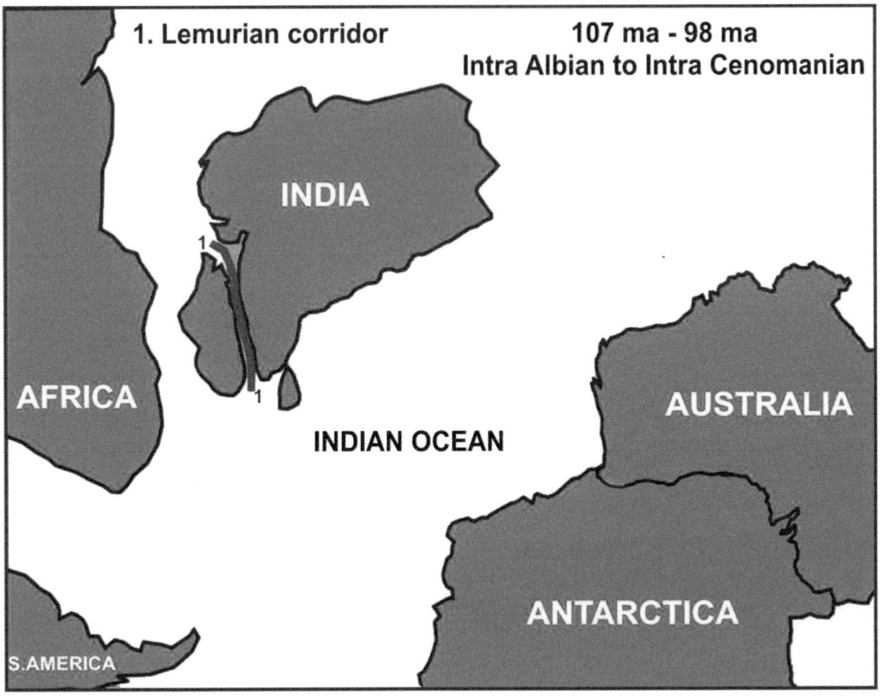

Fig. 7.38 Oceanic isolation of inner from outer Gondwana and Lemurian corridor between Madagascar and India prior to their oceanic separation

7.6.7 Igneous Evidence of Rifting / Island-arc Framework / Subduction in Other Sectors of the Indian Plate Margin Near the First-Order Barremian/Aptian Boundary SB up to the mid-Middle Albian MFS

7.6.7.1 North Sector

Surprisingly, in these 16 years of the new millennium, maximum geological data has been generated in Tibet inclusive of innumerable radiometric dates. The principal phase of Kerguelen plume-genic Late Aptian Rajmahal igneous event in the east sector between Greater India and Australo-Antarctica has had widespread echoeing in the north sector in South Tibet, Nepal, Spiti–Zanskar, Ladakh and other areas/basins (Figs. 7.28, 7.35 and 7.36). The Bangong–Co-Nujiang suture is realized at the said first-order SB age (Krishna 1987b and many others) with the closure of the Bangong–Co-Nujiang Ocean that had originated earlier near the first-order intra-Oxfordian MFS of $\sim$159 ma. Evidence of the Early Jurassic magmatic event also has strengthened along the two sutures (Zhu et al. 2016). The interesting aspect in this data is near echo of the radiometric age ranges between the Indian Ocean and South Tibet and also from the north, south, east, and west sectors all round the Indian plate. The radiometric ages seem to cluster around the ages of the first and second-order sequence surfaces (~205 to ~174 ma, ~170 to ~136 ma, and ~136 to ~100 ma, ~183 to ~100 ma).

A recent editorial to a volume focused on the evolution of Tibet in 2016 includes a good summary. The BNSZ is interpreted to include two Jurassic–Cretaceous magmatic arcs. Double subduction of the BNO is proposed (Zhu et al. 2016) due north beneath Quiangtang and due south beneath Lhasa during ~140 to ~130 ma, i.e. during the terminal Middle Oxfordian–Barremian first-order RST age bracket. Wang et al. (2011) based on zircon ages from rocks associated with ophiolites along the BNSZ constrain the ocean crust in the ~187–164 ma range, i.e. around the first-order intra-Jurassic SB to first-order intra-Jurassic MFS. Zhang et al. (2012), based on zircon ages interpret the ophiolite formation along the YZSZ between ~134 and ~128, i.e. between ~136 ma intra-Valanginian second-order MFS and ~126 ma first-order intra-Cretaceous SB. Luobusa ophiolites in the east YZSZ formed during the Jurassic, while in the central and western YZSZ during the ~130 to ~120 ma interval.

The multiple subductions proposed are age-constrained at the successive first-order surfaces near Pliensbacian/Toarcian SB, intra-Oxfordian MFS, and Barremian/Aptian boundary SB. Zhang et al. (2012) recorded ~205–174 ma magmatism in the southern most Lhasa, ~170 to ~137 ma magmatism in central Lhasa, and ~135 to ~100 ma widespread magmatism over entire Lhasa, and Late Cretaceous ~90 to ~78 ma magmatism sporadically in South Tibet which correspond well to first and second-order sequence surfaces. The ages from east to west and from north to south permit somewhat progressive magmatism from east to west and north to south. Quite similar to the the east sector (~187 to ~159 to ~105 ma volcanics), the Central Lhasa Gangdese region also witnessed long duration magmatism (~143 to ~102 ma) as in the India–Australo-Antarctica divergent region. Magmatism of ~120 to ~100 ma, ~126 to ~123 ma from Yarlung–Psangpo also echoes the same age range. The ~123–124 ma age of the granodiorite porphyry and the associated copper mineralization further strengthen the tectonic-cum-igneous event in the north sector along the Bangong–Co-Nujiang region near the first-order Barremian/Aptian boundary SB. Gibbons et al. (2015) provided radiolarian, foraminifer data and evidence of widespread convergent magmatism along the Bangong–Co-Nujiang suture and neighbourhood (Karakorum, KKBSZ, KLA, ISZ, YTSZ and Himalaya) (Fig. 7.36). The Ladakh batholith started growing at ~105 to ~100 ma due to melting of the subducted oceanic lithosphere, and associated origin of the Shyok–Dras volcanic arc. The localized oceanisation may have started earlier as part of the ~159 ma intra-Oxfordian MFS event with evidences of localized spreading in north, south, east, and west although having principal expression only in the east sector. The subduction event is also suggested at the mid-Middle Albian Dentatus Zone first/second-order MFS. The Early to Middle Albian forams from the Shyok suture (Ehiro et al. 2007) also fall in the same age bracket. The age of the Dras volcanics is suggested in the ~159 to ~91 ma age bracket from the first-order intra-Oxfordian MFS (origin of spreading in the east sector between NW Australia and Greater India précised in Kachchh) to the first/second-order intra-Turonian MFS (start of spreading between India and Madagascar determined in Madagascar, Narmada, and Mannar). The oceanic lithosphere in the region is ranged from ~159 to ~126 ma from the first-order intra-Oxfordian MFS to first-order intra-Cretaceous SB. The said younger age of oceanisation otherwise also marks the initiation of the Rajmahal/Kerguelen volcanism, ceasure of rifting west of Madagascar and initiation east of Madagascar, and start of widespread subduction in the north sector.

7.6.7.2 West Sector

The evidences of the rift-genic igneous event have also emerged in the west sector in Kachchh (Kalmalkar et al. 2014), Jaisalmer (Dasgupta 1976), Pakistan, Afganistan, Madagasar (Valdiya 2010; Okimura and Fatmi 1988; Durr and Gibling 1994; Gaina et al. 2015; Hu et al. 2015) through either igneous rocks or volcanoclasts in sedimentary rocks. Radiometric ages in the said age bracket have also emerged from Nidar, Karzog, Sapat localities of Pakistan (Ahmad et al. 2008). Also, the Shyok volcanics (Gergan and Pant 1983 in Raju 2011; Jain 2014), Dras volcanics at Sapat (U/Pb ages of ~103 and ~101 ma) fall in the same age range. A distinct volcanic event of the said first-order intra-Cretaceous SB age is also recognized in the west sector of the Indian Ocean in Leg 25 site 249 (Vaillier DSDP reports 1974). Thick dolerite intervals have also been observed in a few Kachchh wells (Sanadra 1, GK-22-C-1, and SP-1-1) which are here linked to the first-order intra-Cretaceous SB igneous event, so also ~126 ma lamproite intrusive in the Jhuran Formation at the junction of the Median High and the Kachchh Mainland fault (Kalmalkar et al. 2014). The igneous event in Kachchh, thus, appears widespread at least west of the Median High.

7.6.7.3 Indirect Evidences of the East Sector Magmatism on the Entire GTM Through Transported Volcanoclasts, and Other Biostratigraphic Constraints

Valdiya (2010, 2016) interpreted volanoclasts of the Rajmahal volcanics in sediments of Aptian–Albian age between South Tibet and Baluchistan, which he interpreted as carried along with terrestrial plant debris by the due north flowing rivers to the Neotethys from the relatively higher topographic region in the east and southeast. Alternatively, is suggested widespread submarine volcanic activity on GTM linked to the east sector igneous event suggestively earlier or coeval to

Rajmahal around intra-Cretaceous first-order SB of Barremian/Aptian boundary age. Hu et al. (2015), Dai et al. (2013) and others have recorded U/Pb zircon, and Ar40/Ar39 ages from volcanoclasts of the distal most North Tethys Himalaya of ~142 to ~103 ma, and from South Tethys Himalya of ~146 to ~116 ma, and interpreted major intra-plate volcanism during the terminal Jurassic–Early Cretaceous interval. Guo et al. (2007) considered start of magmatism at ~153 ma, and subduction during Early Cretaceous. They also inferred that ~60 % of the crustal shortening was already realized in the Early Cretaceous subduction event. Leier et al. (2007) interpreted formation of Tibet much earlier in Early Cretaceous, and interpreted Tibet already above sea level at that time. Volkmer (2010) considered the initiation of subduction of the Neotethys in Late Jurassic. Recent record of the mid Aptian radiolarian chert from south of the Indus suture provides credence to the widespread magmatsim up to Desang in High Himalaya and up to Rajmahal in Jharkhand. Zhu et al. (2009) interpreted Tithonian volcanism through basal volcanoclasts at Volong, as part of Middle Jurassic–Early Cretaceous volcanism in Southeast Tibetan basins. The Barremian/Aptian ages of radiolarites, ~109 ma from Jidula Formation at Tingri, Aptian/Albian nannoplanktons from the basal part of Shadu Formation, start of volcanism prior to the Takena Formation of the Xigaze Group (Coulon et al. 1986), ~127 ma of the glauconitc base of the youngest unit the glauconitc Gucuo Group/Formation, also support the widespread expression of the igneous event. The volcanism is considered older to the Late Albian Subticinensis Zone of ~107 to ~104 ma (Silva and Garzanti 1991). Wiedmann and Durr (1995) based on ammonoids of the Albian Mortoniceras inflatum Zone and M. princei Zone interpreted volcanism of Aptian/Albian age. Mafic protoliths in Lesser Himalaya, and eclogites of Aptian–Albian age are found in Nepal (Groppo et al. 2007). The said diverse evidences suggest the initiation of the volcanism in Late Jurassic at the first-order MFS with culmination in Middle Albian MFS. Also is interpreted progression of volcanism from east to west, Tithonian in Tibet, Valanginian–Hauterivian in Nepal and Aptian–Albian in Spiti–Zanskar.

7.6.7.4 Somewhat Similar Age Framework of the Volcanism and Associated Spreading History in the Africa–Antarctica–Patagonia

The oceanisation in the South Africa–Mozambique–Antarctica region is also ranged between ~160 to ~154 to ~107 ma Late Oxfordian–Early Kimmeridgian to Middle Albian bracket (Coffin and Rabinowitz 1992; Segoufin and Patriat 1980; Gaina et al. 2015; Pankhrust et al. 1999). The realization of the South Atlantic Ocean is also placed around ~126 ma. This is preceded by an intial phase of extension in the ~190 to ~167 ma age bracket (Reeves and Vit 2000). There is near continuum of volcanism between ~145 and ~105 ma during the Early

Cretaceous in SE Africa (Moore et al. 2008), so also in the India–Australo-Antarctica region. The initiation and completion of South Atlantic Ocean, like the Indian Ocean between India and Australo-Antarctica, is also similarly ranged between ~126 and ~105 ma volcanism with peak near ~117 ma. The terminal phase continues up to ~110 to ~100 ma. The said deep water marine communication is ranged between ~110 and ~100 ma (Lawver et al. 1992), which is in good accord with appreciable spreading between India and Australo-Antarctica only after Early Albian, i.e. after ~107 ma with the Indian Ocean assuming true deep water oceanic dimensions (Konig and Jokat 2006). The progressive spreading between Africa and South America, thus, is also found similarly age-constrained between the Late Oxfordian first-order MFS–mid-Middle Albian second-order MFS interval.

7.6.7.5 Ammonoid Evidence of Only Restricted/Limited Spreading in the India–Australo-Antarctica Sector up to mid-Middle Albian

The first ammonoids after the widespread second-order mid-Late Valanginian to Barremian sand dominating ammonoid devoid second-order RST are the Early Aptian *Deshayesites* in west and north (Jaisalmer, Kachchh, Spiti, Indus basins) and the Late Aptian *Australiceras* in the east (Cauvery, K-G basin exposed sediments, and even in well cores of Vriddhachalam). The Late Aptian *Australiceras* is also known in Kachchh, and the Late Aptian *Epicheloniceras* in Jaisalmer. The Early to early Middle Albian *Lemuroceras, Douvulleiceras, Hamites, Brancoceras* are known in Jaisalmer and the Indus basins, while *Cleoniceras* and *Lemuroceras* from the youngest bed of the Ukra Member in Kachchhh. These testify the age of the first significant TST from Early Aptian to mid-Middle Albian. The above and/or coeval ammonoids inclusive of ancycloceratins are widely distributed in Australia, Antarctica, east and west sectors of the Indian subcontinent, Madagascar, and E. Africa. The ancycloceratin ammonoids are considered relatively deeper depth constrained up to outer shelf–upper bathyal depths, i.e. up to around ~500 km or more distanting apart of the spreading units. Development of greater depths in excess of upper bathyal would not let the said ammonoids expand from Australo-Antarctica to India or vice versa. This in turn insured that the spreading across India–Australo-Antarctica having begun in Early Aptian would not exceed ~500 km of distance across the two spreading blocks. The ancycloceratin ammonoid evidence, also, indirectly, implied that the spreading exceeded ~500 km only after the mid-Middle Albian MFS event. It is also supported by the culmination of the igneous activity in the Indian east sector around the same time near ~107 ma. The interpreted slow spreading of India vis-à-vis Australia until the mid-Middle Albian MFS is indirectly supported by about tenfold increase in the spreading rate after ~93 ma in Bastia et al. (2010).

7.6.8 *Early Early Aptian–mid-Middle Albian First/Second-Order TST in Kachchh of the First-Order Terminal Mesozoic Sequence*

The first TST of the youngest first-order sequence, let the Tethys transgress only the Mainland Kachchh west of MH as per its farthest east evidence of the Albian (Rai 2006) in the Yaksha temple road side section on the Jhura–Nakhatrana road. This delimited the due east transient extension of the Ukra Member of the Umia Formation. However, the maximum depth realized in this sequence in Kachchh did never equal the depth achieved at the earlier first-order intra-Jurassic MFS (Figs. 7.32 and 7.36), suggestively, so on account of the graded uplift during the terminal Middle Oxfordian–Barremian RST interval in the west and north sectors of the Indian margin.

7.6.8.1 Precise Determination of the mid-Middle Albian MFS in Jaisalmer

The MFS on the GTM is précised in the Jaisalmer basin, only a few hundred kilometers away from Kachchh. The tops of the Abur (Jaisalmer) and Ukra (Kachchh) units on the basis of the included guide ammonoids are dated in the mid-Middle Albian Dentatus Zone based on the occurrence of a late representative of the *Douvilleiceras* lineage, and *Hamites rotundus, Hamites* spp, *Hoplites dentatus*. The youngest exposed levels below the disconformably overlying Sanu Formation include *Hamites rotundus* and indeterminate *Hamites* (Krishna 1980). In Kachchh the youngest Ukra bed—a coquina limestone—yielded a single *Lemuroceras*. Rich ammonoid assemblage of late Early Albian Mammilatum Zone has been recently described from Cauvery (Gautam et al. 2015), which supports the determination in Jaisalmer, however, the Tethyan *Douvilleiceras* is not recorded in the just originated Indian Ocean in the east sector inclusive of the Cauvery basin.

7.6.8.2 Nannoplankton Support to the Ammonoid-Based Determination of the Albian MFS in Kachchh, Jaisalmer, and Cauvery

Jafar and Rai (1989) and Rai et al. (2013a, b) subsequently recorded Middle Albian nannoplaktons in Cauvery, Kachchh, and Jaisalmer. The Kachchh record is in the middle part of the undifferentiated Umia Formation as already mentioned above in a roadside section immediately west of the Median High near the Jakh temple in a thin carbonaceous shale bed in Kachchh. Rai only a few years later also interpreted the same nannoplankton zone in Jaisalmer at the base of the Abur unit in a calcareous sandstone bed. Jafar and Rai (1989) also based on nannoplanktons interpreted ~200 m depth in the grey shale unit of the Kallakuddy quarry (Dalmiapuram) in the Cauvery basin.

7.6.8.3 Extension Towards the Margin in Kachchh of the mid-Middle Albian MFS

In a gradually increasing bathymetric framework, the shoreline is interpreted to have reached up to Jakh temple near the margin in Kachchh. Much unlike the comment of Biswas (2007), it need not at all be considered anomalous. Nevertheless, the duration of the transient marine incursion may have been much shorter only around mid-Middle Albian instead of the Early Aptian–Middle Albian duration of the second-order TST. The second-order mid-Middle Albian MFS event is evidenced widely, also in the Indian Ocean region, and even elsewhere all round the Indian plate incluse of both north and south parts of High Himalaya. Jakh and neighbourhood at the said MFS formed the extreme margin of the basin, since the Neotethys had already withdrawn from the East Mainland by the close of the Barremian during the RST interval of the preceding Toarcian–Barremian first-order sequence.

7.6.8.4 Marine Depositional Environment of the Parihar Formation

Much unlike the understanding of Rai et al. (2013b), the essential marine origin of the Parihar Formation has been repetitively suggested since Krishna (1983) and Krishna et al. (1983) when the Late Valanginian–Barremian sands of the Indian plate margin were interpreted as formed essentially by the near shore marine depositional processes. Moreover, the said calcareous sand level marks the base of the Abur Formation instead of inclusion in the terminal part of the underlying Parihar Formation as made out by Rai et al. (2013b).

7.6.9 Intra-Cretaceous Intra/Inter Plate Reorganization Event

The Cretaceous interval of earth history has been the most eventful of the Mesozoic in terms of the frequency, magnitude and geographical expanse of the events, particularly in context of India and neighbourhood. During the Cretaceous, multi-faceted geological events occurred all round the Indian plate in east, west, north and south. Matthews et al. (2012) has dwelt on the intra-Cretaceous plate reorganization. Within the Cretaceous, the end Cretaceous event attracted more attention than the mid-Cretaceous or the early Cretaceous events. Nevertheless, the mid-Cretaceous happenings did fascinate the geoscientists through a few symposia on the mid-Cretaceous events or otherwise. The evidences in the east and west sectors have been found quite revealing. Extensive volcanism is witnessed in the Cretaceous, in the east sector in the Early Cretaceous, and in the west sector in Late Cretaceous. The volcanism in the Indian east sector ceased at the mid-Middle Albian ~107 ma major MFS with separation and the transfer of the Elan Bank to Antarctica. The subduction is interpreted in the north sector along the Bangong–Co-Nujiang suture since the 1980s. Igneous events are also recorded in the west sector of the Indian Ocean, Madagascar, Pakistan, Kachchh, Jaisalmer, and Barmer. Extensional tectonics is interpreted in Barmer and Bengal. The said event apart from volcanism is also known in the west and north sectors through diverse manifestations. Perhaps, there was a large jump of the spreading axis from east of India to west of India.

The oldest on-land localized volcanism in the west sector in Madagascar is suggestively dated ~102 ma in the terminal Albian (Vaillier, 1974), while somewhat wider expanse is known from mid-Middle Cenomanian onwards to Early and Middle Turonian between ~98 and ~91 ma with interpreted peaking at the ~91 ma major MFS. The transfer of the intra-Oxfordian oceanic crust from the drifting India (Gibbons et al. 2012, 2013) to its present location on the Investigator ridge is linked to the ~98 ma initiation of volcanism and initiation of India/Madagascar separation at the intra-Cenomanian second-order MFS as determined here in Madagascar. The culmination occurred at the second-order Santonian/Campanian boundary at ~84 ma. In the graded separation process, significant steps are interpreted at ~98, ~91, and ~84 ma sequence surfaces. Post-subduction radiolarian cherts above the ophiolites are known in Oman (Warren et al. 2005) of ~98–94 ma, and Cyprus of ~92 ma (Mukasa and Ludden 1987). There is also realized the WBFZ bend during the ~107 to ~98 ma span along with the origin of Wharton basin. The ceasure of spreading in the Wharton basin may be linked to the start of spreading between Madagascar and India, and also the start of spreading between Australia and Antarctica. The major intra-Cretaceous reorganization event is thus age-constrained in the said time span between the cessation of volcanism in the east and start of volcanism in the west along with bending of the WBFZ which coincides with widespread subduction related igneous expression in the north sector along the Bangong–Co-Nujiang suture zone. At this MFS the Indian Ocean also acquired oceanic framework with separation of Elan Bank from India along with its transfer to the Australo-Antarctica plate. Yateesh et al. (2013) have interpreted the intra-Cretaceous plate reorganization during the ~100 to ~90 ma span. During the the pre-plate reorganization span the theatre of principal volcanism is in east and northeast, while in the post plate reorganization time, the prime volcanism shifted to locii west and north of the Indian plate.

7.6.10 Counterclockwise Rotation of India

The spreading between India and Madagascar suggestively initiated around the intra-Cenomanian second-order ~98 ma MFS with the onset of volcanism in Madagascar both in the east and west. The said initiation is additionally supported by significant paleontological/lithological/sedimentological/environmental change with the onset of volcanism. It suggestively peaked at the ~91 ma MFS, and culminated at ~84 ma Santonian/Campanian boundary SB. The extensional tectonics along the India–Madagascar fracture zone until the onset of volcanism remained restricted to mere rifting since the Barremian/Aptian boundary first-order SB. The spreading due northwest of India away from Australo-Antarctica even during Aptian to mid-Middle Albian was space constrained to only few hundred km. The same space constraints in between the WZFZ on the east and the east margin of Madagascar in the west supposedly prompted shifting of the rift axis from west to east of Madagascar. The space constraints also may have forced the counterclockwise rotation on account of the mutually opposed transform sliding of India along parallel to subparallel axes due south in the west and due north in the east between ~136 ma intra-Valanginian second-order MFS—~107 to 98 ma Middle Albian–Middle Cenomanian second-order MFSs, more so after complete separation from Antarctica by mid-Middle Albian MFS at ~107 ma. The said reorganization of the spreading direction from northwest to north is traced on the ocean floor in the form of the curved WBFZ–WZFZ indent on the ocean floor (Gibbons et al. 2013). The counterclockwise rotation of India, thus, is interpreted to have been begun at ~136 ma, however, the bulk of it was realized during the ~107 to ~98 ma span prior to the initiation of separation of Madagascar from India with maximum at ~107 ma through transfer of volcanism from east to west sector.

7.6.11 On the Possible Middle Albian Gap in the Indian Basins

Prasad and Phor 2009 based on their palynological studies interpreted Middle Albian as a gap in the Cauvery and K-G basins respectively between the Andimadam Formation and Sattapadi Shale and between the Raghavapuram Shale and Gajulapadu Shale. Ayyasami (2011) also indicated absence of the Early and Middle Albian ammonoids in Cauvery basin, although rich terminal Early Albian/basal Middle Albian ammonoids have been recently recorded in the Cauvery basin in the early part distinctly well above the base (Gautam et al. 2015). Even earlier Krishna (1983) had recorded Early Albian *Puzosia* nucleus from the Kalakuddy quarry. Thus, the said gap, however, is not evidenced in Kachchh, Jaislamer, Pakistan, and Madagascar, not even in all the Cauvery sections in view of the presence of Albian ammonoids listed elsewhere. In the proximal exposed sections of the Dalmiapuram Formation in the Cauvery basin, the coralline limestones of possible Early Cenomanian age instead suggest distinct shallowing above the relatively ammonoid and plankton bearing Early to mid-Middle Albian grey shales. In Kachchh, Jaisalmer, and Pakistan too, the mid-Middle Albian MFS is followed up by the guide fossil-devoid coarse RST sands of the early part of the Upper Member of Umia Formation in Kachchh and its coeval units, and of Upper Goru Member of Goru Formation in Jaisalmer and Pakistan.

7.7 Intra-Turonian MFS in the Indian Basins

In the ammonoid devoid post Albian exposed succession in Jaisalmer, the said younger MFS (Fig. 7.39) is included in the gap between the Habur and Sanu units in outcrops, and between the Goru and Parh units in subsurface wells. In Kachchh, two alternatives are suggested for the said MFS—near the top of the middle shaly part, and the other immediately below the traps (Figs. 2.42). In Saurashtra, the said MFS is included in the thin Middle Turonian ammonoid bearing middle part of the Wadhwan Formation. In Pakistan, the MFS is determined between the Goru and Parh units. In Narmada, it is placed in the youngest part of the Nodular Limestone, and dated by ammonoids. In Spiti and Janda in West/Central Tibet, the intra-Turonian MFS demarcation is based on the planktonic foraminiferal record (Bertle and Suttner 2005). In Madagascar, the mid-Middle Cenomanian to mid-Late Turonian span is either absent or in non-ammonoid facies. The major part of the Turonian corresponds to volcanism in all the Madagascar basins, although localized older volcanism is suspected even earlier in Middle Cenomanian. There is mention of volcanism even of ~102 ma age in Coffin and Rabinowitz (1992). A significant second-order MFS in Madagascar is suggestively interpreted in the Middle

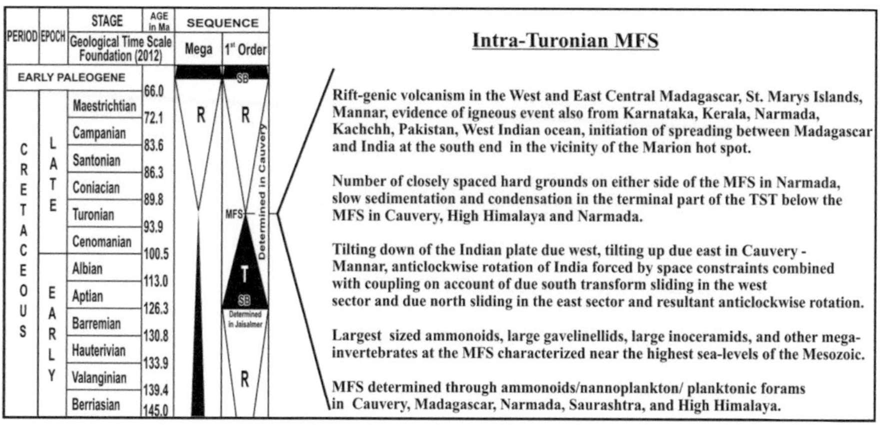

Fig. 7.39 Intra-Turonian MFS

Cenomanian sediments. In the subsurface/offshore wells in the Cauvery and Jaisalmer subbasins, the MFS is précised on the planktonic foraminifer and dinocyst record. In the Mannar subbasin wells, also, the MFS is placed immediately below the volcanic intervals.

7.7.1 Slow Somewhat Condensed Early Aptian–mid-Middle Albian TST of Ukra and Coeval Units Compared to the Overlying Cenomanian in Kachchh, Jaisalmer, and Pakistan

The crudely determined rate of sedimentation in the $\sim$19 my long ammonoid rich Early Aptian–mid-Middle Albian interval in the said basins is found distinctly less compared to that of the overlying $\sim$6.6 my long Cenomanian. This in turn implies either distinct decrease in bathymetry in the otherwise rapid phase of eustatic rise across the globe or uplift/rejuvenation of the provenance. So also is the situation in Spiti (north sector) and Cauvery (east sector).

7.7.2 Turonian Transgression in Narmada and Saurashtra

The early and middle Late Cretaceous Narmada succession of the Bagh Group/Formation includes significant ammonoids and nannoplanktons near the hardgrounded contact of the Nodular Limestone below and the Coralline Limestone above. The inclusion of Albian although surmised in regional context is not confirmed. In Saurashtra, the otherwise ammonoid and other guide fossil-devoid Wadhwan Sandstone unit of Saurashtra in its middle part includes a thin interval of Turonian ammonoids. The included ammonoid/nannoplankton/foram record, thus, evidences merely the Turonian transgression, and therefore any comparison with Pakistan, Jaisalmer, and Kachchh is not possible.

7.7.3 Similar Pattern of Possible Decrease of Bathymetry Across the mid-Middle Albian MFS in Spiti High Himalaya

The Turonian record is ammonoid devoid in Kachchh, Jaisalmer, Pakistan, High Himalaya (Spiti and others), and K-G basins while ammonoids are present in Cauvery, Narmada, Saurashtra, and Madagascar. Good planktonic foraminiferal control is available in Cauvery, Spiti, South Tibet, and Jaisalmer. The ammonoid rich Early Aptian–mid-Middle Albian is only $\sim$10 m thick compared to the overlying ~45 m thick ammonoid devoid, yet planktonic foram bearing, Cenomanian–Early Turonian part in Spiti and South Tibet which suggests similar decrease of bathymetry as in the basins of the west sector.

7.7.4 Late Cenomanian–Early Middle Turonian Condensation in High Himalaya Down the Dip from South to North

Also, the post-Giumal part of the stratigraphic columns in High Himalaya has highly varying thickness of ~40 m at proximal Chichhim and only ~1.6 m at distal Zanskar which strongly suggests increasingly slower rate of sedimentation down the basin under increasing bathymetric framework. The condensation down the dip has perhaps taken place on account of the prevention in supply of terrigenous matter in the distal sites under rapid increase of bathymetry through the combined influence of tectonics and eustatic rise (Figs. 4.6 and 4.14).

7.7.5 Not that Explicit Picture in the Cauvery Basin

In Cauvery, the ammonoid, foram, and nannoplankton rich Cretaceous stratigraphic column owing to bad land topography not only lacks the desired degree of lithostratigraphic differentiation but also is wanting in measured thicknesses of the beds. The subsurface and offshore wells alone provide distinctly impoved differentiation and measurement. On a relative basis, the rate of sedimentation in the Aptian–mid-Middle Albian interval is considered decidedly slower than in the overlying remainder of the Uttatur Group which mostly includes Cenomanian. It is rather difficult to infer relative bathymetry at the mid-Middle Albian and early Middle Turonian MFSs in the Cauvery basin, although the maximum depths suggested in offshore wells are outer shelf to upper bathyal in the early Middle Turonian Helvetica Zone (Chidambaram 2000), and ~200 m at the mid-Middle Albian MFS in the Kalakuddy quarry (Jafar and Rai 1989). Both the MFSs are considered equally significant without differentiation of the first-order, however, preference is given to the intra-Turonian one. May be the increase in bathymetry was faster up to the mid-Middle Albian MFS, and therefore it considerably slowed down and remained somewhat stable between mid-Middle Albian to Early Turonian. The Cenomanian in all the sectors is found thicker under somewhat shallower framework compared to the underlying Albian and overlying Early Turonian.

In the K-G basin, the top of the Raghvapuram shale is interpreted as the said MFS in the Late Cenomanian–Turonian interval. In the outcrops, the Aptian–

Albian transgression is also evidenced through rare ammonoids (Prasad and Pundeer 1999), but comparison between the Albian and Turonian in the K-G basin in the present state of information is not possible.

7.7.6 Ambiguity in the Determination of the First-Order Intra-Cretaceous MFS

The MFSs in Albian and Turonian or near Cenomanian/Turonian appear to include between them a few hundred meters thick distinctly shallower Cenomanian interval in Madagascar, Pakistan, Jaisalmer, Cauvery, and other basins. In context of the bathymetric pattern in the west and north sectors as discussed above, the placement of the intra-Cretaceous first-order MFI (prolonged relatively stable bathymetry) on the Indian plate is suggested between mid-Middle Albian to Early or early Middle Turonian in comparison to the globally accepted graded unidirectional eustatic rise in the Aptian–Turonian interval. On one hand the mid-Middle Albian MFS event is considered more important and widespread, while on the otherhand in eustatic context, the intra-Turonian is considered having slightly better sea level than at the older mid-Middle Albian MFS. The absence of the post Albian ammonoids, coupled with distinctly greater rate of sedimentation in the west and north sectors during the intervening Cenomanian, seem to suggest tectonic uplift and/or eustatic fall with rejuvenation of the provenace for increase sediment supply.

Thus, in Spiti, North Pakistan, Jaisalmer, and Kachchh, the intra-Albian MFS appears more significant, while in Saurashtra, and Narmada, only the intra-Turonian MFS is explicitly evidenced. In Cauvery, both the said MFSs are considered significant. The explicit and decisive differentiation of the first-order one is left for a later occasion although the intra-Turonian is favoured. Alternatively, the MFI as an exception may be ranged though rather a long Middle Albian–Middle Turonian duration. The graphics on the first-order T/R cycles, however, irrespective of the above indecisive discussion continue to depict the first-order MFS in Turonian as per author's earlier studies (Krishna 2012).

7.7.7 Anoxic Events in the Mega-Sequence

Lately, the OAEs have assumed particular significance in context of their association with volcanism, organic enrichment, hydrocarbon source rocks, short duration age bracketing in the extreme terminal part of the TSTs close to the MFSs, ocean water chemistry, condensation/starvation, glauconitization/phosphatisation, body fossil richness, etc. On the GTM, the OAEs have evidenced repetitively in the intra-Permian–intra-Paleogene mega-sequence from the very start. Evidences have surfaced at the start of Late Permian, basal Toarcian, intr-Oxfordian, intra-Valanginian, Late Hauterivian, intra-Barremian, basal Aptian, mid-Middle Albian, Cenomanian/Turonian boundary, intra-Turonian, etc. (Fig. 7.2). The frequency is, particularly, increased during the second-order Late Hauterivian MFS–mid-Middle Albian second-order MFS. The organically rich black shales are considered the best lithological clues for the possible recognition of OAEs in the succession. There are few such dark shales that are age-constrained in the Hauterivian–Barremian of Kachchh. In the Late Barremian, the carbonaceous shales are also associated with a ~3 m thick impure coal seam with significant sulphur content. Dark shales are also recorded in the basal part of the intertrappean succession in Bengal subsurface and Rajmahal in the east sector (Ghose et al. 1996), respectively, of possible early Early Aptian and intra-Valanginian ages, however, their proximal locations do not favour oceanic anoxism but for an extreme transient interlude. Even in the south part of High Himalaya, the oceanic anoxism is similarly not favoured except near the second-order MFSs of the first-order TST in the late Early and Late Cretaceous. In Cauvery, because of the proximity of the nascent/juvenile ocean, the signatures of Mid Albian OAE are somewhat explicit. Nevertheless, the OAE coeval levels are precisely determined in many sections in different Indian sectors for mild/distant ocean anoxic signals than the classic explicit OAE signatures. The best bets are the mid-Middle Albian, and terminal Cenomanian/basal Turonian. Jenkyns (1980, 1985) have discussed in detail the Jurassic–Cretaceous anoxic events across the world. Pandey and Pathak (2016) have also outlined the possibilities of identifying oceanic anoxic events in the Indian Cretaceous geological record. Bertle and Suttner in Spiti have suggested only weak signals of the Cenomanian/Turonian boundary OAE. Wang et al. (2001) have discussed the recognition of the said OAE in South Tibet.

7.7.8 Albian Succession and the Evidences of the Middle Albian Anoxic Event in India

The focus here is on the globally expressed mid-Middle Albian OAE. The start of the last first-order sequence on the Indian margin in basal Aptian is evidenced through the Early Aptian–Middle Albian ammonoids, and planktonic forams, coccoliths and dinocysts in Pakistan, Kachchh, Jaisalmer, Spiti, Cauvery and K-G basins during its first TST. The said ammonoid record required minimum depth of ~several tens of metres near Abur and Ukra localities, and yet more in other basins.

These and coeval levels down the basin in subsurface require geochemical studies to strengthen the paleontological evidences of the globally recognized OAE. The deeper distal facies down the basin hold good promise as hydrocarbon source rock interval. The volcanic connection is explicit. It is no coincidence that the later phase of volcanism in the east divergent sector (Rajmahal LIP province) in Jharkhand, Bengal, Bangladesh, Meghalaya, Naturalist Plateau, Bunbury related to the Kerguelen hot spot ranged through the second-order Early Aptian–Middle Albian TST duration with youngest ages around ~107 ma. It is also noteworthy that the basin margins were reached by the Neotethys only at the peak of the TST and not earlier in Aptian in Kachchh.

7.7.9 OAE Determination in Spiti

In the Chichhim section of Spiti, Bertle and Suttner suspected OAE 1d higher up near the Albian/Cenomanian boundary. It could also be interpreted as close to the Middle/Late Albian transition immediately above the Giumal Formation, about ~8 m above the ammonoid rich part of the Giumal Formation. The juvenile Indian Ocean in spite of initiation at the Barremian/Aptian boundary grew only slowly until the mid-Middle Albian second-order MFS. The intermittent incoming, exchange and mingling of the cooler South Pacific currents is obvious. In spite of the newly developed intercommunication with South Pacific, the communication with Neotethys was never lost. These intermittent communications have influenced the fine cyclic stratigraphy in the regional successions of the second-order TST. The microfloral spatio-temporal distribution, also, seem to support similar anoxic scenario. The South Atlantic like the Indian Ocean after origin near the Barremian–Aptian boundary under similar episodic volcanic time frame, supposedly, also grew slowly.

7.7.10 Mid-Middle Albian MFS Anoxic Event in Cauvery

The Grey Shale / Reefal Limestone contact in the Kalakuddy quarry in the proximal part of Cauvery basin suggests MFS of mid-Middle Albian age. The Grey Shale nannoflora is ecologically controlled in stilled basin conditions with periodic enrichment of nutrients with resultant alternation of organically rich dark shales and nannoplankton rich carbonates. The composition is close to that found elsewhere in the region of the global mid-Middle Albian anoxic event 1c (Jafar and Rai 1989; Bralower et al. 1993; Kale et al. 2000). Kale et al. (Op Cit) based on nannoflora also interpreted change of Cauvery basin latitude from high cool temperate to somewhat warmer lower latitudes. The said change of latitude is linked to the appreciable due northwest/north spreading along with anticlockwise rotation of India inclusive of Madagascar after separation of Elan Bank and South Kerguelen block. Thus, the mid-Middle Albian MFS represents an equally or even more important tectonic event than the subsequent separation of Madagascar and India.

The Karai Shale in outcrops includes nodular/concretionary intervals almost through out, and the nodules often include ammonoids in their nucleus. The *Mortoniceras* bearing nodules in the middle part are noteworthy, and are interpreted as terminal part of the next second-order TST. The top of another nodular interval and stratigraphically associated dark shales in the upper part of the Karai unit seem to mark the significant MFS of Middle to Late Cenomanian or even Early Turonian age. The benthic forams in wells suggest maximum bathymetry near the close of the Sattapadi Shale (equivalent of Karai Formation in wells) in the mid Early Turonian Helvetica Zone (Chidambaram 2000). In the Kunnam section, Ayyasami (2011) and Ayyasami and Bannerjee (1984) have placed the Cenomanian/Turonian boundary in the upper part of the Karai Formation on the basis of the Late Cenomanian index ammonoids below and Early Turonian ammonoids above. In the relatively proximal Kallakuddy quarry, the thick carbonates are again followed up by shales, and Nagendra et al. (2002, 2011) have recorded Early Turonian planktonic forams. They also suggested MFS at the close of the shales. The Karai Shale unit based on included shales does not suggest any major bathymetric changes. Most of the upper part of the Karai Shale unit but for the top few meters is considered of Cenomanian age. On the other hand, the global eustatic framework marks steady eustatic rise in the Cretaceous up to close of Turonian. Ayyasami and Rao (1978) interpreted faunal break at the contact of the Trichnopoli and Ariyalur units.

7.7.11 Asymmetric Inferior Fourth- to Sixth-Order Sequences in the Kachchh Mesozoic

The marine mega-guide-body fossil-devoid and mega-plant fossil-rich Ghuneri Member and Upper Member of the Kachchh Cretaceous sediments were for over a century since Wynne (1872) considered non-marine. These and similar sediments on the Indian plate margin were reinterpreted as essentially marine, and formed by shallow marine nearshore depositional processes (Krishna et al. 1983, Krishna 1987c, 2000 and in between). The individual inferior-order cycles included marker/distinct thin maroon coloured densely bioturbated pebble/concretion/nodule bearing conglomeratic beds as TSTs and ten or more fold thicker RSTs. The RSTs comprised shale/silt/fine sand in the early part followed up by coarse sand during somewhat stabilized to decreasing depths in a regressive framework. The disparity in the sedimentation rate and thickness of the TSTs and RSTs of the individual sequences was several fold and the sequences represented highly skewed and asymmetrical features. The early part of the RSTs featured wavy laminations, smaller tabular cross-bedding, ripple-drift cross-bedding, flaser bedding etc. The later part of the coarse sands was characterized by bigger sized planar trough cross-stratification, hummocky cross-stratification, large tidal channels and inlets, herring bone cross-bedding and sigmoidal bundles. The RSTs have been interpreted as high energy, above wave base, rapidly sedimented in protected coastal prograding tidal complex of tidal flats, lagoons, estuaries, etc. The physically near identical multistoried development suggests uniformity of environment, however, in a fast receding shoreline, the individual sequences cannot be correlated from one section to another, and are highly diachronous, and rapidly younging down the basin. The asymmetric sequences are quite revealing in context of ichno-fauna as well. There is increasing dominance of Skolithos ichnofacies temporally upward and spatially margin ward and of Cruziana ichnofacies down the basin. On the other hand the occurrence of thin condensed densely bioturbated marine body fossil-rich coeval TSTs are known in other basins as well on the Indian plate margin (Jaisalmer, Pakistan, Spiti, Kumaon, Nepal, K-G, Cauvery).

In the author's study of the Kachchh Mesozoics, similar asymmetric sequences have been recognized in the Aalenian (Figs. 2.25–2.26), Early Bathonian (Figs. 2.25–2.26), Middle Callovian (Fig. 2.29), terminal Late Callovian–?early Early Oxfordian (Fig. 2.33), terminal Kimmeridgian (FPs 2.6-7, Figs. 2.36–2.37), Early Tithonian (Fig. 2.39), Hautervian–Barremian (Fig. 2.41), post-Albian part of the Cretaceous (Fig. 2.41). These hold significance and lately have also proven as hydrocarbon reservoirs.

7.7.12 *Possible Gap Between the Umia Formation Below and the Mundra/Naliya Units Above in Offshore Wells*

Both the shallow marine Mundhra Formation in well SP-1-1 of suggestive Santonian age, and the middle to outer neritic Naliya unit in wells of the Coniacian–Santonian age have been interpreted to conformably overly the Umia Formation (ONGC 1993) of suggestive Neocomian–Santonian age. However, Raju (2011) has suggested a break in resistivity and gamma ray logs between the Umia and Naliya units. All the three offshore units (Mundhra, Naliya, and Kori) are of mid to late Late Cretaceous ages in the Coniacian–Maestrichtian age bracket which is included in the Cretaceous part of the terminal first-order late Middle Turonian–Early Paleogene RST. In this RST interval, the Indian plate is spreading due north, and the plate is surmised to have maintained its tilted down posture of its north margin during the early Middle Turonian–Early Paleogene interval with significant change in the direction of drift in steps during the first-order Aptian–early Middle Turonian TST. The Cauvery basin post intra-Cretaceous MFS succession provides support to the surmise. However, the net effect of the mutually opposing eustatic fall and tectonic submergence is not explicitly clear in the Kachchh basin. The well K-1-1 is about 50 km down the dip from well SP-1-1, and distinctly greater bathymetry is expected in K-1-1 well. Nevertheless, tectonic submergence at the second-order MFS is not ruled out. The picture underneath the Kori and Naliya units about the presence or absence of Turonian is not clear, since the wells perhaps did not penetrate below the Santonian.

Based on subsurface microfossil studies, The Naliya and Kori units have been suggested middle shelf to upper bathyal bathymetry respectively in offshore well SP-1-1, KD-1, and K-1-1A while the somewhat older or coeval Mundhra unit in SP-1-1 is considered marginal marine (Raju 2011). The said three units are age ranged in the Coniacian–Maestrichtian part of the first-order RST. Deepening is conceived at least in offshore at the second-order MFS Santonian/Campanian boundary SB, and then shallowing again in Maestrichtian.

In the Banni well, the relatively fine textured subsurface Bibat Formation of Kimmeridgian–Valanginian age is unconformably overlain by the Neogene units, and the vast gap in between is decidedly subareal. However, the stratigraphic framework in the Banni graben does not provide any clues to happenings in the offshore wells ~250 km down the dip.

7.7.13 *Presence of Thin Basalt Layers in a Few Offshore Wells*

The Mundhra Formation inclusive of thin basalt layers is unconformably overlain by the Deccan traps. The basal part of the unit based on coccoliths is of Santonian/Campanian age. Thus, the thin basalt layers included in the unit may be Santonian. If so, these are distinctly older to the unconformably overlying Deccan basalts proper. In context of the intermittent volcanism in Madagascar up to Late Santonian / Early Campanian, these two basalt intervals present in Kachchh suggest two distinct igneous events. The mafic gabbro intrusive in Kachchh basin 'island belt' also have yielded ages (Early Campanian ~81 ma to Paul et al. 2012) way older to the Deccan volcanism, so also the St Marys and other igneous expressions of Karnataka, Kerala, and Tamilnadu on one hand and Mananjari volcanics of East Central Madagascar, and Morondova, Majunga, and Diego volcanics on West Madagascar in the ~94 to ~84 ma age bracket. The older one may be related to the rift-related event marking the initiation of spreading between Madagascar and India, while the later event—the principal one of India—riding over the Reunion plume. Thus, the presence of volcanic flows in the Mundhra Formation in SP-1-1 well of Kachchh offshore of nannplankton based Santonian age is considered significant in a regional frame. Similar basalts of presumable Turonian–Santonian ages are also known in a few other Kachchh offshore wells (Lakhpat–1, Suthri, GK-22-C-1, GK-20-1, GK-36-1 and GK-1-2) (www.dghindia.org/17aspx post DGH 2003). Intra-Turonian volcanics are also recorded in Mannar, and Kerala offshore as evidence of the second-order Turonian sequence (Fig. 7.40). The Santonian volcanism is also known in Narmada basin (Biswas 2005) between the marine Bagh succession below and the largely non-marine Lameta succession above. A few Kachchh wells are mentioned to include even Early Cretaceous volcanics stringers intercalated in the sediments, suggestively of Aptian – mid-Middle Albian age.

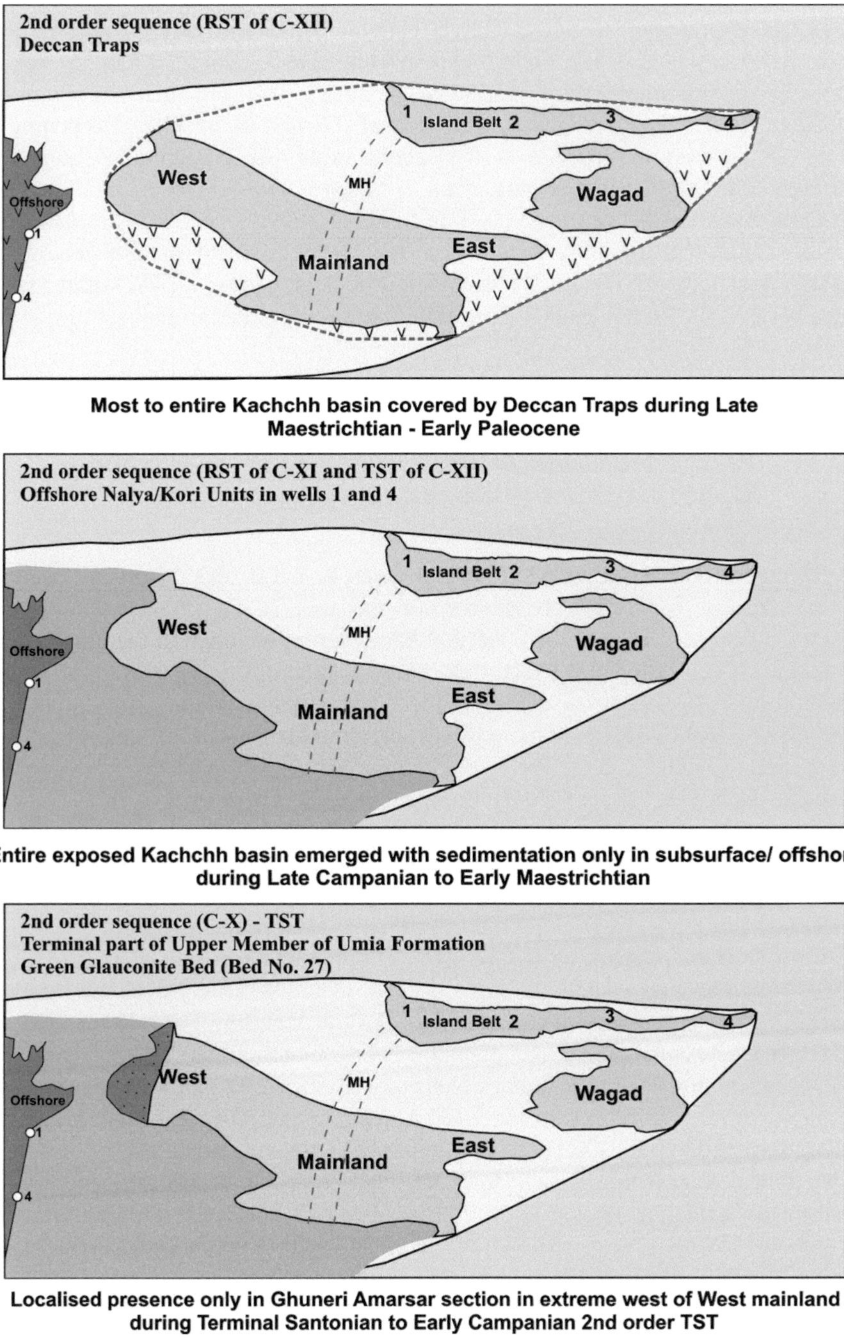

Fig. 7.40 Graphics from bottom to top respectively refer to terminal Santonian to Early Campanian, Late Campanian to Early Maestrichtian, and Late Maestrichtian to Early Paleocene

7.7.14 Possible Convergence of NW India to Either SE Arabia or Kohistan–Ladakh Arc

The said second-order SB igneous manifestation widely in the west sector as above, in addition of marking the oceanic spreading between India, and Madagascar, also, dates the possible convergence event and the resultant folding/doming of the Kachchh Mesozoics. Chatterjee et al. 2013 have invoked the initial collision of India with the Kohistan–Ladakh arc at ~84 ma, i.e. soon after the final separation of Madagascar from India to allow the large similarity of the terminal Cretaceous Campanian–Maastrichtian land vertebrate similarity and intermigration which would not have been otherwise possible had India been oceanically isolated after separation from West Gondwana as also Australo-Antarctica. Gaina et al. (2015) also have interpreted convergence between India and Arabia during ~90 to ~83 ma span. It does not seem improbable that the oceanically isolated India during its due north drifting may have converged either to Southeast Arabia or to the Kohistan–Ladakh arc framework, although evidence to this effect is lacking apart from the vertebrate fossil intermigration constraints. The said constraints are otherwise explained through land bridges in more than one sector, and do not necessitate any convergence. The uplift of the Gondwana basins and the Satpura belt (Biswas 1999) appears to be linked to the widespread geological event in this interval, so also the extensional event (Bladon Andrew et al. 2015) in Barmer, and resumption of non-marine sedimentation after rift-related submergence in Bengal basin (Roy and Chatterjee 2015). The Narmada (Biswas 1999) and Deccan (Chandrasekhar et al. 2011) basins also include stratigraphic gaps with or without igneous expressions, which are here linked to the ~84 ma multifaceted geological event.

7.7.15 Protracted Duration of the Igneous Activities During the Late Permian–intra-Paleocene Sequence

The well known individual and well differentiated igneous activities around the Indian plate in the intra-Permian–intra-Paleogene mega-sequence in this work have been ranged through two or more stages, i.e. ~10 my or more in duration. The separation of Madagascar from India, and associated ride over Marion hot spot, followed by separation of Seychelles and associated ride over Reunion plumes range in age between ~94 and ~60 ma. Even short compressive/convergence events between the spreading India and the Arabian plate have been suggested during the Cretaceous prior to the India–Asia suturing in Paleogene (Gaina et al. 2015). The Mundhra volcanics are considered related to the culmination of the India–Madagascar separation event.

7.7.16 Suggestive Evidence of the Coniacian–Intra-Campanian Second-Order Sequence

The sand dominated Mundhra Formation is interspersed with shales including carbonaceous ones, and suggests formation in shallow marine paralic environment. The coccolith based Santonian–Campanian age (Raju 2011) suggests its broad equivalence to the Sillakudi Formation of the Cauvery basin, which there represents the Late Santonian–Campanian second-order sequence (Fig. 7.42). The late Early Campanian MFS is suggested in the Cauvery and Spiti basins. In the Chichhim section of the Spiti basin (Bertle and Suttner 2005), the said MFS is placed above the marker hard ground between the Lower Limestone and Upper Shale units. In the Cauvery basin, it is determined in the terminal fossiliferous part of the Sillakuddi Formation underlying the Kaller conglomerate. Campanian nannoplanktons have also been recorded in the Chikkim Shale of Spiti (Rai et al. 2015), and are considered here to mark the intra-Campanian second-order MFS. A rich succession of ~15 planktic foram assemblages and planktic zones have found excellent documentation in the relatively distal and deeper Zanda section of Southwest Tibet (Li et al. 2009) from Late Albian to close of Maestrichtian, yet radiolarians are absent. The associated record of rich radiolarian is recorded in the Gamba section farther east in the anoxic organically rich black shales.

7.7.17 Presumable Record of the Terminal Cretaceous Second-Order Sequences

The Kori and Naliya units, both of mid-outer shelf to upper bathyal environment and Campanian–Maestrichtian age, unconformably overly the Upper Member of the Umia Formation. These represent the youngest, the deepest and the farthest offshore sediments of the Kachchh Mesozoic. The early parts are Early Campanian planktonic foram bearing (Biswas 1990), while the upper part is extremely scarce in such forams presumably in the second-order Late Campanian RST. These sediments in the K-1 and K-2 wells make the successive two (Campanian and Maastrichtian) second-order sequences that are in sequence stratigraphic context correlatable to the two successive sequences of Cauvery basin of the Ariyalur Group of late Late Cretaceous age. The marked shallowing in the Kachchh Maastrichtian is ascribed to eustatic fall during the later part of the first-order RST interval. The Kachchh offshore units in turn are unconformably overlain by the Late Maestrichtian Deccan traps. In Zanskar, and several sections of South Tibet close to the Indus–Psangpo suture zone, the Early Paleogene is also preserved with signatures of the initial impingement of the drifting India to Asia.

7.7.18 On the Shallow Marine Deccan Seaway Along the Narmada Fracture from East to West

Linear corridor seaways contextual to India across the Deccan have been postulated in recent decades without clinching evidence (intra-Maestrichtian, Sahni 1984; basal Danian, Keller et al. 2009).The subject is examined in context of the sequence stratigraphic framework developed in the present work. Also are considered the inherent paleogeographic constraints through the Phanerozoic, particularly, during the span of the Late Permian–intra-Paleogene mega-sequence. It is also desirable to differentiate between arms of the seas and seaways. A corridor seaway essentially implies a shallow marine transient corridor across the land from one shallow sea to another, instead of being closed at one end like the interpreted transient Deccan seaway in the basal Danian, which requires revised address only as the Deccan arm of the Neotethys. The short duration corridors are mostly precursors to later origin of the new oceans. Such corridors during their entire short duration existence could be interrupted either by sea level falls or tectonic uplift in between the two ends of the corridor seaway. The arms of the oceans/seas may or may not progress to shallow marine corridor seaways. The time involved in such transformation also is variable, so also the duration of the corridors prior to termination through the origin of a new ocean. The maximum possibilities for such corridors occur during the terminal part of the TSTs around the first/second-order MFSs, invariably in combination with tectonic submergence. In context of the graded increase of sea levels during the Mesozoic up to mid-Late Cretaceous, the chances of such corridor seaways are better in the Cretaceous than either in Jurassic or Triassic below and Paleogene above (Fig. 7.41).

7.7.19 On the Shallow Transient Corridor Seaways

Krishna (1987b, 1996, 2005, 2009, 2012) demonstrated such true shallow marine transient/intermittent seaways (particularly at eustatically influenced highstands at first/second-order MFSs) across the Gondwana. These are the Gondic and Dravidian corridors. The Gondic corridor ran from the IEAP to the Southeast Pacific Gondwanian margin along the divergent margin between East and West Gondwana via the East and Southeast African coasts (Krishna 1983; Krishna 1987b; 1996 and many later papers) during the terminal Late Tithonian. The Dravidian corridor during Early Aptian–mid-Middle Albian extended along the divergent region between India and Australo-Antarctica from the Neotethyan to the Pacific margins of the East Gondwana (Krishna 2005, 2009, 2012). These interpretations have withstood the test of time, and appear even vindicated through the recent geographic reconstructions based on highly constrained voluminous

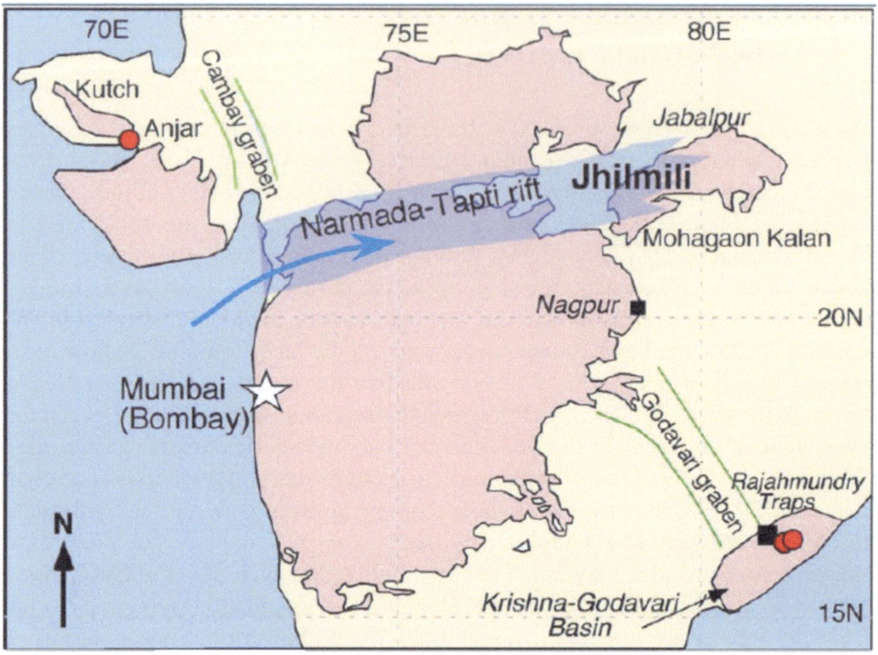

Fig. 7.41 Deccan Seaway of Keller et al. 2009

geophysical and other data (Gibbons et al. 2012, 2013; Alberti et al. 2012b; Gaina et al. 2015 and relevant references therein).

7.7.20 Origin and Progression of the Marine Arms in the Divergent Sectors West and East of India

The Mesozoic rifting along the divergent zones began first in the west sector at the first-order intra-Jurassic SB at the start of Toarcian, i.e. ~183 ma with initiation of marine transgression on the GTM in IEAP as also on the South Pacific margin of the Gondwana in the South Africa–Antarctica–South America region. These shallow marine arms did progress, respectively, due south and due north mostly by way of transform sliding of India against Africa along the Davies fracture, yet could not unite until the terminal Late Tithonian on account of the Mozambique ridge land barrier in between. Any appreciable or desirable spreading across the blocks supposedly did not take place. Even the initiation of the interpreted spreading since the first-order intra-Oxfordian MFS did not catalyze the meeting of the said marine arms into a corridor seaway.

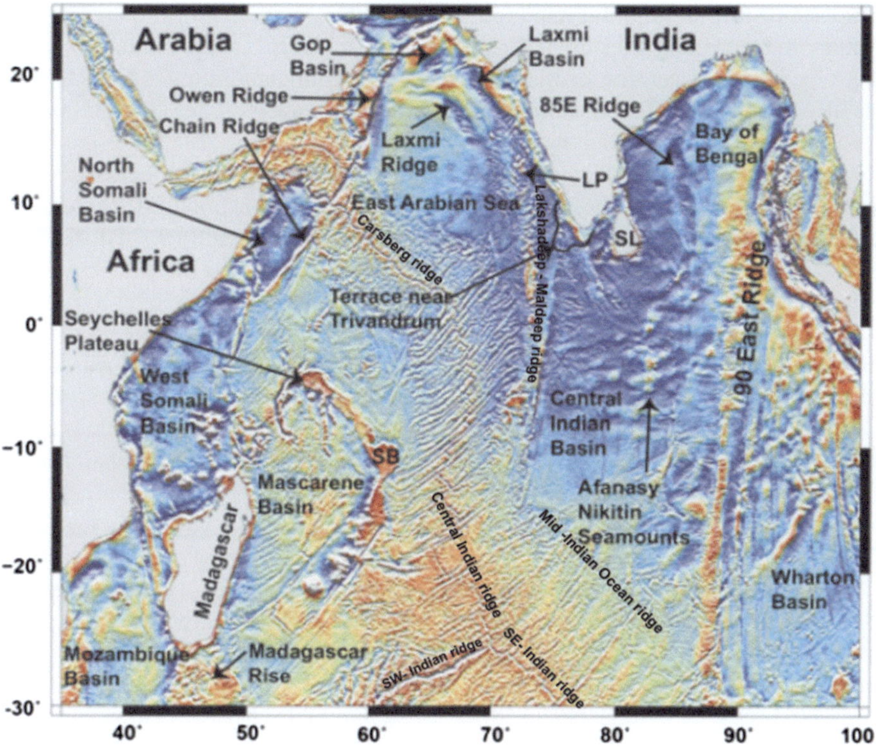

Fig. 7.42 Indian Ocean in East and West of India modified after Gibbons et al. 2013

7.7.21 Gondic Corridor During the Second-Order Terminal Late Tithonian–Early Early Berriasian TST

The rift-related partial/total submergence of the Mozambique ridge allowed the first shallow water ammonoid intercommunication between the two arms in the terminal Late Tithonian second-order TST of common *Argentiniceras, Corongoceras, Durangites* and others in the Argentinian SE Pacific and Neotethyan IEAP, and thus originated the Gondic corridor seaway (Fig. 1.3) as already detailed in Chap. 3. The recent reconstructions (Seton et al. 2012) at ~160 ma, i.e. Middle Oxfordian did not initiate any spreading across the rift axis, but did show slow spreading at ~140 ma, i.e. during the basal Cretaceous Berriasian stage. The two arms did progress mutually towards each other, yet, the respective spreading axes are not shown united. Konig and Jokat (2006) interpreted initiation of the true oceanic Weddell Sea in the South Africa–Antarctica–South America region near ~147 ma, i.e. late Late Tithonian. Thus, the Gondic corridor may have lasted until the early Early

Berriasian MFS through the initiation of the Weddell Sea. At ~120 ma, i.e. Early Aptian, the spreading axes are indicated to have united up to the Southeast Pacific, there by implying ceasure of the shallow marine ammonoid communication from the IEAP to SE Pacific, and consequential disappearance of the Gondic corridor seaway. The author's long held contention (Krishna since 1983) of the space constrained denial of any appreciable/desirable spreading/distanting across the said divergent region west of India is vindicated in recent reconstructions (Seton et al. 2012; Gaina et al. 2015; Gibbons et al. 2015) during the ~160 to ~80 ma span. The distanting across the spreading blocks in the east divergent sector is more than twice over a period of ~40 my in contrast to that of the west sector over a period of ~80 my, i.e. the net crudely estimated average spreading rate in the east sector is four times that of the west sector. In fact, most of the spreading in the west sector has resulted from transform sliding, and much less on account of actual spreading of the mid-oceanic ridge across India and Africa. Even the slow spreading in the west sector is interpreted to have ceased at the close of the terminal Middle Oxfordian–Barremian first-order RST interval at ~126 ma or even earlier at the intra-Valanginian second-order ~136 ma MFS. Gaina et al. (2015) connected the mid-oceanic ridge system in the West Somali basin, and Mozambique basin–Riiser Larson sea region to the Southeast Africa–East Antarctica spreading axis via the Davies Fracture transform fault in their ~130 ma reconstruction. Thus, the spreading is implied to have ceased during the ~136 to ~127 ma span, since spreading is not indicated in the ~120 ma reconstruction. Gaina et al. (2015) have even interpreted an extremely short/transient convergence between Arabia and NW India around ~130 ma which they have related to the uplift of the Masriha ophiolites (Immenhauser 1996). The uplift related convergence, if at all possible, may be related to either the ~136 or ~126 ma sequence surface events.

7.7.22 *Dravidian Corridor During the Second-Order Basal Aptian–mid-Middle Albian TST*

The Dravidian corridor originated in the east sector between India as a constituent of Axial/Inner Gondwana and Australo-Antarctica as a constituent of Outer Gondwana from the Gondwana's Neotethyan margin to its South Pacific margin round the SE Indian (Dravidian) margin. The said origin is evidenced through intermigration of the geographically restricted shallow marine ancycloceratin ammonoids, *Australiceras* and others, from its India–Australia Neotethyan end to the South Pacific Antarctica and Argentina (Fig. 1.3). The shallow marine *Australiceras* with upper bathyal as its maximum depth tolerance is otherwise present both in India and Australia. This in turn attests the existence of the shallow marine Dravidian corridor immediately prior to the initiation of the appreciable rate of spreading between the Axial Gondwana and Outer Gondwna during the Late Aptian to mid-Middle Albian period. The transient existence of the Dravidian corridor appears linked to the initiation and closure of the Rajmahal volcanism. The Dravidian corridor may even have transiently reached the Rajmahal basin.

In case of the Dravidian corridor, two shallow marine arms, respectively, of the South Pacific and Neotethys progressively moved due north and due south towards each other since the late Middle Oxfordian first-order MFS. At the first-order Barremian/Aptian boundary SB after the closure of spreading in the west sector, the unzipping in the sector began its final leg. It prompted the progression/revival of spreading between India and Antarctica in the Enderby–Elan-Bank–South Kerguelen region. This final assault was triggered by the initiation of the Rajmahal/Kerguelen volcanism as evidenced at Bunbury offshore Perth on the West Australian and Enderby on the Antarctica on the Australo-Antarctica side and Meghalaya–Bengal–Jharkhand on the Indian side. The igneous event may have also progressed due east in Southeast Tibet and High Himalaya. At the start of the said final assault, also, may have originated the shallow marine Dravidian corridor seaway. The origin of the corridor appears facilitated by the transform sliding due north-northwest along the east margin of Sri-Lanka–Cauvery region somewhat parallel to the transform margins east and west of Madagascar. This is attested through the distribution of the common Late Aptian *Australiceras* and others on either side of the east divergent sector, i.e. the Indian east and west coast basins on one side and Australo-Antarctica on the other side. The spreading was slow and mostly by way of transform sliding until the mid-Middle Albian second-order MFS. The Aptian ammonoid evidence is present only up to the Godavari basin, and not any farther north in Mahanadi, Bengal and Northeast India or in the West and Northwest Australia. However, diverse other geological evidences are quite convincing. Also, is, extremely, significant, that the Gondic and Dravidian shallow marine corridors were precursors to the subsequent origin and growth of the Indian Ocean in the west and east sectors along with the associated volcanism and sequential oceanic spreading among the Gondwana constituents. The said final leg of spreading is begun at ~126 in Gibbons et al. (2013) and extended up to ~120 ma, however, even at ~120 ma the said spreading is yet not connected to the spreading between South Africa and East Antarctica, which is realized yet later at ~107 ma. Such a spreading framework accords well with the duration of the Dravidian corridor seaway during the Early Aptian, i.e. ~126 ma to ~107 ma, i.e. mid-Middle Albian. Seton et al. (2012) indicate the same in their reconstruction at ~120 ma, while Gaina et al. (2015) between ~130 and ~120 ma. The volcanism and associated oceanic spreading conceptually are, however, not considered obligatory to the origin of such shallow marine corridors. The only pre-requisite to such transient shallow marine corridors is extensional tectonics, and high sea levels at first/second-order MFSs all along the future corridors that could be crustal sagging/thinning/attenuation/stretching/rifting without that is rather impossible to create the corridors. The disappearance of the corridors follows the spreading across the blocks at appreciable spreading rates and associated development of intervening oceanic barriers.

7.7.23 Possibilities of the Deccan Seaway Through the Phanerozoic

The possibilities of the Deccan Sea along the NSL and Godavari fracture are examined through the Phanerozoic in some detail in the backdrop of the earlier discussion on the Gondic and Dravidian corridors.

7.7.24 Paleozoic (Early Permian)

The Paleotethyan GTM witnessed major widespread revival of rifting at the Carboniferous/Permian boundary second-order SB, which on expected lines as a major event also affected the ancient Precambrian inland Gondwana fracture framework, particularly, along the Narmada–Son–Damodar, Son–Mahanadi, and Pranhita–Godavari lineaments. The Narmada lineament had` earlier during the long Cambrian–Carboniferous span remained a positive/uplifted region, away from the reach of the Paleotethys. The said rifting event in combination with post-glaciation melting was followed by a major transgression in Early Permian, which for the first time not only invaded the west sector up to Pakistan and Jaisalmer, but also made longer and multiple inroads in the east sector up to Umaria through Daltonganj, Manaendragarh, West Bengal, Arunachal, etc., from the northeast in the inland rifted basins.

In reconstructions, Krishna (Krishna 1986, 1987b, and later) suspected marine connection with question mark between the marine Early Permian Paleotethyan arm from the northwest in West Rajasthan to the coeval marine arm from the northeast in Umaria, Manendragarh and Daltonganj through the Sanchore, Cambay, and Narmada grabens. However, there is, as yet, no conclusive evidence available of the union or mingling of the said reaches of the marine arms due east from the west and due west from the east. In context of the suggested transient/ephemeral extension of an arm of the Neotethys along the Narmada up to the Lameta Formation in Jabalpur during Campanian/Maestrichtian (Singh 1981; Rai et al. 2002; Shukla and Srivastava 2008), the early Early marine Permian of Umaria might have been, if at all, as well connected to the west as part of the Narmada rift, as to Bihar/Bengal/Sikkim/Bhutan in the east. Such a connection with east is also indicated in Chatterjee and Hotton (1986). The Neotethys originated at the start of the Early to Late Permian, however, nothing to speak of the evidence of marine Early Permian in the Barmer, Sanchore, Cambay, and Narmada basins, these basins even have lacked any evidence of the suggested transgression any time later in the younger Late Permian–late Middle Oxfordian interval. The earliest suggestive yet unconfirmed suggestion of marine conditions is provided by thin marine fossil bearing Jurassic within the Upper Member of the Dhrangadhra Formation in the Lodhika well of Saurashtra basin, and suspected marine strand line tidal flat origin of the otherwise marine body fossil-devoid Nimar Formation sensu stricto in the Narmada graben basin also best age bracketed in the terminal Middle Oxfordian–Barremian. In the Lodhika well are also suspected, non-marine sediments of the Late Permian–Early Jurassic first-order sequence. Reworked Late Permian are also known in Pakistan, Kachchh, and Kerala–Konkan. Even indirect geophysical surveys have surmised non-marine Permian sediments immediately south of the Narmada graben in the northwest part of the Deccan syneclise (DGH docket). Thus, the reactivation of the Narmada rift as early as Permian cannot be ruled out. However, until clinching evidence emerges, the early Early Permian Deccan seaway along the Narmada rift would only remain doubtful and inconclusive, nothing more than a mere possibility as has been the case all these last decades.

7.7.25 Mesozoic

The Mesozoic possibilities of the Deccan Seaway are discussed during Pre-Aptian Mesozoic and post-Barremian Mesozoic.

7.7.25.1 Pre-Aptian Mesozoic

Marine Triassic is not known anywhere in the vicinity of the Narmada-Son-Damodar fracture either in the east or west, and even remote question of the Deccan corridor at the mid-Triassic first-order MFS simply does not arise. Around this time the Narmada fracture is interpreted as emerged. Even non-marine sedimentation is exclusive to the Satpura region. Much to the contrary, marine conditions existed in the west sector in Jaisalmer and Pakistan, perhaps, also in Saurashtra, and may be a transient one even in Narmada at the intra-Oxfordian 1st order MFS. In the east sector, marine conditions are altogether non-existent in the Jurassic but for unconfirmed terminal Jurassic in the newly opened coastal basins. Any corridor sea along Narmada is out of question.

7.7.25.2 Post-Barremian Mesozoic (Early Aptian)

The next significant rifting of consequence in this region is interpreted at the first-order intra-Cretaceous Barremian/Aptian boundary SB which possibly also reactivated the Narmada fracture zone in the west. The Indian East coast K-G and other basins underwent further deepening along with more open sea and widespread marine sedimentation in the Aptian–intra-Paleogene first-order sequence age bracket. However, the said east coast basins do not provide any evidence of the extension of marine sediments in the old Gondwana rifts, since the same probably had by then uplifted, beyond the reach of the sea. Even non-marine sedimentation had ceased in the said basins after the uplift at the first-order intra-Jurassic SB. Casshyap and Khan (2000) have suggested reversal of paleoslope from due north of the Denwa Formation of Ladinian–Pliensbachian span to due south in the disconformably overlying Bagra unit of the terminal Middle Oxfordian–Barremian age bracket. A remote possibility of the tansient reach of the nascent Indian Ocean to the Rajmahal basin is already discussed earlier. The Early Aptian, thus, does not offer any concrete possibility of the Deccan seaway between the Narmada and the old NW–SE striking K-G rift. Moreover, definite marine Aptian is also unknown along the Narmada basin.

The mid-Middle Albian MFS is known to mark widespread rifting and magmatism all round India in the east, north and west, and both tectonic uplifts and submergences along ancient weak zones are possible. Marine Late Aptian/Early Albian seems to be present in the K-G basin through rare guide ammonoids and dinocysts. Cauvery includes rich Albian ammonoid succession. However, any marine signatures are absent in the old Gondwana rift basins in the east sector. Narmada and Konkan–Kerala also lack definite marine Aptian–Albian, and thus a corridor seaway is ruled out.

The unsubstantiated postulations (Sahni 1984; Rai et al. 2002; Keller et al. 2009) relate to the first-order RST interval from the Middle Turonian Helvetica Zone to the intra-Thanetian P4 CP I / CP 2 boundary RST of the Late Permian–intra-Paleogene mega-sequence. The first to second-order framework developed here includes plus five second-order sequences, plus three sequences in the Cretaceous part and the remaining two in the Paleocene part of the first-order RST. The second-order eustatic highs are known in the early Middle Turonian ($\sim$250 m), intra-Coniacian, Early Campanian ($\sim$230 m), Early Maestrichtian, early Early Danian ($\sim$200 m), and mid-Thanetian $\sim$185 m) (Raju 2011) which signal the anticipated eustatically influenced decrease in bathymetry during the first-order RST. Additionally, the extensional tectonics at sequence surfaces, also, influenced increase or decrease of bathymetry in the basins. Among these, the intra-Turonian is the major MFS, while the intra-Campanian, intra-Danian and intra-Thanetian make the second-order MFSs during the first-order RST span of the graded eustatic fall. During the Cretaceous, the best possibility in eustatic context is at the intra-Turonian MFS. However, the absence of the Cenomanian–Turonian ammonoids around the intra-Turonian first-order MFS in the east sector, except Cauvery, does not favour, any ammonoid exchange through the supposed seaway with Narmada. The rich Cauvery basin Cenomanian–Maestrichtian ammonoid succession perhaps is prevented even in the neighbouring K-G basin by a possible bathymetry barrier at the intra-Turonian MFS. Even a transient penetration of the shore line seems unlikely in the Narmada and Godavari rifts, respectively, from the west and southeast, particularly in view of the interpreted pre-Turonian uplift in steps.

A short transient above wave base seaway, if at all present, is still best favoured in Early Campanian second-order MFS in context of a possible rift-genic submergence, and associated widespread transgression in Narmada up to Chakrud, K-G, Mahanadi, Bengal and Assam (Garo hills). The interpretation of lagoonal framework (Singh 1981; Shukla and Srivastava 2008) and associated marine depositional processes of a near shore marine depositional system in the Lameta Formation (Singh 1981) seem to support the Early Campanian nannoplankton evidence in the Lameta Formation at Chakrud in the neighbourhood (Rai et al. 2002), but not without transient intra-Campanian rift episode. It need be born in composite interpretations, that the infra-trappean Lameta Formation may have had diachronous base and top, respectively; the base is somewhat older in areas that escaped arrival of the last flows of the Turonian–Santonian volcanics, and the top is somewhat younger in areas that did not receive the oldest Deccan flows.

7.7.26 Cenozoic (Early Danian)

The basal Early Danian seaway has yet lesser chance in view of the inclusion of the interval in the terminal or later part of the first-order RST. Marine basal Danian P0-P1a Zones are neither recognized in the Assam–Arakan–Bengal–Mahanadi, nor in Kachchh, Narmada, Saurashtra, Jaisamer of the west sector. In fact, there is absence of the underlying Late Maestrichtian as well in the west sector, particularly in Narmada, and Kachchh. Jafar (2015) has also discounted the basal Danian Deccan seaway. In the K-G basin the evidence is mixed in favour and against, however, the regional framework does not seem to favour a widespread Early Danian transgression. It may be restricted to Cauvery and Godavari coastal near-shore belt, and certainly not up to Jhilmili. However, the Early Campanian nannoplankton record from the infra-trappeans and Early Danian planktonic forams from the intertrappeans are evidences that require a marine passage, most probably from the west, which in case of Early Campanian makes a better proposition. Govindan and Ravindran (1996), and Ramkumar et al. (2004) also recognized the Early Campanian Ventricosa Zone MFS in the Ariyalur–Pondicherry subbasin.

The early Danian incursion necessarily also has to be from the west, although evidence at present is lacking. The widespread Campanian transgression is of $\sim$2 my compared to the supposed basal Danian of $\sim$200 ky, which is quite in tune with anticipation in a phase of a rapid eustatic fall immediately after the second-order MFS, and associated fast receding shore line. Moreover, the Santonian/Campanian boundary second-order SB extensional event is evidenced widely on the Indian margin in the west sector in the Kachchh offshore relatively deeper Campanian Kori unit over the Turonian–Santonian shallower Mundhra Formation, so also in Narmada. The upper limit of the Bagh Formation is interpreted up to Santonian. The infra-trappean Lameta Formation is ranged through the Campanian–Early Maestrichtian with possible episodic transient east ward progression of the shore-line even up to Jabalpur. The intertrappean Deccan Formation is ranged up to the terminal Danian second-order MFS of Trinidadensis/Uncinata zonal boundary. Being away from the source of the eruption, the flows encountered in the Jhilmili section may neither include the oldest nor the youngest ones.

7.7.27 Duration of the Deccan Volcanism in Sequence Stratigraphic Context

In context of the rather protracted nature of the principal igneous activities in the intra-Permian–intra-Paleogene mega-sequence in Permo-Triassic, Late Jurassic–Early Cretaceous, mid-Late Cretaceous on the north, east and south margin of the Indian plate, the Deccan volcanism for its short duration terminal Maestrichtian peak is also suggested to similarly have started distinctly earlier and closed distinctly later to the said climax. There is also the aspect that all the flows do not reach the peripheral extent of the spread, particularly so, the oldest and the youngest ones. There is unanimity with regard to the principal phase of Deccan volcanism of terminal Late Maestrichtian of $\sim$1–2 my. The said volcanic events are here interpreted to have begun, climaxed, and closed at first or second-order sequence surfaces. The Deccan volcanism thus may have initiated latest at C33/C32 second-order MFS, and closed at the earliest at the second-order intra-Paleocene MFS. Accordingly, may also be ranged the Lameta Formation from Campanian to basal Maestrichtian, at the most up to early Late Maestrichtian. This is in tune with the ammonoid-based upper limit of the underlying Bagh Formation in early Middle Turonian or Late Santonian/Early Campanian as per possible duration of the pre-Deccan volcanism between the Bagh and Lameta units in the Narmada basin. In view of the recent record of Early Campanian nannoplanktons from Chakrud, the Lameta Formation there should have an older at least Early Campanian base.

7.7.28 Stratigraphic Framework in the Narmada Basin

The first-order intra-Jurassic SB $\sim$183 ma rifting was widespread in the west sector with resultant shallow marine sedimentation in the Toarcian–late Middle Oxfordian age bracket possibly up to the Saurashtra basin as suggested earlier. Contextually, the revival of rifting along the Narmada fracture may have occurred later at the first-order intra-Oxfordian MFS or during the first-order RST interval up to the intra-Cretaceous first-order Barremian/Aptian boundary SB, or even later during the following first-order TST. The oldest Mesozoic sedimentary unit in the basin is the unfossiliferous Nimar Sandstone sensu stricto as defined originally in Bose (1884) of the terminal Middle Oxfordian–Barremian age bracket, and considered shallow marine in a regional tidal flat and associated marine depositional framework. The Nimar Sandstone sensu stricto is assigned to the first-order RST age bracket of Late Oxfordian–Barremian. The upper part, intercalated with 3–5 relatively thin bivalve rich but ammonoid devoid intervals, may be best interpreted within the Aptian–early Middle Turonian first-order TST as a possible facies variant of the Nodular Limestone with a subaerial disconformity between the non-marine lower and marine upper parts. The post-Nimar bivalve rich intercalations are interpreted as transgressive lags at the second/third-order MFSs.

The initiation of the transgression was facilitated through revival of rifting along the ancient Narmada fracture as above. The maximum bathymetry registered near the contact of the bioturbated Nodular Limestone (Chirakhan Member) and the overlying bryozoan rich Coralline Member in the Hardground II of Tripathi and Lahiri (2000). There is interpreted distinct decrease of bathymetry above the Hardground II MFS. The basin registered increase in bathymetry during the first-order TST, and around the first-order intra-Turonian MFS were recovered the Late Turonian nannoplanktons (Jafar 1982), and ammonoids and other marine mega/microfossils in the Bagh successon. As anticipated, the intra-Turonian interval close to the major MFS is slow-sedimented to condensed inclusive of a few hard grounds.

In all probability, either the sea may not have reached Narmada or the rifting revived not earlier than Late Cenomanian. In a regional frame, an older transgression in the Aptian–Turonian cannot be ruled out. Any firm guide fossil evidence older than Turonian has so far not emerged but for the claimed Albian–Cenomanian ammonoids through an abstract (Pal 1971). The entire area supposedly received shallow marine sedimentation during either the first-order eustatic rise from the basal Aptian to Early or early Middle Turonian or during the later part of the first-order TST. In such a framework, the Nodular Limestone along with its suggestive facies variant of fossiliferous marine upper 'Nimar sands' is assigned to the Aptian–early Middle Turonian age bracket, although it may be mostly of Late Cenomanian–Turonian duration in the vertical proximity of the first-order MFS. This is disconformably followed up by the phase of pre-Deccan volcanics in the Bagh area (Biswas 1990) of possible Late Turonian–Santonian ~94–84 ma age bracket, that at the first-order intra-Cretaceous MFS triggered the spreading between India and Madagascar. The post Nodular Limestone part of the Bagh succession (Chirakhan Limestone/Marl) is considered to have been deposited during the LateTuronian second-order RST. The Coniacian Navgam Limestone, so age-constrained through planktonic forams is interpreted as TST of the next terminal Turonian–Coniacian second-order sequence somewhat coeval to its better and fuller expression in Cauvery, Spiti, and Jaisalmer. Yet later, followed the discoformably superposed Lameta succession of Campanian–Early Maestrichtian. The Lameta succesion in turn is terminally and disconformably overlain by the Late Maastrichtian–Danian Deccan traps and associated intertrappeans.

7.7.29 Sequence Stratigraphy-Based Suggestive Correlation of the Narmada and Cauvery Succession

The Nimar unit sensu stricto is corresponded to the Sivaganga unit of Cauvery basin in the terminal Middle Oxfordian–Barremian first-order RST age bracket. The disconformably overlying Nodular Limestone–Coralline Limestone units are similarly coevalled to the Uttatur Group in the Early Aptian–Early Turonian first-order TST age bracket. The Navgam Limestone along with the overlying volcanics is equated to the Trichinopoly Group as the first second-order sequence of the first-order RST. The Lameta succession is age-constrained in the Campanian–Early Maastrichtian age bracket, and along with the overlying Deccan volcanic is correlated to the Sillakudi–Kallamedu units of the Ariyalur Group in the Campanian–Maestrichtian age bracket as successive second-order sequences.

7.7.30 Igneous Events in Kachchh and Neighbourhood in the West Sector During the Late Permian–Early Paleogene Mega-Sequence

The principal igneous event is the Deccan volcanism with climax in Late Maestrichtian. The event has been invariably interpreted to be related to the hot spot volcanism, of the drifting Indian plate riding over the Reunion plume. Many workers have also interpreted pre-Deccan igneous activity of uncertain ages. Vaillier (1974) provided a detailed account of the successive Late Jurassic–Cretaceous volcanic episodes from Leg 25 of the West Indian Ocean at sites 239, 241, and 249. Recently determined ages of the Sadhara carbonatite sill of Patcham region range between ~91 and ~85 ma (Ray et al. 2006; Paul et al. 2012). In context of the ages of the intra-Cretaceous sequence surfaces, the Sadhara event is related to the oceanic separation between India inclusive of Seychelles and Madagascar near the early Middle Turonian Helvetica Zone first-order MFS of ~91–92 ma age. The ~84 ma event is found close to the second-order sequence surfaces near the Santonian/Campanian boundary. Igneous events of the same age range are also known from central-east Madagascar, offshore NW Karnataka, also coeval igneous expressions in South Karnataka, Kerala and recently even in Tamilnadu (Rao et al. 2015), which may also be related to the ride of the drifting India over the Marion plume. The said ~94 to ~84 ma igneous event is also evidenced through the thin basaltic layers in the Mundhra Formation in offshore well SP-1-1 and few other wells, and also in the Narmada basin between the Bagh and Lameta successions. A much older ~124 ma igneous intrusion in the Katrol Formatiom coeval to the Barremian/Aptian boundary first-order SB initiation of the Rajmahal/Kerguelen volcanics is also recorded recently from the Kachchh Mainland Fault and Median High intersection at Jhura (Kalmalkar et al. 2014), so also in the west sector of the Indian Ocean and a few Kachchh wells. In Kachchh and neighbouring Saurashtra are additionally suggested yet older intra-Jurassic first-order MFS and SB igneous events of ~159 and ~183 ma. Permo-Triassic igneous activity in Kashmir and Pakistan not far from Kachchh are also included in the west sector. The igneous plug at Nara may also be older of late Early Jurassic age as suspected in a few wells.

In nutshell, all the successive igneous events of the mega-sequence in spite of their principal plume-related expressions in different sectors around the Indian plate in east, west, north and south appear to have found expression in Kachchh and elsewhere in the west sector. Radiometric age determinations are required on high priority basis of these igneous expressions that appear linked to major sequence surfaces to strengthen or even discard the genetic links among extensional tectonics, igneous events on one side, and their produce—the major sequence surfaces on the other side.

7.7.31 Intra-Cretaceous Origin of the Indian West Coast/Shelf

The Indian West coast/shelf at present stretches from Kachchh–Saurashtra to Cape Comorin. It is neigboured by number of marine sedimentary basins along the coast starting with Kachchh in the north to Kerala in the south. These basins originated at different times during the Mesozoic. In a way, thus, the Kachchh–Saurashtra shelf initiated at the first-order intra-Jurassic Pliensbachian/Toarcian SB initiation of rifting between the West Gondwana and East Gondwana. This is followed by graded growth of the Neotethyan Somali arm due south at the north extreme, and of the Pacific Mozambique arm due north at the south extreme in the west sector during the Toarcian–late Middle Oxfordian first-order TST interval. The emerged Mozambique ridge prevented mingling and communication between the two marine arms. The Konkan–Kerala may have continued with non-marine intra-cratonic regime. The next stage of shaping of the Kachchh–Saurashtra shelf was the initiation of spreading between the two Gondwana units during the terminal Middle Oxfordian–Barremian first-order RST interval through distanting-cum-differentiation of the said shelf. Spreading across India inclusive of Madagascar and Africa is considered slow because of acute space constraints, as also the start of oceanisation in the northeast with spreading away of the Aargo block. Another significant step was the late Late Tithonian origin of the Gondic corridor seaway. Only limited spreading could take place across India–Africa during Berriasian–Early Valanginian. Mumbai–Kerala part of the Indian West coast/shelf initiated with the initiation of rifting between India and Madagascar at the Barremian/Aptian boundary first-order SB. The initiation of the spreading between India and Madagascar started providing it its present configuration since the intra-Cretaceous MFS event above the early Middle Turonian Turoniense Zone of ~92ma, with origin and growth of the Mascarenes basin. The India–Madagascar separation was fully realized by the Santonian/Campanian second-order SB of ~84 ma. The hard rock terrace like feature offshore Trivendrum (Yateesh et al. 2006) evidences the delinking/unzipping at the south extreme of the India/

Madagascar divergence, which is linked to the passing of India over the Marion plume near ~84 ma Santonian/Campanian boundary SB. The north extreme Mumbai shelf was finally sculpted with the spreading away of Seychelles from India at the riding of India over the Reunion plume near the terminal Maestrichtian second-order SB ~65 ma with separation of Seychelles and origin of the Laxmi basin. In some way, the two main triggers to the carving out of the Indian West coast/shelf to its present configuration were provided by the riding over of the fast drifting Indian plate over the Marion and Reunion hotspots in succession over the ~91 to ~65 ma span. Thus, the composite Indian West coast/shelf evolved through a span of well over 100 my between ~183 ma intra-Jurassic SB, and ~65 ma second-order SB.

7.7.32 Intra-Mesozoic Shaping of the Indian East Coast/Shelf

The Indian East coast started taking shape with the commencement of rifting between India and Australo-Antarctica near the base of Permian along with origin of the Indian Gondwana basins. The Indian East coast extends from Bengal basin in the north to the Gulf of Mannar. The Bengal basin started opening at the intra-Cretaceous first-order Barremian/Aptian boundary SB with the oceanisation all along the east divergent sector, even earlier at ~136 ma intra-Valanginian second-order MFS. The final complete distanting was realized by the mid-Middle Albian second-order MFS when Elan Bank broke apart as also opened the Gulf of Mannar between India and Sri Lanka. Desa et al. (2006) recognized anomalies M11 (~134 ma) to M0 (~121 ma) and interpreted break-up in the Sri Lankan region prior to M 11. Coffin et al. (2002) recorded the ~115 ma lamprophyres from the East Antarctic margin. South Kerguelen plateau is held as continental crust (Gibbons et al. 2013, and others). Until the mentioned ages of the origin of the Indian East coast during Early Aptian–mid-Middle Albian, it was preceded by the Dravidian corridor.

Thus, the initiation of the Indian West coast occurred earlier than that of the east coast at ~183 ma nearly 57 to ~47 my earlier. However, the East coast was fully realized by mid-Middle Albian at ~107 ma, distinctly earlier than the West coast by ~40 my. The West coast, thus, took ~100 my more to complete realization compared to East coast, and nearly fourfold time compared to East coast.

7.7.33 The Magnetic Anomaly Evidence from the Indian Ocean in the West (Arabian Sea and Neighbourhood in the India–Africa Divergent Region)

The origin of the Arabian Sea on the magnetic anomaly evidence has been interpreted during ~84 to ~64 ma by Chaube et al. (1993), so also the origin of the

Mascarenes basin by Gaedicke et al. (2002). The igneous events in this divergent sector also have been ranged between ~94 and ~84 my as earlier discussed (Valsangkar et al. 1981; Radhakrishna et al. 1999; Kumar et al. 2001). The same event based on the succession of sequence surfaces peaked at ~91 ma in early Middle Turonian above the Turoniense (ammonoid) Zone/Helvetica (foraminifer) Zone/Gartneri (nanoplankton) Zone first-order MFS. This spreading at ~91 ma is considered consequential to transform sliding due north of India vis-à-vis Madagascar. The ~84 ma igneous event seems related to the passing over of the drifting India (inclusive of Seychelles) over the Marion hot spot of the oceanically isolated Indian plate near Santonian/Campanian boundary second-order sequence surfaces. The age of the new Mascarenes ocean crust of ~84 ma corroborates the interpretation.

7.7.34 Intra-Turonian Initiation of Decline of Temperatures in Early Late Cretaceous

Most of the Cretaceous is claimed as the duration of global warmth—one of the warmest ever of the Phanerozoic. Andrews et al. (1995) ranged the concentration of CO_2 in Early Cretaceous between 1500 and 3000 ppm that dropped significantly in the terminal Cretaceous Maestrichtian stage. Valdiya (2010) interpreted the start of the decline in temperature during the Cenomanian–Turonian interval. The said global decline in temperature appears linked to the mega-order intra-Turonian MFS of the Late Permian–Early Paleogene mega-sequence.

7.7.35 Intra-cretaceous Sequence Surfaces and the Ridge Framework in the Indian Ocean

The Indian Ocean ridge framework mainly includes the 90 E ridge, 85 E ridge, Laccadive ridge, Carlsberg ridge and Laxmi ridge. The 90 E ridge is interpreted to be the resultant of Kerguelen igneous activity and is ranged between ~90 and ~38 ma (Valdiya 2010). The initiation of 90 E ridge in context of sequence surfaces is interpreted at the ~91 my early Middle Turonian Turoniense Zone intra-Cretaceous MFS. The 85 E ridge has been related to the plate reorganization at ~118 to ~107 ma in late Early Cretaceous, as also to the younger ~35 ma hot spot event. On the sequence surface evidence, it may be as well linked to the multifaceted geological event at the intra-Cretaceous first-order SB near the Barremian/Aptian boundary of ~126 ma age. The origin of the Carlsberg ridge on similar lines is interpreted at the first-order intra-Cretaceous MFS in early Middle Turonian of ~91 ma age. The origin of the Laccadive ridge is considered younger to that of Carlsberg, and may be linked to the second-order Santonian/Campanian boundary sequence surface of ~84 ma age.

7.7.36 Spreading Between India and Madagascar

The Early to early Middle Turonian intra/post Helvetica Zone MFS/MFI is readily interpreted in the Mannar, Cauvery, K-G, Spiti, Jaisalmer, Kachchh, Madagascar, Saurashtra, Narmada, and Kerala basins, i.e. almost all around the Indian plate. It provides the precise age of the climax of spreading between India and Madagascar. The rifting between India and Madagascar initiated at the intra-Cretaceous first-order SB at the Barremian/Aptian boundary, and is linked to the ceasure of spreading between India (inclusive of Madgascar) and Africa. There is interpreted jump of the rift axis eastward from west of Madagascar to east of Madagascar at the first-order SB. The said spreading is linked to the igneous activity in the vicinity of the intra-Turonian MFS on both the west and east margins of Madagascar, Pakistan, Kachchh, Narmada, Karnataka, Kerala, Tamilnadu, Mannar, and even in Tethys Himalaya. The initiation of the 90E ridge is also linked to the spreading between India and Madagascar at the early Middle Turonian MFS. The oldest rocks of 90E buried underneath the Bengal basin also range from ~94 to ~84 ma. The event may also be related to the cessation of spreading in the Wharton basin. Most workers have linked the Madagascar volcanism to the Marion plume, however a few (Bardintzeff et al. 2009; Duncan et al. 1997) have contested the same. The radiometric ages of the relevant igneous rocks range from ~94 to ~84 (Valsangkar et al. 1981 ~93.1 ma; Pande et al. 2001 ~85.2 ma; Kumar et al. 2001 ~87.5 ma; Radhakrishna et al. 1999 ~85 ma; Looseveld et al. 1996 ~84 ma; Torsvik et al. 2000 at ~90 ma; Parson and Evans 2005 at ~90 ma; Mellusoe et al. 2009 at ~88 ma; Gibbons et al. 2012, 2013, 2015 between ~94 and ~84 ma; Yateesh et al. 2013 at ~85 ma; Rao et al. 2015, and many other dates in the ~98 to ~84 ma bracket) in Madagascar, Kachchh, Karnataka, Kerala, and Tamilnadu. Yateesh et al. (2006) constrained the India–Madagascar through the terrace like feature off Trivendrum to a conjugate notch in Madagascar, and interpreted the separation around ~86.5 to ~83.6 ma. Biostratigraphically, the volcanics in well sections suggest initiation in Middle Cenomanian around ~98 ma in Madagascar, peak in Middle Turonian at ~91 ma, and prolongation at least up to Coniacian, may be yet longer up to Santonian/Campanian boundary. Even in SE Africa, a significant peak of volcanism is interpreted around ~92 ma near the said first-order MFS (Moore et al. 2008). The subduction of Lhasa and Karakorum terrains is also interpreted between ~94 and ~84 ma interval, so also the onset of Karakorum and Gangdse batholiths in the same interval. The revival of rifting and associated origin of the marine sedimentation of the planktonic foram bearing Campanian–Maastrichtian Dhananjaypur Formaion unconformably above the Rajmahal traps in the Bengal basin is also related to the widespread extensional tectonics with or without volcanism all round the Indian plate around this event. The uplift and associated shallowing or emergence in Madagascar, Pakistan, Jaisalmer, Kachchh, Saurashtra, Narmada, Kerala, Mannar, Cauvery, K-G, and others, and associated drastic geological changes are resultants of the same somewhat Pan-Indian major geological event at the intra-Turonian MFS. The pre-MFS Toarcian–Early Middle Turonian ammonoids based on old (Besaire 1973) and recent (Walaszczyk et al. 2014) data suggest strong similarity between India and Madagascar, while the post-MFS mid-Middle Turonian to end Maastrichtian ammonoids exhibit drastically reduced commonality between the two. The sudden drastic reduction in ammonoid similarity is causally linked for the development of oceanic barrier between the two regions at the spreading away of India from Madagascar that initiated around the early Middle Turonian Turoniense Zone MFS. In most of the Indian basins, the maximum bathymetry and transgression is interpreted at the said MFS. The volcanic flows encountered in the Mannar (Rana et al. 2008), and Kachchh offshore (ONGC 1993; Raju 2011); Narmada (Biswas 1990), appear age-constrained within Turonian. The spreading is evidenced by volcanic igneous event at Mananjari in central-east Madagascar and at St. Mary's Island off the Udipi coast in NW Karnataka along with associated intrusives in Karnataka Cauvery and Kerala in the ~94 ma to ~84 ma age bracket of Turonian to Late Santonian / Early Campanian (Valsangkar et al. 1981; Radhakrishna et al. 1999; Kumar et al. 2001) The spreading and divergence proceeded from north to south, and is interpreted to have culminated at the ~84 ma SB. The second-order SB is interpreted between the Trichonopoly Group and Ariyalur Group in outcrops. In this context, volcanic intercalations in the Parh Formation of Pakistan (Singh 2006) also appear to be part of the widespread event in the west sector. The ammonoid-based age of the intra-Turonian igneous event in particular, and of the Gondwana dismemberment chronicle in general, is more precise and better resolved compared to the tentative radiometry based ages of the recent extremely comprehensive and high quality ocean floor and other geophysical studies (Gaina et al. 2015; Yateesh et al. 2006; Seton et al. 2012; Gibbons et al. 2015; Reeves 2014; Bastia et al. 2010, and references included therein). The ~91–92 ma initiation of oceanic separation of India and Madagascar makes the point of time of sudden rapid drift of India, seemingly unparalleled in the global geological record. Multiple explanations are made like complete emancipation from Gondwana, pull exterted by the subducting Neotethyan slab underneath the Eurasian plate, double or multiple subduction zones (Jagoutz et al. 2015) and others.

7.7.36.1 On the Interpreted Short Duration Convergences Between India and Arabia During the Cretaceous

Gaina et al. (2015) have interpreted two short duration convergence events between the drifting India, and Arabia of ~125 to ~132 ma (terminal second-order RST interval), and then later again of ~90 to ~83 ma (another RST interval). The first convergence was consequented by the separation of India from West Australia and associated Bunbury volcanism). If so, the short-lived convergence would have been caused in the backdrop of acute space constraints in the India–Arabia divergent sector, which also did not permit any appreciable rate of initial spreading until quite some time during the second-order TST from 126 ma base of Aptian to ~107 ma mid-Middle Albian. The space constraints in east and west of India could mostly allow transform related spreading in both the sectors. The three transforms that played cardinal role were oriented along the north-northwest–south-southeast strike.

These transforms were located between West and East Gondwana, Madagascar and India, and India and Antarctica (Figs. 1.1–1.3, 7.11, 7.12 and 7.13). The first was active during the first-order intra-Jurassic TST, and let India move due South-southeast, The other two moved India due north-northeast. The second functioned during the middle and later part of the first-order intra-Cretaceous TST, and the third in the second-order Aptian–mid-Middle Albian TST. The later convergence event is attributed to the counterclockwise rotation of India vis-a-vis Arabia. Again the interpreted counterclockwise rotation and the resultant convergence might have been forced by the transform movement in mutually opposing directions. However, the two short convergences, having occurred, if at all, are considered good possibilities during the early Late Valanginian ~136 to basal Aptian ~126 ma to early Middle Turonian ~91–84 ma Santonian/Campanian boundary. Chatterjee and Scotese (1999) and Chatterjee et al. (2013), also, have necessitated contiguity of India and Arabia through land bridges in late Late Cretaceous for explaining the strong similarity of the post-Turonian Late Cretaceous land vertebrates, which would not have resulted without free intermigration and intermingling otherwise possible only between landmasses connected at least through land bridges.

7.7.37 *Out of India Hypothesis*

At the first thoughts, the hypothesis of the possible ancestry of the north hemisphere mammalian vertebrate life, we see today, on the oceanically isolated Indian plate during the Late Cretaceous, appears improbable yet can not be ruled out. The said mammalian land vertebrate diversity along with its distribution in India, China, Mangolia, Siberia, Central Asia provide testimony to its possible origin during India's supposedly solo venture across the equator for a minimum of ~10 my. The vertebrate similarity during the Late Maestrichtian is otherwise explained to formation of land bridges in the northwest sector, even in other sectors, prior to the time India docked first into Asia (Sahni 1984), short duration convergence between India and Arabia during the mid-Late Cretaceous (Gaina et al. 2015), intermittent availability of the route between the South Europe and North Africa and proximity of the drifting India to Greater Somalia (Chatterjee and Scotese 1999). The Muslimbag ophiolites of ~71–68 ma Ar40/Ar39 age range (Mahmood et al. 1995) also suggest subduction of the Neotethyan floor in the Pakistan sector around the said interval.

7.7.37.1 Terminal Intra-Cretaceous Second-Order late Late Campanian–Early Late Maestrichtian Sequence

The exposed Kallankurichi Limestone (TST) and the Kallamedu Sandstone (RST) of the late Late Campanian to the early Late Maestrichtian are interpreted to constitute the terminal Cretaceous second-order sequence in Cauvery basin with

MFS in early Early Maestrichtian. The said MFS near the top of the Kallankuruchi unit may also mark the start of the Deccan outpouring at the time of Indian plate passing over the Reunion hot spot. The underlying disconformable contact between the Sillakudi and Kallankurichchi units is interpreted as the basal second-order SB of mid/late Late Campanian age.

7.7.38 *Spreading Between Axial and Outer Gondwana*

The post Middle Albian westward spreading of India inclusive of Madagascar within the rather loosely held together Axial Gondwana (India with Madagascar and Africa) occurred without any appreciable drift among its constituents. It had its trailing east and southeast margin progressively submerged in steps during the Aptian–early Middle Turonian first-order TST interval. The westward spreading is interpreted as quite slow, in view of space constraints between Africa and India (inclusive of Madagascar). The space constraints also reactivated the rifting between India and Madagascar at the first-order SB near the Barremian/Aptian boundary. By the mid-Middle Albian second-order MFS, the loosely held together Axial Gondwana without any active appreciable spreading among its three constituents (India, Madagascar and Africa) had oceanically separated from the three member Outer Gondwana (Australia, Antarctica, and South America), since the basal Aptian first-order SB. The Axial Gondwana at this point of time was surrounded by the Indian Ocean in the east, the South Atlantic Ocean in the west, and the Neotethys in the north. In this geographic framework, started the dismemberment separately among the Axial and Outer Gondwana constituents at the ~92 ma early Middle Turonian MFS between India and Madagascar, and among South America, Antarctica, and Australia.

7.7.39 *Change in the Direction of Drift Prior to the Intra-Turonian MFS*

The evidence of magnetic anomalies (Ramana et al. 1994, 1997) interpreted the change in the direction of India's plate movement during ~118 to ~84 ma or near ~84 ma. Desa et al. (2006) suggested the said change in direction at ~50 ma. Gibbons et al. (2015) ranged the anticlockwise rotation in the ~126 to ~84 ma age bracket. They also attributed the said rotation to developments in extreme north. The change in direction of movement is considered to have occurred between the mid-Middle Albian ~107 ma separation of Elan Bank from India along with ceasure of volcanism in the east sector, and the intra-Cenomanian initiation of separation of Madagascar from India along with onset of volcanism in the west sector. The said rotation in the present work is based on the succession of sequence stratigraphic surfaces interpreted between the mid-Middle Albian and intra-Cenomanian MFSs. It is alternatively held that the severe space constraints between the westward spreading India and relatively stable Africa, had made imperative the eventual due north escape of India. The almost simultaneous transform sliding in opposite somewhat parallel directions of India due south-southeast in the west and due north-northwest in the east during ~136 to 107 ma may have forced the initiation of counterclockwise roation on the slow moving Indian plate. The initiation of the curved fracture zone in the Wharton basin may also have been influenced by the start of counterclockwise rotation and associated change in the direction of drifting India in the interpreted time span or vice versa.

7.7.40 The Break-up Among the Axial Gondwana Constituents

Spreading among India, Madagascar and Africa initiated in appreciable manner much later only at the first-order intra-Cretaceous MFS at the early Middle Turonian. It also triggered India's classic due north-northeast rapid drift at a distinctly faster rate than any time earlier in the Mesozoic prior to the intra-Cretaceous MFS. The counterclockwise rotation was by this time already finalized with the resultant distinct change in the direction of drift. Presumably, the much slower pre-MFS drift was rather west-northwest directed while at the MFS event, it became parallel to 90° E ridge. During the Toarcian–Barremian first-order sequence, there was mostly transform sliding between the two Gondwana constituents along the Davies fracture. No appreciable separation occurred across India and Africa. India's west continental margin also did not shape at the commencement of the T/J boundary as suggested recently by Dave (2011, 2012).

7.7.41 Precise Age of the Break-up Unconformity

The complete break-up and beginning of the oceanic spreading was initiated in the east sector at the first-order upper sequence boundary (break-up uncorformity) near the Barremian–Aptian transition. The space constraints did not permit appreciable growth until the mid-Middle Albian. The age of the India–Australo-Antarctica break-up unconformity interpreted as above differs, and is older vis-à-vis other recent assertions (Dave 2011; Raju 2011; Prasad and Phor 2009).

7.7.42 Deccan and Coeval Volcanism, and Oceanic Separation of Seychelles from India

Seychelles may have started drifting away from India through shift/jump of the spreading axis west of India inclusive of Seychelles to between Seychelles and

India. Initially, it only evidenced a precursor rift-related volcanic event prior to the main Deccan volcanism. The massive Deccan outpouring peaked when India passed over the Reunion hot spot at the late Late Maestrichtian second-order SB. The Ar40/Ar39 ages of the associated basalts between ~71 and ~58 ma support the above model. It was a time lag of ~13 my similar to such lag of ~10 my between the successive events of rifting-cum-spreading igneous peak at ~91 ma of the separation of Madagascar from India, and the drifting India's subsequent riding over the Marion plume. The Deccan activity, however, continued into early Paleocene.

7.7.43 Subaereal Stratigraphic Gaps in the Studied Spiti, Kachchh, Cauvery and Other Indian Basins Contextual to the Studied Mega-Sequence

In context of the studied mega-sequence, there are interpreted several subaerial stratigraphic gaps, which are here related to the first and second-order RST intervals from any time between the MFSs to terminal SBs. The subaerial gaps in the early part of the mega-sequence in Spiti are at the basal mega SB within the Kuling Group, at the P/T boundary second-order SB between the Kuling Group and Mikin Formation in Spiti. In Kachchh, the subaerial gaps are inferred between the Precambrian and the Nirona Formation at the first-order intra-Triassic MFS, and between the Precambrian and the Patcham Formation at the first-order intra-Jurassic SB, and between the Umia and Mundhra/Naliya/Kori units in Kachchh around the first-order intra-Cretaceous MFS in Kachchh. In Cauvery, subaereal gaps are interpreted between the Precambrian and the Sivaganga Formation, between the Sivaganga Formation (=Terani Formation) and the Dalmiapuram Formation at the first-order intra-Cretaceous SB, between the Utattur and Trichinopoly, between the Trichinopoly and Ariyalur groups, between the Sillakudi and Kallankuruchhi units, and above the Kallamedu unit at second-order SBs in the Cauvery basin.

7.7.44 Submarine Stratigraphic Gaps

The submarine gaps are interpreted at the first-order MFSs in the Middle Triassic in Spiti and other High Himalaya basins, in the Late Jurassic all over the GTM, while in the Cretaceous only strong condensation is observed in the first TST of Early Aptian–mid-Middle Albian almost in all Indian basins, and again during the Late Cenomanian–Early Turonian succession in distal High Himalaya at Zanskar, and yet again during Campanain inclusive of possible submarine gaps, although as yet no such submarine gaps could be confirmed.

7.7.45 Intertrappean Sediments and Volcanoclasts as Age Evidence of Deccan Igneous Activity

Intra-Maestrichtian intertrappean beds from Anjar (Gujarat) in the west to volcanoclast bearing Mahadek Formation of Assam and Rajahmundry traps in east and K-G wells in southeast indicate either the wide expanse of the Deccan magmatism or individual events not physically connected yet genetically related to the Deccan volcanic event.

7.8 Terminal SB of the Mega-Sequence in Intra-Paleogene

7.8.1 Conceptual Closure of the Mega-Sequence

The mega-sequence has been conceived from the origin to the closure of the Neotethys. The origin is fairly well age-constrained within the Permian between the base of Artinskian and base of Ufimian/Wordian/Guadalopian, and there is no dispute about the base being an SB. However, the major marine/nonmarine change across the closure of the Neotethys is interpreted at a significant MFS, preceded as also followed by significant second-order SBs. Thus, there is diachrony between the conceptual closure of the mega-sequence at an SB and closure of the Neotethys better defined and constrained by a significant MFS. The terminal SB of the mega-sequence need be primarily based upon mega-change of sea level. The closure of the Neotethys is considered protracted over a time in multiple steps from close proximity to initial impingement to final amalgamation depending upon which parameter/s is/are given priority. The major change of sea level as interpreted through facies framewok is observed in Late Paleocene within the Paleogene basal/intra P3 Zone, which suggests conceptual closure of the mega-sequence at that point of time. The P3 Zone event is Pan-Indian in extent with explicit micropaleontological evidence in both, east and west even in Kerala–Konkan and Cauvery. The early part of the disconformably overlying Ranikot and coeval units include volcanic flows/tuffs in Pakistan, Kachchh, Himachal in the west and Myanmar in the east which seem to imply a distinct level and event in the basal most part possibly coeval to the terminal phase of the Deccan volcanism.

Only a few workers lay their stakes in such early closure of the Neotethys, since it makes only the early, if so, a pre-collision proximity instead of the initial well-marked impingement.

7.8.2 *Varying Ages of the Docking of India into Asia Soon After the Conceptual Intra-Paleocene Closure of the Mega-Sequence SB*

The liberation of India from its Gondwana captors initiated in Permian, and culminated with India's complete oceanic isolation at the ~84 ma Late Cretaceous MFS when it became surrounded by Neotethys in north and Indian Ocean in south, east and west. Even during the preceding intra-Devonian–intra-Permian mega-sequence microblocks are known to have spread away from the GTM. The so drifting India travelled across the equator to dock, and assimilate into Asia in multiple steps during the Paleogene. During a significant part of the mega-sequence, almost since mid-Middle Albian, the Indian plate inclusive of Madagscar received marine sedimentation all around it in different sectors. After, the complete separation of Madagascar at ~84 ma, the present Indian west coast began its evolution in a marine regime. Bhatia et al. (2013), Valdiya (2010) have lucidly surveyed in particular the Paleogene geological framework in NW Himalaya as also all around the Indian plate from Indus basin to Meghalaya and Myanmar in the northwest in the Himalayan Foreland (Sirmaur) super-basin with intra-Paleocene individualization at the start of a new mega-sequence. In the said account, a few widespread Pan-Indian intra-Paleogene extensional tectonic events seem to stand out. The age range of ~67 to ~45 ma of the Ladakh batholith determined by Upadhyay et al. (2008) also provides credence to a protracted amalgamation in multiple steps mostly during the Paleogene.

7.8.3 *Initial Impingement*

The initial impingement of the drifting India on stable Asia to final amalgamation of the two is ranged from Late Cretaceous (Chatterjee et al. 2013; Gaina et al. 2015 and others) to Oligocene (Aitchison et al. 2007 and others) in the ~92 to ~34 ma age bracket. Most of the recent multidisciplinary evidence has been contributed from sections in South Tibet close to the Indus-Psangpo suture zone, and most studies now restrict the initial impingement from different angles in Early Eocene in a much smaller range between ~56 and ~50 ma (Zhu et al. (2005), Najman et al. (2010), Meng et al. (2012), Wang et al. (2011), Yang et al. (2015), Jagoutz et al. (2015), Chatterjee et al. (2013), Jain (2014), Gaina et al. (2015), Gibbons et al. (2015), Seton et al. (2012), Bhargava (2015), Bhatia et al. (2013), Valdiya (2016) and numerous other recent works). Diverse aspects of plaeomagnetism, paleolatitude, biostratigraphy, land vertebrate biogeography, size of the Indian plate, marine and non-marine sedimentology, magmatism, spreading rate, metamorphism, change of paleoslope/current/drainage, provenance of deep water sediments, cessation of marine sedimentation, start of mixed detritus, zircon radiometry, chromium spinel provenance and others from different sectors of South Tibet inclusive of localities close to the suture have been applied to interpret the initial impingement. Sound single, double and multiple parameters have been applied to the problem. Also, single, double, and even multiple subduction/collision are envisaged, yet a single initial impingement date remains a far cry.

7.8.4 Convergence of the Initial Impingement in a Short Range

Zhu et al. (2005) interpret the initial impingement prior or within P8 in the Zhepure Shan region of South Tibet of East Central High (Tethys) Himalaya. Hu et al. (2012) also discussed the India Asia collision in the Zhepure region. Najman et al. (2010) in the Qimiba section (Zhepure Shan range) of South Tibet south of the Indus-Psangpo suture on multiple evidences interpret cessation of the marine sediments in P8. In the context of the initial impingement, it is quite logical to consider ceasure of marine sedimentation in different sectors of the drifting India, either isochronous or earliest in the north sector close to the Indus-Psangpo suture, and in quick succession in basins located in the east and west sectors away from the zone of the earliest impingement. Bhatia et al. (2013) suggested the collision at the SBZ 9/10 zonal boundary. Tian et al. (2015) suggested marine framework up to the close of Eocene while Singh et al. (2015) interpreted coastal marine influence even up to Early Miocene. A few models even envisage spreading between the Indian craton and Tethys Himalaya (Gibbons et al. 2013) irrespective of the ground geological evidences to the contrary. Meng et al. (2012) based on paleomagnetic evidence interpret the initial impingement around 24 N latitude at 50 ma. Some models based on palmag data interpret a belt of oceanic island arc framework in the Late Jurassic–Early Cretaceous interval in near equatorial to low latitudinal setting, a few others locate South Tibet in high latitude south hemisphere setting. Jagoutz et al. (2015) in a sensational manner introduced an additional/new Kshiroda oceanic plate between India and Eurasia along with interpretation of double subduction zones on both the edges of the said oceanic plate each ~10000 km long. The said double subduction on north and south of the Kshiroda oceanic plate resulted in shortening of ~3000 km on either side. The first subduction was on the south edge of Kshiroda at ~50 ma and again later at the north edge at ~40 ma. The isotope evidence produced is of a fresh binary system, while the model and dual collision not very different from other recent models. Yang et al. (2015) maintained the unity of the Indian plate until at least intra-Valanginian in high southerly latitude. Thus, even the high precision multidisciplinary voluminous data generated in the present millennium, has failed to reduce the extreme divergence of views on the subject of India–Eurasia collision, neither about the date nor with regard to modalities.

7.8.5 Possible Relevance of Sequence Surfaces to the Initial Impingement in Particular and the Protracted Amalgamation in General

Having hardly ever had even peripheral research interest in the Indian Paleogene development but for occasional visits to the sections and outcrops in different sectors in course of studies in the Mesozoics, the present interpretations are exclusively based on a limited sequence stratigraphic context through scanning of the recently published literature on the subject, particularly in the east and west, and north sectors relatively located away from the theatre of collision. Citing precedents, a few major geological events that occurred during the span of the mega-sequence in geologically complex regions have been assigned precise ages in sequence stratigraphic context in basins located far away, e.g. the initiation of the oceanisation in the northeast with spreading away of Argo block from the Indo-Australian margin at the ~159 ma late Middle Oxfordian orientalis Zone Schilli Subzone first-order intra-Jurassic MFS determined in Kachchh located ~3000 km away. Even a limited sequence stratigraphy based chronicle of Cenozoic events is out of scope in the present work. As such, the collision of India is summarized through a significant initial impingement of the drifting India on either Eurasia or Kohistan Ladakh arc framework near the Early/Middle Eocene boundary within the Cuisian stage, further constrained by planktic zones P8/P9 major MFS of ~51 to ~50 ma age as interpreted in the east, north, and west sectors with focus on Kachchh, inclusive of the action theatre (Fig. 7.43). This initial impingement is preceded by the infra-collision proximity at the intra-Paleocene terminal SB event of the studied mega-sequence, and followed by at least one significant uplift at the intra-Bartonian second-order MFS, prior to the complete continent–continent amalgamation at the terminal Eocene ~34 ma major sequence surface.

The approach is simple, well defined, and satisfactorily applied throughout the Mesozoic. It is emphasized upon once again that the developed first and second-order sequence stratigraphic framework in the Indian Mesozoic validates the governance of the geological happenings round the Indian plate all through the mega-sequence by the same tectono-igneous events, that are near synchronously reflected in the geological record of the individual basins in different ways but for occasional localized small magnitude nuances, noises or distortions, including those of the tectono-igneous events of the suture zone. The second-order Late Cuisian surface is interpreted to mark the most significant widespread geological change in the ~51 to ~50 ma interval, which can only be caused by the significant initial impingement. The compressional/collisional events of the north sector are manifested through extensional tectonics and/or differential submergence in the east, north, and west sectors, e.g. in Shillong (Jauhari and Agrawal 2001) in the Himalayan foreland basins (Bhatia et al. 2013; Bhargava 2015; Valdiya 2016), in Kachchh and neighbourhood (Saraswati 2011; Saraswati et al. 2014 and many others). As a corollary, the sequence surfaces can be well determined in relatively

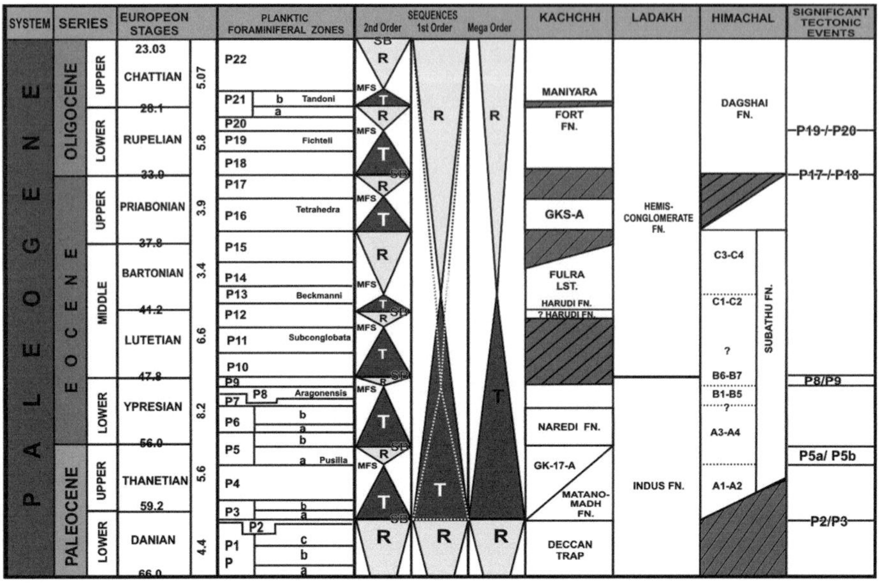

Fig. 7.43 Mega to 2nd order sequence framework and Tectonic events in the Indian Paleogene Geological Record with mega order MFS alternatives in Late Ypressian and Intra-Bartonian

undisturbed basins irrespective of their locations far away, but within the much larger broadly homogenous tectono-stratigraphic region with copy book style sedimento-stratigraphy. Surfaces so determined, in turn, provide reliable ages to events that occurred in geologically complex sectors of the suture or rift zones (within the said region), where most of the evidences are lost while the remainder quite difficult to retrieve.

7.8.5.1 Echoeing of the Same Succession of Refined Radiometric Ages of Magmatic Events All Around and Inside the Indian Plate, in the Collision Theatre as also in the Indian Ocean Region

It is interesting to note that the ages of the major sequence stratigraphic surfaces along with the geological events that these are resultants of like ~370, ~289, ~272, ~243, ~183, ~126, ~107, ~92, ~84, ~78–75, ~68–63, 60, ~51–50, ~40, and ~34 ma events are found echoed in the Indian Ocean, east, north and west sectors, in the inland Gondwana basins, in Lesser Himalaya, in High Himalaya (both proximal and distal sites), as also in the South Tibetan collision theatre between the ITSZ in the south and Bangong–Co-Nujiang in the north. The said correspondence cannot be considered mere coincidence, but is the resultant of a single interrelated melodramatic tell-tale.

7.8.5.2 Same Framework in Other Basins

In Kachchh the Naredi/Harudi unit contact suggests second-order SB at or near the initial impingement, so also in other basins of west, north and east sectors inclusive of transects in South Tibet close to the Indus suture. Thus, the said sequence surfaces in Kachchh and Indus of the west sector can be extended in the north sector close to the suture in South Tibet. The formation of lignite in the west sector (Indus basin, Bikaner, Kachchh, Cambay is interpreted to start at different times under the fast changing climatic influence on the drifting Indian plate, however, always only above the intra-Paleocene SB, and close to the successive climatic/thermal optima in Early Eocene in Cambay and Indus basin, and in Middle Eocene (Late Lutetian–Early Bartonian–Sharma and Saraswati 2015; Saraswati et al. 2014; Punekar and Saraswati 2010) in Kachchh. Saraswati et al. (1993) have suggested temperatures around 24–28 °C after the Bartonian MECO, which may hover in the same range near the EECO within the the Naredi unit. It may also be interpreted that the lignite formed in response to a prolonged but graded thermal build-up that began above the intra-Paleocene SB with optimization in Early Bartonian of P13 Zone age.

7.8.5.3 East Sector

In the northeast in Shillong plateau, the oldest Cenozoic SB can be placed between Therria Formation and Lakadong Formation of intra P3 Zone age; however this SB has hardly any connection to do with the India–Asia collision but for yet closer proximity. The initial impingement related extensional tectonic event in Shillong is interpreted as the tectonically influenced submergence near the close of the Umlatdoh Formation. The Lakadong Formation and Umlatdoh Formation appear to make either two separate third-order sequences or one second-order TST of P3–P8 duration up to the terminal part of the Umlatdoh Limstone Member of Umlatdoh Formation with MFS between the Umlatdoh Limestone Member and Narpuh Sandstone Member of Cuisian P8/P9/SBZ10/11/NP13 Zone age. This east sector event is broadly corresponded to the intra-Naredi event in Kachchh and similarly in Himachal, Jaisalmer and Indus basins.

7.8.5.4 Widespread Recognition of the P8 MFS

In context of ages, the P8/P9 MFS event of the east, north and west sectors is interpreted as a major shallowing/uplift event in South Tibet as also elsewhere around the Indian plate. The terminal Early Eocene mid/late Cuisian P8/P9 significant MFS marks the initial impingement of the drifting India on the stable Asia. A few on the contrary extend the coastal marine influence even up to close of Oligocene in specific sectors e.g. in Ladakh (Singh et al. 2015). The collision in the Indo-Myanmar sector not far from the Shillong plateau is suggested either isochronous or later or earlier to the said impingement compared to South Tibet.

7.8.5.5 Implied Diverse Interpretations

It is interpreted from the above limited discussion that (1) the extensional tectonic event between the lower and upper limestone units in Shillong plateau makes the resultant extensional response to the initial impingement of the drifting India and stable Asia in South Tibet close to the suture, (2) the disconformable contact includes a significant MFS, (3) similar event is also signaled in the HFB, in the west sector basins of Kachchh, Cambay, Bikaner and Indus, (4) The widespread event corresponds to the EECO near P8 Zone, (5) the lignites of the Early Eocene in Cambay and Indus basins correspond to the graded rise in temperature from Late Paleocene Thanetian to Early Eocene Cuisian interval, while of Kachchh correspond to the MECO of Late Lutetian–Early Bartonian span and (6) the said event is best constrained in P8 of an extended P7-P9 range included in the Cuisian stage.

7.9 Terminal Priabonian First-Order ~34 ma SB in the Indian Cenozoic

The first major uplift of the Himalaya, with resultant change of paleoslope from due north to due south along with cessation of marine sedimentation are cited as evidences in support of the amalgamation/collision process at the terminal Priabonian $\sim$34 ma first-order SB by way of the final continent to continent collision as the drop of the curtain to the geodynamic melodrama. The said SB is determined in Kachchh in the Paleogene sequence. The intra-Rupelian first-order MFS is delineated within the unit above the P19 Zone. The terminal SB is determined at the base of Aquitanian stage between P22 and N4 Zones. During the RST between the MFS and the terminal SB is interpreted the phase of major uplift of Himalaya that began at the MFS along with resultant formation of the Siwalik foredeep south of the rising Himalaya. It was in this foredeep that thenceforth initiated the rapid deposition of the non-marine Siwalik group of mammalian vertebrate fossil-rich sedimentary succession. The Siwalik sedimentation began at this early Early Miocene first-order MFS. The major eustatic fall and associated decrease in bathymetry of the order of 200 m is interpreted during the P20–N5 first-order RST interval. There is interpreted change of slope from due west and north to due south on account of the uplift of Himalaya which from the start of the RST started to provide most of the sediments to the foredeep immediately after the MFS. The explosive volcanism in Andaman near the Oligocene/Miocene boundary (Srinivasan 1986) is also linked to the early phase of the eustatic rise which may have begun at the preceding first-order basal SB. The geographically restricted last vestiges of the marine Neotethyan regime also vanished during the preceding intra-Oligocene RST. Thus, finally disappeared the Neotethys after a long well over 200 my spectacular inning that in its own imperceptible fashioned India and the globe at large quite close to what it is at present.

7.10 Summarised Re-enaction of the Succession of Multifaceted Geological Events

Here follows the revised enaction of the principal acts that occurred at the said sequence surfaces (Figs. 7.44 and 7.45). The first major change/reversal of paleoslope and associated igneous event near the Middle/Late Cambrian boundary first-order MFS is followed by first-order Cambrian/Ordovician boundary SB igneous event. The Paleotethys originated at the intra-Devonian, and Neotethys at the intra-Permian mega-order SBs. The widespread basal Permian second-order SB extensional tectonics resulted in the inland Gondwana rift basins. At the Sakmarian/Artinskian MFS initiated the Panjal and coeval volcanism in the north sector. The Neotethys originated with the spreading away of the Cimmerian continent, along with widespread marine transgression over the Panjal and coeval volcanics at the start of Late Permian near the origin of the mega-sequence. The P/T boundary second-order SB event marked the global mass extinctions and widespread cessation of coal formation. The Kachchh, and possibly also the Saurashtra basin formed in the terminal Anisian, and soon after witnessed the start of marine sedimentation in Early Toarcian around the time of the Karoo and coeval volcanism. Rifting in the basin aborted at its climax in intra-Oxfordian MFS. The Proto-Indian Ocean initiated locally with spreading away of AAP at this point of time between Greater India and Australo-Antarctica. See-saw tilted block tectonics caused reversal of paleoslopes to due east and southeast at the said surface. The

Fig. 7.44 Clade diagram for precise chronicle of separation of Successive micro/mega plates/blocks during the dismemberment of Gondwana

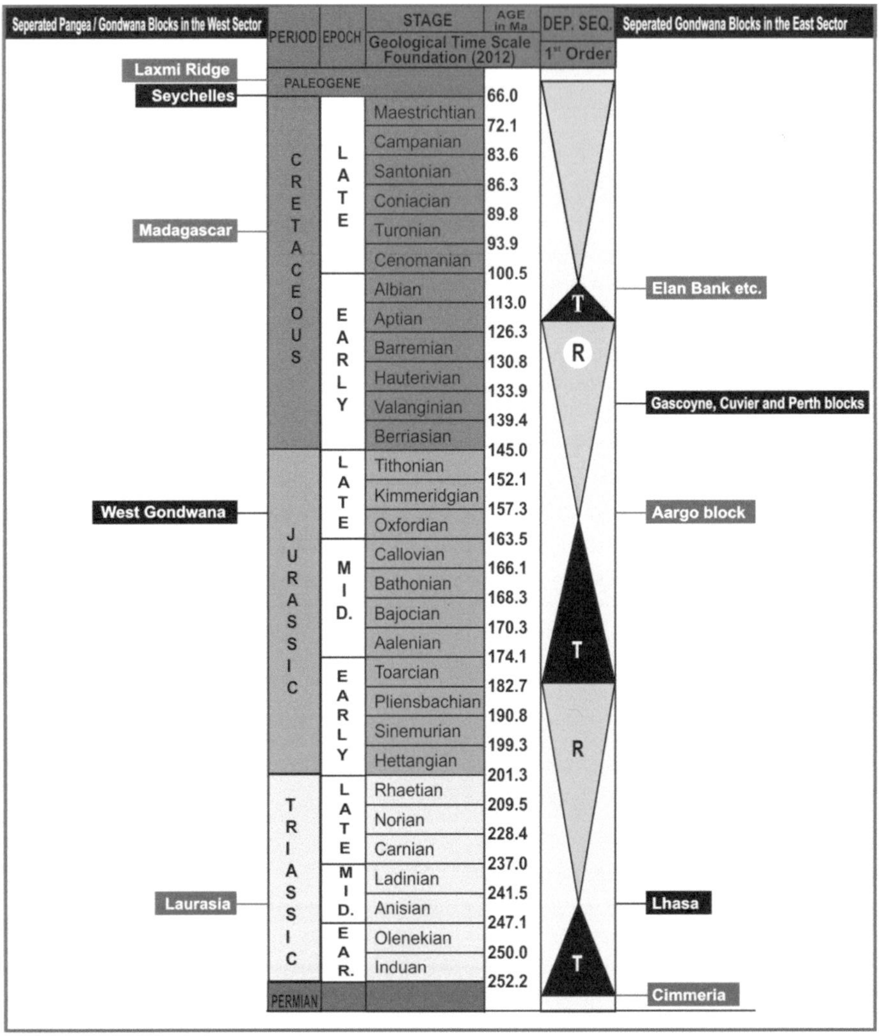

Fig. 7.45 Sequence stratigraphy based chronicle of Indian Phenerozoic geodynamic evolution

transient Gondic corridor was realized in the terminal Late Tithonian. The volcanism, once begun in Oxfordian, continued episodically in the east as also in the west, and so with that, advanced the unzipping due south between India and Australo-Antarctica with culmination at the mid-Middle Albian major MFS. The transient Dravidian corridor signaled the start of the youngest first-order sequence. It operated during Aptian–Early Albian, and precursored the rapid growth of the Indian Ocean. Australo-Antarctica was thus shed away from the Gondwana assembly. In the ~63 my long late Middle Oxfordian–mid-Middle Albian dismemberment interval in the east sector, India in succession shed AAP, GAP, CAP, PAP, and multiple small blocks with origin of abyssal plains/basins offshore NW and West Australia, and Antarctica in as many steps. Finally, broke off the Elan bank and South Kerguelen and associated microblocks from India at the mid-Middle Albian MFS. India moved due west, and faced space constraints in the west between India inclusive of Madagascar and Africa. Thus occurred the inevitable counterclockwise rotation of India during the span of mid-Middle Albian to intra-Cenomanian MFSs. Thereafter followed the spreading between Madagascar and India in Turonian–Santonian and with that began the rift/plume-genic volcanics in the west sector. The oceanically isolated Indian plate rode over the Marion plume as evidenced through plume-genic Late Santonian volcanism. So began in Late Santonian its dream romanceful travel across the equator. Even after getting across the equator, the questful journey of India went on, and then it rode over the Reunion plume that consequented the most spectacular ever igneous outpouring—the Deccan basalts across the K/T along with spreading away of Seychelles. The ~95 my long early intra-Oxfordian–Early Paleogene fragmentation in the west sector also was realized in multiple steps with phased shedding of Africa, Madagascar, Seychelles and other micro-lands. The geodynamic melodrama ended with the docking of India into Asia in Paleogene in multiple steps in Early Eocene and near the close of Oligocene, even in between.

References

Ahmad T, Tanaka T, Sachan HK, Asahara Y, Islam R, Khanna PP (2008) Geochemical and isotopic constraints on the age and origin of the Nidar ophiolitic complex, Indus suture zone, ladakh, India. Tectonophysics 451(1–4):206–224

Ahmad Z (1991) Ultra mafic–mafic alkaline rocks from Spagar, Pichin district Pakistan, magnetism from the waning Gondwanaland. Acta Min Pak 5:25–45

Aitchison JC, Mcdermit IRC, Ali JR, Davis AM, Zyabrev SV (2007) Shoshonite in south Tibet record Late Jurassic melting of a Tethyan intra-oceanic arc. J Geol 115:197–213

Alberti M, Fursich FT, Pandey DK (2013) Deciphering condensed sequences: a case study from the Oxfordian (Upper Jurassic) Dhosa Oolite Member of the Kachchh basin, western India. Sed Geol 60:574–598

Alberti M, Fursich FT, Pandey DK, Ramkumar M (2012a) Stable isotope analysis of belemnites from the Kachchh basin, western India: paleoclimatic implications for the Middle to Late Jurassic transition. Facies 58:261–268

Alberti M, Pandey DK, Fursich FT (2012b) The Oxfordian stable isotope record (d18O, d13C) of belemnoids, brachiopods and oysters from the Kachchh basin (western India) and its potential for paleoecologic, paleoclimatic and paleogeographic reconstructions. Paleogeography Paleoclimateology Paleoecology 344–345:49–68

Andrews JE, Tandon SK, Dennis PF (1995) Concentration of carbon dioxide in the Late Cretaceous atmosphere. J Geol Soc Lond 152:1–3

Arkell WJ (1956) Jurassic geology of the world. Oliver and Boyd, London, pp 1–806

Audley-charles MG, Ballantyne PD, Hall R (1988) Mesozoic–Cenozoic rift-drift sequence of Asian fragments from Gondwanaland. Tectonophysics 155:317–330

Awatar R (2011) The plant megafossils from Indian Gondwana and stratigraphic enigma of supra-Barakar rocks in south Rewa basin, M.P., India, Palaeontology and Stratigraphy: Basics to Application, Jaitly AK, Singh AD, Pandey B, Nath S (ed) BHU, Varanasi, pp 70–77

Ayyasami K (2011) Recognition of the Cretaceous stage boundaries in Southern Indian Cretaceous based on invertebrate fauna, In: Jaitly AK, Singh AD, Pandey B, Nath S (ed) Palaentology and stratigraphy: basics to applications. Banaras Hindu University, Varanasi, pp 78–87

Ayyasami K, Bannerjee RK (1984) Cenomanian – Turonian transition in the Cretaceous of southern India. Bull Geol Soc Den 33:21–30

Ayyasami K, Rao BRJ (1978) On the possible faunal break at the contact of Trichonopoly and Ariyalur groups in the Cretaceous succession of Tiruchirapalli district, Tamilnadu, proceedings of the VII Indian colloquium micropalaeontology and stratigraphy, pp 1–5

Badsah MS, Gnos MQ and Afridi MI (2000) Stratigraphic and tectonic evolution of the north west Indian plate and Kabul block. Geol Soc London Spec Paper 467–475

Banerjee B, Sengupta BJ, Banerjee PK (1995) Signals of Barremian (116 Ma) or younger oceanic crust beneath the Bay of Bengal along 14°N latitude between 81°E and 93°E. Mar Geol 128(1–2):17–23

Bardintzeff JM, Liegeois JP, Bonnin B, Bellon H, Rasaminana G (2009) Madagascar volcanic province linked to the Gondwana break-up, geochemical and isotopic evidences for contrasting mantle sources. Gond Res 18:295–314

Bastia R, Reeves C, Rao P, D'Silva K and Radhakrishnan M (2010) Palaeogeographic reconstruction of East Gondwana and evolution of the Indian continental margin, DCS–DST News, India, pp 1–8

Baud A, Bhat GM, Bachmann GH, Kular-Jurkowest T, Chen ZQ (2014) Report on the first IGCP 630 field workshop 2014. Kashmir, India, Permophiles 60:26–30

Baudin F, Bulot LG, Cecca F, Coccioni R, Gardin S, Renard M (1999) Un equivalent du 'Niveau Faraoni'dans les basin du Sud-Est de la France indice possible d'un evenement anoxique fini hauterivien etendu a la Tethys Mediterraneenne. Bull Soc Geol Fr 170(4):487–498

Beerling DJ, Harfoot M, Lomax B, Pyle JA (2007) The stability of the stratospheric ozone layer during the end-Permian eruption of the Siberian Traps, Phil. Trans. Roy. Soc. London 365:1843–1866

Berner RA (2002) Examination of hypotheses for the Permo-Triassic boundary extension by carbon cycle modeling. Proc Nat Acad Sci, USA 99:4172–4177

Berner RA, Vandenbrooks JM, Ward PD (2007) Oxygen and evolution. Science 316:557–558

Bertle RJ, Suttner TJ (2005) New biostratigraphic data from the Chikkim Formation (Cretaceous Tethys Himalaya, India). Cret Res 26:882–894

Besaire H (1973) Precis de Geologies Malgache. Annales Geologiques de Madagascar, Tananarive 36:1–141

Bhargava ON (2008) Paleozoic succession of the Indian plate. Mem Geol Soc Ind 74:209–244

Bhargava ON (2015) Evolution of the Tethyan and Karewa successions in Kashmir: a synthesis. J Pal Soc Ind 60(1):51–72

Bhargava ON, Krystin L, Balini H, Lein R, Nicora A (2004) Revised litho and biosequence stratigraphy of the Spiti Triassic. Albertiana Suppl 30:21–38

Bhatia SB, Bhargava ON, Singh BP, Bagi H (2013) Sequence stratigraphic framework of the Paleogene succession of the Himalayan Foreland basin: a case study from the Shimla hills. J Pal Soc Ind 58(1):21–38

Biswas SK (1990) Correlation of nonmarine and estuarine Cretaceous of Western India, Pres. Address, Sem-cum-Workshop IGCP 216 and 245, Chandigarh, pp 1–13

Biswas SK (1999) A review of the evolution of rift basins of India during Gondwana with special reference to Western Indian basins and their hydrocarbon prospects. PINSA 65(3):261–283

Biswas SK (2005) Tectonic style and sediment dynamics of rifted basins, their bearing on petroleum habitat—examples from Satpura and Kutch, Basins, India. Ind J Petr Geol 14(1):1–29

Biswas SK (2007) Nannofossil assemblage in Kutch. Curr Sci 92(5)

Bladon Andrew J, Stuart M, Clarke, Stuart D, Burley, Hazel B, Whiteley N, Kothari V, Mohapatra P (2015) previously unrecognised extensional tectonics exposed in the Barmer basin, Northwest India. Search and Discovery Article #30426

Bodin S, Godet A, Follmi K B, Vermuelen J, Arnaud H, Strasser A, Fiet N, Adatte T (2006) The Late Hauterivian Faraoni oceanic event in the Western Tethys; evidence from phosphorus burial rates, vol 235, pp 245–264

Bomfleur B, Schoner R, Schneider JW et al (2014) From the transantarctic basin to the Ferrar large igneous province—new palynostratigraphic age constraints for Triassic—Jurassic sedimentation and magmatism in East Antarctica. Rev Palaeobot Palynol 207:18–37

Bond PG, Wignall PB (2014) Large igneous provinces and mass extinctions: an update. Geol Soc Am Spl Pap 505:29–57

Bose MN (1884) Geology of the Lower Narmada valley between Nimawar and Karwant. Mem Geol Surv Ind 21(1):1–72

Bralower TJ, Sliter WV, Arthur MA, Leckie RM, Allard D, Schlanger SO (1993) Dysoxic/anoxic episodes in the Aptian/Albian (Early Cretaceous). Geophys Monogr 77:5–37

Bronnimann NP, Whittaker JE, Zaninetti L (1975) Triassic forminiferal biostratigraphy of the Kyaukme–Langtaw–Kno area, Norther Shan Straits. Burma Riv Ital Paleont 81:1–30

Brookfield ME, Algo TJ, Hanningon R, Williams J, Bhat GM (2013) Shaken and stirred: seismites and psunamites at the Permian/Triassic boundary, Guryal ravine, Kashmir, India. Palaios 28:568–582

Brookfield ME, Twichet RJ, Goodings C (2003) Paleoenvironments of the Permo-Triassic transition sections in Kashmir, India. Palaeogeogr. Palaeoclimat Palaeoecol 198:353–371

Calkins JA, Offield TW, Abdullah KM, Ali ST (1975) Geology of the Southern Himalaya in Hazara Pakistan and adjacent area, U.S.G.S. Professional Paper, 716 C, 29

Casshyap SM, Khan A (2000) Tectono-sedimentary evolution of the Gondwanian Satpura basin of Central India: evidence of pre-trap doming, rifting, and paleoslope reversal. J Afr Earth Sci 31(1):65–76

Chandrasekhar B, Martha TR, Venkateswarlu N, Subramaniam SK, Kamaraju MUV (2011) Regional geological studies over parts of Deccan syneclise using remotesensing and geophysical data for understanding hydrocarbon prospects. Curr Sci 100(1):10

Chatterjee S, Goswami A, Scotese CR (2013) The longest voyage: tectonic, magmatic, and paleoclimatic evolution of the Indian plate during its northward flight from Gondwana to Asia. Gond Res 23:238–267

Chatterjee S, Hotton N (1986) The palaeoposition of India. J SE Asian Earth Sci 1:145–189

Chatterjee S, Scotese CR (1999) The break-up of Gondwana and the evolution and biogeography of the Indian plate. PINSA 65(A3):397–425

Chaube AK, Bhattacharya GC, Murty GPS, Desa M (1993) Spreading history of the Arabian sea: some new constraints. Mar Geol 112:343–352

Chidambaram L (2000) Middle Turonian—Santonian foraminiferal biostratigraphy of Trichinopoly group, south India. Mem Geol Soc Ind 46:173–204

Coffin MF, Rabinowitz (1992) The Mesozoic East African and Madagascan conjugate margins; stratigraphy and tectonics, Mem. AAPG 53:207–240

Coffin MF, Pringle MS, Duncan RA, Gladczenko TP, Storey M, Muller RD, Gahagan LA (2002) Kerguelen hot spot magma output since 130 ma. J Petrol 43(7):1121–1139

Cogne JP, Humler E (2004) Temporal variation of oceanic spreading and crustal production rates during the last 180 ma. Ear Planet Sci Lett 227:427–439

Coulon C, Maluski H, Bollinger C (1986) Mesozoic and Cenozoic volcanic rocks from central and southern Tibet:39Ar/40Ar dating petrological characteristics, and geodynamic significance. E Planet Sci Lett 79(30):281–302

D G H (2003) Docket for Deccan syneclise, 30 p

Dai JG, Wang C, Li Y (2012) Relicts of early Cretaceous sea mounts in the central western Yarlung—Tsangpo suture zone southern Tibet. J As Earth Sci 53:25–37

Dai JG, Wang C, Polat A, Santosh M, Li Y, Ge Y (2013) Rapid fore arc spreading between 130 and 120 ma; evidence from geochronology and geochemistry of the Xigaze Ophiolite, southern Tibet. Lithos 172–173:1–16

Dasgupta SK (1976) The stratigraphy of the West Rajasthan Shelf, Proc. 4th Ind. Coll Micropal Strat 2:19–223

Datta AK, Poddar BC, Kolhapur VK, Bhattacharya BK, Sanval R (1996) Geochronological studies of Mesozoic igneous rocks from Eastern India. J SE Asian E Sci 21:188–193
Dave A (2011) Recent Developments in sequence stratigraphy and depositional sequences of Indian petroliferous basins. In: Jaitly AK, Singh AD, Pandey B and Nath S (ed) Palaeontology and stratigraphy: basics to applications. Banaras Hindu University, Varanasi, pp 88–112
Dave A (2012) Depositional sequence stratigraphy and petroleum systems of Kachchh basin—Status Talk. In: Thakkar MG (ed) Geology of Kachchh basin. Western India, Bhuj–Kachchh, pp 94–97
Dera G, Brigaud B, Monna F, Laffont R, Pucéat E, Deconinck JF, Pellenard P, Joachimski MM, Durlet C (2011) Climatic ups and downs in a disturbed Jurassic world. Geology 39:215–218
Desa M, Ramana MV, Ramaprasad T (2006) Sea-floor spreading anomalies south of Shrilanka. Mar Geol 229:227–240
Dromart G, Garcia JP, Picard S, Atrops F, Lecuyer C, Sheppard SMF (2003) Ice age at the Middle/Late Jurassic transition: geological and geochemical evidence. Am J Sci 303:667–707
Duncan RA, Hooper PRH, Rehace, Arsh SM, Duncan AR (1997) The timing and duration of the Karoo igneous event, southern Gondwana. J Geophys Res 102:B8, 18, 127–18, 138A
Durr SB, Gibling MR (1994) Early Cretaceous volcanoclastics and quartzose sandstones from north-central Nepal. Geol Rundsch 83:62–75
Ehiro M, Kojima S, Sato T, Ahmad T and Ohtani T (2007) Discovery of Jurassic ammonoids from the Shyok suture zone to the northeast of Chang La pass, vol 16. Ladakh, Northwest India and its Tectonic Significance, Island Arc, pp 124–132
Fatmi AN (1986) Lower and Middle Jurassic ammonites from Windar Group of Phusi Jhal (Kanrach) and Sand (Windar Nai) sections of Las Bella, Baluchistan, Pakistan. Mem Geol Surv Pak, Pal Pak 16:1–17
Gaedicke C, Schluter HU, Roeser HA, Prexl A, Shrekenperger B, Meyer H, Reichert C, Clift P, Amjad S (2002) Origin of the northern Indus fan and Murray ridge, northern Arabian sea: in terpretations from seismic and magnetic imaging. Tectonophysics 355:123–143
Gaina C, Hinsbergen DJJ, Spakman W (2015) Tectonic interactions between India and Arabia since the Jurassic reconstructed from marine geophysics, ophiolite geology, and seismic tomography. Tectonics 34:875–906
Garzanti E (1999) Stratigraphy and sedimentation history of the Nepal Tethys Himalaya passive margin. J.As. Earth Sci 17:805–827
Gautam J, Pandey B, Pathak DB, Jaitly AK (2015) Recognition of the Early Albian Douvilleiceras mammillatum Zone in the Cauvery Basin. SE India. J Earth Sci Ind 8(4):100–111
Geary E, Kay R (1989) Identification of an Early Cretaceous ophiolite in the Camarines Norte-Calaguas Islands basement complex, eastern Luzon. Philippines, Tectonophysics 168 (109–126):1989. doi:10.1016/0040-1951(89)90371-5
Geary E, Harrison TM, Heizler M (1988) Diverse ages and origins of basement complexes. Luzon, Philippines, Geol 16:341–344. doi:10.1130/0091-7613(1988)016<0341:DAAOOB>2.3.CO;2
Geiger M (2004) Sedimentary and stratal patterns in Jurassic successions of Western Madagascar–facies, stratigraphy, and architecture of Gondwana break up and drill sequences, Ph.D thesis, Uni. Bremen
Ghose NC, Mitra I, Singh BP (1996) Volcanoclastic and related rocks of Rajmahal traps, Eastern India, Sp. Publ. 42, Geol. Surv. Ind., pp 133–142
Ghose NC, Shmakin BM, Smirnov BN (1973) Some geochronological observations on the Precambrians of Chotanagpur, Bihar. Geol Mag 110:481–484
Ghosh N, Basu AR, Bhargava ON, Shukla UK, Ghatak AR, Garzione CN and Ahluwalia AD (2014) Catastrophic environmental transition at the Permian/Triassic Neotethyan margin of Gondwanaland: geochemical, isotopic, and sedimentological evidences in the Spiti valley, India. Gond Res. doi:10.1016/j.gr.2015.04.006
Gibbons AD, Zahirovic S, Muller RD, Whittaker JM, Yattesh V (2015) A tectonic model reconciling evidence for the collisions between India, Eurasia, and intra-oceanic arcs of the central-eastern Tethys. Gond Res 28(2):451–492

Gibbons AD, Barckhausen U, Bogaard VDPK, Hoemler K, Werner R, Whittaker JM, Muller RD (2012) Constraining the Jurassic extent of Greater India: tectonic evolution of the West Australian margin, Geochem. Geophys Geosyst 13:Q05W13. doi:10.1029/2011GC003919

Gibbons AD, Whittakar JM, Mueller RD (2013) The break-up of East Gondwana: assimilating constraints from Cretaceous ocean basins and India into a best fit tectonic model. J Geophys Res Solid Earth 118:808–822

Gibling MR, Gradstein FM, Kristianen IL, Nagy J, Sarti M, Wiedmann J (1994) Early Cretaceous strata of Nepal Himalaya: conjugate margins and rift volcanism during Gondwana break-up. J Geol Soc Lond 151:269–290

Gnos E, Immenhauser A, Peters T (1997) Late Cretaceous / Early Tertiary convergence between the Indian and Arabian plates recorded in ophiolites and related sediments. Tectonophysics 271:1–19

Govindan A, Ravindran CN (1996) Cretaceous biostratigraphy and sedimentation history of Cauvery basin, India. In: Pandey et al (ed) Contribution to XV Indian Colloquium on micropaleontology and stratigraphy, 19–32

Gradstein FM, Gibling MR, Sarti M, Rad UV, Thruro JW, Ogg JG, Jansa LF, Kaminski MA, Westermann GEG (1991) Mesozoic Tethyan strata of Thakkhola, Nepal: evidence for the drift and break-up of Gondwana. Paleogeogr Paleoclimat Paleoecol 88:193–218

Groppo C, Lombardo B, Rolfo F, Pertusati P (2007) Clockwise exhumation path of granulitized eclogites from the Ama Drime range (eastern Himalayas). J Metamorph Geol 25:51–75

Guo Z, Wilson M, Liu J (2007) Post-collisional adakites in south Tibet: products of partial melting of subduction-modified lower crust. Lithos 96:205–224

Haggerty B (1996) Episodes of flood basalt volcanism correlated with mass extinctions. J Undergrad Sci 3:155–164

Haq BU, Hardenbol J, Vail PR (1987) Chronology of fluctuating sea-level since the Triassic. Science 235:1156–1167

Harris LB (1996) Polyphase faulting in the Perth Basin and Yilgarn Craton marin, Western Australia: Relationship to Upper Palaeozoic–Mesozoic Tectonics of Gondwanaland. Gondwana 9(2):1051–1060

Hein C (2002) The tectonic evolution of the Northwest Shelf of Australia and southern Southeast Asia. Unpublished Thesis

Hill KC, Hall R (2003) Mesozoic-Cenozoic evolution of Australia's New Guinea margin in a west pacific context. Geol Soc Aust Spl Publ 22; Geol Soc Am Spl Pap 372:266–289

Hornung T, Krystyn L and Brandner R (2007) A Tethys-wide mid-Carnian (Upper Triassic) carbonate productivity crisis: evidence for the alpine Reingraben event from Spiti (Indian Himalaya)?. J Asian Earth Sci doi:10.1016/j.jseaes.2006.10.001

Hu XM, Sinclair HD, Wang JG, Jiang HH, Wu FY (2012) Late Cretaceous-Palaeogene stratigraphic and basin evolution in the Zhepure Mountain of southern Tibet: implications for the timing of India-Asia initial collision. Basin Res 24:520–543

Hu X, Garzanti E, An W (2015) Provenance and drainage system of the Early Cretaceous volcanic detritus in the Himalaya as constrained by detrital zircon geochronology. J Paleogeogr 4(1):85–98

Immenhauser A (1996) Cretaceous sediments on the Masirah Island ophiolite (Sultanate of Oman): Evidence for an unusual bathymetric history. J Geol Soc London 153:539–551

Jafar SA (2015) Episodes of subaerial large igneous provinces (LIPs) link to late Turonian/late Maestrichtian deep incursion of the sea on Indian continental block. (abstract), XII Symposium on Mesozoic terestrial systems, Shenyang, China

Jafar SA (1982) Nannoplankton evidence of the Turonian transgression during Narmada valley, India, and Turonian–Coniacian boundary problem. J Pal Soc Ind 27:17–30

Jafar SA, Rai J (1989) Discovery of Albian nannoflora from type Dalmiapuram Formation, Cauvery basin, India, and palaeooceanographic remarks. Curr Sci 58(7):358–363

Jagoutz O, Royden L, Holt AF, Becke W (2015) Anomalously fast convergence of India and Eurasia by double subduction. Nat Geosci 8:475–478

Jain AK (2014) When did India–Asia collide and make the Himalaya. Curr Sci, Spl Sec Sci Himalaya 106(2):254–266

Jain AK, Talat Ahmad, Sandeep Singh, Ghosh SK, Patel RC, Kumar Rohtash KK, Agarwal J, Perumal Islam R, Bhargava ON (2012) Evolution of the Himalaya, Proc Indian Nat Sci Acad 78(3):259–275

Jauhari AK, Agrawal KK (2001) Early Paleogene in south Shillong plateau Northeast India: local biostratigraphic signals of global oceanic and tectonic changes. Palaeogeogr Palaeoclimatol Palaeoecol 168:187–203

Jenkyns HC (1980) Cretaceous anoxic events: from continents to oceans. J Geol Soc, London 137:171–188

Jenkyns HC (1985) The early Toacian and Cenomanian – Turonian anoxic events in Europe: Comparisons and contrasts. Geol Rdsch 74:505–518

Johnson SY, Alam AMN (1991) Sedimentation and tectonics of the Sylhet trough, Bangladesh. Geol Soc Am Bull 103:1513–1527

Jourdan F, Feraud G, Bertrand H, Watkeys MK, Renne PR (2007) Distinct brief major events in the Karoo large igneous province clarified by new Ar40/Ar 39 ages on the Lesotho basalts. Lithos 98(124):195–209

Kale AS, Lotfalikani A, Phansalkar VG (2000) Calcareous nannofossils from the Uttatur group of Trichonopoly Cretaceous, South India. Mem Geol Soc Ind 46:213–227

Kalmalkar NR, Duraiswami RA, Jonalagadda MK, Griffin WL (2014) Mid-Cretaceous lamproite from the Kutch region, Gujarat, India. Gond Res 26(3–4):942–956

Kanjilal S, Singh SR (2012) Arrival of Quenstedtoceras Hyatt: a late Callovian ammonite in Kachchh and its significance, Geology of Kachchh basin, p 27

Keller G, Adatte T, Bajpai S, Mahobay DM, Widdowson M, Khosla A, Sharma R, Khosla SC, Gertsch B, Fleitmann D, Sahni A (2009) K/T transition in Central India marks major marine seaways across India. Earth Planet Sci Lett 282:10–23

Kent RW, Pringle MS, Muller RD, Saunders AD, Ghose NC (2002) 40Ar/39Ar Geochronology of the Rajmahal Basalts, India, and their Relationship to the Kerguelen Plateau. J Petrol 43(7):1141–1153

Khan AA (1989) Stratigraphy and hydrocarbon potentian of Permo-Triassic sequence of rocks in the state of Kuwait, Proceedings of the deep formation, organization of Arab paetroleam exporting countries, Kuwait, E3–E29

Konig M, Jokat W (2006) The mesozoic breakup of the Weddell sea. J Geophys Res 111:B 12102 (2006). doi:10.1029/2005JB004035

Korte C, Kozur HW (2011) Temperature changes across the Perman-Triassic boundary and observations pertaining to some geological; geochemical features around this level. Mem Geol Soc Ind. 78:100–139

Korte C, Hesselbo SP, Ullmann CV, Dietl G, Ruhl M, Schweigert G, Thibault N (2015) Jurassic climate mode governed by ocean gateway. Nat Commun 6. doi:10.1038/ncomms10015

Krishna J (1980) Uncoiled ammonites of middle Albian (Lower Cretaceous) age from Habur Series, Jaisalmer, Rajasthan. J Pal Soc Ind 23, 24:49–54

Krishna J (1983) Callovian – Albian ammonoid stratigraphy and paleobiogeography in the Indian subcontinent with special reference to the Tethys Himalaya. Him Geol 11:43–72

Krishna J (1986) Bearing of paleontological record on plate tectonics with reference to Indian plate. Short Term Course in Paleontology and Stratigraphy. COSIP-ULP in Geol, 257–269

Krishna J (1987a) Biological evidence for better appreciation of Indian Gondwana. Palaeobotanist 36:268–284

Krishna J (1987b) The Jurassic-Cretaeous ammonoid geography in the Indian subcontinent vis-à-vis marine seaways and plate tectonics in the Indian occean region, 3 Decades of Development in Paleontology and Stratigraphy in India. Spl Publ Geol Surv Ind 11(1):453–482

Krishna J (1987c) An overview of the mesozoic stratigraphy of Kachchh and Jaisalmer. J Pal Soc Ind 32:136–149

Krishna J (1994) Environmental reconstruction studies of the Mesozoic sedimentary sequences in Western Kachchh (ONGC Sponsored Project), (Technical Report), 1–143, pts, 1–10

Krishna J (1996) Origin of the Gondic (Trans-Gondwana) corridor: the ammonoid evidence. Gondwana Nine, Oxford and IBH Publishing Company, Delhi, 2, 1091–1099

Krishna J (2000) An overview of the recent progress in the Cretaceous stratigraphy of Kachchh basin, India. Mem Geol Surv India 46:399–401

Krishna J (2005) Update on Kachchh Jurassic ammonoid zonation: summarised intra-basinal to global perspectives. Spl. Publ.-1, Ass. Petrol. Geol., 63–68

Krishna J (2009) Genetic relationships among extensional tectonics, regional geology, sequence stratigraphy and biotic evolution: example of the Gondwanian Tethyan Mesozoic contextual to hydrocarbon exploration as modeled for the Kachchh sector, India. J Ind Geol Cong 1(1):30–34

Krishna J (2012) Litho-bio-chrono-magneto-igno-sequento-stratigraphic integration in the Kachchh Mesozoic, Status Talk–Mesozoic Biostratigraphy. In: Thakkar MG (ed) Abstract Volume, Geology of Kachchh basin, Western India, pp 21–25

Krishna J (2014) Ammonoid–sequence- magneto stratigraphic integration in the Kachchh Jurassic. Unpublished DST project report, 60 p

Krishna J (2016) Initiation of Indian ocean at ~159 ma late Middle Oxfordian Orientalis Zone Schilli Subzone MFS event (abstr), 35th Internat. Geol Congr, Capetown

Krishna J, Cariou E (1993) The Tethyan Macrocephalitinae: evolutionary, environmental and dispersal strategies. Geobios 15:227–236

Krishna J, Pathak B (1994) Stratigraphic, biogeographic and environmental signatures in the amonoid bearing Jurassic- Cretaceous in the Himalaya on the south margin of the Tethys. J Him Geol 4(2):151–167

Krishna J, Pandey B, Pathak DB (2010) Update on the Jurassic stratigraphy of the Indian subcontinent with focus on the tectonically controlled regional transgressive–regressive couplets. Earth Sci Frontiers, China 17:138–140

Krishna J, Pandey B, Pathak DB (2011) Current status of the Jurassic ammonoid stratigraphic framework in the Indian subcontinent with focus on the tectonically controlled regional transgressive—regressive couplets. Geol Soc Ind Mem 78:140–176

Krishna J, Pandey B, Ojha JR, Pathak DB (2000) Transgressive sediment intervals in the Late Jurassic of Kachchh. India, GeoRes Forum 6:321–332

Krishna J, Pathak DB, Pandey B (1998) Develoment of Oxfordian (early Upper Jurassic) in the most proximal exposed part of Kachchh basin at wagad outside the mainland Kachchh. J Geol Soc Ind 52:513–522

Krishna J, Pathak DB, Pandey B (1994) Discovery of middle and late Oxfordian in Kachchh (India) with first ever approximation of the Oxfordian/Kimmeridgian boundary in the Indian subcontinent (abstr) Oxfordian—Kimmeridgian Joint Working Groups Meet, Lyon, 10

Krishna J, Singh IB, Howard JD, Jafar SA (1983) Implications of new data on the Mesozoic rocks of Kachchh. India, Nat 305:790–792

Krishna KS, Abraham H, Sager WW, Pringle MS, Frey F, Gopala Rao D, Levchenko OV (2012) Tectonics of the Ninetyeast Ridge derived from spreading records in adjacent oceanic basins and age constraints of the ridge. J Geophys Res: Solid Earth 1978–2012:117

Krystyn L, Bhargava ON, Richoz S (2007) A candidate GSSP for the base of the Olenekian Stage: Mud at Pin Valley; district lahul & Spiti, Himachal Pradesh (Western Himalaya). India, Albertiana 35:5–29

Kumar A, Pande K, Venkatesan TR, Bhaskar-Rao YJ (2001) The Karnataka Late Cretaceous dykes as products of Marion hot spot at the Madagascar–India break-up event; evidence from Ar40-Ar39 geochronology and geochemistry. Geophys Res Lett 28:2715–2718

Lakshminarayana (2002) Evolution in basin fill style during the Mesozoic Gondwana continental breakup in the Godavari triple junction, southeastern India. Gondawana Res 5:227–244

Lawver LA, Gahagan LM, Coffin MF (1992) The development of palaeoseaways around Antarctica. Antarc Res Ser 60:145–174

Leier AL, DeCelles PG, Kapp P, Gehrels GE (2007) Lower Cretaceous strata in the Lhasa terrane, Tibet, with implications for understanding the early tectonic history of the Tibetan Plateau. J Sed Res 77:809–825

Li G, Jiang G, Hu X, Wan X (2009) New biostratigraphic data from the Cretaceous Bolinxiala Formation in Zanda, southwestern Tibet of China, and their paleogeographic and paleoceanographic implications. Cret Res 30:1005–1018

Looseveld RJH, Bell A, Terken JJM (1996) The tectonic evolution of Oman. Geoarabia 1:28–51

Mahmood K, Boudier F, Gnos E, Monie P, Nicolas A (1995) 40Ar/39Ar dating of the emplacement of the Muslim Bagh ophiolite. Pakistan, Tectonophysics 250:169–181

Malkowski MA, Grove M, Graham SA (2016) Unzipping the Patagonia–Andies long lived influence of rifting history on foreland basin evolution. Lithosphere. doi:10.1130/L489.1

Manninen T, Eerola T, Makiti H, Uvori S, Luttinen A, Senvano A, Manhika V (2008) The Karoo volcanic rocks and related intrusive in souther central Mozambic. J Geol Surv Finl, Spec Pap 48:211–250

Matthews KJ, Seton M, Müller RD (2012) A global-scale plate reorganization event at 105–100 ma. Ear Planet Sci Lett 355:283–298

Mehrotra NC, Aswal HS (2003) Late Jurassic–Cretaceous Dinoflagellage Cysts. Pal Ind 7, ONGC, Dehradoon, 1–108

Mellusoe L, Sheth SC, Mahoney JH, Morra V, Petrone CM, Storey M (2009) Correlations between silicic volcanic rocks of the St Mary's Islands (southwestern India) and eastern Madagascar: implications for Late Cretaceous, India-Madagascar reconstructions. J Geol Soc Lond 166:283–294

Meng J, Wang C, Zhao X, Coe R, Li, Finn D (2012) India-Asia collision was at 24°N and 50 ma: palaeomagnetic proof from southernmost Asia, Scientific reports 2:925, PMC 3514643

Metcalfe I (2011) Tectonic framework and Phanerozoic evolution of Sundaland. Gond Res 19:3–21

Meyer KM, Kump LR (2008) Oceanic euxinia in earth history: causes and consequences. Annu Rev Earth Planet Sci 36:251–288

Mitra ND, Raja Rao CS (1987) Recent advances in the study of the Gondwana Stratigraphy of India. Geo Sur Ind Spec Pub 2:136–145

Moore A, Blenkinsop T, Cotterill F (2008) Controls on post-Gondwana alkaline volcanism in Southern Africa. Ear Planet Sci Lett 268:151–164

Mukasa SB, Ludden JN (1987) Uranium–lead isotopic ages of plagiogranites from the Troodos ophiolite, Cyprus, and their tectonic significance. Geology 15:825–828

Müller RD, Roest WR, Royer J-Y, Gahagan LM, Sclater JG (1997) A digital age map of the ocean floor, SIO Reference Series 93–30, Scripps Institute of Oceanography. http://www.ngdc.noaa.gov/mgg/global/crustage.HTML

Mutterlose J (1986) Upper Jurassic belemnites from the Orville coast, Western Antarctica, and their palaeobiogeographical significance. Brit Antarct Surv Bull 70:1–22

Nagendra R, Kamalnarayanan BV, Gargi S, Gilbert H, Bakkalraj D, Nallapa-Reddy A, Jaiprakash B (2011) Sequence surfaces and paleo-bathymetric trends in Albian to Maestrichtian sediments of Ariyalur area, Cauvery basin. J Mar Petr Geol 28:895–905

Nagendra R, Nagendran G, Narsimha K, Jaiprakash BC, Nallapa-Reddy A (2002) Sequence stratigraphy of Dalmiapuram Formation, Kalakuddy quarry II, South India. J Geol Soc Ind 59:249–258

Najman Y, Appel E, Boudagher-Fadel M, Bown P, Carter A, Garzanti E, Godin L, Han J, Liebke U, Oliver G, Parrish R, Vezzoli G (2010) Timing of India-Asia collision: Geological, biostratigraphic, and palaeomagnetic constraints. J Geophys Res 115:B12. doi:10.1029/2010JB007673

Nathaniel DM (2013) Hydrocarbon potential of sub-basalt Mesozoics of deep water Kerala basin, 10th biennial International Conference and Exposition. Kochi 428:6

Singh SR (1992) Ammonite biostratigraphy of the Callovian—Oxfordian (Jurassic) rocks of Nara dome, district Kachchh, Gujarat. Unpublished PhD thesis BHU, 328p

Nurnburg D, Muller RD (1991) The tectonic evolution of the south Atlantic from late Jurassic to Present. Tectonophysics 191:27–33

Ogg JG, Gutowski J (1996) Oxfordian and Lower Kimmeridgian magnetic polarity time scale. GeoRes Forum 1–2:406–414

Okimura and Fatmi (1988) Tectonics and sedimentation of the Indo-Eurasian colliding plate boundary region and its influence on the mineral development in Pakistan. Jour, Overseas Scientific Researchy, Hiroshima University, 32:1–32

ONGC (1993) Lithostratigraphy of Indian petroliferous basins documents. KDMIPE, Dehradun

ONGC (2008) Pteroleum systems sequence stratigraphy. KDMIPE, Dehradun

Pal AK (1971) A note on the biozonation of the Bagh Group of Madhya Pradesh, Abstract. Proc Ind Sci Congr 333

Pande K, Sheth HC, Bhutani R (2001) Ar40/Ar39 age of the St. Mary's Islands volcanic Southern India. Records of India Madagascar breakup of the Indian Sub-continent. Earth Planet Sci Lett 193:39–46

Pandey B, Pathak DB (2015a) Record of Early Bathonian ammonoids from Kachch, India: biostratigraphic and paleobiogeographic implications. J Pal Soc Ind 60(1):33–44

Pandey B, Pathak DB (2015b) Status of the Indian Early Cretaceous ammonoid record in light of recent observations in the Spiti Valley. Himachal Himalaya, Himalayan Geol 36(1):1–8

Pandey B, Pathak DB (2016) The possibility of the oceanic anoxic events (OAEs) study in the Indian marine Jurassic–Cretaceous outcrops. J Geol Soc Ind 87:261–267

Pankhrust RJ, Weaver SD, Herve F, Larrondo P (1999) Mesozoic–Cenozoic evolution of the Northern Patagonian Batholith in Aysen, southern Chile. J Geol Soc London 156:673–694

Parson LM, Evans AJ (2005) Seafloor topography and tectonic elements of the Western Indian Ocean. Phil Trans Roy Soc Lond Ser A 363:15–24

Pathak DB, Krishna J, Pandey B (2011) Differentiation of the Significant Late Valanginian (Early Cretaceous) Transgressive Event in the Spiti Himalaya, India. J Asian Earth Sci 42:1226–1231

Paul DK, Ray A, Biswas SK, Parh DSM (2012) Petrological and geochemical heterogeneity of the magmatic rocks of Kachchh, Gujarat, in relation to tectonic setting and source characteristics, Status Talk–Magmatism in Kachchh. In: (ed.) Thakkar MG (ed) Abstract Volume, Geology of Kachchh basin, Western India, pp 80–82

Pearce CR, Hesselbo SP, Coe AL (2005) The mid-Oxfordian (Late Jurassic) positive carbon-isotope excursion recognised from fossil wood in the British Isles. Palaeogeogr Palaeoclimatol Palaeoecol 221(3–4):343–357

Pedersen RB, Searle MP, Corfield RI (2001) U-Pb zircon ages from the Spongtang ophiolite, Ladakh Himalaya. J Geol Sco London 158:513–520

Plummer PhS, Belle ER (1995) Mesozoic tectono-stratigraphic evolution of the Seychelles microcontinetn. Sed Geol 96:73–91

Poblete F, Roperch P, Arriagada C, Ruffet G et al (2016) Dike systems and their volcanic host rocks on King George Island. Antartica, Tectonophysics 668–679:15–34

Polvé M, Maury R, Bellon H, Rangin C, Priadi B, Yuwono S, Joron J, Atmadja RS (1997) Magmatic evolution of Sulawesi (Indonesia): constraints on the Cenozoic geodynamic history of the Sundaland active margin. Tectonophysics 272:69–92. doi:10.1016/S0040-1951(96)00276-4

Potonie R, Sah SCD (1960) Sporae dispersae of the lignites from Cannanore beach on the Malabar coast of India. Palaeobotanist 7(2):121–135

Prasad B (2011a) Palynology: Principles, methodology and applications in biostratigraphy and oil exploration. In: Palaeontology and stratigraphy: basics to applications. Jaitly AK, Singh AD, Pandey B, Nath S (eds), BHU Varanasi, pp 16–35

Prasad B, Phor (2009) Palynostratigraphy of the subsurface Gondwana and post Gondwna Mesozoics of the Cauvey Basin, India. J Pal Soc Ind 54(1):41–71

Prasad B, Pundeer BS (1999) Biostratigraphy of the exposed Gondwana and Cretaceous rocks of the Krishna–Godavari basin, India. J Pal Soc Ind 44:91–117

Prasad B, Jain AK, Mathur YK (1995) A standard palynozonation scheme for Cretaceous and Pre Cretaceous subsurface sediments of Krishna-Godavar Basin, India. Geo Sci J 16(2):151–232

Prasad B, Pundeer BS, Swami SN (2000) Age, depositional environment, source potential and distribution of pay sands in Mandapeta subbasin with remarks on the biostratigraphic classification of Gondwana sediments in Krishna-Godavari basin. Ind J Petrol Geol 9(1):1–19

Prasad GVR (2011) Vertebrate Palaeontological research in India: recent advances and future prospects. In: Jaitly AK, Singh AD, Pandey B, Nath S (eds) Palaeontology and stratigraphy: basics to applications, pp 49–58

Prinz P, Legler C, Legler G (1993) Marine Jurassic sediments in Mozambique. Neues Jahbuch der Geologie und Palaeontologie 8:489–496

Punekar J, Saraswati PK (2010) Age of the Vastan Lignite in context of Some Oldest Cenozoic Fossil Mammals from India. J Geo Soc Ind 76:63–68

Qi M, Xiang H, Zhong ZQ, Qiu HN, Wang H, Sun XL, Xu B (2013) Ar40/Ar39 geochronology constraints on the formation age of Myanmar Jadeitite. Lithos 162–163:107–114

Qiu BB, Zhu DC, Zhao ZD, Wang LQ (2010) The westward extension of Comei fragmented large igneous province in southern Tibet and its implications. Acta Petrologica Sinica 26(7):2207–2216 (in Chinese with English abstract)

Qiu Z, Wu F, Yang S, Zhu M, Sun J, Yang P (2009) Age and genesis of Myanmar Jadeitite constraints from U-Pb ages and H F Isotopes of Zircon inclusions. Bull Chinese Sci 54:658–668

Rabinowitz PD, Coffin MF, Falvey D (1983) The separation of Madagascar and Africa. Science 220:67–69

Radhakrishna T, Maluski H, Mitchell JG, Joseph M (1999) Ar40/Ar39 and K/Ar geochronology of the dykes from the south Indian granulite terrain. Tectonophysics 304:109–129

Rai J (2006) Discovery of nannofossils in a plant bed of the Bhuj Member, Kutch and its significance. Curr Sci 91(4):519–526

Rai J, Jain S (2012) Early jurassic gondwana break-up—a nanno-fossil story, in Geology of Kachchh basin, western India (abstract), pp 31–32

Rai J, Jain S (2013) Pliensbachian nanno-fossils from Kachchh: implications on the earliest Jurassic transgressive event on the western Indian margin. Zitteliana, A 53:105–120

Rai J, Garg R, Kumar S (2002) Early campanian nannofossils from chakrud, near Zeerabad, Bagh area: stratigraphi and palaeoenvironmental implications. National Conference on Biodiversity-Past and Present, The palaeobotanical Society, Lucknow, pp 53

Rai J, Garg S, Gupta M, Singh A, Pandey DK, Fursich FT, Alberti M, Garg R (2015a) Integrated biostratigraphy of the Jurassic strata of the Wagad uplift, Kachchh, Western India. Vol Jur, XIII(2):55–80

Rai J, Pathak DB, Pandey B (2015) Early Campanian calcareous nannofossils from Chikkim Formation, Spiti Valley, Tethys Himalaya, India, 30th Himalaya-Karakoram-Tibet Workshop Dehradun, India, pp 32–33

Rai J, Singh A, Garg R (2013a) Calcareous nannofossils of Albian age from Tanot well 1, Jaisalmer basin, Rajasthan, and its palaeobiogeographic significance. J Pal Soc Ind 58(1):67–73

Rai J, Singh A, Pandey DK (2013b) Early to Middle Albian age calcareous nannofossils from Pariwar Formation of Jaisalmer basin. Rajasthan, Western India, Significance 105(11): 1604–1611

Raina BK, Bhattacharya DP (1977) The Geology of a part of the Charap and Sarchu valley, Lahul and Spiti district, Himachal Prades. Quat J Geol Min Met Soc Ind 129–142

Rais P, Louis-Schmid B, Bernasconi SM, Weissert H (2007) Paleoceanographic and paleoclimatic reorganization around the Middle/Late Jurassic transition. Paleogeogr Paleoclimat Paleoecol 251:527–546

Rajaram M, Anand SP, Balakrishna TS (2006) Composite magnetic anomaly map of India and its contiguous regions. J Geol Soc Ind 68:569–576

Rajnath (1934) Detailed stratigraphy of the Jumara area, Cutch (abstr). In: Proceedings of the 21 st Ind Sci Congr, pp 346

Raju DSN (2011) Cretaceous stratigraphic record on Indian plate: an overview. Mem Geol Soc Ind 78:177–228

Ramana MV, Nair RR, Sarma KVLNS, Ramaprasad T, Krishna KS, Subrahmanyam V, D'Cruz M, Subrahmanyam C, Paul J, Subrahmanyam AS, Chandrasekhar DV (1994) Mesozoic anomalies in the Bay of Bengal, Earth Planet. Sci Lett 121:469–475

Ramana MV, Ramprasad T, Desa M (2001) Sea floor spreading magnetic anomalies in the Enderby basin, east Antarctica, Earth Planet. Sci Lett 191:241–255

Ramana MV, Subrahmanyam B, Chaubey AK et al (1997) Structure and origin of the 85° E. ridge. J Geophys Res 102:17995–18102

Ramkumar M, Steuben D, Berner Z (2004) Lithostraigraphy, depositional history and sea-level changes of the Cauvery basin, South India. Ann Geol Penins Balk 65:1–27

Rana MS, Chakrobarty C, Sharma R, Girdhar M (2008) Mannar volcanic–implications for Madagascar breakup, P-358, 7th International confrence and exposition on petroleum geophysics, Hyderabad, p 6

Rao NVC, Dongreb R, Wu FU, Lehmannd B (2015) A Late Cretaceous (ca. 90 ma) kimberlite event in southern India: implication for sub-continental lithospheric mantle evolution and diamond exploration, Gond. Res., GR-01463, p 12

Rao TR, Rao YV, Prasad GJS, Rao PT (1996) Tectonics of chintalapudi and the adjoining subbasins of gondwanas of godavari valley and the east coast of India, Gondwana. Spl Publ Geol Surv Ind 9(2):755–782

Rawal U (1995) Geodynamics of the tectonomagnetic and geophysical signatures with in mobile part of the transact. Mem Geol Soc Ind 31:37–61

Ray A, Patil SK, Paul DK, Biswas SK, Das B, Pant NC (2006) Petrology, geochemistry and magnetic properties of Sadara sill: evidence of rift related magmatism from Kutch basin, northwestern India. J As Earth Sci 27:907–921

Reeves CV (2014) The position of Madagascar within Gondwana and its movement during Gondwana dispersal. J Afr Earth Sci 94:45–57

Reeves CV, Vit De (2000) Making ends meet in Gondwana; retracing the transforms of the Indian ocean and reconnecting the continental shear zones. Terranova 12:272–280

Reichow MK, Pringle MS, Al'Mukhamedov AI, Allen MB, Andreichev VI, Buslov MM, Davies CE, Fedoseev GS, Fitton JG, Inger S, Medevedev AY, Mitchell C, Puchkov VN, Safanova IY, Scott RA, Saunders AD (2009) The timing and extent of the eruption of the Siberian traps large igneous province implications for the end Permian environmental crisis. Ear Planet, Sci Lett 277:9–20

Retallack GJ, Smith R, Ward P (2003) Vertebrate extinction across the Permian Triassic boundary in karoo Basin, South Africa, Bulletin. Geo Soc Am 115:1133–1152

Riccardi AC (1991) Jurassic and Cretaceous marine connections between southeast Pacific and Tethys. Paleogeogr Paleoclimat Paleoecol 87:155–189

Robertson A, Sharp I (1998) Mesozoic deep water slope/rise sedimentation and volcanism along the north Indian passive margin. J As Earth Sci 16:195–215

Robinson P, Morrison K, Fittal M (2012) Seychelles microcontient, an integration of regional tectonics and stratigraphy to unlock the hydrocarbon potential (presentation) to Geol Soc Lond 52 p

Roy AB, Chatterjee A (2015) Tectonic framework and evolutionary history od the Bengal basin in the Indian subcontinent. Curr Sci 109(2):271–279

Ruban DA (2014) Mesozoic long-term eustatic cycles and their uncertain hierarchy, Geosci Front 07/2014; 82(4). doi:10.1016/j.gsf.2014.06.001

Sahni A (1984) Cretaceous–Paleocene terrestrial faunas of India: lack of endemism during drifting phase of the Indian plate. Science 226:441–443

Sakai H (1984) Stratigraphy of Tansen area in the Nepal Lessere Himalaya. J Nepal Geol Soc 4:41–52

Saraswati PK (2011) Paleogene foraminiferal biostratigraphy of India in reference to continuity and gaps at epoch boundaries. Mem Geol Soc Ind 78:227–244

Saraswati PK, Khanolkar S, Raju DSN, Dutta S, Banerjee S (2014) Foraminiferal biostratigraphy of lignite mines of Kutch, India: Age of lignite and fossil vertebrates. J Palaeogeogr 3(1):90–98

Saraswati PK, Ramesh R, Navada SV (1993) Paleogene isotopic temperatures of Western India. Lethaia 26:89–98

Sarkar A, Datta AK, Poddar BC, Kolhapur VK, Bhattacharya BK, Sanval R (1996) Geochronological studies of Mesozoic igneous rocks from Eastern India. J SE Asian E Sci 21:188–193

Segoufin, Patriat (1980) Existence d'anomolies mesozoicques dans le basin do somalie; implecation pour le relations Afrique- Antarcticques-Madagascar. C.R Acad Sci Peris 291:85–88

Seton M, Muller RD, Zahirovic S, Gaina C, Torsvik T, Shepard G, Talsma A, Gurnis M, Turner M, Maus S, Chandler M (2012) Global continental and ocean basin constructions since 200 ma. Earth Sci Rev 113:212–270

Shankaran AV (1997) Polar wander and the Cambrian biological leap. Curr Sci 73:636–639

Shanker R, Singh G, Kumar G (1996) Sequence stratigraphy in Himalaya and its reflections in Gondwana sedimentation. In: Proceedings of the 9th Gondwana Symposium, Spl Publ Geol Surv Ind, Hyderabad, India. 1159–1173.

Sharland P, Archer R, Casey DM et al (2001) Arabian plate sequence stratigraphy, GeoArabia. Spe Pub 2:371

Sharma J, Saraswati PK (2015) Lignites of Kutch, western India: Dinoflagellate biostratigraphy and palaeoclimate. Revue de Micropaleont 58:107–119

Shellnut JG, Bonit GM, Brookfield ME, Jaba BM (2009) No link between the Panjal traps (Kashmir) and Permian mass extinction. Geophys Res Lett 38:219308. doi:10.1829/2011GL059052

Shellnut JG, Lee TY, Chiu HY, Lee YH, Wong J (2015) Evidence of middle Jurassic magmatism within the seychelles microcontinent: implications for the break-up of Gondwana. Geophys Res Lett 42:10 pp

Shi GH, Cui WY, Cao SM, Jiang N, Jian P, Lui DY, Miao LC, Chu BB (2008) Ion microprobe Zircon U-Pb age and Geochronology of the Mynmar Jaedeite. J Geol Soc Lond 166:221–234

Shi GH, Yeiyan L, Huayu H, Yi NN, Yan L, Yingxin L, Ye Y (2014) Superimposed tectonometamorphic episodes of Jurassic and Eocene age in the Jadeite uplift as revealed by Ar40/Ar39 dating. Gond Res 26:464–474

Shukla UK, Srivastava R (2008) Lizard eggs from the lateCretaceous Lameta Formation of Jabalpur, central India with interpretation of depositional environments of nest-bearing horizons. Cretac Res 29:674–686

Silva PI, Garzanti E, Gaetani M (1991) Stratigraphy of the Chikkim and Fatu La Formations in the Zangla and Zumlung units (Zanskar Range, India) with comparisons to the Thakkhola region (central Nepal). Mid- Cretaçeous evolution of the Indian passive margin, Rivista Italiana di Paleontologia e Stratigrafia 97(3–4):511–564

Singh IB (1981) Paleoenvironment and paleogeography of the Lameta Group sediments (Late Cretaceous) in Jabalpur area, India. J Pal Soc Ind 26:38–53

Singh NP (2006) Mesozoic lithostratigraphy of the Jaisalmer basin, Rajasthan. J Pal Soc Ind 51(2):1–25

Singh NP, Singh D, Bharktya DK, Srivastava S (2000) Subsurface stratigraphy, sedimentaion model and hydrocarbon prospects of Saurashtra basin, India. Bull ONGC 37(1):167–179

Singh RK, Agarwalla RC, Verma DP, Goel AK, Gupta SK (2006) Hydrocarbon exploration of Mesozoics in Kutch offshore area. In: 6th international conference and exposition on petroleum geophysics, 49–54

Sinha AK, Bagdasarian GP (1976) K/Ar dating of some magmatic and metamorphic rocks from Tethyan and Lesser Zones of Kumaon and Garhwal and Indian Himalaya and its implication in the Himalayan tectogenesis. Colloq Int CNRS Ecol Et Geol De Himalaya 268:387–394

Srinivasan MS (1986) neogene reference sections of Andaman-Nicobar: their bearing on volcanis, Sea floor tectonism and global sea-level changes. In: Ghose NC, Varadarajan S (ed) Ophiolites and Indian Plate Margin. Patna, pp 295–308

Srivastava RK, Heaman LM, Sinha A, Shihua S (2005) Emplacement, age and isotope geochemistry of Sung valley alkaline carbonatite complex. Shilong plateau, northeastern India: implications for primary melt and genesis of the associated silicate rocks, Lithos 81:33–54

Stampfli GM, Borel GD (2002) A plate tectonic model for the Paleozoic and Mesozoic constrained by dynamic plate boundaries and restored synthetic oceanic isochrons, Earth Planet. Sci Lett 196:17–33. doi:10.1016/S0012-821X(01)00588-X

Storey M, Kent RW, Saunders AD, Hergt J, Salters VJM, Whitechurch H, Sevigny JH, Thiwall NF, Leat P, Ghose NC, Gifford M (1992) Lower Cretaceous rocks on continental margins and their relationship to Kerguelen plateau. Proc Ocean Drilling Programme, Sci Results 120:33–53

Symonds P, Planke S, Frey O, Skogseid J (1998) Volcanic evolution of the western Australia continental margin and its implication for basin development. Proc. Petrol. Expl. Soc. Australia, 33–34

Tewari RC (1998) Permian Gondwana sedimentation in Yellandu (Singareni) coalfield, Andhra Pradesh, India, with notes on regional palaeogeography and tectonic history. J Geol Soc Ind 23:67–79

Tewari RC, Casshyap SM (1996) Mesozoic tectonic events including rifting in peninsular India and their bearing on Gondwana sedimentation, Gondwana. Spl Publ Geol Surv Ind 9(2): 865–881

Thanh NX, Rajesh VJ, Itaya T, Windley B, Kwon S, Park CS (2012) A Cretaceous forearc 104 ma ophiolite in the Shyok suture zone, Ladakh, NW India, Implications for the tectonic evolution of the NW Himalaya. Lithos 155:81–93

Torsvik TH, Tucker RD, Ashwal LD, Carter LM, Jamtveit B, Vidyadharan KT, Venkataraman P (2000) Late Cretaceous India–Madagascar fit and timing of break-up related magmatism. Terra Nova 12:220–224

Tripathi SC, Lahiri TC (2000) Marine oscillation event stratification: an example from Late Cretaceous Bagh carbonate sequence of Narmada valley India. Mem Spl Geol Soc Ind 46:15–24

Upadhyay R, Frisch W, Siebel W (2008) Tectonic implications of new U-Pb zircon ages of the Ladakh batholith. Indus Suture Zone, NW Himalaya, India, Terra Nova 20:309–317

Vaillier TL (1974) Volcanogenic sediments and their relation to landmass volcanism and sea-floor-continent movements, Western Indian ocean, leg 25. Deep Sea Drilling Proj 21:515–542

Valdiya KS (2010) The making of India: Geodynamic evolution, 1st edn. McMillan, India, 816 p

Valdiya KS (2016) The making of India: geodynamic evolution (2nd edn). Springer, 924 p

Valsangkar AB, Radhakrishnamuri C, Subbarao KV, Beckinsale RD (1981) Palaeomagnetism and K/Ar age studies of acid igneous rocks from the St Marys Island. Mem Geol Soc Ind 3:265–275

Van Vagoner JC, Posamentier HM, Mitchum RM (1988) An overview of the fundamentals of sequence stratigraphy and definitions. In: Wilgus CK et al (eds) Sea-level changes: an integrated approach, SEPM Sp. Publ, vol 42, pp 39–45

Veevers JJ (ed) (1984) Phanerozoic Earth history of Australia, Clarendon press, Oxford, 418 pp

Veevers JJ, Tiwari RC (1995) Gondwana master basin of peninsular India. Geol Soc Am, Boulder 72p

Veevers JJ, Tewari RC, Mishra HK (1996) Aspects of Late Triassic to Early Cretaceous disruption of the Gondwana coal bearing fan of East-Central Gondwanaland, Gondwana. Spl Publ Geol Surv Ind 9(2):637–646

Venkatachala BS (1970) Palynology of the Mesozoic sediments of Kachchh, West India; Reworked Permian Pollens from the upper Jurassic sediments-a discussion. Palaeobotanist 18(1):45–49

Vijaya (1997) Benchmark Permian–Triassic palynoevents in peninsular India and northwest Himalaya. J Pal Soc Ind 42:117–126

Vijaya (2009) Palynofloral changes in the Upper Paleozoic and Mesozoic of the Deocha-Pachami area, Birbhum Coalfield, West Bengal. Sci China Ser D-Earth Sci 52, 12:1932–1952

Volkmer JE (2010) the Cretaceous-Tertiary tectonic evolution of the Lhasa terrane, Tibet Ph.D. Thesis, University of Arizona, Tucson, p 213

Walaszczyk I, Kennedy WJ, Dembiez K, Gale AS, Praszkier T, Rasomiaramanana AH, Radrianaly H (2014) Ammonite and inoceramid biostratigraphy and biogeography through basal Middle Campanian (Upper Cretaceous) of the Morondova basin, Western Madagascar. J Afr Earth Sci 89:79–132

Wang CS, Hu XM, Jansa L, Wan XQ, Tao R (2001) The Cenomanian – Turonian anoxic event in Southern Tibet. Cret Res 22:481–490

Wang JG, Hu XM, Jansa L, Huang ZC (2011) Provenance of the Upper Cretaceous-Eocene deep-water sandstones in Sangdanlin, Southern Tibet: Constraints on the timing of initial India-Asia collision. J Geol 119(3):293–309

Warren CJ, Parrish RR, Waters DJ, Searle MP (2005) Dating the geologic history of Oman's Semail ophiolite: Insights from U-Pb geochronology. Contr Mineral Petr 150:403–422

Westphal H, Aignar T (1993) Seismic stratigraphy and subsidence analysis in the Barrow–Dampier basin, Northwest Australia. Am Ass Petrol Geol B V 81:1721–1749

Wiedmann J, Durr SB (1995) First ammonites from the Mid to Upper Cretaceous Xigaze Group, South Tibet, and their significance. Newslet Stratigr 32:17–26

Williams SE, Whttaker JM, Granot R, Muller RD (2013) Early India–Australia spreading history by newly detected magnetic anomalies in the Perth abyssal plain. J Geophy Res 118: 3275–3284

Wynne AB (1872) Memoir on the Geology of Kutch, to accompany the map compiled by Wynne, A.B. Fedden, F., during the session 1867–1869. Mem Geol Surv Ind 9:1–293

Xiachi L, Grant-Mackie JA (1993) Jurassic sedimentary cycles and sea level changes in Southern Tibet. Palaeogeogr Palaeoclimateol Palaeoecol 101:27–48

Yang T, Ma Y, Bian W, Ding J (2015) Paleomagnetic results from the Early Cretaceous Lakang Formation lavas; constraints on the paleolatitude of the Tethyan Himalaya and the India–Asia collision, E Planet Sci Lett, vol 428, pp 122–133

Yateesh V, Bhattacharya GC, Mahender K (2006) The terrace like feature in the mid central slope region off Trivendrum and a plausible model for India–Madagascar juxtaposition in immediate pre-drift scenario. Gond Res 10(1–2):179–185

Yateesh V, Dyment J, Bhattacharya GC, Muller RD (2013) Deciphering detailed plate kinematics of the Indian ocean and developing a unified model for East Gondwana reconstruction: an India–Australia–French venture, DCS Newsletter

Yin A, Dubey CS, Kelty TK, Webb AAG, Harrison TM, Chou CY, Célérier J (2011) Geologic correlation of the Himalayan orogen and Indian craton: part 2. structural geology, geochronology, and tectonic evolution of the Eastern Himalaya. Geol Soc Am Bull 122:360–395

Yui TF, Fukyama M, Ezuka Y, Wu Y, Wu TW, Liou JG, Grove M (2013) Is Mynmar jadeite of Jurassic age? A result from incompletely recrystallised in heritage zircon. Lithos 160–161: 268–282

Zahirovic S, Seton M, Muller RD (2014) The Cretaceous and Cenozoic tectonic evolution of Southeast Asia. Solid Earth 5:227–273

Zhang K, Zhang Y, Tang X, Xu B (2012) Late Mesozoic tectonic evolution and growth of the Tibetan plateau prior to the Indo-Asian collision. Ear Sci Lett 114:236–249

Zhu B, Kidd SF, Rowley DB, Currie BS, Shafique N (2005) Age of initiation of India-Asia collision in the East-Central Himalaya. J Geol 113:265–285

Zhu DC, Chung S, Mo X, Zhao Z, Yaoling N, Song B, Yang Y (2009) The 132 ma Comei-Bunbury large igneous province: Remnants identified in present-day southeastern Tibet and southwestern Australia. Geology 37(7):583–586

Zhu DC, Chung S, Niu Y (2016) Recent advances on the tectonic and magmatic evolution of the Greater Tibetan plateau; a special issue in honour of Prof. Guitang Pan, Lithos 245:1–6

Zhu DC, Pan GT, Chung SL, Liao ZL, Wang LQ, Li GM (2008) SHRIMP zircon age and geochemical constraints on the origin of lower Jurassic volcanic rocks from the Yeba formation, Southern Gangdese, south Tibet, Int Geol Rev 50:442–471

MIX
Papier aus verantwortungsvollen Quellen
Paper from responsible sources
FSC® C105338

If you have any concerns about our products,
you can contact us on
ProductSafety@springernature.com

In case Publisher is established outside the EU,
the EU authorized representative is:
Springer Nature Customer Service Center GmbH
Europaplatz 3, 69115 Heidelberg, Germany

Printed by Libri Plureos GmbH
in Hamburg, Germany